ELECTRONIC DEVICES
A Design Approach

Ali Aminian
DeVry University—Columbus, OH

Marian K. Kazimierczuk
Wright State University

Upper Saddle River, New Jersey
Columbus, Ohio

Library of Congress Cataloging in Publication Data

Aminian, Ali.
   Electronic Devices: a design approach / Ali Aminian, Marian K. Kazimierczuk.
     p. cm.
   Includes index.
   ISBN 0-13-013560-7
    1.Electronic apparatus and appliances. 2. Electronic circuit design. 3. Electronics. I. Kazimierczuk, Marian. II. Title.

TK7870 .A527 2004
621.381—dc21   2002035536

Editor in Chief: Stephen Helba
Editor: Dennis Williams
Development Editor: Kate Linsner
Production Editor: Rex Davidson
Design Coordinator: Diane Ernsberger
Cover Designer: Ali Mohrman
Cover Art: Digital Vision
Production Manager: Pat Tonneman
Marketing Manager: Ben Leonard

Copyright © 2004 by Pearson Education, Inc., Upper Saddle River, New Jersey 07458.
Pearson Prentice Hall. All rights reserved. Printed in the United States of America. This publication is protected by Copyright and permission should be obtained from the publisher prior to any prohibited reproduction, storage in a retrieval system, or transmission in any form or by any means, electronic, mechanical, photocopying, recording, or likewise. For information regarding permission(s), write to: Rights and Permissions Department.

Pearson Prentice Hall™ is a trademark of Pearson Education, Inc.
Pearson® is a registered trademark of Pearson plc
Prentice Hall® is a registered trademark of Pearson Education, Inc.

Pearson Education Ltd.   Pearson Education Australia Pty. Limited
Pearson Education Singapore Pte. Ltd.   Pearson Education North Asia Ltd.
Pearson Education Canada, Ltd.   Pearson Educación de Mexico, S.A. de C.V.
Pearson Education—Japan   Pearson Education Malaysia Pte. Ltd.

ISBN: 0-13-013560-7

This book is dedicated to my wife Mary and my son Ashley

Ali Aminian

This book is dedicated to my wife Alicja and my children Anna and Andrzej

Marian K. Kazimierczuk

PREFACE

This textbook presents a comprehensive coverage of electronic devices, discrete and integrated, with real-world applications. It is intended for use in four-year engineering technology and engineering programs, and was written with several objectives in mind, some of which are presented below:

Preciseness: One of the objectives is to make available a textbook that is technically correct and academically precise in its presentation of theory, algebraic derivations, and the analysis and design of electronic circuits and systems. As you will notice, most algebraic solutions, which are based on algebraic derivations, are tested and verified by simulation.

Completeness: The second objective is to provide complete coverage of the topics. That is, the text begins with the basic theory and ends with real-world applications of each device, while covering all that is needed in between without overwhelming the student with extraneous information. The authors believe that a book must be complete enough to be used as a textbook and as a reference. However, the instructor is not obliged to teach every single chapter or every single page in a chapter. The instructor decides the breadth and depth of coverage based on his or her own timetable and instructional objectives. In some cases, though, only the end results in a chapter may be presented and utilized for the analysis and design of circuits.

Design approach: The third objective is to be practical in coverage of the course. That is, in addition to offering complete coverage of theory and analysis, the textbook presents a real-world application of the device by designing a system or circuit for some given practical specifications.

Use of real devices: The fourth objective is to make use of real devices with actual characteristics based on device specifications provided by the manufacturer. To promote and facilitate the use of data sheets, device specifications are included at the end of the chapter in which the device is introduced.

Simulation: In addition to algebraically proving the analysis and design processes, some analysis-type and most design-type examples are tested and verified by simulation using Electronics Workbench, the most popular and user-friendly simulation software. With the availability of the latest presentation equipment in most classrooms, these simulations can be carried out live in the classroom following the board work, which adds more variety to the lecture and captures students' attention.

Use of MATLAB: Another objective is to expose the students to the most popular computation and visualization software in academia and industry. Beginning in Chapter 1, MATLAB is used for plotting the diode equation, and is used more extensively in Chapters 7 and 9 for precise graphical solutions and frequency response plots. However, beginning with Laboratory 1 in the lab manual, students are encouraged/required to use MATLAB for plotting the collected data of device characteristics.

ACKNOWLEDGMENTS

We would like to extend our gratitude and appreciation to Professor Aminian's dear friend and colleague Professor Carol Dietrich who, despite her very busy schedule, volunteered to proofread every chapter as it was written.

We would also like to thank the reviewers and our colleagues at different DeVry Universities for taking time to review the manuscript and provide answers to an extensive questionnaire:

Michael Miller, DeVry University, Phoenix, AZ

Mostafa Mortezaie, DeVry University, Fremont, CA

Shoja Mazidi, DeVry University, Calgary, Canada

Ahmad Ibrahim, DeVry University, Toronto, Canada

Their comments, suggestions, and recommendations were given serious consideration and most of them were implemented in the final version of the manuscript.

We also thank to the Prentice Hall editors and staff: Dennis Williams, Rex Davidson, and Lara Dimmick for all their help in getting this manuscript published.

Ali Aminian
Marian K. Kazimierczuk

CONTENTS

Chapter 1   Semiconductor Theory and the *P-N* Junction Diode

    1.1   Introduction   1
    1.2   A Brief Review of Atomic Theory   2
    1.3   The *P-N* Junction   7
    1.4   The *P-N* Junction Diode   12
          - Diode I-V Characteristics   12
          - Temperature Effects   14
          - The Diode Equation   15
          - DC and AC Resistances of the Diode   16
          - Basic Diode Circuits   20
          - Diode Switching Circuits   23
    1.5   Zener Diode   24
    1.6   Light-Emitting Diode   28
    1.7   Summary   30
    1.8   Device Specifications and Data Sheets   32
         Problems   38

Chapter 2   Diode Applications

    2.1   Introduction   43
    2.2   The Basic Power Supply   44
          - The Transformer   44
          - Half-Wave Rectifier   45
          - Full-Wave Rectifier   47
          - Filtering   53
          - Regulation   56
          - Regulated Power Supply   58
    2.3   Summary   62
         Problems   63

Chapter 3   Bipolar Junction Transistors and DC Biasing the BJT Amplifier

    3.1   Introduction   67
    3.2   Basic Theory of BJT Operation   68
    3.3   Common-Emitter Characteristics   71
    3.4   Biasing the Common-Emitter Amplifier   76
          - The Base Bias   77
          - Emitter Bias   81
          - Voltage-Divider Bias   84
    3.5   Biasing the Common-Collector Amplifier   88
    3.6   Biasing the Common-Base Amplifier   89
    3.7   DC Bias Analysis of the Cascode Amplifier   91
    3.8   Reading the Transistor Data Sheets   93

3.9 DC Bias Design of BJT Amplifiers   94
3.10 The Transistor Switch   100
3.11 Summary   105
3.12 Device Specifications and Data Sheets   107
Problems   116

Chapter 4   Transistor Modeling

4.1 Introduction   121
4.2 The r-parameter Model   122
4.3 The Hybrid or h-parameter Model   131
Problems   149

Chapter 5   Small-Signal Operation of the BJT Amplifier

5.1 Introduction   155
5.2 The BJT Amplifier Fundamentals   157
5.3 Analysis of the BJT Amplifier   163
5.4 The Cascode Amplifier   183
5.5 Troubleshooting   187
5.6 Summary   191
Problems   193

Chapter 6   Design of the Small-Signal BJT Amplifier

6.1 Introduction   199
6.2 The AC Load Line and the Output Signal Swing   200
6.3 A Brief Introduction to Frequency Response   210
6.4 Design of the Common-Emitter Amplifier   212
6.5 Design of the Cascode Amplifier   219
6.6 Summary   224
Problems   225

Chapter 7   Field-Effect Transistors

7.1 Introduction   227
7.2 Junction Field-Effect Transistors (JFET)   228
- JFET Structure and Characteristics   228
- DC Biasing the JFET   235
- Self-Bias   237
- Voltage-Divider Bias   246
- DC Bias Design of the JFET   254
7.3 Metal-Oxide Semiconductor FET (MOSFET)   261
- Depletion MOSFET   261
- Enhancement MOSFET   264
- DC Biasing the Enhancement MOSFET   267
- DC Bias Design of the MOSFET Amplifier   275
7.4 Summary   279

7.5   Device Specifications and Data Sheets   280
      Problems   284

Chapter 8   Small-Signal Operation of the FET Amplifier

8.1   Introduction   289
8.2   Small-Signal Circuit Model of FET   290
8.3   Analysis of the Common-Source JFET Amplifier   294
8.4   Analysis of the Common-Drain JFET Amplifier   303
8.5   Analysis of the Enhancement MOSFET Amplifier   310
8.6   Analysis of the Common-Gate MOSFET Amplifier   323
8.7   Analysis of the FET Cascode Amplifier   327
8.8   Design of Small-Signal JFET and MOSFET Amplifiers   330
8.9   Summary   340
      Problems   343

Chapter 9   Frequency Response of BJT and FET Amplifiers

9.1   Introduction   349
9.2   Frequency Response Fundamentals   349
9.3   Frequency Response of the CE BJT Amplifier   359
9.4   Frequency Response Plot of the CE BJT Amplifier   375
9.5   Frequency Response of the CB BJT Amplifier   377
9.6   Frequency Response of the BJT Cascode Amplifier   380
9.7   Frequency Response of the FET Amplifier   384
9.8   Frequency Response of the FET Cascode Amplifier   391
9.9   Summary   394
      Problems   396

Chapter 10   Current-Mirror Current Sources and Differential Amplifiers

10.1   Introduction   401
10.2   Current-Mirror Current Sources   402
  - Basic Current Mirror   402
  - Wilson Current-Mirror   404
  - MOSFET Current-Mirror   408
10.3   Differential Amplifier 412
10.4   The Basic Differential Amplifier Circuit   415
10.5   Diff-Amp with Current-Mirror Current Source   424
  - Diff-Amp with Basic Current-Mirror   425
  - Diff-Amp with Wilson Current-Mirror   428
10.6   MOSFET Differential Amplifier   439
  - Diff-Amp with Basic MOSFET Current-Mirror   440
  - Diff-Amp with Cascode MOSFET Current-Mirror   444
10.7   Summary   446
       Problems   447

Chapter 11   Operational Amplifiers

- 11.1 Introduction   453
- 11.2 Operational Amplifier   454
- 11.3 Open-Loop Operation   456
- 11.4 Closed-Loop Operation   464
- 11.5 Basic Op-Amp Configurations   465
  - Inverting Amplifier   465
  - Non-Inverting Amplifier   471
  - Unity-Gain Buffer   472
  - Compensation Resistor   476
- 11.6 Summing Amplifiers   479
  - Inverting Summer   479
  - Non-Inverting Summer   481
- 11.7 Difference Amplifier   486
  - Noise Suppression   488
  - Instrumentation Amplifier   490
- 11.8 Integrator and Differentiator   494
- 11.9 Summary   504
- 11.10 Device Specifications and Data Sheets   506
- Problems   516

Chapter 12   Negative Feedback, Op-Amp Characteristics, and Single-Supply Operation

- 12.1 Introduction   523
- 12.2 Negative Feedback   524
  - Stability   524
- 12.3 Basic Feedback Topologies   526
  - Series-Shunt Feedback   526
- 12.4 Op-Amp Characteristics   539
  - DC Characteristics   539
  - AC Characteristics   541
- 12.5 Single-Supply Operation   545
- 12.6 Summary   553
- 12.7 Device Specifications and Data Sheets   554
- Problems   561

Chapter 13   Power Amplifiers and Output Stages

- 13.1 Introduction   565
- 13.2 Classes of Operation   566
- 13.3 Class A Power Amplifier   567
- 13.4 Class B Push-Pull Power Amplifier   568
- 13.5 Class AB Push-Pull Power Amplifier   577
- 13.6 Heat Sinks and Thermal Considerations   585
- 13.7 Summary   588

Problems   593

Chapter 14   Active Filters

14.1   Introduction   595
14.2   Active Filters   596
14.3   First-Order Low-Pass Active Filter   596
14.4   First-Order High-Pass Active Filter   600
14.5   Design of the First-Order Active Filter   604
14.6   Second-Order Active Filters   608
- Second-Order Low-Pass Active Filter   610
- Second-Order High-Pass Active Filter   612
- Sallen-Key Unity-Gain Active Filter   613
- Sallen-Key Equal Component Active Filter   615
- Design of the Second-Order Active Filter   619

14.7   Higher Order Active Filters   625
14.8   Band-Pass Active Filter   633
14.9   Band-Stop Active Filter   639
14.10   Summary   644
        Problems   645

Chapter 15   IC Regulators and Switching Power Supplies

15.1   Introduction   649
15.2   Integrated Circuit Regulators   650
15.3   Switching Power Supplies   658
- Buck (Step-Down) DC-DC Converter   659
- Boost (Step-Up) DC-DC Converter   669
- Buck-Boost (Inverting Step-Down/Up) DC-DC Converter   678

15.4   Summary   688
15.5   Device Specifications and Data Sheets   689
        Problems   702

Chapter 16   Oscillators and Waveform Generators

16.1   Introduction   705
16.2   Oscillator Principles of Operation   706
16.3   RC Phase-Shift Oscillator   707
16.4   Wien-Bridge Oscillator   711
16.5   Colpitts Oscillator   715
16.6   Hartley Oscillator   717
16.7   Square-Wave Generator   720
16.8   The 555 Timer   722
16.9   Function Generator   726
- Sine-Wave and Square-Wave Generation   726
- Pulse and Ramp Generation   727
- FSK Generation   729

	16.10	Summary 730	
		Problems 732	

Chapter 17 Digital-to-Analog and Analog-to-Digital Conversion

- 17.1 Introduction 735
- 17.2 Digital-to-Analog Conversion 736
  - R-2R Ladder DAC 736
  - Binary-Weighted Resistor DAC 740
  - Monolithic Integrated Circuit DAC 742
- 17.3 Analog-to-Digital Conversion 743
  - Successive Approximation ADC 743
  - Monolithic Integrated Circuit ADC 744
  - Mixed Signal Design Example 746
- 17.4 Summary 749
  - Problems 751

Chapter 18 Four-Layer Electronic Devices and Optoelectronic Devices

- 18.1 Introduction 753
- 18.2 Four-Layer Devices 754
  - Silicon Controlled Rectifier (SCR) 754
  - Silicon Controlled Switch (SCS) 757
  - DIACs and TRIACs 758
- 18.3 Optoelectronic Devices 763
  - Photoresistors 763
  - Photodiodes 764
  - Phototransistors 766
  - Optocouplers 766
- 18.4 Summary 768
- 18.5 Device Specifications and Data Sheets 770
  - Problems 789

INDEX 791

Chapter 1

SEMICONDUCTOR THEORY AND THE *P-N* JUNCTION DIODE

1.1 INTRODUCTION

Semiconductor materials have played a significant role in the development of modern electronic devices such as *diodes*, *transistors*, and *integrated circuits*, which are used in the design and construction of many useful electronic products and equipment. In this chapter, we will present a brief theory of the semiconductor *p-n* junction, which is a fundamental component of many electronic devices. We will also study the characteristics, biasing, and operation of the *p-n* junction diode, Zener diode, and light-emitting diode (*LED*). Further applications of these devices will be covered in the following chapter.

Chapter 1

1.2 A BRIEF REVIEW OF ATOMIC THEORY

Because all matter is made of atoms, and the configuration of certain electrons in an atom is the key factor in determining if an element is a conductor, insulator, or semiconductor, we will begin with a brief introduction to atomic theory.

Atomic Structure

An atom is the smallest particle of an element that retains the characteristics of that element. The structure of an atom basically consists of a *nucleus* at the center surrounded by orbiting *electrons*:
- The nucleus carries almost the total mass of the atom and consists of:
  - *Neutrons* that are neutral and carry no charge
  - *Protons* that carry positive charges
- The orbiting electrons carry negative charges.

The number of positively charged protons is equal to the number of negatively charged electrons in an atom, which makes it electrically neutral or balanced.

Atomic Number

The number of protons in the nucleus or the number of orbiting electrons in a neutral atom determines the atomic number of an atom.

Atomic Weight

The atomic weight of an atom equals the number of neutrons plus the number of protons in the nucleus of the atom.

The hydrogen atom is the simplest atom with an atomic number equal to 1, and the next simplest is the helium atom with an atomic number of 2, as illustrated in Figure 1-1 below:

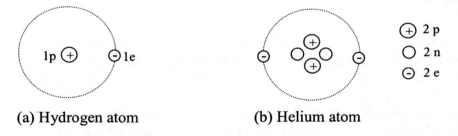

(a) Hydrogen atom   (b) Helium atom

Figure 1-1: Bohr model of hydrogen and helium atoms

The atomic weight of hydrogen atom is 1 and the atomic weight of helium is 4.

Chapter 1

Atomic Shells and Subshells

The orbiting electrons of an atom are aligned in a structured manner consisting of *shells* and *subshells*. Each shell, which is located at a different level, can contain a maximum number of electrons. In general, the nth shell can contain a maximum of $2n^2$ electrons, where n is the shell number, and shell number 1 is the innermost and closest to the nucleus. Let us now consider the structure of the copper atom, whose atomic number is 29, which means there are a total of 29 orbiting electrons.

1st shell (K): $2n^2 = 2(1)^2 = 2$ electrons

2nd shell (L): $2n^2 = 2(2)^2 = 8$ electrons

3rd shell (M): $2n^2 = 2(3)^2 = 18$ electrons

4th shell (N): 1 electron

Total: 29 electrons

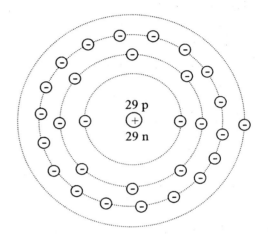

Figure 1-2:
Bohr model of copper atom (*Cu*)

Furthermore, each shell is divided into *subshells*, the nth shell having n subshells. Shells are generally designated with upper case letters *K, L, M, N, . . .*, and subshells are designated with lowercase letters *s, p, d, f*. The first subshell (*s*) can have a maximum of two electrons, and the succeeding subshells can have four electrons more than the preceding subshell.

Shell	Subshells	Capacity ($2n^2$)	Content
K ($n = 1$)	s	2	2
L ($n = 2$)	s	2	2
	p	6	6
M ($n = 3$)	s	2	2
	p	6	6
	d	10	10
N ($n = 4$)	s	2	1
	p	6	0
	d	10	0
	f	14	0

Table 1-1: Electron contents of shells and subshells of the copper atom

Let us now consider the atomic structure of two popular semiconductors, *silicon* and *germanium*, and find out what they have in common. The atomic numbers for silicon and germanium are 14 and 32, respectively.

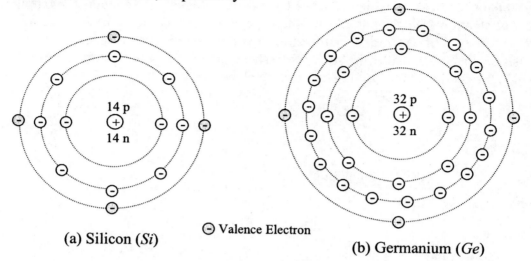

(a) Silicon (*Si*)   ⊖ Valence Electron   (b) Germanium (*Ge*)

Figure 1-3: Shell structures of silicon and germanium atoms

The electron contents of shells and subshells of silicon and germanium atoms are shown in Tables 1-2 and 1-3 below.

Shell	Subshells	Capacity ($2n^2$)	Content
K ($n = 1$)	s	2	2
L ($n = 2$)	s	2	2
	p	6	6
M ($n = 3$)	s	2	2
	p	6	2
	d	10	0

Table 1-2: Electron contents of shells and subshells of the **silicon** atom

Shell	Subshells	Capacity ($2n^2$)	Content
K ($n = 1$)	s	2	2
L ($n = 2$)	s	2	2
	p	6	6
M ($n = 3$)	s	2	2
	p	6	6
	d	10	10
N ($n = 4$)	s	2	2
	p	6	2
	d	10	0
	f	14	0

Table 1-3: Electron contents of shells and subshells of the **germanium** atom

Chapter 1

Valence Shell and Valence Electrons

The outermost shell in an atom is called the *valence shell* and the electrons in the valence shell are called *valence electrons*. The copper atom, as shown in Figure 1-2, has only one valence electron; however, silicon and germanium both have four valence electrons, as shown in Figure 1-3 above. The number of valence electrons in an atom has significant influence on the electrical properties of an element. The fewer the valence electrons and the more distant the valence shell is from the nucleus, the more likely it is that the valence electrons will break away from the parent atom and become *free* electrons with a minimum amount of external energy (such as heat). For example, the single valence electron in the fourth shell of the copper atom can break away easily at room temperature (25°C) and thus create an abundance of free electrons in copper, which makes it a good conductor. Free electrons are abundant in good conductors such as copper, silver, and gold because of the following two reasons:
1. The single valence electron is in a higher level shell farther away from the nucleus; hence, it is not tightly bound to the parent atom by the positive force of attraction by the nucleus.
2. The atom tends to complete its outer shell by freeing the loosely bound valence electron.

Semiconductors and Covalent Bonding

Silicon and germanium are the two most widely used semiconductors in the fabrication of electronic devices. Silicon, however, is the most popular because of its tolerance to higher temperatures. As Figure 1-3 depicts, both silicon and germanium atoms have four valence electrons. Furthermore, referring to Tables 1-2 and 1-3, we notice that both atoms have two electrons in the outermost *p* subshell, which is four electrons short of being complete. Because of this unique atomic structure, atoms in semiconductor materials tend to share their four valence electrons with the neighboring atoms, which results in a complete *p* subshell for every atom. This sharing of valence electrons in semiconductors is called *covalent bonding*, which produces a stable, tightly bound, lattice structure called a *crystal*. For the sake of simplicity, we will just show the four valence electrons of each semiconductor atom and the symbolic covalent bonding between one and its four neighboring atoms, as illustrated in Figure 1-4 below:

⊕ Nucleus

⊖ Valence Electron

(⊖ ⊖) Covalent Bond

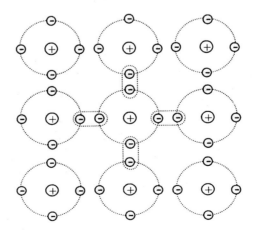

Figure 1-4:
Covalent bonding in a semiconductor crystal

Energy Bands

In the atomic structure, shells and subshells represent distinct energy levels. The more distant the shell or subshell is from the nucleus, the higher the energy level of the orbiting electrons. Hence, valence electrons of an atom possess the highest level of energy, and thus, can break away from the parent atom by acquiring a sufficient amount of external energy and becoming free electrons. When atoms bond together to form molecules of matter, the energy levels, which existed before for the single atoms, form bands of energy or *energy bands*. To better visualize this process, an *energy band* diagram is generally used to represent different energy levels to which the orbiting electrons belong, as illustrated in Figure 1-5. The amount of energy that the valence electrons must attain to be elevated to the next level (*conduction band*) is measured in *electron volts* (1 eV = 1.6 × 10^{-19} joules), which is the *energy gap* between *valence band* and *conduction band*. This energy gap is generally referred to as the *forbidden band*, because the valence electrons can leave the valence band only when they have acquired a sufficient amount of energy to jump across the forbidden band. As shown in Figure 1-5(a), the width of the forbidden band for a typical insulator is relatively wide; that is, the valence electrons need to acquire a large amount of energy in order to make the transition from valence band to conduction band. In a semiconductor, however, the width of the forbidden band is relatively narrow, as seen in Figure 1-5(b). For example, the energy gaps for silicon and germanium at room temperature (25°C) are about 1.1 eV and 0.67 eV, respectively. Moreover, the energy gap for conductors is almost nonexistent; the valence and conduction bands are generally considered to be overlapping, as shown in Figure 1-5(c).

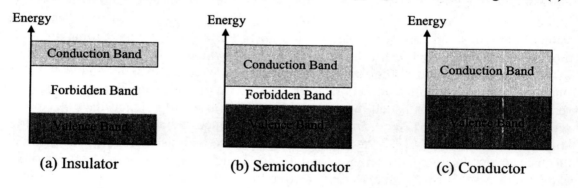

Figure 1-5: Energy band diagrams for three different materials

Electron-Hole Pair and Hole Current

Recall that the flow of negatively charged electrons constitutes the electrical current in conductors. In semiconductors, however, there is also another kind of charge flow, referred to as *hole current*, in addition to the electron flow. As illustrated in Figure 1-6, whenever a valence electron attains sufficient energy to make the transition to the conduction band, a *hole* or absence of an electron is left behind in the crystal structure. As a result of the loss of this negatively charged electron, the atom is no longer neutral, but has a net positive charge, and is called a positive *ion*. Now, if an electron that is released from the neighboring covalent bond fills the hole, leaving behind a freshly produced hole, then the net effect is that a *hole* or *positive charge* has moved from the neighboring atom to the first. This phenomenon can continue among the remaining atoms and the flow of these holes constitutes what is called the *hole current* in the semiconductor.

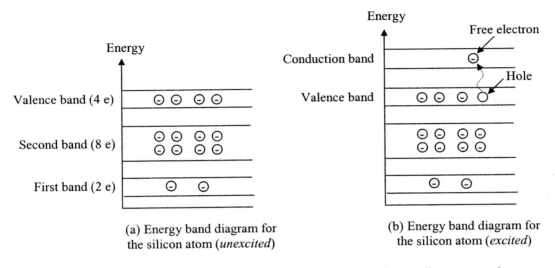

(a) Energy band diagram for the silicon atom (*unexcited*)

(b) Energy band diagram for the silicon atom (*excited*)

Figure 1-6: Creation of electron-hole pair in silicon crystal

1.3 THE *P-N* JUNCTION

Before we discuss the semiconductor *p-n* junction, let us define a few terms and processes that are frequently used and referred to in *p-n* junction theory:

Intrinsic semiconductor is the pure semiconductor, in which the number of free electrons equals the number of holes in the crystal structure.

Doping is the process by which impurity atoms are introduced to the intrinsic semiconductor in order to alter the balance between holes and electrons.

N-type impurities (donors) are the type of impurities that add (donate) electrons to intrinsic semiconductors, when combined.

P-type impurities (acceptors) are the type of impurities that produce holes (accept electrons) in intrinsic semiconductors, when combined.

Extrinsic semiconductor is the impure semiconductor that has been doped with *n*-type or *p*-type impurity atoms, resulting in an imbalance between the hole and electron densities.

Ionization is the process of losing or gaining a valence electron. If a neutral atom loses a valence electron, it is no longer neutral and is called a *positive ion*. On the other hand, if a neutral atom gains a valence electron, it is called a *negative ion*.

Diffusion current results when there is a non-uniform concentration of charge carriers (electrons or holes) in the semiconductor; that is, if there is a higher density of carriers in one region and lower density in another, carriers start migrating from the region of higher density to the region with lower density until a fairly uniform concentration is established in the semiconductor. The flow of these charge carriers during migration constitutes a current flow called *diffusion current*, and the carriers are said to diffuse from one region to another.

N-type Semiconductor

An *n*-type semiconductor is produced when the intrinsic semiconductor is doped with *n*-type impurity atoms that have five valence electrons (*pentavalent*), such as *arsenic*, *antimony*, and *phosphorus*. When the pentavalent (donor) impurity is introduced to the intrinsic semiconductor, it becomes an integral part of the crystal structure; four of its five valence electrons bond covalently with the surrounding semiconductor atoms; the fifth valence electron, which is an excess electron, becomes a free electron. The net effect is that each pentavalent impurity atom donates an electron to the semiconductor; hence, it is called *donor*, and the semiconductor is then an *n*-type semiconductor because it contains excessive electrons. A graphical illustration of this process is presented in Figure 1-7.

P-type Semiconductor

A *p*-type semiconductor is produced when the intrinsic semiconductor is doped with *p*-type impurity atoms that have three valence electrons (*trivalent*), such as *aluminum*, *boron*, and *gallium*. When the trivalent (acceptor) impurity is introduced to the intrinsic semiconductor, it also becomes an integral part of the crystal structure and its three valence electrons bond covalently with three neighboring semiconductor atoms. However, it is short of one electron when it tries to establish a covalent bond with the fourth neighboring atom; hence, a hole is created within the crystal structure. The net effect is that each trivalent impurity atom creates a *hole* in the semiconductor, which is ready to accept a wandering electron and fill the covalent bond. Hence, it is called *acceptor*, and the semiconductor is then a *p*-type semiconductor because it contains excessive holes.

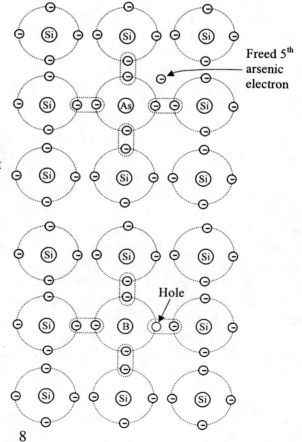

(Si) Nucleus of silicon atom
(As) Nucleus of arsenic atom
⊖ Valence Electron
⊖ ⊖ Covalent Bond

Figure 1-7: *N*-type semiconductor. Creation of a free electron by doping the semiconductor with a pentavalent (*donor*) impurity

(Si) Nucleus of silicon atom
(B) Nucleus of boron atom
⊖ Valence Electron
⊖ ⊖ Covalent Bond

Figure 1-8: *P*-type semiconductor. Creation of a hole by doping the semiconductor with a trivalent (*acceptor*) impurity

Chapter 1

Formation of the *p-n* Junction

The *p-n* junction is a fundamental component of many electronic devices and is formed by joining, through a certain manufacturing process, a block of *p*-type semiconductor to a block of *n*-type semiconductor. Although the process is not as simple as it sounds, we can, however, view the formation of the junction in terms of charge redistribution, as illustrated in Figures 1-9 through 1-11 below. Figure 1-9 shows isolated blocks of *p*- and *n*-type semiconductors that are both electrically neutral. Figure 1-10 shows the two semiconductor blocks at the instant they are joined, where the free electrons at the edge of the *n*-type semiconductor are anxious to diffuse and recombine with the holes across the junction.

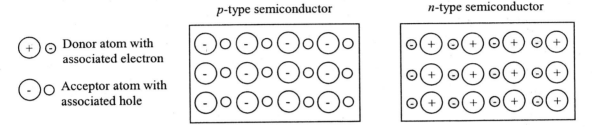

Figure 1-9: Blocks of *p*-type and *n*-type semiconductors before they are joined

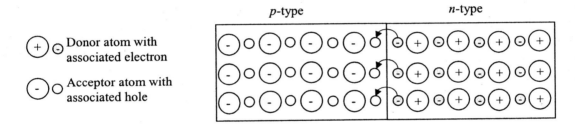

Figure 1-10: Blocks of *p*-type and *n*-type semiconductors at the instant they are joined

V_b = Barrier potential develops as both sides of the junction are depleted of holes and electrons.

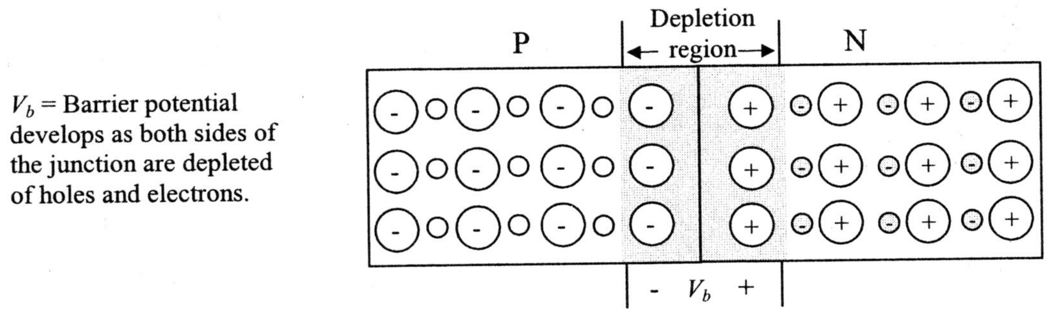

Figure 1-11: The *p-n* junction after recombination of electron-hole pairs

Figure 1-11 shows the formation of the *p-n* junction after the first set of electrons have diffused and recombined with the first set of available holes across the junction. The recombination of these electron-hole pairs leaves behind a *depletion region* at both sides of the junction, where a portion of the *n*-type semiconductor is depleted of electrons and a portion of the *p*-type semiconductor is depleted of holes. Also note that as a result of the recombination process, the donor atoms lose their associated electrons and the acceptor atoms lose their associated holes and thus become positive and negative ions, respectively. The accumulation of these positive and negative ions builds up a potential difference across the junction known as *barrier potential* (V_b), which prohibits the flow of further electrons. The barrier potential is so named because it acts as a barrier to the flow of further diffusion current. The magnitude of the barrier potential depends on the doping levels of the *p* and *n* materials, the type of material (*Si* or *Ge*), and the temperature. For silicon diodes the barrier potential at room temperature (25°C) is approximately 0.7 V, and for germanium diodes it is approximately 0.3 V. This potential, however, cannot be measured unless an appropriate external source is applied to forward bias the *p-n* junction.

Forward-Bias

Consider the following *p-n* junction, which is biased by the external DC source *E*. To forward-bias the *p-n* junction, the positive terminal of the voltage source is connected to the *p*-type material and the negative terminal is connected to the *n*-type material, as shown in Figure 1-12 below:

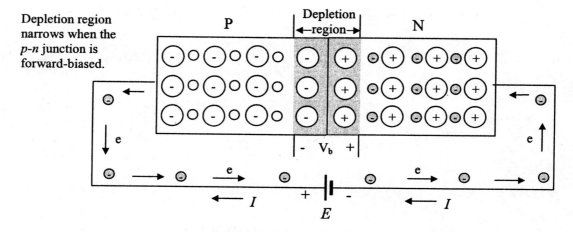

Figure 1-12: Forward biasing the *p-n* junction with an external source

Since like charges repel each other, the negatively charged electrons in the *n*-type semiconductor are repelled by the negative terminal of the voltage source and the holes in the *p*-type are repelled by the positive terminal toward the junction. As a result of this repulsion of charge carriers, the depletion region narrows and electrons are forced to diffuse across the junction and recombine with holes. For every electron that recombines with a hole, an electron leaves the covalent bond in the *p*-region and enters the positive terminal of the source, thus the amount of current (the number of electrons in a given time) entering and leaving the external source is always equal. Recall that the direction of the conventional current flow (*I*) is considered opposite to the electron flow, hence the current *I* leaves the positive terminal of the DC source and returns to the negative terminal, as shown in Figure 1-12 above.

Reverse-Bias

Consider the following *p-n* junction, which is biased by the external DC source *E*, but with an opposite polarity compared to forward-bias. That is, the positive terminal of the voltage source is connected to the *n*-type material and the negative terminal is connected to the *p*-type material, as shown in Figure 1-13 below:

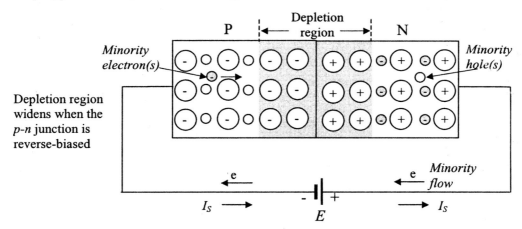

Figure 1-13: Reverse biasing the *p-n* junction with an external source

In order to concentrate on the main aspects of the *p-n* junction and to minimize the confusion, thus far we have ignored the presence of the minority carriers. However, there are generally very few electrons present in the *p*-type semiconductor, in addition to the abundantly present holes. Similarly, there are very few holes in the *n*-type semiconductor, in addition to the abundantly present *majority carrier* electrons. The *minority carriers*, electrons in the *p*-type semiconductor, are referred to as the *minority electrons*, and the holes in the *n*-type are called *minority holes*. When the *p-n* junction is reverse-biased, the negatively charged majority electrons in the *n*-type semiconductor are attracted by the positive terminal of the voltage source and the holes in the *p*-type are attracted by the negative terminal away from the junction. As a result of this attraction of majority charge carriers, the depletion region widens, and thus, majority electrons cannot diffuse across the junction and recombine with holes. In other words, reverse biasing inhibits the flow of majority carriers in a *p-n* junction. The minority carriers, which are actually forward-biased, establish a minority current flow that is much smaller than the majority current flow under forward-bias and is referred to as the *reverse saturation current* I_S, also denoted in some diode data sheets as I_R.

Reverse Breakdown

If the source voltage *E* is increased in the reverse-bias circuit of Figure 1-13 above, a point will be reached that the *p-n* junction will allow a substantial current flow. The voltage at which this reverse current flow occurs is called the *reverse breakdown voltage* V_{BR}.

1.4 THE *P-N* JUNCTION DIODE

One practical application of the semiconductor *p-n* junction is the *p-n* junction diode. The semiconductor diode functions like a unidirectional switch for the current flow; that is, it behaves like a closed switch when forward-biased and like an open switch when reverse-biased. The typical appearance, terminal identification, and electronic symbol for the *p-n* junction diode are shown in Figure 1-14(a) below:

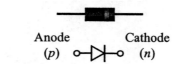

(a) Diode terminal identification and symbol

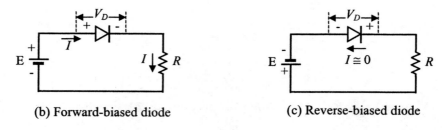

(b) Forward-biased diode   (c) Reverse-biased diode

Figure 1-14: Diode symbol and two different bias circuits

Figure 1-14(b) above shows a forward-biased diode circuit in which the positive terminal of the voltage source (E) is connected to the anode or *p* region of the *p-n* junction diode. As mentioned previously, the barrier potential or the forward-biased diode voltage V_D, which is generally referred to as V_F, is nominally 0.7 V for a silicon diode and 0.3 V for a germanium diode. The reverse breakdown voltage V_{BR} is usually quite high (hundreds of volts) for general-purpose diodes; however, special *Zener* diodes are designed to have a V_{BR} of a few volts. Despite the name *breakdown*, the diode is not necessarily damaged at the breakdown voltage; however, like any other electronic device, the diode is susceptible to damage by overheating as a result of excessive current flow. Hence, a current limiting resistor must be placed in series with the diode (forward-biased or reverse-biased) to avoid excessive current and overheating. The maximum diode current or power is generally given in diode data sheets (included at the end of the chapter). The reverse saturation current I_S is usually very small. For example, it is 50 nA for the 1N4001 silicon diode; however, it can be a few μA for germanium diodes. This is one of the reasons that germanium diodes are not as popular and widely used as their silicon counterparts.

Diode I-V Characteristics

Assume that the forward-bias circuit of Figure 1-14(b) is set up in the lab. The diode current I_D and diode voltage V_D are measured and recorded as the source voltage E is increased beyond V_F. Then the same procedure is repeated with the reverse-bias circuit of Figure 1-14(c) while the voltage E is increased beyond V_{BR}. The plot of the diode current I_D versus diode voltage V_D, which is called diode *I-V* characteristic, will look like the plot shown in Figure 1-15.

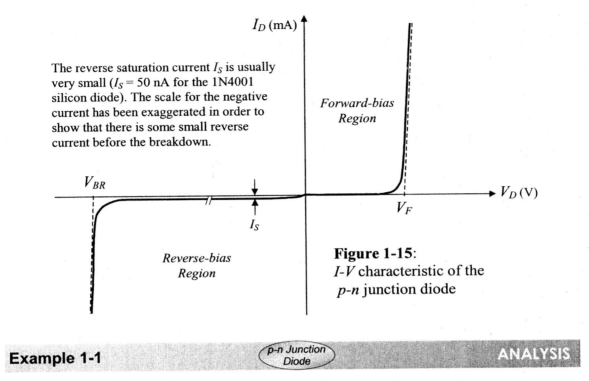

The reverse saturation current I_S is usually very small ($I_S = 50$ nA for the 1N4001 silicon diode). The scale for the negative current has been exaggerated in order to show that there is some small reverse current before the breakdown.

Figure 1-15: I-V characteristic of the p-n junction diode

Example 1-1 — p-n Junction Diode — ANALYSIS

Let us determine the diode current I_D and the diode voltage V_D for the forward-biased circuit of Figure 1-14(b), assuming that the diode is silicon, $E = 10$ V, and $R = 1$ kΩ.

Solution:
The amount of current that flows through the forward-biased diode depends on the magnitude of the source voltage E, the type of the diode (Si, Ge), and the value of the resistor R. That is, considering that the diode is silicon, there will be a 0.7 V drop across the diode ($V_D = V_F = 0.7$ V), and the remainder of the source voltage ($E - V_F$) will be across the resistor R. Since the circuit is a simple series one, current through the diode equals the current through the resistor; hence, the diode current can be determined by applying Ohm's law as follows:

$$I_D = \frac{E - V_F}{R} \qquad (1\text{-}1)$$

$$I_D = \frac{10\text{ V} - 0.7\text{ V}}{1\text{ k}\Omega} = 9.3\text{ mA}$$

Example 1-2 — p-n Junction Diode — ANALYSIS

Let us now determine the diode current I_D and the diode voltage V_D for the reverse-biased circuit of Figure 1-14(c), assuming that:
    (a) The diode $V_{BR} = 90$ V, $E = 10$ V, and $R = 1$ kΩ.
    (b) The diode $V_{BR} = 90$ V, $E = 100$ V, and $R = 1$ kΩ.

Solution:

(a) Since $E < V_{BR}$, the reverse-biased diode behaves like an open-circuit; hence the only current flow is the reverse saturation current I_S, which is very small, a few nA for a silicon diode and a few μA for a germanium diode.

$$I_D = I_S \cong 0$$

$$V_D = E = -10 \text{ V}$$

(b) Since $E > V_{BR}$, reverse breakdown occurs, V_{BR} drops across the diode and the remainder of the source voltage $(E - V_{BR})$ drops across the resistor R.

$$V_D = V_{BR} \cong -90$$

$$I_D = \frac{E - V_{BR}}{R} \tag{1-2}$$

$$I_D = \frac{100 \text{ V} - 90 \text{ V}}{1 \text{ k}\Omega} = 10 \text{ mA}$$

Temperature Effects

The change in temperature can have a significant effect on the diode characteristics, as shown in Figure 1-16. As the temperature increases, the forward-bias voltage V_F of the p-n junction decreases, and as a result, the I-V characteristic curve in the forward-bias region moves to the left. In fact, for every 1°C increase in temperature V_F decreases by approximately 2 mV.

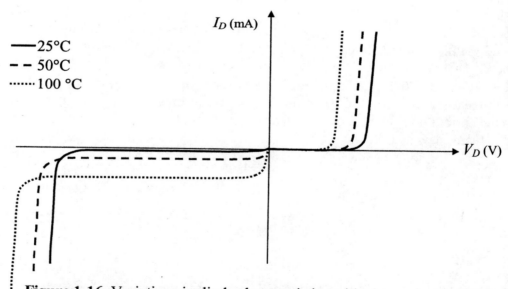

Figure 1-16: Variations in diode characteristics with change in temperature

Practice Problem 1-1 — p-n Junction Diode — ANALYSIS

Determine the diode current I_D and the diode voltage V_D for the diode circuits of Figure 1-17, assuming that the diode is silicon and $V_{BR} = 60$ V.

Figure 1-17: Diode bias circuits

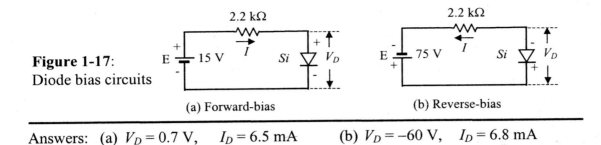

(a) Forward-bias   (b) Reverse-bias

Answers: (a) $V_D = 0.7$ V, $I_D = 6.5$ mA   (b) $V_D = -60$ V, $I_D = 6.8$ mA

The Diode Equation

General characteristics of the *p-n* junction can be described by the following equation known as the *diode equation*, which is valid at both forward- and reverse-bias regions.

$$I_D = I_S(e^{kV_D/T} - 1) \qquad (1\text{-}3)$$

I_S = *reverse saturation current*
$k = 11,600/\eta$
V_D = *diode voltage* (positive for forward-bias and negative for reverse-bias)
η = *emission coefficient*, $\eta = 1$ for *Ge* and *Si* diodes at the rapidly increasing sections of the curve (relatively higher diode current)
T = *temperature* in °K = °C + 273°

Expanding the diode equation results in the following:

$$I_D = I_S e^{kV_D/T} - I_S \qquad (1\text{-}4)$$

In the forward-bias region where V_D is positive, the first part of the equation grows, rapidly nullifying the effect of the second term, which is very small in magnitude. In the reverse-bias region (up to the breakdown) where V_D is negative, the first term decreases rapidly, resulting in $I_D = -I_S$, which is a straight line at the I_S level. The plot of the *diode equation* for a silicon diode is exhibited in Figure 1-18 ($I_S = 10$ nA).

Note that the plot of the *diode equation* shows $V_F \cong 0.5$ V. However, the characteristics curve of the commercially available silicon diode is shifted to the right by about 0.2 V, resulting in $V_F \cong 0.7$ V. This additional voltage drop is due to the internal *DC* resistance and external *contact* resistance of the actual diode.

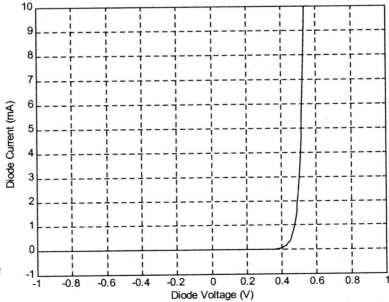

Figure 1-18:
Plot of the *diode equation* with MATLAB

DC and AC Resistances of the Diode

DC Resistance R_D:

The *DC resistance* of a diode, which is also referred to as *static resistance*, is determined simply by direct application of Ohm's law. The diode DC resistance is the ratio of the diode DC voltage to the diode DC current

$$R_D = \frac{V_D}{I_D} \qquad (1\text{-}5)$$

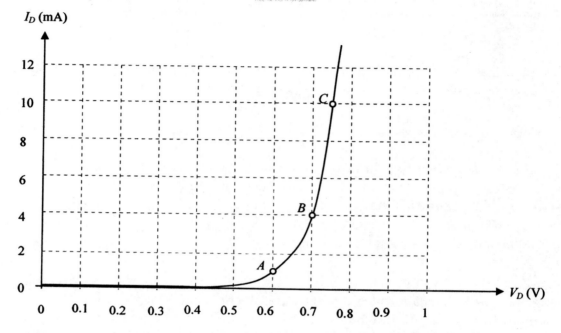

Figure 1-19: Diode *I-V* characteristic in the forward-bias region

Given the diode characteristic shown in Figure 1-19, let us determine the diode DC resistance at the operating points *A*, *B*, and *C*. The diode current I_D at point *A* is 1 mA, the diode voltage V_D is 0.6 V, and the diode DC resistance R_D is determined as follows:

$$R_D = \frac{V_D}{I_D} = \frac{0.6\,\text{V}}{1\,\text{mA}} = 600\,\Omega$$

The diode current I_D at point *B* is 4 mA, the diode voltage V_D is 0.7 V, and the diode DC resistance R_D is calculated to be

$$R_D = \frac{V_D}{I_D} = \frac{0.7\,\text{V}}{4\,\text{mA}} = 175\,\Omega$$

The diode current I_D at point *C* is 10 mA, the diode voltage V_D is 0.75 V, and the diode DC resistance R_D is

$$R_D = \frac{V_D}{I_D} = \frac{0.75\,\text{V}}{10\,\text{mA}} = 75\,\Omega$$

Hence, the DC resistance drops from 600 Ω to 75 Ω as the current rises from 1 mA to 10 mA. The change in the diode voltage is small, especially in the linear region of the curve

($V_D \geq 0.7$ V); however, the change in the diode current, which can be significant, has the most influence in determining the DC resistance of the diode. The higher the diode current, the lower the diode DC resistance.

Small-Signal (AC) Resistance r_d:

The *small-signal* (AC) or the *dynamic resistance* of a diode is defined as the ratio of the change in diode voltage ΔV_D to the change in diode current ΔI_D, which is the reciprocal of the slope of the diode *I-V* characteristic at a given operating DC diode current I_D

$$r_d = \frac{\Delta V_D}{\Delta I_D} \qquad (1\text{-}6)$$

Let us determine the AC resistance of the diode at the same operating points *A*, *B*, and *C*. The slope of the curve at each point is exhibited in Figure 1-20.

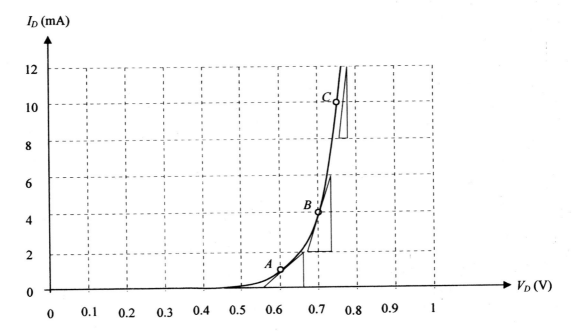

Figure 1-20: Diode *I-V* characteristics in the forward-bias region

The change in current at point *A* is 2 mA, the change in voltage is approximately 0.1 V, and the small-signal resistance is calculated to be

$$r_d = \frac{\Delta V_D}{\Delta I_D} = \frac{0.1 \text{ V}}{2 \text{ mA}} = 50 \, \Omega$$

The change in current at point *B* is 4 mA, the change in voltage is approximately 0.05 V, and the small-signal resistance is given by

$$r_d = \frac{\Delta V_D}{\Delta I_D} = \frac{0.05 \text{ V}}{4 \text{ mA}} = 12.5 \, \Omega$$

The change in current at point C is 4 mA, the change in voltage is approximately 0.025 V, and the small-signal resistance is determined as follows:

$$r_d = \frac{\Delta V_D}{\Delta I_D} = \frac{0.025 \text{ V}}{4 \text{ mA}} = 6.25 \, \Omega$$

The graphical determination of the dynamic resistance of the diode is not a practical one because drawing the tangent line at the operating point is an approximation and the diode characteristics may not always be available. There is, however, a more convenient approach, which is derived from the *diode equation*. Recall that the dynamic resistance r_d of the *p-n* junction diode is the reciprocal of the slope of the curve $\Delta V_D/\Delta I_D$ at the operating point. Furthermore, recall from your differential calculus that the slope of the curve at any point is the instantaneous rate of change or derivative, which is, in this case, the derivative of the diode voltage with respect to the diode current. Since the diode equation states the diode current as a function of the diode voltage, we will take the derivative of the diode current with respect to the diode voltage, then after some simplification, will invert the result. Thus,

$$I_D = I_S e^{V_D/V_T} - I_S \tag{1-7}$$

$$\frac{dI_D}{dV_D} = \frac{I_S e^{V_D/V_T}}{V_T} \tag{1-8}$$

Also, note that solving for $I_D + I_S$ in Equation 1-7 results in the following:

$$I_S e^{V_D/V_T} = I_D + I_S \tag{1-9}$$

Substituting for $I_S e^{V_D/V_T}$ in Equation 1-8, we obtain the diode dynamic conductance

$$g_d = \frac{dI_D}{dV_D} = \frac{I_D + I_S}{V_T} \cong \frac{I_D}{V_T} \tag{1-10}$$

At room temperature $T_K = 25°C + 273° = 298°K$, the diode small-signal conductance is

$$g_d = \frac{dI_D}{dV_D} = \frac{I_D}{V_T} = \frac{qI_D}{kT} = \frac{11,600}{298} I_D = 38.926 I_D \tag{1-11}$$

The diode small-signal or dynamic resistance is described by

$$r_d = \frac{dV_D}{dI_D} = \frac{kT}{qI_D} = \frac{V_T}{I_D} = \frac{0.0257}{I_D} \tag{1-12}$$

$$r_d = \frac{V_T}{I_D} \cong \frac{26 \text{ mV}}{I_D} \tag{1-13}$$

Practice Problem 1-2 ANALYSIS

Determine the DC resistance of the diode at points A, B, and C. The diode I-V characteristic is shown in Figure 1-21. Also, determine the small-signal resistance of the diode at points A, B, and C using Equation 1-13.

Chapter 1

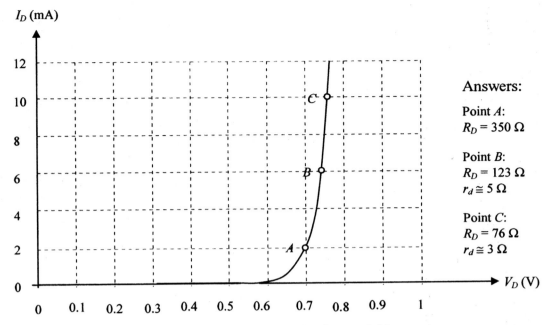

Answers:

Point A:
$R_D = 350\ \Omega$

Point B:
$R_D = 123\ \Omega$
$r_d \cong 5\ \Omega$

Point C:
$R_D = 76\ \Omega$
$r_d \cong 3\ \Omega$

Figure 1-21: Diode *I-V* characteristic in the forward-bias region

Example 1-3 *p-n Junction Diode* **ANALYSIS**

Determine the following for diode circuits of Figure 1-22, assuming that the diode is silicon with $V_F = 0.7$ V, $V_{BR} = 75$ V, and $I_S = 10$ nA at 25°C.
(a) Diode voltage V_D for the circuit of Figure 1-22(a) at 75°C.
(b) Diode current I_D for the circuit of Figure 1-22(b) at 75°C.

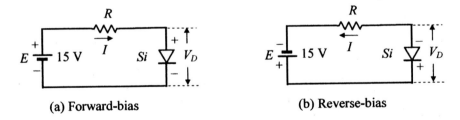

(a) Forward-bias (b) Reverse-bias

Solution: **Figure 1-22**: Diode bias circuits

(a) For every 1°C increase in temperature, V_F decreases by approximately 2 mV.

$$\frac{\Delta V_D}{\Delta T} = -\frac{2\ \text{mV}}{1°C} \tag{1-14}$$

$$\Delta V_D = -\frac{2\ \text{mV}}{1°C} \times \Delta T = -\frac{2\ \text{mV}}{1°C}(T_2 - T_1)°C = -\frac{2\ \text{mV}}{1°C}(75 - 25)°C = -100\ \text{mV}$$

$$V_D = 700\ \text{mV} - 100\ \text{mV} = 0.60\ \text{V}$$

(b) The reverse saturation current I_S doubles for every 10°C increase in temperature.

$$\frac{T_2 - T_1}{10°C} = \frac{75°C - 25°C}{10°C} = 5$$

Thus, I_S doubles 5 times and becomes at $T_2 = 75°C$

$$I_S(T_2) = I_S(T_1) \times 2^{\frac{T_2 - T_1}{10}} = 10 \text{ nA} \times 2^5 = 320 \text{ nA}$$

$$I_D = -I_S = -320 \text{ nA}$$

Practice Problem 1-3 ANALYSIS

Determine the diode voltage V_D for the circuit of Figure 1-22(a) and the diode current I_D for the circuit of Figure 1-22(b) at 50°C, assuming that the diode is germanium with $V_F = 0.3$ V and $I_S = 1$ µA at 25°C.

Answers: (a) $V_D = 0.25$ V   (b) $I_D = -5.656$ µA

Basic Diode Circuits

Having learned the fundamentals of the *p-n* junction diode, we will now present a few examples involving the analysis of basic diode circuits.

Example 1-4 ANALYSIS

Let us now determine the diode current I_D and the diode voltage V_D for the two circuits of Figure 1-23.

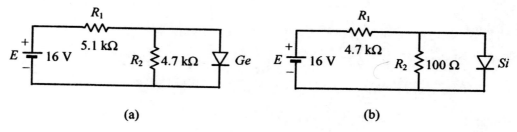

Solution: **Figure 1-23**: Diode bias circuits

We can convert the above circuits to a simple series one by Thevenizing the circuit across the diode, as shown in Figure 1-24.

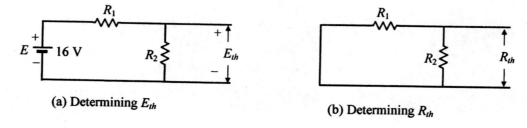

Figure 1-24: Thevenin's equivalent circuits for determining E_{th} and R_{th}

Circuit of Figure 1-23(a):

Thevenin's equivalent voltage source E_{th} is the voltage across R_2, which is determined simply by applying the voltage divider rule

$$E_{th} = E\frac{R_2}{R_1 + R_2} = 16\,\text{V}\,\frac{4.7\,\text{k}\Omega}{5.1\,\text{k}\Omega + 4.7\,\text{k}\Omega} = 7.67\,\text{V}$$

Thevenin's equivalent resistance R_{th} is the parallel combination of R_1 and R_2 given by

$$R_{th} = \frac{R_1 R_2}{R_1 + R_2} = \frac{5.1\,\text{k}\Omega \times 4.7\,\text{k}\Omega}{5.1\,\text{k}\Omega + 4.7\,\text{k}\Omega} = 2.45\,\text{k}\Omega$$

Inserting the diode back into the circuit, Thevenin's equivalent circuit is shown in Figure 1-25.

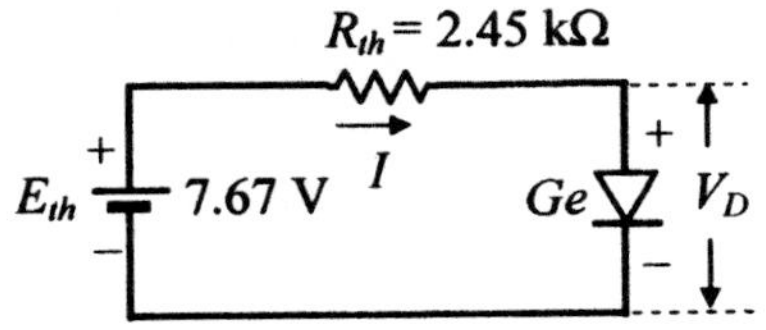

Figure 1-25: Thevenin's equivalent circuit of Figure 1-23(a)

$$V_D = V_F = 0.3\,\text{V}$$

$$I_D = \frac{E_{th} - V_F}{R_{th}} = \frac{7.67\,\text{V} - 0.3\,\text{V}}{2.45\,\text{k}\Omega} = 3\,\text{mA}$$

Circuit of Figure 1-23(b):

$$E_{th} = E\frac{R_2}{R_1 + R_2} = 16\,\text{V}\,\frac{100\,\Omega}{4.7\,\text{k}\Omega + 100\,\Omega} = 0.333\,\text{V}$$

$$R_{th} = \frac{R_1 R_2}{R_1 + R_2} = \frac{4.7\,\text{k}\Omega \times 100\,\Omega}{4.7\,\text{k}\Omega + 100\,\Omega} = 98\,\Omega$$

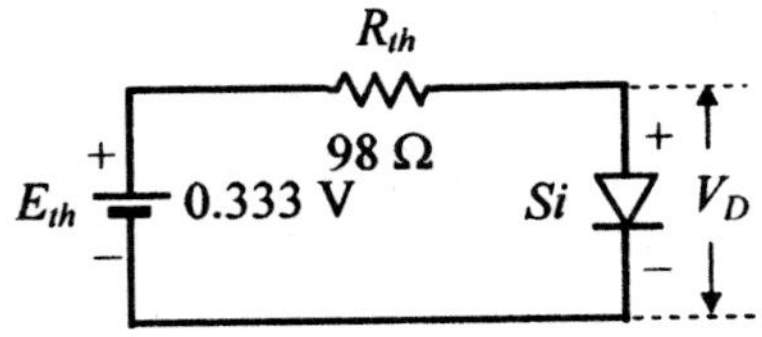

Figure 1-26: Thevenin's equivalent circuit of Figure 1-23(b)

$$V_D = 0.333\,\text{V} < 0.7\,\text{V}$$
$$I_D = 0\,\text{mA}$$

Note that when $E_{th} \leq V_F$, there will be no voltage drop across the resistor R_{th}, and thus, no current flows through the diode ($I_D = 0$). The diode is neither forward-biased nor reverse-biased; it is unbiased (equivalent to an open circuit), and V_D equals E_{th} ($V_D = E_{th}$).

Practice Problem 1-4 ANALYSIS

Determine the diode voltage V_D and diode current I_D for the circuit of Figure 1-23:
(a) Assume that the diode is silicon with $E = 15$ V, $R_1 = 5.1$ kΩ, and $R_2 = 7.5$ kΩ.
(b) Assume that the diode is silicon with $E = 15$ V, $R_1 = 15$ kΩ, and $R_2 = 510$ Ω.

Answers: (a) $V_D = 0.7$ V, $I_D = 2.71$ mA   (b) $V_D = 0.49$ V, $I_D = 0$ mA

Example 1-5 ANALYSIS

Determine the diode current I_D and the output voltage V_O for the two circuits (a) and (b) of Figure 1-27.

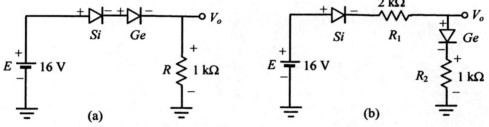

Solution:

Figure 1-27: Diode bias circuits

Both circuits are simple series circuits with both diodes forward-biased. Hence, the voltage drop across both diodes is a total of 1 V.

Circuit of Figure 1-27(a):

$$V_o = 16 \text{ V} - 0.7 \text{ V} - 0.3 \text{ V} = 15 \text{ V}$$
$$I_D = I_R = \frac{V_o}{R} = \frac{15 \text{ V}}{1 \text{ k}\Omega} = 15 \text{ mA}$$

Circuit of Figure 1-27(b):

$$I_D = \frac{E - (0.7 \text{ V} + 0.3 \text{ V})}{R_1 + R_2} = \frac{15 \text{ V}}{3 \text{ k}\Omega} = 5 \text{ mA}$$
$$V_o = I_D R_2 + 0.3 \text{ V} = (5 \text{ mA} \times 1 \text{ k}\Omega) + 0.3 \text{ V} = 5.3 \text{ V}$$

Practice Problem 1-5 ANALYSIS

Determine the output voltage V_o and the diode current I_D for the circuits of Figure 1-28.

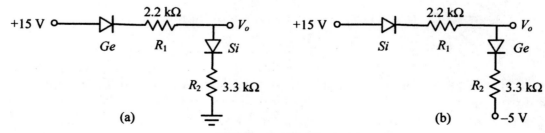

Figure 1-28: Diode bias circuits

Answers: (a) $V_o = 9.1$ V, $I_D = 2.545$ mA   (b) $V_o = 6.7$ V, $I_D = 3.4545$ mA

Diode Switching Circuits

It is possible to design simple switching circuits such as *AND* gates and *OR* gates with two diodes and a resistor, although the integrated circuit logic gates are made of diodes and transistors.

Example 1-6 — Diode Circuits — ANALYSIS

Let us now determine the output voltage levels V_{o1} and V_{o2} for the two circuits of Figure 1-29 and verify that they are indeed logic gates. Assume that the diodes are silicon.

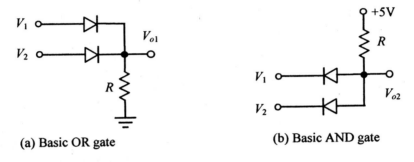

(a) Basic OR gate   (b) Basic AND gate

Figure 1-29: Diode logic circuits

Table 1-4: Logic Table for Circuits of Figure 1-29(a) and (b)

V_1		V_2		V_{o1}		V_{o2}	
0 V	L	0 V	L	0 V	L	0.7 V	L
0 V	L	5 V	H	4.3 V	H	0.7 V	L
5 V	H	0 V	L	4.3 V	H	0.7 V	L
5 V	H	5 V	H	4.3 V	H	5 V	H

Circuit of Figure 1-29(a):

With both inputs V_1 and V_2 set to 0 V (*low*), both diodes are unbiased (*open*), resulting in a *low* output.

$$V_{o1} = 0 \text{ V } (low)$$

With one of the inputs V_1 or V_2 set to 0 V (*low*), and the other set to 5 V (*high*), the diode with 5 V input will be forward-biased and the other will be unbiased resulting in a *high* output:

$$V_{o1} = 5 \text{ V} - 0.7 \text{ V} = 4.3 \text{ V } (high)$$

With both inputs V_1 and V_2 set to 5 V (*high*), both diodes are forward-biased resulting again in a *high* output given by

$$V_{o1} = 5\text{ V} - 0.7\text{ V} = 4.3\text{ V } (high)$$

Hence, the output is *high* when either *or* both inputs are *high*. Thus, the circuit is an *OR* gate.

<u>Circuit of Figure 1-29(b):</u>

With both inputs V_1 and V_2 set to 0 V (*low*), both diodes are forward-biased with 0.7 V dropped across each diode. The output, which is measured from the anode of either diode to ground, is

$$V_{o2} = 0.7\text{ V } (low)$$

With one of the inputs V_1 or V_2 set to 0 V (*low*), and the other set to 5 V (*high*), the diode with 0 V input will be forward-biased and the other will be unbiased. The output is measured from the anode of the forward-biased diode to ground and is

$$V_{o2} = 0.7\text{ V } (low)$$

With both inputs V_1 and V_2 set to 5 V (*high*), both diodes are unbiased (*open*) resulting in a *high* output

$$V_{o2} = 5\text{ V } (high)$$

Hence, the output is *high* only when both inputs V_1 and V_2 are *high*. Thus, the circuit is an *AND* gate. The above results have been entered in Table 1-4 for both V_{o1} and V_{o2}.

Practice Problem 1-6 — Diode Circuits — **ANALYSIS**

Draw the circuit diagrams of a three-input *AND* gate and a three-input *OR* gate, and show the truth table with inputs V_1, V_2, and V_3, and outputs V_{o1} and V_{o2}, similar to Table 1-4.

1.5 ZENER DIODE

Zener diode is a *p-n* junction diode that is designed for specific *reverse breakdown* voltage. For *Zener* diodes, the reverse breakdown voltage is much smaller in magnitude than the reverse breakdown voltage for regular diodes and is referred to as the *nominal Zener voltage* V_Z. The breakdown voltage level V_Z is controlled by the doping levels. Increasing the doping, that is, increasing the number of added impurities, decreases the breakdown voltage V_Z. Zener diodes are operated only in the reverse-bias region and are mainly used as constant DC voltage sources or reference voltage sources. Some practical applications of regular diodes and Zener diodes are presented in the following chapter. The symbol and terminal identification of the Zener diode and the *I-V* characteristic of a 3.3 V Zener diode are displayed in Figure 1-30.

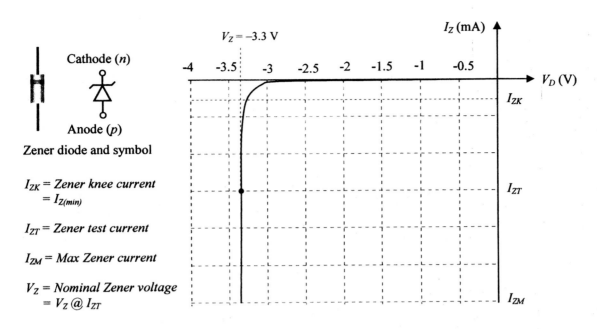

Figure 1-30: *I-V* characteristic of a 3.3 V Zener diode

Four of the main characteristics of the Zener diode are I_{ZK}, I_{ZT}, I_{ZM}, and V_Z. I_{ZK} is the designation for the *knee current*, which is the minimum Zener current to keep the diode operating in the Zener region (breakdown region). Note that at or near I_{ZK}, the Zener voltage is less than the nominal Zener voltage V_Z. I_{ZT} is the notation used for the *Zener test current*, which corresponds to the rated nominal Zener voltage V_Z. I_{ZM} is the designation for the *maximum Zener diode current*, which may not be exceeded. V_Z is the notation used for the rated nominal Zener voltage at I_{ZT}. Another important characteristic of the Zener diode is its *AC* or dynamic resistance r_z, which is generally given in data sheets for two current levels I_{ZT} and I_{ZK}, labeled in some data sheets as z_{ZT} and z_{ZK}. Since the dynamic resistance r_z is directly proportional to the slope of the characteristic, which is almost vertical, r_z is approximately constant around I_{ZT}.

For the 1N5226B Zener diode, for example, $V_Z = 3.3$ V, $I_{ZK} = 0.25$ mA, $I_{ZT} = 20$ mA, $I_{ZM} = 151.5$ mA, r_z at I_{ZT} is 28 Ω, and r_z at I_{ZK} is 1600 Ω. The maximum Zener diode current I_{ZM} is determined from the given power rating of $P_{D(max)} = 500$ mW, as follows:

$$P_{D(max)} = V_Z I_{ZM} \qquad (1\text{-}15)$$

$$I_{ZM} = \frac{P_{D(max)}}{V_Z} = \frac{500 \text{ mW}}{3.3 \text{ V}} = 151.5 \text{ mA}$$

Note that the Zener diode data sheets are included at the end of the chapter.

Example 1-7 — Zener Diode — ANALYSIS

Let us determine V_Z and I_Z for the circuit of Figure 1-31.

Figure 1-31: Basic Zener diode bias circuit

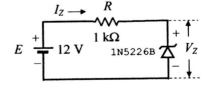

Solution:

Since $V_Z = 3.3$ V for 1N5226B, and E is higher than V_Z, the Zener diode is operating in the Zener region (breakdown region). Hence, 3.3 V is dropped across the Zener diode and the remaining source voltage appears across the resistor R. Thus,

$$I_Z = \frac{V_R}{R} = \frac{E - V_Z}{R} \tag{1-16}$$

$$I_Z = \frac{12\,\text{V} - 3.3\,\text{V}}{1\,\text{k}\Omega} = 8.7\,\text{mA}$$

Practice Problem 1-7 — Zener Diode — ANALYSIS

Determine I_Z for the circuit of Figure 1-31 if $E = 20$ V and the Zener is 1N5240B. Also, determine I_{ZK}, I_{ZM}, and r_z @ I_{ZT} for this Zener diode.

Answers: $I_Z = 10$ mA, $I_{ZK} = 0.25$ mA, $I_{ZM} = 50$ mA, and $r_z = 17\,\Omega$

Example 1-8 — Zener Circuits — ANALYSIS

Let us determine the Zener diode current I_Z and the output voltage V_o for the two circuits of Figure 1-32. Assume the Zener diode is a 1N5231B with $V_Z = 5.1$ V and $I_{ZM} = 98$ mA.

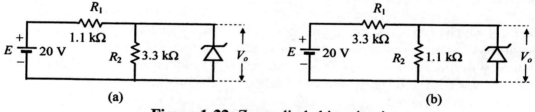

Figure 1-32: Zener diode bias circuits

Solution:

As we have done in Example 1-4, we will convert the above circuits to a simple series one by Thevenizing the circuit across the diode, as shown in Figure 1-33.

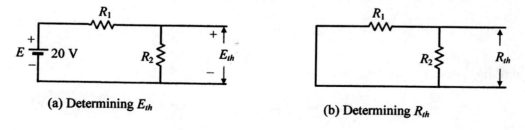

Figure 1-33: Thevenin's equivalent circuits for determining E_{th} and R_{th}

Circuit of Figure 1-32(a):

Circuit of Figure 1-32(a):

Thevenin's equivalent voltage source E_{th} is the voltage across R_2, which is determined simply by applying the voltage divider rule

$$E_{th} = E\frac{R_2}{R_1 + R_2} = 20\text{ V}\frac{3.3\text{ k}\Omega}{1.1\text{ k}\Omega + 3.3\text{ k}\Omega} = 15\text{ V}$$

Thevenin's equivalent resistance R_{th} is the parallel combination of R_1 and R_2

$$R_{th} = \frac{R_1 R_2}{R_1 + R_2} = \frac{1.1\text{ k}\Omega \times 3.3\text{ k}\Omega}{1.1\text{ k}\Omega + 3.3\text{ k}\Omega} = 825\text{ }\Omega$$

Inserting the diode back into the circuit, Thevenin's equivalent circuit is obtained.

Figure 1-34:
Thevenin's equivalent circuit of Figure 1-32(a)

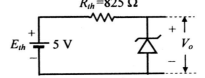

$$V_o = V_Z = 5.1\text{ V}$$

$$I_Z = \frac{E_{th} - V_Z}{R_{th}} = \frac{15\text{ V} - 5.1\text{ V}}{825\text{ }\Omega} = 12\text{ mA}$$

Circuit of Figure 1-32(b): $E_{th} = 5$ V, $R_{th} = 825$ Ω

Figure 1-35:
Thevenin's equivalent circuit of Figure 1-32(b)

Since $E_{th} < V_Z$, the diode is reverse biased, but is not operating in the Zener region (breakdown region). Hence, the Zener diode is approximately equivalent to an open circuit, and thus, no significant current flows. Thus,

$$V_o = E_{th} = 5\text{ V}$$

$$I_Z \cong 0$$

Practice Problem 1-8 — Zener Circuits — ANALYSIS

Determine the output voltage V_o and the Zener diode current I_Z for the two circuits of Figure 1-36 (a) and (b). The Zener diode is 1N5236B.

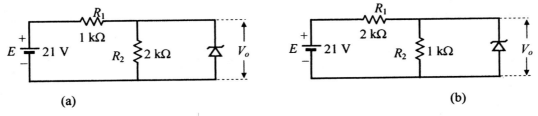

Figure 1-36: Zener diode bias circuits

Answers: (a) $V_o = 7.5$ V, $I_Z = 9.75$ mA (b) $V_o = 7$ V, $I_Z \cong 0$ mA

1.6 LIGHT-EMITTING DIODE

Recall that when a *p-n* junction is forward-biased, free electrons in the *n* side cross the junction and recombine with the holes in the *p* side. Since the migrating free electrons are in the *conduction* band with higher energy level than the holes in the *valence* band with lower energy level, electrons release some energy during the recombination process. For silicon and germanium most of this released energy is in the form of heat. However, in other semiconductor materials, such as *gallium arsenide* and *gallium phosphide*, most of this energy is released in the form of light. If the semiconductor material is translucent, the emitted light is visible, and the *p-n* junction is called a *light-emitting diode* (*LED*). This process of converting electrical energy to light energy is called *electroluminescence*. The forward voltage drop V_F for an LED is approximately 2 V, although it can be as high as 3 V, depending on the type and forward current. For adequate brightness of emitted light, an LED requires between 10 to 20 mA of current. The terminal identification and the symbol of an LED are shown in Figure 1-37(a) below. Data sheets for the TLHK4200 LED are included at the end of the chapter.

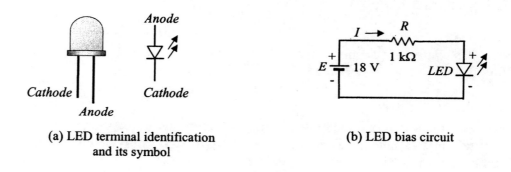

(a) LED terminal identification and its symbol

(b) LED bias circuit

Figure 1-37: LED symbol and bias circuit

Example 1-9 — LED Bias Circuit — ANALYSIS

Let us determine the LED current (I_{LED}) for the circuit of Figure 1-37 above.

Solution:

Assuming the typical 2 V drop across the LED, the rest of the source voltage will drop across the resistor *R*.

$$I_{LED} = I_R = \frac{E - V_F}{R}$$

$$I_{LED} = \frac{18\,\text{V} - 2\,\text{V}}{1\,\text{k}\Omega} = 16\,\text{mA}$$

Practice Problem 1-9 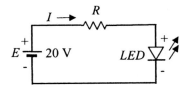 DESIGN

Determine the value of the resistor R for the circuit of Figure 1-38 below so that the LED current is approximately 20 mA for full brightness. Assume $V_F = 2$ V, and use the commercially available standard value resistor.

Figure 1-38: LED bias circuit (LED driver)

Answer: $R = 910\,\Omega$

Practice Problem 1-10 LED Bias Circuit DESIGN

Determine the value of the resistor R for the circuit of Figure 1-39 below so that the LED current is approximately 20 mA when ON. Assume $V_F = 2$ V, and use the commercially available standard value resistor. Also construct a truth table showing when the LED is ON. Assume that $V_{1(H)} = V_{2(H)} = 5.7$ V, and $V_{1(L)} = V_{2(L)} = 0$ V.

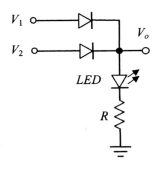

Figure 1-39: Diode logic circuit with LED

Answer:

$R = 150\,\Omega$

Example 1-10 ANALYSIS

With reference to the data sheets at the end of the chapter, let us now determine the following specifications for the TLHK4200 LED:
a) Relative luminous intensity (RLI) at an angle θ of 0° and 30°.
b) Relative luminous intensity (RLI) at forward current $I_F = 10$ mA and 5 mA.
c) Forward voltage drop V_F at $I_F = 20$ mA and 5 mA.

Solution:

a) With reference to Figure 3 of the data sheets for the TLHK4200 LED, the following observations are made: RLI = 1 at $\theta = 0°$ and RLI = 0.35 at $\theta = 30°$.
b) With reference to Figure 7 of the data sheets for the TLHK4200 LED, the following observations are made: RLI = 1 at $I_F = 10$ mA and RLI = 0.5 at $I_F = 5$ mA.
c) With reference to Figure 4 of the data sheets for the TLHK4200 LED, the following observations are made: $V_F = 2$ V at $I_F = 20$ mA and $V_F = 1.7$ V at $I_F = 5$ mA.

Practice Problem 1-11 ANALYSIS

With reference to the data sheets at the end of the chapter, determine the following specifications for the TLHK4200 LED:
a) Relative luminous intensity (RLI) at an angle θ of 40° and 90°.
b) Relative luminous intensity (RLI) at forward current $I_F = 20$ mA and 50 mA.
c) Forward voltage drop V_F at $I_F = 10$ mA and 50 mA.

1.7 SUMMARY

- An atom is the smallest particle of an element that retains the characteristics of that element.

- The structure of an atom consists of a *nucleus* at the center surrounded by orbiting *electrons*. The nucleus consists of *neutrons* that are neutral and *protons* that carry positive charges. The orbiting electrons carry negative charges.

- The number of positively charged protons is equal to the number of negatively charged electrons in an atom, which makes it electrically neutral or balanced.

- The number of protons in the nucleus or the number of orbiting electrons in a neutral atom determines the atomic number of an atom.

- The atomic weight of an atom equals the number of neutrons plus the number of protons in the nucleus of the atom.

- The orbiting electrons of an atom are aligned in a structured manner consisting of *shells* and *subshells*. The *n*th shell can contain a maximum of $2n^2$ electrons, where *n* is the shell number, shell number 1 being the innermost and closest to the nucleus.

- The outermost shell in an atom is called the *valence shell,* and the electrons in the valence shell are called *valence electrons*.

- Because of a unique atomic structure, atoms in semiconductor materials tend to share their four valence electrons with the neighboring atoms, which results in a complete *p* subshell for every atom. The sharing of valence electrons in semiconductors is called *covalent bonding*, which produces a stable, tightly bound, lattice structure called a *crystal*.

- An *n*-type semiconductor is produced when the intrinsic (pure) semiconductor is doped with *n*-type impurity atoms that have five valence electrons (*pentavalent*), such as *arsenic*, *antimony*, and *phosphorus*.

- A *p*-type semiconductor is produced when the intrinsic semiconductor is doped with *p*-type impurity atoms that have three valence electrons (*trivalent*), such as *aluminum*, *boron*, and *gallium*.

- The *p-n* junction is a fundamental component of many electronic devices and is formed by joining, through a certain manufacturing process, a block of *p*-type semiconductor to a block of *n*-type semiconductor.

- A diode is a two-electrode device that is a one-way conductor.

- To forward-bias the *p-n* junction, the positive terminal of the voltage source is connected to the *p*-type material and the negative terminal is connected to the *n*-type material. When the *p-n* junction is forward-biased, a substantial forward current can flow. In the forward-biased region, the diode current increases exponentially with increase in the diode voltage. An external resistor, however, can control the amount of diode current.

- The threshold voltage is 0.7 V for silicon diodes and 0.2 V for germanium diodes.

- A change in temperature can have a significant effect on the diode characteristics. For every 1°C increase in temperature, V_F decreases by approximately 2 mV, and V_{BR} doubles for every 10°C increase in temperature. The reverse saturation current I_S doubles for every 10°C increase in temperature.

- A *Zener* diode is a *p-n* junction diode that is designed for a specific *reverse breakdown* voltage. For *Zener* diodes, the reverse breakdown voltage is referred to as the *nominal Zener voltage* V_Z. The breakdown voltage level V_Z, which is much smaller in magnitude than the reverse breakdown voltage for regular diodes, is controlled by the doping levels. Zener diodes are operated only in the reverse-bias region and are mainly used as constant DC voltage sources or reference voltage sources.

- When a *p-n* junction is forward-biased, free electrons in the *n* side cross the junction and recombine with the holes in the *p* side. Since the migrating free electrons are in the *conduction* band with higher energy level than the holes in the *valence* band with lower energy level, electrons release some energy during the recombination process.

- For certain semiconductor materials, such as *gallium arsenide* (GaAs), *gallium phosphide* (GaP), and *gallium arsenide phosphide* (GaAsP) most of this energy is released in the form of light in the visible or infrared spectrum range, and the *p-n* junction is called a *light-emitting diode* (LED), discovered in 1904.

- The forward voltage drop V_F for an LED is approximately 1.5 to 3.6 V, depending on the emitted light color and the forward current. For adequate brightness of emitted light, an LED requires between 10 to 20 mA of current.

- LEDs are used for emergency exit lights, traffic lights, and displays because of their low-power consumption, low maintenance, ultra-long lifetime, and different colors.

- There are metal-semiconductor diodes formed by bonding a metal, such as aluminum or platinum to an *n*-type semiconductor. These devices are called Schottky diodes and have a threshold voltage of 0.3 V. The recombination time is very short, on the order of 10 ps. Schottky diodes are used in high-speed switching applications.

- There are high temperature diodes made of wide bandgap semiconductor meterial, such as silicon carbide diode (SiC), used in high temperature electronics.

- There are *varactor diodes* operated in the reverse-bias mode, where the *p-n* junction capacitance depends on the diode reverse voltage. They are used as variable capacitors in applications, such as *voltage-controlled oscillators* (VCO).

- There are *PIN* diodes with an intrinsic region (*i*-region) placed between the *p*- and *n*-regions. PIN diodes have low capacitance and find applications at high frequencies.

1.8 DEVICE SPECIFICATIONS AND DATA SHEETS

Data sheets of the following devices discussed in this chapter are included for your reference and convenience.
- 1N4001 through 1N4007 P-N Junction Diodes (Rectifiers)
  Copyright of Semiconductor Component Industries, LLC. Used by permission.

- 1N5221B through 1N5267B, 500 mW Zener Diodes
  Courtesy of Vishay Intertechnology

- TLHK4200 Light Emitting Diode (LED)
  Courtesy of Vishay Intertechnology

Chapter 1

MOTOROLA
SEMICONDUCTOR TECHNICAL DATA

Order this document
by 1N4001/D

Axial Lead
Standard Recovery Rectifiers

This data sheet provides information on subminiature size, axial lead mounted rectifiers for general–purpose low–power applications.

Mechanical Characteristics

- Case: Epoxy, Molded
- Weight: 0.4 gram (approximately)
- Finish: All External Surfaces Corrosion Resistant and Terminal Leads are Readily Solderable
- Lead and Mounting Surface Temperature for Soldering Purposes: 220°C Max. for 10 Seconds, 1/16" from case
- Shipped in plastic bags, 1000 per bag.
- Available Tape and Reeled, 5000 per reel, by adding a "RL" suffix to the part number
- Polarity: Cathode Indicated by Polarity Band
- Marking: 1N4001, 1N4002, 1N4003, 1N4004, 1N4005, 1N4006, 1N4007

**1N4001
thru
1N4007**

1N4004 and 1N4007 are
Motorola Preferred Devices

**LEAD MOUNTED
RECTIFIERS
50–1000 VOLTS
DIFFUSED JUNCTION**

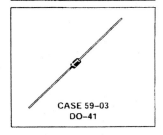

CASE 59–03
DO–41

MAXIMUM RATINGS

Rating	Symbol	1N4001	1N4002	1N4003	1N4004	1N4005	1N4006	1N4007	Unit
*Peak Repetitive Reverse Voltage Working Peak Reverse Voltage DC Blocking Voltage	V_{RRM} V_{RWM} V_R	50	100	200	400	600	800	1000	Volts
*Non–Repetitive Peak Reverse Voltage (halfwave, single phase, 60 Hz)	V_{RSM}	60	120	240	480	720	1000	1200	Volts
*RMS Reverse Voltage	$V_{R(RMS)}$	35	70	140	280	420	560	700	Volts
*Average Rectified Forward Current (single phase, resistive load, 60 Hz, see Figure 8, T_A = 75°C)	I_O	1.0							Amp
*Non–Repetitive Peak Surge Current (surge applied at rated load conditions, see Figure 2)	I_{FSM}	30 (for 1 cycle)							Amp
Operating and Storage Junction Temperature Range	T_J T_{stg}	– 65 to +175							°C

ELECTRICAL CHARACTERISTICS*

Rating	Symbol	Typ	Max	Unit
Maximum Instantaneous Forward Voltage Drop (i_F = 1.0 Amp, T_J = 25°C) Figure 1	v_F	0.93	1.1	Volts
Maximum Full–Cycle Average Forward Voltage Drop (I_O = 1.0 Amp, T_L = 75°C, 1 inch leads)	$v_{F(AV)}$	—	0.8	Volts
Maximum Reverse Current (rated dc voltage) (T_J = 25°C) (T_J = 100°C)	I_R	0.05 1.0	10 50	µA
Maximum Full–Cycle Average Reverse Current (I_O = 1.0 Amp, T_L = 75°C, 1 inch leads)	$I_{R(AV)}$	—	30	µA

*Indicates JEDEC Registered Data

Preferred devices are Motorola recommended choices for future use and best overall value.

Rev 5

Ⓜ **MOTOROLA**

© Motorola, Inc. 1996

1N5221B...1N5267B 500 mW Zener Diodes

Vishay Telefunken

Type								
1N5221B	2.4	20	< 30	< 1200	0.25	< 100	1.0	< −0.085
1N5222B	2.5	20	< 30	< 1250	0.25	< 100	1.0	< −0.085
1N5223B	2.7	20	< 30	< 1300	0.25	< 75	1.0	< −0.080
1N5224B	2.8	20	< 30	< 1400	0.25	< 75	1.0	< −0.080
1N5225B	3.0	20	< 29	< 1600	0.25	< 50	1.0	< −0.075
1N5226B	3.3	20	< 28	< 1600	0.25	< 25	1.0	< −0.070
1N5227B	3.6	20	< 24	< 1700	0.25	< 15	1.0	< −0.065
1N5228B	3.9	20	< 23	< 1900	0.25	< 10	1.0	< −0.060
1N5229B	4.3	20	< 22	< 2000	0.25	< 5	1.0	< +0.055
1N5230B	4.7	20	< 19	< 1900	0.25	< 5	2.0	< +0.030
1N5231B	5.1	20	< 17	< 1600	0.25	< 5	2.0	< +0.030
1N5232B	5.6	20	< 11	< 1600	0.25	< 5	3.0	< +0.038
1N5233B	6.0	20	< 7	< 1600	0.25	< 5	3.5	< +0.038
1N5234B	6.2	20	< 7	< 1000	0.25	< 5	4.0	< +0.045
1N5235B	6.8	20	< 5	< 750	0.25	< 3	5.0	< +0.050
1N5236B	7.5	20	< 6	< 500	0.25	< 3	6.0	< +0.058
1N5237B	8.2	20	< 8	< 500	0.25	< 3	6.5	< +0.062
1N5238B	8.7	20	< 8	< 600	0.25	< 3	6.5	< +0.065
1N5239B	9.1	20	< 10	< 600	0.25	< 3	7.0	< +0.068
1N5240B	10	20	< 17	< 600	0.25	< 3	8.0	< +0.075
1N5241B	11	20	< 22	< 600	0.25	< 2	8.4	< +0.076
1N5242B	12	20	< 30	< 600	0.25	< 1	9.1	< +0.077
1N5243B	13	9.5	< 13	< 600	0.25	< 0.5	9.9	< +0.079
1N5244B	14	9.0	< 15	< 600	0.25	< 0.1	10	< +0.082
1N5245B	15	8.5	< 16	< 600	0.25	< 0.1	11	< +0.082
1N5246B	16	7.8	< 17	< 600	0.25	< 0.1	12	< +0.083
1N5247B	17	7.4	< 19	< 600	0.25	< 0.1	13	< +0.084
1N5248B	18	7.0	< 21	< 600	0.25	< 0.1	14	< +0.085
1N5249B	19	6.6	< 23	< 600	0.25	< 0.1	14	< +0.086
1N5250B	20	6.2	< 25	< 600	0.25	< 0.1	15	< +0.086
1N5251B	22	5.6	< 29	< 600	0.25	< 0.1	17	< +0.087
1N5252B	24	5.2	< 33	< 600	0.25	< 0.1	18	< +0.088
1N5253B	25	5.0	< 35	< 600	0.25	< 0.1	19	< +0.089
1N5254B	27	4.6	< 41	< 600	0.25	< 0.1	21	< +0.090
1N5255B	28	4.5	< 44	< 600	0.25	< 0.1	21	< +0.091
1N5256B	30	4.2	< 49	< 600	0.25	< 0.1	23	< +0.091
1N5257B	33	3.8	< 58	< 700	0.25	< 0.1	25	< +0.092
1N5258B	36	3.4	< 70	< 700	0.25	< 0.1	27	< +0.093
1N5259B	39	3.2	< 80	< 800	0.25	< 0.1	30	< +0.094
1N5260B	43	3.0	< 93	< 900	0.25	< 0.1	33	< +0.095
1N5261B	47	2.7	< 105	< 1000	0.25	< 0.1	36	< +0.095
1N5262B	51	2.5	< 125	< 1100	0.25	< 0.1	39	< +0.096
1N5263B	56	2.2	< 150	< 1300	0.25	< 0.1	43	< +0.096
1N5264B	60	2.1	< 170	< 1400	0.25	< 0.1	46	< +0.097
1N5265B	62	2.0	< 185	< 1400	0.25	< 0.1	47	< +0.097
1N5266B	68	1.8	< 230	< 1600	0.25	< 0.1	52	< +0.097
1N5267B	75	1.7	< 270	< 1700	0.25	< 0.1	56	< +0.098

1) Based on dc-measurement at thermal equilibrium; lead length = 9.5mm (3/8"); thermal resistance of heat sink = 30K/W

TLHK4200
Vishay Telefunken

High Intensity LED in ø 3 mm Tinted Non–Diffused Package

Color	Type	Technology	Angle of Half Intensity
Red	TLHK4200	AlInGaP on GaAs	22°

Description

This device has been designed to meet the increasing demand for AlInGaP technology.
It is housed in a 3 mm clear plastic package. The small viewing angle of these devices provides a high brightness.
All LEDs are categorized in luminous intensity groups. That allows users to assemble LEDs with uniform appearance.

Features

- AlInGaP technology
- Standard ø 3 mm (T-1) package
- Small mechanical tolerances
- Suitable for DC and high peak current
- Very small viewing angle
- Very high intensity
- Luminous intensity categorized

Applications

Status lights
OFF / ON indicator
Background illumination
Readout lights
Maintenance lights
Legend light

Document Number 83059
Rev. A1, 04-Feb-99

TLHK4200
Vishay Telefunken

Absolute Maximum Ratings
T_{amb} = 25°C, unless otherwise specified
TLHK4200

Parameter	Test Conditions	Symbol	Value	Unit
Reverse voltage		V_R	5	V
DC forward current	$T_{amb} \leq 60°C$	I_F	30	mA
Surge forward current	$t_p \leq 10$ μs	I_{FSM}	0.1	A
Power dissipation	$T_{amb} \leq 60°C$	P_V	80	mW
Junction temperature		T_j	100	°C
Operating temperature range		T_{amb}	−40 to +100	°C
Storage temperature range		T_{stg}	−55 to +100	°C
Soldering temperature	$t \leq 5$ s, 2 mm from body	T_{sd}	260	°C
Thermal resistance junction/ambient		R_{thJA}	400	K/W

Optical and Electrical Characteristics
T_{amb} = 25°C, unless otherwise specified
Red (TLHK4200)

Parameter	Test Conditions	Type	Symbol	Min	Typ	Max	Unit
Luminous intensity	I_F = 10 mA		I_V	25	50		mcd
Dominant wavelength	I_F = 10 mA		λ_d		630		nm
Peak wavelength	I_F = 10 mA		λ_p		643		nm
Angle of half intensity	I_F = 10 mA		φ		±22		deg
Forward voltage	I_F = 20 mA		V_F		1.9	2.6	V
Reverse voltage	I_R = 10 μA		V_R	5			V
Junction capacitance	V_R = 0, f = 1 MHz		C_j		15		pF

Typical Characteristics (T_{amb} = 25°C, unless otherwise specified)

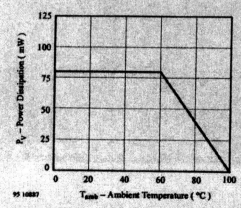

Figure 1 Power Dissipation vs. Ambient Temperature

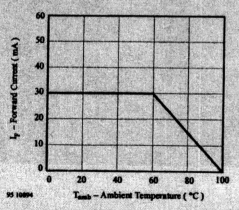

Figure 2 Forward Current vs. Ambient Temperature

TLHK4200
Vishay Telefunken

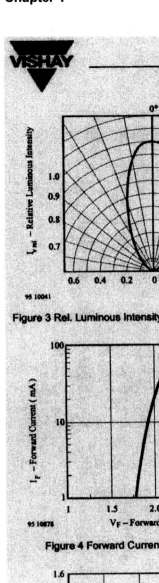
Figure 3 Rel. Luminous Intensity vs. Angular Displacement

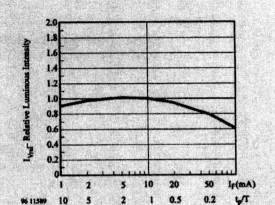
Figure 6 Rel. Lumin. Intensity vs. Forw. Current/Duty Cycle

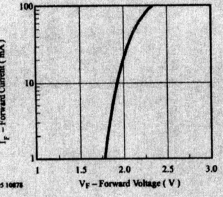
Figure 4 Forward Current vs. Forward Voltage

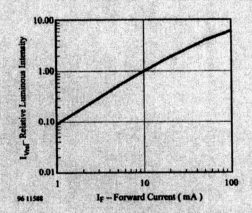
Figure 7 Relative Luminous Intensity vs. Forward Current

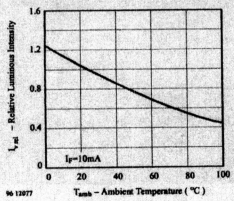
Figure 5 Rel. Luminous Intensity vs. Ambient Temperature

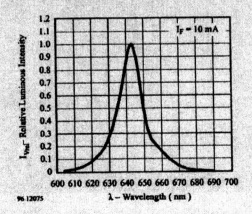
Figure 8 Relative Luminous Intensity vs. Wavelength

Document Number 83059
Rev. A1, 04-Feb-99

www.vishay.de • FaxBack +1-408-970-5600

Chapter 1

PROBLEMS

Section 1.2 Atomic Theory

1.1 State the reason(s) why is it useful to briefly review *atomic theory* before discussing the *p-n* junction and electronic devices.

1.2 Define the *atom* and its structure.

1.3 Define the *atomic number* and *atomic weight*.

1.4 Define the *valence* electron(s).

1.5 Explain what is common in the atomic structure of (a) *conductors*, (b) *semiconductors*.

Section 1.3 The P-N Junction

1.6 Explain (a) *p*-type semiconductor, (b) *n*-type semiconductor.

1.7 Explain how the *p-n* junction is connected to an external DC source, so that the junction is (a) forward-biased, (b) reverse-biased.

Section 1.4 The P-N Junction Diode

1.8 Draw the schematic diagram, including the DC source and a current limiting resistor, for a (a) forward-biased diode, (b) reverse-biased diode.

1.9 Carry out Practice Problem 1-1.

1.10 Determine the diode current I_D and the diode voltage V_D for the diode circuits of Figure 1-1P below, assuming that the diode $V_{BR} = 80$ V.

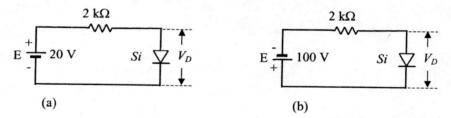

Figure 1-1P: Diode bias circuits

1.11 How do the diode characteristics change with variations in temperature? What is the rule of thumb for changes in V_F and I_S with temperature?

1.12 Carry out Practice Problem 1-2.

1.13 Determine the *DC* resistance of the diode at points *A, B,* and *C*. The forward characteristic curve of the diode is shown in Figure 1-2P. Also, determine the *AC* resistance of the diode at all three operating points using Equation 1-12.

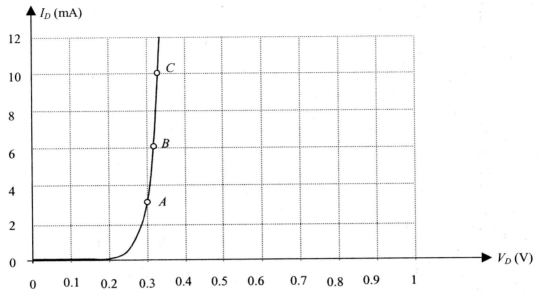

Figure 1-2P: Diode *I-V* characteristic in the forward-bias region

1.14 Carry out Practice Problem 1-3.

1.15 Determine the following for diode circuits of Figure 1-3P, assuming that the diode is silicon with $V_F = 0.7$ V, $V_{BR} = 100$ V, and $I_S = 50$ nA at 25°C.
  (a) Diode voltage V_D for the circuit of Figure 1-3P(a) at 100°C.
  (b) Diode current I_D for the circuit of Figure 1-3P(b) at 100°C.

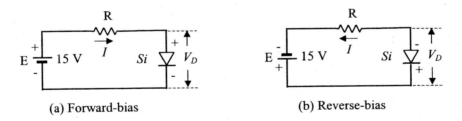

(a) Forward-bias (b) Reverse-bias

Figure 1-3P: Diode bias circuits

1.16 Carry out Practice Problem 1-4.

1.17 Determine the diode current I_D and the diode voltage V_D for the circuits of Figure 1-4P as shown below:

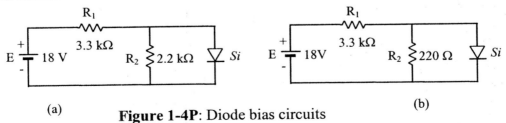

(a) (b)

Figure 1-4P: Diode bias circuits

Chapter 1

1.18 Carry out Practice Problem 1-5.

1.19 Determine the output voltage V_o and the diode current I_D for the circuit of Figure 1-5P.

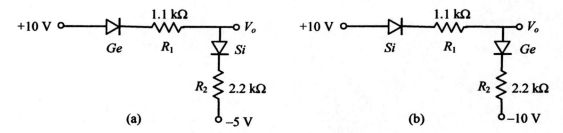

Figure 1-5P: Diode bias circuits

1.20 Carry out Practice Problem 1-6.

1.21 Determine the output voltages V_{o1} and V_{o2} for the two circuits of Figure 1-6P for all input combinations of V_1 and V_2. Assume that $V_1(H) = V_2(H) = 4.7$ V, $V_1(L) = V_2(L) = 0$ V. Assume that diodes are germanium.

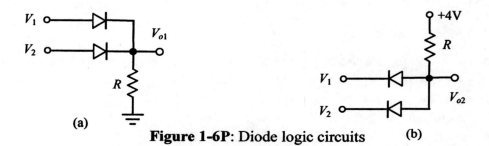

Figure 1-6P: Diode logic circuits

Section 1.5 Zener Diode

1.22 Explain the difference between the Zener diode and the regular p-n junction diode.

1.23 Explain the following Zener diode parameters: I_{ZK}, I_{ZT}, I_{ZM}, V_Z, r_Z, and $P_{Z(max)}$.

1.24 Carry out Practice Problem 1-7.

1.25 Determine V_o and I_Z for the circuit of Figure 1-7P. Also, determine I_{ZM} for the Zener diode.

Figure 1-7P:
Basic Zener diode bias circuit

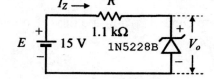

1.26 Carry out Practice Problem 1-8.

1.27 Determine the Zener diode current I_Z and the output voltage V_o for the circuit of Figure 1-8P. Also, determine I_{ZM} for the Zener diode.

Figure 1-8P: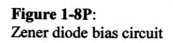
Zener diode bias circuit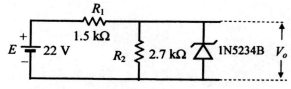

1.28 Determine the Zener diode current I_Z and the output voltage V_o for the circuit of Figure 1-9P.

Figure 1-9P:
Zener diode bias circuit

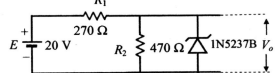

Section 1.6 Light-Emitting Diode

1.29 Explain the fundamental differences in the operation of LED diodes and ordinary diodes.

1.30 What kind of semiconductor materials are used in fabrication of light-emitting diodes and why?

1.31 Carry out Practice Problem 1-9.

1.32 Carry out Practice Problem 1-10.

1.33 Design a simple LED driver circuit with $E = 20$ V and $I_{LED} = 18$ mA.

1.34 Carry out Practice Problem 1-11.

Chapter

DIODE APPLICATIONS
Rectification, Filtering, and Regulation

2.1 INTRODUCTION

Having introduced and discussed *p-n* junction theory and the *p-n* junction diode and its characteristics in the previous chapter, we are now ready to explore the practical applications of the diode. One major area of the diode application is in electronic power supply systems, in which the regular diode is used for rectification and the Zener diode is used for regulation. A thorough analysis and design of basic DC power supplies, which consists of rectification, filtering, and regulation, will be carried out in this chapter.

2.2 THE BASIC POWER SUPPLY

A basic power supply, which converts the 110 V AC utility line voltage to a desired DC level, comprises a *transformer*, a *rectifier*, a *filter*, and a *regulator*. The ordinary p-n junction diode, also referred to as *rectifier diode*, is the principal component of the rectifier section, and the Zener diode is the principal component in the regulator section of the power supply.

2.2.1 The Transformer

The transformer section of the power supply generally down-converts the 110 V utility line voltage. A brief overview of the transformer principles is presented in Figure 2-1.

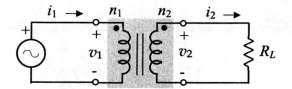

Figure 2-1: Schematic diagram of a loaded transformer

A transformer generally comprises a *primary winding* with n_1 turns, a *secondary winding* with n_2 turns, and an *iron core*. The voltage v_1 applied at the primary is magnetically induced into the secondary winding, and the iron core facilitates the induction process. The voltage ratio (v_1/v_2) is directly proportional to the *turns ratio* (n_1/n_2).

$$\frac{v_1}{v_2} = \frac{n_1}{n_2} \tag{2-1}$$

$$v_2 = v_1 \frac{n_2}{n_1} = \frac{v_1}{\text{turns ratio}} \tag{2-2}$$

For an ideal (*lossless*) transformer, the power delivered to the load at the secondary equals the power supplied at the primary. That is,

$$p_1 = p_2 \tag{2-3}$$

$$v_1 i_1 = v_2 i_2 \tag{2-4}$$

$$\frac{i_1}{i_2} = \frac{v_2}{v_1} \tag{2-5}$$

Hence, the current ratio (i_1/i_2) is inversely proportional to the turns ratio.

$$\frac{i_1}{i_2} = \frac{n_2}{n_1} \tag{2-6}$$

$$i_1 = i_2 \frac{n_2}{n_1} = \frac{i_2}{\text{turns ratio}} \tag{2-7}$$

Example 2-1 — Transformer ANALYSIS

Assuming that the input v_i is a 60 Hz 110 V utility line voltage, let us determine the output $v_{o(rms)}$, $v_{o(p)}$, i_1, i_2, and the power p_i supplied by the source for the circuit of Figure 2-2.

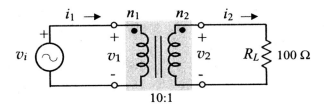

Figure 2-2: Schematic diagram of the loaded transformer

Solution:

The nominal 110 V AC is actually the rms value of the utility line voltage, and since the turns ratio is 10:1, the rms value of the secondary voltage v_2 is determined as follows:

$$v_2 = \frac{v_1}{turns\ ratio} = \frac{110\ V(rms)}{10} = 11\ V(rms)$$

$$v_{2(p)} = \sqrt{2}\, v_{2(rms)} \tag{2-8}$$

$$v_{o(p)} = v_{2(p)} = \sqrt{2}(11\ V_{rms}) = 15.55\ V$$

$$i_{o(rms)} = \frac{v_{o(rms)}}{R_L} = \frac{11\ V}{100\ \Omega} = 110\ mA$$

$$i_1 = \frac{i_2}{turns\ ratio}$$

$$i_{1(rms)} = \frac{110\ mA}{10} = 11\ mA$$

$$p_o = i_{o(rms)} v_{o(rms)} = (i_{o(rms)})^2 R_L = \frac{v_{o(rms)}^2}{R_L} \tag{2-9}$$

$$p_i \cong p_o = i_{o(rms)} v_{o(rms)} = 110\ mA \times 11\ V = 1.21\ W$$

2.2.2 Half-Wave Rectifier

The *half-wave rectifier* is simply a basic series-clipper circuit connected to the secondary of the step-down transformer, as shown in Figure 2-3.

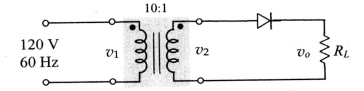

Figure 2-3: Half-wave rectifier

Since the turn ratio is 10-to-1, v_2 is 12 V(rms) with a peak value of approximately 17 V.

$$v_{2(p)} = \sqrt{2}\, v_{2(rms)} = 1.414 \times 12\,\text{V} = 16.97 \cong 17\,\text{V}$$

$$v_{o(p)} = v_{2(p)} - 0.7\,\text{V} = 17\,\text{V} - 0.7\,\text{V} = 16.3\,\text{V}$$

Applying the *express* analysis procedure for the series clipper circuit, the clipping level for v_2 is +0.7 V, which becomes the 0 V level for the output v_o. Since this circuit is a negative clipper, the portion of the waveform below the clipping level is clipped off, producing the output waveform shown in Figure 2-4.

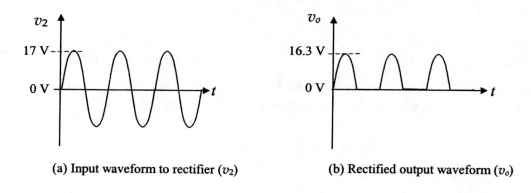

(a) Input waveform to rectifier (v_2)   (b) Rectified output waveform (v_o)

Figure 2-4: Input and output waveforms of the half-wave rectifier

The input to the rectifier, which is a full sinusoid, has an *average* or *DC* value of 0 V. However, the output of the half-wave rectifier has a DC value of $0.3183 v_o(p)$, as shown below.

Average or DC value at the output of the half-wave rectifier $V_{o(DC)}$:

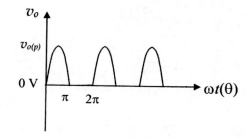

Figure 2-5: Half-wave rectifier input and output waveforms

$$v_{o(DC)} = \frac{1}{T} \int_T v_o\, d\theta \qquad (2\text{-}10)$$

$$v_{o(DC)} = \frac{1}{2\pi} \int_0^\pi v_{o(p)} \sin\theta\, d\theta \qquad (2\text{-}11)$$

Chapter 2

$$v_{o(DC)} = \frac{v_{o(p)}}{2\pi}[-\cos\theta]_0^\pi \qquad (2\text{-}12)$$

$$v_{o(DC)} = -\frac{v_{o(p)}}{2\pi}[\cos(\pi)-\cos(0)] \qquad (2\text{-}13)$$

$$v_{o(DC)} = -\frac{v_{o(p)}}{2\pi}[-2] = \frac{v_{o(p)}}{\pi} \qquad (2\text{-}14)$$

$$v_{o(DC)} = \frac{v_{o(p)}}{\pi} = 0.3183 v_{o(p)} \qquad (2\text{-}15)$$

Hence, the $V_{o(DC)}$ of the half-wave rectifier of Figure 2-5 is approximately 5.2 V.

Practice Problem 2-1 — Half-Wave Rectifier — ANALYSIS

Determine the $V_{o(DC)}$ for the half-wave rectifier of Figure 2-6.

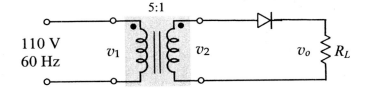

Figure 2-6: Half-wave rectifier

Answer: $V_{o(DC)} = 9.67$ V

2.2.3 Full-Wave Rectifier

There are two versions of the full-wave rectifier (a) full-wave center-tapped rectifier, and (b) full-wave bridge rectifier.

Full-Wave Center-Tapped Rectifier

The *full-wave center-tapped* rectifier is a combination of two *half-wave* rectifiers with a center-tapped transformer, as illustrated in Figure 2-7.

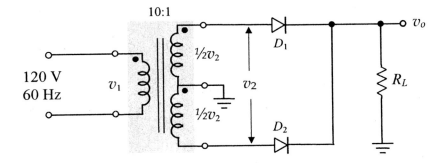

Figure 2-7: Full-wave center-tapped rectifier

Again, since the turns ratio is 10-to-1, v_2 is 12 V(rms) with a peak of approximately 17 V.

$$v_{2(p)} = \sqrt{2}v_{2(rms)} = 1.414 \times 12 \text{ V} = 16.97 \cong 17 \text{ V}$$

In order to analyze the operation of the above full-wave rectifier, we will redraw the above circuit and consider the positive and negative half-cycles separately.

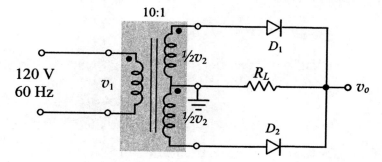

Figure 2-8: Full-wave center-tapped rectifier redrawn

Positive half-cycle:

For the positive half-cycle of the input signal, diode D_1 is forward-biased and diode D_2 is reverse-biased. The resulting equivalent circuit and the corresponding output are shown in Figure 2-9.

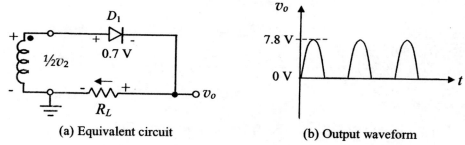

(a) Equivalent circuit

(b) Output waveform

Figure 2-9: Equivalent circuit of the center-tapped rectifier and the corresponding output for the *positive half-cycle*

$$v_{o(p)} = 0.5v_{2(p)} - 0.7 \text{ V} \qquad (2\text{-}16)$$

$$v_{o(p)} = 8.5 \text{ V} - 0.7 \text{ V} = 7.8 \text{ V}$$

Negative half-cycle:

For the negative half-cycle of the input signal, diode D_2 is forward-biased and diode D_1 is reverse-biased. The resulting equivalent circuit and the corresponding output are shown in Figure 2-10.

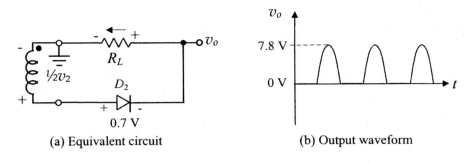

(a) Equivalent circuit (b) Output waveform

Figure 2-10: Equivalent circuit of the center-tapped rectifier and the corresponding output for the *negative half-cycle*

$$v_{o(p)} = 0.5v_{2(p)} - 0.7\text{ V} = 8.5\text{ V} - 0.7\text{ V} = 7.8\text{ V}$$

Note that in both cases, positive and negative half-cycles, the output current flows in the same direction through the R_L, and v_o is positive with respect to the common ground. By combining the two outputs for positive and negative cycles, we obtain the full-wave output waveform depicted in Figure 2-11.

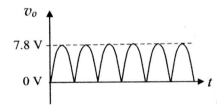

Figure 2-11: Output waveform of the full-wave center-tapped rectifier

Luckily, we do not have to go through the derivation process again to find the average or DC value of the voltage at the output of the full-wave center-tapped rectifier; it is simply twice that of the half-wave rectifier, because the area under the curve has doubled.

$$v_{o(DC)} = \frac{2v_{o(p)}}{\pi} = 0.6366 v_{o(p)} \qquad (2\text{-}17)$$

$$V_{o(DC)} = 0.6366 \times 7.8\text{ V} = 4.965\text{ V}$$

Hence, $V_{o(DC)}$ of the full-wave rectifier of Figure 2-11 is approximately 5 V.

Practice Problem 2-2 (Full-Wave Rectifier) ANALYSIS

Draw the output waveform and determine the $V_{o(DC)}$ for the full-wave rectifier of Figure 2-12.

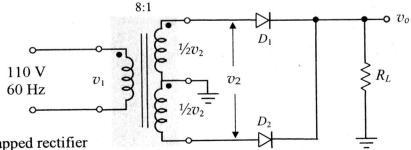

Figure 2-12: Full-wave center-tapped rectifier

49

Answer: $V_{o(DC)} = 5.74$ V

Full-Wave Bridge Rectifier

Four diodes are used in constructing the *full-wave bridge* rectifier. In the traditional schematic diagram of the full-wave bridge rectifier, the four diodes are arranged in a diamond shape as illustrated in Figure 2-13 below:

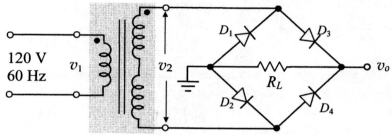

Figure 2-13:
Traditional schematic diagram of the full-wave bridge rectifier

However, the diamond-shaped arrangement of the diodes is neither easy nor convenient to draw. Hence, we will adopt an alternate way of drawing the same schematic of the full-wave bridge rectifier, which is much easier to draw and more convenient to analyze and explain its operation.

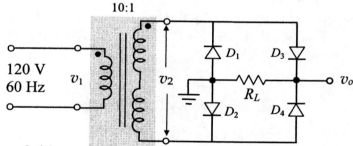

Figure 2-14:
Alternate schematic diagram of the full-wave bridge rectifier

We will hold on to the same turns ratio of the transformer of 10:1, so that a fair comparison can be made of the output DC levels of all three rectifiers with the same input line voltage of 120 V.

$$v_{2(p)} = \sqrt{2}v_{2(rms)} = 1.414 \times 12 \text{ V} = 16.97 \cong 17 \text{ V}$$

In order to analyze the operation of the full-wave rectifier, we will consider the positive and negative half-cycles separately.

Positive half-cycle:

For the positive half-cycle of the input signal, diodes D_2 and D_3 are forward-biased and diodes D_1 and D_4 are reverse-biased. The resulting equivalent circuit and the corresponding output are shown in Figure 2-15.

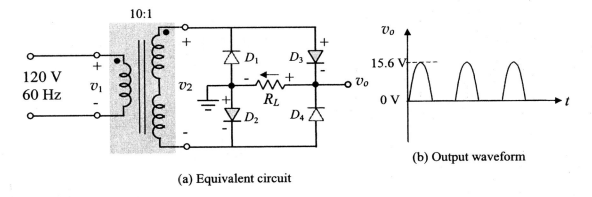

(a) Equivalent circuit

(b) Output waveform

Figure 2-15: Equivalent circuit of the full-wave bridge rectifier and the corresponding output for the *positive half-cycle*

$$v_{o(p)} = v_{2(p)} - 1.4\,\text{V} \tag{2-18}$$

$$v_{o(p)} = 17\,\text{V} - 1.4\,\text{V} = 15.6\,\text{V}$$

Negative half-cycle:

For the negative half-cycle of the input signal, diodes D_1 and D_4 are forward-biased and diodes D_2 and D_3 are reverse-biased. The resulting equivalent circuit and the corresponding output are shown in Figure 2-16 below:

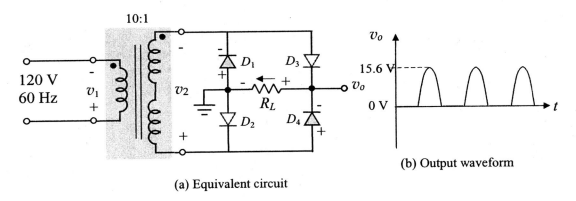

(a) Equivalent circuit

(b) Output waveform

Figure 2-16: Equivalent circuit of the full-wave bridge rectifier and the corresponding output for the *negative half-cycle*

Chapter 2

Combining the above two half-wave outputs for positive and negative half-cycles, we obtain the *full-wave* output waveform shown in Figure 2-17.

Figure 2-17:
Overall output waveform of the full-wave bridge rectifier

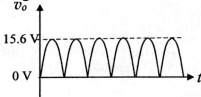

The average or DC value of the output waveform is determined with Equation 2-17:

$$v_{o(DC)} = \frac{2v_{o(p)}}{\pi} = 0.6366 v_{o(p)}$$

$$V_{o(DC)} = 0.6366 \times 15.6 \text{ V} = 9.93 \text{ V} \cong 10 \text{ V}$$

Hence, $V_{o(DC)}$ of the *full-wave bridge* rectifier is twice that of the *full-wave center-tapped* rectifier, and that is why the full-wave bridge is the most widely used rectifier circuit.

Practice Problem 2-3 Full-Wave Rectifier ANALYSIS

Determine the $V_{o(DC)}$ for the full-wave bridge rectifier of Figure 2-18 as shown below:

Figure 2-18:
Full-wave bridge rectifier

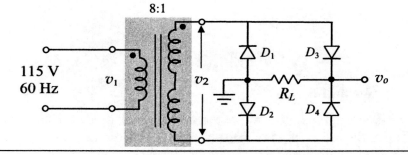

Answer: $V_{o(DC)} \cong 12$ V

Practice Problem 2-4 Full-Wave Rectifier ANALYSIS

Determine the $V_{o(DC)}$ for the full-wave bridge rectifier of Figure 2-19 as shown below:

Figure 2-19:
Full-wave bridge rectifier

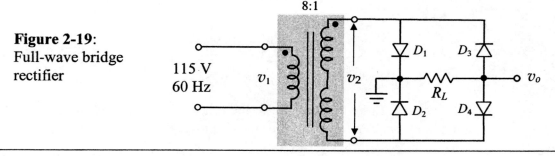

Answer: $V_{o(DC)} \cong -12$ V

2.2.4 Filtering

Although the output of the full-wave bridge rectifier, as shown in Figure 2-17, has a DC value of $2v_{o(p)}/\pi$, it is a pulsating DC. The plot of a pure DC would be a fixed horizontal line. We can, however, achieve something close to a pure DC and increase the magnitude of the $V_{o(DC)}$ at the same time by inserting a suitable filter capacitor C at the output, as shown in Figures 2-20 and 2-21.

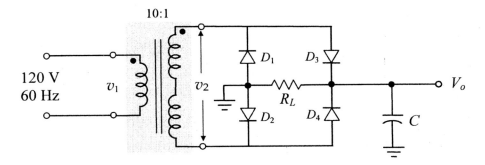

Figure 2-20: Full-wave bridge rectifier with filter capacitor

Figure 2-21: Filtered output of the full-wave bridge rectifier

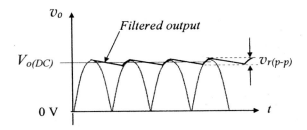

Obviously, the filtered output is not exactly a straight line, as the pure DC would be, but it is more like a triangular waveform. However, compared to the non-filtered output, it is much more desirable and much closer to an ideal DC. Furthermore, its DC value $V_{o(DC)}$ is larger than before and the *ripple voltage* ($v_{r(p-p)}$) is much smaller. The magnitude of the peak-to-peak ripple voltage $v_{r(p-p)}$ is directly proportional to the output current $I_{o(DC)}$, and inversely proportional to the output frequency and the capacitor value C.

$$v_{r(p-p)} = \frac{I_{o(DC)}}{f_o C} = \frac{I_{o(DC)}}{2 f_{in} C} \tag{2-19}$$

where

$$I_{o(DC)} = \frac{V_{o(DC)}}{R_L} \tag{2-20}$$

$$V_{o(DC)} = v_{o(p)} - \frac{v_{r(p-p)}}{2} \tag{2-21}$$

With a suitable filter capacitor C, the ripple voltage can be minimized so that $V_{o(DC)}$ is approximately equal to $v_{o(p)}$.

Example 2-2 — Filtering — ANALYSIS

Assuming that $R_L = 100\ \Omega$ and $C = 1000\ \mu F$, let us determine the output $v_{r(p\text{-}p)}$ and $V_{o(DC)}$ for the full-wave bridge circuit of Figure 2-20.

Solution:
Initially, we assume that $V_{o(DC)}$ is approximately equal to $v_{o(p)}$, then determine $v_{r(p\text{-}p)}$.

$$V_{o(DC)} \cong v_{o(p)} = 15.6\ \text{V}$$

$$I_{o(DC)} = \frac{V_{o(DC)}}{R_L} = \frac{15.6\ \text{V}}{100\ \Omega} = 156\ \text{mA}$$

$$v_{r(p\text{-}p)} = \frac{I_{o(DC)}}{2 f_{in} C} = \frac{156\ \text{mA}}{2 \times 60\ \text{Hz} \times 1\ \text{mF}} = 1.3\ \text{V}$$

Having determined the ripple voltage, we can now determine the actual $V_{o(DC)}$, as follows:

$$V_{o(DC)} = v_{o(p)} - \frac{v_{r(p\text{-}p)}}{2} = 15.6\ \text{V} - 0.65\ \text{V} = 14.95\ \text{V} \cong 15\ \text{V}$$

Practice Problem 2-5 — Full-Wave Rectifier — ANALYSIS

Determine the $v_{r(p\text{-}p)}$ and $V_{o(DC)}$ for the full-wave bridge rectifier of Figure 2-22 as shown below. Assume $C = 2000\ \mu F$ and $R_L = 120\ \Omega$.

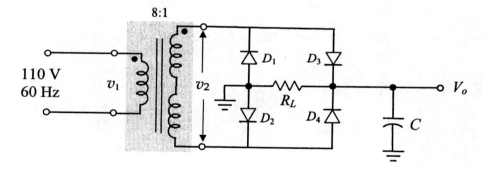

Figure 2-22: Full-wave bridge rectifier with filter capacitor

Answers: $v_{r(p\text{-}p)} = 0.625\ \text{V}$, $V_{o(DC)} = 17.7\ \text{V}$

Chapter 2

Section Summary — Rectification — ANALYSIS

The Transformer:

$$v_2 = v_1 \frac{n_2}{n_1} = \frac{v_1}{turn\ ratio}$$

$$v_{2(p)} = \sqrt{2}\, v_{2(rms)} = 1.414\, v_{2(rms)}$$

Half-wave Rectifier:

$$v_{o(p)} = v_{2(p)} - 0.7\ \text{V}$$

$$v_{o(DC)} = \frac{v_{o(p)}}{\pi} = 0.3183\, v_{o(p)}$$

Full-wave Center-tapped Rectifier:

$$v_{o(p)} = 0.5\, v_{2(p)} - 0.7\ \text{V}$$

$$v_{o(DC)} = \frac{2 v_{o(p)}}{\pi} = 0.6366\, v_{o(p)}$$

Full-wave Bridge Rectifier:

$$v_{o(p)} = v_{2(p)} - 1.4\ \text{V}$$

$$v_{o(DC)} = \frac{2 v_{o(p)}}{\pi} = 0.6366\, v_{o(p)}$$

Filtering:

$$v_{r(p-p)} = \frac{I_{o(DC)}}{f_o C} = \frac{I_{o(DC)}}{2 f_{in} C}$$

where

$$I_{o(DC)} = \frac{V_{o(DC)}}{R_L}$$

Initially consider $V_{o(DC)} \cong v_{2(p)}$ and solve for $I_{o(DC)}$ and $v_{r(p-p)}$, then recalculate $V_{o(DC)}$

$$V_{o(DC)} = v_{o(p)} - \frac{v_{r(p-p)}}{2}$$

2.2.5 Regulation

The nominal 110 V utility line voltage is not necessarily fixed at 110 V; it may fluctuate approximately between 90 to 130 V, depending on the consumer demand for power and some other factors. However, your electronic equipment powered by a DC power supply may not tolerate that kind of voltage level fluctuations. Hence, a voltage regulator is needed in every power supply to keep the output DC voltage fixed at a desired level, irrespective of the input fluctuations. The *voltage regulation* of a power supply can be accomplished simply by employing the services of a Zener diode, which is one of its major applications. Consider the following DC circuit with a variable input voltage E, in which a Zener diode is used to achieve a reasonably fixed voltage at the output.

Example 2-3 — Zener Regulator — ANALYSIS

Assuming that $R_S = 100\ \Omega$ and $R_L = 330\ \Omega$, let us determine the output V_o, I_Z, and I_L.

1N5236B:
$V_Z = 7.5$ V
$I_{ZT} = 20$ mA
$I_{ZK} = 0.25$ mA
$P_{Z(max)} = 500$ mW
$r_Z = 6\ \Omega\ @\ I_{ZT}$

Figure 2-23: Zener diode used as voltage regulator

Solution:

Thevenizing the circuit external to the Zener, we obtain the equivalent circuit shown in Figure 2-24.

$$E_{th} = E\frac{R_L}{R_S + R_L} \tag{2-22}$$

$$E_{th} = 14\ \text{V}\frac{330\ \Omega}{100\ \Omega + 330\ \Omega} = 10.744\ \text{V}$$

$$R_{th} = R_S \| R_L \tag{2-23}$$

$$R_{th} = 100\ \Omega \| 330\ \Omega = 76.7\ \Omega$$

Figure 2-24: Thevenin's equivalent circuit

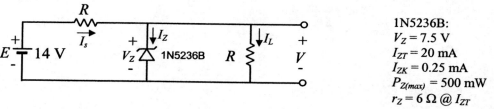

$$I_Z = \frac{E_{th} - V_Z}{R_{th}} \tag{2-24}$$

$$I_Z = \frac{10.744\text{ V} - 7.5\text{ V}}{76.7\text{ }\Omega} = 42.3\text{ mA}$$

Having determined the Zener current I_Z, we now return to the original circuit and determine I_L and I_S, as follows:

$$V_o = V_Z = 7.5\text{ V}$$

$$I_L = \frac{V_o}{R_L} \qquad (2\text{-}25)$$

$$I_L = \frac{7.5\text{ V}}{330\text{ }\Omega} = 22.7\text{ mA}$$

$$I_S = I_Z + I_L$$

$$I_S = 42.29\text{ mA} + 22.72\text{ mA} = 65\text{ mA}$$

Alternately, we can also determine I_S and I_L as follows:

$$I_S = \frac{E - V_Z}{R_S}$$

$$I_S = \frac{14\text{ V} - 7.5\text{ V}}{100\text{ }\Omega} = 65\text{ mA}$$

$$I_L = I_S - I_Z \qquad (2\text{-}26)$$

$$I_L = 65\text{ mA} - 42.3\text{ mA} = 22.7\text{ mA}$$

Practice Problem 2-6 — Zener Regulator — ANALYSIS

Determine the output V_o, I_Z, I_L, and I_S for the circuit of Figure 2-23 with the following supply voltage levels, and then comment on why the circuit of Figure 2-23 above qualifies as a *voltage regulator*.

(a) $E = 16$ V,   (b) $E = 12$ V

All other circuit parameters remain unchanged.

Answers:

(a) $V_o = V_Z = 7.5$ V, $I_Z = 62.3$ mA, $I_L = 22.7$ mA, $I_S = 85$ mA

(b) $V_o = V_Z = 7.5$ V, $I_Z = 22.3$ mA, $I_L = 22.7$ mA, $I_S = 45$ mA

2.2.6 Regulated Power Supply

Consider the complete power supply circuit, including *rectification*, *filtering*, and *regulation*, as exhibited in Figure 2-25 below:

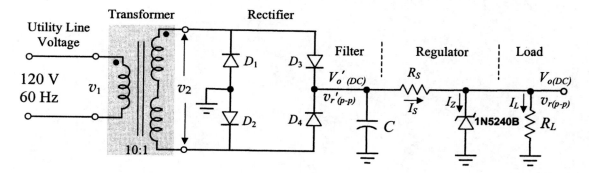

Figure 2-25: Regulated power supply

Assuming that $R_L = 100\ \Omega$, and $C = 1000\ \mu F$, let us pick a value for R_S such that $I_Z = I_{ZT}$, and then determine the output $v_{r(p-p)}$ and $V_{o(DC)}$ for the regulated power supply circuit as shown above. Zener characteristics are as given below:

1N5240B: $V_Z = 10\ V$, $I_{ZT} = 20\ mA$, $I_{ZK} = 0.25\ mA$, $I_{ZM} = 50\ mA$, $r_Z = 17\ \Omega$ @ I_{ZT}

Solution:

$$v_{2(p)} = \sqrt{2}\,v_{2(rms)} = 1.414 \times 12\ V = 16.97 \cong 17\ V$$

$$v_{o\,(p)}' = v_{2(p)} - 1.4\ V = 15.6\ V$$

With $I_Z = I_{ZT}$, V_o equals V_Z.

$$V_o = V_Z = 10\ V$$

$$I_L = \frac{V_{o(DC)}}{R_L} = \frac{10\ V}{100\ \Omega} = 100\ mA$$

$$I_S = I_Z + I_L$$

$$I_S = 20\ mA + 100\ mA = 120\ mA$$

Initially, we assume that $V_{o(DC)}'$ is approximately equal to $v_{o(p)}'$, then determine $v_{r(p-p)}'$.

$$V_{o(DC)}' \cong v_{o(p)}' = 15.6\ V$$

$$v_{r(p-p)}' = \frac{I_S}{2 f_{in} C} = \frac{120\ mA}{2 \times 60\ Hz \times 1\ mF} = 1\ V$$

Having determined the ripple voltage $V_{r(p-p)}'$, we can now determine the actual $V_{o(DC)}'$, as follows:

$$V_{o(DC)}' = v_{o(p)}' - \frac{v_{r(p-p)}'}{2} = 15.6\ V - 0.5\ V = 15.1\ V$$

$$R_S = \frac{V_{o(DC)}' - V_{o(DC)}}{I_S} = \frac{15.1\ V - 10\ V}{120\ mA} = 42.5\ \Omega$$

Use $R_S = 43\,\Omega$

The ripple voltage, which is a triangular waveform, is treated like an AC signal; thus, the Zener diode appears like a resistor with a value of r_Z. However, the load resistor in parallel with r_Z produces an equivalent resistance R_L'.

$$R_L' = r_Z \parallel R_L \tag{2-27}$$

$$R_L' = 17\,\Omega \parallel 100\,\Omega = 14.5\,\Omega$$

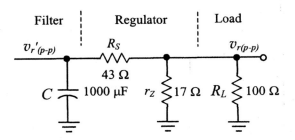

Figure 2-26: Equivalent circuit for analysis of the ripple voltage

Applying the voltage divider rule, we can now determine the ripple voltage $v_{r(p-p)}$ at the output, as follows:

$$v_{r(p-p)} = v_r'{}_{(p-p)} \frac{R_L'}{R_S + R_L'} \tag{2-28}$$

$$v_{r(p-p)} = 1\,\text{V}\,\frac{14.5\,\Omega}{43\,\Omega + 14.5\,\Omega} = 252\,\text{mV}$$

If this amount of ripple is considered to be too high for certain applications, then the ripple at the filter section $v_{r\,(p-p)}$ can be reduced by increasing the capacitor value, which will decrease the ripple $v_{r(p-p)}$ at the output. Let us now find the minimum capacitor value so that the output ripple is no more than 1% of the output DC voltage, which is the industry standard.

$$v_{r(p-p)} \leq 0.01\,V_{o(DC)} = 0.01 \times 10\,\text{V} = 100\,\text{mV}$$

Solving for $v_r'{}_{(p-p)}$ in Equation 2-28 yields the following:

$$v_r'{}_{(p-p)} = v_{r(p-p)} \frac{R_S + R_L'}{R_L'} \tag{2-29}$$

$$v_r'{}_{(p-p)} \leq 100\,\text{mV}\,\frac{43\,\Omega + 14.5\,\Omega}{14.5\,\Omega} = 396.55\,\text{mV}$$

also,

$$v_r'{}_{(p-p)} = \frac{I_S}{2 f_{in} C}$$

$$C \geq \frac{I_S}{2 f_{in} v_r'{}_{(p-p)}} = \frac{120\,\text{mA}}{2 \times 60\,\text{Hz} \times 396.55\,\text{mV}} = 2.52\,\text{mF}$$

Use $C = 2.7\,\text{mF} = 2{,}700\,\mu\text{F}$

Practice Problem 2-7   *Power Supply*   ANALYSIS

Determine I_L, I_Z, $v_{r(p-p)}$, and $V_{o(DC)}$ for the regulated power supply of Figure 2-27.
Assume $C = 2000$ μF, $R_S = 47$ Ω, and $R_L = 120$ Ω.
1N5242B: $V_Z = 12$ V, $I_{ZT} = 20$ mA, $I_{ZK} = 0.25$ mA, $I_{ZM} = 41.66$ mA, and $r_z = 30$ Ω @ I_{ZT}.

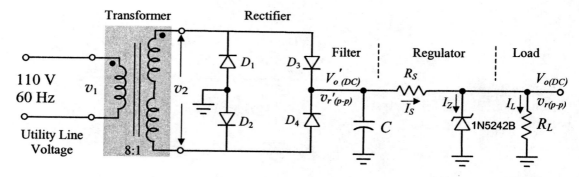

Figure 2-27: Regulated power supply

Hint: Initially, assume $V_o'{(DC)} = v_o'{(p)}$, determine I_S and $v_r'{(p-p)}$, and then determine $V_o'{(DC)}$. Having determined $V_o'{(DC)}$, recalculate I_S and $v_r'{(p-p)}$, then determine $v_{r(p-p)}$.

Answers: $V_{o(DC)} = 12$ V, $I_Z = 22$ mA, $I_S = 122$ mA, $v_{r(p-p)} = 169$ mV

Example 2-4   *Power Supply*   DESIGN

Let us now design a regulated power supply with $V_{o(DC)} = 10$ V and $I_L = 250$ mA.

Solution:
Since $V_{o(DC)} = 10$ V, we pick the 1N5240B Zener diode with the following specifications:
1N5240B: $V_Z = 10$ V, $I_{ZT} = 20$ mA, $I_{ZK} = 0.25$ mA, $I_{ZM} = 50$ mA, and $r_z = 17$ Ω @ I_{ZT}.
We will assume the nominal utility line voltage $v_1 = 115$ V, and pick a transformer with turns ratio of 8:1.

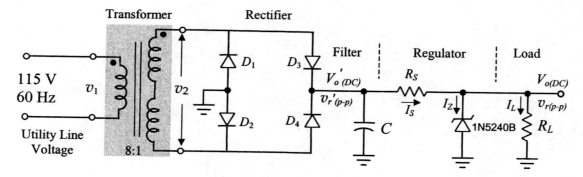

Figure 2-28: Regulated power supply

Chapter 2

$$v_{2(rms)} = \frac{v_{1(rms)}}{8} = \frac{115 \text{ V}}{8} = 14.375 \text{ V}$$

$$v_{2(p)} = \sqrt{2}\, v_{2(rms)} = 1.414 \times 14.375 \text{ V} = 20.3 \text{ V}$$

$$v_o'{}_{(p)} = v_{2(p)} - 1.4 \text{ V} = 18.9 \text{ V}$$

Let $I_Z = I_{ZT} = 20$ mA

$$R_{L(min)} = \frac{V_{o(DC)}}{I_{L(max)}} = \frac{10 \text{ V}}{250 \text{ mA}} = 40\ \Omega$$

$$I_S = I_Z + I_L$$

$$I_S = 20 \text{ mA} + 250 \text{ mA} = 270 \text{ mA}$$

Initially, we assume that $V_o'{}_{(DC)}$ is approximately equal to $v_o'{}_{(p)}$, then determine R_S.

$$V_o'{}_{(DC)} \cong v_o'{}_{(p)} = 18.9 \text{ V}$$

$$R_S = \frac{V_o'{}_{(DC)} - V_{o(DC)}}{I_S} = \frac{18.9 \text{ V} - 10 \text{ V}}{270 \text{ mA}} = 32.96\ \Omega$$

Let $R_S = 33\ \Omega$

$$R_L' = r_Z \| R_L$$

$$R_L' = 17\ \Omega\ \|\ 40\ \Omega = 12\ \Omega$$

The ripple at the output cannot be more than 1% of the output DC voltage $V_{o(DC)}$.

$$V_{r(p\text{-}p)} \leq 0.01\, V_{o(DC)} = 0.1 \text{ V} = 100 \text{ mV}$$

$$v_r'{}_{(p\text{-}p)} = v_{r(p\text{-}p)} \frac{R_S + R_L'}{R_L'}$$

$$v_r'{}_{(p\text{-}p)} \leq 100 \text{ mV}\, \frac{33\ \Omega + 12\ \Omega}{12\ \Omega} = 375 \text{ mV}$$

$$C \geq \frac{I_S}{2 f_{in} v_r'{}_{(p\text{-}p)}} = \frac{270 \text{ mA}}{2 \times 60 \text{ Hz} \times 375 \text{ mV}} = 6 \text{ mF}$$

Let $C = 6{,}800\ \mu\text{F}$

Having determined the ripple $V_r'{}_{(p\text{-}p)}$ at the filter, we can now determine the actual $V_{o(DC)}$ at the output, and then determine a new value for R_S, as follows:

$$V_o'{}_{(DC)} = v_o'{}_{(p)} - \frac{v_r'{}_{(p\text{-}p)}}{2} = 18.9 \text{ V} - 0.1875 \text{ V} = 18.71 \text{ V}$$

$$R_S = \frac{V_o'{}_{(DC)} - V_{o(DC)}}{I_S} = \frac{18.71 \text{ V} - 10 \text{ V}}{270 \text{ mA}} = 32.27\ \Omega$$

Use $R_S = 33\ \Omega$

As you can see, neither $V_o'{}_{(DC)}$ nor R_S is much different from the previously calculated values, because the ripple at the output is so small.

Practice Problem 2-8 DESIGN

Design a regulated power supply with $V_{o(DC)} = 12$ V and $I_L = 240$ mA. Assume the nominal line voltage $v_1 = 110$ V, and use a transformer with turns ratio of 8:1.

Answer: Zener is 1N5242B, $R_S = 23 \, \Omega$, $C \geq 6{,}800 \, \mu F$

2.3 SUMMARY

- A basic power supply, which converts the 110 V AC utility line voltage to a desired DC level, comprises a *transformer*, a *rectifier*, a *filter*, and a *regulator*. The ordinary p-n junction diode, also referred to as a *rectifier diode*, is the principal component of the rectifier section, and the Zener diode is the principal component in the regulator section of the power supply.

- The output of the half-wave rectifier has an average or DC value of $0.3183 v_o(p)$.

- There are two versions of the full-wave rectifier: (a) full-wave center-tapped rectifier, and (b) full-wave bridge rectifier.

- The average or DC value of the voltage at the output of the full-wave rectifier is simply twice that of the half-wave rectifier. That is, $v_{o(DC)} = 0.6366 v_o(p)$.

- The output of the full-wave bridge rectifier is a pulsating DC. The plot of a pure DC, however, is a fixed horizontal line. We can achieve something close to a pure DC with some ripple and increase the magnitude of the $V_{o(DC)}$ at the same time by inserting a suitable filter capacitor C at the output.

- The magnitude of the ripple voltage $v_{r(p-p)}$ is directly proportional to the output current $I_{o(DC)}$, and inversely proportional to the capacitor value C.

- The nominal 110 V utility line voltage is not necessarily fixed at 110 V; it may fluctuate approximately between 90 to 130 V. Hence, a voltage regulator is needed in every power supply to keep the output DC voltage fixed at a desired level, irrespective of the input fluctuations.

- The *voltage regulation* of a power supply can be accomplished simply by employing the services of a Zener diode, which is one of its major applications.

- The industry standard for maximum ripple voltage $v_{r(p-p)}$ at the output of a power supply is 1% of the output DC voltage ($0.01 V_{o(DC)}$).

Chapter 2

PROBLEMS

Section 2.2.2 Half-Wave Rectifier

2.1 Carry out Practice Problem 2-1.

2.2 Determine the $V_{o(DC)}$ for the half-wave rectifier of Figure 2-1P as shown below:

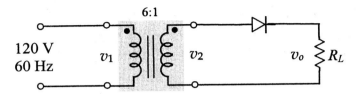

Figure 2-1P: Half-wave rectifier

Section 2.2.3 Full-Wave Rectifier

2.3 Carry out Practice Problem 2-2.

2.4 Determine the $V_{o(DC)}$ and $I_{o(DC)}$ for the full-wave rectifier of Figure 2-2P as shown below:

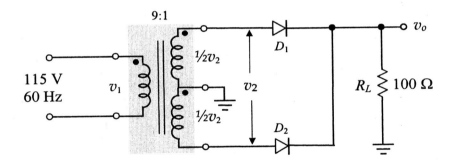

Figure 2-2P: Full-wave center-tapped rectifier

2.5 Carry out Practice Problem 2-3.

2.6 Determine the $V_{o(DC)}$ and $I_{o(DC)}$ for the full-wave bridge rectifier of Figure 2-3P as shown below:

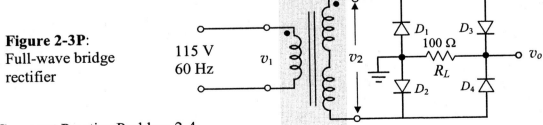

Figure 2-3P: Full-wave bridge rectifier

2.7 Carry out Practice Problem 2-4.

Chapter 2

Section 2.2.4 Filtering

2.8 Determine the $V_{o(DC)}$, $I_{o(DC)}$, and $v_{r(p-p)}$ for the full-wave bridge rectifier of Figure 2-4P. Assume $R_L = 100\ \Omega$, and $C = 3{,}300\ \mu F$.

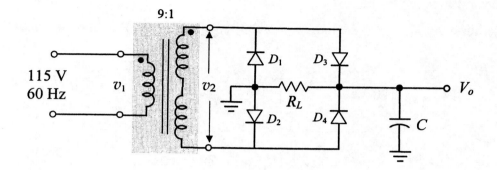

Figure 2-4P: Full-wave bridge rectifier with filter capacitor

2.9 Repeat problem 2.8 above with $C = 4{,}700\ \mu F$.

2.10 Carry out Practice Problem 2-5.

2.11 Carry out Practice Problem 2-6.

Section 2.2.5 Regulation

2.12 Assuming that $R_S = 100\ \Omega$ and $R_L = 200\ \Omega$, determine the output V_o, I_Z, and I_L for the circuit of Figure 2-5P.

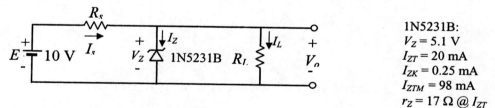

1N5231B:
$V_Z = 5.1$ V
$I_{ZT} = 20$ mA
$I_{ZK} = 0.25$ mA
$I_{ZTM} = 98$ mA
$r_Z = 17\ \Omega\ @\ I_{ZT}$

Figure 2-5P: Zener diode used as voltage regulator

2.13 Assuming that $R_S = 200\ \Omega$ and $R_L = 100\ \Omega$, determine the output V_o, I_Z, and I_L for the circuit of Figure 2-6P.

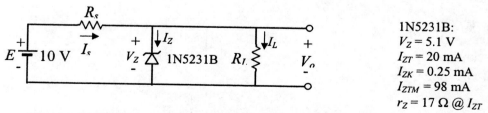

1N5231B:
$V_Z = 5.1$ V
$I_{ZT} = 20$ mA
$I_{ZK} = 0.25$ mA
$I_{ZTM} = 98$ mA
$r_Z = 17\ \Omega\ @\ I_{ZT}$

Figure 2-6P: Zener diode used as voltage regulator

Chapter 2

Section 2.2.6 Regulated Power Supply

2.14 Carry out Practice Problem 2-7.

2.15 Determine the I_L, I_Z, $v_{r(p-p)}$, and $V_{o(DC)}$ for the regulated power supply of Figure 2-7P as shown below. Assume $C = 3{,}300$ μF, $R_S = 120$ Ω, and $R_L = 75$ Ω.
1N5231B: $V_Z = 5.1$ V, $I_{ZT} = 20$ mA, $I_{ZK} = 0.25$ mA, $P_{ZM} = 500$ mW, $r_Z = 17$ Ω @ I_{ZT}

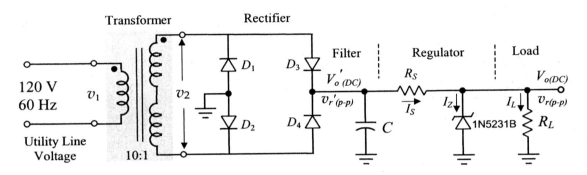

Figure 2-7P: Regulated power supply

2.16 Carry out Practice Problem 2-8.

2.17 Design a regulated power supply with $V_{o(DC)} = 7.5$ V and $I_L = 200$ mA. Assume the nominal line voltage $v_1 = 115$ V, and use a transformer with turns ratio of 10:1.

Chapter

BIPOLAR JUNCTION TRANSISTORS AND DC BIASING THE BJT AMPLIFIER

Analysis & Design

3.1 INTRODUCTION

Having explored and studied the *p-n* junction diode and its applications in the previous two chapters, we now turn our attention to a three-layer device (*pnp* or *npn*) known as the *bipolar junction transistor* (BJT). A BJT is basically a current-controlled device and may be looked at as a current-controlled current source. The *bipolar junction transistor* is so named (*bipolar*) because its operation depends on both types of charge carriers: holes and electrons. Other devices that operate with only one type of charge carrier such as *field-effect transistors* (FET) are unipolar. This chapter covers the basic theory of transistor operation, the transistor characteristics, and DC biasing methods. We will examine several different biasing methods and discuss their advantages and disadvantages. Once we are thoroughly familiar with the popular biasing methods, we will focus our attention on the DC bias design of BJT amplifiers. The AC operation of a BJT amplifier, commonly referred to as the small-signal operation, depends on the DC bias design with a properly selected stable operating point. In this chapter, we will carry out a thorough and in-depth coverage of the analysis and design of several DC biasing methods. The following three chapters will be dedicated to the study of the small-signal operation: modeling, analysis, and design of BJT amplifiers.

3.2 BASIC THEORY OF BJT OPERATION

The bipolar junction transistor is a three-layer, three-terminal device. The three layers consist of two layers of the same type of semiconductor material (*n*-type or *p*-type) and a single layer of the opposite type of semiconductor (*p*-type or *n*-type). The single layer of the opposite type is sandwiched between the two layers of the same type, making the transistor either *npn* or *pnp,* as illustrated in Figure 3-1. In either case, there are two *p-n* junctions as the semiconductor type changes from *n* to *p* and back to *n*, or from *p* to *n* and back to *p*. The back-to-back diodes below are approximate equivalent representations of the BJT, and are presented to help visualize the two junctions of the transistor.

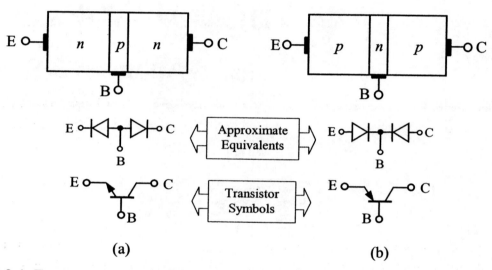

Figure 3-1: Two types of BJTs with corresponding approximate equivalents and transistor symbols: (a) *npn*, (b) *pnp*

The three layers, as shown in Figure 3-1, are termed **emitter**, **base**, and **collector** and labeled *E, B*, and *C*, respectively. The base region, which is very lightly doped compared to the other two regions, is sandwiched between the emitter and the collector layers. The emitter region is heavily doped, and the collector region is moderately doped. It must be stressed that the BJT is much more than two diodes connected back-to-back as illustrated in Figure 3-1. In fact, two diodes connected in such a fashion will not function as a BJT. These sketches, referred to as approximate equivalents, are presented to help visualize the two junctions of the BJT and will be particularly useful in transistor biasing. Figure 3-1 also displays the corresponding transistor symbols. Note that in both cases the lead with the arrow is the emitter terminal and that the tip of the arrow points to the *n*-type semiconductor. Therefore, when the arrow is pointing out, the transistor is *npn*, and when the arrow is pointing in, the transistor is *pnp*. Here is an easy way to remember all this: "*npn* stands for *not pointing in!*"

To properly bias the transistor for normal operation in the active region, one junction must be forward-biased while the other junction is reverse-biased. One such bias arrangement of a *npn* transistor is shown in Figure 3-2, where the emitter-base junction is forward-biased and the collector-base junction is reverse-biased.

Chapter 3

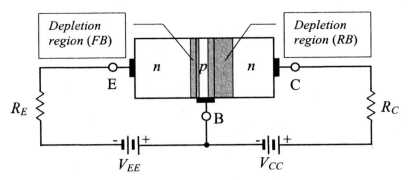

Figure 3-2: Biasing an *npn* transistor

For the sake of clarity, let us consider the biasing of each junction separately and then recombine the results.

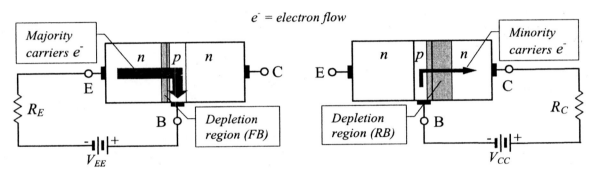

Figure 3-3(a): Forward biasing the emitter-base junction

Figure 3-3(b): Reverse biasing the collector-base junction

The forward-biased emitter-base junction of Figure 3-3(a) is similar to a forward-biased *p-n* junction diode. As a result of the forward bias, the depletion region at this junction narrows, allowing the majority carriers (electrons) to diffuse into the base.

The reverse-biased collector-base junction of Figure 3-3(b) is similar to a reverse-biased *p-n* junction diode. As a result of the reverse bias, the depletion region widens, inhibiting the flow of majority carriers. However, a limited number of minority carriers (electrons) will drift through the junction, establishing what is known as the reverse current or the leakage current. Let us now apply both the V_{CC} and the V_{EE} simultaneously and combine the results.

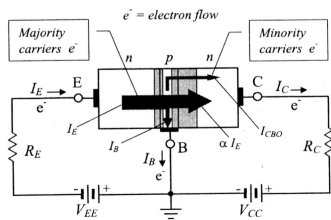

Figure 3-4: Electron flow of majority and minority carriers in a biased *npn* transistor

With the emitter-base junction forward-biased and the collector-base junction reverse-biased as illustrated in Figure 3-4, we are now looking at the whole picture. Remember that the base region is very thin and is very lightly doped, which means there are very few holes in that region compared to the number of electrons injected from the emitter. Thus, very few of the injected electrons will recombine with the holes. Instead, a majority of them will diffuse to the collector-base junction where they will be swept across that junction by the electric field set up by the positive V_{CC}. In addition to the majority carrier electrons, there are also the minority carrier electrons of the reverse-biased collector-base junction that have already established a path of their own. As mentioned earlier, the flow of these minority carrier electrons is actually the leakage current under reverse bias and is called I_{CBO} or I_{CO} (*collector-to-base current with the emitter open*). For an ideal transistor, this leakage current would have a value of zero. In a general-purpose transistor, I_{CBO} could be a few μA or even nA, while the collector current I_C and emitter current I_E would be in the mA range. Treating the transistor of Figure 3-4 as a current node, where a single current (I_E) enters and the two currents (I_C and I_B) exit the transistor, we can write the following equations using Kirchhoff's current law:

$$I_E = I_B + I_C \tag{3-1}$$

where

$$I_C = I_{C(majority)} + I_{C(minority)} = I_{C(maj)} + I_{CBO} \tag{3-2}$$

Since the I_{CBO} is quite small compared to $I_{C(maj)}$, I_{CBO} can be ignored in most applications. However, it must be noted that I_{CBO} is very sensitive to variations in temperature and can approximately double for every 10°C rise in temperature. Thus, it can become a concern when high temperature variations and large power dissipations are expected.

You have probably been wondering about the addition of the ground connection in Figure 3-4. Frankly, introduction of the ground connection in Figure 3-4 does not make it any different from the circuit of Figure 3-2 where there is no ground connection. It only helps establish a common reference point that happens to be the *base* terminal. With the introduction of this common ground, it is now clear that for a *npn* transistor, the *emitter* terminal is at a negative potential and the *collector* terminal is at a positive potential with respect to the *base* terminal where the voltage level is defined to be zero (*ground level*). These are the conditions required for the transistor to operate in the active region. See Figure 3-9 for a graphical description of the *active region* where the input current (I_E) and the output voltage (V_{CB}) are both positive. Let us repeat this in another form: *To properly bias the BJT for normal operation in the active region, **the emitter-base junction must be forward-biased, and the collector-base junction must be reverse-biased.***

Remember that the directions of currents I_E, I_C, and I_B, as shown in Figure 3-4, are the direction of the *electron flow*. For the *conventional flow* that we will be using throughout the text, the current directions will have to be reversed. Therefore, the schematic diagram for the setup of Figure 3-4 and the current directions therein will be redrawn as illustrated in Figures 3-5(a) and (b).

Chapter 3

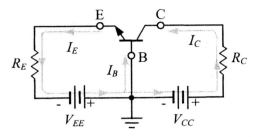

Figure 3-5(a): Direction of *conventional* current flow in a biased *npn* transistor

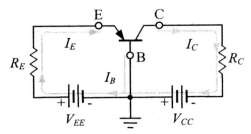

Figure 3-5(b): Direction of *conventional* current flow in a biased *pnp* transistor

Notice that with *conventional flow* the emitter current I_E always flows in the direction of the emitter arrow, and the other two currents, I_B and I_C, join together and make up the I_E. This fundamental current relationship is always in force for both *npn* and *pnp* transistors. Let us repeat this in mathematical notation.

$$I_E = I_B + I_C \qquad (3\text{-}3)$$

So far, we have discussed the transistor theory mainly in terms of the *npn* type transistors. However, the underlying theory presented thus far is equally applicable to the *pnp* type transistors. The only difference will be reversing the polarity of the supply voltages V_{CC} and V_{EE} as illustrated in Figure 3-5(b). We will, however, examine a circuit with a *pnp* transistor later in the chapter.

3.3 COMMON-EMITTER CHARACTERISTICS

In general, the BJT may be configured in one of the three bias arrangements: *common-base*, *common-emitter*, and *common-collector*. However, as you will see in the following chapters, common-emitter is the most widely used and frequently encountered configuration of the three bias arrangements. Hence, we will only examine the common-emitter characteristics of the BJT. With the common-emitter, the emitter will be the common reference or the common ground to both the base and the collector terminals.

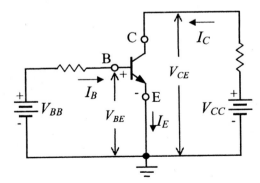

Figure 3-6(a):
npn transistor biased in the *common-emitter* configuration

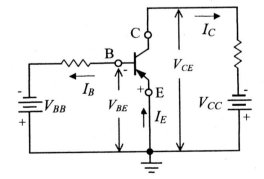

Figure 3-6(b):
pnp transistor biased in the *common-emitter* configuration

As illustrated in Figures 3-6(a) and 3-6(b), the base-emitter junction is forward-biased, while the collector-base junction is reverse-biased. Apparently, the base is the input and the collector is the output in both the *npn* and the *pnp* bias circuits.

Note that the fundamental current relation that was true with the common-base is also true with the common-emitter, that is:

$$I_E = I_C + I_B \qquad (3\text{-}4)$$

Before we start examining the input and output characteristics of the common-emitter configuration, let us introduce an important transistor parameter called β (*beta*). β is the forward current gain of the common-emitter configuration. In other words, β is the ratio of the output current I_C to the input current I_B.

$$\beta = \frac{I_C}{I_B} \qquad (3\text{-}5)$$

A typical β of general-purpose transistors such as 2N3904 or 2N2222 would be in the range of 100 to 300. β is also referred to as h_{FE} in some data sheets. Solving for I_C in the above equation, we have

$$I_C = \beta I_B \qquad (3\text{-}6)$$

also

$$I_E = I_B + I_C = I_B + \beta I_B = (\beta + 1)I_B \qquad (3\text{-}7)$$

Therefore,

$$I_E = (\beta + 1)I_B \qquad (3\text{-}8)$$

Also, note that the ratio of I_C to I_E is referred to as α (alpha).

$$\alpha = \frac{I_C}{I_E} \qquad (3\text{-}9)$$

Substituting for I_C and I_E, we will have

$$\alpha = \frac{\beta I_B}{(\beta + 1)I_B} \qquad (3\text{-}10)$$

Therefore,

$$\alpha = \frac{\beta}{\beta + 1} \qquad (3\text{-}11)$$

Proofs of the following two equations (3-12) and (3-13) are left to the student as an exercise at the end of this chapter.

$$\beta = \frac{\alpha}{1 - \alpha} \qquad (3\text{-}12)$$

$$\beta + 1 = \frac{1}{1 - \alpha} \qquad (3\text{-}13)$$

Example 3-1 — COMMON-EMITTER

An *npn* transistor with $\alpha = 0.995$ is biased in a circuit and is allowed a base current of 25 µA. Answer the following:
  a) Determine the collector current I_C and the emitter current I_E.
  b) Determine the approximate input voltage V_{BE}.

Solution:

a) $\beta = \dfrac{\alpha}{1-\alpha} = \dfrac{0.995}{1-0.995} = 199$

$I_C = \beta I_B = 199 \times 25\ \mu A = 4.975\ mA$

$I_E = (\beta + 1) I_B = (199 + 1) \times 25\ \mu A = 5\ mA$

b) V_{BE}, which is the voltage across the forward-biased junction, is approximately 0.7 V.

3.3.1 Input Characteristics of the Common-Emitter Configuration

Since the input of the common-emitter configuration is the forward-biased base-emitter junction, its characteristics are similar to those of a *p-n* junction diode. Hence, the plot of the input current I_B versus the input voltage V_{BE}, for different values of V_{CE} generates a family of curves as illustrated in Figure 3-7.

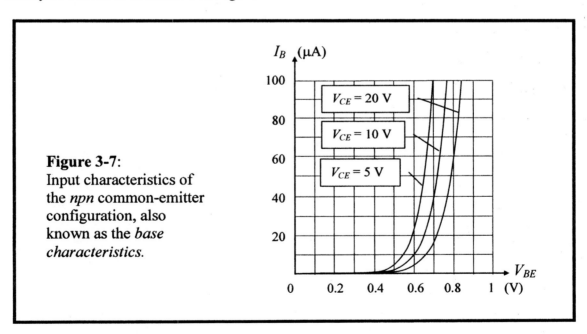

Figure 3-7: Input characteristics of the *npn* common-emitter configuration, also known as the *base characteristics*.

Notice that the base current I_B decreases with increased V_{CE} for a fixed V_{BE}. This is due to widening of the reverse-biased collector-emitter depletion region as voltage V_{CE} increases. The widening of the depletion region makes the base region narrower. With the narrowed base, fewer of the injected majority carriers go through the recombination

Chapter 3

process, resulting in a smaller base current. We will revisit this phenomenon while examining the output characteristics. The input characteristics of the common-emitter are also referred to as the *base characteristics*.

3.3.2 Output Characteristics of the Common-Emitter Configuration

To determine the output characteristics of any configuration, we need to examine the relationship between the output current and the output voltage, and the effect of the input voltage or current on this relationship. Referring to Figure 3-8, the output current is the I_C, the output voltage is V_{CE}, and the input current is I_B. Thus, imagine a test circuit as illustrated in Figure 3-8 with adjustable supply voltages V_{BB} and V_{CC}, so that we can control the input current I_B and the output voltage V_{CE}.

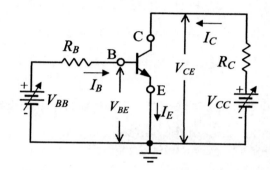

Figure 3-8:
A test setup for measuring the output characteristics of the *common-emitter* configuration, also known as the *collector characteristics*.

In addition to the test circuit of Figure 3-8, we also need Table 3-1 in order to collect the resulting data, as follows:

1. With $V_{CC} = 0$, set the I_B to 90 µA, by adjusting the V_{BB}. Then set V_{CE} to 0 V, 1 V, 2 V, 3 V, and so on by adjusting the V_{CC}, while measuring and recording the I_C.
2. Repeat step 1 for different values of I_B as listed in Table 3-1.

Table 3-1: I_C (mA) to be measured and recorded at the corresponding I_B and V_{CE}

$I_B \downarrow$ $V_{CE} \rightarrow$	0 V	1 V	2 V	3 V	4 V	5 V	6 V	7 V	8 V	9 V	10 V
90 µA											
80 µA											
70 µA											
60 µA											
50 µA											
40 µA											
30 µA											
20 µA											
10 µA											
0 µA											

Plotting the 10 sets of collected data, I_C versus V_{CE} for 10 different I_B settings, will produce the output characteristic curves of Figure 3-9.

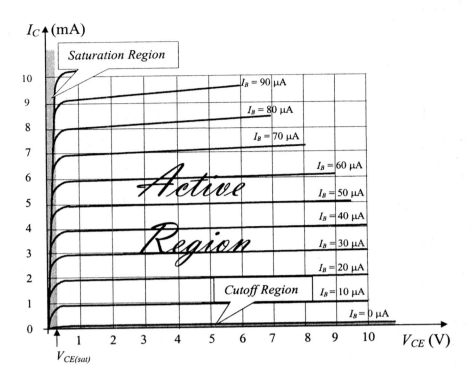

Figure 3-9: Common-emitter output characteristics

Examining the output characteristic curves of Figure 3-9, we note that the curves get steeper as the current level increases. In fact, if these lines are extended to the left, they are supposed to meet and intersect the V_{CE} axis at exactly the same point, as illustrated in Figure 3-10. This point, commonly denoted as V_A, is referred to as the *Early voltage* after J. M. Early, who discovered this and other characteristics of the transistor. The Early voltage is a useful transistor parameter that will be utilized later in the analysis and design of transistor circuits.

Figure 3-10: The Early voltage V_A

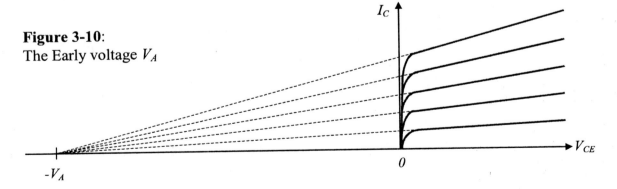

As in the case of the common-base, if the V_{CE} is increased beyond the reverse breakdown voltage V_{BR}, the *punch through* phenomenon will take place and the breakdown will occur, resulting in a large and sudden current flow. The common-emitter output characteristics are also referred to as the *collector characteristics*.

An important reminder:
Note that the linear relationship of the collector current I_C to the base current I_B ($I_C = \beta I_B$) is limited to the active region only, and is not valid in the saturation region. That is, I_B may continue to increase, but the collector current will not increase beyond $I_{C(sat)}$.

Example 3-2 — COMMON-EMITTER

Assuming that Figure 3-9 shows the output characteristics of a certain transistor, determine the transistor β at the following voltages and currents:

I. With $V_{CE} = 6$ V, and:
   a. $I_B = 10$ μA
   b. $I_B = 30$ μA
   c. $I_B = 50$ μA

II. With $I_B = 60$ μA, and:
   a. $V_{CE} = 2$ V
   b. $V_{CE} = 4$ V
   c. $V_{CE} = 6$ V

I. The intersection of the vertical line at $V_{CE} = 6$ V with the I_B curves corresponds to the following collector currents, from which the β is calculated using Equation 3-8.
   a. $I_B = 10$ μA $I_C = 0.95$ mA $\beta = 95$
   b. $I_B = 30$ μA $I_C = 3.0$ mA $\beta = 100$

II. The intersection of the 60 μA I_B curve with the vertical lines at each of the given V_{CE} corresponds to the following collector currents, from which the β is calculated using Equation 3-8.
   a. $V_{CE} = 2$ V $I_C = 5.95$ mA $\beta = 99$
   b. $V_{CE} = 4$ V $I_C = 6.0$ mA $\beta = 100$

3.4 BIASING THE COMMON-EMITTER AMPLIFIER

As mentioned previously, the common-emitter is the most popular and the most frequently encountered configuration. One of the reasons is that the common-emitter can be biased from a single DC source. A familiar common-emitter DC bias circuit is illustrated in Figure 3-11(a). Since both DC sources V_{BB} and V_{CC} are +10 V, both R_B and R_C can be attached to the same +10 V DC source, as illustrated in Figure 3-11(b).

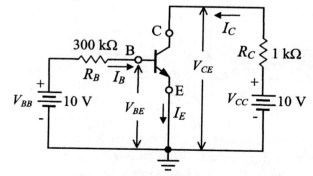

Figure 3-11(a): Dual-supply common-emitter DC bias circuit

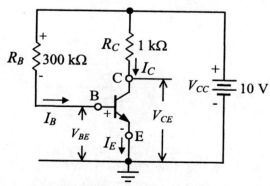

Figure 3-11(b): Single-supply common-emitter DC bias circuit

Chapter 3

One obvious question is *"What if V_{BB} is not equal to V_{CC}?"* The answer is that a different value of V_{BB}, in conjunction with the R_B, produces a certain I_B. When the V_{BB} is made equal to V_{CC} by using a single DC source, R_B can be adjusted to achieve the same I_B. In other words, there is no reason for V_{BB} to be different from V_{CC}. Hence, the common-emitter amplifier can run from a single DC source, because both sources are of the same value and polarity.

With the common-emitter configuration, several biasing methods can be employed. We will now explore these biasing methods.

3.4.1 Base Bias   ANALYSIS

The circuit of Figure 3-11(c) and its equivalent Figure 3-11(b) are biased in the simplest of all biasing methods, known as the *base bias* or *fixed bias*. Figure 3-11(c) is the exact equivalent of Figure 3-11(b) but with a simpler schematic diagram, which is the preferred style of drawing a common-emitter. Let us now analyze the DC bias of Figure 3-11(c).

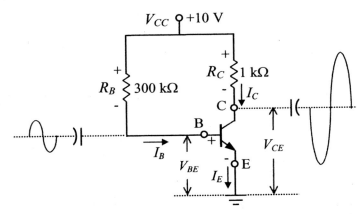

Figure 3-11(c): Base-biased common-emitter amplifier circuit.

Figure 3-11(c) represents the circuit of a basic common-emitter amplifier, biased in the simple biasing method of *base bias*. As an amplifier, this circuit can receive a small signal at the input and produce a large signal at the output. At the same time, this circuit will take a small DC current at the input (I_B) and produce a large DC current at the output (I_C). As we have seen previously, the ratio of the output current I_C to the input current I_B is the DC current gain of the common-emitter, referred to as the DC β or h_{FE}.

The purpose of the two capacitors at the input and the output is to isolate the DC operation of the amplifier from the signal source at the input side and from the load at the output side. In other words, the DC operation is contained between the two capacitors.

To begin the DC analysis of Figure 3-11(c), we will start with the input loop or the input path. The input path starts at the V_{CC} and ends at the ground, with the input current I_B passing through R_B and the forward biased base-emitter junction. The voltage drop at the base-emitter junction (V_{BE}) is assumed to be the average of 0.7 V.

The Input Path

The input path consists of the DC source V_{CC}, the resistor R_B, and the forward-biased emitter-base junction. The source V_{CC} must equal the sum of the voltage drops along the input path. Therefore, we can write the following KVL equation.

$$V_{CC} = I_B R_B + V_{BE} \tag{3-14}$$

Rearranging the above equation results in the following:

$$V_{CC} - V_{BE} = I_B R_B \tag{3-15}$$

Solving for the input current, we obtain the following equation for I_B:

$$I_B = \frac{V_{CC} - V_{BE}}{R_B} \tag{3-16}$$

Substituting the values for all variables, we get the exact value for I_B.

$$I_B = \frac{10\,\text{V} - 0.7\,\text{V}}{300\,\text{k}\Omega} = 31\,\mu\text{A}$$

Assuming a typical β of 150, we can solve for the output current I_C.

$$I_C = \beta I_B = 150 \times 31\,\mu\text{A} = 4.65\,\text{mA}$$

The Output Path

Similarly, applying the KVL to the output path will result in the following equation. That is, the source V_{CC} must equal the sum of the voltage drops around the output path.

$$V_{CC} = I_C R_C + V_{CE} \tag{3-17}$$

Solving for V_{CE} results in the following.

$$V_{CE} = V_{CC} - I_C R_C \tag{3-18}$$

Substituting the values for all variables, we obtain the value for V_{CE}, the output voltage.

$$V_{CE} = 10\,\text{V} - 4.65\,\text{mA} \times 1\,\text{k}\Omega = 5.35\,\text{V}$$

Hence, the operating point for the circuit is: $I_C = 4.65$ mA, and $V_{CE} = 5.35$ V.

The Operating Point and the DC Load Line

As mentioned previously, the AC operation or the amplifying action of the BJT amplifier depends heavily on the DC operation and the properly selected DC operating point. The DC operating point is actually a point on the output characteristic curves of the transistor that corresponds to a specific input current, output current, and output voltage. Let us take another look at the common-emitter output characteristics.

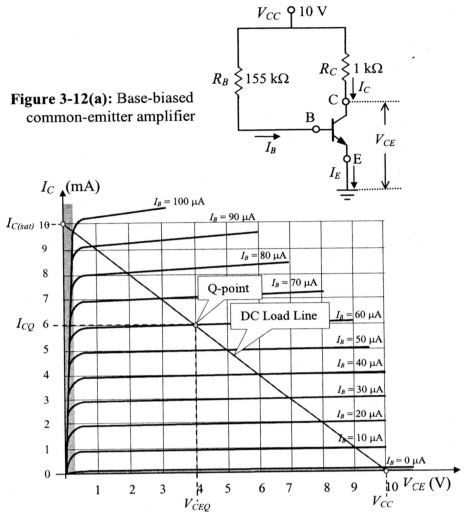

Figure 3-12(a): Base-biased common-emitter amplifier

Figure 3-12(b) Output characteristics, DC load line and the Q-point

Any point in the active region will correspond to some specific input current I_B, output current I_C, and output voltage V_{CE}. Consider the amplifier circuit of Figure 3-12(a). The DC input and output currents and the output voltage are as follows:

$$I_B = \frac{V_{CC} - V_{BE}}{R_B} = \frac{10\text{ V} - 0.7\text{ V}}{155\text{ k}\Omega} = 60\text{ μA}$$

The average β at I_B of 60 μA is about 100, therefore

$I_C = \beta I_B = 100 \times 60\text{ μA} = 6\text{ mA}$, and

$V_{CE} = V_{CC} - I_C R_C = 10 - 6\text{ V} = 4\text{ V}$.

The Q-point will be at the intersection of the I_B curve (60 μA) and the DC load line. The DC load line is simply the plot of the output equation (Equation 3-18, I_C versus V_{CE}), which is a linear equation. To construct a line, two points are needed. One convenient set of these two points are the x and y intercepts. In our case, these two points would be the

Chapter 3

V and I intercepts. To be more specific, they would be the V_{CE} intercept and the I_C intercept. These two intercepts are referred to as the *cutoff voltage* and the *saturation current*.

The output equation (Equation 3-18):

$$V_{CE} = V_{CC} - I_C R_C$$

Cutoff Voltage $V_{CE(cutoff)}$:

Setting $I_C = 0$, in the above equation, we obtain the V_{CE} intercept.

$$V_{CE(cutoff)} = V_{CE}\Big|_{I_C = 0} = V_{CC} \tag{3-19}$$

Saturation Current $I_{C(sat)}$:

Setting $V_{CE} = 0$, in the output equation, we obtain the I_C intercept.

$$I_{C(sat)} = I_C\Big|_{V_{CE} = 0} = \frac{V_{CC}}{R_C} \tag{3-20}$$

$$V_{CE(cutoff)} = V_{CC} = 10 \text{ V}, \quad \text{and} \quad I_{C(sat)} = \frac{10 \text{ V}}{1 \text{ k}\Omega} = 10 \text{ mA}$$

A line drawn from the $I_{C(sat)}$ (10 mA) to the V_{CC} (10 V) is the DC load line. The Q-point, which is the intersection of the load line and the I_B curve (60 µA), corresponds to an I_C of 6 mA and a V_{CE} of 4 V. These Q-point values are in full agreement with the algebraically obtained I_C and V_{CE}.

Practice Problem 3-1 ANALYSIS

Figure 3-12(b) represents the collector characteristics of a generic transistor that is used in a base-biased common-emitter amplifier with the following parameters:

$$V_{CC} = 9 \text{ V}, \quad R_B = 270 \text{ k}\Omega, \quad R_C = 1.2 \text{ k}\Omega$$

a. Determine the I_B from the given values.
b. Draw the DC load line and locate the Q-point graphically. Record I_{CQ} and V_{CEQ}.
c. Determine the transistor β in the active region of operation.
d. Using the β found in step c, determine the theoretically expected I_{CQ} and V_{CEQ}.

Answers: $I_B \cong 30$ µA, $I_{CQ} \cong 3$ mA, $V_{CEQ} \cong 5.3$ V, $\beta \cong 100$

3.4.2 Emitter Bias — ANALYSIS

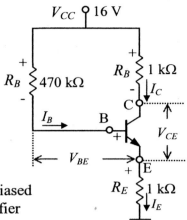

Figure 3-13: Emitter-biased common-emitter amplifier

The *emitter bias*, also referred to as the *emitter-stabilized bias*, is obtained by adding an emitter resistor to the *base bias* configuration, as illustrated in Figure 3-13. The emitter resistor gives some stability to the bias circuit by providing a current-series feedback. The topic of negative feedback is explored in Chapter 12.

The approach to analyze the emitter-biased amplifier will be the same approach we have used for the base bias. That is, we will start with the input path, find the input current I_B, and then analyze the output path to solve for V_{CE}.

The Input Path

Again, the sum of the voltage drops around the input path must equal the DC source V_{CC}.

$$V_{CC} = I_B R_B + V_{BE} + I_E R_E \tag{3-21}$$

Separating the variables and substituting for I_E in terms of I_B results in the following:

$$V_{CC} - V_{BE} = I_B R_B + I_B(\beta + 1) R_E \tag{3-22}$$

Factoring out the I_B, then solving for I_B, we obtain the following:

$$I_B = \frac{V_{CC} - V_{BE}}{R_B + (\beta + 1) R_E} \tag{3-23}$$

Assuming a β of 200 and substituting the values for all parameters yields the following:

$$I_B = \frac{V_{CC} - V_{BE}}{R_B + (\beta + 1) R_E} = \frac{16\text{ V} - 0.7\text{ V}}{470\text{ k}\Omega + (201)1\text{ k}\Omega} = 22.8\ \mu\text{A}$$

Having found I_B, we can then solve for I_C and I_E.

$$I_C = \beta I_B = 200 \times 22.8\ \mu\text{A} = 4.56\text{ mA}$$

$$I_E = (\beta + 1) I_B = 201 \times 22.8\ \mu\text{A} = 4.58\text{ mA}$$

The Output Path

Again, the DC source V_{CC} must be equal to the voltage drops in the output path.

$$V_{CC} = I_C R_C + V_{CE} + I_E R_E \tag{3-24}$$

Separating the variables and solving for V_{CE} results in the following equation:

$$V_{CE} = V_{CC} - I_C R_C - I_E R_E \tag{3-25}$$

Substituting the values for all components determines the V_{CE}.

$$V_{CE} = 16 - (4.56 \text{ mA} \times 1 \text{ k}\Omega) - (4.58 \text{ mA} \times 1 \text{ k}\Omega) = 6.86 \text{ V}$$

Since $I_C \cong I_E$, especially with high β, Equation 3-25 may be revised as follows:

$$V_{CE} = V_{CC} - I_C(R_C + R_E) \tag{3-26}$$

Saturation Current

By definition, the saturation current is the collector-emitter current with $V_{CE} = 0$.

$$I_{C(sat)} = \frac{V_{CC}}{R_C + R_E} \tag{3-27}$$

Example 3-3 — Emitter Bias — ANALYSIS

Carry out the following for the emitter-biased amplifier circuit of Figure 3-14.
a. Draw the DC load line on the collector characteristic curves of Figure 3-15.
b. Algebraically determine the Q-point values.
c. Locate the Q-point on the load line.
d. Compare the algebraically determined and graphically located Q-points.

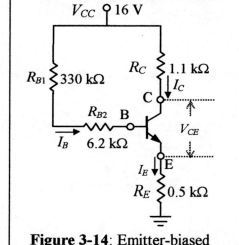

Figure 3-14: Emitter-biased common-emitter amplifier

Solution:

Load line will be drawn from $V_{CC} = 16$ V to $I_{C(sat)}$.

$$I_{C(sat)} = \frac{V_{CC}}{R_C + R_E} = \frac{16 \text{ V}}{1.6 \text{ k}\Omega} = 10 \text{ mA}$$

After drawing the load line on the characteristics of Figure 3-15, we look for an average β value in that region. The average β in that region of the characteristics is about 200. The base current I_B is determined with Equation 3-23, as follows:

$$I_B = \frac{V_{CC} - V_{BE}}{R_B + (\beta+1)R_E} = \frac{16 \text{ V} - 0.7 \text{ V}}{(330 \text{ k}\Omega + 6.2 \text{ k}\Omega) + (201)(0.5 \text{ k}\Omega)} = 35 \mu A$$

Therefore, $I_C = \beta I_B = 200 \times 35 \mu A = 7$ mA.
Using Equation 3-26, we can solve for V_{CE}.

$$V_{CE} = V_{CC} - I_C(R_C + R_E) = 16 \text{ V} - (7 \text{ mA} \times 1.6 \text{ k}\Omega) = 4.8 \text{ V}$$

Hence, the algebraically calculated Q-point is at $I_{CQ} = 7$ mA, and $V_{CEQ} = 4.8$ V.

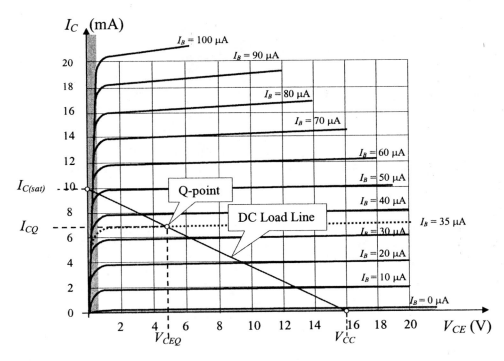

Figure 3-15: DC load line and the Q-point for the amplifier of Figure 3-14

The graphical Q-point is at I_{CQ} of approximately 7 mA, and V_{CEQ} of about 4.8 V. Thus, the algebraically determined Q-point values of 7 mA and 4.8 V agree very closely with the graphically found Q-point.

Practice Problem 3-2   (Emitter Bias)   ANALYSIS

Determine the I_{CQ} and V_{CEQ} for the amplifier of Figure 3-16. Assume $\beta = 180$.

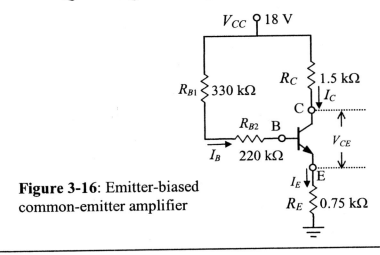

Figure 3-16: Emitter-biased common-emitter amplifier

Answers: $I_{CQ} = 4.54$ mA,   and $V_{CEQ} = 7.78$ V

Voltage-Divider Bias — ANALYSIS

The voltage-divider bias is an improved version of the emitter bias, which is obtained by the addition of another resistor to the emitter bias configuration. The advantage of the voltage-divider bias over the emitter bias is that if, and only if, it is designed correctly, the voltage-divider biased circuit becomes β-*independent*. The transistor β is sensitive to variations in temperature, and it can vary as the temperature changes. Hence, a circuit that is β-dependent will not be very stable, and the established Q-point will vary with changes in temperature. Although the emitter resistor in the emitter-biased configuration does provide some stability, the circuit still is β-dependent. Aside from the sensitivity of the transistor β to variations in temperature, it would not be uncommon for the same type of general-purpose transistors to have β in the range of 100 to about 300. Thus, if a circuit is designed using a transistor with a certain β, there is no guarantee that all transistors will have the same β. The voltage-divider bias, if designed correctly, takes care of this problem, and is therefore the most popular biasing method. A voltage-divider bias circuit is illustrated in Figure 3-17.

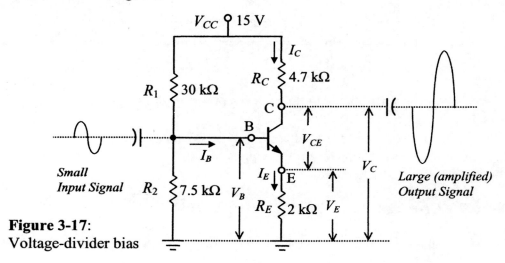

Figure 3-17: Voltage-divider bias

There are two approaches to analyzing the voltage-divider bias: the exact approach, and the approximate approach. If designed correctly, the result of the approximate analysis will be very close to the exact analysis, and the circuit will be β-independent. We will try both approaches with the circuit of Figure 3-17.

The Approximate Method — ANALYSIS

Applying the voltage-divider rule at the base, we can solve for the approximate V_B.

$$V_B = V_{CC} \frac{R_2}{R_1 + R_2} \qquad (3\text{-}28)$$

Substituting all component values in the above equation yields the base voltage.

$$V_B = 15\,\text{V} \frac{7.5\,\text{k}\Omega}{30\,\text{k}\Omega + 7.5\,\text{k}\Omega} = 3\,\text{V}$$

The emitter voltage is found by subtracting the 0.7 V drop across the base-emitter junction from the base voltage.

$$V_E = V_B - V_{BE} \tag{3-29}$$

$$V_E = 3\text{ V} - 0.7\text{ V} = 2.3\text{ V}$$

Having found the emitter voltage, we can readily solve for the emitter current.

$$I_E = \frac{V_E}{R_E} \tag{3-30}$$

$$I_E = \frac{V_E}{R_E} = \frac{2.3\text{ V}}{2\text{ k}\Omega} = 1.15\text{ mA}$$

Since I_C and I_E are usually very close in value, it is permissible to approximate $I_C = I_E$; therefore,

$$I_C \cong I_E = 1.15\text{ mA} \tag{3-31}$$

Having solved for I_E or I_C, we move to the output path, as usual. That is, the source V_{CC} must be equal to the voltage drops around the output path.

$$V_{CC} = I_C R_C + V_{CE} + I_E R_E \tag{3-32}$$

Replacing I_E with I_C in Equation 3-25 and solving for V_{CE} yields the following:

$$V_{CE} = V_{CC} - I_C (R_C + R_E) \tag{3-33}$$

$$V_{CE} = 15 - 1.15\text{ mA }(4.7\text{ k}\Omega + 2\text{ k}\Omega) = 7.3\text{ V}$$

The Exact Method — ANALYSIS

The exact method requires Thevenizing the base circuit.

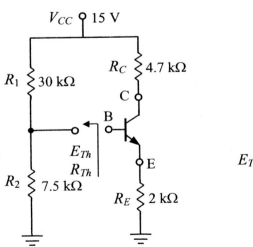

Figure 3-18: Thevenizing Figure 3-17

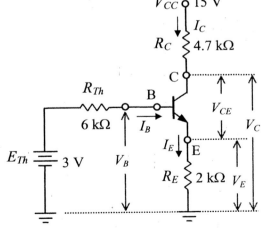

Figure 3-19: Thevenized equivalent of Figure 3-17

$$E_{th} = V_{CC}\frac{R_2}{R_1 + R_2} \qquad (3\text{-}34)$$

$$R_{Th} = R_1 \parallel R_2 \qquad (3\text{-}35)$$

$$E_{Th} = 15\text{ V}\frac{7.5\text{ k}\Omega}{30\text{ k}\Omega + 7.5\text{ k}\Omega} = 3\text{ V, and} \qquad R_{Th} = 7.5\text{ k}\Omega \parallel 30\text{ k}\Omega = 6\text{ k}\Omega$$

Having determined the E_{Th} and R_{Th}, we then concentrate on the Thevenin's equivalent circuit of Figure 3-18, and determine I_B by applying KVL to the input loop.

$$E_{Th} = (I_B \times R_{Th}) + V_{BE} + (I_E \times R_E) \qquad (3\text{-}36)$$

Substituting for I_E in terms of I_B, yields the following equation.

$$E_{Th} = (I_B \times R_{Th}) + V_{BE} + [I_B(\beta + 1) \times R_E] \qquad (3\text{-}37)$$

Factoring out the I_B, separating the variables, and solving for I_B results in the following:

$$I_B = \frac{E_{Th} - V_{BE}}{R_{Th} + (\beta + 1)R_E} \qquad (3\text{-}38)$$

Apparently, we need a value for β. Let $\beta = 199$ and substitute for all variables.

$$I_B = \frac{3\text{ V} - 0.7\text{ V}}{6\text{ k}\Omega + (200)(2\text{ k}\Omega)} = 5.665\ \mu\text{A}$$

$I_C = \beta I_B = 199 \times 5.665\ \mu\text{A} = 1.127335$ mA, $I_E = I_B(\beta + 1) = 5.665\ \mu\text{A} \times 200 = 1.133$ mA

Then we can solve for V_{CE}.

$$V_{CE} = V_{CC} - I_C R_C - I_E R_E \qquad (3\text{-}39)$$

$V_{CE} = 15\text{ V} - (1.127335\text{ mA} \times 4.7\text{ k}\Omega) - (1.133\text{ mA} \times 2\text{ k}\Omega) = 7.4355$ V

Notice that with the approximate method β was never used. The absence of β in the whole process verifies the independence of the circuit on β.

Now, let us compare the Q-point values determined using the approximate method with those determined using the exact method.

Q-point	Approx. Method	Exact Method	%Difference
I_E	1.15 mA	1.133 mA	1.5%
V_{CE}	7.3 V	7.4355 V	1.8%

Table 3-2: Approximate solutions versus exact solutions

As depicted in Table 3-2, the approximate solutions are fairly close to the exact solutions, with a maximum difference of no more than 1.8%. But remember that this is not always the case with every voltage-divider bias circuit. *The approximate solutions will be close to the exact solutions only if the circuit is correctly designed.* The approximate solutions are based on the assumption that the resistance R_B seen from the base, as shown in Figure 3-20, is much higher than the resistance R_2 so that R_2 in parallel with $R_B \cong \beta R_E$ is approximately equal to R_2. This assumption allows us to use Equation 3-34 as a voltage divider between R_1 and R_2 to determine V_B without the need to use the transistor β.

$$R_B = R_{BE} + (\beta+1)R_E \cong \beta R_E \qquad (3\text{-}40)$$

Figure 3-20:
Clarification of the base resistance R_B, which is in parallel with R_2 in a voltage-divider bias circuit

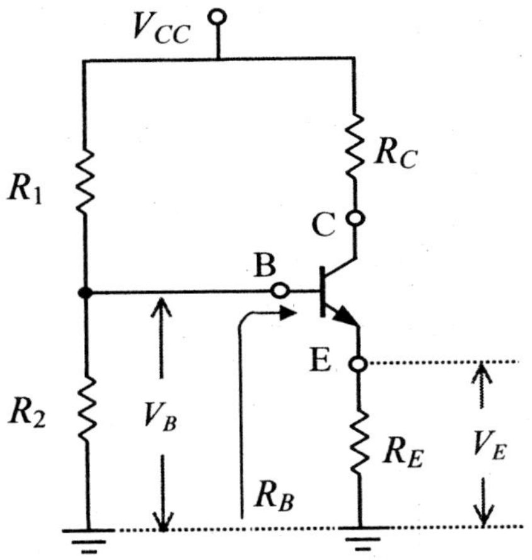

In order to have $R_2 \cong R_2 \| R_B$, $R_B \gg R_2$ or

$$R_2 \ll R_B \cong \beta R_E \qquad (3\text{-}41)$$

The larger the β, the larger is the βR_E than R_2.
Hence, let us assume a typical minimum β of 100 for a general-purpose transistor, and let R_B be larger than $10R_2$, or $10R_2$ smaller than βR_E.

$$10R_2 < \beta R_E = 100R_E \qquad (3\text{-}42)$$

Hence, the rule of thumb for a properly designed β-independent voltage-divider bias configuration is as follows:

$$R_2 < 10R_E \qquad (3\text{-}43)$$

Otherwise, the circuit will not be β-independent, which is the principal objective of the voltage-divider bias.

Practice Problem 3-3   *Voltage-divider Bias*   **ANALYSIS**

Determine the I_{EQ} and V_{CEQ} for a voltage-divider biased BJT amplifier with the following parameters: $R_1 = 39\ k\Omega$, $R_2 = 11\ k\Omega$, $R_C = 2.7\ k\Omega$, $R_E = 1.3\ k\Omega$, $V_{CC} = 18\ V$
  a) Using the approximate method.
  b) Using the exact method, $\beta = 160$

Answers:

Approximate: $I_{CQ} = 2.5\ mA$, $V_{CEQ} = 8\ V$ Exact: $I_{CQ} = 2.4\ mA$, $V_{CEQ} = 8.4\ V$

3.5 BIASING THE COMMON-COLLECTOR AMPLIFIER

The practical common-collector configuration is very similar to the emitter-biased common-emitter configuration, as shown in Figure 3-21 below. Hence, the DC bias analysis of the common-collector will be similar to that of the common-emitter amplifier.

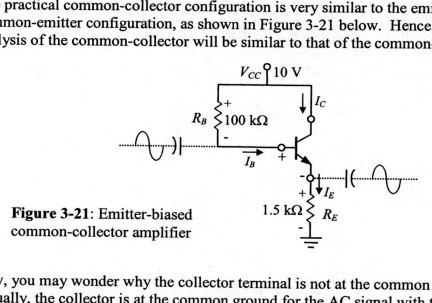

Figure 3-21: Emitter-biased common-collector amplifier

Now, you may wonder why the collector terminal is not at the common ground. Actually, the collector is at the common ground for the AC signal with the base being the input and the emitter being the output as shown in Figure 3-21.

Example 3-4 — Common Collector — ANALYSIS

Let us determine I_C and V_{CE} for the emitter-biased circuit of Figure 3-21 with $\beta = 199$.

Solution: Approach to analyzing this circuit will be the same approach we have used to analyze the emitter-biased common-emitter amplifier, except that there is no R_C. Using Equation 3-23, we can solve for the base current.

$$I_B = \frac{V_{CC} - V_{BE}}{R_B + (\beta+1)R_E} = \frac{10\text{ V} - 0.7\text{ V}}{100\text{ k}\Omega + (200)(1.5\text{ k}\Omega)} = 23.25\text{ }\mu\text{A}$$

$$I_E = (\beta+1)I_B = 200 \times 23.25\text{ }\mu\text{A} = 4.65\text{ mA}$$

$$V_{CE} = V_{CC} - I_E R_E = 10\text{ V} - (4.65\text{ mA} \times 1.5\text{ k}\Omega) = 3\text{ V}$$

Practice Problem 3-4 — Common Collector — ANALYSIS

Determine I_E and V_{CE} for the emitter-biased circuit of Figure 3-21 with $\beta = 99$.

Answers: $I_E = 3.72$ mA, $V_{CE} = 4.4$ V

3.6 BIASING THE COMMON-BASE AMPLIFIER

The common-base amplifier does not have much practical application, because it suffers from a very low input resistance; however, its advantage is the superior high-frequency response. Its practical application is in the cascode amplifier, which is a cascade of the common-emitter and the common-base amplifiers. Hence, in order to be prepared for the cascode amplifier, we will take a quick look at the common-base DC bias analysis.

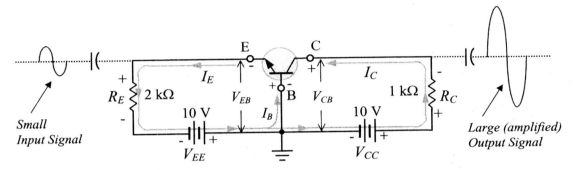

Figure 3-22: Common-base bias amplifier circuit

Recall that when analyzing the DC bias of a BJT amplifier, there will always be two loops or current paths to be considered for analysis, the *input loop* and the *output loop*. Each loop must be considered separately, the input loop being the first to be analyzed.

The Input Loop

The input loop is defined as the loop or the current path that contains the forward biased base-emitter junction. In Figure 3-22, the input loop comprises a negative DC source V_{EE}, a resistor R_E, and the forward-biased base-emitter junction. The voltage drop across the forward biased base-emitter junction will be assumed to be 0.7 V. Since the DC source in the input loop is negative, direct application of Kirchhoff's voltage law (KVL) to the input loop will result in a negative current. To avoid the negative current, let us apply the KVL in a different style, without worrying about the polarities. That is, in a closed loop, the source voltage (absolute value) must equal the sum of the individual voltage drops.

$$V_{EE} = V_{BE} + I_E R_E$$
$$V_{EE} - V_{BE} = I_E R_E \tag{3-44}$$

Solving for the input current I_E results in the following:

$$I_E = \frac{V_{EE} - V_{BE}}{R_E} \tag{3-45}$$

$$I_E = \frac{10\,\text{V} - 0.7\,\text{V}}{2\,\text{k}\Omega} = 4.65\,\text{mA}$$

$$V_{EB} = -V_{BE} = -0.7\,\text{V} \tag{3-46}$$

Chapter 3

The Output Loop

The output loop consists of a DC source voltage V_{CC}, a resistor R_C, and the reverse-biased collector-base junction. Applying KVL to the output loop results in the following:

$$V_{CC} = I_C R_C + V_{CB} \qquad (3\text{-}47)$$

Solving for the junction voltage V_{CB} yields

$$V_{CB} = V_{CC} - I_C R_C \qquad (3\text{-}48)$$

Assuming α of 0.99, I_C is given by

$$I_C = \alpha I_E = 0.99 \times 4.65 \text{ mA} \cong 4.6 \text{ mA}$$

Substituting for all parameters in Equation 3-48, we obtain

$$V_{CB} = 10 \text{ V} - (1 \text{ k}\Omega \times 4.6 \text{ mA}) = 5.4 \text{ V}$$

This concludes the DC bias analysis of the common-base amplifier of Figure 3-22.

Example 3-5 ANALYSIS

Determine the input and output currents and voltages for the common-base amplifier of Figure 3-23, if $V_{EE} = +5$ V, $V_{CC} = -10$ V, $R_E = 1$ kΩ, $R_C = 1$ kΩ, and $\beta = 125$.

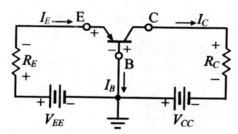

Figure 3-23:
Common-base amplifier (*pnp*)

Solution:

Recall that for a *pnp* transistor, all polarities and current directions are reversed. Thus,

$$I_E = \frac{V_{EE} - V_{EB}}{R_E} = \frac{5 \text{ V} - 0.7 \text{ V}}{1 \text{ k}\Omega} = 4.3 \text{ mA}$$

$$\alpha = \frac{\beta}{\beta + 1} = 0.992 \qquad V_{EB} = +0.7 \text{ V}$$

$$I_C = \alpha I_E = 0.992 \times 4.3 \text{ mA} = 4.27 \text{ mA}$$

$$V_{CB} = V_{CC} + I_C R_C$$

$$V_{CB} = -10 \text{ V} + 4.27 \text{ mA} \times 1 \text{ k}\Omega = -5.73 \text{ V}$$

$$V_{CE} = V_C - V_E = V_{CB} - V_{EB} = -5.73 - 0.7 = -6.43 \text{ V}$$

3.7 DC BIAS ANALYSIS OF THE CASCODE AMPLIFIER

The cascode amplifier is a cascade of a common-emitter and a common-base amplifier, which will be discussed and clarified in more detail when analyzing the AC operation. It has all the characteristics of the common-emitter amplifier, but possesses the superior high frequency response chracteristics of the common-base amplifier. A cascode amplifier circuit is shown in Figure 3-24 below.

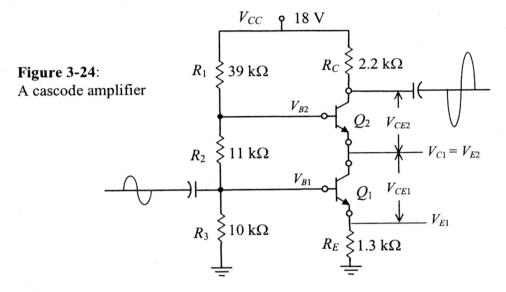

Figure 3-24: A cascode amplifier

Just like the voltage-divider biased common-emitter amplifier, if designed correctly, the result of the approximate analysis of the cascode amplifier will be very close to the exact analysis, and the circuit will be β-independent.

Applying the voltage-divider rule at the base of Q_1, we can solve for V_{B1}, as follows:

$$V_{B1} = V_{CC} \frac{R_3}{R_1 + R_2 + R_3} \qquad (3\text{-}49)$$

Substituting all component values in the above equation yields the base voltage V_{B1}.

$$V_{B1} = 18\,\text{V} \frac{10\,\text{k}\Omega}{39\,\text{k}\Omega + 11\,\text{k}\Omega + 10\,\text{k}\Omega} = 3\,\text{V}$$

The emitter voltage is found by subtracting the 0.7 V drop across the base-emitter junction from the base voltage.

$$V_{E1} = V_{B1} - V_{BE} \qquad (3\text{-}50)$$

$V_{E1} = 3\,\text{V} - 0.7\,\text{V} = 2.3\,\text{V}$

Having found the emitter voltage, we can readily solve for the emitter current.

$$I_{E1} = \frac{V_{E1}}{R_E} \qquad (3\text{-}51)$$

$$I_{E1} = \frac{V_E}{R_E} = \frac{2.3\,\text{V}}{1.3\,\text{k}\Omega} = 1.77\,\text{mA}$$

Chapter 3

$I_{C1} \cong I_{E1} = I_{C2} \cong I_{E2} = 1.77$ mA

Applying the voltage-divider rule one more time at the base of Q_2, we can solve for V_{B2}, as follows:

$$V_{B2} = V_{CC}\frac{R_2 + R_3}{R_1 + R_2 + R_3} \quad (3\text{-}52)$$

Substituting all component values in the above equation yields the base voltage V_{B2}.

$$V_{B2} = 18\text{ V}\frac{11\text{ k}\Omega + 10\text{ k}\Omega}{39\text{ k}\Omega + 11\text{ k}\Omega + 10\text{ k}\Omega} = 6.3\text{ V}$$

The emitter voltage of Q_2, which is also equal to the collector voltage of Q_1, is found by subtracting the 0.7 V drop across the base-emitter junction from the base voltage.

$$V_{E2} = V_{C1} = V_{B2} - V_{BE} \quad (3\text{-}53)$$

$V_{E2} = V_{C1} = 6.3 - 0.7\text{ V} = 5.6\text{ V}$

Voltage at the collector of Q_2 is found by subtracting the voltage drop across the R_C from the supply voltage V_{CC}, as follows:

$$V_{C2} = V_{CC} - I_C R_C$$

$V_{C2} = 18 - (1.77\text{ mA} \times 2.2\text{ k}\Omega) = 14.1\text{ V} \quad (3\text{-}54)$

Having determined the collector and emitter voltages V_C and V_E for both transistors, the collector-emitter voltages V_{CE1} and V_{CE2} are determined by subtracting the emitter voltage V_E from the collector voltage V_C for both transistors.

$$V_{CE1} = V_{C1} - V_{E1} \quad (3\text{-}55)$$

$V_{CE1} = 5.6 - 2.3\text{ V} = 3.3\text{ V}$

$$V_{CE2} = V_{C2} - V_{E2} \quad (3\text{-}56)$$

$V_{CE2} = 14.1\text{ V} - 5.6\text{ V} = 8.5\text{ V}$

Practice Problem 3-5 (Cascode Amplifier)   ANALYSIS

Determine the DC voltages V_{CE1} and V_{CE2} for a cascode amplifier with the following parameters:

$V_{CC} = 20$ V, $R_1 = 43$ kΩ, $R_2 = 15$ kΩ, $R_3 = 12$ kΩ, $R_E = 1.5$ kΩ, $R_C = 2$ kΩ

Answers: $I_E = 1.82$ mA, $V_{CE1} = 4.3$ V, $V_{CE2} = 9.4$ V

Chapter 3

Section Summary — DC Bias — ANALYSIS

Emitter bias:

$$I_B = \frac{V_{CC} - V_{BE}}{R_B + (\beta + 1)R_E}$$

$$I_E = (\beta + 1)I_B, \quad I_C = \beta I_B$$

$$V_{CE} = V_{CC} - I_C(R_C + R_E)$$

Voltage-divider bias:

$$V_B = V_{CC}\frac{R_2}{R_1 + R_2}$$

$$V_E = V_B - V_{BE}$$

$$I_E = \frac{V_E}{R_E}, \quad I_C \cong I_E$$

$$V_{CE} = V_{CC} - I_C(R_C + R_E)$$

Cascode amplifier:

$$V_{B1} = V_{CC}\frac{R_3}{R_1 + R_2 + R_3}$$

$$V_{E1} = V_{B1} - V_{BE}$$

$$I_E = \frac{V_E}{R_E}, \quad I_C \cong I_E$$

$$V_{C2} = V_{CC} - I_C R_C$$

$$V_{B2} = V_{CC}\frac{R_2 + R_3}{R_1 + R_2 + R_3}$$

$$V_{E2} = V_{C1} = V_{B2} - V_{BE}$$

$$V_{CE1} = V_{C1} - V_{E1}$$

$$V_{CE2} = V_{C2} - V_{E2}$$

3.8 READING THE TRANSISTOR DATA SHEETS

Data sheets or the specification sheets of the transistor contain a considerable amount of information regarding the DC and AC characteristics and the operation of the transistor. For the time being, we will examine some of the DC characteristics and specifications, leaving the AC characteristics for later chapters. In our examination of the transistor data sheets, we will review the specifications of Motorola's 2N2222A, which are included at the end of the chapter.

Maximum Ratings

Maximum ratings are usually the first series of specifications listed in the transistor data sheets. The first 3 entries in this list are V_{CEO}, V_{CBO}, and V_{EBO}. The letter "O" stands for "*open*", as was the case with I_{CEO}. For example, V_{CEO} is the collector-to-emitter voltage with the base open. Yes, you guessed it right, the missing terminal is the one that is open. Therefore, the maximum DC voltage that may be applied across the collector-emitter terminals (V_{CE}) of this transistor, with the base open, is 40 V. The next entry in the list is the *continuous collector current*, which is a maximum of 600 mA. Next in the list is the *total device dissipation* (P_D) @ 25°C, with a derating factor above 25°C. This is actually the power rating of the device, but note that there are two different P_D ratings. The 625 mW is the military rating, and the 1.5 W is the manufacturer's rating for commercial use, each with its own derating factor. That is, if the device is operated in an environment higher than 25°C, one must derate the given P_D by 12 mW for each 1°C, for the commercial use. The last entry is the *operating and storage junction temperature range*, which is from −55°C to +150°C.

Thermal Characteristics

Thermal ratings are primarily intended for the design engineer, to be considered in the circuits and systems design. We will review the thermal ratings in later chapters when analyzing or designing power amplifiers.

Electrical Characteristics

Under the *electrical characteristics,* there are the *off characteristics,* and the *on characteristics*, each listed separately.

- Off Characteristics

   The first three in the list are the minimum breakdown voltages $V_{(BR)CEO}$, $V_{(BR)CBO}$, and $V_{(BR)EBO}$ which are the same parameters listed in the maximum ratings list as V_{CEO}, V_{CBO}, and V_{EBO}. These are simply repeated here for the convenience of the reader. The following entries are the maximum cutoff currents such as I_{CBO}, I_{EBO}, and I_{CEO}, which are the *collector-to-base current with the emitter open, emitter-to-base current with the collector open*, and the *collector-to-emitter current with the base open*, respectively. For example, the I_{CEO} of 10 nA is the maximum collector-to-emitter current with the base current cutoff (open). There are also two more maximum cutoff currents in the list: I_{CEX} of 10 nA, and I_{BEX} of 20 nA. The I_{CEX} is the maximum collector-to-emitter current with the emitter-base junction reverse-biased with a 3 V potential difference, while the V_{CE} is kept at 60 V.

- On Characteristics

   The first item in the list is the DC current gain h_{FE} or β_{dc}. Recall that in the examination of the common-emitter output characteristics we found out that the value of β was changing with I_C and V_{CE}. Hence, instead of a single β listing, several minimum values of the h_{FE} are listed at different I_C values with a V_{CE} of 10 V. The next item in the list is the collector-emitter saturation voltage $V_{CE(sat)}$, which is a maximum of 0.3 V with an I_C of 150 mA, but it can be as high as 1 V if I_C is increased to 500 mA. The last item in the list is the base-emitter saturation voltage $V_{BE(sat)}$, which is a maximum of 1.2 V with an I_B of 15 mA, but it can be as high as 2 V if I_B is increased to 50 mA. This means that if the transistor is driven into saturation with a base current of 15 mA, a maximum of 1.2 V can be expected across the base-emitter junction instead of the usual 0.7 V.

Looking up the transistor β or h_{FE}:

As illustrated in Figure 3 of the data sheet for 2N2222A, the transistor β or DC current gain varies with I_C and with temperature. For instance, at room temperature (25°C) and with $V_{CE} = 10$ V, β for this transistor varies from a minimum of approximately 100 at $I_C = 0.1$ mA to a maximum of approximately 260 at $I_C = 50$ mA.

3.9 DC BIAS DESIGN OF BJT AMPLIFIERS

Since the DC operating point sets the limits of the signal swing at the output of the amplifier, the proper DC bias design is of considerable importance in the overall design process of the BJT amplifier of any configuration. The objective of an optimum DC bias is placing the operating point on the load line where a maximum signal swing can be achieved at the output. This optimum operating point is usually at the center of the AC load line. Since the AC analysis and design are explored in the following chapters, we will, for the time being, practice centering the Q-point on the DC load line. This procedure will be best understood and appreciated by going over the design process of several different bias arrangements, as follows.

Let's design and test the DC bias of a common-emitter BJT amplifier for a centered Q-point, with $V_{CC} = 16$ V, and $I_{CQ} = 2$ mA. We will begin with the emitter bias.

Example 3-6: Emitter Bias Design   DESIGN

Given: $V_{CC} = 16$ V, and $I_{CQ} = 2$ mA.
For a centered Q-point on the DC load line, $V_{CE} = \frac{1}{2} V_{CC} = 8$ V
The remaining 8 V will be distributed such that $0.1 V_{CC} \leq V_E \leq 0.2 V_{CC}$
Let $V_E = 2$ V, and $V_{RC} = 6$ V, as shown in Figure 3-25.

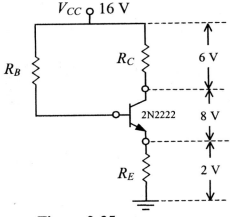

Figure 3-25: DC voltage allocation of the design example

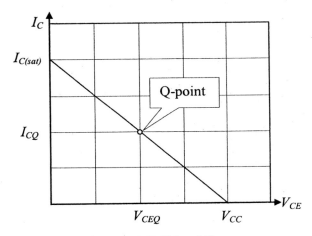

Figure 3-26: DC load line and the centered Q-point

From the allocated voltages and currents, we can now determine the resistor values that will produce the intended Q-point.

$$R_E = \frac{V_E}{I_E} = \frac{2\text{ V}}{2\text{ mA}} = 1 \text{ k}\Omega, \quad R_E = 1 \text{ k}\Omega$$

$$R_C = \frac{V_{RC}}{I_C} = \frac{6\text{ V}}{2\text{ mA}} = 3 \text{ k}\Omega, \quad R_C = 3 \text{ k}\Omega$$

The average β for the 2N2222 transistor will be assumed to be 200.

$V_B = V_E + 0.7 \text{ V} = 2.7 \text{ V}$

$$I_B = \frac{I_C}{\beta} = \frac{2\text{ mA}}{200} = 10 \text{ μA}$$

$$R_B = \frac{V_{CC} - V_B}{I_B} = \frac{(16 - 2.7)\text{ V}}{10\text{ μA}} = 1.33 \text{ M}\Omega \quad \text{Use } R_B = 1 \text{ M}\Omega + 330 \text{ k}\Omega$$

Chapter 3

This concludes the design of the DC bias circuit of the emitter-bias design example. To test the design, the circuit will be simulated using the Electronics Workbench (EWB).

The results of the simulation are presented in Figure 3-27 below. The slight difference in I_C and V_{CE} is the result of slightly different β of the transistor (202.6 rather than 200).

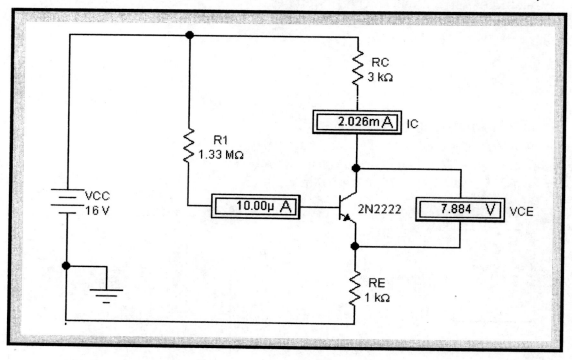

Figure 3-27: Circuit setup with simulation results of Design Example 3-6 created in Electronics Workbench

Example 3-7: Voltage-Divider Bias Design   DESIGN

Given: $V_{CC} = 16$ V, and $I_{CQ} = 2$ mA.

The voltage distribution will be the same as in Design Example 3-6, therefore

$R_E = 1$ kΩ, and $R_C = 3$ kΩ.

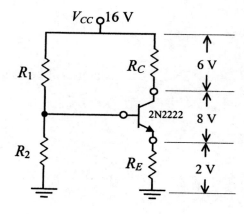

Figure 3-28: Circuit diagram and DC voltage allocation of the voltage-divider bias circuit

Again, $V_B = V_E + 0.7 \text{ V} = 2.7 \text{ V}$, also $V_B = V_{CC} \dfrac{R_2}{R_1 + R_2}$

Hence, $\dfrac{V_{CC}}{V_B} = \dfrac{R_1 + R_2}{R_2} = \dfrac{R_1}{R_2} + 1$, or $\dfrac{V_{CC}}{V_B} - 1 = \dfrac{R_1}{R_2}$

Solving for R_1 yields the following in terms of R_2:

$$R_1 = \left(\dfrac{V_{CC}}{V_B} - 1\right) R_2$$

Recall that the rule of thumb for a properly designed β-independent voltage-divider bias is to have $R_2 < 10 R_E$.

Let **$R_2 = 7.5 \text{ k}\Omega$**

$R_1 = 4.926$, $R_2 = 36.9 \text{ k}\Omega$, use **$R_1 = 36 \text{ k}\Omega$** standard value resistor.

If the above-mentioned relation between R_2 and R_E is not implemented, the circuit will still function, but we will not be utilizing the desirable and advantageous characteristics of the voltage-divider bias, which makes the circuit β-independent, and, as a result, more stable.

This concludes the DC bias design of the voltage-divider bias design example. The next step is to test the design by simulation. The results of the simulation are presented in Figure 3-29.

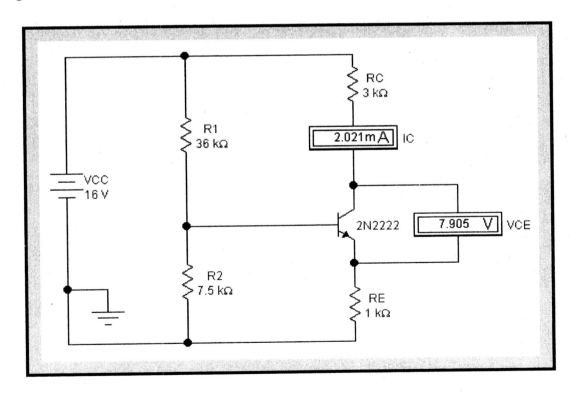

Figure 3-29: Circuit setup with simulation results of Design Example 3-7 created in Electronics Workbench

Practice Problem 3-6 **DESIGN**

Design the DC bias of a common-emitter amplifier in the following bias configurations, and with the specifications given below.
Centered Q-point, $V_{CC} = 15$ V, $I_{CQ} = 1.5$ mA,
a) Emitter bias.
b) Voltage-divider bias

After completing the design for each bias type, test your design by simulation.

Example 3-8   (Cascode)   **DESIGN**

DC Bias Design of the Cascode Amplifier:

First, we need to make some informed decisions regarding the DC voltage distribution. As you will see later in the AC analysis of the cascode amplifier, the signal swing at the output of the first stage (v_{ce1}) is very small, but the second stage has a much larger output signal swing. Hence, the DC voltage allocation of V_{CE2} will have to be larger compared to V_{CE1}, in order to support the corresponding signal swings. In addition, just like any other voltage-divider bias design, to make the circuit most stable and β-independent, the rule of thumb is that R_3 must be smaller than ten times R_E. We start the DC voltage allocation by following the established guidelines, as follows:

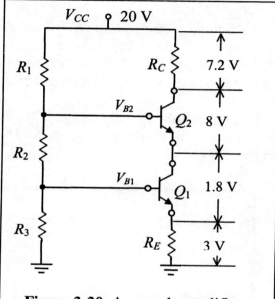

Figure 3-30: A cascode amplifier

$$0.1 V_{CC} \leq V_{E1} \leq 0.15 V_{CC}, \text{ and } V_{CE1} + V_{CE2} \cong \tfrac{1}{2} V_{CC}$$

DC bias decisions: Let $V_{CC} = 20$ V, $I_C = 2$ mA.

Let $V_{E1} = 3$ V, $V_{CE1} = 1.8$ V, $V_{CE2} = 8$ V, as shown in Figure 3-30.

Let $Q_1 = Q_2 = $ 2N2222 transistors.

$$R_E = \frac{V_{E1}}{I_E} = \frac{3\text{ V}}{2\text{ mA}} = 1.5\text{ k}\Omega \qquad R_E = 1.5\text{ k}\Omega$$

$$R_C = \frac{V_{RC}}{I_C} = \frac{7.2\text{ V}}{2\text{ mA}} = 3.6\text{ k}\Omega \qquad R_C = 3.6\text{ k}\Omega$$

$$V_{B1} = V_{E1} + 0.7\text{ V} = 3.7\text{ V}$$

Hence, $R_2 = \dfrac{V_{B2} - V_{B1}}{I_2} \cong \dfrac{5.5\text{ V} - 3.7\text{ V}}{0.336\text{ mA}} = 5.36\text{ k}\Omega$. Use $R_2 = 5.6\text{ k}\Omega$.

$R_1 = \dfrac{V_{CC} - V_{B2}}{I_1} = \dfrac{20\text{ V} - 5.5\text{ V}}{0.336\text{ mA}} = 43.2\text{ k}\Omega$ Use $R_1 = 43\text{ k}\Omega$.

Test Run and Design Verification

Let us now test and verify the DC bias design of the above cascode amplifier by simulation. The result of simulation by EWB is depicted in Figure 3-31.

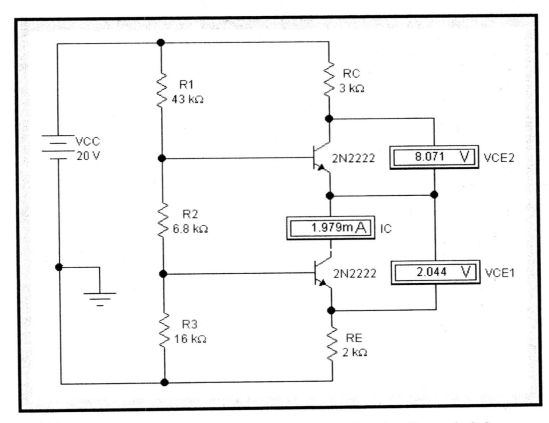

Figure 3-31: Circuit setup with simulation results of Design Example 3-8 created in Electronics Workbench

As you can see, the Q-point I_C, V_{CE1}, and V_{CE2} agree very closely with the design specifications of 2 mA, 1.8 V, and 8 V, respectively.

Practice Problem 3-7   *Cascode*   **DESIGN**

Design DC bias of a cascode amplifier for the following specifications:

$V_{CC} = 22\text{ V}$, $I_C = 3\text{ mA}$, $V_{CE1} = 2\text{ V}$, $V_{CE2} = 8\text{ V}$, and $V_{E1} = 4\text{ V}$.

Answers: $R_E = 1.3\text{ k}\Omega$, $R_C = 2.7\text{ k}\Omega$, $R_3 = 10\text{ k}\Omega$, $R_2 = 4.3\text{ k}\Omega$, $R_1 = 33\text{ k}\Omega$

$V_{CC} = 22$ V, $I_C = 3$ mA. Let $V_{CE1} = 2$ V, $V_{CE2} = 8$ V, and $V_{E1} = 4$ V.

Answers: $R_E = 1.3$ kΩ, $R_C = 2.7$ kΩ, $R_3 = 10$ kΩ, $R_2 = 4.3$ kΩ, $R_1 = 33$ kΩ

3.10 THE TRANSISTOR SWITCH

For the normal operation in the analog applications, the transistor operation is mainly restricted to the active region of its collector characteristics. In switching applications, however, the regions of operation are the cutoff and saturation regions, where the transistor switches from cutoff to saturation and vice-versa. Consider the circuit of Figure 3-32, which is basically a simple base-biased common-emitter configuration. However, since the R_B is not connected to the V_{CC}, the transistor is not readily biased. Hence, with no signal applied at the input terminal V_{in}, there will be no current flowing at the base or at the collector, and as a result, there will be no voltage drop across the R_C. Thus, as illustrated in Figure 3-32(a), with no signal applied at the input ($V_{in} = 0$), the current is *cutoff* ($I_C = 0$) and the output voltage V_{CE} is equal to V_{CC}. In other words, using digital terms, when the input is *low*, the output is *high*. Hence, for apparent reasons, this circuit is commonly called the *BJT inverter*.

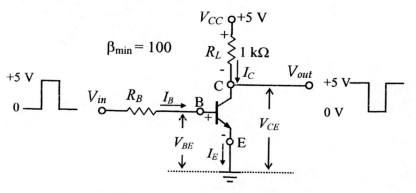

Figure 3-32: The BJT Inverter

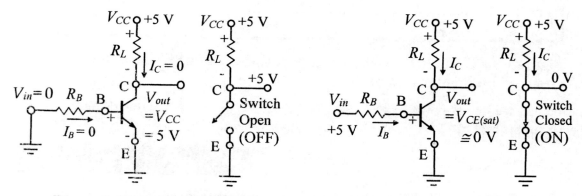

Figure 3-32(a): Transistor OFF   **Figure 3-32(b)**: Transistor ON

The transistor will be biased only when a signal of sufficient amplitude appears at the input. But, for switching applications, base current I_B must be high enough so that the transistor is driven into saturation. As the transistor saturates, the output voltage V_{CE} becomes approximately zero, although up to about 0.1 V, 0.2 V, or even 0.3 V could be measured at higher current levels. But, for practical purposes, we will assume that $V_{CE(sat)}$ is approximately zero. Also, note that to ensure hard saturation, β_{min} is generally used in switching applications. That is, with increased β, the transistor will be driven further into saturation.

Now, assuming that $R_B = 100$ kΩ, let us find out if the transistor of Figure 3-32 will be driven into saturation when the input goes *high*. Apparently, when the input goes *high* ($V_{in} = 5$ V), as shown in Figure 3-32(b), the base-emitter junction is forward biased, dropping approximately 0.7 V at the junction. Thus, the base current will be as follows:

$$I_B = \frac{V_{in} - V_{BE}}{R_B} \tag{3-57}$$

$$I_B = \frac{5\text{ V} - 0.7\text{ V}}{100\text{ k}\Omega} = 43\text{ }\mu\text{A}, \qquad I_C = \beta I_B = 4.3\text{ mA}$$

However, is this collector current high enough to cause saturation? To find that out, we can determine $I_{C(sat)}$, as follows:

$$I_{C(sat)} = \frac{V_{CC}}{R_C} = \frac{V_{CC}}{R_L} \tag{3-58}$$

$$I_{C(sat)} = \frac{5\text{ V}}{1\text{ k}\Omega} = 5\text{ mA}$$

Evidently, 4.3 mA is not large enough to drive the transistor into full saturation; a minimum of 5 mA is needed to start the saturation. Now the question is *what is the maximum R_B to start saturation?* To answer this question, first we need to know the minimum I_B that initiates the saturation $I_{B(sat)min}$.

$$I_{B(sat)min} = \frac{I_{C(sat)}}{\beta_{min}} \tag{3-59}$$

$$I_{B(sat)min} = \frac{5\text{ mA}}{100} = 50\text{ }\mu\text{A}$$

$$R_{B(max)} = \frac{V_{in} - 0.7}{I_{B(sat)min}} \tag{3-60}$$

$$R_{B(max)} = \frac{5 - 0.7\text{ V}}{50\text{ }\mu\text{A}} = 86\text{ k}\Omega$$

The R_B of 86 kΩ will just start the saturation; however, if the β changes downward, I_C will drop and the transistor will no longer be in saturation. Therefore, to take some precautionary measures, it is advisable to use a smaller R_B, about one-half the $R_{B(max)}$, such as 43 kΩ. Using R_B of 43 kΩ will result in the following I_B:

$$I_B = \frac{5\text{ V} - 0.7\text{ V}}{43\text{ k}\Omega} = 100\text{ }\mu\text{A}$$

Recall that $I_C = \beta I_B$ is no longer valid at saturation. That is, I_B may increase beyond $I_{B(sat)min}$, but the collector current will not increase beyond $I_{C(sat)}$.

Chapter 3

$$I_{C(max)} = I_{C(sat)} \qquad (3\text{-}61)$$

Therefore, with an I_B of 100 µA, the collector current will be no more than 5 mA, and the saturation will be guaranteed even with changes in transistor β.

Design of the transistor switch

Example 3-9 — BJT Switch — DESIGN

Let us design a BJT inverter, with a collector current of 10 mA when saturated (ON), and a V_{CE} of 5 V when cutoff (OFF). Assume V_{in} is either 0 V or 6 V, and $\beta_{min} = 80$.

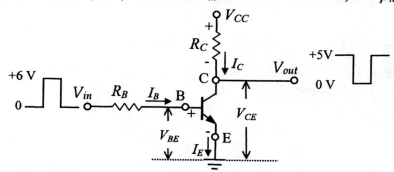

Figure 3-33: The BJT inverter

Solution: Let $V_{CC} = 5$ V, so that $V_{CE} = 5$ V when *cutoff*.

$$I_{C(sat)} = \frac{V_{CC}}{R_C}, \qquad R_C = \frac{V_{CC}}{I_{C(sat)}}, \qquad R_C = \frac{5\,\text{V}}{10\,\text{mA}} = 500\,\Omega$$

May use 510 Ω standard resistor, or two 1 kΩ resistors in parallel.

$$I_{B(sat)min} = \frac{I_{C(sat)}}{\beta_{min}} \qquad I_{B(sat)min} = \frac{10\,\text{mA}}{80} = 125\,\mu\text{A}$$

$$R_{B(max)} = \frac{V_{in} - 0.7}{I_{B(sat)min}} = \frac{6 - 0.7\,\text{V}}{125\,\mu\text{A}} = 42.4\,\text{k}\Omega \qquad \text{Use } R_B = 22\,\text{k}\Omega$$

Practice Problem 3-8 — DESIGN

Design a BJT switch, with a collector current of 5 mA when saturated (ON), and a V_{CE} of 6 V when cutoff (OFF). Assume V_{in} is either 0 V or 4 V, and $\beta_{min} = 66$.

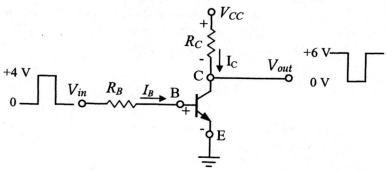

Figure 3-34: The BJT switch

Answers: $V_{CC} = 6$ V, $R_C = 1.2$ kΩ, $R_{B(max)} = 43.5$ kΩ   Use $R_B = 22$ kΩ

Example 3-10: Design of LED driver   **DESIGN**

Let us design a BJT switch that turns on an LED when 5 V DC is applied at the input. Assume that the LED requires about 10 mA for full brightness and drops about 1.8 V. Also assume that $\beta_{min} = 50$.

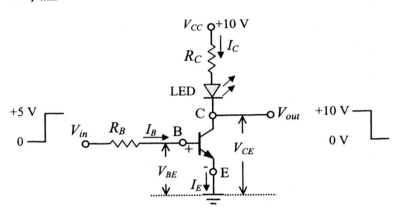

Figure 3-35: BJT as LED driver

Solution:

Let $V_{CC} = 10$ V. With the application of 5 V at the input, the transistor must saturate and allow 10 mA of I_C to flow through the LED, which will cause a drop of 1.8 V across the LED and 8.2 V across the R_C. Hence, we can determine the R_C and R_B as follows:

$$R_C = \frac{V_{CC} - V_{LED}}{I_{C(sat)}} = \frac{10\text{ V} - 1.8\text{ V}}{10\text{ mA}} = 820\text{ Ω}$$

With $\beta_{min} = 50$, we can determine the $I_{B(sat)min}$ and $R_{B(max)}$.

$$I_{B(sat)min} = \frac{I_{C(sat)}}{\beta_{min}}$$

$$I_{B(sat)min} = \frac{10\text{ mA}}{50} = 200\text{ μA}$$

$$R_{B(max)} = \frac{V_{in} - 0.7}{I_{B(sat)min}} = \frac{5\text{ V} - 0.7\text{ V}}{200\text{ μA}} = 21.5\text{ kΩ} \qquad \text{Use } R_B = 11\text{ kΩ}$$

Switching Times (ON and OFF Delay Times)

The transistor switch of Figure 3-32, like any other switching device, does not change instantaneously from cutoff to saturation or from saturation to cutoff, as the input pulse switches from *low* to *high* or *vice versa*. A timing diagram of the transistor switch is shown in Figure 3-36.

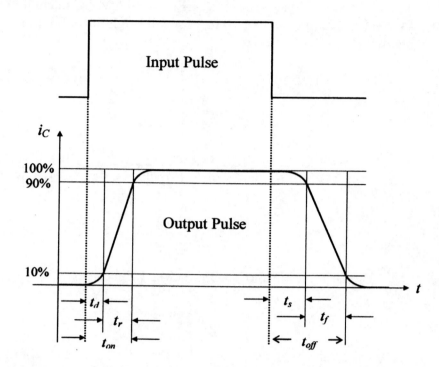

Figure 3-36: Timing diagram of the transistor switch

The time it takes to switch from the *off* state to the *on* state is known as the *on delay time* t_{on}, and the time it takes to make the transition from *on* state to *off* state is known as the *off delay time* t_{off}. As illustrated in Figure 3-36, the *on delay time* t_{on} is the sum of the *delay time* t_d and the *rise time* t_r. The *delay time* t_d is the time interval it takes for the collector current to rise from 0% to 10% of its final value. The *rise time* t_r is the time interval it takes for the collector current to rise from 10% to 90% of its final value. The on delay time is

$$t_{on} = t_d + t_r \qquad (3\text{-}62)$$

Likewise, the *off delay time* t_{off} is the sum of the *storage time* t_s and the *fall time* t_f. The *storage time* t_s is the time interval it takes for the collector current to drop from 100% to 90% of its final value, during which the excess charge stored in the base is depleted. The greater the degree of saturation, the greater is the storage time. The *fall time* t_f is the time interval it takes for the collector current to fall from 90% to 10% of its final value. The off delay time is

$$t_{off} = t_s + t_f \qquad (3\text{-}63)$$

Example 3-11

Let us determine the on delay time and the off delay time for the 2N2222A general-purpose transistor.
From the data sheets we get the following information.

delay time	_rise time_	_storage time_	_fall time_
$t_d = 10$ ns	$t_r = 25$ ns	$t_s = 225$ ns	$t_f = 60$ ns

Therefore, the *on delay time* is: $t_{on} = t_d + t_r = 10 + 25 = 35$ ns

and the *off delay time* is: $t_{off} = t_s + t_f = 225 + 60 = 285$ ns

---

Switching Transistors:

For switching applications, specially designed switching transistors are much faster than the general-purpose transistors. Here are two examples of switching transistors with their corresponding switching speeds:

MSC1621T1   switching transistor $t_{on} = 20$ ns, and $t_{off} = 40$ ns

BSV52LT1   switching transistor $t_{on} = 12$ ns, and $t_{off} = 18$ ns

3.11 SUMMARY

- The *bipolar junction transistor* (BJT) is a three-layer, three-terminal device with two *pn* junctions. The BJT is a current-controlled device and may be looked at as a current-controlled current source, where a small amount of input current can produce a larger amount of output current.

- There are two types of BJT: *npn* and *pnp*. Recall that *npn* stands for *not pointing in!*

- The three terminals of the BJT are called *emitter* (E), *base* (B), and *collector* (C).

- A BJT may be biased in one of three bias arrangements: *common-base, common-emitter,* and *common-collector.* Common-emitter is the most widely used and the most frequently encountered configuration of the three bias arrangements.

- For proper biasing of the BJT in the active region, the emitter-base junction must be forward-biased, while the collector-base junction is reverse-biased.

- In digital applications where the transistor is used primarily as a switch, saturation and cutoff regions are the appropriate regions of operation.

- The ratio of the output current to the input current in the *common-base* is called α (*alpha*), and the same ratio in the *common-emitter* is called β (*beta*) or h_{FE}. For a general-purpose transistor, typical value of α is in the range of 0.99 to 0.995, and the typical value of β is in the range of 100 to 200.

- For the conventional current flow that we will be using throughout the text, the direction of the emitter current is always in the direction of the emitter arrow, and the other two currents (I_C and I_B) join together and make up I_E.

- The emitter current I_E is always equal to the sum of the collector current I_C and the base current I_B, although I_B is much smaller compared to I_C and I_E, making $I_E \cong I_C$.

- Input characteristics of the common-base, common-emitter, and common-collector bias configurations resemble the characteristics of a forward-biased diode.

- Output characteristics of the *CE* and the *CC* bias configurations are almost identical.

- The I_{CBO} is the collector-to-base current with the emitter open, and the I_{CEO} is the collector-to-emitter current with the base open.

- The relation $I_C = \beta I_B$ is valid only up to, not including, saturation. That is, I_B may increase beyond $I_{B(sat)min}$, but the collector current will not increase beyond $I_{C(sat)}$.

- The BJT amplifier may be biased in one of the three bias configurations: common-base, common-emitter, and common-collector. However, the common-emitter, which is the most widely used configuration, can be biased using several different methods: base bias, emitter bias, and voltage-divider bias.

- The voltage-divider bias is the most popular of all biasing methods because, if designed correctly, the circuit can be β-independent, and as a result, more stable.

- The operating point, quiescent point, or simply the Q-point, is a point on the transistor collector characteristics that corresponds to certain input current, output current, and output voltage at which the transistor is biased to operate.

- The DC load line is actually the plot of the output equation of the bias configuration, which extends from the saturation current $I_{C(sat)}$ to the cutoff voltage V_{CC}, while passing through the Q-point.

- When analyzing a DC bias configuration, there are always two loops or current paths to be considered for analysis: the input loop and the output loop; the input loop is the first to be analyzed.

- The input loop is the one that contains the forward-biased base-emitter junction and not necessarily the loop that happens to be to the left.

Chapter 3

- The objective of an optimum DC bias is placing the Q-point on the load line where a maximum signal swing can be achieved at the output. This optimum Q-point is usually at the center of the AC load line.

- DC bias of an amplifier sets the limits on the output signal swing. Hence, a proper bias design is of considerable importance in the overall design process.

- It is a good idea to test your bias design by simulation in order to verify the accuracy of the design.

3.12 DEVICE SPECIFICATIONS AND DATA SHEETS

Data sheets of the following devices discussed in this chapter are included for your reference and convenience.
- 2N2222A Bipolar Junction Transistor
  Copyright of Semiconductor Component Industries, LLC. Used by permission.
- 2N3904 Bipolar Junction Transistor (Courtesy of Fairchild Semiconductor)

MOTOROLA
SEMICONDUCTOR TECHNICAL DATA

Order this document by P2N2222A/D

Amplifier Transistors
NPN Silicon

P2N2222A

CASE 29–04, STYLE 17
TO–92 (TO–226AA)

MAXIMUM RATINGS

Rating	Symbol	Value	Unit
Collector–Emitter Voltage	V_{CEO}	40	Vdc
Collector–Base Voltage	V_{CBO}	75	Vdc
Emitter–Base Voltage	V_{EBO}	6.0	Vdc
Collector Current — Continuous	I_C	600	mAdc
Total Device Dissipation @ T_A = 25°C Derate above 25°C	P_D	625 5.0	mW mW/°C
Total Device Dissipation @ T_C = 25°C Derate above 25°C	P_D	1.5 12	Watts mW/°C
Operating and Storage Junction Temperature Range	T_J, T_{stg}	–55 to +150	°C

THERMAL CHARACTERISTICS

Characteristic	Symbol	Max	Unit
Thermal Resistance, Junction to Ambient	$R_{\theta JA}$	200	°C/W
Thermal Resistance, Junction to Case	$R_{\theta JC}$	83.3	°C/W

ELECTRICAL CHARACTERISTICS (T_A = 25°C unless otherwise noted)

Characteristic	Symbol	Min	Max	Unit
OFF CHARACTERISTICS				
Collector–Emitter Breakdown Voltage (I_C = 10 mAdc, I_B = 0)	$V_{(BR)CEO}$	40	—	Vdc
Collector–Base Breakdown Voltage (I_C = 10 µAdc, I_E = 0)	$V_{(BR)CBO}$	75	—	Vdc
Emitter–Base Breakdown Voltage (I_E = 10 µAdc, I_C = 0)	$V_{(BR)EBO}$	6.0	—	Vdc
Collector Cutoff Current (V_{CE} = 60 Vdc, $V_{EB(off)}$ = 3.0 Vdc)	I_{CEX}	—	10	nAdc
Collector Cutoff Current (V_{CB} = 60 Vdc, I_E = 0) (V_{CB} = 60 Vdc, I_E = 0, T_A = 150°C)	I_{CBO}	— —	0.01 10	µAdc
Emitter Cutoff Current (V_{EB} = 3.0 Vdc, I_C = 0)	I_{EBO}	—	10	nAdc
Collector Cutoff Current (V_{CE} = 10 V)	I_{CEO}	—	10	nAdc
Base Cutoff Current (V_{CE} = 60 Vdc, $V_{EB(off)}$ = 3.0 Vdc)	I_{BEX}	—	20	nAdc

© Motorola, Inc. 1996

Chapter 3

P2N2222A

ELECTRICAL CHARACTERISTICS (T_A = 25°C unless otherwise noted) (Continued)

Characteristic	Symbol	Min	Max	Unit
ON CHARACTERISTICS				
DC Current Gain (I_C = 0.1 mAdc, V_{CE} = 10 Vdc) (I_C = 1.0 mAdc, V_{CE} = 10 Vdc) (I_C = 10 mAdc, V_{CE} = 10 Vdc) (I_C = 10 mAdc, V_{CE} = 10 Vdc, T_A = −55°C) (I_C = 150 mAdc, V_{CE} = 10 Vdc)(1) (I_C = 150 mAdc, V_{CE} = 1.0 Vdc)(1) (I_C = 500 mAdc, V_{CE} = 10 Vdc)(1)	h_{FE}	35 50 75 35 100 50 40	— — — — 300 — —	—
Collector–Emitter Saturation Voltage(1) (I_C = 150 mAdc, I_B = 15 mAdc) (I_C = 500 mAdc, I_B = 50 mAdc)	$V_{CE(sat)}$	— —	0.3 1.0	Vdc
Base–Emitter Saturation Voltage(1) (I_C = 150 mAdc, I_B = 15 mAdc) (I_C = 500 mAdc, I_B = 50 mAdc)	$V_{BE(sat)}$	0.6 —	1.2 2.0	Vdc
SMALL–SIGNAL CHARACTERISTICS				
Current–Gain — Bandwidth Product(2) (I_C = 20 mAdc, V_{CE} = 20 Vdc, f = 100 MHz)	f_T	300	—	MHz
Output Capacitance (V_{CB} = 10 Vdc, I_E = 0, f = 1.0 MHz)	C_{obo}	—	8.0	pF
Input Capacitance (V_{EB} = 0.5 Vdc, I_C = 0, f = 1.0 MHz)	C_{ibo}	—	25	pF
Input Impedance (I_C = 1.0 mAdc, V_{CE} = 10 Vdc, f = 1.0 kHz) (I_C = 10 mAdc, V_{CE} = 10 Vdc, f = 1.0 kHz)	h_{ie}	2.0 0.25	8.0 1.25	kΩ
Voltage Feedback Ratio (I_C = 1.0 mAdc, V_{CE} = 10 Vdc, f = 1.0 kHz) (I_C = 10 mAdc, V_{CE} = 10 Vdc, f = 1.0 kHz)	h_{re}	— —	8.0 4.0	$\times 10^{-4}$
Small–Signal Current Gain (I_C = 1.0 mAdc, V_{CE} = 10 Vdc, f = 1.0 kHz) (I_C = 10 mAdc, V_{CE} = 10 Vdc, f = 1.0 kHz)	h_{fe}	50 75	300 375	—
Output Admittance (I_C = 1.0 mAdc, V_{CE} = 10 Vdc, f = 1.0 kHz) (I_C = 10 mAdc, V_{CE} = 10 Vdc, f = 1.0 kHz)	h_{oe}	5.0 25	35 200	µmhos
Collector Base Time Constant (I_E = 20 mAdc, V_{CB} = 20 Vdc, f = 31.8 MHz)	$rb'C_c$	—	150	ps
Noise Figure (I_C = 100 µAdc, V_{CE} = 10 Vdc, R_S = 1.0 kΩ, f = 1.0 kHz)	N_F	—	4.0	dB
SWITCHING CHARACTERISTICS				
Delay Time	(V_{CC} = 30 Vdc, $V_{BE(off)}$ = −2.0 Vdc, I_C = 150 mAdc, I_{B1} = 15 mAdc) (Figure 1)	t_d	—	10	ns
Rise Time		t_r	—	25	ns
Storage Time	(V_{CC} = 30 Vdc, I_C = 150 mAdc, I_{B1} = I_{B2} = 15 mAdc) (Figure 2)	t_s	—	225	ns
Fall Time		t_f	—	60	ns

1. Pulse Test: Pulse Width ≤ 300 µs, Duty Cycle ≤ 2.0%.
2. f_T is defined as the frequency at which $|h_{fe}|$ extrapolates to unity.

P2N2222A

SWITCHING TIME EQUIVALENT TEST CIRCUITS

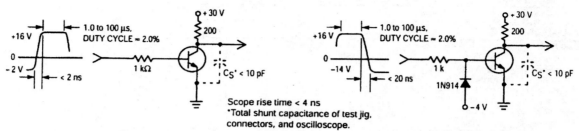

Figure 1. Turn–On Time

Figure 2. Turn–Off Time

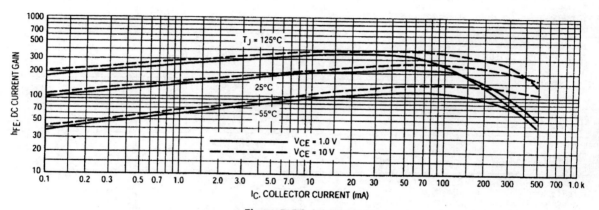

Figure 3. DC Current Gain

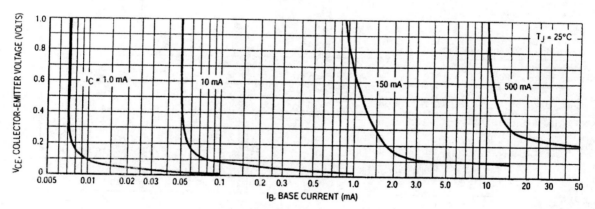

Figure 4. Collector Saturation Region

Motorola Small–Signal Transistors, FETs and Diodes Device Data

Chapter 3

P2N2222A

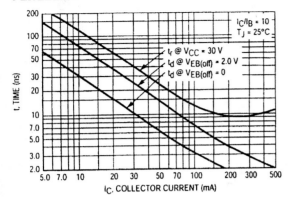

Figure 5. Turn–On Time

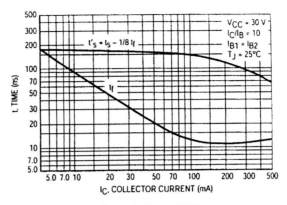

Figure 6. Turn–Off Time

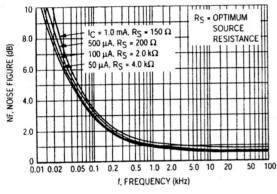

Figure 7. Frequency Effects

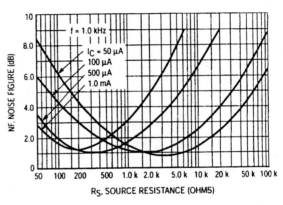

Figure 8. Source Resistance Effects

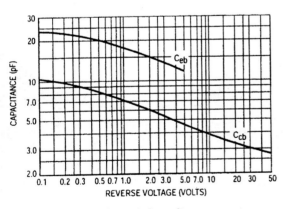

Figure 9. Capacitances

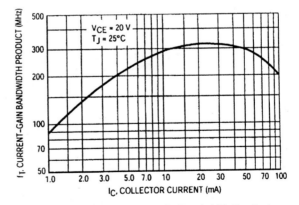

Figure 10. Current–Gain Bandwidth Product

Discrete POWER & Signal Technologies

2N3904
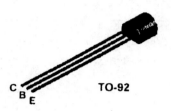
TO-92

MMBT3904
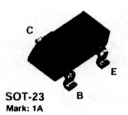
SOT-23
Mark: 1A

MMPQ3904
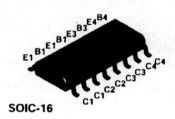
SOIC-16

PZT3904
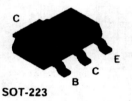
SOT-223

NPN General Purpose Amplifier

This device is designed as a general purpose amplifier and switch. The useful dynamic range extends to 100 mA as a switch and to 100 MHz as an amplifier. Sourced from Process 23.

Absolute Maximum Ratings* $T_A = 25°C$ unless otherwise noted

Symbol	Parameter	Value	Units
V_{CEO}	Collector-Emitter Voltage	40	V
V_{CBO}	Collector-Base Voltage	60	V
V_{EBO}	Emitter-Base Voltage	6.0	V
I_C	Collector Current - Continuous	200	mA
T_J, T_{stg}	Operating and Storage Junction Temperature Range	-55 to +150	°C

*These ratings are limiting values above which the serviceability of any semiconductor device may be impaired.

NOTES:
1) These ratings are based on a maximum junction temperature of 150 degrees C.
2) These are steady state limits. The factory should be consulted on applications involving pulsed or low duty cycle operations.

© 1997 Fairchild Semiconductor Corporation

NPN General Purpose Amplifier
(continued)

Electrical Characteristics
T_A = 25 °C unless otherwise noted

Symbol	Parameter	Test Conditions	Min	Max	Units
OFF CHARACTERISTICS					
$V_{(BR)CEO}$	Collector-Emitter Breakdown Voltage	I_C = 1.0 mA, I_B = 0	40		V
$V_{(BR)CBO}$	Collector-Base Breakdown Voltage	I_C = 10 µA, I_E = 0	60		V
$V_{(BR)EBO}$	Emitter-Base Breakdown Voltage	I_E = 10 µA, I_C = 0	6.0		V
I_{BL}	Base Cutoff Current	V_{CE} = 30 V, V_{EB} = 0		50	nA
I_{CEX}	Collector Cutoff Current	V_{CE} = 30 V, V_{EB} = 0		50	nA
ON CHARACTERISTICS*					
h_{FE}	DC Current Gain	I_C = 0.1 mA, V_{CE} = 1.0 V	40		
		I_C = 1.0 mA, V_{CE} = 1.0 V	70		
		I_C = 10 mA, V_{CE} = 1.0 V	100	300	
		I_C = 50 mA, V_{CE} = 1.0 V	60		
		I_C = 100 mA, V_{CE} = 1.0 V	30		
$V_{CE(sat)}$	Collector-Emitter Saturation Voltage	I_C = 10 mA, I_B = 1.0 mA		0.2	V
		I_C = 50 mA, I_B = 5.0 mA		0.3	V
$V_{BE(sat)}$	Base-Emitter Saturation Voltage	I_C = 10 mA, I_B = 1.0 mA	0.65	0.85	V
		I_C = 50 mA, I_B = 5.0 mA		0.95	V
SMALL SIGNAL CHARACTERISTICS					
f_T	Current Gain - Bandwidth Product	I_C = 10 mA, V_{CE} = 20 V, f = 100 MHz	300		MHz
C_{obo}	Output Capacitance	V_{CB} = 5.0 V, I_E = 0, f = 1.0 MHz		4.0	pF
C_{ibo}	Input Capacitance	V_{EB} = 0.5 V, I_C = 0, f = 1.0 MHz		8.0	pF
NF	Noise Figure (except MMPQ3904)	I_C = 100 µA, V_{CE} = 5.0 V, R_S = 1.0 kΩ, f = 10 Hz to 15.7 kHz		5.0	dB
SWITCHING CHARACTERISTICS (except MMPQ3904)					
t_d	Delay Time	V_{CC} = 3.0 V, V_{BE} = 0.5 V,		35	ns
t_r	Rise Time	I_C = 10 mA, I_{B1} = 1.0 mA		35	ns
t_s	Storage Time	V_{CC} = 3.0 V, I_C = 10 mA		200	ns
t_f	Fall Time	I_{B1} = I_{B2} = 1.0 mA		50	ns

*Pulse Test: Pulse Width ≤ 300 µs, Duty Cycle ≤ 2.0%

Spice Model

NPN (Is=6.734f Xti=3 Eg=1.11 Vaf=74.03 Bf=416.4 Ne=1.259 Ise=6.734 Ikf=66.78m Xtb=1.5 Br=.7371 Nc=2 Isc=0 Ikr=0 Rc=1 Cjc=3.638p Mjc=.3085 Vjc=.75 Fc=.5 Cje=4.493p Mje=.2593 Vje=.75 Tr=239.5n Tf=301.2p Itf=.4 Vtf=4 Xtf=2 Rb=10)

NPN General Purpose Amplifier
(continued)

Thermal Characteristics
TA = 25 °C unless otherwise noted

Symbol	Characteristic	Max		Units
		2N3904	*PZT3904	
P_D	Total Device Dissipation Derate above 25°C	625 5.0	1,000 8.0	mW mW/°C
$R_{\theta JC}$	Thermal Resistance, Junction to Case	83.3		°C/W
$R_{\theta JA}$	Thermal Resistance, Junction to Ambient	200	125	°C/W

Symbol	Characteristic	Max		Units
		**MMBT3904	MMPQ3904	
P_D	Total Device Dissipation Derate above 25°C	350 2.8	1,000 8.0	mW mW/°C
$R_{\theta JA}$	Thermal Resistance, Junction to Ambient Effective 4 Die Each Die	357	 125 240	°C/W °C/W °C/W

*Device mounted on FR-4 PCB 36 mm X 18 mm X 1.5 mm; mounting pad for the collector lead min. 6 cm².

**Device mounted on FR-4 PCB 1.6" X 1.6" X 0.06."

Typical Characteristics

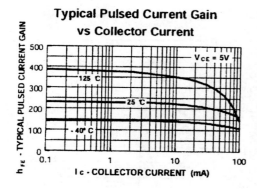

Typical Pulsed Current Gain vs Collector Current

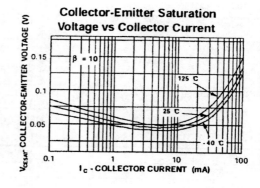

Collector-Emitter Saturation Voltage vs Collector Current

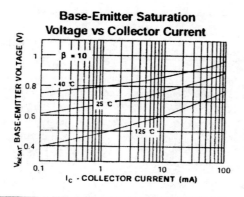

Base-Emitter Saturation Voltage vs Collector Current

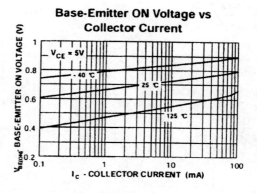

Base-Emitter ON Voltage vs Collector Current

NPN General Purpose Amplifier (continued)

Typical Characteristics (continued)

Collector-Cutoff Current vs Ambient Temperature

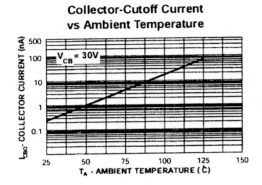

Capacitance vs Reverse Bias Voltage

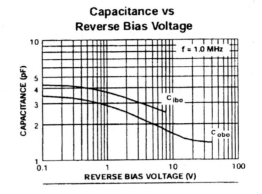

Noise Figure vs Frequency

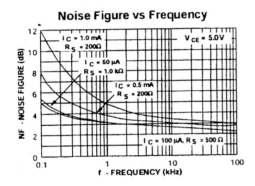

Noise Figure vs Source Resistance

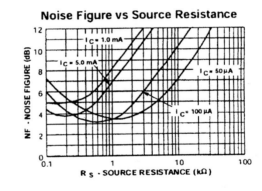

Current Gain and Phase Angle vs Frequency

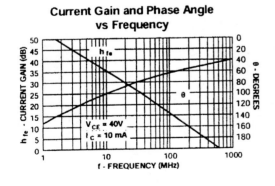

Power Dissipation vs Ambient Temperature

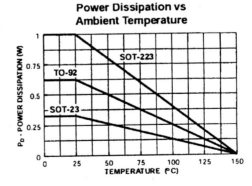

PROBLEMS

Section 3.2 Basic Theory of BJT Operation

3.1 Why is the BJT so named, *bipolar* junction transistor?

3.2 Draw the symbols and the approximate equivalents for the *npn* and *pnp* transistors and label each terminal.

3.3 Which layer of the BJT is the most heavily doped and which layer is the most lightly doped?

Section 3.3 Common-Emitter Characteristics

3.4 Prove Equation (3-12). That is, show that $\beta = \alpha/(1-\alpha)$.

3.5 Prove Equation (3-13). That is, show that $\beta + 1 = 1/(1-\alpha)$.

3.6 Draw the bias circuit in common-emitter configuration (a) with an *npn* transistor, and (b) with a *pnp* transistor. Show all current directions and label all components.

3.7 A *pnp* transistor with an α of 0.9925 is biased in the circuit of Figure 3-6(b) with a base current of 40 µA. Determine the transistor β, I_C, and the emitter current I_E.

3.8 Assuming that Figure 3-9 represents the output characteristics of a certain *npn* transistor, determine the I_C of the transistor with the $I_B = 80$ µA and with V_{CE} of:
a) 1 V b) 3 V c) 5 V d) 7 V

3.9 Define the Early voltage, in words.

Section 3.4.1 CE Base-Bias

3.10 Draw the circuit diagram of Figure 3-11(c) with the following parameters.
$V_{CC} = +12$ V, $R_C = 1.5$ kΩ, $R_B = 330$ kΩ, $\beta = 120$
Determine: I_B, I_C, and V_{CE} a) algebraically b) by simulation

3.11 Draw the *pnp* version of Figure 3-11(c) with the following parameters.
$V_{CC} = -15$ V, $R_C = 1.8$ kΩ, $R_B = 330$ kΩ, $\beta = 120$
Determine: a) V_{BE} b) I_B c) I_C d) V_{CE}

3.12 Draw the schematic diagram of a *base-biased* CE amplifier (*npn*), with the following parameters: $V_{CC} = +15$ V, $R_C = 1.5$ kΩ, $R_B = 360$ kΩ, $\beta = 202$
   a) Algebraically, determine I_B, I_C, and V_{CE}
   b) Draw the DC load line on a copy of the collector characteristics of Figure 3-2P and locate the Q-point, graphically.
   c) How do the two Q-points compare?

3.13 Carry out Practice Problem 3-1.

3.14 Determine R_B for the circuit of Figure 3-1P, so that $V_{CE} = 0.5 V_{CC}$.

3.15 Referring to Figure 3-11(c), determine $I_{C(sat)}$, $I_{B(sat)}$, and $R_{B(sat)}$, the R_B that will initiate the saturation of the transistor. Let $\beta = 100$.

3.16 Let $R_B = 220\ k\Omega$ in Figure 3-1P and determine I_B, I_C, and V_{CE}.

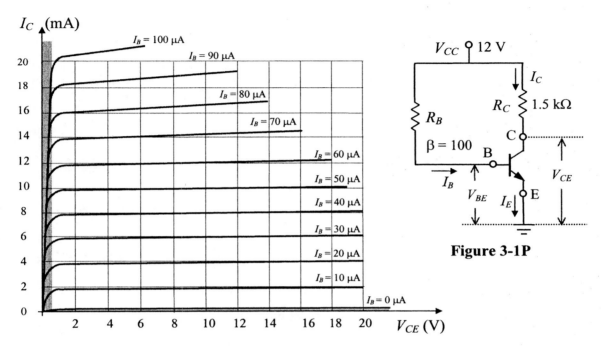

Figure 3-2P: Transistor collector characteristic curves

Section 3.4.2 CE Emitter-Bias

3.17 Draw the circuit diagram of an emitter-biased common-emitter amplifier with the following parameters.
$V_{CC} = 14\ V$, $R_B = 470\ k\Omega$, $R_C = 1\ k\Omega$, $R_E = 0.5\ k\Omega$, $\beta = 199$

   a) Determine I_B, I_C, and V_{CE} algebraically.
   b) Determine I_B, I_C, and V_{CE} by simulation.

3.18 Determine R_C for the amplifier circuit of Figure 3-3P for a centered Q, $\beta = 199$.

3.19 Carry out Practice Problem 3-2.

3.20 Determine the value of R_B in Figure 3-4P so that $V_{CE} = 0.5 V_{CC}$. Let $\beta = 100$.

3.21 Draw the circuit diagram of an emitter-biased common-emitter amplifier (*npn*), with the following parameters: $V_{CC} = +15\ V$, $R_C = 1\ k\Omega$, $R_E = 0.5\ k\Omega$, $R_B = 510\ k\Omega$, and $\beta = 204$.
   a) Algebraically, determine I_B, I_C, and V_{CE}.
   b) Draw the DC load line on a copy of the collector characteristics of Figure 3-2P, and locate the Q-point, graphically.
   c) How do the two Q-points compare?

Chapter 3

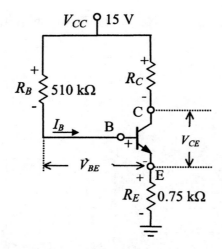

Figure 3-3P: Problem 3.18   **Figure 3-4P**: Problem 3.20

Section 3.4.3 CE Voltage-Divider Bias

3.22 Carry out Practice Problem 3-3.

3.23 Draw the circuit diagram of a common-emitter voltage-divider bias with the following parameters.
$V_{CC} = 16$ V, $R_1 = 30$ kΩ, $R_2 = 10$ kΩ, $R_C = 3.3$ kΩ, $R_E = 1.5$ kΩ, $\beta = 199$

   a) Determine I_{CQ} and V_{CEQ} using the exact method.
   b) Determine I_{CQ} and V_{CEQ} using the approximate method.
   c) Determine the same by simulation.

3.24 Determine R_C in Figure 3-5P, if $V_{CE} = 6$ V.

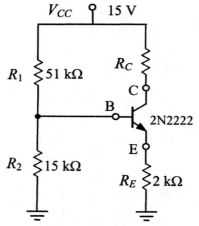 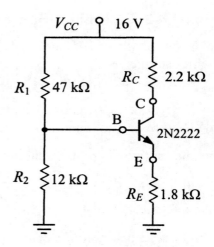

Figure 3-5P   **Figure 3-6P**

3.25 In Figure 3-5P, determine I_{CQ} and V_{CEQ}, if $R_C = 3.3$ kΩ.

   a) Algebraically
   b) By simulation

Chapter 3

3.26 In Figure 3-6P, determine I_{CQ} and V_{CEQ}.

  a) Using the approximate method
  b) Using the exact method

Section 3.5 Common-Collector Bias

3.27 In Figure 3-7P, determine I_{BQ}, I_{EQ}, V_{ECQ}, and V_{BCQ}, if $\beta = 199$.

3.28 In Figure 3-7P, determine I_{BQ}, I_{EQ}, V_{ECQ}, and V_{BCQ}, if $\beta = 99$.

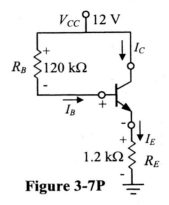

Figure 3-7P

Section 3.7 Cascode Configuration

3.29 Carry out Practice Problem 3-5.

3.30 Determine the voltages V_{CE1} and V_{CE2} for the cascode amplifier of Figure 3-8P.

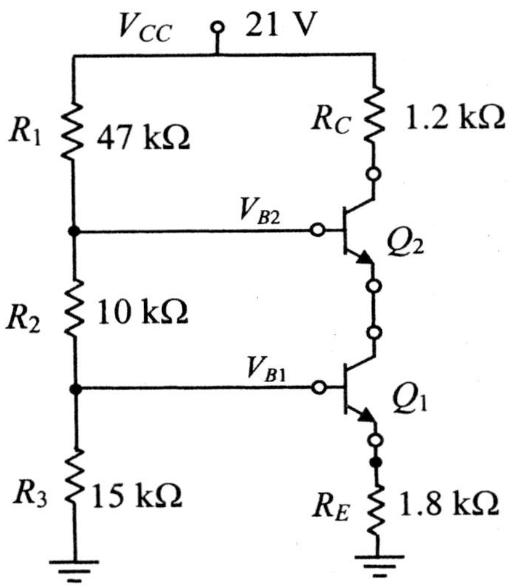

Figure 3-8P: Cascode amplifier

Section 3.9 DC Bias Design

3.31 Design the DC bias circuit of an emitter-biased common-emitter amplifier with the specifications given below.
Centered Q-point, $V_{CC} = 16$ V, $I_{CQ} = 2.2$ mA. Assume $\beta = 200$.

3.32 Design the DC bias circuit of a voltage-divider biased common-emitter amplifier with the specifications given below.
Centered Q-point, $V_{CC} = 16$ V, $I_{CQ} = 2.2$ mA

3.33 Carry out Practice Problem 3-6.

3.34 Design the DC bias circuit of a voltage-divider biased common-emitter amplifier with the specifications given below.
$V_{CC} = 16$ V, $I_{CQ} = 1.5$ mA, $V_{CEQ} = 6$ V

3.35 Design the DC bias circuit of an emitter-biased common-emitter amplifier with the specifications given below.
$V_{CC} = 16$ V, $I_{CQ} = 1.5$ mA, $V_{CEQ} = 6$ V. Assume $\beta = 150$.

Chapter 3

3.36 Design the DC bias circuit of a voltage-divider biased common-emitter amplifier with the specifications given below.
$V_{CC} = 14$ V, $I_{CQ} = 1$ mA, $V_{CEQ} = 6$ V

3.37 Design the DC bias circuit of an emitter-biased common-collector amplifier with the specifications given below.
$V_{CC} = 12$ V, $I_{CQ} = 1.5$ mA, $V_{CEQ} = 6$ V. Assume $\beta = 200$.

3.38 Carry out Practice Problem 3-7.

3.39 Design the DC bias circuit of a cascode amplifier with the specifications given below.
$V_{CC} = 20$ V, $I_{CQ} = 2$ mA, $V_{CEQ1} = 2$ V, $V_{CEQ2} = 8$ V

3.40 Design the DC bias circuit of a cascode amplifier with the specifications given below.
$V_{CC} = 22$ V, $I_{CQ} = 2.5$ mA, $V_{CEQ1} = 2$ V, $V_{CEQ2} = 8$ V

Section 3.10 The Transistor Switch

3.41 Design the transistor switch of Figure 3-9P. Assume $\beta_{min} = 100$. Verify your design by simulation.

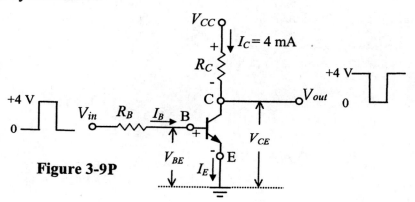

Figure 3-9P

3.42 Design the transistor switch of Figure 3-9P for $I_C = 6$ mA, and V_{out} from 0 V to 6 V, if V_{in} is from 0 to 5 V. Assume $\beta_{min} = 77$. Verify your design by simulation.

3.43 Design the transistor switch of Figure 3-9P for $I_C = 10$ mA, and V_{out} from 0 to 10 V, if V_{in} is from 0 to 5 V. Assume $\beta_{min} = 85$.

3.44 Refer to the 2N3904 transistor data sheets and determine the t_{on} and t_{off} for the transistor.

3.45 Design a BJT switch that turns on an LED when 5 V DC is applied at the input. Assume that the LED requires about 15 mA for full brightness and drops about 2 V. Also assume that $\beta_{min} = 60$, and $V_{CC} = 10$ V. Verify your design by simulation.

3.46 Design a BJT switch that turns on an LED when 5 V DC is applied at the input. Assume that the LED requires about 18 mA for full brightness and drops 1.8 V. Also assume that $V_{CE(sat)} = 0.2$ V, $\beta_{min} = 80$, and $V_{CC} = 10$ V.

Chapter

TRANSISTOR MODELING
r-parameter model
h-parameter model
universal r/h model

4.1 INTRODUCTION

To study and analyze the small-signal operation of the BJT amplifier, it would be helpful and convenient to use a transistor model, instead of using the transistor symbol, in the amplifier circuit in which the transistor is utilized. A model would be an equivalent circuit, usually a combination of simple electronic symbols and components, to be substituted for the transistor symbol, wherever it appears in the circuit. The model must be an equivalent substitute and true representative of the transistor characteristics under the small-signal[1] operation. Several different models are available, each being more suitable for a certain type of analysis. The two popular models that are better suited to our purpose and our level of analysis are the ones known as the *hybrid model* or the *h-parameter model*, and the *r-parameter model*. Most authors use one or the other and some devote a section to each model. We will present the theory involved in developing each model, then in conclusion, we will utilize a *universal* model that can be interpreted as the *h-parameter* model or the *r-parameter* model.

---

[1] Refer to section 5.1 for definition of the small-signal.

Chapter 4

4.2 THE *r-parameter* MODEL

Figure 4-1 is the exact and familiar common-base circuit of Figure 3-27, with the addition of two new labels r_e and r_c. These two parameters represent the small-signal (AC) input and output resistance of the BJT, in the common-base configuration.

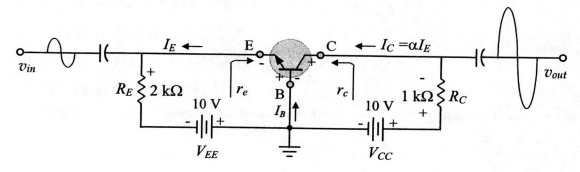

Figure 4-1: Common-base amplifier circuit (*npn*)

The r_e, which is the input resistance of the BJT, stands for the *emitter-to-base resistance*, and the r_c, which is the output resistance of the BJT, stands for the *collector-to-base resistance*. Since the base is the common ground in the common-base configuration, r_e and r_c are referred to as the *emitter resistance* and the *collector resistance*, respectively. The small-signal *r-parameter* model of the BJT is shown in Figure 4-2. Note that the following figure shows the transistor models only. The rest of the amplifier circuit is not included in this figure. We will be looking at the whole picture shortly.

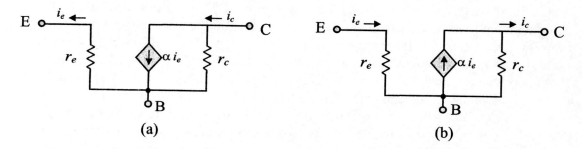

Figure 4-2: The CB *r-parameter* model of the BJT: (a) *npn*, (b) *pnp*

4.2.1 Determining the *r-parameters* in the Common-Base Configuration

Apparently, the *r-parameters* are to be defined and determined under small-signal operating conditions. One familiar parameter that appears in the common-base model is α, and is defined as follows under the above-mentioned conditions. Notice that in the following equations, holding the DC quantity constant ensures that the corresponding AC quantity equals zero. For example, keeping the DC voltage V_{CB} constant ensures that the small-signal v_{cb}, which is actually the deviation from V_{CBQ}, is kept equal to zero.

Chapter 4

$$\alpha = \left.\frac{i_c}{i_e}\right|_{v_{cb} = 0} \quad (4\text{-}1)$$

$$\alpha = \left.\frac{\Delta I_C}{\Delta I_E}\right|_{V_{CB} = constant} \quad (4\text{-}2)$$

Since $\Delta I_C \cong \Delta I_E$, $\alpha \cong 1$

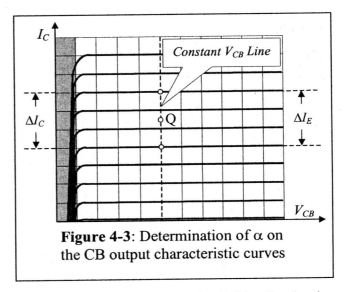

Figure 4-3: Determination of α on the CB output characteristic curves

Of course, we could have predicted that $\alpha \cong 1$ without going through all this. But having gone through this simple procedure, we are now familiar with the graphical determination of a transistor parameter in its simplest form, which can be useful for determining other more complex parameters in this or other transistor models. In addition, we have verified the validity of the parameter definition and the method of evaluation.

Another familiar parameter that will appear in the *r-parameter* model of the common-emitter configuration is the small-signal β or $\beta_{(ac)}$, and is defined as follows.

$$\beta = \left.\frac{i_c}{i_b}\right|_{v_{ce} = 0} \quad (4\text{-}3)$$

$$\beta = \left.\frac{\Delta I_C}{\Delta I_B}\right|_{V_{CE} = constant} \quad (4\text{-}4)$$

As mentioned previously, $\beta_{(dc)}$ and $\beta_{(ac)}$ are usually close enough to be considered equal. Hence, we will be using the notation β to represent the average $\beta_{(dc)}$ and $\beta_{(ac)}$.

The next parameter r_e, the CB input resistance of the BJT, is defined as follows:

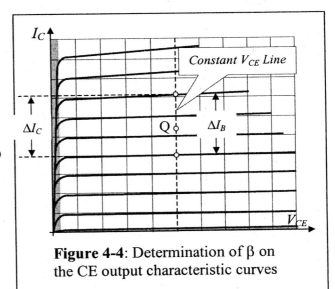

Figure 4-4: Determination of β on the CE output characteristic curves

$$r_e = \left.\frac{v_{be}}{i_e}\right|_{v_{cb}=0} \quad (4\text{-}5)$$

$$r_e = \left.\frac{\Delta V_{BE}}{\Delta I_E}\right|_{V_{CB}=constant} \quad (4\text{-}6)$$

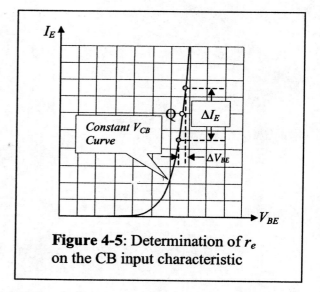

Figure 4-5: Determination of r_e on the CB input characteristic

Recall that the common-base input characteristics are very similar to the *pn-junction* diode characteristics. Hence, the small-signal input resistance of the BJT, which is basically the reciprocal of the slope of the constant V_{CB} curve, can be approximated according to Equation 1-13 with the exception that the diode current I_D will be replaced by the junction current I_E.

$$r_e \cong \frac{26\,mV}{I_{EQ}} \quad (4\text{-}7)$$

The last parameter to be determined is r_c, the CB output resistance of the BJT, which is defined as follows:

$$r_c = \left.\frac{v_{cb}}{i_c}\right|_{i_e=0} \quad (4\text{-}8)$$

$$r_c = \left.\frac{\Delta V_{CB}}{\Delta I_C}\right|_{I_E=constant} \quad (4\text{-}9)$$

Evidently, due to the fact that ΔI_C is very small, r_c will be very large, about several MΩ. This will not be the case with the CE configuration, as the output characteristic curves of the CE are much more inclined, and are not as flat as the CB characteristic curves.

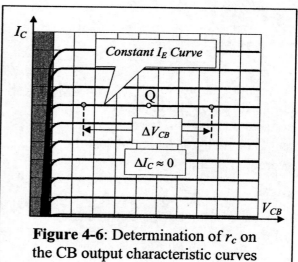

Figure 4-6: Determination of r_c on the CB output characteristic curves

4.2.2 Determining the *r-parameters* in the Common-Emitter Configuration

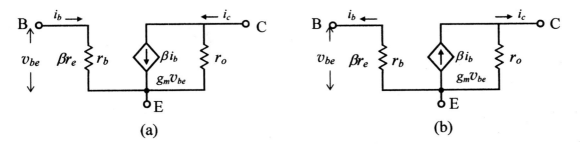

Figure 4-7: The CE *r-parameter* model of the BJT. (a) *npn*. (b) *pnp*

We have already defined and determined the parameter β, which is the transistor current gain in the common-emitter configuration.

The parameter r_b, being the resistance measured from the base to the common-emitter terminal, is defined as follows:

$$r_b = \left.\frac{v_{be}}{i_b}\right|_{v_{ce}=0} \tag{4-10}$$

$$r_b = \left.\frac{\Delta V_{BE}}{\Delta I_B}\right|_{V_{CE}=constant} \tag{4-11}$$

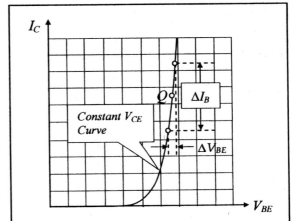

Figure 4-8: Determination of r_b, the CE *input resistance* of BJT

The common-emitter input characteristics are also very similar to the *pn-junction* diode characteristics. Hence, the small-signal input resistance of the BJT, which is the reciprocal of the slope of the constant V_{CE} curve, can be approximated according to Equation 1-12 with the exception that the diode current I_D will be replaced by the base current I_B

$$r_b = \frac{V_T}{I_{BQ}} \cong \frac{26\,\text{mV}}{I_{BQ}} \tag{4-12}$$

where $V_T = kT/q \approx 26$ mV at $T = 25°C$. Substituting for I_B in terms of I_E yields

$$r_b = \frac{V_T}{I_{BQ}} \cong \frac{26\,\text{mV}}{I_{BQ}} = \frac{26\,\text{mV}}{\dfrac{I_E}{\beta+1}} = (\beta+1)\frac{26\,\text{mV}}{I_E} = (\beta+1)r_e$$

Therefore, the CE input resistance of the BJT can be expressed as follows:

$$r_b = (\beta+1)r_e \cong \beta r_e \tag{4-13}$$

The last *r-parameter* of the common-emitter configuration is the output resistance r_o, which is defined as follows:

$$r_o = \left.\frac{v_{ce}}{i_c}\right|_{i_b=0} = \left.\frac{\Delta V_{CE}}{\Delta I_C}\right|_{I_B = constant} \tag{4-14}$$

Note that the Early voltage (V_A), as presented in Chapter 3, is repeated below in Figure 4-9. The Early voltage was defined as the point on the negative V_{CE} axis where the extensions of all I_C curves meet.

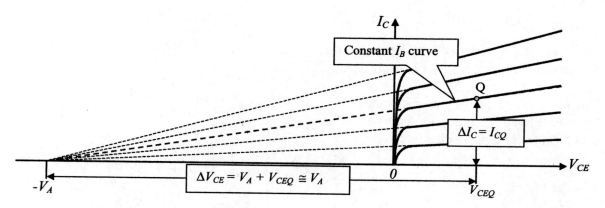

Figure 4-9: Determining r_o, the CE output resistance of the BJT

As illustrated in Figure 4-9, the above expression (4-14) is basically the reciprocal of the slope of the base current line I_B upon which the Q-point is residing. Since the extensions of all I_C curves intercept the V_{CE} axis at the negative Early voltage $-V_A$, r_o can be defined and determined as follows.

$$r_o = \left.\frac{\Delta V_{CE}}{\Delta I_C}\right|_{I_B = constant} = \left.\frac{\Delta V_{CE}}{\Delta I_C}\right|_{I_{BQ}} = \frac{V_A + V_{CEQ}}{I_{CQ}} \cong \frac{V_A}{I_{CQ}}$$

Therefore, r_o can be determined simply by knowing I_{CQ} and the Early voltage.

$$r_o = \frac{V_A}{I_{CQ}} \text{ ohms} \tag{4-15}$$

The typical value of the *Early voltage* V_A may range from 50 V to 150 V, and the typical value of r_o can be in the range of 40 kΩ to 200 kΩ. The typical r_e can be from 10 Ω to 50 Ω, and r_c is usually in the range of 4 MΩ to 6 MΩ. Of course, these figures are the average or typical values. The actual parameter value, which depends largely on the DC bias Q-point, could be different from a typical value.

Chapter 4

Table 4-1: Summary of the *r-parameters*

Configuration →	Common-Base		Common-Emitter		Common-Collector	
Parameter ↓	Notation	Typical Value	Notation	Typical Value	Notation	Typical Value
Current Gain	α	$\cong 1$	β	150	β	150
Input Resistance	r_e	15 Ω	r_b	1.5 kΩ	βr_e	1.5 kΩ
Output Resistance	r_c	5 MΩ	r_o	50 kΩ	r_o	50 kΩ

where $r_e = \dfrac{V_T}{I_E} = \dfrac{26 \text{ mV}}{I_E}$ $r_b = \beta r_e$ $r_c = \beta r_o \cong \infty$ $r_o = \dfrac{V_A}{I_C}$ $V_A \cong 100$ V

Example 4-1 — ANALYSIS

Determine the *r-parameters* and draw the small-signal equivalent circuit for the amplifier circuit of Figure 4-10. Assume $\beta = 120$ and $V_A = 120$ V.

Solution:

First, we need to determine I_{BQ}.

$$I_B = \frac{V_{CC} - V_{BE}}{R_B + (\beta+1)R_E} = 17.9 \text{ μA}$$

$I_C = \beta I_B = 2.15$ mA

$I_E = (\beta+1)I_B = 2.166$ mA $r_e = \dfrac{V_T}{I_E} = \dfrac{26 \text{ mV}}{I_E} = 12$ Ω $r_o = \dfrac{V_A}{I_C} = \dfrac{120 \text{ V}}{2.15 \text{ mA}} = 55.8$ kΩ

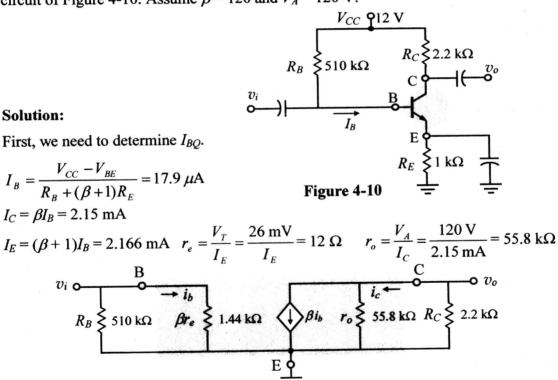

Figure 4-10

Figure 4-11: Small-signal equivalent circuit of Figure 4-10

Note that the colored area in Figure 4-11 is the r-parameter model and the small-signal equivalent of the BJT.

Example 4-2 ANALYSIS

Determine the *r-parameters* and draw the small-signal equivalent circuit for the CE amplifier circuit of Figure 4-12.

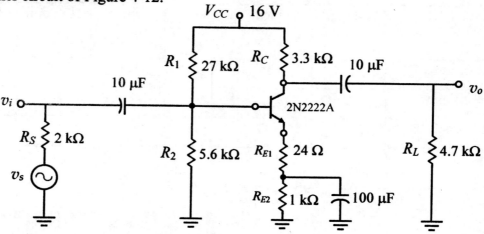

Figure 4-12: A common-emitter amplifier circuit

Solution:
First, we need to determine I_{EQ}. Since $R_2 < 10R_E$, the approximate method of analysis will suffice. But in order to determine I_E, we need to find V_B and then V_E. Thus,

$$V_B = V_{CC} \frac{R_2}{R_1 + R_2} = 16 \times \frac{5.6}{27 + 5.6} = 2.75 \text{ V}, \quad V_E = V_B - 0.7 \text{ V} = 2.05 \text{ V}$$

$$I_E = \frac{V_E}{R_{E1} + R_{E2}} = 2 \text{ mA}, \quad r_e = \frac{V_T}{I_E} = \frac{26 \text{ mV}}{I_E} = 13 \text{ }\Omega$$

Next, we need to determine the transistor β. Referring to the data sheets of 2N2222A, I_C of 2 mA corresponds to $\beta = 160$. Therefore, $r_b = \beta r_e = 2080 \text{ }\Omega \cong 2 \text{ k}\Omega$. Let us assume a typical voltage V_A of 100 V. Hence,

$$r_o = \frac{V_A}{I_C} = \frac{100 \text{ V}}{2 \text{ mA}} = 50 \text{ k}\Omega \qquad \text{Let } R_B = R_1\|R_2 = 27\|5.6 = 4.638 \text{ k}\Omega.$$

Having determined all of the *r-parameters*, we can now draw the small-signal equivalent circuit for the amplifier of Figure 4-12. Notice that the signal source v_s, its associated resistance R_s, and the load resistor R_L are all external to the amplifier circuit. Also, note that R_{E2} is bypassed by the 100 µF capacitor for the AC signal, but R_{E1} is not bypassed.

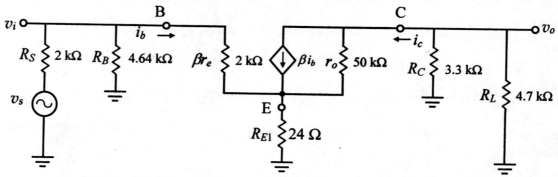

Figure 4-13: Small-signal equivalent circuit of Figure 4-12

Chapter 4

Practice Problem 4-1 — Modeling — ANALYSIS

Determine the *r-parameters* for the amplifier of Figure 4-14 and draw the small-signal equivalent circuit with all parameter values.

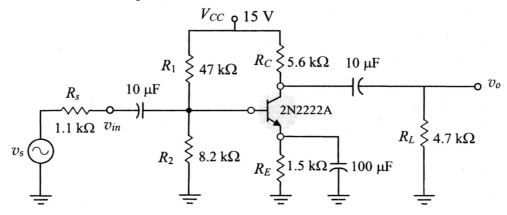

Figure 4-14: A common-emitter amplifier

Answer:

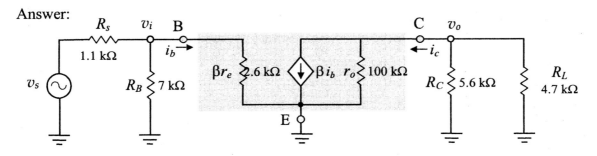

Figure 4-15: Small-signal equivalent circuit of Figure 4-14

Example 4-3 — Modeling — ANALYSIS

Determine the *r-parameters* and draw the small-signal equivalent circuit for the amplifier circuit of Figure 4-16. Assume $\beta = 120$, and $V_A = 100$ V.

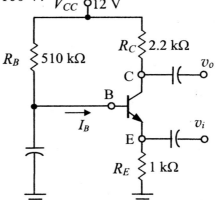

Figure 4-16: A CB amplifier circuit

Chapter 4

Solution:

Note that the emitter is the input, the collector is the output, and the base is the common ground for the AC signal. Therefore, this circuit is a CB amplifier. It is also DC equivalent to the CE amplifier of Example 4-1 (Figure 4-10). Hence, the DC analysis will be exactly the same.

$$I_B = \frac{V_{CC} - V_{BE}}{R_B + (\beta+1)R_E} = 17.9 \, \mu A$$

$$I_C = \beta I_B = 2.15 \text{ mA}$$

$$I_E = (\beta + 1)I_B = 2.167 \text{ mA}$$

$$r_e = \frac{26 \text{ mV}}{I_E} = 12 \, \Omega$$

$$r_o = \frac{V_A}{I_C} = \frac{100 \text{ V}}{2.15 \text{ mA}} = 46.5 \text{ k}\Omega$$

$$r_c = \beta r_o = 120 \times 46.5 \text{ k}\Omega = 5.58 \text{ M}\Omega \cong \infty.$$

This agrees with the typical value of approximately 5 MΩ, and if desired, can be replaced with an open circuit.

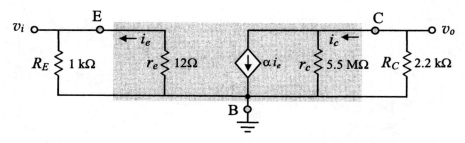

Figure 4-17: Small-signal equivalent circuit of Figure 4-16

Example 4-4 — Modeling — ANALYSIS

Determine the *r-parameters* for the transistor and draw the small-signal equivalent circuit for the amplifier of Figure 4-18 below.

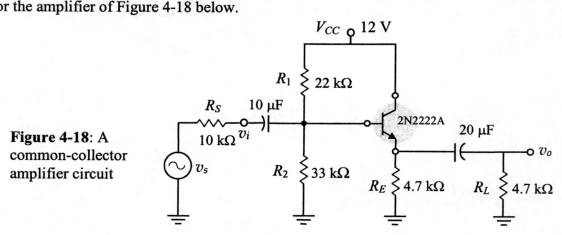

Figure 4-18: A common-collector amplifier circuit

Chapter 4

Solution:

$$V_B = V_{CC}\frac{R_2}{R_1+R_2} = 7.2 \text{ V} \qquad V_E = V_B - 0.7 \text{ V} = 6.5 \text{ V}$$

$$I_E = \frac{V_E}{R_E} = 1.383 \text{ mA} \qquad r_e = \frac{26 \text{ mV}}{I_E} = 18.8 \text{ }\Omega$$

Referring to the data sheets of 2N2222A, I_C of 1.4 mA corresponds to $\beta \cong 160$.
Therefore, $r_b = \beta r_e \cong 3$ kΩ. Assuming $V_A = 100$ V, $r_o = \frac{V_A}{I_E} = = 72$ kΩ

Let $R_B = R_1||R_2 = 13.2$ kΩ

Results: $\beta \cong 160$, $r_b = \beta r_e \cong 3$ kΩ

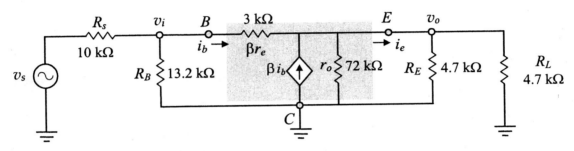

Figure 4-19: The *r-parameter* model and small-signal equivalent circuit of Figure 4-18

4.3 THE HYBRID OR *h-parameter* MODEL

The hybrid model is a general two-port model, which can be utilized for modeling any two-port network. Consider the following two-port system as illustrated in Figure 4-20.

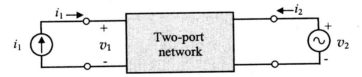

Figure 4-20: A two-port network under test

In the above network, both v_1 and i_2 are considered to be functions of i_1 and v_2, where i_1 and v_2 are the test signals. That is, the voltage v_1, developed at the input port depends on both the current i_1 applied at the input port, and the voltage v_2 applied at the output port.

$$v_1 = f(i_1, v_2) \qquad (4\text{-}16)$$

$$i_2 = f(i_1, v_2) \qquad (4\text{-}17)$$

Thus, the following two network equations can be written for v_1 and i_2 in terms of some imaginary *h*-parameters (h_{11}, h_{12}, h_{21}, and h_{22}).

$$v_1 = h_{11}i_1 + h_{12}v_2 \qquad (4\text{-}18)$$

$$i_2 = h_{21}i_1 + h_{22}v_2 \qquad (4\text{-}19)$$

Chapter 4

In the above two equations, both v_1 and i_2 are functions of i_1 and v_2, and if the values of the h-parameters are known, both v_1 and i_2 can be easily determined. In addition, by examining these equations, we can determine the identity of each parameter.

4.3.1 Definitions and Evaluation of the *h-parameters*

Notice that setting $v_2 = 0$ in Equation 4-18 will result in the following equation, which can be solved for the parameter h_{11}. Thus,

$$v_1 = h_{11} i_1 \qquad (4\text{-}20)$$

$$h_{11} = \left. \frac{v_1}{i_1} \right|_{v_2 = 0} \quad \text{Ohms} \qquad (4\text{-}21)$$

The parameter h_{11}, which is the ratio of the input voltage v_1 to the input current i_1 with v_2 set to 0, is the resistance seen by the input source, and is defined as the *short-circuit input impedanc*.

Setting $i_1 = 0$ in Equation 4-18 will result in the following equation, which can be solved for the parameter h_{12}

$$v_1 = h_{12} v_2 \qquad (4\text{-}22)$$

$$h_{12} = \left. \frac{v_1}{v_2} \right|_{i_1 = 0} \quad \text{no unit} \qquad (4\text{-}23)$$

The parameter h_{12}, which is the ratio of the input voltage v_1 to the output voltage v_2 with the input current i_1 set to 0, is defined as the *open-circuit reverse voltage gain*. This gain is a dimensionless quantity. However, if one insists on having a unit for this parameter, the term volt per volt (V/V) would be appropriate and academically acceptable.

Setting $v_2 = 0$ in Equation 4-19 results in the following equation, which can be solved to yield the parameter h_{21}. Thus,

$$i_2 = h_{21} i_1 \qquad (4\text{-}24)$$

$$h_{21} = \left. \frac{i_2}{i_1} \right|_{v_2 = 0} \quad \text{no unit} \qquad (4\text{-}25)$$

The parameter h_{21}, which is the ratio of the output current i_2 to the input current i_1 with v_2 set to 0, is also a dimensionless quantity, or in A/A, and is simply the current gain of the network. It is formally defined as the *forward current gain*.

To evaluate the last parameter h_{22}, the current i_1 will be set equal to 0 in Equation 4-19, resulting in the following, which can be solved to yield the parameter h_{22}. Thus,

$$i_2 = h_{22} v_2 \qquad (4\text{-}26)$$

$$h_{22} = \left. \frac{i_2}{v_2} \right|_{i_1 = 0} \quad \text{Siemens} \qquad (4\text{-}27)$$

Chapter 4

The parameter h_{22}, being the ratio of the output current i_2 to the output voltage v_2 with i_1 set to 0 is the output conductance and its reciprocal is the output resistance. This parameter is usually referred to as the *open-circuit output admittance*.

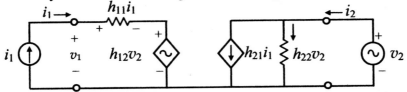

Figure 4-21: The *h-parameter* model of a two-port network

The above model completely satisfies the two-port network equations. In addition, it reveals the identity of the *h*-parameters. Evidently, the voltage v_1 is the sum of the two voltages $h_{11}i_1$ and $h_{12}v_2$, which satisfies Equation 4-18, and the current i_2 is the sum of the two currents $h_{21}i_1$ and $h_{22}v_2$, satisfying Equation 4-19.

Example 4-5 — Hybrid Model — **ANALYSIS**

Determine the *h-parameters* for the resistive network given in Figure 4-21. Then draw the *hybrid* equivalent circuit.

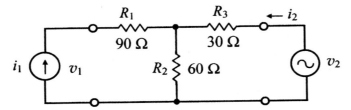

Figure 4-22: A simple two-port network

Solution:

a) Set $v_2 = 0$, as illustrated in Figure 4-22(a), and solve for the parameters h_{11} and h_{21}

$$h_{11} = \frac{v_1}{i_1}\bigg|_{v_2=0}$$

$$v_1 = i_1[R_1 + (R_2 \| R_3)] = i_1 (110\, \Omega)$$

$$h_{11} = \frac{i_1(110\, \Omega)}{i_1} = 110\, \Omega$$

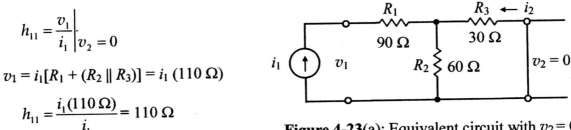

Figure 4-23(a): Equivalent circuit with $v_2 = 0$

Applying the current divider rule, we will solve for i_2 in Figure 4-22(a) in terms of i_1 and substitute the result in the equation for the h_{21} parameter (Equation 4-24)

$$i_2 = -i_1 \frac{R_2}{R_2 + R_3} = -i_1 \frac{60\, \Omega}{90\, \Omega} = -i_1 \frac{2}{3} \qquad h_{21} = \frac{i_2}{i_1}\bigg|_{v_2=0} = \frac{-\frac{2}{3}i_1}{i_1} = -\frac{2}{3}$$

b) Set $i_1 = 0$, as illustrated in Figure 4-22(b), and solve for the two parameters h_{12} and h_{22}

$$h_{12} = \left.\frac{v_1}{v_2}\right|_{i_1 = 0}$$

$$v_1 = v_2 \frac{R_2}{R_2 + R_3} = v_2 \frac{60\,\Omega}{90\,\Omega} = v_2 \frac{2}{3}$$

$$h_{12} = \frac{v_2 \frac{2}{3}}{v_2} = \frac{2}{3}$$

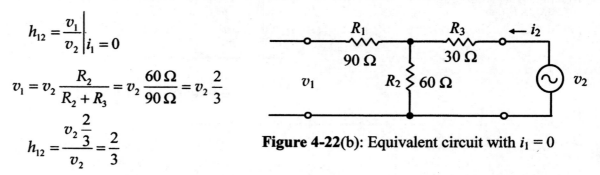

Figure 4-22(b): Equivalent circuit with $i_1 = 0$

To evaluate the parameter h_{22}, we will first solve for current i_2 in Figure 4-22(b) and substitute the result into the equation for h_{22} (Equation 4-27). Thus,

$$i_2 = \frac{v_2}{R_2 + R_3} = \frac{v_2}{90}$$

$$h_{22} = \left.\frac{i_2}{v_2}\right|_{i_1 = 0} = \frac{v_2}{90\,v_2} = \frac{1}{90}\ \text{S}$$

or

$$\frac{1}{h_{22}} = 90\,\Omega$$

Summarizing the results: $h_{11} = 110\,\Omega$, $h_{21} = -2/3$, $h_{12} = 2/3$, and $1/h_{22} = 90\,\Omega$. Hence, the two-port network equations can be written as follows:

$$v_1 = h_{11} i_1 + h_{12} v_2 = 110\,i_1 + (2/3)\,v_2$$

$$i_2 = h_{21} i_1 + h_{22} v_2 = (-2/3)\,i_1 + (1/90)\,v_2$$

The following equivalent circuit is the implementation of the above two equations, which represents the *h-parameter model* of the resistive two-port network given in Example 4-5.

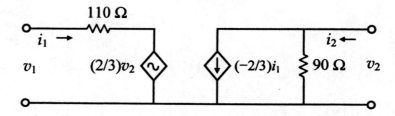

Figure 4-23: Model for Example 4-5

Having identified the *h-parameters*, we can now realize the two-port network equations. This realization is known as the hybrid equivalent circuit or *h-parameter model* of a two-port network, as illustrated in Figure 4-23.

4.3.2 The *Hybrid (h-parameter)* Model of the BJT:

The *h-parameter* model of the BJT is based on the *hybrid* equivalent circuit of the two-port network, and likewise, it comprises a resistance in series with a dependent voltage source, and a dependent current source in parallel with a conductance, as shown in Figure 4-24.

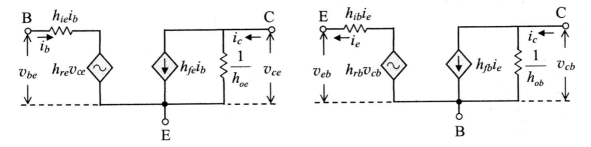

Figure 4-24: *h-parameter* model of the BJT in common-emitter configuration

Figure 4-25: *h-parameter* model of the BJT in common-base configuration

For the purpose of convenience and clarity, the numeric subscripts are replaced with appropriate notations representing the identity of the parameter. For example, in the common-emitter configuration, the h_{11} is replaced with h_{ie}, and h_{12} with h_{re}. The subscript *e* represents the common-emitter configuration, the *i* in h_{ie} stands for the *input resistance*, and the *r* in h_{re} signifies the *reverse voltage ratio*. Similarly, the *f* in h_{fe} stands for the *forward current ratio*, and the *o* in h_{oe} symbolizes the *output admittance*. Apparently, if the subscript *e* represents the common-emitter, then the subscript *b* will represent the common-base, and *c* will stand for the common-collector. The *h-parameter designations* of the three configurations are listed in Table 4-2.

Configuration	h_i	h_r	h_f	h_o
Common Base	h_{ib}	h_{rb}	h_{fb}	h_{ob}
Common Emitter	h_{ie}	h_{re}	h_{fe}	h_{oe}
Common Collector	h_{ic}	h_{rc}	h_{fc}	h_{oc}

Table 4-2: *h-parameter* designations of the BJT configurations

Because the common-emitter is the most frequently encountered configuration, only the common-emitter *h*-parameters are usually listed in most data sheets.

4.3.3 Graphical Determination of the *h-parameters*

The *h-parameters* of the BJT, like the *r-parameters*, can be determined graphically from the transistor base and collector characteristics. We will define and graphically determine the *h-parameters* for the CB and CE configurations.

Determining the *h-parameters* in CE configuration:

Note that when evaluating h_{ie} we must let $v_{ce} = 0$, and when evaluating h_{re}, i_b must be set equal to zero. These conditions can be achieved by keeping the DC V_{CE} and I_B constant.

$$h_{11} = \left.\frac{v_1}{i_1}\right|_{v_2 = 0}$$

$$h_{ie} = \left.\frac{v_{be}}{i_b}\right|_{v_{ce} = 0}$$

$$h_{ie} = \left.\frac{\Delta V_{BE}}{\Delta I_B}\right|_{V_{CE} = constant} \quad (4\text{-}28)$$

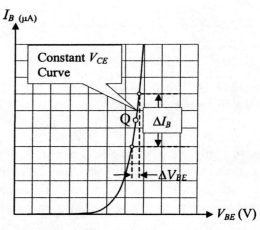

CE Input Characteristics

Figure 4-26: Graphical determination of the *h-parameter* h_{ie}

$$h_{12} = \left.\frac{v_1}{v_2}\right|_{i_1 = 0}$$

$$h_{re} = \left.\frac{v_{be}}{v_{ce}}\right|_{i_b = 0}$$

$$h_{re} = \left.\frac{\Delta V_{BE}}{\Delta V_{CE}}\right|_{I_B = constant} \quad (4\text{-}29)$$

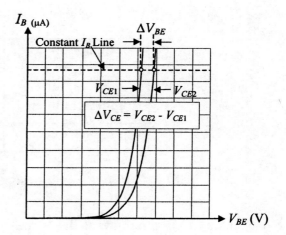

CE Input Characteristics

Figure 4-27: Graphical determination of the *h-parameter* h_{re}

Note that when evaluating h_{fe}, we must let $v_{ce} = 0$, and when evaluating h_{oe}, i_b must be set equal to zero. These conditions can be achieved by keeping the DC V_{CE} and I_B constant.

$$h_{21} = \left.\frac{i_2}{i_1}\right|_{v_2 = 0}$$

$$h_{fe} = \left.\frac{i_c}{i_b}\right|_{v_{ce} = 0}$$

$$h_{fe} = \left.\frac{\Delta I_C}{\Delta I_B}\right|_{V_{CE} = constant} \quad (4\text{-}30)$$

$$= \beta$$

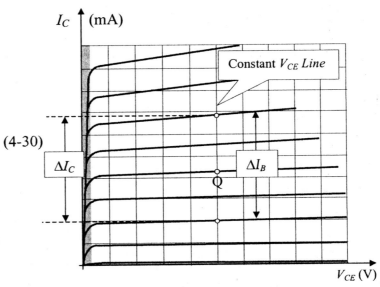

Figure 4-28: Graphical determination of the *h*-parameter h_{fe}

CE Output Characteristics

$$h_{22} = \left.\frac{i_2}{v_2}\right|_{i_1 = 0}$$

$$h_{oe} = \left.\frac{i_c}{v_{ce}}\right|_{i_b = 0}$$

$$h_{oe} = \left.\frac{\Delta I_C}{\Delta V_{CE}}\right|_{I_B = constant} \quad (4\text{-}31)$$

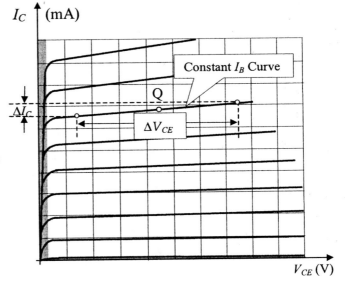

Figure 4-29: Graphical determination of the *h*-parameter h_{oe}

CE Output Characteristics

137

Chapter 4

Determining the *h-parameters* in CB configuration:

Note that when evaluating h_{ib} and h_{fb}, we must let $v_{cb} = 0$, and when evaluating h_{rb} and h_{ob}, i_e must be set equal to zero. These conditions can be achieved by keeping the DC V_{CB} and I_E constant. Having gained some experience in this area, let us now evaluate two parameters on each characteristic's curve.

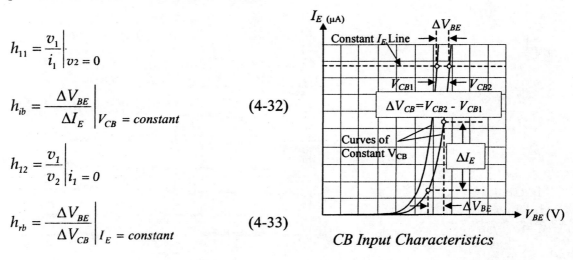

$$h_{11} = \left.\frac{v_1}{i_1}\right|_{v_2 = 0}$$

$$h_{ib} = \left.\frac{\Delta V_{BE}}{\Delta I_E}\right|_{V_{CB} = constant} \quad (4\text{-}32)$$

$$h_{12} = \left.\frac{v_1}{v_2}\right|_{i_1 = 0}$$

$$h_{rb} = \left.\frac{\Delta V_{BE}}{\Delta V_{CB}}\right|_{I_E = constant} \quad (4\text{-}33)$$

Figure 4-30: Graphical determination of the *h-parameters* h_{ib} and h_{rb}

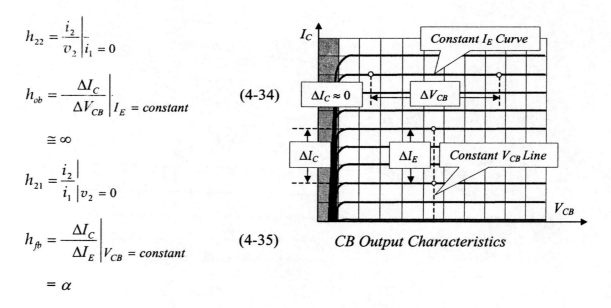

$$h_{22} = \left.\frac{i_2}{v_2}\right|_{i_1 = 0}$$

$$h_{ob} = \left.\frac{\Delta I_C}{\Delta V_{CB}}\right|_{I_E = constant} \quad (4\text{-}34)$$

$$\cong \infty$$

$$h_{21} = \left.\frac{i_2}{i_1}\right|_{v_2 = 0}$$

$$h_{fb} = \left.\frac{\Delta I_C}{\Delta I_E}\right|_{V_{CB} = constant} \quad (4\text{-}35)$$

$$= \alpha$$

Figure 4-31: Graphical determination of the *h-parameters* h_{fb} and h_{ob}

Chapter 4

The transistor characteristic curves are not always readily available for the purpose of determining the *h-parameters*, although one can generate both the input and output characteristics by utilizing the methods described in Chapter 3, or using an instrument known as the *curve tracer*. Generally, transistor data sheets do not offer characteristic curves, although the minimum, maximum, and average values of the *h-parameters* are usually given in most data sheets. The h_{fe}, being the forward current gain of the CE configuration, is actually the AC β, which is usually not much different from the DC β or h_{FE}. Hence, a single average β obtained from the data sheet may be used in both the DC and AC analysis. Likewise, h_{fb}, being the forward current gain of the CB configuration, is actually the AC α, which can be considered to have a value of approximately 1. As we will see shortly, the parameters h_{ie}, h_{ib}, h_{oe}, and h_{ob} are directly related to the DC bias current, and both can be determined knowing I_{CQ}. That leaves only h_{re} and h_{rb}, which can be totally ignored in most applications or, if necessary, either an average value from the data sheet may be used, or the parameter can be measured by generating a set of CE or CB input characteristic curves, respectively. As shown in Figures 4-27 and 4-30, both of these parameters can be easily determined from such characteristics. Recall the dynamic or AC resistance of the *p-n* junction diode as discussed in Chapter 2 and concluded with Equation 1-13, which is repeated below.

$$r_d = \frac{26\,\text{mV}}{I_D}$$

Now, let us consider the parameter h_{ib}, the BJT input resistance of the CB configuration. With reference to the input characteristics of the BJT in the common-base configuration, h_{ib} is determined as follows.

$$h_{ib} = \frac{\Delta V_{BE}}{\Delta I_E}\bigg|_{V_{CB}\,=\,constant} = r_e \qquad (4\text{-}36)$$

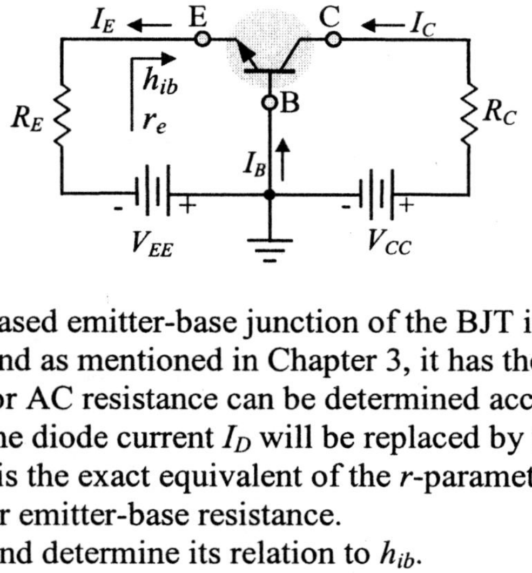

Figure 4-32: *npn* transistor biased in the *CB* configuration

Referring to Figure 4-32, the forward-biased emitter-base junction of the BJT is similar to the forward-biased *p-n* junction diode, and as mentioned in Chapter 3, it has the same characteristics. Therefore, its dynamic or AC resistance can be determined according to Equation 1-13, with the exception that the diode current I_D will be replaced by the junction current I_E. The *h-parameter* h_{ib} is the exact equivalent of the *r-parameter* r_e, which stands for the emitter resistance or emitter-base resistance.

Now, let us examine the parameter h_{ie}, and determine its relation to h_{ib}.

$$h_{ie} = \frac{\Delta V_{CE}}{\Delta I_B} = \frac{\Delta V_{CE}}{\Delta \frac{I_E}{\beta+1}} = (\beta+1)\frac{\Delta V_{CE}}{\Delta I_E} = (\beta+1)h_{ib}$$

$$h_{ie} = (\beta+1)h_{ib} = (\beta+1)r_e \cong \beta r_e \quad \text{ohms} \qquad (4\text{-}37)$$

The h-parameter to be evaluated is h_{oe}, which is defined as follows:

$$h_{oe} = \left.\frac{i_c}{v_{ce}}\right|_{i_b=0} = \left.\frac{\Delta I_C}{\Delta V_{CE}}\right|_{I_B = constant} \qquad (4\text{-}38)$$

The reciprocal of the above equation (4-38) is the exact equivalent of Equation 4-14 that was used earlier to define and evaluate the r-parameter r_o. Evaluation of r_o involved using the CE output characteristics and the Early voltage, as illustrated in Figure 4-9. To evaluate h_{oe} we would utilize the same technique and procedure, which will be briefly repeated here for convenience.

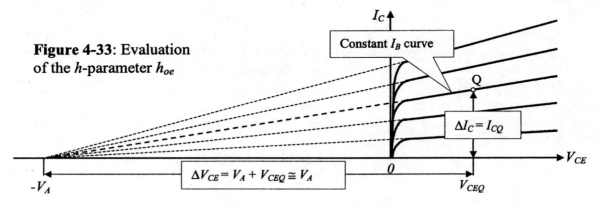

Figure 4-33: Evaluation of the h-parameter h_{oe}

With reference to Figure 4-33, the above expression (Equation 4-38) is basically the slope of the base current line I_B upon which the Q-point is residing. Since the extensions of all I_B curves intercept the V_{CE} axis at the Early voltage V_A, h_{oe} can be defined and determined as follows:

$$h_{oe} = \left.\frac{\Delta I_C}{\Delta V_{CE}}\right|_{I_B = constant} = \left.\frac{\Delta I_C}{\Delta V_{CE}}\right|_{I_{BQ}} = \frac{I_{CQ} - 0}{V_A + V_{CEQ}} \cong \frac{I_{CQ}}{V_A} \qquad \text{A/V or siemens} \qquad (4\text{-}39)$$

The reciprocal of the above expression yields the output resistance, which is the r-parameter r_o.

$$\frac{1}{h_{oe}} = \frac{V_A}{I_{CQ}} = r_o \quad \text{ohms} \qquad (4\text{-}40)$$

A typical value of the Early voltage V_A would be in the range of about 50 V to 150 V. The last h-parameter to be evaluated is h_{ob}, the CB output resistance of the BJT, which is defined as follows:

$$h_{ob} = \left.\frac{i_c}{v_{cb}}\right|_{i_e=0} \qquad \text{A/V or siemens} \qquad (4\text{-}41)$$

$$h_{ob} = \left.\frac{\Delta I_C}{\Delta V_{CB}}\right|_{I_E = constant} \qquad (4\text{-}42)$$

$$\frac{1}{h_{ob}} = \left.\frac{v_{cb}}{i_c}\right|_{i_e=0} = r_c \quad \text{ohms} \qquad (4\text{-}43)$$

Chapter 4

This is the exact definition of the r-parameter r_c, which was evaluated from the CB output characteristics, and is briefly repeated here for convenience. Due to the fact that ΔI_C is very small, h_{ob} will be very large, several MΩ. As we saw earlier, this is not the case with the CE configuration, as the output characteristic curves of the CE are much more inclined, and are not as horizontal as the CB characteristic curves.

Table 4-3: Summary of the *h-parameters*

Configuration →	Common-Base		Common-Emitter		Common-Collector	
Parameter ↓	Notation	Typical Value	Notation	Typical Value	Notation	Typical Value
Current Gain	h_{fb}	$\cong 1$	h_{fe}	150	h_{fc}	150
Input Resistance	h_{ib}	15 Ω	h_{ie}	1.5 kΩ	h_{ic}	1.5 kΩ
Output Resistance	$1/h_{ob}$	5 MΩ	$1/h_{oe}$	50 kΩ	$1/h_{oc}$	50 kΩ
Feedback Ratio	h_{rb}	4×10^{-5}	h_{re}	2×10^{-4}	h_{rc}	$\cong 1$

$$h_{ib} = \frac{26\,\text{mV}}{I_E} = r_e, \quad h_{ie} = \frac{26\,\text{mV}}{I_B} = h_{fe}\,h_{ib} = \beta r_e, \quad \frac{1}{h_{ob}} = \frac{h_{fe}}{h_{oe}} \cong \infty, \quad \frac{1}{h_{oe}} = \frac{V_A}{I_C}, \quad V_A \cong 100\,\text{V}$$

Example 4-6 (Hybrid Model — ANALYSIS)

Determine the *h*-parameters h_{fe}, h_{ie}, and h_{oe} for the transistor of Figure 4-34, and draw the small-signal equivalent of the amplifier circuit.

Solution:

First, we need to determine I_{EQ}.

$$V_B = V_{CC}\frac{R_2}{R_1 + R_2} = 2.7\,\text{V}$$

$$V_E = V_B - 0.7\,\text{V} = 2\,\text{V}$$

$$I_E = \frac{V_E}{R_E} = 2\,\text{mA}, \quad r_e = \frac{26\,\text{mV}}{I_E} = 13\,\Omega$$

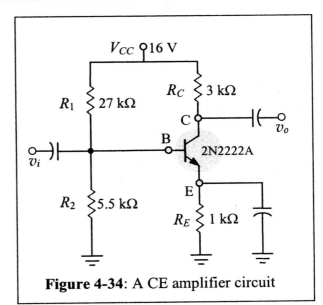

Figure 4-34: A CE amplifier circuit

Next, we need to determine the transistor β. Referring to the data sheets of 2N2222A, I_C of 2 mA corresponds to $\beta = 160$. Therefore,

$$h_{ie} = r_b = \beta r_e \cong 2.1 \text{ k}\Omega$$

Next, $R_B = R_1 \| R_2 = 27\|5.5 = 4.57$ kΩ. Assuming a typical voltage $V_A = 100$ V, we get

$$\frac{1}{h_{oe}} = \frac{V_A}{I_{CQ}} = 50 \text{ k}\Omega$$

Results: $h_{fe} = \beta = 160$, $h_{ie} = r_b = \beta r_e = 2.1$ kΩ, $1/h_{oe} = 50$ kΩ

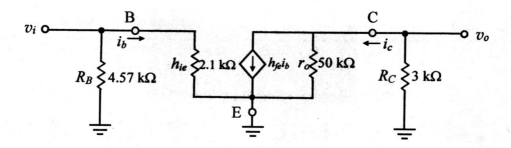

Figure 4-35: Small-signal equivalent circuit of Figure 4-34

Reviewing the typical *h-parameter* values listed above, it is apparent that h_{re} and h_{rb} are very small quantities with typical values of 2×10^{-4} and 4×10^{-5}, respectively. Hence, in ordinary applications such as systems analysis and design, the h_r parameter may be omitted from the transistor model by assuming that $h_r \cong 0$. This assumption replaces the dependent voltage source $h_{re}v_{ce}$ or $h_{rb}v_{cb}$ at the input side with a short circuit.

Example 4-7   Hybrid Model   **ANALYSIS**

Determine the *h-parameters* and draw the small-signal equivalent circuit for the amplifier circuit of Figure 4-36.

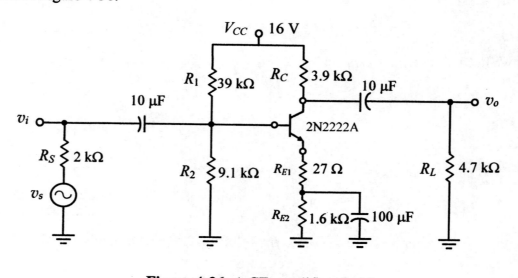

Figure 4-36: A CE amplifier circuit

Chapter 4

Solution: First, we need to determine I_{EQ}. Since $R_2 < 10R_E$, the approximate method of analysis will suffice. But in order to determine I_E, we need to find V_B and then V_E. Thus,

$$V_B = V_{CC}\frac{R_2}{R_1+R_2} = 16 \times \frac{9.1}{39+9.1} = 3.03 \text{ V} \qquad V_E = V_B - 0.7 \text{ V} = 3.03 - 0.7 \text{ V} = 2.33 \text{ V}$$

$$I_E = \frac{V_E}{R_{E1}+R_{E2}} = \frac{2.33}{0.027+1.6} = 1.43 \text{ mA} \qquad r_e = \frac{V_T}{I_E} = \frac{26 \text{ mV}}{1.43 \text{ mA}} = 18.18 \text{ }\Omega$$

Next, we need to determine the transistor β. Referring to the data sheets of 2N2222A, I_C of 1.43 mA corresponds to $\beta = h_{fe} \cong 155$. Therefore, $h_{ie} = r_b = \beta r_e = 155 \times 18.18 \cong 2.82 \text{ k}\Omega$. Let us assume a typical voltage V_A of 100 V. Hence,

$$r_o = \frac{1}{h_{oe}} = \frac{V_A}{I_C} = \frac{100 \text{ V}}{1.43 \text{ mA}} = 69.93 \text{ k}\Omega \qquad R_B = R_1 \| R_2 = 39 \| 9.1 = 7.38 \text{ k}\Omega$$

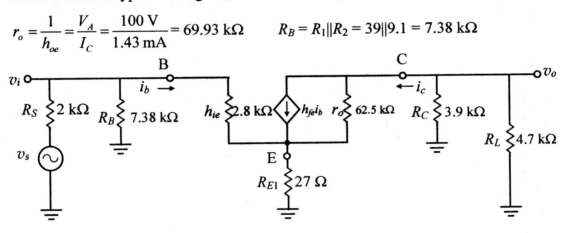

Figure 4-37: Small-signal equivalent circuit of Figure 4-36

Practice Problem 4-2 — Hybrid Model — ANALYSIS

Determine the *h-parameters* for the amplifier of Figure 4-38 and draw the small-signal equivalent circuit with all parameter values.

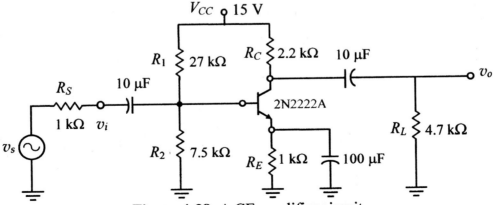

Figure 4-38: A CE amplifier circuit

Chapter 4

Answer:

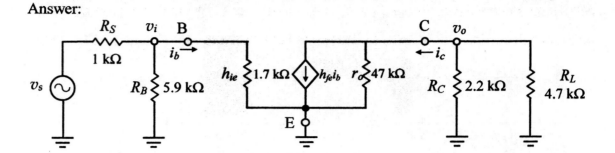

Figure 4-39: Small-signal equivalent circuit of Figure 4-38

Example 4-8 — Hybrid Model — ANALYSIS

Determine the *h-parameters* and draw the small-signal equivalent circuit for the amplifier circuit of Figure 4-40. Assume $\beta = 160$ and $V_A = 90$ V.

Solution:

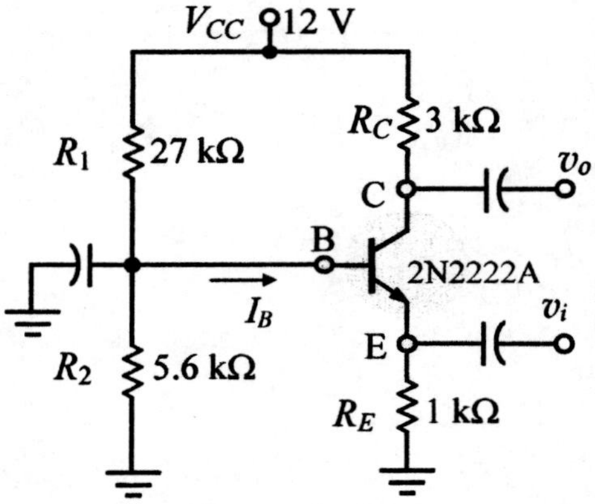

Notice that this circuit, with the emitter being the input, the collector being the output, and the base being the common ground, is a CB amplifier. It is also DC equivalent to the CE amplifier of Example 4-6 (Figure 4-34). Hence, the DC analysis will be exactly the same.

$$V_B = V_{CC}\frac{R_2}{R_1+R_2} = 12\frac{5.6}{27+5.6} = 2.06 \text{ V}$$

Figure 4-40: A CB amplifier circuit

$V_E = V_B - 0.7 \text{ V} = 2.06 - 0.7 = 1.36 \text{ V}$

$I_E = \dfrac{V_E}{R_E} = \dfrac{1.36}{1} = 1.36 \text{ mA}, \quad r_e = \dfrac{V_T}{I_E} = \dfrac{26 \text{ mV}}{1.36} = 19.118 \text{ }\Omega, \quad h_{ib} = r_e = 19.118 \text{ }\Omega$

Referring to the data sheets of 2N2222A, I_C of 1.36 mA corresponds to $h_{fe} = \beta = 160$.

$h_{fb} = \alpha = \dfrac{\beta}{\beta+1} = \dfrac{160}{160+1} = 0.9938$. Assuming a typical voltage $V_A = 90$ V, we obtain

$\dfrac{1}{h_{oe}} = \dfrac{V_A}{I_{CQ}} = \dfrac{90 \text{ V}}{1.36 \text{ mA}} = 67.18 \text{ k}\Omega \qquad \dfrac{1}{h_{ob}} = \dfrac{h_{fe}}{h_{oe}} = 160 \times 67.18 \text{ k}\Omega = 10.75 \text{ M}\Omega \cong \infty.$

$\dfrac{1}{h_{ob}}$ may be dropped from the model for being practically equivalent to an open circuit.

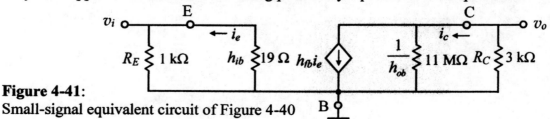

Figure 4-41: Small-signal equivalent circuit of Figure 4-40

144

Having completed the definitions and evaluation of the r and h parameters, we can conclude that, neglecting h_r, both models are so close that we can consider them nearly identical. Omitting h_r from the *hybrid* model, let us compare the two models side-by-side and see if we can combine the two in one *universal* model.

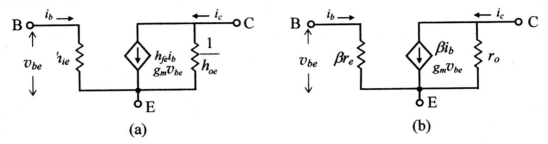

Figure 4-42: The CE models of the BJT: (a) *h-parameter*. (b) *r-parameter*

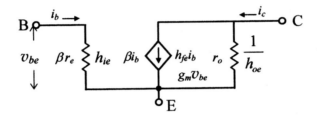

Figure 4-43: The universal h and r parameter model of BJT in CE configuration

Example 4-9 — Universal Model — ANALYSIS

Draw the universal h and r parameter model for the circuit of Figure 4-36 (Example 4-7).

Solution: The *h-parameters* as determined in Example 4-7 are as follows:

$$h_{fe} = \beta \cong 155, \qquad h_{ie} = r_b = \beta r_e \cong 2.5\,k\Omega, \qquad r_o = \frac{1}{h_{oe}} = \frac{V_A}{I_C} = \frac{120\,V}{1.6\,mA} = 75\,k\Omega$$

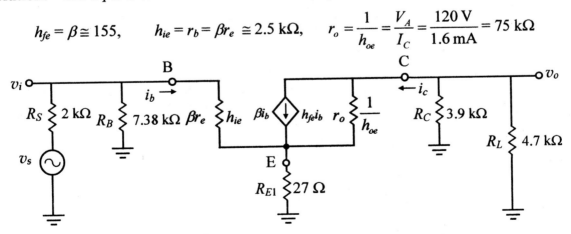

Figure 4-44: The universal h and r parameter model of Figure 4-36

4.3.4 The Approximate BJT Model

Because the *hybrid* model accounts for the voltage feedback effect by introducing the dependent voltage source $h_{re}v_{ce}$ at the input side, it represents a more accurate equivalent circuit of the BJT and must be used as it appears without modification in certain applications, such as research and development of new devices. However, some approximations to the model are permissible in ordinary and less sophisticated applications. As mentioned earlier, the h_r parameter can be omitted in most applications, such as systems analysis and design, without significantly affecting the outcome. The second approximation involves neglecting the effect of the parameter h_o. The parameter h_{ob} in CB configuration, which has a resistance value of several MΩ, may be replaced by an open circuit in almost any ordinary application. Note that $\beta i_b = g_m v_{be}$ and therefore the dependent source can be described by a current-dependent source βi_b or by a voltage dependent source $g_m v_{be}$. The parameter h_{oe} in the CE configuration, which has a resistance ranging from 40 to 100 kΩ, may be omitted from the model if and only if the situation allows. If a typical h_{oe} of 70 kΩ is to be in parallel with R_C of about 2 kΩ, then it seems reasonable to neglect the effect of h_{oe} by replacing it with an open circuit. Therefore, the most simplified small-signal BJT model, which may be considered either as *r-parameter* model or *h-parameter* model, is as depicted in Figure 4-45. The model of Fig. 4-45(a) is most commonly used and can be applied to all configurations: CE, CB and CC.

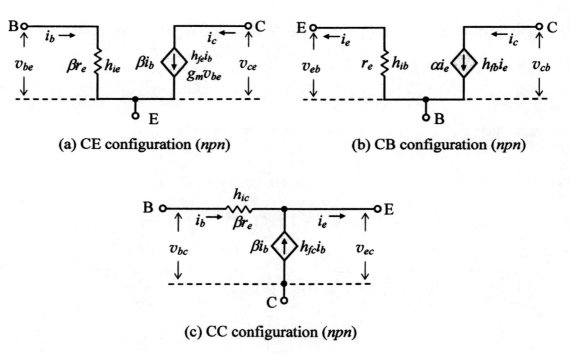

(a) CE configuration (*npn*) (b) CB configuration (*npn*)

(c) CC configuration (*npn*)

Figure 4-45: The approximate h and r parameter model of BJT

Example 4-10 — Approximate Model — ANALYSIS

Determine the approximate h and r model for the transistor of Figure 4-46, and draw the small-signal equivalent of the amplifier circuit.

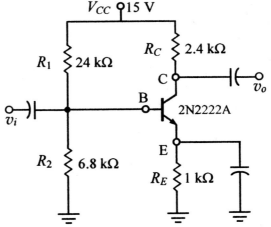

Figure 4-46: A CE amplifier circuit

Solution:

First, we need to determine I_{EQ}. Thus,

$$V_B = V_{CC}\frac{R_2}{R_1+R_2} = 15 \times \frac{6.8}{24+6.8} = 3.3\text{ V}$$

$$V_E = V_B - 0.7\text{ V} = 3.3 - 0.7 = 2.6\text{ V}$$

$$I_E = \frac{V_E}{R_E} = \frac{2.6}{1} = 2.6\text{ mA}$$

$$r_e = \frac{V_T}{I_E} = \frac{26\text{ mV}}{I_E} = \frac{26\text{ mV}}{2.6} = 10\ \Omega$$

Next, we need to determine the transistor β. Referring to data sheets of 2N2222A, I_C of 2.6 mA corresponds to $\beta = 170$. Therefore,

$$h_{ie} = r_b = \beta r_e = 170 \times 10 = 1.7\text{ k}\Omega$$

$R_B = R_1\|R_2 = 24\|6.8 = 5.3\text{ k}\Omega$.

Results: $h_{fe} = \beta = 170$ $h_{ie} = r_b = \beta r_e = 1.77\text{ k}\Omega$

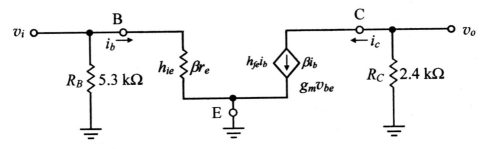

Figure 4-47: Approximate h and r model and small-signal equivalent circuit of Figure 4-46

Example 4-11 — Universal Model — ANALYSIS

Determine the universal h and r model for the transistor and draw the small-signal equivalent circuit for the CE amplifier with *gain stabilization* of Figure 4-48.

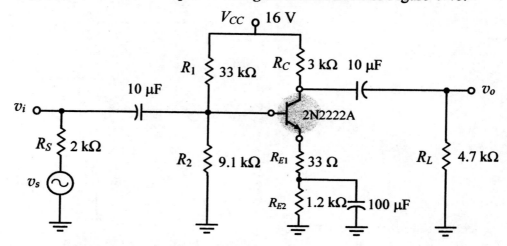

Figure 4-48: A common-emitter amplifier circuit

Solution:

$$V_B = V_{CC} \frac{R_2}{R_1 + R_2} = 16 \times \frac{9.1}{33 + 9.1} = 3.46 \text{ V} \qquad V_E = V_B - 0.7 \text{ V} = 3.46 - 0.7 = 2.76 \text{ V}$$

$$I_E = \frac{V_E}{R_{E1} + R_{E2}} = \frac{2.76}{0.033 + 1.2} = 2.238 \text{ mA} \qquad r_e = \frac{V_T}{I_E} = \frac{26 \text{ mV}}{I_E} = \frac{26 \text{ mV}}{2.238} = 11.6 \text{ }\Omega$$

Referring to the data sheets of 2N2222A, I_C of 2.2 mA corresponds to $\beta \cong 170$. Therefore, $h_{ie} = r_b = \beta r_e = 170 \times 11.6 = 1972 \text{ }\Omega \cong 2 \text{ k}\Omega$. Let $R_B = R_1 | R_2 = 33 \| 9.1 = 7.1 \text{ k}\Omega$.

Results: $h_{fe} = \beta \cong 170$ and $h_{ie} = r_b = \beta r_e \cong 2 \text{ k}\Omega$.

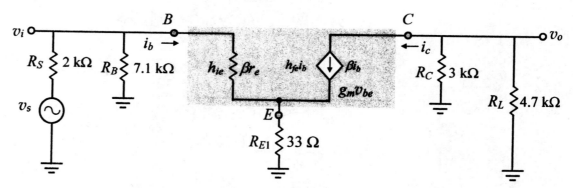

Figure 4-49: Universal h and r model and small-signal equivalent circuit of Figure 4-48

Chapter 4

PROBLEMS

Section 4.2 The *r-parameter* Model

4.1 Determine the *r-parameters* and draw the small-signal equivalent circuit of Figure 4-10, with $R_B = 750$ kΩ. All other circuit parameters remain unchanged.

4.2 Determine the *r-parameters* and draw the small-signal equivalent circuit of Figure 4-10, with $R_E = 2$ kΩ. All other circuit parameters remain unchanged.

4.3 Determine the *r-parameters* and draw the small-signal equivalent circuit of Figure 4-12, with $R_1 = 39$ kΩ, and $R_2 = 7.5$ kΩ. All other circuit parameters remain unchanged.

4.4 Determine the *r-parameters* and draw the small-signal equivalent circuit of Figure 4-12, with $R_1 = 47$ kΩ, $R_2 = 10$ kΩ, $R_{E1} = 30$ Ω, and $R_{E2} = 1.2$ kΩ. All other circuit parameters remain unchanged.

4.5 Determine the *r-parameters* and draw the small-signal equivalent circuit of Figure 4-14, with $R_1 = 39$ kΩ, $R_2 = 10$ kΩ, and $R_C = 4.7$ kΩ. All other circuit parameters remain unchanged.

4.6 Determine the *r-parameters* and draw the small-signal equivalent circuit of Figure 4-14, with $R_1 = 33$ kΩ, $R_2 = 7.5$ kΩ, and $R_C = 5.1$ kΩ. All other circuit parameters remain unchanged.

4.7 Determine the *r-parameters* for the transistor of Figure 4-1P below, and draw the small-signal equivalent circuit.

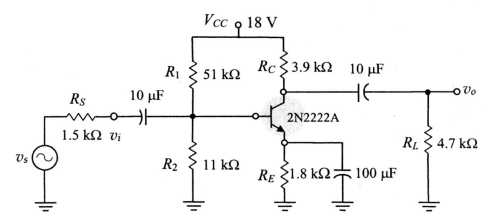

Figure 4-1P: CE amplifier (problems 4.7 & 4.8)

4.8 Determine the *r-parameters* and draw the small-signal equivalent circuit of Figure 4-1P, with $R_1 = 39$ kΩ, $R_2 = 9.1$ kΩ, $R_E = 1.5$ kΩ, and $R_C = 4.7$ kΩ. All other circuit parameters remain unchanged.

parameters and draw the small-signal equivalent circuit of Figure 4-16, ... kΩ, R_E = 1.6 kΩ, and R_C = 2.7 kΩ. All other circuit parameters remain ...

4.9 ...ine the *r-parameters* and draw the small-signal equivalent circuit of Figure 4-18, R_1 = 39 kΩ, R_2 = 51 kΩ, R_E = 7.5 kΩ, and V_{CC} = 15 V. All other circuit parameters ...nain unchanged.

Section 4.3.1 The *Hybrid (h-parameter)* Model

4.11 Determine the *h-parameters* for the circuit of Figure 4-21. Write the network equations, and draw the *hybrid* equivalent circuit, with R_1 = 3.3 kΩ, R_2 = 2.2 kΩ, and R_3 = 1.1 kΩ.

4.12 Determine the *h-parameters* for the circuit of Figure 4-21. Write the network equations, and draw the *hybrid* equivalent circuit, with R_1 = 1.1 kΩ, R_2 = 2.2 kΩ, and R_3 = 3.3 kΩ.

4.13 Determine the *h-parameters* for the circuit of Figure 4-2P below. Write the network equations, and draw the *hybrid* equivalent circuit.

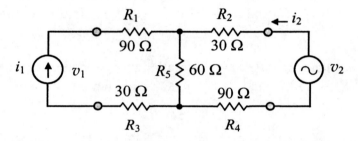

Figure 4-2P: Resistive two-port network (problems 4.13 & 4.14)

4.14 Determine the *h-parameters* for the circuit of Figure 4-2P above. Write the network equations, and draw the *hybrid* equivalent circuit, with R_1 = 1.1 kΩ, R_2 = 2.2 kΩ, R_3 = 2.2 kΩ, R_4 = 1.1 kΩ, and R_5 = 3.3 kΩ.

Section 4.3.2 & 4.3.3 The *Hybrid (h-parameter)* Model of the BJT

4.15 Determine the *h-parameters* for the transistor of Figure 4-34 and draw the small-signal equivalent circuit, with R_1 = 33 kΩ, R_2 = 6.8 kΩ, and R_C = 3.3 kΩ. All other circuit parameters remain unchanged.

4.16 Determine the *h-parameters* for the transistor of Figure 4-34 and draw the small-signal equivalent circuit, with R_1 = 39 kΩ, R_2 = 9.1 kΩ, R_E = 1.5 kΩ, and R_C = 3.3 kΩ. All other circuit parameters remain unchanged.

4.17 Determine the *h-parameters* for the transistor of Figure 4-3P and draw the small-signal equivalent circuit.

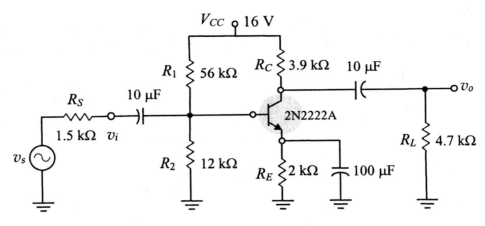

Figure 4-3P: CE amplifier (problems 4.17 & 4.18)

4.18 Determine the *h-parameters* for the transistor of Figure 4-3P and draw the small-signal equivalent circuit, with $R_1 = 43$ kΩ, $R_2 = 10$ kΩ, $R_E = 1.5$ kΩ, and $R_C = 3.6$ kΩ. All other circuit parameters remain unchanged.

4.19 Determine the *h-parameters* for the transistor of Figure 4-36 and draw the small-signal equivalent circuit, with $R_1 = 47$ kΩ, $R_2 = 11$ kΩ, $R_E = 1.5$ kΩ, and $R_C = 3.3$ kΩ. All other circuit parameters remain unchanged.

4.20 Determine the *h-parameters* for the transistor of Figure 4-36 and draw the small-signal equivalent circuit, with $R_1 = 51$ kΩ, $R_2 = 12$ kΩ, $R_E = 1.8$ kΩ, and $R_C = 3.6$ kΩ. All other circuit parameters remain unchanged.

4.21 Determine the *h-parameters* for the transistor of Figure 4-40 and draw the small-signal equivalent circuit, with $R_1 = 39$ kΩ, $R_2 = 10$ kΩ, $R_E = 1.5$ kΩ, and $R_C = 5.1$ kΩ. All other circuit parameters remain unchanged.

4.22 Determine the *h-parameters* for the transistor of Figure 4-40 and draw the small-signal equivalent circuit, with $R_1 = 43$ kΩ, $R_2 = 11$ kΩ, $R_E = 1.6$ kΩ, and $R_C = 4.7$ kΩ. All other circuit parameters remain unchanged.

4.23 Determine the *h-parameters* for the transistor of Figure 4-4P below, and draw the small-signal equivalent circuit.

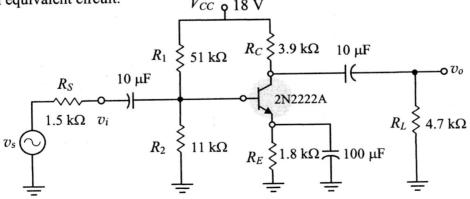

Figure 4-4P: CE amplifier (problems 4.23 & 4.24)

Chapter 4

4.24 Determine the *h-parameters* for the transistor of Figure 4-4P above, and draw the small-signal equivalent circuit, with $R_1 = 39$ kΩ, $R_2 = 9.1$ kΩ, $R_E = 1.5$ kΩ, and $R_C = 3.9$ kΩ. All other circuit parameters remain unchanged.

4.25 Determine the combined *h* and *r* model of the transistor and draw the small-signal equivalent circuit for the amplifier of Figure 4-4P, with $R_1 = 39$ kΩ, $R_2 = 9.1$ kΩ, $R_E = 1.5$ kΩ, and $R_C = 3.9$ kΩ. All other circuit parameters remain unchanged.

4.26 Determine the combined *h* and *r* model of the transistor and draw the small-signal equivalent circuit for the amplifier of Figure 4-4P, with $R_1 = 33$ kΩ, $R_2 = 8.2$ kΩ, $R_E = 1.2$ kΩ, and $R_C = 3.3$ kΩ. All other circuit parameters remain unchanged.

4.27 Determine the combined *h* and *r* model of the transistor and draw the small-signal equivalent circuit for the amplifier of Figure 4-40, with $R_1 = 33$ kΩ, $R_2 = 8.2$ kΩ, $R_E = 1.2$ kΩ, and $R_C = 3.3$ kΩ. All other circuit parameters remain unchanged.

Section 4.3.4 The Approximate BJT Model

4.28 Determine the approximate *h* and *r parameters* for the transistor and draw the small-signal equivalent circuit for the amplifier of Figure 4-5P below.

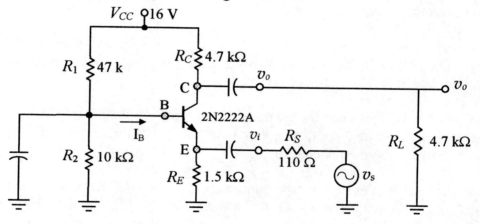

Figure 4-5P: A common-base amplifier circuit

4.29 Determine the approximate *h* and *r* model for the transistor and draw the small-signal equivalent circuit for the amplifier of Figure 4-5P above, with $R_1 = 33$ kΩ, $R_2 = 8.2$ kΩ, $R_E = 1.2$ kΩ, and $R_C = 3.3$ kΩ. All other circuit parameters remain unchanged.

4.30 Determine the approximate *h* and *r* model for the transistor and draw the small-signal equivalent circuit for the amplifier of Figure 4-46, with $R_1 = R_2 = 27$ kΩ, and $R_E = 3.3$ kΩ. All other circuit parameters remain unchanged.

4.31 Determine the approximate *h* and *r* model for the transistor and draw the small-signal equivalent circuit for the amplifier of Figure 4-46, with $R_1 = 24$ kΩ, $R_2 = 36$ kΩ, and $R_E = 5.1$ kΩ. All other circuit parameters remain unchanged.

Chapter 4

4.32   Determine the approximate h and r model for the transistor and draw the small-signal equivalent circuit for the amplifier of Figure 4-48, with $R_1 = 27$ kΩ, $R_2 = 7.5$ kΩ, $R_{E2} = 1.1$ kΩ, and $R_C = 2.7$ kΩ. All other circuit parameters remain unchanged.

4.33   Determine the approximate h and r model for the transistor and draw the small-signal equivalent circuit for the amplifier of Figure 5-48, with $R_1 = 47$ kΩ, $R_2 = 11$ kΩ, and $R_{E2} = 1.6$ kΩ. All other circuit parameters remain unchanged.

Chapter

SMALL-SIGNAL OPERATION OF THE BJT AMPLIFIER

5.1 INTRODUCTION

Having discussed the basic theory of the BJT operation, an in-depth and comprehensive analysis and design of the DC bias of the BJT amplifier in Chapter 3, followed by transistor modeling for small-signal operation in Chapter 4, we now wish to explore the small-signal operation of the BJT amplifier. The AC operation is generally classified in two categories: *small-signal operation* and *large-signal operation*. Small-signal operation is the analysis and design of amplifiers that take signals of very small amplitude in the range of a few millivolts, and produce at the output large signals in the range of several volts. Hence, the focus of attention in the *small-signal operation* is mainly the AC voltage and current gains or the amplifying action of the amplifier. Other parameters such as input resistance, output resistance, and the frequency response of the amplifier are also of great concern.

Chapter 5

For example, the unit called *preamp* in your audio system is a small-signal amplifier; it takes the small-signal from a variety of input transducers, such as CD player, tape player, receiver, or a microphone, and amplifies it to a level that is acceptable by the large-signal amplifier. The large-signal operation, which is discussed in later chapters, is the analysis and design of *large-signal amplifiers*, commonly known as power amplifiers or current drivers. In fact, your audio system is a familiar and good example of an electronic system that utilizes both the small-signal and large-signal amplifiers, whose block diagram is shown in Figure 5-1 below:

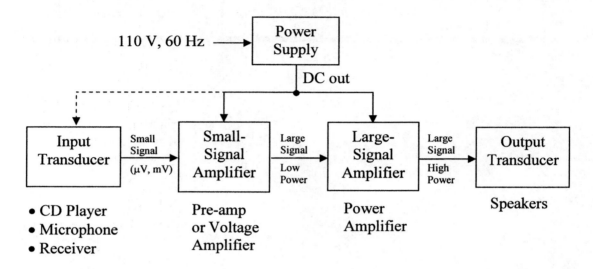

Figure 5-1: Block diagram of an electronic system (audio system)

In case of other applications, the input transducer could be, for example, a temperature sensor, pressure sensor, humidity sensor, or some other small-signal device. The output transducer could be a motor, a solenoid, or some other high-power device. In either case, the small signal from the input transducer is amplified and fed to the large signal amplifier, which puts power behind the large signal and delivers it to the output transducer. The power that is delivered to the output transducer is drawn from the power supply by the large-signal amplifier, which is also referred to as a current driver.

The analysis of the small-signal amplifier is covered in this chapter and its design is covered in the following chapter. Since the operation of the large-signal amplifier is perfected by the use of the operational amplifier, its analysis and design will be covered in Chapter 13, after the introduction of the operational amplifier.

5.2 THE BJT AMPLIFIER FUNDAMENTALS

In the previous chapter, we have seen a variety of BJT amplifier circuits and have learned that an amplifier is a two-port system with input and output terminals. Consider the following block diagram of a BJT amplifier, as illustrated in Figure 5-2 below.

Figure 5-2: Block diagram of an amplifier

Assuming a common reference ground for the input and output ports, the above block diagram may be redrawn as illustrated in Figure 5-3 below.

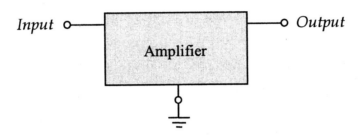

Figure 5-3: An amplifier with a common reference ground

We will now place this amplifier under test by applying a signal from the signal source v_s at the input terminal and connecting a load resistor R_L at the output terminal. Note that these additions are all external to the amplifier circuit. As a result of applying a small signal at the input, a large signal will develop at the output, as shown in Figure 5-4 below. This amplifying action is called the gain in voltage level or simply the *voltage gain* A_v.

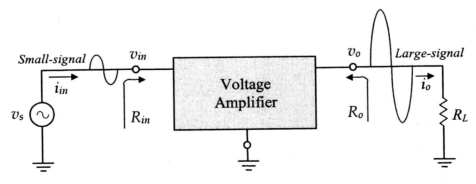

Figure 5-4: An amplifier under test

The amplifier will draw an input current i_{in} from the signal source. According to Ohm's law, the amount of this input current will depend on the signal level v_s and the input resistance R_{in} of the amplifier. Hence, at the input side, the amplifier will look like a resistor with a value of R_{in} to the signal source. Likewise, an equivalent circuit can be obtained at the output terminal by applying Thevenin's theorem, which will result in a dependent voltage source with a series resistor. The amplitude of this voltage source will depend on the applied signal v_{in} and the voltage gain or amplification factor (A_v) of the amplifier. Hence, the Thevenin's equivalent circuit at the output, which is the open-circuit voltage and resistance measured at the output terminal with no load, will be as follows. The amplifier's equivalent circuit is shown in Figure 5-5 below.

$$E_{th} = v_{o(NL)} = v_{in} A_v \qquad (5\text{-}1)$$

$$R_{th} = R_o \qquad (5\text{-}2)$$

where,

$$A_v = \frac{v_o}{v_{in}} \qquad (5\text{-}3)$$

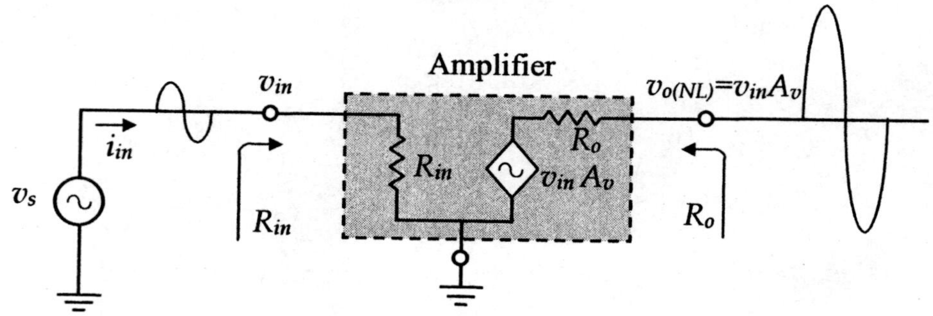

Figure 5-5: Equivalent circuit of an amplifier and the output voltage with no load

The output signal is shown to be in-phase with the input signal in the above diagram. This is not necessarily the case with all types of amplifiers. As we will see shortly, the common-emitter amplifier's output will be out-of-phase with the input signal and the direction of the output current will be opposite to what is shown in Figure 5-5. Application of the load resistor R_L at the output terminal will draw an output current i_o. The amount of the output current i_o and the output voltage with load $v_{o(WL)}$ will depend on the no-load output voltage $v_{o(NL)}$, the output resistance R_o, and the load resistor R_L.

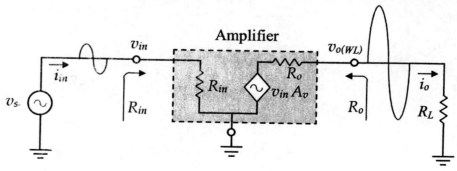

Figure 5-6: Equivalent circuit of an amplifier and the output voltage with load

$$i_o = \frac{v_{in} A_v}{R_o + R_L} \quad (5\text{-}4)$$

$$v_{o(WL)} = i_o R_L \quad (5\text{-}5)$$

Also, the output voltage with load $v_{o(WL)}$ can be determined directly by applying the voltage divider rule at the output, as follows:

$$v_{o(WL)} = v_{in} A_v \frac{R_L}{R_o + R_L} = v_{o(NL)} \frac{R_L}{R_o + R_L} \quad (5\text{-}6)$$

where A_v is the voltage gain with no load.

$$A_v = \frac{v_{o(NL)}}{v_{in}} \quad (5\text{-}7)$$

The Amplifier Parameters

Having developed an equivalent circuit for the amplifier, and having introduced some amplifier terminology such as *voltage gain*, *input resistance*, and *output resistance*, we now wish to discuss and evaluate some important amplifier parameters. Note that the input and output resistances, labeled R_{in} and R_o, are not meant to give the impression that these parameters are DC resistances. These parameters are AC resistances and certainly not reactances. The academically correct labeling would have been the lowercase r with corresponding subscripts designating AC resistance. However, since lowercase r labels (r_e, r_o, r_b, r_c) have already been used to represent the input and output resistance of the transistor, and in order to use meaningful labels that closely describe the parameter, we will be using uppercase R labels with lowercase subscripts, such as R_{in} and R_o, for the input and output resistance of the amplifier, and other notations, such as R_b and R_e, to designate the equivalent AC resistances seen from the base and emitter of the transistor.

The input resistance R_{in}:

The input resistance of an amplifier is actually the resistance seen by the signal source as shown in Figure 5-6. It is also the resistance measured from the input terminal to the common ground under live conditions, using Ohm's law, not an ohmmeter, as follows:

$$R_{in} = \frac{v_{in}}{i_{in}} \quad (5\text{-}8)$$

In the laboratory, it is more convenient to measure the input current i_{in} indirectly by measuring the voltage across a current-sense resistor R_s of known value, as shown in Figure 5-7, and then computing the current according to Equation 5-9.

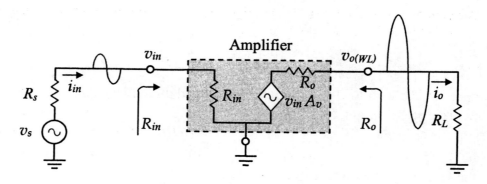

Figure 5-7: Equivalent circuit of an amplifier under test

$$i_{in} = \frac{v_s - v_{in}}{R_s} \tag{5-9}$$

Substituting Equation 5-9 for i_{in} in Equation 5-8 yields the following.

$$R_{in} = \frac{v_{in}}{v_s - v_{in}} R_s \tag{5-10}$$

Note that Equation 5-8, which is the definition of the input resistance, will be used when analyzing amplifier circuits. In contrast, Equation 5-10 is to be used when measuring the amplifier's input resistance in a laboratory situation. In order to measure the input resistance R_{in} of the amplifier in the laboratory, a current-sense resistor R_s is placed in series with the signal source. But from the analysis point of view, R_s may account for the internal resistance of the signal source v_s. In either case, R_s is external to the amplifier and is not to be considered part of the input resistance of the amplifier.

The output resistance R_o:

The output resistance of an amplifier is actually the resistance seen by the load as shown in Figure 5-7 above. It is also the resistance measured from the output terminal to the common ground under live conditions, using Ohm's law, not an ohmmeter, as follows:

$$R_o = \frac{v_{o(NL)} - v_{o(WL)}}{i_o} \tag{5-11}$$

where,

$$i_o = \frac{v_{o(WL)}}{R_L} \tag{5-12}$$

Substituting Equation 5-12 for i_o in Equation 5-11 results in the following.

$$R_o = \frac{v_{o(NL)} - v_{o(WL)}}{v_{o(WL)}} R_L \tag{5-13}$$

The above equation for R_o is more useful in the laboratory when measuring the output resistance of an amplifier. But in order to determine the output resistance R_o of the amplifier, analytically, a test voltage is applied at the output terminal while the input is

Chapter 5

grounded. We have actually used this concept in Chapter 4 (Equation 4-14) in order to determine the output resistance r_o of the BJT.

Voltage gain with and without load:

The voltage gain of an amplifier, as mentioned previously, is the ratio of the output voltage v_o to the input voltage v_{in}. Apparently, the output voltage may be measured with or without load. Recall that the output voltage drops according to Equation 5-6 when the amplifier is loaded. Hence, the voltage gain drops accordingly, when the load is applied.

$$A_{v(NL)} = \frac{v_{o(NL)}}{v_{in}} \tag{5-14}$$

$$A_{v(WL)} = \frac{v_{o(WL)}}{v_{in}} \tag{5-15}$$

The current gain A_i:

The current gain of an amplifier is the ratio of the output current i_o to the input current i_{in}.

$$A_i = \frac{i_o}{i_{in}} \tag{5-16}$$

where the input and output currents can be defined as follows:

$$i_{in} = \frac{v_{in}}{R_{in}} \tag{5-17}$$

$$i_o = \frac{v_{o(WL)}}{R_L} \tag{5-18}$$

Substituting Equations 5-17 and 5-18 for Equation 5-16 results in the following:

$$A_i = \frac{i_o}{i_{in}} = \frac{v_{o(WL)}}{R_L} \times \frac{R_{in}}{v_{in}} = \frac{v_{o(WL)}}{v_{in}} \times \frac{R_{in}}{R_L} \tag{5-19}$$

Since the ratio of the output voltage with load to the input voltage is the *voltage gain* with load, the *current gain* can be expressed as follows:

$$A_i = \frac{i_o}{i_{in}} = A_{v(WL)} \frac{R_{in}}{R_L} \tag{5-20}$$

The power gain A_p:

The power gain of an amplifier is defined as the ratio of the output power P_o to the input power P_{in}. The input and output powers are simply the product of the *rms* value of the corresponding current and voltage, as follows:

$$P_{in} = i_{in(rms)} \times v_{in(rms)}$$

$$P_o = i_{o(rms)} \times v_{o(rms)}$$

161

$$A_p = \frac{p_o}{p_{in}} = \frac{i_{o(rms)} \times v_{o(rms)}}{i_{in(rms)} \times v_{in(rms)}} = \frac{i_o}{i_{in}} \times \frac{v_o}{v_{in}} \tag{5-21}$$

Hence, the *power gain* is the product of the *current gain* and the *voltage gain*.

$$A_p = A_i \times A_{v(WL)} \tag{5-22}$$

The output voltage with and without load:

Referring to Equations 5-14 and 5-15, we can solve for the output voltage v_o, as follows:

$$A_{v(NL)} = \frac{v_{o(NL)}}{v_{in}} \qquad\qquad A_{v(WL)} = \frac{v_{o(WL)}}{v_{in}}$$

$$v_{o(NL)} = A_{v(NL)} \, v_{in} \tag{5-23}$$

and,

$$v_{o(WL)} = A_{v(WL)} \, v_{in} \tag{5-24}$$

where v_{in} can be determined by applying a voltage divider rule at the input side.

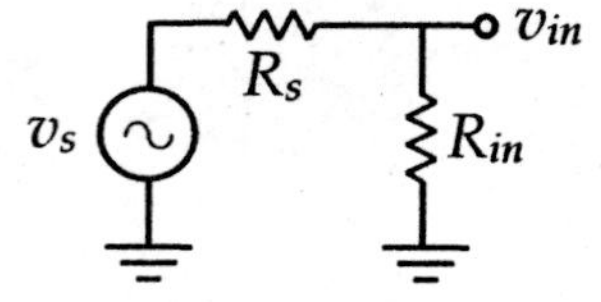

$$v_{in} = v_s \frac{R_{in}}{R_s + R_{in}} \tag{5-25}$$

Figure 5-8: Computing v_{in}

Example 5-1 — ANALYSIS

Recall the amplifier that was placed under test a while back in Figure 5-4. Assuming that the test result revealed the following data, determine the amplifier parameters R_{in}, R_o, $A_{v(NL)}$, $A_{v(WL)}$, A_i, and A_p.

Collected data:

$v_s = 50$ mV, $v_{in} = 40$ mV, $v_{o(NL)} = 5$ V, $v_{o(WL)} = 3$ V, $R_s = 1$ kΩ, $R_L = 2$ kΩ

Solution:

Input resistance can be determined from Equation 5-10,

$$R_{in} = \frac{v_{in}}{v_s - v_{in}} R_s = \frac{40 \text{ mV}}{50 \text{ mV} - 40 \text{ mV}} \times 1 \text{ k}\Omega = 4 \text{ k}\Omega$$

Output resistance can be determined from Equation 5-13.

$$R_o = \frac{v_{o(NL)} - v_{o(WL)}}{v_{o(WL)}} R_L = \frac{5 \text{ V} - 3 \text{ V}}{3 \text{ V}} \times 2 \text{ k}\Omega = 1.33 \text{ k}\Omega$$

No-load voltage gain can be determined from Equation 5-14.

$$A_{v(NL)} = \frac{v_{o(NL)}}{v_i} = \frac{5\text{ V}}{50\text{ mV}} = 100$$

Voltage gain with load can be determined from Equation 5-15.

$$A_{v(WL)} = \frac{v_{o(WL)}}{v_i} = \frac{3\text{ V}}{50\text{ mV}} = 60$$

Current gain can be determined from Equation 5-20.

$$A_i = A_{v(WL)} \frac{R_{in}}{R_L} = 60 \times \frac{4\text{ k}\Omega}{2\text{ k}\Omega} = 120$$

Power gain can be determined from Equation 5-22.

$$A_p = A_i \times A_{v(WL)} = 120 \times 60 = 7200$$

5.3 ANALYSIS OF THE BJT AMPLIFIER

In analyzing any amplifier circuit, our intention is to determine all the amplifier parameters, some of which were discussed in the previous section. Upon completion of this chapter, you are expected to have a comprehensive knowledge and understanding of the theory underlying the small-signal operation of the BJT amplifier. In order to achieve that goal, we will discuss at length the small-signal analysis of all three amplifier configurations, one configuration at a time. Some of the examples that are introduced throughout the chapter will be tested by simulation in order to verify the accuracy of the algebraic solutions and to validate our method of analysis. A number of problems will be given at the end of the chapter as homework assignments that will require verification of the algebraic solutions by simulation. As you have already seen in Chapter 3, we have been using Electronics Workbench for simulation of electronic circuits. The Electronics Workbench was chosen as the simulation software because it produces simulation results that are more visual, more realistic, and easily recognizable.

5.3.1 The Common-Emitter Amplifier

Analysis of an amplifier begins with the DC analysis in order to determine the I_{CQ}, upon which depend some of the small-signal parameters of the transistor. Having determined the I_{CQ}, we then determine the V_{CEQ}, which gives us an idea about the Q-point location, which reveals the limitations on the output signal swing. Next, we begin the AC analysis by drawing the small-signal equivalent circuit with the transistor model substituted for the transistor. From the equivalent circuit, we can determine all the small-signal parameters such as the input resistance R_{in}, output resistance R_o, voltage gain A_v with and without load, current gain A_i, and the power gain A_p. Having determined the voltage gain, we can then determine the expected output signal level for a given small-signal v_s, although the expected signal may or may not be achieved at the output, depending on the Q-point location. We have already discussed the importance of the Q-point location and its relation to the output signal swing in Chapter 4. This topic will be revisited when we introduce the AC load line shortly.

Chapter 5

Consider the familiar emitter-biased common-emitter amplifier of Figure 5-9. Let us determine the following parameters for this amplifier. Assume $\beta = 120$ and $V_A = 120$ V.

a) I_{CQ} and V_{CEQ}

b) R_{in}

c) R_o

d) $A_{v(NL)}$

e) $A_{v(WL)}$

f) A_i

g) A_p

h) $v_{o(WL)}$

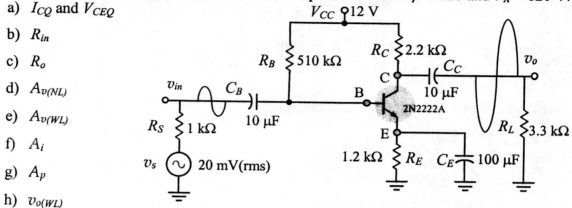

Figure 5-9: An emitter-biased common-emitter amplifier

For the DC analysis, we first need to determine I_{BQ}. Thus,

$$I_B = \frac{V_{CC} - V_{BE}}{R_B + (\beta+1)R_E} = \frac{12 - 0.7}{5.1 + (120+1)1.2} = 17.25 \, \mu A$$

$I_C = \beta I_B = 120 \times 0.01725 = 2.07$ mA $\quad I_E = (\beta + 1)I_B = (120 + 1)0.01725 = 2..09$ mA

$V_{CE} = V_{CC} - (I_C R_C + I_E R_E) \cong V_{CC} - I_C (R_C + R_E)$

$= 12$ V $- 2.07$ mA$(2.2$ k$\Omega + 1.2$ k$\Omega) = 4.96$ V

Therefore, the Q-point is at I_C of approximately 2 mA and V_{CE} of 5.17 V. With V_{CC} of 12 V, this is a fairly well placed Q-point. Next, we need to decide on a model and determine the small-signal parameters. The model we have adopted is the universal *h* and *r* model, which can be interpreted as *r-parameter* or *h-parameter* model. Hence, we will now evaluate the *r-parameters* of the transistor to obtain

$$r_e = \frac{V_T}{I_E} = \frac{26 \, mV}{2.09} = 12.44 \, \Omega \, , \, r_b = \beta r_e = 1.5 \, k\Omega = h_{ie}, \, r_o = \frac{V_A}{I_C} = \frac{120 \, V}{2.07 \, mA} = 58 \, k\Omega = \frac{1}{h_{oe}}$$

Having determined the *r-parameters*, we can draw the small-signal equivalent circuit. The coupling capacitors C_B and C_C and the emitter bypass capacitor C_E have been selected so that they will have a very low and negligible reactance in the frequency range at which the amplifier will operate. This band of frequency can range typically from 20 Hz to 100 kHz at which the capacitors can be replaced by a short circuit in the small-signal model of an amplifier.

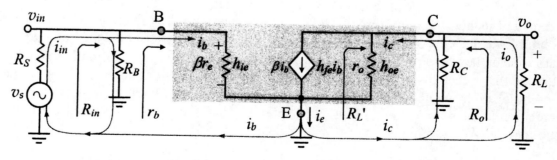

Figure 5-10: Small-signal equivalent circuit of Figure 5-9

For the sake of convenience, let us refer to these R labels (R_{in}, R_o, R_L', etc.) as the *R-parameters*, not to be mistaken with the *r-parameters* of the transistor.
The next step is to evaluate the R-parameters R_{in}, R_L', and R_o.

The input resistance:

The input resistance R_{in} is the parallel combination of R_B and r_b.

$$R_{in} = R_B \parallel r_b \tag{5-26}$$

$R_{in} = 510 \text{ k}\Omega \parallel 1.5 \text{ k}\Omega \cong 1.5 \text{ k}\Omega$

The total AC load R_L' at the output is the parallel combination of the r_o, R_C, and R_L.

$$R_L' = r_o \parallel R_C \parallel R_L \tag{5-27}$$

$R_L' = 58 \text{ k}\Omega \parallel 2.2 \text{ k}\Omega \parallel 3.3 \text{ k}\Omega = 1.29 \text{ k}\Omega$

However, with no load, that is, without external load R_L, the internal load of the amplifier will be the r_o in parallel with R_C, and we will designate this internal load as R_L'', in order to determine the no-load voltage gain.

$$R_L'' = r_o \parallel R_C$$

$R_L'' = 58 \text{ k}\Omega \parallel 2.2 \text{ k}\Omega = 2.12 \text{ k}\Omega \tag{5-28}$

The output resistance:

The output resistance R_o seen by the load is simply the parallel combination of R_C and r_o, as the resistance of the current source is ideally equal to an open circuit.

$$R_o = r_o \parallel R_C \tag{5-29}$$

$R_o = 58 \text{ k}\Omega \parallel 2.2 \text{ k}\Omega = \mathbf{2.12 \text{ k}\Omega}$

Having determined the R-parameters, we are now ready to determine the voltage and current gains. Recall that the voltage gain is defined as the ratio of the output voltage v_o to the input voltage v_{in}. Also, notice that when the input voltage at the base is going positive, the output voltage at the load is going negative. This indicates that the signal at the output is 180° out-of-phase with the input signal. Thus, both the voltage gain and the current gain are expected to be negative, where the negative sign indicates inversion. Hence, the common-emitter is an inverting configuration.

Voltage gain with load:

$$A_{v(WL)} = \frac{v_{o(WL)}}{v_{in}} = \frac{-\beta \cdot i_b \cdot R_L'}{+i_b \cdot \beta \cdot r_e} = -\frac{R_L'}{r_e}$$

$$A_{v(WL)} = \frac{v_{o(WL)}}{v_{in}} = -\frac{R_L'}{r_e} \tag{5-30}$$

$$A_{v(WL)} = -\frac{R_L'}{r_e} = -\frac{1290 \text{ }\Omega}{12.5 \text{ }\Omega} = \mathbf{-103.2}$$

Voltage gain without load:

$$A_{v(NL)} = \frac{v_{o(NL)}}{v_{in}} = \frac{-\beta i_b R_L''}{i_b \beta r_e} = -\frac{R_L''}{r_e} = -g_m R_L''$$

$$A_{v(NL)} = \frac{v_{o(NL)}}{v_{in}} = -\frac{R_L''}{r_e} = -g_m R_L'' \tag{5-31}$$

$$A_{v(NL)} = -\frac{R_L''}{r_e} = -\frac{2120\,\Omega}{12.5\,\Omega} = \mathbf{-169.6}$$

The current gain:

The current gain can be determined from Equation 5-20 as

$$A_i = \frac{i_o}{i_{in}} = A_{v(WL)} \frac{R_{in}}{R_L}$$

$$A_i = \frac{i_o}{i_{in}} = -103 \times \frac{1.5\,k\Omega}{3.3\,k\Omega} = \mathbf{-46.8}$$

The power gain:

The power gain is determined by Equation 5-22 as

$$A_p = A_i A_{v(WL)}$$

$$A_p = A_i A_{v(WL)} = (-46.8) \times (-103) = \mathbf{4820.4}$$

The output voltage:

The output voltage with load is obtained from Equations 5-24 and 5-25 as

$$v_{in} = v_s \frac{R_{in}}{R_S + R_{in}} = 20\,mV_{rms} \times \frac{1.5\,k\Omega}{1\,k\Omega + 1.5\,k\Omega} = 12\,mV_{(rms)} = 2 \times \sqrt{2} \times 12 = 33.94\,mV_{(p-p)}$$

$$v_{o(WL)} = A_{v(WL)} v_{in} = 103.2 \times 12\,mV(rms) = 1.238\,V_{(rms)} = 2 \times \sqrt{2} \times 1.238 = 3.48\,V_{(p-p)}$$

Practice Problem 5-1 — CE Amplifier — ANALYSIS

Determine the following parameters for the amplifier circuit of Figure 5-11. Assume $\beta = 140$ and $V_A = 120\,V$.

1) I_{CQ} and V_{CEQ}
2) R_{in}
3) R_o
4) $A_{v(NL)}$
5) $A_{v(WL)}$
6) A_i

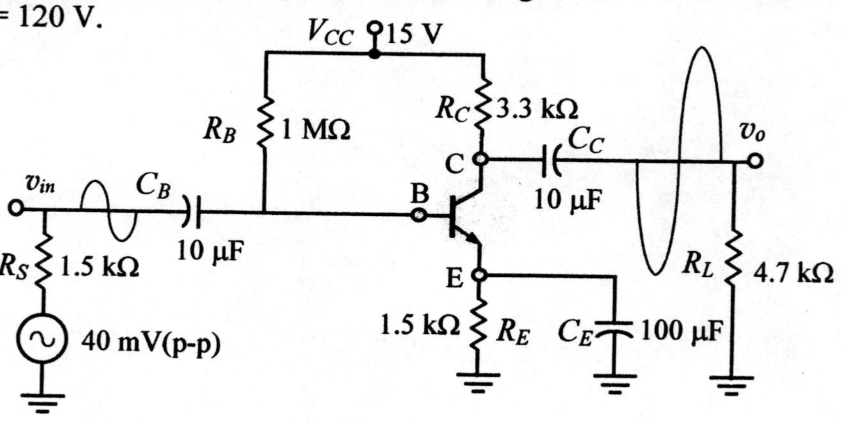

Figure 5-11: An emitter-biased common-emitter amplifier

7) A_p

8) $v_{o(WL)}$

Answers:

1) $I_{CQ} = 1.65$ mA and $V_{CEQ} = 7$ V

2) $R_{in} = 2.2$ kΩ

3) $R_o = 3.16$ kΩ

4) $A_{v(NL)} = -200$

5) $A_{v(WL)} = -120$

6) $A_i = -56$

7) $A_p = 6740$

8) $v_{o(WL)} = 2.85$ V

Example 5-2 ANALYSIS

Analyze and determine all the amplifier parameters for the voltage-divider biased common-emitter amplifier of Figure 5-12.

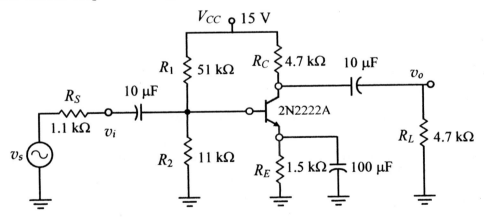

Figure 5-12: A voltage-divider biased common-emitter amplifier

Solution:

DC analysis:

First, we need to determine I_{EQ}. Thus,

$$V_B = V_{CC}\frac{R_2}{R_1 + R_2} = 15 \times \frac{11}{51+11} = 2.66 \text{ V}$$

$V_E = V_B - 0.7\text{ V} = 2.66\text{ V} - 0.7\text{ V} = 1.96\text{ V}$ $\qquad I_E = \dfrac{V_E}{R_E} = \dfrac{1.96\text{ V}}{1.5\text{ k}\Omega} = \mathbf{1.3\text{ mA}}$

$V_{CE} = V_{CC} - I_E(R_C + R_E) = 15\text{ V} - 1.3\text{ mA}(1.5\text{ k}\Omega + 4.7\text{ k}\Omega) \cong \mathbf{7\text{ V}}$

For the AC analysis, again our adopted transistor model is the *h* and *r* model, which can be interpreted as *r-parameter* or *h-parameter* model. Thus, we need to evaluate r_e, r_b, and r_o. Let us assume $V_A = 120$ V. Referring to the 2N2222A data sheets, I_E of 1.3 mA corresponds to β of approximately 150. Next,

$r_e = \dfrac{V_T}{I_E} = \dfrac{26\text{ mV}}{1.3\text{ mA}} = 20\ \Omega$

$r_b = \beta r_e = 150 \times 20\ \Omega = 3\text{ k}\Omega = h_{ie}$

$r_o = \dfrac{V_A}{I_C} = \dfrac{120\text{ V}}{1.3\text{ mA}} = 92.3\text{ k}\Omega = \dfrac{1}{h_{oe}}$

For having such a large value, r_o may be omitted from the model, if desired. At the base, resistors R_1 and R_2 will appear in parallel to the AC signal. Hence, we will designate this parallel combination as R_B.

$R_B = R_1 \parallel R_2 = 51 \parallel 11 = 9.05\text{ k}\Omega$

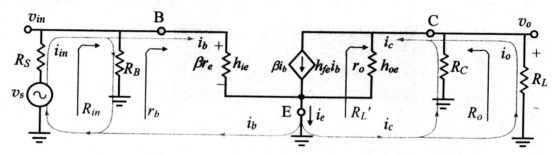

Figure 5-13: Small-signal equivalent circuit of Figure 5-12

Next, we will evaluate the R-parameters R_i, R_L', and R_o. The input resistance R_{in} is the parallel combination of R_B and r_b

$$R_{in} = R_B \parallel r_b \tag{5-32}$$

$R_{in} = 9\text{ k}\Omega \parallel 3\text{ k}\Omega \cong \mathbf{2.25\text{ k}\Omega}$

The total AC load R_L' is the parallel combination of r_o, R_C, and R_L

$$R_L' = r_o \parallel R_C \parallel R_L \tag{5-33}$$

$R_L' = 92\text{ k}\Omega \parallel 4.7\text{ k}\Omega \parallel 4.7\text{ k}\Omega = 2.29\text{ k}\Omega$

Without the external load R_L, the internal load of the amplifier is the parallel combination of r_o and R_C, which is designated as R_L''

$$R_L'' = r_o \parallel R_C \tag{5-34}$$

Chapter 5

$R_L'' = 92\text{ k}\Omega \parallel 4.7\text{ k}\Omega = 4.47\text{ k}\Omega$

The output resistance R_o seen by the load is also the parallel combination of r_o and R_C

$$R_o = r_o \parallel R_C \tag{5-35}$$

$R_o = r_o \parallel R_C = 4.47\text{ k}\Omega$

Voltage gain with and without load:

$$A_{v(WL)} = \frac{v_{o(WL)}}{v_{in}} = \frac{-\beta i_b R_L'}{i_b \beta r_e} = -\frac{R_L'}{r_e} = -g_m R_L'$$

$$A_{v(WL)} = \frac{v_{o(WL)}}{v_{in}} = -\frac{R_L'}{r_e} = -g_m R_L' \tag{5-36}$$

$$A_{v(WL)} = -\frac{R_L'}{r_e} = -\frac{2.29\text{ k}\Omega}{20\text{ }\Omega} = -114.5$$

$$A_{v(NL)} = \frac{v_{o(NL)}}{v_{in}} = \frac{-\beta i_b R_L''}{i_b \beta r_e} = -\frac{R_L''}{r_e} = -g_m R_L''$$

$$A_{v(NL)} = \frac{v_{o(NL)}}{v_{in}} = -\frac{R_L''}{r_e} = -g_m R_L'' \tag{5-37}$$

$$A_{v(NL)} = -\frac{R_L''}{r_e} = -\frac{4.47\text{ k}\Omega}{20\text{ }\Omega} = -223.5$$

The current gain:

From Equation 5-20 the current gain can be determined by

$$A_i = \frac{i_o}{i_{in}} = A_{v(WL)} \frac{R_{in}}{R_L}$$

$$A_i = \frac{i_o}{i_{in}} = -115 \times \frac{2.25\text{ k}\Omega}{4.7\text{ k}\Omega} = -55$$

The power gain:

From Equation 5-22, the power gain is determined as follows:

$$A_p = A_i A_{v(WL)}$$

$A_p = A_i A_{v(WL)} = (-55) \times (-115) = \mathbf{6325}$

The output voltage with load is obtained using Equations 5-24 and 5-25

$$v_{in} = v_s \frac{R_{in}}{R_S + R_{in}} = 100\text{ mV} \frac{2.25\text{ k}\Omega}{1.1\text{ k}\Omega + 2.25\text{ k}\Omega} = 67.164\text{ mV}$$

$v_{o(WL)} = A_{v(WL)} v_{in} = -115 \times 67.164\text{ mV} = \mathbf{-7.723\text{ V}}$

Practice Problem 5-2

ANALYSIS

Draw the small-signal equivalent circuit, and determine the following for the amplifier of Figure 5-14. Assume $\beta = 160$ and $V_A = 100$ V.

a) I_{CQ} and V_{CEQ}
b) R_{in}
c) R_o
d) $A_{v(NL)}$
e) $A_{v(WL)}$
f) A_i
g) A_p
h) $v_{o(WL)}$

Figure 5-14: A voltage-divider biased common-emitter amplifier

Answers:

$I_{CQ} = 1.42$ mA, $\quad V_{CEQ} = 5.63$ V, $\quad R_{in} = 2.13$ kΩ, $\quad R_o = 4.75$ kΩ,

$A_{v(NL)} = -259.5$, $\quad A_{v(WL)} = -176$, $\quad A_i = -37.5$, $\quad A_p = 6600$, $\quad v_{o(WL)} = 4.33$ V.

Example 5-3

ANALYSIS

Determine all the amplifier parameters, including the output voltage for the CE amplifier with *gain stabilization* shown in Figure 5-15, using the approximate transistor model (neglecting the effect of r_o).

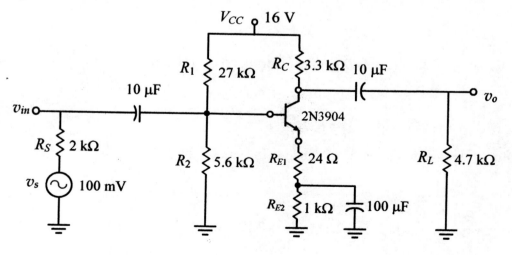

Figure 5-15: A CE amplifier circuit with partially unbypassed R_E

Solution:

DC analysis:

$$V_B = V_{CC} \frac{R_2}{R_1 + R_2} = 16 \times \frac{10}{43 + 10} = 2.75 \text{ V} \qquad V_E = V_B - 0.7 \text{ V} = 2.75 \text{ V} - 0.7 \text{ V} = 2.05 \text{ V}$$

$$I_E = \frac{V_E}{R_{E1} + R_{E2}} = \frac{2.05}{0.025 + 1} = 2 \text{ mA} \qquad r_e = \frac{V_T}{I_E} = \frac{26 \text{ mV}}{2} = 13 \text{ }\Omega$$

$$V_{CE} = V_{CC} - I_C(R_C + R_E) = 16 \text{ V} - 2 \text{ mA}(3.3 \text{ k}\Omega + 1.024 \text{ k}\Omega) = 7.35 \text{ V}.$$

Assuming a typical $\beta = 160$ and V_A of 120 V, $r_o = \frac{V_A}{I_C} = \frac{120 \text{ V}}{2 \text{ mA}} = 60 \text{ k}\Omega$. Therefore, $r_b = \beta r_e = 2080 \text{ }\Omega \cong 2 \text{ k}\Omega$. Next, $R_B = R_1 \parallel R_2 = 27 \parallel 5.6 = 4.638 \text{ k}\Omega$. Since we are utilizing the approximate model, there will be no use for r_o. Thus, we can now draw the small-signal equivalent circuit. Note that R_{E2} is bypassed by the 100 μF capacitor for the AC signal, but R_{E1} is not bypassed. This resistor, although quite small in value, plays an important role in the overall amplifier operation by providing gain control, improved stability, and increased input resistance.

$$R_B = 4.638 \text{ k}\Omega \qquad R_{E1} = 24 \text{ }\Omega \qquad R_C = 3.3 \text{ k}\Omega \qquad R_L = 4.7 \text{ k}\Omega$$
$$v_s = 100 \text{ mV} \qquad R_S = 2 \text{ k}\Omega \qquad \beta = 160 \qquad r_e = 13 \text{ }\Omega$$

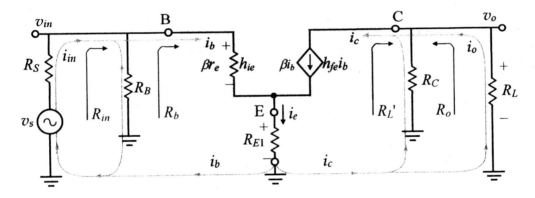

Figure 5-16: Small-signal equivalent circuit of Figure 5-15

Next, we will evaluate the R-parameters R_{in}, R_b, R_L', and R_o. Note that the resistance seen by the signal from base to the common ground is labeled R_b and is determined as follows:

$$R_b = \frac{v_b}{i_b} = \frac{i_b \beta r_e + i_e R_{E1}}{i_b}$$

Substituting for i_e in terms of i_b results in the following:

$$R_b = \frac{i_b \beta r_e + (\beta + 1)i_b R_{E1}}{i_b} = \beta r_e + (\beta + 1) R_{E1}$$

$$R_b \cong \beta r_e + \beta R_{E1} = \beta(r_e + R_{E1}) \tag{5-38}$$

$R_b = 160(13 + 24) = 5.92 \text{ k}\Omega$

Chapter 5

The input resistance R_{in} is the parallel combination of R_B and R_b

$$R_{in} = R_B \parallel R_b \qquad (5\text{-}39)$$

$R_{in} = 4.6 \text{ k}\Omega \parallel 5.92 \text{ k}\Omega \cong \mathbf{2.59 \text{ k}\Omega}$

The total AC load R_L' is the parallel combination of R_C and R_L

$$R_L' = R_C \parallel R_L \qquad (5\text{-}40)$$

$R_L' = 3.3 \text{ k}\Omega \parallel 4.7 \text{ k}\Omega = 1.94 \text{ k}\Omega$

Without the external load R_L, the internal load of the amplifier is R_C, which will be designated as R_L''

$$R_L'' = R_C \qquad (5\text{-}41)$$

$R_L'' = 3.3 \text{ k}\Omega$

The output resistance R_o seen by the load is also R_C

$$R_o = R_C \qquad (5\text{-}42)$$

$R_o = \mathbf{3.3 \text{ k}\Omega}$

<u>Voltage gain with load:</u>

$$A_{v(WL)} = \frac{v_{o(WL)}}{v_{in}} = \frac{v_o}{v_b} = \frac{-\beta i_b R_L'}{i_b R_b} = -\frac{\beta R_L'}{R_b} = -\frac{\beta R_L'}{\beta(r_e + R_{E1})}$$

$$A_{v(WL)} = \frac{v_{o(WL)}}{v_{in}} = -\frac{R_L'}{r_e + R_{E1}} \qquad (5\text{-}43)$$

$$A_{v(WL)} = -\frac{R_L'}{r_e + R_{E1}} = -\frac{1.94 \text{ k}\Omega}{13 \, \Omega + 24 \, \Omega} = \mathbf{-52.4}$$

<u>Voltage gain without load:</u>

Likewise, the voltage gain without load is given by

$$A_{v(NL)} = \frac{v_{o(NL)}}{v_{in}} = -\frac{R_L''}{r_e + R_{E1}} = -\frac{R_C}{r_e + R_{E1}} \qquad (5\text{-}44)$$

$$A_{v(NL)} = -\frac{R_C}{r_e + R_{E1}} = \frac{3.3 \text{ k}\Omega}{(13+24)\,\Omega} = \mathbf{-89.189}$$

<u>The current gain:</u>

From Equation 5-20, the current gain can be determined as follows:

$$A_i = \frac{i_o}{i_{in}} = A_{v(WL)} \frac{R_{in}}{R_L}$$

$$A_i = \frac{i_o}{i_{in}} = -52.4 \times \frac{2.59 \text{ k}\Omega}{4.7 \text{ k}\Omega} = \mathbf{-28.876}$$

<u>The power gain:</u>

From Equation 5-22, the power gain is

$$A_p = A_i A_{v(WL)}$$

$A_p = A_i A_{v(WL)} = (-29) \times (-52.4) = \mathbf{1519.6}$

The output voltage with load is obtained using Equations 5-24 and 5-25 as

$$v_{in} = v_s \frac{R_{in}}{R_s + R_{in}} = 100 \text{ mV} \times \frac{2.59 \text{ k}\Omega}{2 \text{ k}\Omega + 2.59 \text{ k}\Omega} = 56.427 \text{ mV}$$

$v_{o(WL)} = A_{v(WL)} v_{in} = -52.4 \times 56.427 \text{ mV} = \mathbf{-2.957 \text{ V}}$

Practice Problem 5-3  **ANALYSIS**

Determine all the parameters for the CE amplifier with *gain stabilization* shown in Figure 5-17, using the approximate transistor model (neglecting the effect of r_o). Assume $\beta = 160$.

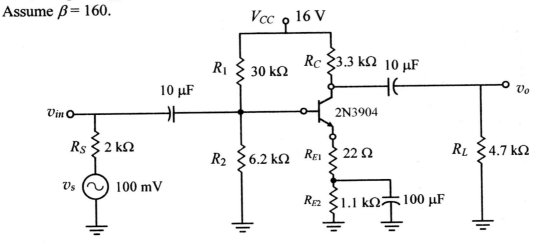

Figure 5-17: A CE amplifier circuit with partially unbypassed R_E

Answers:

a) $I_{CQ} = 1.818$ mA, $V_{CEQ} = 7.96$ V

b) $R_{in} = 2.7$ kΩ

c) $R_o = 3.3$ kΩ

d) $A_{v(NL)} = -91$

e) $A_{v(WL)} = -53.4$

f) $A_i = -31$

g) $A_p = 1651$

h) $v_{o(WL)} = 3$ V

Example 5-4

Let us now analyze the circuit of Figure 5-18 of Example 5-3 with the r_o represented in the transistor model and then compare the results.

$$R_B = 4.6 \text{ k}\Omega \qquad R_{E1} = 24 \text{ }\Omega \qquad R_C = 3.3 \text{ k}\Omega \qquad R_L = 4.7 \text{ k}\Omega$$
$$v_s = 100 \text{ mV} \qquad R_s = 2 \text{ k}\Omega \qquad \beta = 160 \qquad r_e = 13 \text{ }\Omega \qquad r_o = 60 \text{ k}\Omega$$

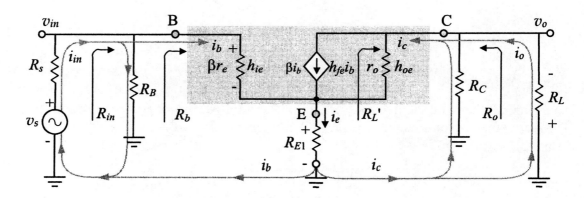

Figure 5-18: Small-signal equivalent circuit of Figure 5-15

For a precise and comprehensive analysis, we will consider the above amplifier as a two-port network and apply the two-port network theory as discussed in the previous chapter.

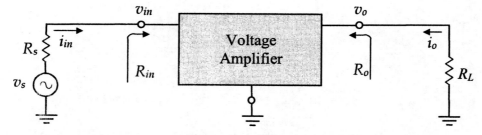

Figure 5-19: Two-port network model of Figure 5-18

According to Equations 4-21 and 4-27, the input resistance and output admittance of the amplifier are designated as parameters h_{11} and h_{22} and are defined as follows:

$$h_{11} = \left.\frac{v_1}{i_1}\right|_{v_2=0}$$

or,

$$R_{in} = \left.\frac{v_{in}}{i_{in}}\right|_{v_o=0} \tag{5-45}$$

and,

$$h_{22} = \left.\frac{i_2}{v_2}\right|_{i_1=0}$$

or,

$$\frac{1}{R_o} = \left.\frac{i_o}{v_2}\right|_{i_{in}=0} \tag{5-46}$$

Inverting both sides of the above equation yields R_o.

$$R_o = \left.\frac{v_o}{i_o}\right|_{i_{in}=0} \tag{5-47}$$

Let us now determine the input resistance R_{in} under the conditions set by Equation 5-45; that is, by grounding the output (set $v_o = 0$).

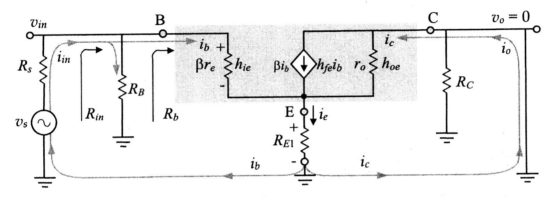

Figure 5-20: Circuit of Figure 5-19 with output grounded (set $v_o = 0$)

The input resistance R_{in} is the parallel combination of R_B and R_b, with R_b to be determined under the above set conditions ($v_o = 0$). Notice that with the output grounded, R_C is shorted out and one side of r_o is at the ground, making it parallel to R_{E1}. Hence, the total resistance at the emitter is the parallel combination of R_{E1} and r_o. Because r_o is 60 kΩ and R_{E1} is only 24 Ω, this parallel combination is practically equal to 24 Ω.

$$R_b = \left.\frac{v_b}{i_b}\right|_{v_o=0} \tag{5-48}$$

$$R_b = \frac{v_b}{i_b} = \frac{i_b \cdot (\beta \cdot r_e) + i_e(R_{E1} \parallel r_o)}{i_b}$$

$$R_b = \frac{v_b}{i_b} = \frac{i_b \cdot (\beta \cdot r_e) + i_e(R_{E1})}{i_b}$$

$$R_b = \frac{v_b}{i_b} = \frac{i_b \cdot (\beta \cdot r_e) + (\beta+1)i_b(R_{E1})}{i_b} \cong \frac{i_b \cdot \beta \cdot r_e + \beta \cdot i_b \cdot R_{E1}}{i_b}$$

$$R_b = \beta(r_e + R_{E1}) \tag{5-49}$$

$R_b = 160(13\ \Omega + 24\ \Omega) = 160 \times 37\ \Omega = 5920\ \Omega$

The input resistance:

The input resistance R_{in} is the parallel combination of R_B and R_b and will have exactly the same value as determined without r_o in Example 5-4.

$$R_{in} = R_B \| R_b \tag{5-50}$$

$R_{in} = 4.6 \text{ k}\Omega \| 5.92 \text{ k}\Omega = \mathbf{2.59 \text{ k}\Omega}$

Next, we will determine the output resistance R_o under the conditions set by Equation 5-47; that is, by setting $v_s = 0$ so that $i_{in} = 0$, and applying a test voltage v_t at the output, as shown in Figure 5-21 below.

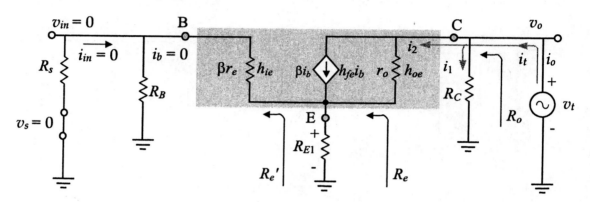

Figure 5-21: Determining the output resistance of Figure 5-20

Although the v_s, v_{in}, and i_{in} are set to zero, the DC bias of the amplifier is fully operational, and the r-parameters r_e and r_o, whose values depend on the Q-point DC current, are undisturbed.

Let us begin the analysis with the resistance labeled R_e, which is the total resistance seen by the test signal v_t at the emitter. Apparently, R_e is the parallel combination of R_{E1} and the resistance labeled R_e', and this parallel combination can never be larger than R_{E1}, which is only 24 Ω. Therefore, the maximum resistance value that R_e can possibly have is no more than 24 Ω.

The output resistance:

Application of the test signal v_t at the output will cause the flow of a test current i_t which will branch out into i_1 and i_2 at the collector terminal, where i_1 is simply the current through R_C, and i_2 is the current through the series combination of r_o and R_e.

$$i_t = i_1 + i_2 \tag{5-51}$$

$$i_1 \frac{v_t}{R_C} = \frac{v_o}{R_C} \tag{5-52}$$

$$i_2 = \frac{v_t}{r_o + R_e} \cong \frac{v_o}{r_o} \tag{5-53}$$

Chapter 5

$$i_t = \frac{v_o}{R_C} + \frac{v_o}{r_o} = v_o\left(\frac{1}{R_C} + \frac{1}{r_o}\right) = v_o\left(\frac{R_C + r_o}{R_C r_o}\right) = \frac{v_o(R_C + r_o)}{R_C r_o} = i_o \quad (5\text{-}54)$$

Having determined that $i_o = i_t$, we can now determine R_o according to Equation 5-47, as follows:

$$R_o = \left.\frac{v_o}{i_o}\right|_{i_{in}=0}$$

$$R_o = \frac{v_o}{i_o} = v_o\left[\frac{R_C r_o}{v_o(R_C + r_o)}\right] = \frac{R_C r_o}{R_C + r_o} \quad (5\text{-}55)$$

The above expression, which is the product over sum, is the parallel equivalent of the two resistors R_C and r_o. Therefore, the output resistance R_o is

$$R_o = R_C \parallel r_o \quad (5\text{-}56)$$

$R_o = 3.3\ \text{k}\Omega \parallel 60\ \text{k}\Omega = \mathbf{3.13\ k\Omega}$

The total AC load R_L' is the parallel combination of r_o, R_C, and R_L

$$R_L' = r_o \parallel R_C \parallel R_L \quad (5\text{-}57)$$

$R_L' = 60\ \text{k}\Omega \parallel 3.3\ \text{k}\Omega \parallel 4.7\ \text{k}\Omega = 1.88\ \text{k}\Omega$

Without the external load R_L, the internal load of the amplifier will be the parallel combination of r_o and R_C, which will be designated as R_L''

$$R_L'' = r_o \parallel R_C \quad (5\text{-}58)$$

$R_L'' = 60\ \text{k}\Omega \parallel 3.3\ \text{k}\Omega = 3.13\ \text{k}\Omega$

Voltage gain with load:

$$A_{v(WL)} = \frac{v_{o(WL)}}{v_{in}} = \frac{v_o}{v_b} = \frac{-\beta i_b R_L'}{+i_b R_b} = -\frac{\beta R_L'}{R_b} = -\frac{\beta R_L'}{\beta(r_e + R_{E1})}$$

$$A_{v(WL)} = \frac{v_{o(WL)}}{v_{in}} = -\frac{R_L'}{r_e + R_{E1}} \quad (5\text{-}59)$$

$$A_{v(WL)} = -\frac{R_L'}{r_e + R_{E1}} = -\frac{1.88\ \text{k}\Omega}{13\ \Omega + 24\ \Omega} = \mathbf{-50.8}$$

Voltage gain without load:

Likewise, voltage gain without load is expressed by

$$A_{v(NL)} = \frac{v_{o(NL)}}{v_{in}} = -\frac{R_L''}{r_e + R_{E1}} \quad (5\text{-}60)$$

$$A_{v(NL)} = -\frac{R_L''}{r_e + R_{E1}} = -\frac{3.13\ \text{k}\Omega}{(13 + 24)\ \Omega} = \mathbf{-84.6}$$

Chapter 5

The current gain:

From Equation 5-20, the current gain can be determined by

$$A_i = \frac{i_o}{i_{in}} = A_{v(WL)} \frac{R_{in}}{R_L}$$

$$A_i = \frac{i_o}{i_{in}} = -50.8 \times \frac{2.59 \text{ k}\Omega}{4.7 \text{ k}\Omega} = -28$$

The power gain:

From Equation 5-22, the power gain is

$$A_p = A_i A_{v(WL)}$$

$A_p = A_i A_{v(WL)} = (-28) \times (-50.8) = \mathbf{1422.4}$

The output voltage with load is obtained using Equations 5-24 and 5-25

$$v_{in} = v_s \frac{R_{in}}{R_S + R_{in}} = 100 \text{ mV} \times \frac{2.59 \text{ k}\Omega}{2 \text{ k}\Omega + 2.59 \text{ k}\Omega} = 56.427 \text{ mV}$$

$$v_{o(WL)} = A_{v(WL)} v_{in} = -50.8 \times 56.427 \text{ mV} = \mathbf{-2.866 \text{ V}}$$

Let us now simulate the circuit and compare the algebraically obtained results to those obtained by simulation. The results of the simulation without load are presented in Figure 5-22 and Figure 5-23.

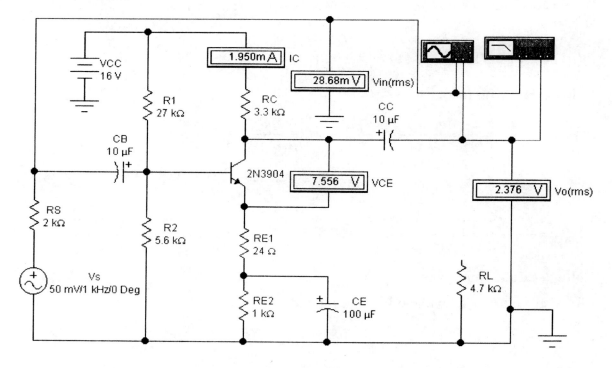

Figure 5-22: Circuit setup with simulation results of Example 5-4 **without load** created in Electronics Workbench

The Q-point: The simulation results of I_{CQ} = 1.95 mA and V_{CEQ} = 7.55 V as depicted on DMMs of EWB agree very closely with the algebraic solutions of 2 mA and 7.35 V.

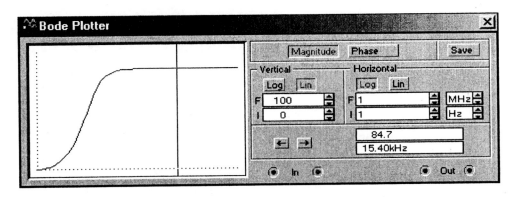

Figure 5-23: Electronics Workbench Bode plotter showing the voltage gain without load of Example 5-4

No-load voltage gain: The simulated voltage gain of 84.7 as depicted on the Bode plotter of Figure 5-23 above is almost identical to the algebraically determined gain of 84.5. The results of simulation with load are presented in Figures 5-24 and 5-25.

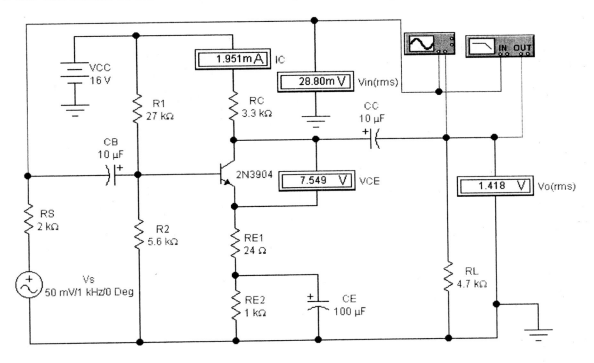

Figure 5-24: Circuit setup with simulation results of Example 5-4 **with load**

Voltage gain with load: Figure 5-24 above depicts the circuit setup with digital readouts of simulation results with load. Figure 5-25 depicts the Electronics Workbench Bode plotter showing the voltage gain of 50.4 with load, which agrees very closely with the algebraically determined voltage gain of 50.8.

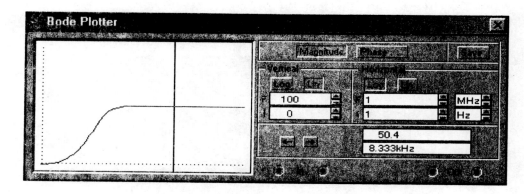

Figure 5-25: Electronics Workbench Bode plotter showing the voltage gain with load for Example 5-4

Input resistance:

The simulation result of the input resistance can be determined from the DMM readings of Figure 5-24 and from Equation 5-10, as follows:

$$R_{in} = \frac{v_{in}}{v_s - v_{in}} R_S = \frac{28.68 \text{ mV}}{50 \text{ mV} - 28.68 \text{ mV}} \times 2 \text{ k}\Omega = \mathbf{2.69 \text{ k}\Omega}$$

Output resistance:

The simulation result of the output resistance can be determined from the DMM readings of Figures 5-22 and 5-24, and from Equation 5-13, as follows:

$$R_o = \frac{v_{o(NL)} - v_{o(WL)}}{v_{o(WL)}} R_L = \frac{2.376 \text{ V} - 1.418 \text{ V}}{1.418 \text{ V}} \times 4.7 \text{ k}\Omega = \mathbf{3.175 \text{ k}\Omega}$$

A summary of the theoretically expected results versus the results obtained by simulation are presented in Table 5-1.

Table 5-1: Results Summary of Example 5-3

Parameter	Algebraic solution	Simulation results
I_{CQ}	2 mA	1.95 mA
V_{CEQ}	7.35 V	7.5 V
$A_{v(NL)}$	−84.6	−84.7
$A_{v(WL)}$	−50.8	−50.4
R_{in}	2.59 kΩ	2.69 kΩ
R_o	3.13 kΩ	3.175 kΩ

As you can see, all algebraic solutions agree very closely with the solutions obtained by simulation. Let us now compare the algebraically obtained results of Example 5-3, in which r_o was omitted, to the results of Example 5-4, in which r_o was accounted for.

Table 5-2: Results of Example 5-3 (without r_o) Versus Example 5-4 (with r_o)

Parameter	Solution without r_o	Solution with r_o
$A_{v(NL)}$	−89.2	−84.6
$A_{v(WL)}$	−52.4	−50.8
R_{in}	2.59 kΩ	2.69 kΩ
R_o	3.3 kΩ	3.13 kΩ
$v_{o(WL)}$	2.96 V	2.86 V

Apparently, using the approximate model (neglecting r_o) does not significantly affect the outcome. The output voltage $v_{o(WL)}$ has a value of 2.86 V with r_o and 2.95 V without r_o, a difference of 3%. Hence, for the sake of convenience and simplicity, we will be using the approximate model unless the situation dictates otherwise.

Practice Problem 5-4 ANALYSIS

Draw the small-signal model for the CE amplifier with *gain stabilization* of Figure 5-26 and determine: (a) with r_o and (b) without r_o. Assume $\beta = 150$ and $V_A = 110$ V.

a) I_{CQ} and V_{CEQ}
b) R_{in}
c) R_o
d) $A_{v(NL)}$
e) $A_{v(WL)}$
f) A_i
g) A_p
h) $v_{o(WL)}$

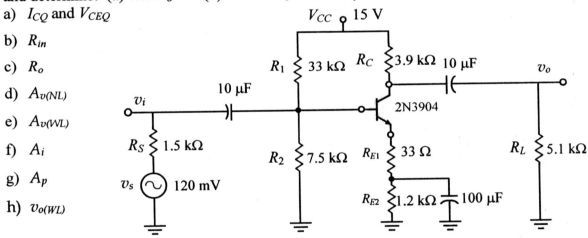

Figure 5-26: A CE amplifier with partially unbypassed R_E

Answers:

With r_o	Without r_o
a) $I_{CQ} = 1.685$ mA and $V_{CEQ} = 6.35$ V	the same
b) $R_{in} = 3.3$ kΩ	$R_{in} = 3.3$ kΩ
c) $R_o = 3.368$ kΩ	$R_o = 3.9$ kΩ
d) $A_{v(NL)} = -76$	$A_{v(NL)} = -80$
e) $A_{v(WL)} = -41$	$A_{v(WL)} = -44$
f) $A_i = -29$	$A_i = -30$
g) $A_p = 1197$	$A_p = 1320$
h) $v_{o(WL)} = 3.4$ V	$v_{o(WL)} = 3.6$ V

Section Summary — CE Amplifier — ANALYSIS

Summary of Equations for the Analysis of the Common-Emitter BJT Amplifier

DC Analysis:

a) Emitter Bias

$$I_B = \frac{V_{CC} - V_{BE}}{R_B + (\beta+1)R_E} \qquad I_C = \beta I_B \qquad r_e = \frac{26\,\text{mV}}{I_C} \qquad r_o = \frac{V_A}{I_C}$$

$$V_{CE} = V_{CC} - (I_C R_C + I_E R_E) \cong V_{CC} - I_C(R_C + R_E)$$

b) Voltage-divider Bias

$$V_B = V_{CC}\frac{R_2}{R_1 + R_2} \qquad V_E = V_B - 0.7\,\text{V} \qquad I_E = \frac{V_E}{R_E} \qquad r_e = \frac{26\,\text{mV}}{I_E}$$

Note if R_E is split, then $R_E = R_{E1} + R_{E2}$ $\qquad r_o = \dfrac{V_A}{I_E}$

$$V_{CE} = V_{CC} - I_C(R_C + R_E)$$

Small-Signal (AC) Analysis:

Referring to the transistor data sheets, the actual β may be found that corresponds to the operating point I_C.

Then, $R_b = \beta(r_e + R_{E1})$ and $R_B = R_1 \parallel R_2$ for voltage-divider bias.

Note, if R_E is not split and there is no R_{E1}, then $R_{E1} = 0$, and if V_A is not known, then r_o may be considered infinite (∞).

$$R_{in} = R_B \parallel R_b \qquad R_L' = R_C \parallel r_o \parallel R_L \qquad R_L'' = R_C \parallel r_o \qquad R_o = R_C \parallel r_o$$

$$A_{v(WL)} = \frac{v_{o(WL)}}{v_{in}} = -\frac{R_L'}{r_e + R_{E1}} \qquad\qquad A_{v(NL)} = \frac{v_{o(NL)}}{v_{in}} = -\frac{R_L''}{r_e + R_{E1}}$$

$$A_i = \frac{i_o}{i_{in}} = A_{v(WL)}\frac{R_{in}}{R_L} \qquad\qquad A_p = A_i \times A_{v(WL)}$$

$$v_{in} = v_s \frac{R_{in}}{R_s + R_{in}} \qquad v_{o(NL)} = A_{v(NL)} v_{in} \qquad v_{o(WL)} = A_{v(WL)} v_{in}$$

5.4 THE CASCODE AMPLIFIER

The DC analysis of the cascode amplifier was presented in Chapter 3. Let us now carry out the complete AC analysis for the cascode amplifier of Figure 5-27 as shown below:

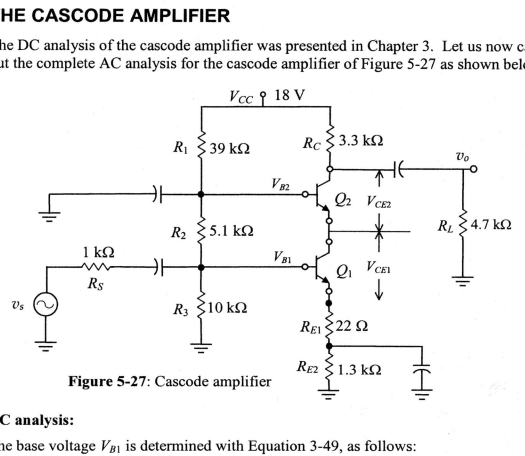

Figure 5-27: Cascode amplifier

DC analysis:

The base voltage V_{B1} is determined with Equation 3-49, as follows:

$$V_{B1} = V_{CC} \frac{R_3}{R_1 + R_2 + R_3} = 18\,\text{V} \frac{10\,\text{k}\Omega}{39\,\text{k}\Omega + 5.1\,\text{k}\Omega + 10\,\text{k}\Omega} = 3.327\,\text{V}$$

$$V_{E1} = V_{B1} - V_{BE} = 3.327\,\text{V} - 0.7\,\text{V} = 2.627\,\text{V}$$

$$I_{E1} = \frac{V_{E1}}{R_E} = \frac{2.627\,\text{V}}{1.322\,\text{k}\Omega} \cong 2\,\text{mA}, \quad I_{C1} \cong I_{E1} = I_{C2} \cong I_{E2} = 2\,\text{mA}, \quad r_e = \frac{26\,\text{mV}}{I_E} = 13\,\Omega$$

For the 2N2222 transistor, the I_C of 2 mA corresponds to $\beta = 160$.

The base voltage V_{B2} is determined with Equation 3-56, as follows:

$$V_{B2} = V_{CC} \frac{R_2 + R_3}{R_1 + R_2 + R_3} = 18\,\text{V} \frac{10\,\text{k}\Omega + 5.1\,\text{k}\Omega}{39\,\text{k}\Omega + 5.1\,\text{k}\Omega + 10\,\text{k}\Omega} = 5\,\text{V}$$

$$V_{E2} = V_{C1} = V_{B2} - V_{BE} = 5 - 0.7\,\text{V} = 4.3\,\text{V}$$

$$V_{C2} = V_{CC} - I_C R_C = 18 - (2\,\text{mA} \times 3.3\,\text{k}\Omega) = 11.4\,\text{V}$$

$$V_{CE1} = V_{C1} - V_{E1} = 4.3 - 2.6\,\text{V} = 1.7\,\text{V}$$

$$V_{CE2} = V_{C2} - V_{E2} = 11.4\,\text{V} - 4.3\,\text{V} = 7.1\,\text{V}$$

Chapter 5

AC analysis:

We will start the AC analysis with the AC equivalent circuit, which will be a cascade of a common-emitter followed by a common-base configuration, as shown in Figure 5-28 below:

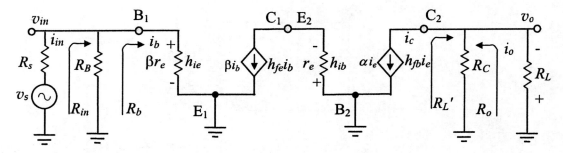

Figure 5-28: Small-signal equivalent circuit of Figure 5-27

<u>First stage (common-emitter):</u>

$$R_b = \beta(r_e + R_{E1}) \tag{5-61}$$

$R_b = 160(13\ \Omega + 22\ \Omega) = 5.6\ \text{k}\Omega$

The resistance R_B is the parallel combination of R_2 and R_3.

$$R_B = R_2 \parallel R_3 \tag{5-62}$$

$R_B = 10\ \text{k}\Omega \parallel 5.1\ \text{k}\Omega = 3.77\ \text{k}\Omega$

The input resistance R_{in} is the parallel combination of R_B and R_b.

$$R_{in} = R_B \parallel R_b \tag{5-63}$$

$R_{in} = 3.77\ \text{k}\Omega \parallel 5.6\ \text{k}\Omega \cong 2\ \text{k}\Omega$

$$R_o = R_C \tag{5-64}$$

$R_o = 3.3\ \text{k}\Omega$

The load for the first stage is the input resistance of the second stage, which is simply r_e. Hence, the gain of the first stage is determined with Equation 5-43 as follows:

$$A_{v1} = \frac{v_{C1}}{v_{in}} = -\frac{R_{L1}'}{r_e + R_{E1}} = -\frac{r_e}{r_e + R_{E1}} \tag{5-65}$$

<u>Second stage (common-base):</u>

The load for the second stage is the parallel combination of R_C and R_L.

$$R_L' = R_C \parallel R_L \tag{5-66}$$

$R_L' = 3.3\ \text{k}\Omega \parallel 4.7\ \text{k}\Omega = 1.94\ \text{k}\Omega$

The gain for the second stage is determined as follows:

$$A_{v2} = \frac{v_o}{v_{C1}} = \frac{\alpha i_e R_L'}{\beta i_b r_e} \cong \frac{\beta i_b R_L'}{\beta i_b r_e} = \frac{R_L'}{r_e} \tag{5-67}$$

Chapter 5

The overall voltage gain is the product of the two gains.

$$A_v = \frac{v_o}{v_{in}} = \frac{v_{C1}}{v_{in}} \cdot \frac{v_o}{v_{C1}} = A_{v1} \cdot A_{v2}$$

$$A_v = -\frac{R_L'}{r_e + R_{E1}} \tag{5-68}$$

$$A_v = -\frac{1.94 \text{ k}\Omega}{35 \, \Omega} = -55.429$$

$$v_{in} = v_s \frac{R_{in}}{R_s + R_{in}} = 100 \text{ mV} \frac{2 \text{ k}\Omega}{3 \text{ k}\Omega} = 66.667 \text{ mV}$$

$$v_o = A_{v(WL)} v_{in} = 55 \times 66.667 \text{ mV} = 3.667 \text{ V}$$

Practice Problem 5-5 ANALYSIS

Determine all the amplifier parameters for the following cascode amplifier:

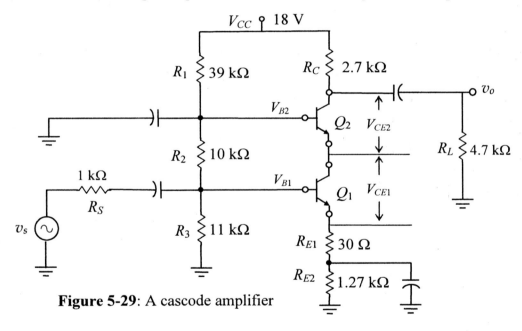

Figure 5-29: A cascode amplifier

Answers:

$I_C = 2$ mA, $V_{CE2} = 7$ V, $R_{in} = 2.975$ kΩ, $R_o = 2.7$ kΩ, $A_v = -40$

185

Chapter 5

Section Summary — Cascode Amplifier — ANALYSIS

Summary of Equations for the Analysis of the Cascode BJT Amplifier

DC analysis:

$$V_{B1} = V_{CC} \frac{R_3}{R_1 + R_2 + R_3}, \qquad V_{E1} = V_{B1} - V_{BE}, \qquad I_{E1} = \frac{V_{E1}}{R_E}$$

$$I_{C1} \cong I_{E1} = I_{C2} \cong I_{E2} \qquad r_e = \frac{26\,\text{mV}}{I_E}$$

$$V_{B2} = V_{CC} \frac{R_2 + R_3}{R_1 + R_2 + R_3}, \qquad V_{E2} = V_{C1} = V_{B2} - V_{BE}$$

$$V_{C2} = V_{CC} - I_C R_C, \qquad V_{CE1} = V_{C1} - V_{E1}, \qquad V_{CE2} = V_{C2} - V_{E2}$$

AC analysis:

If R_E is split; that is, $R_E = R_{E1} + R_{E2}$

$$A_v = -\frac{R_L{'}}{r_e + R_{E1}}$$

where R_{E1} is the unbypassed portion of the R_E, otherwise

$$A_v = \frac{R_L{'}}{r_e}$$

$$R_B = R_2 \parallel R_3, \qquad R_{in} = R_B \parallel R_b, \qquad R_b = \beta(r_e + R_{E1}), \qquad R_o = R_C$$

$$A_i = \frac{i_o}{i_{in}} = A_{v(WL)} \frac{R_{in}}{R_L}$$

$$A_p = A_i \times A_{v(WL)}$$

$$v_{in} = v_s \frac{R_{in}}{R_s + R_{in}}$$

$$v_{o(WL)} = A_{v(WL)} v_{in}$$

5.5 TROUBLESHOOTING

Having gained some experience in troubleshooting the DC bias operation of the BJT amplifier in Chapter 4, we now wish to carry out some troubleshooting exercises regarding the AC operation. Recall that the AC operation of the BJT amplifier depends on a satisfactory DC bias operation. Hence, the cause of unsatisfactory AC operation can very likely be a DC bias problem. Also, note that the following examples are designed to include the errors and problems that occur frequently in the laboratory. Hence, do not rule out the common errors in breadboarding and misread component values.

Example 5-5   TROUBLESHOOTING

Consider the amplifier circuit of Figure 5-30. The application of a $v_s = 50$ mV(p-p) sinusoidal signal with a frequency of 5 kHz results in a $v_o = 0.65$ V(p-p) at the output as shown below. Determine whether there is an operational problem with this circuit. State what the problem is, if there is one, and the cause of it.

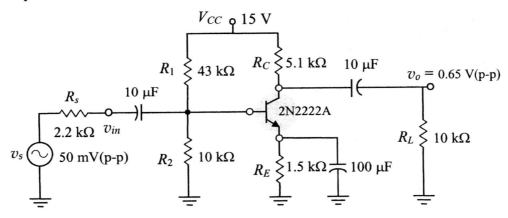

Figure 5-30: A voltage-divider biased common-emitter amplifier

Solution:

From the experience we have gained thus far in analyzing amplifier circuits, we suspect that the output signal of 0.65 V(p-p) is too small for an input of 50 mV(p-p). Hence, we continue checking the circuit as follows:

1. Using an oscilloscope, we check the input voltage v_{in}. Let us assume that a sinusoidal signal of 37 mV(p-p) is observed at the input terminal designated as v_{in}, which is indicative of a voltage gain $A_v = (v_o/v_{in}) \cong 17.5$.
2. The experience tells us that a voltage gain of 17.5 seems to be too low for a CE amplifier with a fully bypassed emitter resistor. To verify our suspicion, we perform a quick analysis of the circuit to determine the theoretically expected DC bias voltages and the voltage gain. The result of the quick analysis shows the following expected values: $V_B \cong 2.8$ V, $V_E \cong 2.1$ V, $r_e \cong 18.5$ Ω, $A_v = v_o/v_{in} \cong 175$.

3. Obviously, there is something wrong with the circuit operation. The signal seems to be going through and is amplified, but with only one-tenth of the expected gain. Hence, we start verifying the DC measurements.
4. The DC voltage readings at the base and at the emitter measure 2.8 V and 2.1 V, respectively. Both of these readings agree with the expected DC values. We then check V_C and V_{CE}, which measure approximately 14.2 V and 12.1 V, respectively. The measurement of $V_C = 14.2$ V implies a collector current of approximately 0.15 mA ($V_{CC} - V_C / R_C \cong 0.15$ mA). The voltage measurement of 2.1 V at the emitter, however, indicates an emitter current of 1.4 mA. Now the question is "Which of the two currents (0.15 mA and 1.4 mA) is the actual current flow?" We can answer this question without breaking the circuit and measuring the I_C, simply by making use of our knowledge about the circuit operation. We know that the expression for the voltage gain is ($v_o/v_{in} = R_L'/r_e = R_L'/(26$ mV$/I_E) = I_E R_L'/26$ mV), which is a function of the bias current I_E. Hence, the reason for the voltage gain being one-tenth of the expected value must be the I_C of 0.15 mA instead of 1.4 mA.
5. Now the question is "What is the cause of such a significant drop in I_C?" To answer this question, we again refer to our knowledge of the circuit operation and find that $I_E = V_E/R_E$. We have already verified that $V_E = 2.1$ V. Thus, the problem must be the R_E; probably its value is not the 1.5 kΩ that we thought it was, which is a common mistake in the laboratory. Hence, we check the R_E, and we observe that its color bands are brown, green, orange, which are the color bands for a 15 kΩ resistor.
6. We replace the 15 kΩ resistor R_E with the correct value of 1.5 kΩ (brown, green, red) and measure an output voltage of approximately 4.3 V(p-p).

Conclusion: Certainly, there was an operational problem with the circuit, because the output voltage was much less than expected. The reason for such a low output was the low voltage gain because of a very low collector current, which was the result of a wrong R_E (15 kΩ instead of 1.5 kΩ) used in the circuit.

Example 5-6   TROUBLESHOOTING

Consider the amplifier circuit of Figure 5-31. The application of a $v_s = 100$ mV(p-p) sinusoidal signal with a frequency of 5 kHz results in a $v_o = 130$ mV(p-p) at the output as shown below. Determine whether there is an operational problem with this circuit. State what the problem is, if there is one, and the cause of it.

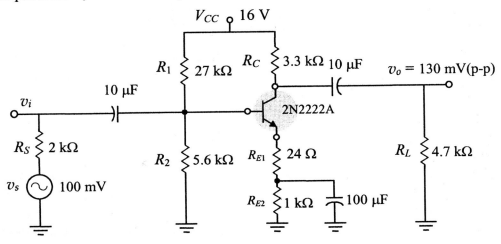

Figure 5-31: A CE amplifier with partially unbypassed R_E

Solution:

Again, having analyzed several amplifiers of this type throughout the chapter, an output signal of 130 mV(p-p) seems too small for an input of 100 mV(p-p). However, in order to begin an effective and efficient troubleshooting process, it is best to have an idea of what is expected from this amplifier circuit. Hence, we begin the process by performing a quick analysis of the circuit operation.

1. The analysis of the circuit produces the following expected results: $V_B = 2.77$ V, $V_E = 2.07$ V, $I_E = 2$ mA, $r_e = 13\ \Omega$, $A_v = v_o/v_{in} = 52$, $v_{in} = 56$ mV, $v_o \cong 2.9$ V

2. Apparently, there is something wrong and we intend to find it. We continue our examination of the circuit with checking the DC bias voltages.

3. The DC voltage readings are as follows: $V_B = 2.7$ V, $V_E = 2.0$ V, $V_{CE} = 7.4$ V. These measurements agree very closely with the theoretically expected bias voltages. Hence, we conclude that there is no problem with the DC operation.

4. Having found no problem with DC operation and the measurement of a stable but small signal at the output, we turn our attention to the coupling and bypass capacitors.

5. The inspection of all three capacitors for correct values and polarities proves satisfactory. To make sure that capacitors are actually passing the signal, we check the signal at both sides of each capacitor with an oscilloscope.

6. The measurement of signals before and after each capacitor proves satisfactory for the coupling capacitors. However, a nonzero signal is measured across the resistor R_{E2}, implying that the resistor is not being bypassed by the capacitor C_E. This discrepancy may be the result of an open bypass capacitor or a poor connection at one end of the capacitor. Most likely the problem is caused by a poor connection.

7. An inspection of connections at both sides of the capacitor proves them to be faulty at the ground side. After the capacitor is firmly grounded, the oscilloscope displays an output signal of approximately 2.9 V(p-p).

Conclusion: The problem was a very low output signal because of a very low gain, which was due to a poor ground connection of the bypass capacitor.

Example 5-7 — TROUBLESHOOTING

Consider the amplifier circuit of Figure 5-32. The application of a $v_s = 50$ mV(p-p) sinusoidal signal with a frequency of 5 kHz results in an output of approximately $v_o = 0.8$ V(p-p) as shown below. Determine whether there is an operational problem with this circuit. State what the problem is, if there is one, and the cause of it.

Solution:

Again, the experience tells us that we can expect an output signal larger than 0.8 V(p-p). Hence, we check the v_{in} with an oscilloscope and find that it measures in the vicinity of 4 to 5 mV(p-p). The ratio of v_o/v_{in} produces a gain of 160 to 200, an average of 180, which seems to be reasonable for an amplifier with a fully bypassed emitter resistor. Thus, we turn our attention to R_s and notice that it is a 20 kΩ resistor instead of a 2 kΩ. We replace the 20 kΩ resistor with a 2 kΩ and observe an output of around 4.3 V(p-p).

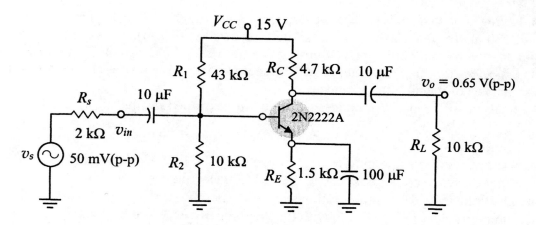

Figure 5-32: Circuit for Example 5-7

Chapter 5

Practice Problem 5-6 — TROUBLESHOOTING

The application of a $v_s = 50$ mV(p-p) with $f = 1$ kHz to Figure 5-33 results in 4.8V(p) half-sinusoid at the output (similar to the output of a half-wave rectifier). State what the problem is and the cause of it.

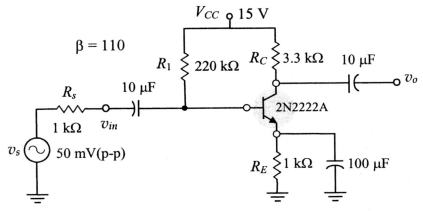

Figure 5-33: Circuit for Practice Problem 5-6

5.6 SUMMARY

- The AC operation is generally classified in two categories: *small-signal operation* and *large-signal operation*.

- Small-signal operation is the analysis and design of amplifiers that take signals of very small amplitude in the range of a few millivolts, and produce at the output large signals in the range of several volts.

- The focus of attention in the *small-signal operation* is mainly the AC voltage and current gains or the amplifying action of the amplifier. Other parameters such as input resistance, output resistance, and the frequency response of the amplifier are also of great concern.

- The large-signal operation, which is discussed in later chapters, is the analysis and design of *large-signal amplifiers*, commonly known as power amplifiers or current drivers.

- Some of the amplifier parameters that have been discussed in this chapter are: input resistance R_{in}, output resistance R_o, no-load voltage gain $A_{v(NL)}$, voltage gain with load $A_{v(WL)}$, current gain A_i, and power gain A_p.

- Analysis of an amplifier begins with the DC analysis in order to determine the I_{CQ}, upon which depend some of the small-signal parameters of the transistor. Having determined the I_{CQ}, we then determine the V_{CEQ}, which gives us an idea about the Q-point location, which reveals the limitations on the output signal swing. Next, we begin the AC analysis by drawing the small-signal equivalent circuit with the transistor model substituted for the transistor. From the equivalent circuit, we can determine all the small-signal amplifier parameters as listed above.

- The common-emitter amplifier is an inverting configuration, which can produce reasonably high voltage gain, has moderately high input and output resistances, and offers moderately high current gain.

- The common-base amplifier is a non-inverting configuration, which can produce reasonably high voltage gain, has moderately high output resistance, but suffers from very low input resistance, and offers no current gain.

- The common-collector amplifier is a non-inverting configuration, which can produce reasonably high current gain, has very low output resistance, can offer very high input resistance, but offers no voltage gain.

- The resistor R_{E1} in a common-emitter amplifier, although quite small in value, plays an important role in the overall amplifier operation by providing gain control, improved stability, and increased input resistance.

- The input resistance of the common-base amplifier may be improved by addition of a base resistor R_B without affecting the DC bias arrangement.

- All three types of amplifier circuits, CE, CB, and CC, may be biased in the voltage-divider bias configuration.

- The DC operating point sets the limits of the signal swing at the output of the amplifier, and the objective of an optimum DC bias is to place the operating point on the load line so that a maximum signal swing can be achieved at the output. This optimum operating point is usually at the center of the AC load line.

- The DC load line has a slope of $m = -1/(R_C + R_E)$, and is constructed by connecting the V_{CC} to $I_{C(sat)}$ on the collector characteristic curves.

- Likewise, the AC load line has a slope equal to the negative reciprocal of the total AC resistance R_L' through which the output signal current i_c flows. To construct the AC load line with this particular slope, a line may be drawn connecting the two intercepts I_o and V_o.

- To avoid or minimize the non-linearity distortion, the amplifier operation must be limited to the active or nearly linear region of the device, where the equal intervals of the base current correspond to nearly equal spacing of the I_B curves on the collector characteristics.

- The non-linearity distortion causes the upper half-cycle of the sinusoidal output to appear compressed (wider and smaller in amplitude) compared to the lower half-

Chapter 5

cycle, which becomes more pronounced with an increased input signal.

PROBLEMS

Section 5.2 Amplifier Fundamentals

5.1 An amplifier was tested in the laboratory and the following data were collected. Determine the amplifier parameters R_{in}, R_o, $A_{v(NL)}$, $A_{v(WL)}$, A_i, and A_p.

Collected data:

$v_s = 100$ mV, $v_{in} = 60$ mV, $v_{o(NL)} = 6$ V
$v_{o(WL)} = 4$ V, $R_s = 1.2$ kΩ, $R_L = 2.2$ kΩ

5.2 The following data were collected from an amplifier in the laboratory. Determine the amplifier parameters R_{in}, R_o, $A_{v(NL)}$, $A_{v(WL)}$, A_i, and A_p.

Collected data:

$v_s = 110$ mV, $v_{in} = 65$ mV, $v_{o(NL)} = 5.5$ V
$v_{o(WL)} = 4.2$ V, $R_s = 1.5$ kΩ, $R_L = 2.2$ kΩ

5.3 The following data were collected from an amplifier. If it is known that $A_{v(NL)} = 100$, determine the amplifier parameters R_{in}, R_o, $A_{v(WL)}$, A_i, and A_p.

Collected data:

$v_s = 100$ mV, $v_{in} = 75$ mV, $v_{o(WL)} = 5$ V, $R_s = 1$ kΩ, $R_L = 2$ kΩ

Section 5.3.1 Analysis of CE Amplifier

5.4 Carry out Practice Problem 5-1.

5.5 Determine the following parameters for the amplifier of Figure 5-1P. Assume $\beta = 150$ and $V_A = 120$ V.

a. I_{CQ} and V_{CEQ}
b. R_{in}
c. R_o
d. $A_{v(NL)}$
e. $A_{v(WL)}$
f. A_i
g. A_p
h. $v_{o(WL)}$

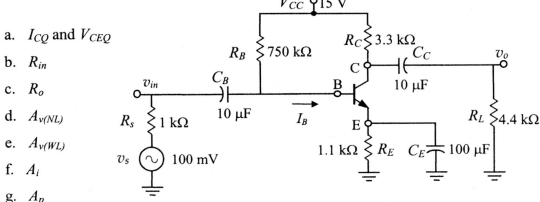

Figure 5-1P: A CE amplifier

5.6 Repeat Problem 5.5 if $R_B = 1$ MΩ, and all other parameters remain unchanged.

5.7 Carry out Practice Problem 5-2.

Chapter 5

5.8 Determine the following parameters for the amplifier of Figure 5-2P.
Assume $\beta = 150$ and $V_A = 120$ V.

a) I_{CQ} b) V_{CEQ} c) R_{in} d) R_o e) $A_{v(NL)}$ f) $A_{v(WL)}$ g) A_i h) A_p i) $v_{o(WL)}$

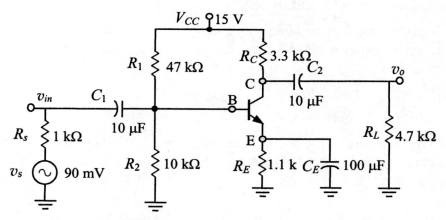

Figure 5-2P: CE amplifier (problems 5.6 through 5.9)

5.9 Repeat problem 5.8 with $R_1 = 33$ kΩ, $R_2 = 7.5$ kΩ, and all other parameters unchanged.

5.10 Repeat problem 5.8 with $R_1 = 27$ kΩ, $R_2 = 5.1$ kΩ, and all other parameters unchanged.

a) Algebraically
b) By simulation (a through f)

5.11 Carry out Practice Problem 5-3.

5.12 Carry out Practice Problem 5-4.

5.13 Determine the following for the amplifier of Figure 5-3P.
Assume $\beta = 175$ and $V_A = 120$ V. Draw the small-signal equivalent circuit.

a) I_{EQ} b) V_{CEQ} c) R_{in} d) R_o e) $A_{v(NL)}$ f) $A_{v(WL)}$ g) A_i h) A_p i) v_o

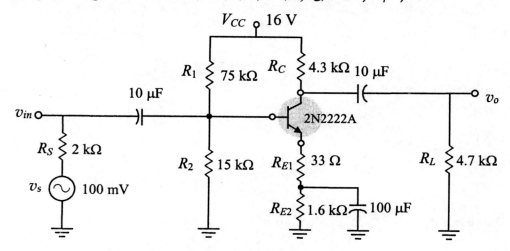

Figure 5-3P: A CE amplifier

Chapter 5

5.14 Repeat problem 5.13 with $R_1 = 56$ kΩ, $R_2 = 11$ kΩ, $R_{E2} = 1.5$ kΩ, and all other parameters unchanged.

5.15 Repeat problem 5.13 with $R_1 = 51$ kΩ, $R_2 = 10$ kΩ, $R_{E2} = 1.5$ kΩ, and all other parameters unchanged.

5.16 Repeat problem 5.13, neglecting the effect of r_o.

5.17 Repeat problem 5.14, neglecting the effect of r_o.

Section 5.4 Cascode Amplifier

5.18 Carry out Practice Problem 5.5.

5.19 Determine the following parameters for the cascode amplifier of Figure 5-4P:
a) I_{EQ} b) V_{CEQ} c) R_{in} d) R_o e) A_v f) v_o. Assume $\beta = 180$ and $V_A = 100$ V.

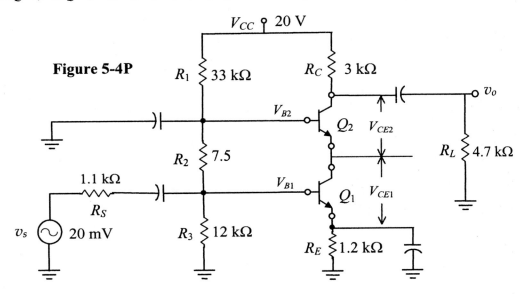

5.20 Determine the following parameters for the cascode amplifier of Figure 5-5P:
a) I_{EQ}
b) V_{CEQ}
c) R_{in}
d) R_o
e) $A_{v(NL)}$
f) $A_{v(WL)}$

Assume $\beta = 180$ and $V_A = 100$ V

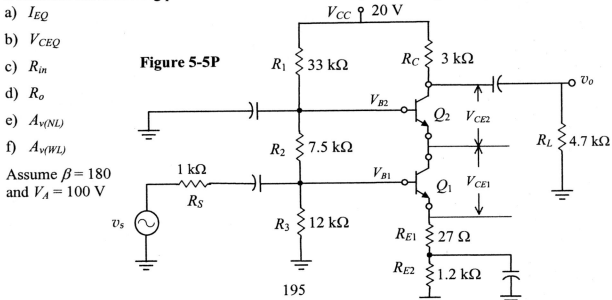

Section 5.5 Troubleshooting

Note that these troubleshooting exercises are designed to include the errors and problems that occur frequently in the laboratory. Hence, do not rule out the common errors in breadboarding and misread component values.

5.21 The application of a $v_s = 20$ mV(p-p), 1 kHz sinusoidal signal at the input of Figure 5-6P results in a sinusoidal signal at the output that has a peak-to-peak amplitude equal to approximately one-half the theoretically expected voltage. Determine what the problem is and the cause of it.

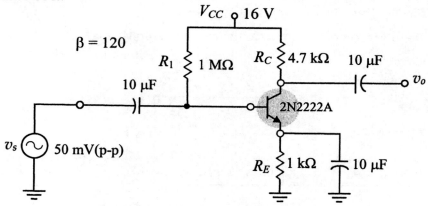

Figure 5-6P

5.22 Application of a $v_s = 100$ mV(p-p) sinusoidal signal with a frequency of 5 kHz at the input of Figure 5-7P results in a $v_o = 220$ mV(p-p) at the output as shown below. Determine whether there is an operational problem with this circuit. State what the problem is, if there is one, and the cause of it.

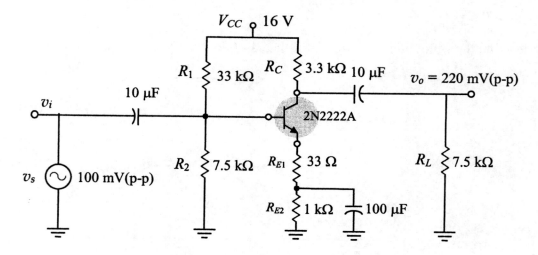

Figure 5-7P

5.23 With the application of a v_s = 100 mV(p-p), 5 kHz sinusoidal signal at the input of Figure 5-8P, an output signal larger than 6 v(p-p) is expected. However, an output signal v_o = 1.4 V(p-p) is measured at the output, as shown below. Determine what the problem is, if there is one, and the cause of it.

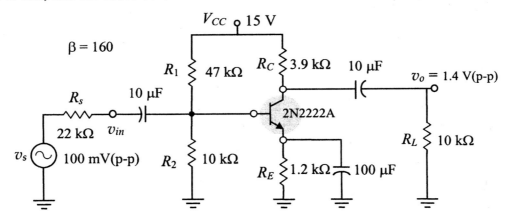

Figure 5-8P

5.24 With the application of a v_s = 100 mV(p-p) sinusoidal signal with a frequency of 1 kHz at the input of Figure 5-9P an output signal larger than 4 V(p-p) is expected. However, a sinusoidal signal of approximately v_o = 3 V(p-p) is measured at the output as shown below. Determine what the problem is, if there is one, and the cause of it.

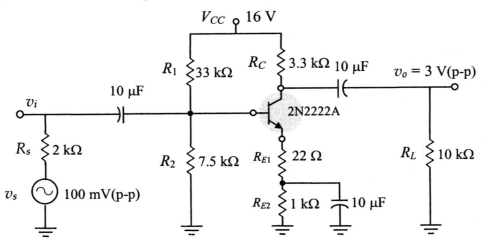

Figure 5-9P

Chapter

DESIGN OF THE SMALL-SIGNAL BJT AMPLIFIER

6.1 INTRODUCTION

Having discussed small-signal operation and the underlying theory, we wish now to present the primary and the terminal objective of the previous three chapters, the design of the BJT amplifier. However, in order to begin the design procedure properly and achieve optimum signal swing at the output, we need to take a close look at the AC load line and the limitations it can impose on the output signal swing. Upon the completion of this chapter you should be able to design amplifiers for a variety of specifications. In order to achieve this goal, the precise and comprehensive design methodology of the two most important configurations will be discussed along with a number of practical design examples. All design examples are tested by simulation in order to verify the accuracy of the design.

6.2 THE AC LOAD LINE AND THE OUTPUT SIGNAL SWING

As mentioned previously, the DC operating point sets the limits of the signal swing at the output of the amplifier, and the objective of an optimum DC bias is to place the operating point on the load line so that a maximum signal swing can be achieved at the output. This optimum operating point is usually at the center of the AC load line.

Recall that the DC load line has a slope of $m = -1/(R_C + R_E)$, and is constructed by connecting the V_{CC} to $I_{C(sat)}$ on the collector characteristic curves. Note that the slope is the negative reciprocal of the total resistance $(R_C + R_E)$ through which the output DC current I_C flows. Likewise, the AC load line has a slope equal to the negative reciprocal of the total AC resistance R_L' through which the output signal current i_c flows. To construct the AC load line with this particular slope, the two intercepts I_o and V_o are determined as follows:

I_C-intercept:

$$I_o = I_{CQ} + \frac{V_{CEQ}}{R_L'} \tag{6-1}$$

V_{CE}-intercept:

$$V_o = V_{CEQ} + I_{CQ} R_L' \tag{6-2}$$

These two intercept notations, I_o and V_o, should not be confused with i_o and v_o, which denote the output signal current and voltage, respectively. Consider the common-emitter amplifier of Figure 6-1 below. Let us construct the DC and AC load lines for this amplifier.

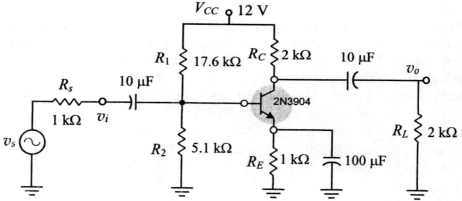

Figure 6-1: A voltage-divider biased common-emitter amplifier

Since the I_{CQ} and V_{CEQ} will be needed for determining the intercepts of the AC load line, let us first determine the Q-point.

$$V_B = V_{CC} \frac{R_2}{R_1 + R_2} = 12\text{V} \frac{5.1\,\text{k}\Omega}{17.6\,\text{k}\Omega + 5.1\,\text{k}\Omega} = 2.7\text{ V}$$

$$V_E = V_B - 0.7\text{ V} = 2\text{ V}$$

$$I_E = \frac{V_E}{R_E} = \frac{2\text{ V}}{1\text{ k}\Omega} = 2\text{ mA}$$

$$V_{CE} = V_{CC} - I_E(R_C + R_E) = 12\text{ V} - 2\text{ mA}(2\text{ k}\Omega + 1\text{ k}\Omega) = 6\text{ V}$$

The DC operating point is at $I_{CQ} = 2$ mA and $V_{CEQ} = 6$ V.

DC load line:

The DC load line will be drawn from $V_{CC} = 12$ V to $I_{C(sat)}$.

$$I_{C(sat)} = \frac{V_{CC}}{R_C + R_E} = \frac{12\text{ V}}{3\text{ k}\Omega} = 4\text{ mA}$$

The AC load line may be determined with or without the external load R_L.

6.2.1 AC Load Line and the Maximum Output Signal Swing with No Load (no R_L)

Since the AC load without R_L is $R_L'' = R_C$, the AC load line intercepts are determined by substituting the R_C for R_L' in Equations 6-1 and 6-2, as follows:

$$I_{o(NL)} = I_{CQ} + \frac{V_{CEQ}}{R_C} = 2\text{ mA} + \frac{6\text{ V}}{2\text{ k}\Omega} = 5\text{ mA}$$

$$V_{o(NL)} = V_{CEQ} + I_{CQ}R_C = 6\text{ V} + (2\text{ mA} \times 2\text{ k}\Omega) = 10\text{ V}$$

Both the DC and AC load lines without external load R_L are depicted in Figure 6-2. Note that both lines must and do pass through the Q-point with I_C of 2 mA and V_{CE} of 6 V. Notice that below the Q-point, the input signal i_b is limited by the cutoff region to 20 μA at the $I_B = 0$ line. That is, if i_b is increased further, it will be clipped off at $I_B = 0$ as shown in Figure 6-3. Hence without being clipped off at one side, the maximum base current is limited to 20 μA below and 20 μA above the Q-point, a total of 40 μA(p-p). The 40 μA of base current produces an output current i_c of 4 mA(p-p) and an output voltage v_{ce} of 8 V(p-p), as shown in Figure 6-2. Further increase in i_b will cause clipping of i_c and v_{ce} at the output, as depicted in Figure 6-3.

Also note that in Figure 6-2, for the 20 μA swing of i_b above the Q-point, the output signal i_c swing is 2.1 mA; but for the other 20 μA swing of i_b below the Q-point, the output signal swing i_c is only 1.9 mA, a total of 4 mA(p-p). Similarly, the output voltage swing v_{ce} below the Q-point is 4.2 V, but it only swings 3.8 V above the Q-point. In other words, the output signal is not vertically symmetrical. A reduced input signal, however, will cause an output signal to be nearly distortion free, with amplitude that is proportionally low. This kind of distortion can be observed in the laboratory on the oscilloscope, when one side of the output waveform, upper or lower half, appears wider and compressed, particularly when the amplifier output is driven to its maximum limit just before being clipped off by the cutoff or saturation region. Simulation results of this amplifier, which are to be presented shortly, accurately depict similar results. Distortion of this kind is due to the nonlinear characteristics of the device; that is, equal increments of the base current do not necessarily correspond to equal spacing of the I_B curves in the collector characteristics.

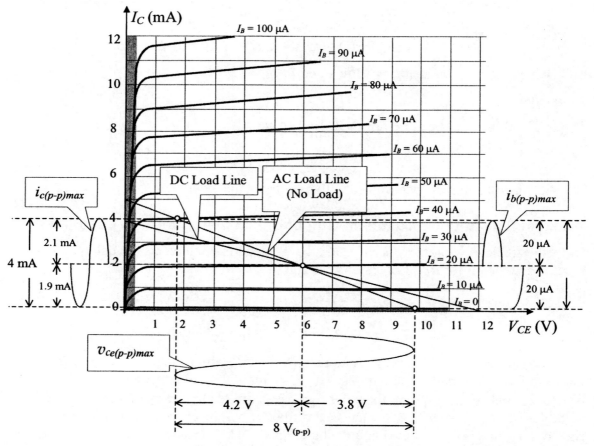

Figure 6-2: DC and AC load lines with the maximum input and output signals for the CE amplifier of Figure 6-1, with no load

As you can see, the spacing between I_B curves gets wider with increased I_B. To minimize the non-linearity distortion, the transistor operation must be limited to the active or linear region where the I_B curves are nearly evenly spaced.

Let us now examine the DC and AC operation of this amplifier (Figure 6-1) by simulation and see if we can verify the non-linearity distortion suggested in Figure 6-2, and the limitations imposed on the output signal by the cutoff region as depicted in Figure 6-3. The Electronics Workbench is used for circuit simulation and the simulation results are presented in Figures 6-4 and 6-5. The circuit setup and simulation results of the DC operation are depicted in Figure 6-4. The current-meter at the collector registers I_C of 1.953 mA and the voltmeter across the collector-emitter terminals registers V_{CE} of 6.13 V. These readings agree very closely with the algebraically determined Q-point values of $I_C = 2$ mA and $V_{CE} = 6$ V.

Chapter 6

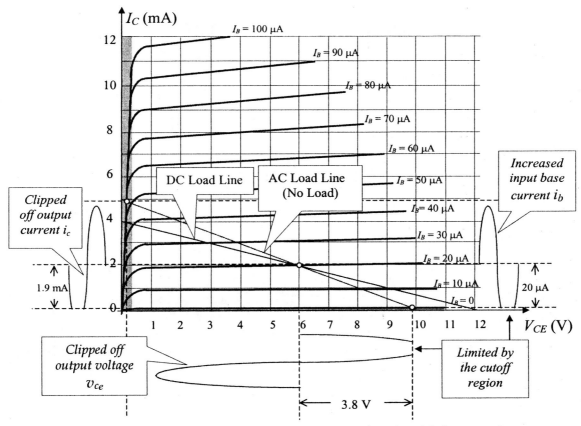

Figure 6-3: Output current and voltage signals with increased input signal i_b for the CE amplifier of Figure 6-1, without load

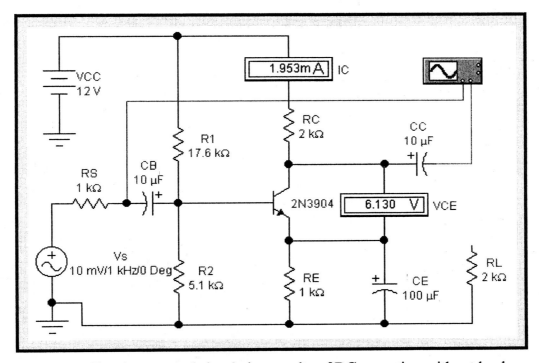

Figure 6-4: Circuit setup and simulation results of DC operation without load created in Electronics Workbench

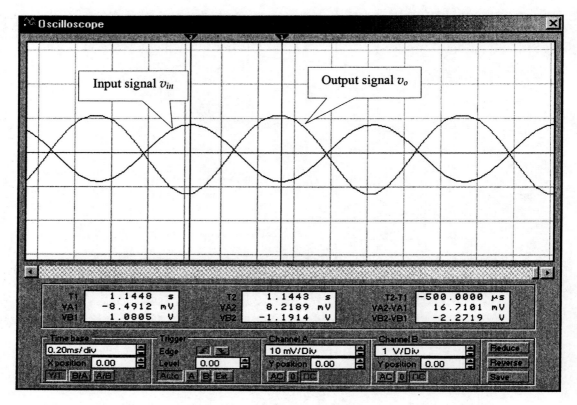

Figure 6-5: Electronics Workbench oscilloscope showing the no-load input and output signals with $v_s = 10$ mV$_{(rms)}$, $v_{in} = 16.71$ mV$_{(p-p)}$, and $v_o = 2.27$ V$_{(p-p)}$

As depicted on the oscilloscope of Figure 6-5, application of a small signal v_s of 10 mV$_{(rms)}$ results in a v_{in} of 16.71 mV$_{(p-p)}$ and v_o of 2.27 V$_{(p-p)}$. The output signal appears fairly distortion free; however, with a closer look, the presence of non-linearity distortion is evident even with such a small input signal. That is, the oscilloscope registers the negative peak of the output signal $V_{B1} = 1.08$ V and the positive peak $V_{B2} = -1.19$ V, a difference of 0.11 V. For a distortion free signal, the positive peak would equal the negative peak in amplitude, and the signal would be vertically symmetrical across the horizontal axis.

In the next simulation, as depicted in Figure 6-6, the source signal v_s is increased to 20 mV$_{(rms)}$ resulting in a v_{in} of 33.7 mV$_{(p-p)}$ and v_o of 4.4 V$_{(p-p)}$. The non-linearity distortion appears a bit more noticeable compared to the previous case and the oscilloscope readings do verify this. The oscilloscope of Figure 6-6 registers the positive peak of the output signal as $V_{B1} = 1.988$ V and the negative peak as $V_{B2} = -2.42$ V, a difference of 0.432. As expected, because of the increased unequal spacing between equal increments of I_B curves, the non-linearity distortion is more pronounced with increased input signal.

Chapter 6

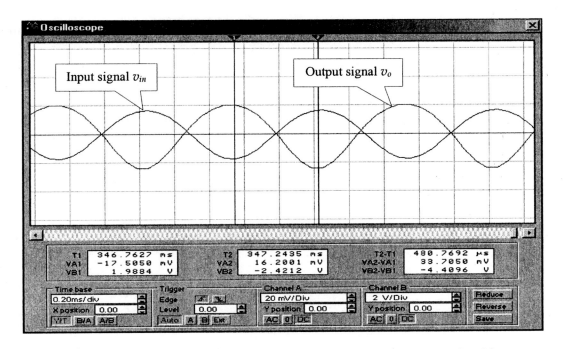

Figure 6-6: Electronics Workbench oscilloscope showing the no load input and output signals with $v_s = 20$ mV$_{(rms)}$, $v_{in} = 33.7$ mV$_{(p-p)}$, and $v_o = 4.4$ V$_{(p-p)}$

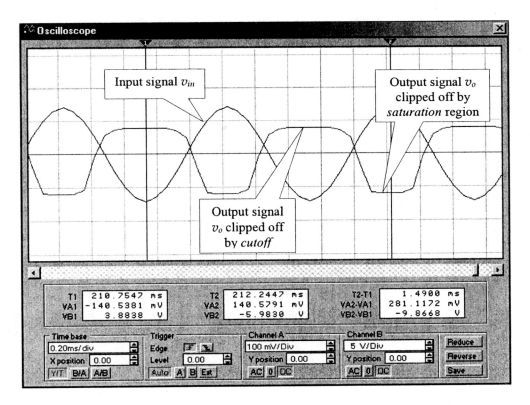

Figure 6-7: Electronics Workbench oscilloscope showing the no-load output signal clipped off by *saturation* and *cutoff* regions with $v_s = 100$

205

The results of the third and the final simulation without load are depicted in Figure 6-7, in which the output is driven to *saturation* and *cutoff* by application of a larger source signal $v_s = 100 \text{ mV}_{(rms)}$, resulting in a $v_{in} = 281 \text{ mV}_{(p-p)}$ and $v_o = 9.86 \text{ V}_{(p-p)}$. As registered by the oscilloscope, the lower half of the output waveform is clipped off by the saturation region at $V_{B2} = -5.983$ V, and the upper half is clipped off by the cutoff region at $V_{B1} = +3.88$ V. These limits set by the cutoff and saturation regions are in close agreement with the cutoff and saturation limits introduced by the AC load line of Figure 6-3, which were approximately 3.8 V above and 5.9 V below the Q-point.

6.2.2 AC Load Line and the Maximum Output Signal Swing with Load

The two intercepts of the AC load line with load are determined by substituting the total AC load R_L in Equations 6-1 and 6-2, as follows:

$$R_L' = R_C \| R_L = 2 \text{ k}\Omega \| 2 \text{ k}\Omega = 1 \text{ k}\Omega$$

$$I_{o(WL)} = I_{CQ} + \frac{V_{CEQ}}{R_L'} = 2 \text{ mA} + \frac{6 \text{ V}}{1 \text{ k}\Omega} = 8 \text{ mA}$$

$$V_{o(WL)} = V_{CEQ} + I_{CQ} \cdot R_L' = 6 \text{ V} + (2 \text{ mA} \times 1 \text{ k}\Omega) = 8 \text{ V}$$

The AC load line, which has a slope $m = -1/R_L'$ or -1 mA per volt, is constructed by connecting the I_o at $I_C = 8$ mA to V_o at $V_{CE} = 8$ V, as shown in Figure 6-8 below. As expected, this line intersects the DC load line at the Q-point.

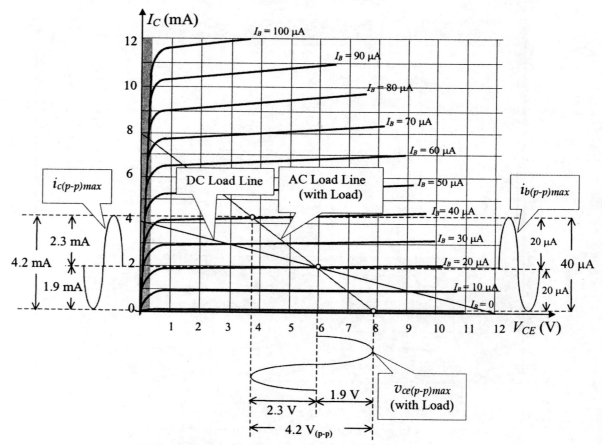

Figure 6-8: Maximum output current and voltage signals for the CE amplifier of Figure 6-1, with load

Below the Q-point, the swing of the base current i_b is again limited by the cutoff region to a maximum of 20 μA, which corresponds to a swing of 1.9 mA in i_c and a swing of 1.9 V in v_{ce}. Above the Q-point, however, the swing of 20 μA in i_b corresponds to a change of 2.3 mA in i_c and a change of 2.3 V in v_{ce}, a maximum signal swing of 4.2 V(p-p) without clipping at the output. Apparently, this difference in the upper and lower swings of the output signal, which results in a vertically nonsymmetrical waveform at the output, is due to the inherent nonlinear characteristics of the transistor. As mentioned earlier, to avoid or minimize the non-linearity distortion, the amplifier operation must be limited to the active or nearly linear region of the device, where the equal intervals of the base current correspond to nearly equal spacing of the I_B curves on the collector characteristics.

The above observations are verified when the amplifier circuit is simulated with load. Since the load resistor R_L is equal to R_C, the total AC load resistance at the output and the voltage gain are halved when the amplifier is loaded. Thus, application of a $v_s = 20$ mV$_{(rms)}$ with load is comparable to $v_s = 10$ mV$_{(rms)}$ without load, since the output signal swing in both cases is expected to be nearly equal. Two simulations are carried out to examine the amplifier with load, one with $v_s = 20$ mV$_{(rms)}$ and the other with $v_s = 100$ mV$_{(rms)}$. The results of the first simulation are depicted in Figures 6-9 and 6-10, in which the application of $v_s = 20$ mV$_{(rms)}$ results in a v_{in} of 33.7 mV$_{(p-p)}$ and v_o of 2.25 V$_{(p-p)}$. The non-linearity distortion appears more noticeable compared to the case without load, although the output signal swing in both cases is about 2.5 V$_{(p-p)}$. The oscilloscope of Figure 6-10 registers the negative peak of the output signal as $V_{B1} = -1.24$ V and the positive peak as $V_{B2} = +1.01$ V, a difference of 0.23 V, nearly twice the difference without load. Hence, we can conclude that the increased non-linearity distortion is due to the increase in the amplitude of the input signal, not the output signal.

The results of the second and the final simulation with load are depicted in Figures 6-11 and 6-12, in which the application of a $v_s = 100$ mV$_{(rms)}$ results in the output being clipped off by the cutoff region at $V_{B1} = +1.99$ V. This limit set by the cutoff region is again in close agreement with the cutoff limit introduced by the AC load line of Figure 6-8, which is approximately 1.9 V above the Q-point.

Chapter 6

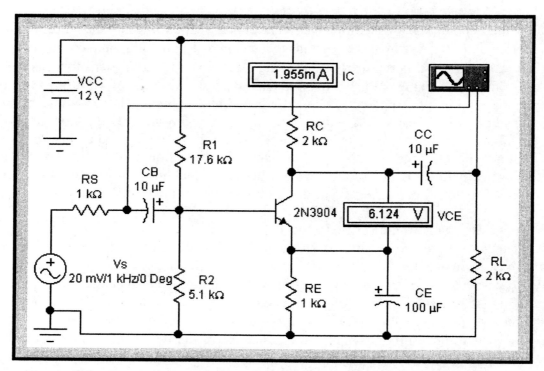

Figure 6-9: Circuit setup for simulation with $v_s = 20$ mV, with load created in Electronics Workbench

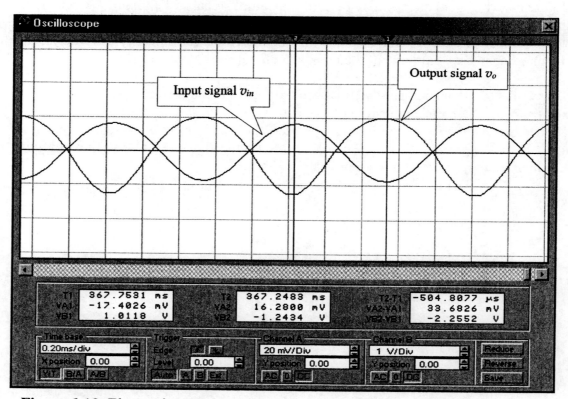

Figure 6-10: Electronics Workbench oscilloscope showing the output signal with load distorted by the nonlinear characteristic of the BJT

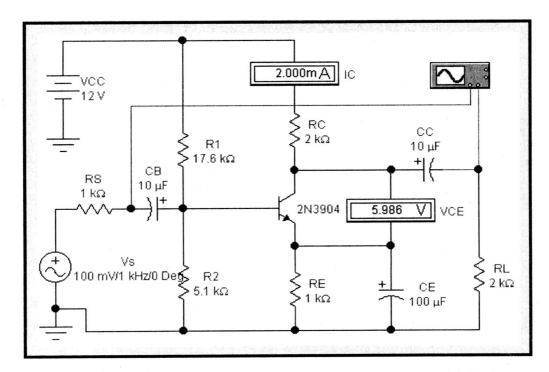

Figure 6-11: Circuit setup for simulation with $v_s = 100$ mV, with load created in Electronics Workbench

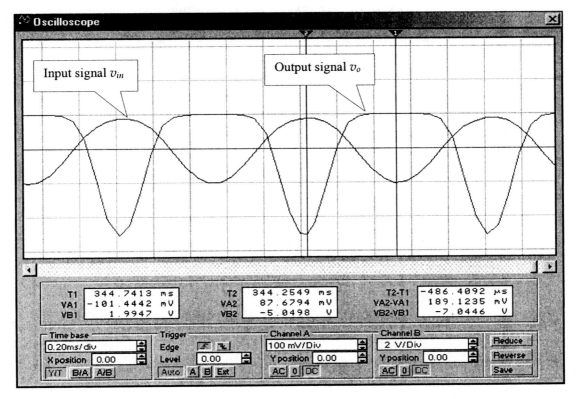

Figure 6-12: Electronics Workbench oscilloscope showing the loaded output signal clipped off by the *cutoff* region with increased v_s

6.3 A BRIEF INTRODUCTION TO FREQUENCY RESPONSE

Although the frequency response of BJT and FET amplifiers will be covered in Chapter 9, a brief introduction to the frequency response is useful in the design of the BJT amplifiers. In our analysis of the BJT amplifier, we have assumed a range of operating frequency in which the coupling and bypass capacitors were replaced by a short circuit; that is, the operating frequency was considered high enough so that the reactance of the capacitors would be very low, nearly zero. Consider the common-emitter amplifier of Figure 6-13.

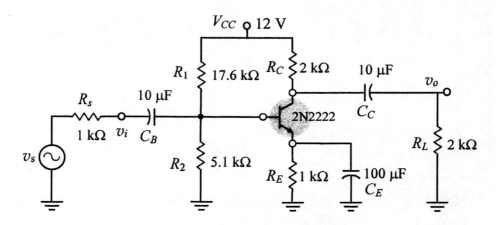

Figure 6-13: A voltage-divider biased common-emitter amplifier

This is the same amplifier circuit we have been analyzing and simulating with and without load, for exploring the role of the AC load line. The frequency response of this amplifier, that is, the voltage gain A_v versus frequency, will be similar to the response illustrated in Figure 6-14. Notice that at zero frequency (DC), the voltage gain A_v is zero because capacitors are an open circuit for DC and as a result the output voltage v_o is zero. Then the gain rises as the frequency increases, and levels off for further increases in frequency, and starts dropping again at higher frequencies. The range of frequency over which the voltage gain $A_{v(mid)}$ is nearly constant is called the *midband frequency range*. The two frequencies at which the gain drops to $0.707 A_{v(mid)}$ are called the lower and upper cutoff frequencies, f_L and f_H, respectively. The band of frequency between the two cutoff frequencies is called the *bandwidth* of the amplifier $BW = f_H - f_L$.

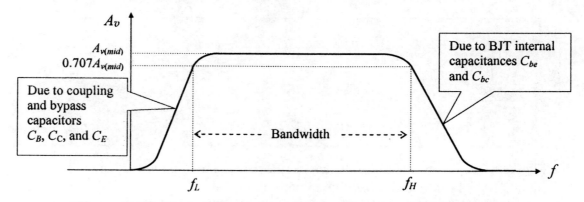

Figure 6-14: Typical frequency response diagram of a BJT amplifier

The upper cutoff frequency f_H is introduced by the internal parasitic or junction capacitances of the transistor. The coupling capacitors C_B and C_C and the bypass capacitor C_E introduce the lower cutoff frequency f_L. In fact, each of these three capacitors along with their associated resistance introduces a cutoff frequency. The highest of the three cutoffs could be dominant and is equal to the lower cutoff frequency f_L of the amplifier. Because of the very low resistance associated with the emitter bypass capacitor, generally the cutoff frequency introduced by C_E is the dominant low cutoff f_L.

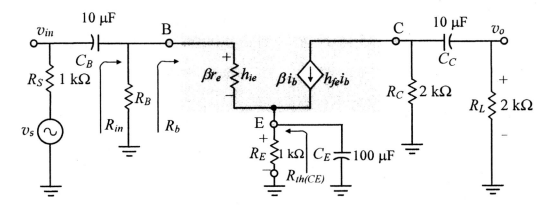

Figure 6-15: Small-signal low-frequency equivalent circuit of Figure 6-13

The Lower Cutoff Frequency

The full description and derivations regarding the general frequency response, the lower cutoff frequency and the upper cutoff frequency are given elsewhere in the text. However, the end result of those derivations regarding the lower cutoff frequency is presented below for the amplifier circuit of Figure 6-13. Since the dominant lower cutoff frequency is introduced by the emitter bypass capacitor C_E, the associated resistance and the resultant cutoff frequency are evaluated according to the following two equations:

$$R_{thCE} = R_E \| \left[r_e + \frac{R_S \| R_B}{\beta} \right] \cong r_e + \frac{R_S \| R_B}{\beta} \quad (6\text{-}3)$$

$$R_{thCE} = 13\,\Omega + \frac{1\,\text{k}\Omega \| 3.94\,\text{k}\Omega}{160} = 17.99\,\Omega$$

$$f_L = f_{CE} = \frac{1}{2\pi(R_{thCE})C_E} \quad (6\text{-}4)$$

$$f_L = \frac{1}{2\pi(17.99\,\Omega)(100\,\mu\text{F})} \cong 88.5\,\text{Hz}$$

Thus, the lower cutoff frequency for the amplifier of Figure 6-13 is 88.5 Hz. However, if this cutoff frequency proves to be too high, an increase in the emitter bypass capacitance C_E will reduce the cutoff frequency to a desired value.

6.4 DESIGN OF THE COMMON-EMITTER AMPLIFIER

As you must have already noticed, the *common-emitter* amplifier is the most versatile circuit of all three BJT amplifier configurations. It can have moderately high input resistance while allowing both voltage and current gains. Although the DC bias design has been covered in Chapter 3, we will cover the whole design procedure of a common-emitter amplifier and a cascode amplifier, for a number of given specifications, in the following two design examples.

Example 6-1 — DESIGN

Let us design a *common-emitter* amplifier to meet the specifications given below.

$15 \text{ V} \leq V_{CC} \leq 20 \text{ V}$ \hspace{2em} $1.5 \text{ mA} \leq I_C \leq 2.5 \text{ mA}$
Input resistance $R_{in} \geq 3 \text{ k}\Omega$ \hspace{2em} Output resistance $R_o \leq 4 \text{ k}\Omega$
Output signal swing $V_{o(p-p)} \geq 6 \text{ V}$ \hspace{2em} No load voltage gain $A_{v(NL)} = -60 \pm 5$
Lower cut-off frequency $f_L \leq 35 \text{ Hz}$ \hspace{2em} Assume $R_S = 1 \text{ k}\Omega$ and $R_L = 4.7 \text{ k}\Omega$

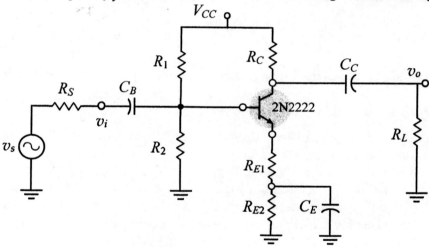

Figure 6-16: Circuit diagram of the common-emitter amplifier

DESIGN STRATEGY

I. General Considerations:

We will employ a voltage-divider bias so that the circuit is β-independent. We will also split R_E into two parts, R_{E1} and R_{E2}. The resistance value of R_{E1} will be relatively small, having very little or no effect on the DC bias. However, it will have a considerable effect on the gain and input resistance. By properly selecting R_{E1}, one can control the gain to a desired value while improving the input resistance. In addition, R_{E1} will contribute significantly to the overall stability of the system by providing a current-series feedback. A series resistor R_S will be used at the input side in order to facilitate the measurement of the input resistance. Similarly, a load resistor R_L will be used at the output in order to facilitate the measurement of the output resistance.

II. DC Bias Decisions:

In order to achieve a maximum signal swing at the output, we will try to center the Q-point on the AC load line. This can be accomplished by keeping R_C much higher than R_E, and by making V_{CEQ} slightly less than ½V_{CC}. Hence, the next step in the design process is to make some informed decisions within the specified boundaries, as follows.

Let V_{CC} = 16 V, I_C = 2 mA, V_{CE} = 6.7 V, and $A_{v(NL)}$ = –60.
Use a 2N2222 transistor.

Of the remaining 9.4 V, we will allocate 10 to 25 percent of V_{CC} to R_E and the rest to R_C. Let V_{RE} = 2.1 V and V_{RC} = 7.2 V, as displayed in Figure 6-17.

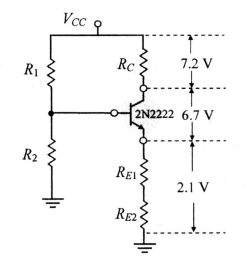

Figure 6-17: DC voltage allocation of the *common-emitter* amplifier

III. Design Computations:

From the allocated voltages and currents, we will now determine the resistor values.

$R_E = \dfrac{V_E}{I_E} = \dfrac{2.1\,\text{V}}{2\,\text{mA}} = 1.05\,\text{k}\Omega$ Let R_{E2} = 1 kΩ. R_{E1} to be determined shortly.

$R_C = \dfrac{V_{RC}}{I_C} = \dfrac{7.2\,\text{V}}{2\,\text{mA}} = 3.6\,\text{k}\Omega$ R_C = **3.6 kΩ**

$V_B = V_E + 0.7\,\text{V} = 2.8\,\text{V}$, also $V_B = V_{CC}\dfrac{R_2}{R_1 + R_2}$

$\dfrac{V_{CC}}{V_B} = \dfrac{R_1 + R_2}{R_2} = \dfrac{R_1}{R_2} + 1$, $\dfrac{V_{CC}}{V_B} - 1 = \dfrac{R_1}{R_2}$, $\dfrac{16}{2.8} - 1 = \dfrac{R_1}{R_2}$, $\dfrac{R_1}{R_2} = 4.714$

Let R_2 = 7.5 kΩ. R_1 = 4.714 × 7.5 kΩ = 35.35 kΩ, R_1 = **36 kΩ**

The no-load voltage gain is $A_{v(NL)} = -\dfrac{R_C}{r_e + R_{E1}}$, where $r_e = \dfrac{V_T}{I_E} = \dfrac{26\,\text{mV}}{2\,\text{mA}} = 13\,\Omega$

$r_e + R_{E1} = \dfrac{R_C}{|A_{v(NL)}|} = \dfrac{3.6\,\text{k}\Omega}{60} = 60$, $R_{E1} = 60 - 13 = 47$, R_{E1} = **47 Ω**

Let R_S = 1 kΩ and R_L = 4.7 kΩ, $R_L' = R_C \parallel R_L$ = 3.6 kΩ ∥ 4.7 kΩ = 2 kΩ

The expected voltage gain with load will be:

$$A_{v(WL)} = -\dfrac{R_L'}{r_e + R_{E1}} = -\dfrac{2\,\text{k}\Omega}{60\,\Omega} = -33.3$$

Chapter 6

Capacitors:

The dominant cutoff frequency will be introduced by the emitter bypass capacitor C_E. Let $C_E = 100$ μF and coupling capacitors $C_B = C_C = 10$ μF. We will evaluate the dominant cutoff frequency. If it is found to be higher than the specified $f_L = 35$ Hz, we will then increase the size of the capacitor C_E. Assuming an average β of 200 for the 2N2222 transistor, the Thevenin's equivalent resistance seen by C_E and the resultant cutoff frequency is determined according to Equations 6-3 and 6-4, as follows.

$$R_{th_{CE}} = R_{E2} \| \left[R_{E1} + r_e + \frac{(R_1 \| R_2 \| R_S)}{\beta} \right]$$

$$R_{th_{CE}} = 1\,k\Omega \| \left[47\,\Omega + 13\,\Omega + \frac{(34.7\,k\Omega \| 7.5\,k\Omega \| 1k\Omega)}{200} \right] = 60.417\,\Omega$$

$$f_L = \frac{1}{2\pi R_{th_{CE}} C_E} = \frac{1}{2\pi \times 60 \times 100 \times 10^{-6}} = 26.342\,Hz$$

Expected Input and Output Resistances:

$$R_{in} = (R_1 \| R_2) \| \beta(r_e + R_{E1}) \quad (6-5)$$

$R_{in} = (35.1\,k\Omega \| 7.5\,k\Omega) \| 200(13\,\Omega + 47\,\Omega) = 6.18\,k\Omega \| 12\,k\Omega = 4.08\,k\Omega$

$$R_o = R_C \| r_o \cong R_C$$

$R_o \cong 3.6\,k\Omega$

IV. Test Run and Design Verification:

Now that the design considerations and computations are complete, we will take it for a test drive and evaluate its performance using the Electronics Workbench. The results of simulation are presented in Figures 6-18 through 6-19.

The Quiescent Point:

Notice the DC bias current I_C of 2 mA and the bias voltage V_{CE} of 6.7 V as displayed on the DMMs of Figure 6-18. The almost exact match of these measurements with the expected design data verifies the accuracy of the DC bias design.

Voltage Gain with and without Load:

The Bode plotter of Figure 6-19 captures the no-load voltage gain of 58.5, and the loaded gain of 33.2 is depicted on the Bode Plotter of Figure 6-20. These results agree very closely with the expected/intended design specifications.

Input Resistance:

The input resistance can be calculated from the AC measurements displayed on the DMMs of Figure 6-18, as follows: $v_s = 100$ mV(rms), $v_{in} = 80.43$ mV(rms),

$$R_{in} = \frac{v_{in}}{v_s - v_{in}} R_S = \frac{80.43}{100 - 8.43} \times 1 = 4.1\,k\Omega$$

Chapter 6

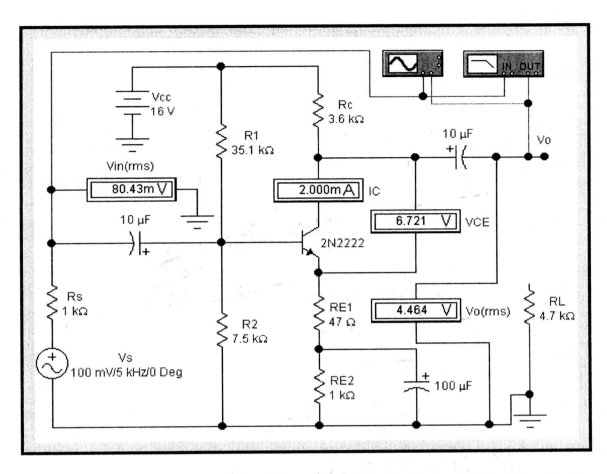

Figure 6-18: Circuit setup with simulation results of Design Example 6-1 without load created in Electronics Workbench

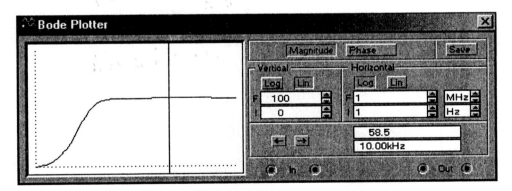

Figure 6-19: Electronics Workbench Bode plotter showing the no-load voltage gain at midband

Chapter 6

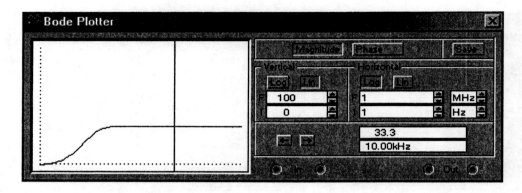

Figure 6-20: Electronics Workbench Bode plotter showing the voltage gain with load at midband

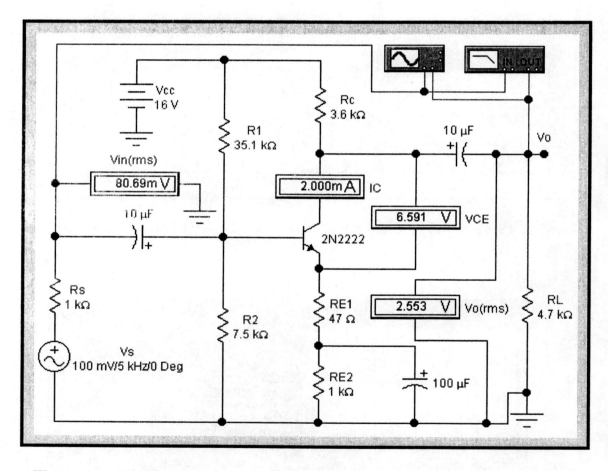

Figure 6-21: Circuit setup with simulation results of Design Example 6-1 with load created in Electronics Workbench

Chapter 6

Output Resistance:

From the data depicted on the DMMs of Figures 6-18 and 6-21, we have the following:

$$v_{o(NL)} = 4.464 \text{ V} \qquad v_{o(L)} = 2.553 \text{ V}$$

$$R_o = \frac{v_{o(NL)} - v_{o(WL)}}{v_{o(WL)}} R_L = \frac{1.911 \text{ V}}{2.553 \text{V}} 4.7 \text{ k}\Omega = 3.518 \text{ k}\Omega$$

Inversion and Output Signal Swing:

The signals displayed on the Electronics Workbench oscilloscope of Figure 6-22 verify the inversion of the output with respect to the input signal. In addition, the undistorted output signal swing of 7 volts with load is not only in full compliance with design specifications, but is also indicative of a well-placed Q-point.

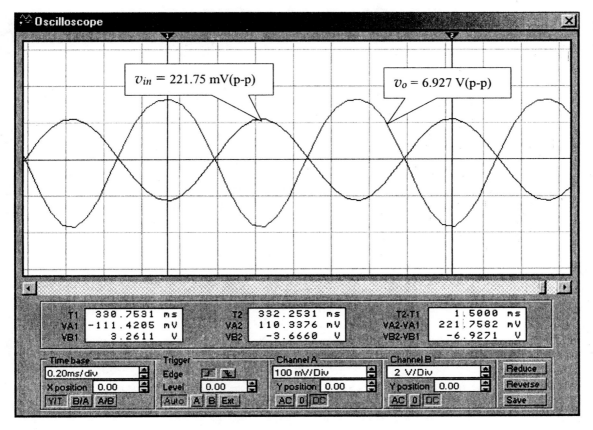

Figure 6-22: Electronics Workbench oscilloscope displaying the input and output signals of Design Example 6-1 with load

Lower Cutoff Frequency:

Figure 6-23 below captures the low cutoff frequency at 28.5 Hz, which is also in full compliance with the design specifications.

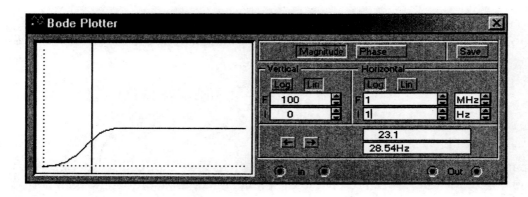

Figure 6-23: Electronics Workbench Bode plotter showing the lower cutoff frequency at 28.5 Hz

Summary of the simulation results versus analytically expected results are displayed in Table 6-1 below:

Table 6-1: Summary of Results of Design Example 6-1

Parameter	Specifications	Expected	Measured
I_{CO}	1.5 to 2.5 mA	2.0 mA	2.0 mA
V_{CEQ}		6.7 V	6.72 V
R_{in}	≥ 3.6 kΩ	4.08 kΩ	4.1 kΩ
R_o	≤ 3.6 kΩ	3.6 kΩ	3.52 kΩ
$A_{v(NL)}$	60 ± 5	60	58.5
$A_{v(WL)}$		33.3	33.3
$v_{o(p\text{-}p)}$	≥ 6 V	≥ 6 V	7 V
f_L	≤ 35 Hz	25 Hz	28.5 Hz

Practice Problem 6-2 DESIGN

Design a common-emitter amplifier to meet the specifications given below:

$V_{CC} = 15$ V, $V_{CE} = 6.6$ V, $V_E = 2.1$ V, $I_C = 1.9$ mA

Input resistance $R_{in} \geq 3.3$ kΩ Output resistance $R_o \leq 3.3$ kΩ

Output signal swing $v_{o(p\text{-}p)} \geq 6$ V   No-load voltage gain $A_{v(NL)} = 75$

Lower cut-off frequency $f_L \leq 20$ Hz $R_s = 1$ kΩ, $R_L = 4.7$ kΩ

Chapter 6

6.5 DESIGN OF THE CASCODE AMPLIFIER

As mentioned previously, the cascode amplifier is a cascade of a common-emitter and a common-base amplifier. It has all the characteristics of the common-emitter amplifier, and possesses the superior high-frequency response characteristics of the common-base amplifier. Hence, it is generally used in high-frequency applications, such as *RF* (radio frequency) amplifiers.

Example 6-2   DESIGN

Let's now design a cascode amplifier to meet the specifications given below.

Design specifications:

$I_C = 2$ mA,

$A_{v(NL)} = 45$,

$f_L \leq 30$ Hz.

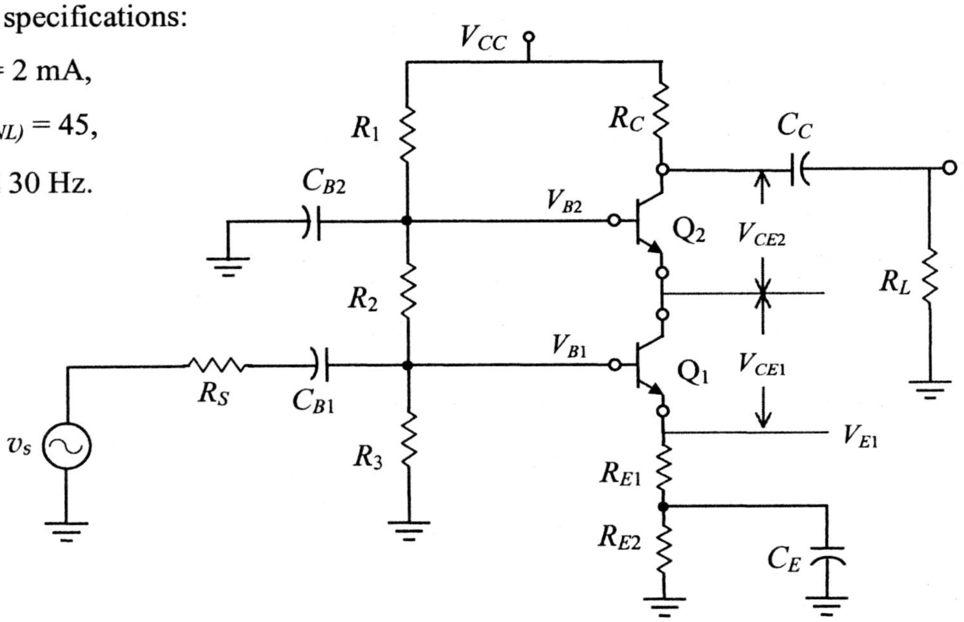

Figure 6-24: Circuit diagram of a cascode amplifier

DC Bias Decisions:

Recall from the analysis of the cascode amplifier that the voltage gain of the first stage $A_{v1} \leq 1$. Hence, there is no large signal swing at the output of the first stage; thus, we don't need to assign a large DC voltage to V_{CE1}.

Let $V_{CC} = 20$ V, $V_{CE2} = 8$ V, and $V_{CE1} = 1$ V. Of the remaining 11 volts, we will allocate 10 to 15 percent of V_{CC} to R_E, and the rest to R_C. Let $V_{RE} = 2.4$ V and $V_{RC} = 8.6$ V.

Design Computations:

From the allocated voltages and currents, we will now determine the resistor values.

$R_E = \dfrac{V_E}{I_E} = \dfrac{2.4 \text{ V}}{2 \text{ mA}} = 1.2 \text{ k}\Omega$ Let $R_{E2} = 1.1$ kΩ, R_{E1} to be determined shortly

$$R_C = \frac{V_{RC}}{I_C} = \frac{8.6\,\text{V}}{2\,\text{mA}} = 4.3\,\text{k}\Omega \qquad \boldsymbol{R_C = 4.3\,\text{k}\Omega}$$

$V_{B1} = V_{E1} + 0.7\,\text{V} = 2.4\,\text{V} + 0.7\,\text{V} = 3.1\,\text{V}$

$V_{B2} = V_{E2} + 0.7\,\text{V} = V_{E1} + V_{CE1} + 0.7\,\text{V} = 4.1\,\text{V}$

To comply with the proper voltage divider rule $R_3 \leq 10 R_E$; let $\boldsymbol{R_3 = 10\,\text{k}\Omega}$. Ignoring the very low base currents, the currents I through R_1, R_2, and R_3 will be approximately the same. Hence,

$$I = \frac{V_{B1}}{R_3} = \frac{3.1\,\text{V}}{10\,\text{k}\Omega} = 0.31\,\text{mA}$$

$$R_2 = \frac{V_{B2} - V_{B1}}{I} = \frac{1\,\text{V}}{0.31\,\text{mA}} = 3.226\,\text{k}\Omega \quad \text{Use } \boldsymbol{R_2 = 3.3\,\text{k}\Omega}.$$

$$R_1 = \frac{V_{CC} - V_{B2}}{I} = \frac{15.9\,\text{V}}{0.31\,\text{mA}} = 51.3\,\text{k}\Omega \qquad \text{Use } \boldsymbol{R_1 = 51\,\text{k}\Omega}.$$

The no-load voltage gain is $A_{v(NL)} = -\dfrac{R_C}{r_e + R_{E1}}$, where $r_e = \dfrac{V_T}{I_E} = \dfrac{26\,\text{mV}}{2\,\text{mA}} = 13\,\Omega$

$$r_e + R_{E1} = \frac{R_C}{|A_{v(NL)}|} = \frac{4.3\,\text{k}\Omega}{45} = 95\,\Omega \qquad R_{E1} = 95 - 13 = 82\,\Omega, \qquad \boldsymbol{R_{E1} = 82\,\Omega}$$

Design Summary:

$I_C = 2\,\text{mA}$, $R_C = 4.3\,\text{k}\Omega$, $R_1 = 51\,\text{k}\Omega$, $R_2 = 3.3\,\text{k}\Omega$, $R_3 = 10\,\text{k}\Omega$, $R_{E1} = 82\,\Omega$, $R_{E2} = 1.1\,\text{k}\Omega$,

$A_{v(NL)} = -45$, let $C_{B1} = C_{B2} = C_C = 10\,\mu\text{F}$, and $C_E = 100\,\mu\text{F}$.

Let $R_S = 1\,\text{k}\Omega$ and $R_L = 4.7\,\text{k}\Omega$. $A_{v(WL)} = (R_L\|R_C)/(r_e + R_{E1}) = 2.345/0.095 = 24.68$.

<u>Expected input and output resistances:</u>

$$R_{in} = (R_3 \| R_2) \| \beta(r_e + R_{E1})$$

$R_{in} = (10\,\text{k}\Omega \| 3.3\,\text{k}\Omega) \| 160(13\,\Omega + 82\,\Omega) = 2.48\,\text{k}\Omega \| 15.2\,\text{k}\Omega = 2.13\,\text{k}\Omega$

$$R_o = R_C \| r_o \cong R_C = 4.3\,\text{k}\Omega$$

<u>Dominant low cutoff frequency:</u>

$$R_{th_{CE}} = R_{E2} \| \left[R_{E1} + r_e + \frac{(R_2 \| R_3 \| R_S)}{\beta} \right]$$

$$R_{th_{CE}} = 1\,\text{k}\Omega \| \left[43\,\Omega + 13\,\Omega + \frac{(10\,\text{k}\Omega \| 3.3\,\text{k}\Omega \| 1\,\text{k}\Omega)}{160} \right] = 57\,\Omega$$

$$f_L = \frac{1}{2\pi R_{th_{CE}} C_E} = 28\,\text{Hz}$$

Test Run and Design Verification

The results of simulation created in Electronics Workbench are presented in Figures 6-25 through 6-29.

Chapter 6

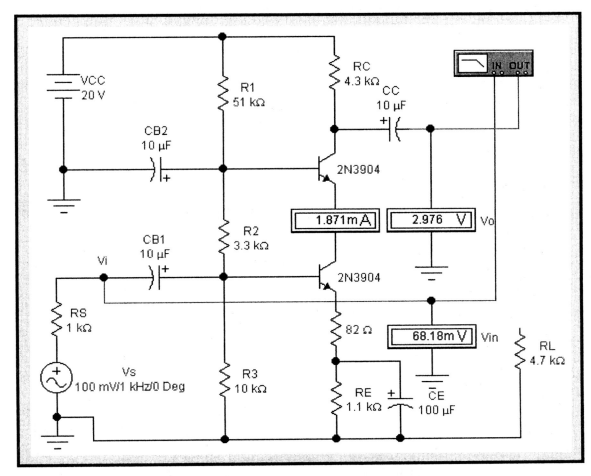

Figure 6-25: Circuit setup for simulation with no load created in Electronics Workbench

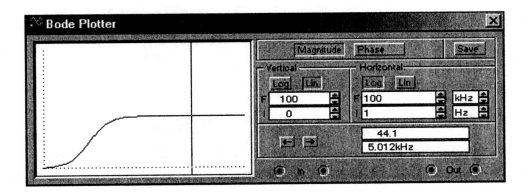

Figure 6-26: Electronics Workbench Bode plotter showing the no-load voltage gain at midband

Chapter 6

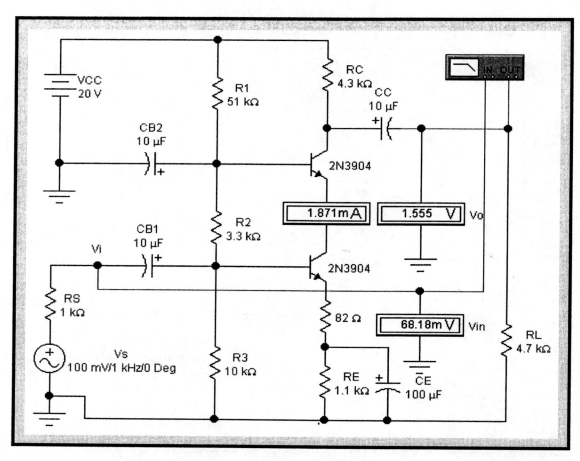

Figure 6-27: Circuit setup for simulation with load created in Electronics Workbench

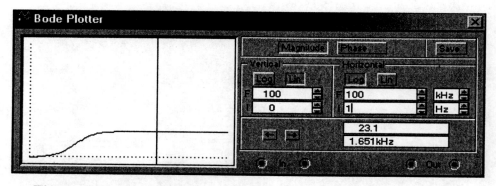

Figure 6-28: Electronics Workbench Bode plotter showing the voltage gain with load at midband

Input Resistance:

The input resistance can be calculated from the AC measurements displayed on the DMMs of Figure 6-25 or Figure 6-27 as follows: $v_s = 100$ mV, $v_{in} = 68.18$ mV,

Chapter 6

$$R_{in} = \frac{v_{in}}{v_s - v_{in}} R_s = \frac{68.18 \text{ mV}}{(100 - 68.18) \text{ mV}} 1 \text{ k}\Omega = 2.14 \text{ k}\Omega$$

Output Resistance:

From the data depicted on the DMMs of Figures 6-25 and 6-27, we have the following:

$$v_{o(NL)} = 2.976 \text{ V} \qquad v_{o(L)} = 1.555 \text{ V}$$

$$R_o = \frac{v_{o(NL)} - v_{o(WL)}}{v_{o(WL)}} R_L = \frac{1.421 \text{ V}}{1.555 \text{ V}} 4.7 \text{ k}\Omega = 4.3 \text{ k}\Omega$$

Lower Cutoff Frequency:

Figure 6-29 below captures the low cutoff frequency at 22.39 Hz, which is also in full compliance with the design specifications.

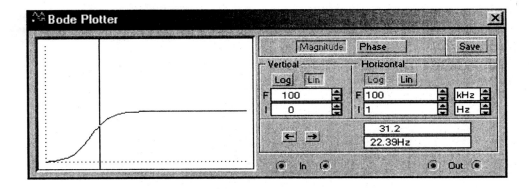

Figure 6-29: Electronics Workbench Bode plotter showing the low cutoff frequency

Table 6-2: Summary of Results of Design Example 6-2

Parameter	Expected	Measured
I_{CQ}	2.0 mA	1.871 mA
R_{in}	2.14 kΩ	2.14 kΩ
R_o	4.3 kΩ	4.3 kΩ
$A_{v(NL)}$	45	44.1
$A_{v(WL)}$	23.6	23.1
f_L	28 Hz	22.4 Hz

As you can see in Table 6-2 above, the measured results of simulation agree very closely with the theoretically expected results.

6.6 SUMMARY

- The DC operating point sets the limits of the signal swing at the output of the amplifier, and the objective of an optimum DC bias is to place the operating point on the load line so that a maximum signal swing can be achieved at the output. This optimum operating point is usually at the center of the AC load line.

- The DC load line has a slope of $m = -1/(R_C + R_E)$, and is constructed by connecting the V_{CC} to $I_{C(sat)}$ on the collector characteristic curves.

- Likewise, the AC load line has a slope equal to the negative reciprocal of the total AC resistance R_L' through which the output signal current i_c flows. To construct the AC load line with this particular slope, a line may be drawn connecting the two intercepts I_o and V_o as given by Equations 6-1 and 6-2.

- To avoid or minimize the non-linearity distortion, the amplifier operation must be limited to the active or nearly linear region of the device, where the equal intervals of the base current correspond to nearly equal spacing of the I_B curves on the collector characteristics.

- The non-linearity distortion causes the upper half-cycle of the sinusoidal output to appear compressed (wider and smaller in amplitude) compared to the lower half-cycle, which becomes more pronounced with an increased input signal.

- The range of frequency over which the gain is nearly constant is called the midband gain $A_{v(mid)}$. The two frequencies at which the gain drops to $0.707 A_{v(mid)}$ are called the lower and upper cutoff frequencies, respectively. The band of frequency between the two cutoff frequencies is called the bandwidth of the amplifier.

- By properly selecting the R_{E1}, one can control the gain to a desired value while improving the input resistance of the CE amplifier. In addition, R_{E1} contributes significantly to the overall stability of the system by providing a current-series feedback.

- In order to achieve a maximum signal swing at the output, try to center the Q-point on the AC load line. This can be achieved by keeping R_C much larger than R_E, and by making V_{CEQ} slightly less than ½VCC.

- Use of a series resistor R_S at the input side facilitates the measurement of the input resistance. Similarly, the use of a load resistor R_L at the output facilitates the measurement of the output resistance.

Chapter 6

PROBLEMS

Section 6.2 AC Load Line and the Output Signal Swing

6.1 Draw the DC and AC load lines for the amplifier of Figure 6-1P on a copy of the collector characteristics of Figure 6-2P and determine the maximum output signal swing without clipping.

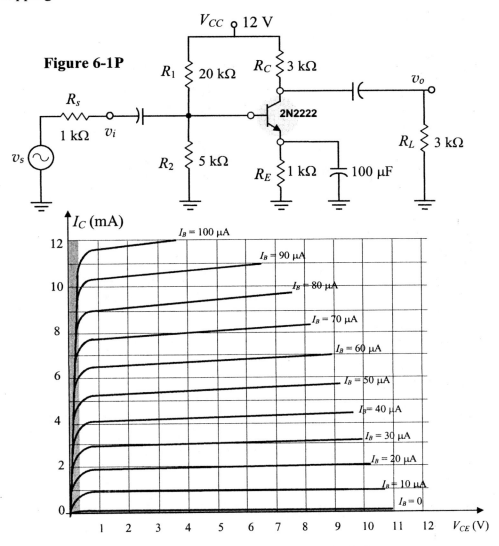

Figure 6-2P: Collector characteristics

Section 6.3 Frequency Response

6.2 Carry out Practice Problem 6-1.

6.3 Determine the low cutoff frequency for the amplifier of Figure 6-1P.

Chapter 6

Section 6.4 Design of CE Amplifier

6.4 Carry out Practice Problem 6-2.

6.5 Design a *common-emitter* amplifier to meet the specifications given below.

$V_{CC} = 14$ V
Input resistance $R_i \geq 3.3$ kΩ
Output signal swing $v_{o(p-p)} \geq 5$ V
Lower cutoff frequency $f_L \leq 40$ Hz

$I_C = 1.5$ mA
Output resistance $R_o \leq 3.3$ kΩ
No-load voltage gain $A_{v(NL)} = 55$
Assume $R_s = 1$ kΩ, $R_L = 10$ kΩ

6.6 Carry out Practice Problem 6-3.

6.7 Design a *common-emitter* amplifier to meet the specifications given below.

$V_{CC} = 15$ V
Input resistance $R_i \geq 3.3$ kΩ
Output signal swing $v_{o(p-p)} \geq 6$ V
Lower cutoff frequency $f_L \leq 40$ Hz

$I_C = 1.65$ mA
Output resistance $R_o \leq 3.9$ kΩ
No-load voltage gain $A_{v(NL)} = 55$
Assume $R_s = 1$ kΩ, $R_L = 10$ kΩ

6.8 Design a *common-emitter* amplifier to meet the specifications given below.

$V_{CC} = 16$ V
Input resistance $R_i \geq 3.3$ kΩ
Output signal swing $v_{o(p-p)} \geq 6$ V
Lower cutoff frequency $f_L \leq 40$ Hz

$I_C = 1.8$ mA
Output resistance $R_o \leq 3.6$ kΩ
No-load voltage gain $A_{v(NL)} = 75$
Assume $R_s = 1$ kΩ, $R_L = 10$ kΩ

6.9 Design a *common-emitter* amplifier to meet the specifications given below.

$V_{CC} = 18$ V
Input resistance $R_i \geq 3.3$ kΩ
Output signal swing $v_{o(p-p)} \geq 6$ V
Lower cutoff frequency $f_L \leq 40$ Hz

$I_C = 2$ mA
Output resistance $R_o \leq 3.3$ kΩ
No-load voltage gain $A_{v(NL)} = 96$
Assume $R_s = 1$ kΩ, $R_L = 10$ kΩ

Section 6.5 Design of Cascode Amplifier

6.10 Design a *cascode* amplifier to meet the specifications given below.

$V_{CC} = 18$ V
Input resistance $R_i \geq 3.6$ kΩ
Output signal swing $v_{o(p-p)} \geq 5$ V
Lower cutoff frequency $f_L \leq 40$ Hz

$I_C = 1.6$ mA
Output resistance $R_o \leq 5.1$ kΩ
No-load voltage gain $A_{v(NL)} = 60$
Assume $R_s = 1$ kΩ, $R_L = 10$ kΩ

6.11 Design a *cascode* amplifier to meet the specifications given below.

$V_{CC} = 22$ V
Input resistance $R_i \geq 3.6$ kΩ
Output signal swing $v_{o(p-p)} \geq 5$ V
Lower cutoff frequency $f_L \leq 40$ Hz

$I_C = 2$ mA
Output resistance $R_o \leq 5.1$ kΩ
No-load voltage gain $A_{v(NL)} = 80$
Assume $R_s = 1$ kΩ, $R_L = 10$ kΩ

Chapter 7

FIELD-EFFECT TRANSISTORS

7.1 INTRODUCTION

Field-effect transistors (FETs) are the next generation of transistors after the bipolar junction transistors (BJTs). Like the BJT, the field-effect transistor is a three-terminal device, but it operates under different principles than the BJT. As mentioned previously, the BJT is a bipolar semiconductor device because its current flow depends on both types of charge carriers: majority and minority carriers, or holes and electrons. The FET, however, is a unipolar semiconductor device in which the current flow depends on only one kind of charge carrier: the majority carrier, either holes or electrons. The field-effect transistor (FET) is so named because its output current depends on the electric field that is set up within the device by an external voltage applied at the input terminal. Hence, the FET is a voltage-controlled device, whereas the BJT is a current-controlled device in which the output current depends on the input current ($I_C = \beta I_B$). Because of the flow of only one type of charge carrier, the FET produces less noise compared to BJT. There are two FET types: junction field-effect transistor (JFET), and metal-oxide-semiconductor field-effect transistor (MOSFET). Both types have applications as discrete components and as building blocks of integrated circuits; however, MOSFET is becoming more popular in both areas. MOSFET is currently the device of choice in very-large-scale integrated (VLSI) circuits such as microprocessors and memory chips. Two chapters have been devoted to exploring the FET characteristics and its applications. This chapter covers the basic theory of the FET operation, DC biasing methods, and the DC bias design. Transistor modeling, small-signal operation, analysis, and design of the FET amplifiers will be covered in the following chapter.

7.2 JUNCTION FIELD-EFFECT TRANSISTORS (JFET)

7.2.1 JFET Structure and Characteristics

There are two types of junction FETs: *n*-channel and *p*-channel. The structure and the corresponding symbol of an *n*-channel JFET are displayed in Figure 7-1 below. The *n*-type channel, which is formed from a lightly doped semiconductor material, is sandwiched between the two *p*-type gate regions. Two metal ohmic contacts connect both ends of the channel to the *drain* and the *source* leads of the transistor. The *gate* regions are formed from a heavily doped *p*-type semiconductor material, and are connected together to the gate terminal by ohmic metal contacts.

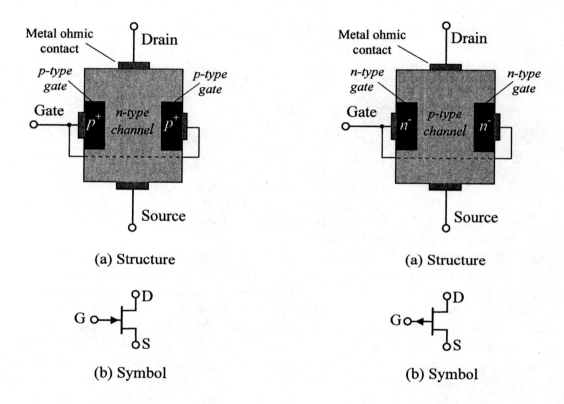

Figure 7-1: An *n-channel* JFET   Figure 7-2: A *p-channel* JFET

In a *p*-channel JFET, the *p*-type channel is sandwiched between the two *n*-type gate regions, as shown in Figure 7-2 above. The symbol for the *p*-channel is similar to the symbol for the *n*-channel JFET, except for the direction of the arrow, which is reversed for the *p*-channel.

The application of a small positive voltage V_{DS}, as shown in Figure 7-3(a) below, causes the flow of current I_D (conventional flow) through the channel from drain to source terminal. The resistance of the channel limits the amount of current I_D. Reverse biasing the gate-source junction by a small negative voltage applied at the gate causes the formation of depletion regions at the two *p-n* junctions. Since the channel is lightly doped compared to the gate regions, the depletions are broader in the channel, narrowing the channel width and as a result, causing an increase in channel resistance, which results in a decrease in the current I_D. The width of the depletion regions is directly proportional to the amount of reverse biasing voltage V_{GS} applied at the gate-source terminals. In other words, the amount of current I_D that can flow through the channel is controlled by the magnitude of the reverse biasing voltage V_{GS}.

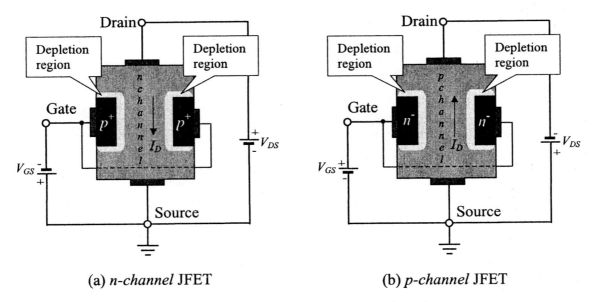

(a) *n-channel* JFET   (b) *p-channel* JFET

Figure 7-3: Reverse biasing the gate-to-source junction causes the narrowing of the channel by formation of depletion regions

Let us consider the case with gate-source terminals shorted ($V_{GS} = 0$) and V_{DS} is gradually increased as shown in Figures 7-4 and 7-5. Initially the current increases linearly with an increased V_{DS}, resulting in the formation of a non-uniform depletion region at both junctions, which is wider toward the drain and narrower toward the source terminal. This non-uniform depletion is caused by the change in voltage along the channel as it decreases from the applied V_{DS} at the *drain* to zero at the *source*. Although there is no external reverse biasing voltage V_{GS}, the *p-n* junctions are still reverse-biased because the *n*-channel is at a higher potential than the *p*-type gate, which is at the ground level. However, the amount of reverse bias and the resultant depletion width are proportional to the voltage level along the channel, which is the highest at the drain (equal to applied V_{DS}) and the lowest at the source (zero).

With further increase in V_{DS}, the increase in current is no longer proportional to the increase in V_{DS}. Instead, the current starts leveling off (Figure 7-6), and there will come a point at which the two depletion regions meet and the current will not increase further with an increased V_{DS}. The voltage V_{DS} at which the current levels off (with $V_{GS} = 0$) is referred to as the *pinch-off* voltage V_P. The maximum current I_D that flows through the channel with $V_{DS} \geq V_P$ and with $V_{GS} = 0$ is called the *drain-source saturation current* (I_{DSS}). Mathematically, I_{DSS} may be described as follows:

$$I_{DSS} = I_D \bigg|_{V_{GS} = 0} \tag{7-1}$$

Chapter 7

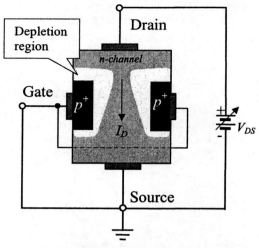

(a) Increased V_{DS} causes the formation of non-uniform depletion regions

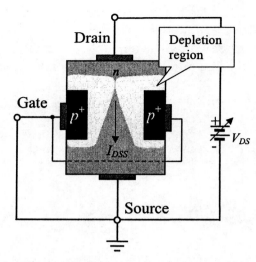

(b) Further increase in V_{DS} ($V_{DS} \geq V_P$) causes the depletion regions to meet and I_D to level off

Figure 7-4: With V_{GS} kept constant (in this case zero), increased V_{DS} causes the formation of non-uniform depletion regions (*n-channel* JFET)

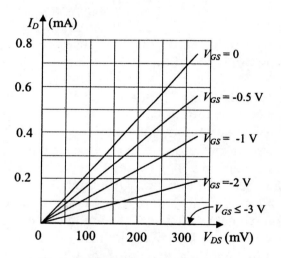

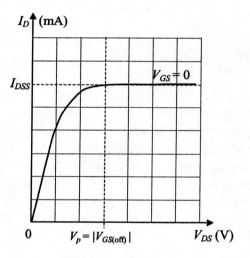

Figure 7-5: With small V_{DS} the current I_D rises linearly with V_{DS} and the device behaves like a *voltage-controlled resistance* (VCR).

Figure 7-6: With increased V_{DS}, I_D levels off and remains at its saturation value I_{DSS} for $V_{DS} \geq V_P$.

Let us now repeat the above procedure with a $V_{GS} = -0.5$ V, as illustrated in Figure 7-6. Because the gate-source junction is reverse-biased with a $V_{GS} = -0.5$ V, the already established depletion regions start getting wider with an increase in V_{DS}, resulting in a narrower channel width with a higher resistance. As the voltage V_{DS} is increased further, the *pinch-off* occurs at a smaller V_{DS} compared to the previous case, and the current starts leveling off early, as shown in Figure 7-7(b).

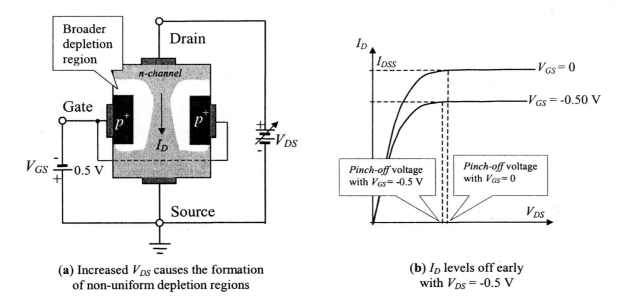

(a) Increased V_{DS} causes the formation of non-uniform depletion regions

(b) I_D levels off early with $V_{DS} = -0.5$ V

Figure 7-7: Application of a negative V_{GS} causes the formation of broader non-uniform depletion regions

If this procedure is repeated with different values of V_{GS}, such as: 0, −0.5 V, −1 V, −1.5 V, −2 V, and so on, there will come a point at which the channel will be totally depleted of the charge carriers, and as a result, the current will cease flowing. The V_{GS} that causes the total depletion of the channel and stops the current flow is referred to as the *gate-to-source cutoff voltage* ($V_{GS(off)}$). Mathematically, $V_{GS(off)}$ may be defined as follows:

$$V_{GS(off)} = V_{GS}\Big|_{I_D = 0} \qquad (7\text{-}2)$$

The voltage $V_{GS(off)}$ is equal to the *pinch-off voltage* V_P in absolute value. That is:

$$|V_{GS(off)}| = V_P \qquad (7\text{-}3)$$

If an experiment is completed as described by the above set of procedures, the plot of the collected data (I_D versus V_{DS}, for different values of V_{GS}) will resemble the family of curves shown in Figure 7-8. This family of curves is referred to as the *drain characteristics* of the JFET. The plot of I_D versus V_{GS}, which is the *transfer characteristics* of the JFET, can be plotted from the drain characteristics, as shown in Figure 7-8 (b).

Joining the points on the family of curves where *pinch-offs* occur results in a parabolic curve known as the pinch-off locus, as shown in Figure 7-8(a). The equation of this parabola and the parabolic transfer characteristic curve of Figure 7-8(b), which is known as *Shockley's equation*, is as follows:

$$I_D = I_{DSS}\left(\frac{V_{GS(off)} - V_{GS}}{V_{GS(off)}}\right)^2 = I_{DSS}\left(1 - \frac{V_{GS}}{V_{GS(off)}}\right)^2 \qquad (7\text{-}4)$$

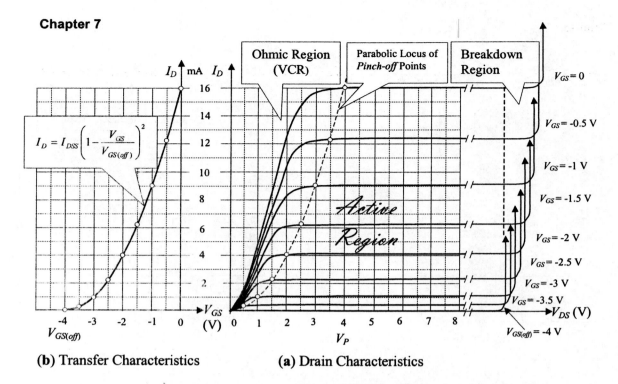

(b) Transfer Characteristics   **(a)** Drain Characteristics

Figure 7-8: Plot of Drain and Transfer Characteristics for an *n*-channel JFET

The region to the left of the *pinch-off* locus parabola on the *drain characteristics* is called the *ohmic region*, where the resistance between the *drain* and *source* can be controlled by the applied *gate-to-source* voltage V_{GS}, thus making the device a *voltage-controlled resistance* (*VCR*), if operated in this region. Note that, in the *ohmic region*, slope of the curve ($\Delta I_D / \Delta V_{DS}$) decreases with increased V_{GS} (absolute value), and thus the resistance of the device, which is the reciprocal of the slope, increases with increased V_{GS} (absolute value). As shown in Figure 7-8(a), the transistor breakdown may occur at excessively large values of V_{DS}, a phenomenon similar to BJT breakdown, causing a large and sudden increase in the drain current. The region between the *ohmic* and the *breakdown* regions is the *active region*, also known as the *saturation region*, where the device is normally operated for amplification of the small signal.

Example 7-1 — n-channel JFET — ANALYSIS

A JFET is known to have a $V_{GS(off)} = -4$ V, and $I_{DSS} = 16$ mA. Determine the drain current I_D with $V_{GS} = -2$ V

Solution:
Knowing the I_{DSS} and $V_{GS(off)}$, drain current can be determined with Shockley's equation, as follows:

$$I_D = I_{DSS}\left(1 - \frac{V_{GS}}{V_{GS(off)}}\right)^2$$

$$I_D = 16\,\text{mA}\left(1 - \frac{-2\,\text{V}}{-4\,\text{V}}\right)^2 = 16\,\text{mA}(0.5)^2 = 4\,\text{mA}$$

Example 7-2 — n-channel JFET — ANALYSIS

Following the first lecture on JFETs, an experimenter student set up the circuit of Figure 7-9 in the laboratory and recorded the following observations:
1. With a $|V_{GG}| = 2.0$ V, the DMM placed across R_D registered 2.5 V.
2. As the $|V_{GG}|$ was slowly stepped up, the voltmeter reading across R_D kept going down until it reached a value of zero volts when $|V_{GG}|$ was 3.6 V. Determine I_{DSS} and V_p of the JFET.

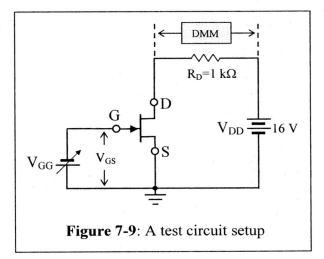

Figure 7-9: A test circuit setup

Solution:

$V_{GS(off)}$ is defined as the voltage that cuts off the I_D. Since the current flow stopped at V_{GS} of -3.6 V, $V_{GS(off)} = -3.6$ V and,
$$V_P = |V_{GS(off)}| = 3.6 \text{ V}.$$
According to the student observations, measurement of 2.5 V across R_D with a $|V_{GG}|$ of 2.0 V corresponds to 2.5 mA of I_D and a V_{GS} of -2 V.
Knowing the $V_{GS(off)}$, I_D, and V_{GS}, I_{DSS} can be determined with Shockley's equation, as follows:

$$I_D = I_{DSS}\left(1 - \frac{V_{GS}}{V_{GS(off)}}\right)^2$$

$$I_{DSS} = \frac{I_D}{\left(1 - \frac{V_{GS}}{V_{GS(Off)}}\right)^2}$$

Substituting for I_D, V_{GS}, and $V_{GS(off)}$ in the above equation results in the following I_{DSS}.

$$I_{DSS} = \frac{2.5 \text{ mA}}{\left(1 - \frac{-2 \text{ V}}{-3.6 \text{ V}}\right)^2} = 12.656 \text{ mA}$$

Practice Problem 7-1 — n-channel JFET — ANALYSIS

In his next set of experiments with another JFET, the experimenter student of Example 7-2 observes the following with the same setup of Figure 7-9:
1. The DMM placed across R_D reads 12.0 V when $|V_{GG}|$ is adjusted to zero V.
2. The DMM placed across R_D reads 3.0 V when $|V_{GG}|$ is adjusted to 1.5 V.
Determine the I_{DSS} and $V_{GS(off)}$ of the JFET.

Answers: $I_{DSS} = 12$ mA, and $V_{GS(off)} = -3$ V

Chapter 7

Plotting the Transfer Characteristics with Shockley's Equation

Given the I_{DSS} and $V_{GS(off)}$, the transfer characteristics curve may be plotted with Shockley's equation. Let us plot the transfer characteristics of the JFET described in Example 7-1, which has the following characteristics: $V_{GS(off)} = -4$ V, and $I_{DSS} = 16$ mA.

Option 1: Plotting the old-fashioned way, by hand

To plot the parabolic curve, we will need several points, each with an I_D and a V_{GS} value. Apparently, the more the number of points, the more accurate the curve will be. A rule of thumb for having sufficient data to do an adequate plot is to construct a two-column data table like the one shown below. The entries in the V_{GS} column go from zero up to $V_{GS(off)}$ in increments of 0.5 V. The entries in the I_D column, which will be the corresponding I_D for each V_{GS}, are to be determined with Shockley's equation and entered accordingly, as shown below.

V_{GS} (V)	I_D (mA)
0	16
−0.5	12.25
−1.0	9.0
−1.5	6.25
−2.0	4
−2.5	2.25
−3.0	1.0
−3.5	0.25
−4.0	0

Table 7-1

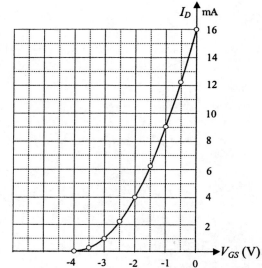

Figure 7-10: Plot of transfer characteristics

Option 2: Plotting with computer using software such as MATLAB

$$I_D = I_{DSS}\left(1 - \frac{V_{GS}}{V_{GS(off)}}\right)^2 = 16\,\text{mA}\left(1 - \frac{V_{GS}}{-4\,\text{V}}\right)^2 = 16\,\text{mA}\left(1 + \frac{V_{GS}}{4\,\text{V}}\right)^2$$

You may ask MATLAB to plot the equation, print a title, and label both axes, as follows:

1. Enter the Shockley's equation with corresponding numerical values. Since the current axis will be labeled in mA, the mA unit may be dropped from the equation.
```
» y ='16*(1+(x/4))^2';
```
2. Ask MATLAB to plot the function from $V_{GS(off)}$ to I_{DSS}.
```
» fplot(y,[-4  0  0  16])
```
3. Define the ranges for x and y axes, and print grid lines.
```
»  axis([-5 1  0 18]), grid;
```
4. Ask MATLAB to print x-label and y-label.
```
» xlabel('Gate-to-Source Voltage (V)')
» ylabel('Drain Current (mA)')
```

Chapter 7

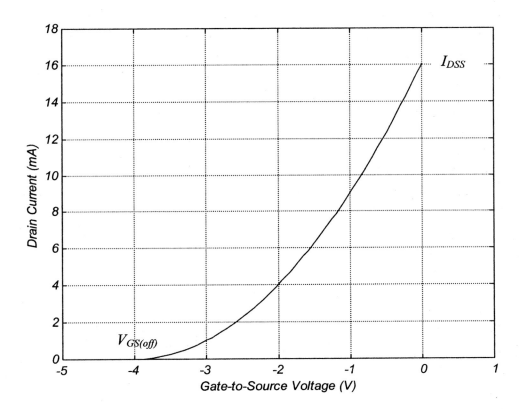

Figure 7-11: MATLAB plot of the transfer characteristics of an *n*-channel JFET

Practice Problem 7-2 — PLOTTING

Plot the transfer characteristics curve of the JFET given in Practice Problem 7-1
a) the old-fashioned way, b) with computer, preferably MATLAB

7.2.2 DC Biasing the JFET

To use the Junction FET for amplification of small signals, it must be biased properly to operate in the active region. Like the BJT, JFET may be biased in one of several different methods.

Fixed Bias Circuit — JFET — ANALYSIS

Figure 7-12 shows the circuit of a fixed bias JFET. Note that since the gate-source junction is reverse-biased, gate current is approximately zero, and, as a result, there is no voltage drop across the R_G. Hence, the gate-to-source voltage V_{GS} equals the gate supply voltage V_{GG}. That is,

$V_{GS} = V_G = V_{GG} = -1$ V

Chapter 7

$I_{DSS} = 12$ mA $V_{GS(off)} = -4$ V

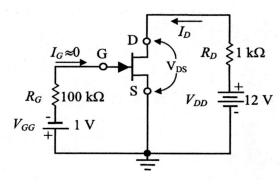

Figure 7-12: A fixed bias circuit

I_D may be determined by substituting for all variables in Shockley's equation.

$$I_D = I_{DSS}\left(1 - \frac{V_{GS}}{V_{GS(off)}}\right)^2$$

$$I_D = 12\text{ mA}\left(1 - \frac{-1\text{V}}{-4\text{V}}\right)^2 = 12\text{ mA }(0.75)^2 = 6.75\text{ mA}$$

According to KVL, the supply voltage V_{DD} must equal the voltage drops around the loop.

$$V_{DD} = I_D R_D + V_{DS} \qquad (7\text{-}5)$$

$$V_{DS} = V_{DD} - I_D R_D \qquad (7\text{-}6)$$

$V_{DS} = 12$ V $-$ (6.75 mA $\times$ 1 kΩ) $= 5.25$ V

Hence, the Q-point voltages and currents are as follows:

$V_{GSQ} = V_G = V_{GG} = -1$ V, $I_{DQ} = 6.75$ mA,   and $V_{DSQ} = 5.25$ V

Example 7-3   ANALYSIS

Determine the Q-point voltages and currents for the fixed bias circuit of Figure 7-13.

Solution:

Since $I_G = 0$, there will be no voltage drop across R_G. Hence,
$V_{GS} = V_{GG} = -1.5$ V

Next, I_{DQ} is determined with Shockley's equation.

$$I_D = I_{DSS}\left(1 - \frac{V_{GS}}{V_{GS(off)}}\right)^2$$

$$I_D = 16\text{ mA}\left(1 - \frac{-1.5\text{V}}{-4\text{V}}\right)^2 = 6.25\text{ mA}$$

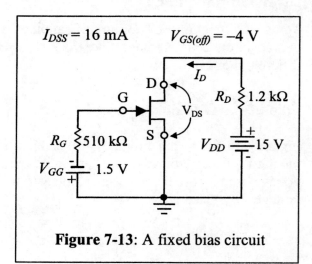

Figure 7-13: A fixed bias circuit

Having solved for I_{DQ}, V_{DSQ} is determined with Equation 8-5, as follows:
$$V_{DS} = V_{DD} - I_D R_D$$
$$V_{DS} = 15\ V - (6.25\ mA \times 1.2\ k\Omega) = 15 - 7.5\ V = 7.5\ V$$

Example 7-4 *(p-channel JFET) — ANALYSIS*

Determine the I_{DQ} and V_{DSQ} for the fixed bias circuit of Figure 7-14.

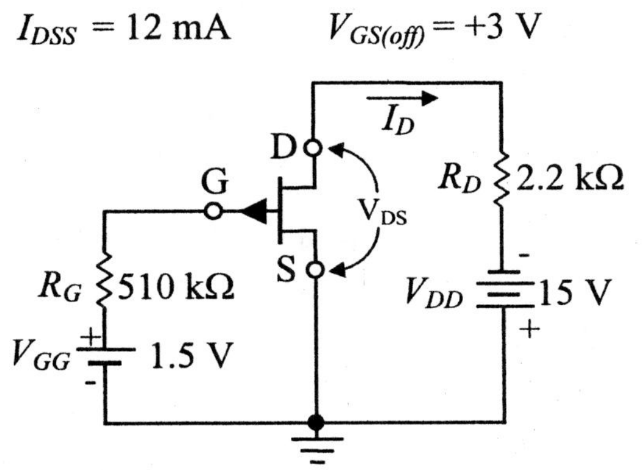

Solution:

$V_{GS} = V_{GG} = +1.5\ V$

Next, I_{DQ} is determined with Shockley's equation.

$$|I_D| = 12\ mA\left(1 - \frac{1.5\ V}{3\ V}\right)^2 = 3\ mA$$

Having solved for I_{DQ}, V_{DSQ} is determined with Equation 7-6, as follows:

$$V_{DS} = V_{DD} - I_D R_D$$

$$V_{DS} = -15\ V - (-3\ mA \times 2.2\ k\Omega) = -15 + 6.6\ V = -8.4\ V$$

$I_{DSS} = 12\ mA$, $V_{GS(off)} = +3\ V$

Figure 7-14: A fixed bias circuit

Self-Bias Circuit *(n-channel JFET) — ANALYSIS*

Self-bias circuit of JFET is similar to the emitter-bias in BJT. As shown in Figure 7-15, a resistor R_S is installed at the source terminal, eliminating the need for a V_{GG}. The negative V_{GS} needed to reverse bias the gate-source junction will be developed as a result of the current I_D flowing through the resistor R_S. Hence, the circuit is called *self-bias*.

Because there is no gate current, there will be no voltage drop across the R_G, which makes the gate terminal to be at the ground potential. That is, $V_G = 0$. Applying KVL around the input loop results in the following:

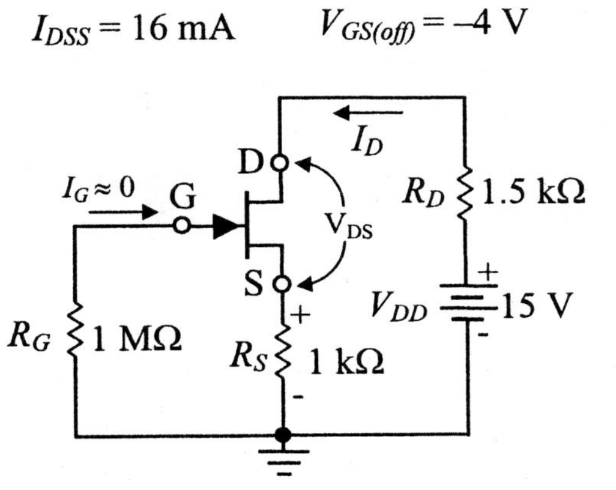

$I_{DSS} = 16\ mA$, $V_{GS(off)} = -4\ V$

Figure 7-15: A self-biased circuit

$$V_G = V_{GS} + I_D R_S = 0$$

Chapter 7

Solving for V_{GS} yields the following:

$$V_{GS} = V_G - I_D R_S = 0 - I_D R_S = -I_D R_S$$

Solving for I_D results in the following:

$$I_D = -\frac{V_{GS}}{R_S} \qquad (7\text{-}7)$$

The plot of the above equation is a straight line; hence it is referred to as the self-bias line. We also know of another equation for I_D, Shockley's equation, whose plot is a parabola. The intersection of the self-bias line and the parabolic transfer characteristics curve is the operating point or the Q-point, which corresponds to I_{DQ} and V_{GSQ}. Hence, the Q-point may be solved for either graphically, or algebraically. The graphical method involves plotting the transfer characteristics curve and the self-bias line, then trying to estimate the point of intersection. Hence, the graphical method is not going to be very accurate. The algebraic solution involves solving a first-order equation and a second-order equation simultaneously. There will be no plotting and the answer will be accurate. For these reasons, the algebraic solution is encouraged. However, for the purpose of demonstration, we will solve the circuit of Figure 7-15 using both methods.

Graphical solution: $I_{DSS} = 16$ mA $V_{GS(off)} = -4$ V

First, we need to plot the transfer characteristics of the JFET, and then the self-bias line. Both equations are repeated below for convenience.

$$I_D = -\frac{V_{GS}}{R_S} = -\frac{V_{GS}}{1\,k\Omega}$$

V_{GS}	I_D (mA)
0	0
−4 V	4

Let $V_{GS} = 0 \Rightarrow I_D = 0$,   Let $V_{GS} = -4$ V $\Rightarrow I_D = 4$ mA
Hence, the two points on the line are (0 V, 0 mA) and (−4 V, 4 mA)

$$I_D = I_{DSS}\left(1 - \frac{V_{GS}}{V_{GS(off)}}\right)^2 = 16\,\text{mA}\left(1 - \frac{V_{GS}}{-4\,\text{V}}\right)^2 = 16\,\text{mA}\left(1 + \frac{V_{GS}}{4\,\text{V}}\right)^2$$

We will now ask MATLAB to plot both equations, as follows:
Since the current axis (I_D) is in mA, we will drop the mA units from both equations.

MATLAB Commands for plotting both equations:
Enter the Shockley's equation (note: $y = I_D$, $x = V_{GS}$).
```
» y ='16*(1+x/4)^2';
```
Ask MATLAB to plot the above equation in blue, and define the x and y ranges.
```
» fplot(y,[-4 0 0 16])
```
Ask MATLAB to hold the current plot and draw the grid lines.
```
» hold, grid;
Current plot held
```
Enter the coordinates of the two points on the line, points (0, 0) and (−4, +4). Note that the first entry x1 is x-coordinates, and y1 is y-coordinates of both points.
```
» x1=[0 -4]; y1=[0 +4];
» axis([-5 0  0 16]);
```
Ask MATLAB to plot the line in red.
```
» plot(x1,y1)
```

Ask MATLAB to print x and y labels, the grid lines, and a title for the plot.
```
» xlabel('Gate-to-Source Voltage (V)')
» ylabel('Drain Current (mA)')
»
```

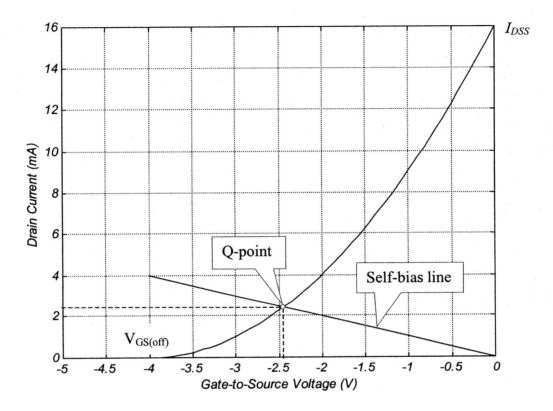

Figure 7-16: MATLAB plot of the transfer characteristics curve and the self-bias line

The graphical solution is approximately at $V_{GSQ} = -2.4$ V and $I_{DQ} = 2.4$ mA. According to KVL, the supply voltage V_{DD} must equal the sum of all voltage drops around the output loop.

$$V_{DD} = I_D R_D + V_{DS} + I_D R_S = I_D(R_D + R_S) + V_{DS}$$

Solving for V_{DS} in the above equation results in the following:

$$V_{DS} = V_{DD} - I_D(R_D + R_S) \qquad (7\text{-}8)$$

$$V_{DS} = 15 \text{ V} - 2.4 \text{ mA} (1.5 \text{ k}\Omega + 1 \text{ k}\Omega) = 15 \text{ V} - 6 \text{ V} = 9 \text{ V}$$

Algebraic solution:

Since the solution is the point of intersection of the self-bias line and the parabolic curve, algebraic solution begins by equating the two equations.

$$-\frac{V_{GS}}{1 \text{ k}\Omega} = 16 \text{ mA}\left(1 + \frac{V_{GS}}{4\text{V}}\right)^2$$

Multiplying both sides by 10^3 and expanding the squared term results in the following:

$$-V_{GS} = 16 \times \left(1 + \frac{1}{2}V_{GS} + \frac{V_{GS}^2}{16}\right)$$

$$-V_{GS} = 16 + 8V_{GS} + V_{GS}^2$$

Collecting the terms results in the following quadratic equation:

$$\boxed{V_{GS}^2 + 9V_{GS} + 16 = 0} \tag{7-9}$$

There will be two solutions for the above equation, and only one of those two will be the valid solution. At this point, you can enter the coefficients of the above polynomial in your calculator and solve for the roots. However, solving the old-fashioned way, we may factor out the equation if factorable, or use the quadratic formula.

For a second-order or quadratic equation in the general form of $(ax^2 + bx + c = 0)$, the solution is:

$$x = \frac{-b \pm \sqrt{b^2 - 4ac}}{2a}$$

Applying the quadratic formula to Equation 7-9 results in the following solutions:

$$V_{GS} = \frac{-9 \pm \sqrt{9^2 - 4 \times 16}}{2} = \frac{-9 \pm \sqrt{17}}{2} = -2.44 \text{ V}, -6.56 \text{ V}$$

Since the V_{GSQ} must be somewhere between 0 and $V_{GS(off)}$, that is: $0 \geq V_{GS} \geq -4$ V, the valid solution is $V_{GS} = -2.44$ V. Note that the valid solution always corresponds to using the plus sign in the numerator of the quadratic formula, as follows:

$$\boxed{V_{GS} = \frac{-b + \sqrt{b^2 - 4ac}}{2a}} \tag{7-10}$$

The current I_{DQ} may be determined by substituting for V_{GS} in the equation of the self-bias line (Equation 7-7), as follows:

$$I_D = \frac{-V_{GS}}{R_S} = -\frac{-2.44 \text{ V}}{1 \text{ k}\Omega} = 2.44 \text{ mA}$$

The voltage V_{DS} can now be determined by substituting for I_D in Equation 7-8, as follows:

$$V_{DS} = V_{DD} - I_D(R_D + R_S)$$

$$V_{DS} = 15 \text{ V} - 2.44 \text{ mA}(1.5 \text{ k}\Omega + 1 \text{ k}\Omega) = 15 \text{ V} - 6.1 \text{ V} = 8.9 \text{ V}$$

Hence, the algebraic solutions for the DC operating point are as follows:

$$V_{GSQ} = -2.44 \text{ V} \qquad I_D = 2.44 \text{ mA} \qquad V_{DSQ} = 8.9 \text{ V}$$

General algebraic solution for a self-biased circuit

The equation of the self-bias line and Shockley's equation are as follows:

$$I_D = -\frac{V_{GS}}{R_S} \qquad I_D = I_{DSS}\left(1 - \frac{V_{GS}}{V_{GS(off)}}\right)^2$$

Equating the two equations results in the following:

Chapter 7

$$-\frac{V_{GS}}{R_S} = I_{DSS}\left(1 - \frac{V_{GS}}{V_{GS(off)}}\right)^2$$

Since $V_{GS(off)}$ is a negative constant (for n-channel) and $V_P = |V_{GS(off)}|$, the above equation can be written as follows, with V_P being the absolute value of $V_{GS(off)}$:

$$-\frac{V_{GS}}{R_S} = I_{DSS}\left(1 + \frac{V_{GS}}{V_P}\right)^2$$

Multiplying both sides by R_S and expanding the squared term results in the following:

$$-V_{GS} = I_{DSS}R_S\left(1 + \frac{2V_{GS}}{V_P} + \frac{V_{GS}^2}{V_P^2}\right)$$

$$-V_{GS} = I_{DSS}R_S + \frac{2I_{DSS}R_S}{V_P}V_{GS} + \frac{I_{DSS}R_S}{V_P^2}V_{GS}^2$$

Collecting the terms we obtain the following quadratic equation:

$$\frac{I_{DSS}R_S}{V_P^2}V_{GS}^2 + \left(\frac{2I_{DSS}R_S}{V_P} + 1\right)V_{GS} + I_{DSS}R_S = 0 \qquad (7\text{-}11)$$

The above equation is now in the form of the following general quadratic equation:

$$ax^2 + bx + c = 0$$

which has the following general solution:

$$V_{GS}\Big|_{n-channel} = \frac{-b + \sqrt{b^2 - 4ac}}{2a} \qquad (7\text{-}12)$$

$$V_{GS}\Big|_{p-channel} = \frac{+b - \sqrt{b^2 - 4ac}}{2a} \qquad (7\text{-}13)$$

where,

$$a = \frac{I_{DSS}R_S}{V_P^2} \qquad b = \frac{2I_{DSS}R_S}{|V_P|} + 1 \qquad c = I_{DSS}R_S$$

Having determined the V_{GS} with the above equations, I_D and V_{DS} can be determined with Equations 7-7 and 7-8, respectively.

$$I_D = \frac{-V_{GS}}{R_S}$$

$$V_{DS} = V_{DD} - I_D(R_D + R_S)$$

Let us now apply the results of the above derivations to the self-bias circuit of Figure 7-15, and compare the results with those already obtained, algebraically.

$$I_{DSS} = 16 \text{ mA} \qquad V_{GS(off)} = -4 \text{ V} \Rightarrow V_P = |V_{GS(off)}| = 4 \text{ V}$$

$$a = \frac{I_{DSS}R_S}{V_P^2} = \frac{16 \text{ mA} \times 1 \text{ k}\Omega}{4^2} = \frac{16}{16} = 1$$

$$b = \frac{2I_{DSS}R_S}{V_P}+1 = \frac{2\times 16\,\text{mA}\times 1\,\text{k}\Omega}{4\text{V}}+1 = \frac{32}{4}+1 = 9$$

$$c = I_{DSS}R_S = 16\,\text{mA}\times 1\,\text{k}\Omega = 16$$

$$V_{GS}\Big|_{n-channel} = \frac{-b+\sqrt{b^2-4ac}}{2a} = \frac{-9+\sqrt{9^2-4\times 16}}{2} = \frac{-9+\sqrt{17}}{2} = -2.438 \cong -2.44\,\text{V}$$

Example 7-5 ANALYSIS

Determine the operating point voltages and currents for the self-bias circuit of Figure 7-17 for a *p*-channel JFET.

Solution:

Using the results of general algebraic solution for the self-bias circuit, we have the following:

$$a = \frac{I_{DSS}R_S}{V_P^2} = \frac{15\,\text{mA}\times 510\,\Omega}{5^2} = 0.306$$

$$b = \frac{2I_{DSS}R_S}{V_P}+1 = \frac{2\times 15\,\text{mA}\times 510\,\Omega}{5\text{V}}+1 = 4.06$$

$$c = I_{DSS}R_S = 15\,\text{mA}\times 0.51\,\text{k}\Omega = 7.65$$

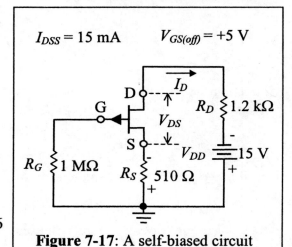

Figure 7-17: A self-biased circuit

Substituting for *a*, *b*, and *c* in Equation 7-13 results in the following solution for V_{GS}.

$$V_{GS}\Big|_{p-channel} = \frac{+b-\sqrt{b^2-4ac}}{2a} = \frac{4.06-\sqrt{4.06^2-4\times 0.306\times 7.65}}{2\times 0.306} = \frac{4.06-2.67}{0.612} \cong 2.27\,\text{V}$$

We can now determine the I_{DQ} and V_{DSQ} with Equations 7-7 and 7-8.

$$I_D = \frac{-V_{GS}}{R_S} = \frac{-2.27\,\text{V}}{510\,\Omega} = -4.45\,\text{mA}$$

$$V_{DS} = V_{DD} - I_D(R_D + R_S)$$

$$V_{DS} = -15\,\text{V} - (-4.45\,\text{mA})(1.2\,\text{k}\Omega + 0.51\,\text{k}\Omega) = -15\,\text{V} + 7.6\,\text{V} = -7.4\,\text{V}$$

Let us now plot the graphical solution with MATLAB and compare the results.
On the MATLAB command window, enter Shockley's equation ($y = I_D$, $x = V_{GS}$):
```
» y='15*(1-(x/5))^2';
```
Ask MATLAB to plot the equation in blue, and define ranges for $x = V_{GS}$ and $y = I_D$.
```
» fplot(y,[0 5 0 16])
» grid
» hold
```

```
Current plot held
```
Enter the two points on the self-bias line (0 V, 0 mA) and (5 V, 9.8 mA).
```
» x1=[0   5];
» y1=[0   9.8];
```
Ask MATLAB to plot the self-bias line in red.
```
» plot(x1,y1)
```
Ask MATLAB to print x label, y label, and a title for the plot.
```
» xlabel('Gate-to-Source Voltage (V)')
» ylabel('Drain Current(mA)')
»
```

Figure 7-18 depicts the two plots and the Q-point, which is the intersection of the line and the parabolic curve. As you can see, the graphical solution plotted with MATLAB matches the algebraically determined solutions of I_D = 4.45 mA and V_{GS} = 2.27 V.

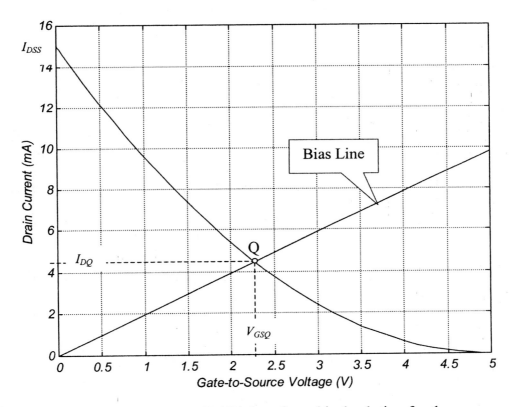

Figure 7-18: MATLAB plot of graphical solution for the self-bias circuit of Example 7-5 with *p*-channel JFET

Practice Problem 7-3 — ANALYSIS

Determine the graphical and algebraic solutions for the circuit of Figure 7-19.

I_{DSS} = 12 mA

$V_{GS(off)}$ = –3.3 V

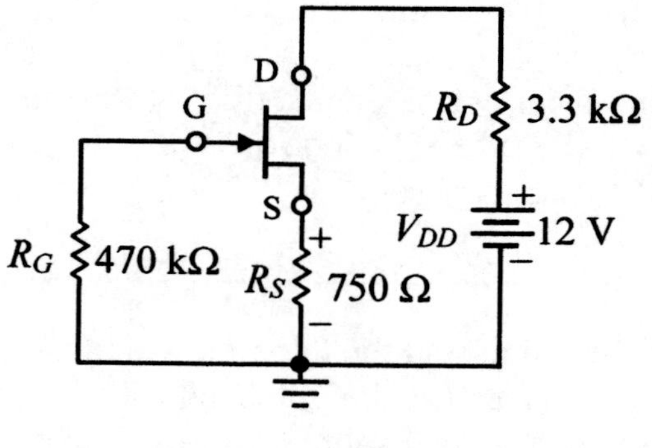

Figure 7-19: A self-biased circuit

Answers: V_{GS} = –1.817V, I_D = 2.422 mA, $V_{DS} \cong$ 2.19 V

Stability Comparison of Fixed Bias Circuit versus Self-Bias Circuit

Let us compare the stability of the fixed bias circuit of Figure 7-20 with the self-bias circuit of Figure 7-21. The JFETs used in both circuits are identical and have the same parameters I_{DSS} = 16 mA and $V_{GS(off)}$ = –4 V.

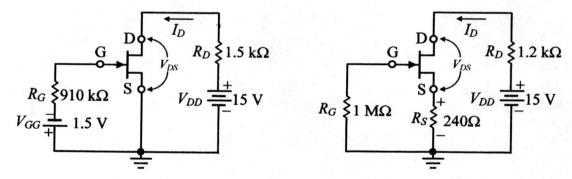

Figure 7-20: A fixed bias circuit   **Figure 7-21**: A self-biased circuit

For a fair comparison, both circuits were designed to operate at the same Q-point with V_{GSQ} = –1.5 V and I_{DQ} = 6.25 mA, as illustrated in Figure 7-22. The other curve is the altered transfer characteristic due to assumed change in the JFET characteristics.

The change in the drain current for the self-bias circuit ΔI_{D1} is 0.78 mA, and the change for the fixed bias circuit ΔI_{D2} is 1.68 mA, more than twice the change in the self-bias circuit. Hence, the self-bias circuit is more stable than the fixed-biased circuit. The severity of deviation from the original Q-point I_{DQ} is directly related to the slope of the bias line. That is, the closer to vertical the slope of the bias line is, the larger the deviation (ΔI_D), and the closer to horizontal the slope is, the smaller will be the deviation. Since the slope of the bias line is the negative reciprocal of the *source* resistor R_S, a smaller slope would require a larger resistor. A larger resistor, however, would limit the current I_D. For example, in the self-bias circuit of Figure 7-18 with a $V_{GS} = -1.5$ V and $R_S = 240\ \Omega$, I_D is 6.25 mA ($I_D = -V_{GS}/R_S$). The use of a larger R_S, say 24 kΩ instead of 240 Ω, would allow only 62.5 µA of drain current.

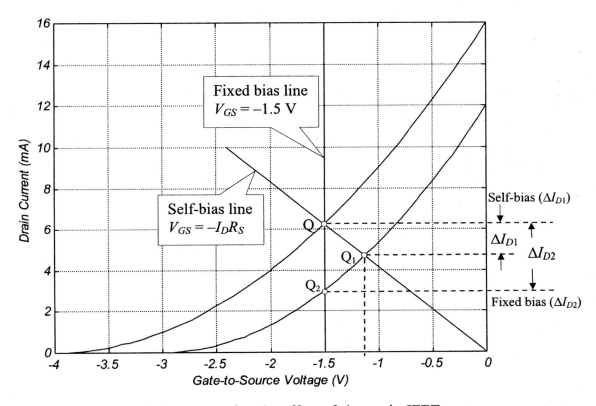

Figure 7-22: Demonstrating the effect of change in JFET characteristics on fixed bias and self-bias circuits

In order to have a reasonable amount of current flow with a fairly large R_S, we will use the familiar bias configuration known as the *voltage-divider* bias.

MATLAB commands for plotting the graphs of Figure 7-22:
```
» y ='16*(1+(x/4))^2';
» z ='12*(1+(x/3))^2';
» x1=[0  -2.4], y1=[0 10];
» x2=[-1.5 -1.5], y2=[0 16];
» fplot(y,[-4 0  0 16])
» hold on, grid
» fplot(z,[-3 0  0 12])
```

```
» axis([-4 0  0 16])
» plot(x1,y1)
» plot(x2,y2)
» xlabel('Gate-to-Source Voltage (V)')
» ylabel('Drain Current (mA)'
```

Practice Problem 7-4 ANALYSIS

Repeat the above procedure of stability comparison for the following two circuits. Initially I_{DSS} = 12 mA and $V_{GS(off)}$ = –3 V, then it changes to 14 mA and –3.5 V.

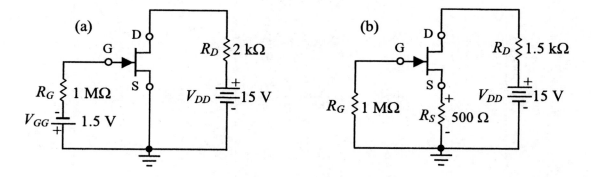

Voltage-Divider Bias ANALYSIS

The gate in a *voltage-divider* bias circuit, instead of being at the ground potential, is kept at a positive potential (for *n*-channel) by means of a voltage-divider network. The effect of keeping the gate at a positive potential V_G rather than at the ground level is to shift the intercept of the bias line with the horizontal axis from (0,0) at the origin to a positive V_G, as shown in Figure 7-23. As you can see, shifting of the intercept from $V_{GS} = 0$ to $V_{GS} = +V_G$ lessens the slope of the bias line and thus improves the stability of the circuit. Recall that the voltage-divider bias was used

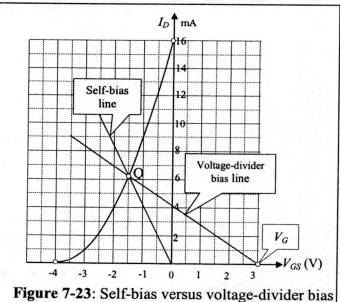

Figure 7-23: Self-bias versus voltage-divider bias

in BJT configurations in order to make the circuit β-independent, and as a result, provided more stability by not responding to changes in the transistor characteristics, such as the change in β. Although there is no β with JFET, the use of a voltage-divider bias serves the same purpose, providing more stable operation. Analysis of the voltage-divider biased circuit is similar to the self-biased circuit, except that $V_G > 0$:

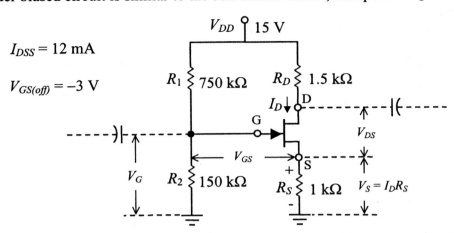

Figure 7-24: A voltage-divider biased JFET amplifier circuit

Analysis of the voltage-divider bias begins with determining the V_G by applying the voltage-divider rule at the gate.

$$V_G = V_{DD} \frac{R_2}{R_1 + R_2} \qquad (7\text{-}14)$$

$$V_G = 15\,\text{V} \frac{150\,\text{k}\Omega}{750\,\text{k}\Omega + 150\,\text{k}\Omega} = 2.5\,\text{V}$$

By definition,

$$V_{GS} = V_G - V_S$$

Substituting for V_S in the above equation yields the following:

$$V_{GS} = V_G - I_D R_S$$

Rearranging the variables and solving for I_D results in the following:

$$I_D = \frac{V_G - V_{GS}}{R_S} \qquad (7\text{-}15)$$

Derivation of a general solution for the voltage-divider bias is similar to the derivation we have done for the self-bias circuit, which begins by equating the above equation for I_D to Shockley's equation. The general solution is given below in Equation 7-16; however, the detail of derivation is left to the student as an exercise.

$$\frac{I_{DSS} R_S}{V_P^2} V_{GS}^2 + \left(\frac{2 I_{DSS} R_S}{|V_P|} + 1 \right) V_{GS} + I_{DSS} R_S - V_G = 0 \qquad (7\text{-}16)$$

The above equation is in the form of the following general quadratic equation:

$$ax^2 + bx + c = 0$$

which has the following general solution:

$$V_{GS}\bigg|_{n-channel} = \frac{-b+\sqrt{b^2-4ac}}{2a} \qquad (7\text{-}17)$$

$$V_{GS}\bigg|_{p-channel} = \frac{+b-\sqrt{b^2-4ac}}{2a} \qquad (7\text{-}18)$$

where,

$$a = \frac{I_{DSS}R_S}{V_P^2} \qquad b = \frac{2I_{DSS}R_S}{|V_P|}+1 \qquad c = I_{DSS}R_S - |V_G|$$

After the V_{GS} is determined with the above equations, I_D and V_{DS} can be determined with Equations 7-15 and 7-8, respectively.

$$I_D = \frac{V_G - V_{GS}}{R_S}$$

$$V_{DS} = V_{DD} - I_D(R_D + R_S)$$

We will now determine the parameters a, b, c, and then V_{GS}, I_D, and V_{DS}.

$$a = \frac{12\,\text{mA} \times 1\,\text{k}\Omega}{3\,\text{V}^2} = \frac{4}{3}$$

$$b = \frac{2 \times 12\,\text{mA} \times 1\,\text{k}\Omega}{3\,\text{V}} + 1 = 9$$

$$c = (12\,\text{mA} \times 1\,\text{k}\Omega) - 2.5 = 9.5$$

$$V_{GS} = \frac{-9 + \sqrt{81 - 4 \times 1.33 \times 9.5}}{\frac{8}{3}} = -1.31\,\text{V}$$

$$I_D = \frac{2.5\,\text{V} - (-1.31\,\text{V})}{1\,\text{k}\Omega} = 3.81\,\text{mA}$$

$V_{DS} = 15\,\text{V} - 3.81\,\text{mA}\,(1.5\,\text{k}\Omega + 1\,\text{k}\Omega) = \mathbf{5.475\,V}$

Let us now plot the graphical solution with MATLAB and verify the algebraic solution. To determine two points for plotting the bias line, the equation of the line is:

$$I_D = \frac{V_G - V_{GS}}{R_S}$$

Let $V_{GS} = V_{GS(off)} = -3\,\text{V} \Rightarrow I_D = 5.5\,\text{mA}$
Let $V_{GS} = V_G = 2.5\,\text{V} \Rightarrow I_D = 0\,\text{mA}$

V_{GS}	I_D (mA)
-3 V	5.5
2.5 V	0

MATLAB commands for plotting the transfer characteristics curve and the bias line:

1. Enter Shockley's equation.
```
» y='12*(1+(x/3))^2';
```
2. Plot the curve.
```
» fplot(y,[-3 0  0 12])
```
3. Draw the grid lines and hold the current plot.

Chapter 7

```
» grid, hold
```
4. Enter the two points on the bias line (−3 V, 5.5 mA) and (+2.5 V, 0 mA).
```
» x1=[-3   2.5], y1=[5.5   0];
```
5. Plot the bias line.
```
» plot(x1,y1)
```
6. Define the ranges of both axes, and print x and y labels.
```
» axis([-4 3  0 12]);
» xlabel('Gate-to-Source Voltage (V)')
» ylabel('Drain Current (mA)')
```

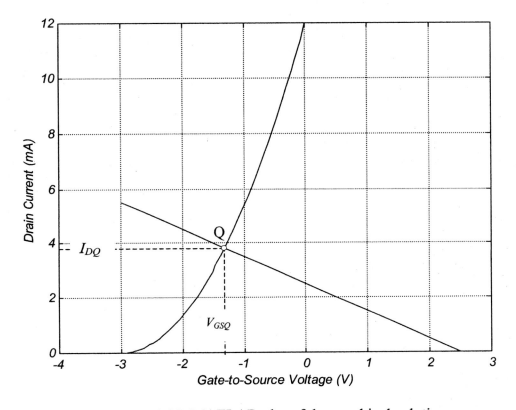

Figure 7-25: MATLAB plot of the graphical solution

Example 7-6 — p-channel JFET — ANALYSIS

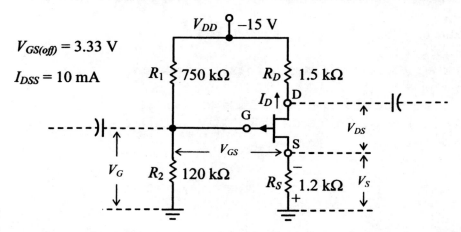

Figure 7-26: Voltage-divider biased *p*-channel JFET amplifier

Once again, analysis of the voltage-divider bias begins with determining V_G, and the rest of the procedure will be similar to the *n*-channel circuit, we just analyzed. The gate voltage V_G is determined from Equation 7-14 as

$$V_G = V_{DD} \frac{R_2}{R_1 + R_2}$$

$$V_G = -15\,\text{V} \frac{120\,\text{k}\Omega}{750\,\text{k}\Omega + 120\,\text{k}\Omega} = -2.07\,\text{V}$$

The voltage V_{GS} will be determined from Equation 7-18

$$V_{GS}\bigg|_{p-channel} = \frac{+b - \sqrt{b^2 - 4ac}}{2a}$$

where

$$a = \frac{I_{DSS} R_S}{V_P^2} \qquad b = \frac{2 I_{DSS} R_S}{V_P} + 1 \qquad c = I_{DSS} R_S - V_G$$

After V_{GS} is determined from the above equations I_D and V_{DS} can be determined from Equations 7-15 and 7-8, respectively. Then,

$$I_D = \frac{V_G - V_{GS}}{R_S}$$

$$V_{DS} = V_{DD} - I_D (R_D + R_S)$$

Applying the results of the above derivations, we can now determine the parameters a, b, c, and then V_{GS}, I_D, and V_{DS}

$$a = \frac{10\,\text{mA} \times 1.2\,\text{k}\Omega}{(3.33\,\text{V})^2} = 1.082$$

$$b = \frac{2 \times 10 \text{ mA} \times 1.2 \text{ k}\Omega}{3.33 \text{ V}} + 1 = 8.2$$

$$c = (10 \text{ mA} \times 1.2 \text{ k}\Omega) - 2.07 = 9.93$$

$$V_{GS} = \frac{+8.2 - \sqrt{(8.2)^2 - 4 \times 1.082 \times 9.93}}{2 \times 1.082} = 1.512 \text{ V}$$

$$I_D = \frac{-2.07 \text{ V} - 1.512 \text{ V}}{1.2 \text{ k}\Omega} = -2.977 \text{ mA} \qquad |I_D| \cong 3\text{mA}$$

$V_{DS} = -15 \text{ V} - [-3 \text{ mA } (1.5 \text{ k}\Omega + 1.2 \text{ k}\Omega)] = \mathbf{-6.938 \text{ V}}$

Let us now verify the algebraically obtained solutions above with the graphical solutions plotted with MATLAB. Note that the absolute value of the current will be considered for plotting, so that the parabola opens upward.

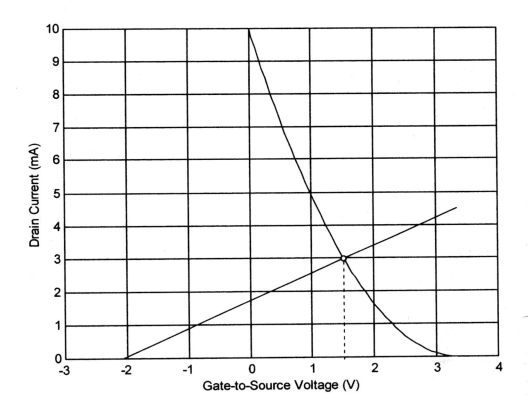

Figure 7-27: MATLAB plot of the graphical solution for Example 7-6

To determine two points for plotting the bias line, the equation of the line is

$$I_D = \frac{V_G - V_{GS}}{R_S}$$

Let $V_{GS} = V_G = -2.07 \text{ V} \Rightarrow I_D = 0 \text{ mA}$.

Let $V_{GS} = V_{GS(off)} = 3.33 \text{ V} \Rightarrow |I_D| = 4.5 \text{ mA}$.

MATLAB commands for plotting the graphical solution for Example 7-6:

Chapter 7

```
» y='10*(1-(x/3.33))^2';
» fplot(y,[0  3.33  0 10])
» grid, hold
» x1=[-2.07  3.33];
» y1=[0  4.5];
» axis([-3 4  0 10]);
» plot(x1,y1)
» xlabel('Gate-to-Source Voltage (V)')
» ylabel('Drain Current (mA)')
»
```

Practice Problem 7-5   ANALYSIS

Determine the V_{GS}, I_D, and V_{DS} for the amplifier circuit of Figure 7-28
a) algebraically
b) graphically, preferably with MATLAB.

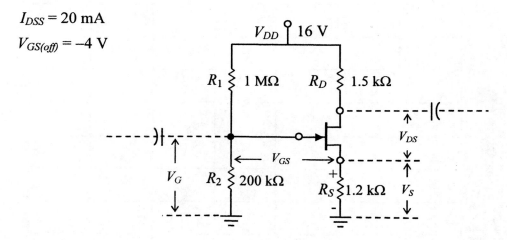

$I_{DSS} = 20$ mA

$V_{GS(off)} = -4$ V

Figure 7-28

Answers:

$V_{GS} = 2.22$ V, $I_D = 4.06$ mA, $V_{DS} = 5.065$ V

Chapter 7

Section Summary — JFET DC-BIAS — ANALYSIS

Summary of Equations for the Analysis of Self-Biased and Voltage-Divider Biased JFET Amplifier

Self-bias:

$$V_{GS}\bigg|_{n-channel} = \frac{-b + \sqrt{b^2 - 4ac}}{2a}$$

$$V_{GS}\bigg|_{p-channel} = \frac{+b - \sqrt{b^2 - 4ac}}{2a}$$

where,

$$a = \frac{I_{DSS} R_S}{V_P^2} \qquad b = \frac{2 I_{DSS} R_S}{|V_P|} + 1 \qquad c = I_{DSS} R_S$$

$$I_D = \frac{-V_{GS}}{R_S}$$

$$V_{DS} = V_{DD} - I_D(R_D + R_S)$$

Voltage-divider bias:

$$V_{GS}\bigg|_{n-channel} = \frac{-b + \sqrt{b^2 - 4ac}}{2a}$$

$$V_{GS}\bigg|_{p-channel} = \frac{+b - \sqrt{b^2 - 4ac}}{2a}$$

where,

$$a = \frac{I_{DSS} R_S}{V_P^2} \qquad b = \frac{2 I_{DSS} R_S}{|V_P|} + 1 \qquad c = I_{DSS} R_S - |V_G|$$

$$I_D = \frac{V_G - V_{GS}}{R_S}$$

$$V_{DS} = V_{DD} - I_D(R_D + R_S)$$

7.2.3 DC Bias Design of the JFET Amplifier

As we have seen in the previous chapters, the BJT amplifier circuit can be designed to be totally β-independent. That is, with a correct design, the changes in the transistor characteristics will not cause significant change in the position of the established operating point. With the FET, however, we can only minimize the shift in the operating point due to changes in transistor characteristics. As demonstrated in Figures 7-22 and 7-23, the stability of the Q-point is directly related to the slope of the bias line, and the optimum slope can be achieved with the voltage-divider bias. Hence, in order to optimize the stability of the circuit, we will employ the voltage-divider bias; and the first step in the design procedure will be to determine the slope of the bias line that will help achieve the intended/specified stability.

Example 7-7 — DESIGN

Let us design the DC bias of a JFET amplifier with the specifications given below:

$$V_{DD} = 16 \text{ V}, \quad I_{DQ} = 3 \text{ mA}, \quad I_{DSS} = 12 \text{ mA}, \quad V_{GS(off)} = -3 \text{ V}$$

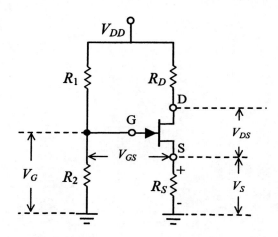

Figure 7-29: Voltage-divider biased JFET amplifier circuit

Stability considerations:
The I_{DQ} is not to shift more than ±10% of its original position for the following changes in the JFET characteristics:
I_{DSS}:   12 mA ± 2 mA
$V_{GS(off)}$: −3 V ± 1 V

Solution:

The transfer characteristics of the JFET and the corresponding variations in the transistor characteristics are plotted with MATLAB and depicted in Figure 7-30.

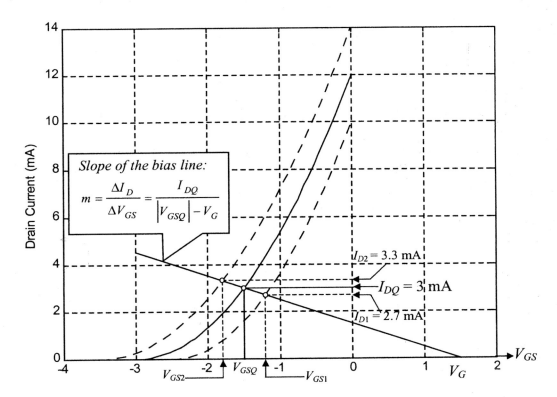

Figure 7-30: MATLAB plot of transfer characteristics curves and the bias line

According to the given specifications, as the transistor characteristics vary from the nominal $I_{DSS} = 12$ mA by ±2 mA, the drain current I_D can vary a maximum of 0.3 mA upward to $I_{D2} = 3.3$ mA or downward to $I_{D1} = 2.7$ mA, a total of 0.6 mA deviation from the $I_{DQ} = 3$ mA. As a result of the changes in I_D, there will be corresponding changes in V_{GS}, which can be determined with Shockley's equation, as follows:

$$I_D = I_{DSS}\left(1 - \frac{V_{GS}}{V_{GS(off)}}\right)^2$$

Dividing both sides of the above equation by I_{DSS} and taking the square root of both sides results in the following:

$$\sqrt{\frac{I_D}{I_{DSS}}} = \left(1 - \frac{V_{GS}}{V_{GS(off)}}\right)$$

Separating the variables and solving for V_{GS} yields the following general solution:

$$V_{GS} = \left(1 - \sqrt{\frac{I_D}{I_{DSS}}}\right) \cdot V_{GS(off)} \qquad (7\text{-}19)$$

With the above equation we can now solve for V_{GSQ}, V_{GS1}, and V_{GS2}, which correspond to I_{DQ}, I_{D1}, and I_{D2}, respectively.

$$V_{GSQ} = \left(1 - \sqrt{\frac{I_{DQ}}{I_{DSS}}}\right) \cdot V_{GS(off)} = \left(1 - \sqrt{\frac{3\,\text{mA}}{12\,\text{mA}}}\right) \times (-3\,\text{V}) = -1.5\,\text{V}$$

$$V_{GS1} = \left(1 - \sqrt{\frac{I_{D1}}{I_{DSS1}}}\right) \cdot V_{GS(off)1} = \left(1 - \sqrt{\frac{2.7\,\text{mA}}{10\,\text{mA}}}\right) \times (-2.5\,\text{V}) = -1.2\,\text{V}$$

$$V_{GS2} = \left(1 - \sqrt{\frac{I_{D2}}{I_{DSS2}}}\right) \cdot V_{GS(off)2} = \left(1 - \sqrt{\frac{3.3\,\text{mA}}{14\,\text{mA}}}\right) \times (-3.5\,\text{V}) = -1.8\,\text{V}$$

As shown in Figure 7-30, the slope of the bias line is:

$$m = \frac{\Delta I_D}{\Delta V_{GS}} = \frac{I_{D2} - I_{D1}}{V_{GS2} - V_{GS1}} \quad (7\text{-}20)$$

$$m = \frac{\Delta I_D}{\Delta V_{GS}} = \frac{3.3\,\text{mA} - 2.7\,\text{mA}}{-1.8\,\text{V} - (-1.2\,\text{V})} = \frac{0.6\,\text{mA}}{-0.6\,\text{V}} = -1\,\frac{\text{mA}}{\text{V}}$$

Hence, the absolute value of the slope is 1 mA per volt.

$$|m| = 1\,\text{mA/V}$$

Returning to Figure 7-30, the slope of the bias line can also be expressed as the following:

$$|m| = \frac{I_{DQ}}{|V_{GSQ}| + V_G} \quad (7\text{-}21)$$

Rearranging the equation and solving for V_G results in the following general solution for V_G:

$$V_G = \frac{I_{DQ}}{|m|} - |V_{GSQ}| \quad (7\text{-}22)$$

Substituting for all variables yields the value for V_G.

$$V_G = \frac{3\,\text{mA}}{1\,\text{mA/V}} - 1.5\,\text{V} = 3\,\text{V} - 1.5\,\text{V} = 1.5\,\text{V}$$

Having determined the V_G and V_{GS}, we can now solve for V_S, as follows:

$$V_{GS} = V_G - V_S$$
$$V_S = V_G - V_{GS} = I_D R_S \quad (7\text{-}23)$$

$$V_S = 1.5\,\text{V} - (-1.5\,\text{V}) = 3\,\text{V}$$

$$R_S = \frac{V_S}{I_D} \quad (7\text{-}24)$$

$$R_S = \frac{3\,\text{V}}{3\,\text{mA}} = 1\,\text{k}\Omega$$

Of the remaining 13 V from the $V_{DD} = 16$ V, we will allocate 7 V to V_{DS} and 6 V to R_D. The general guidelines for allocation of these two voltages are similar to ones we have discussed with BJT bias design.

Hence, the R_D can be determined as follows:

$$R_D = \frac{V_{RD}}{I_D} \quad (7\text{-}25)$$

$$R_D = \frac{6\text{ V}}{3\text{ mA}} = 2\text{ k}\Omega$$

Applying the voltage-divider rule at the gate terminal results in the following equation for the gate voltage

$$V_G = V_{DD} \frac{R_2}{R_1 + R_2}$$

Some algebraic manipulation of the above equation results in the following equation for R_1 in terms of R_2

$$R_1 = \left(\frac{V_{DD}}{V_G} - 1\right) R_2 \qquad (7\text{-}26)$$

$$R_1 = \left(\frac{16\text{ V}}{1.5\text{ V}} - 1\right) R_2 = 9.67 R_2 \text{ or } R_2 = \frac{R_1}{9.67}$$

Next, we will determine R_1 and R_2, by assuming a value for R_2. In contrast to the BJT, there are no restrictions in the value of R_2. Hence, in order to achieve a high input resistance, we can assign a large value for R_2, usually in the range of a few hundred kΩ. Likewise, we can assume a value for R_1, and then determine R_2.

Letting $R_1 = 1.1$ MΩ, $R_2 = 1.1/9.67 = 113.754$ kΩ. Use $R_2 = 115$ k$\Omega \approx 120$ kΩ.

Summary of design computations:

$R_1 = 1.1$ MΩ, $R_2 = 120$ kΩ, $R_S = 1$ kΩ, $R_D = 2$ kΩ,
$I_D = 3$ mA, $V_{DS} = 7$ V, $V_S = 3$ V, $V_G = 1.5$ V, $V_{GS} = -1.5$ V.

Example 7-8 DESIGN

Let us now design the DC bias of another JFET amplifier with somewhat tighter stability specifications as given below:

$V_{DD} = 18$ V
$I_{DQ} = 3$ mA
$I_{DSS} = 14$ mA
$V_{GS(off)} = -3.85$ V
Stability considerations:
I_{DQ} is not to shift more than ± 0.5 mA from its original position for the following changes in the JFET parameters:
$\quad I_{DSS} = 14$ mA ± 3 mA
$\quad V_{GS(off)} = -3.85$ V ± 1 V

Solution:

Let us first clarify the stability specifications, as follows:
Nominal characteristics and the Q-point:

$I_{DSS} = 14$ mA, $V_{GS(off)} = -3.85$ V,

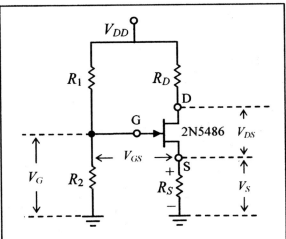

Figure 7-31: Voltage-divider biased JFET amplifier circuit

Chapter 7

$I_{DQ} = 3$ mA

Upward deviation of the characteristics and the limit of the corresponding shift in I_D:

$$I_{DSS2} = 17 \text{ mA}, \quad V_{GS(off)2} = -4.85 \text{ V}, \quad I_{D2} = 3.5 \text{ mA}$$

Downward deviation of the characteristics and the limit of the corresponding shift in I_D:

$$I_{DSS1} = 11 \text{ mA}, \quad V_{GS(off)1} = -2.85 \text{ V}, \quad I_{D1} = 2.5 \text{ mA}$$

The next step in the design process is to determine the slope of the bias line in order to meet the specified stability. Using Equation 7-19, we can determine V_{GSQ}, V_{GS1}, and V_{GS2} that correspond to I_{DQ}, I_{D1}, and I_{D2}, respectively

$$V_{GS} = \left(1 - \sqrt{\frac{I_D}{I_{DSS}}}\right) V_{GS(off)}$$

$$V_{GSQ} = \left(1 - \sqrt{\frac{I_{DQ}}{I_{DSS}}}\right) V_{GS(off)} = \left(1 - \sqrt{\frac{3 \text{ mA}}{14 \text{ mA}}}\right) \times (-3.85 \text{ V}) = -3.025 \text{ V}$$

$$V_{GS1} = \left(1 - \sqrt{\frac{I_{D1}}{I_{DSS1}}}\right) V_{GS(off)1} = \left(1 - \sqrt{\frac{2.5 \text{ mA}}{11 \text{ mA}}}\right) \times (-2.85 \text{ V}) = -1.5 \text{ V}$$

$$V_{GS2} = \left(1 - \sqrt{\frac{I_{D2}}{I_{DSS2}}}\right) V_{GS(off)2} = \left(1 - \sqrt{\frac{3.5 \text{ mA}}{17 \text{ mA}}}\right) \times (-4.85 \text{ V}) = -2.65 \text{ V}$$

The slope of the bias line can be determined from Equation 7-20

$$m = \frac{\Delta I_D}{\Delta V_{GS}} = \frac{I_{D2} - I_{D1}}{V_{GS2} - V_{GS1}}$$

$$m = \frac{\Delta I_D}{\Delta V_{GS}} = \frac{3.5 \text{ mA} - 2.5 \text{ mA}}{-2.65 \text{ V} - (-1.5 \text{ V})} = \frac{1 \text{ mA}}{-1.15 \text{ V}} = -0.87 \frac{\text{mA}}{\text{V}}$$

$$|m| = 0.87 \text{ mA/V}$$

Using Equation 7-22, we can solve for V_G as follows

$$V_G = \frac{I_{DQ}}{|m|} - |V_{GSQ}|$$

$$V_G = \frac{3 \text{ mA}}{0.87 \text{ mA/V}} - 3.025 \text{ V} = 0.423 \text{ V}$$

Next, V_S is determined from Equation 7-23

$$V_S = V_G - V_{GS}$$

$$V_S = 0.423 \text{ V} - (-3.025 \text{ V}) = 3.871 \text{ V}$$

Having determined V_S, R_S can be determined from Equation 7-24

$$R_S = \frac{V_S}{I_D}$$

$$R_S = \frac{3.45 \text{ V}}{3 \text{ mA}} = 1.15 \text{ k}\Omega \quad \text{Pick } R_S = 1.2 \text{ k}\Omega.$$

Of the remaining 14.55 V from $V_{DD} = 18$ V, we will allocate 7 V to V_{DS} and 7.55 V across R_D. Then, R_D can be determined from Equation 7-25

$$R_D = \frac{V_{RD}}{I_D}$$

$$R_D = \frac{7.55 \text{ V}}{3 \text{ mA}} = 2.5 \text{ k}\Omega$$

Pick $R_D = 2.4$ kΩ. Assuming $R_2 = 100$ kΩ, R_1 can be determined from Equation 7-26

$$R_1 = \left(\frac{V_{DD}}{V_G} - 1\right) R_2$$

$$R_1 = \left(\frac{18 \text{ V}}{1.617 \text{ V}} - 1\right) R_2 = 12 R_2 = 1 \text{ M}\Omega$$

Summary of design computations:

$R_1 = 1.2$ MΩ, $R_2 = 100$ kΩ, $R_S = 1.2$ kΩ, $R_D = 2.4$ kΩ,
$I_D = 3$ mA, $V_{DS} = 7$ V, $V_G = 1.617$ V, $V_{GS} = -1.831$ V, $V_S = 3.45$ V.

Test Run and Design Verification:

We will now test the design by simulating the circuit with Electronics Workbench. The result of the simulation is presented in Figure 7-32.

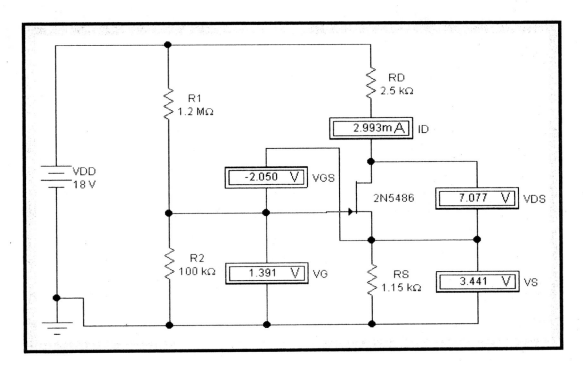

Figure 7-32: Circuit setup with simulation results of Design Example 7-8

Table 7-2: Summary of Results

Parameter	Specification	Expected	Measured
I_{DQ}	3 mA	3 mA	2.99 mA
V_{DSQ}	7 V	7 V	7.07 V
V_{GSQ}	-	−2.06 V	−2.05 V
V_G	-	1.38 V	1.39 V
V_S	-	3.45 V	3.44 V

As you can see, we have practically achieved the specified DC operating point of 3 mA and 7 V. In addition, the specified range of stability is assured because that was the primary objective of the whole design procedure.

Practice Problem 7-6 — n-channel JFET — DESIGN

Design the DC bias of the JFET amplifier of Figure 7-33 to meet the specifications given below:

V_{DD} = 16 V I_{DSS} = 10 mA
I_{DQ} = 3 mA $V_{GS(off)}$ = −3 V
V_{DS} = 7 V

Stability considerations:

The I_{DQ} is not to shift more than ±10% from its original position for the following changes in the JFET characteristics:

I_{DSS} : 10 mA ± 2 mA
$V_{GS(off)}$: −3 V ± 0.5 V

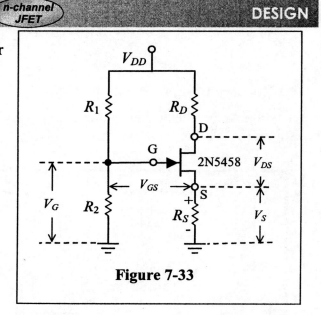

Figure 7-33

Answers: R_1 = 1 MΩ, R_2 = 150 kΩ, R_S = 1 kΩ, R_D = 2 kΩ

Practice Problem 7-7 — n-channel JFET — DESIGN

Design the DC bias of a voltage-divider biased JFET amplifier to meet the specifications given below: V_{DD} = 20 V, I_{DQ} = 4.5 mA, V_{DS} = 7.2 V
   JFET: 2N5670, $V_{GS(off)}$ = −4.5 V, I_{DSS} = 18 mA

Stability considerations:
The I_{DQ} is not to shift more than ±0.5 mA from its original position for the following changes in the JFET characteristics:
   I_{DSS} :   18 mA ± 4 mA $V_{GS(off)}$: −4.5 V ± 1 V

| Answers: | $R_1 = 1.1$ MΩ, | $R_2 = 222$ kΩ, | $R_S = 1.25$ kΩ, | $R_D = 1.6$ kΩ |

7.3 METAL-OXIDE-SEMICONDUCTOR FET (MOSFET)

The principal difference between a metal-oxide-semiconductor FET (MOSFET) and a JFET is that the gate region is physically isolated from the channel region by means of a thin layer of silicon dioxide, eliminating the possibility of a gate current even with a forward biased gate-source junction. For this reason the MOSFET is also referred to as the *insulated-gate* FET or IGFET. There are two types of MOSFETs: the *depletion*-type (D-MOS) and the *enhancement*-type (E-MOS).

7.3.1 Depletion MOSFET

Like the JFET, there are two types of *depletion* MOSFETs: *n*-channel and *p*-channel. The physical structure of an *n*-channel *depletion* MOSFET is illustrated in Figure 7-34. A lightly doped block of *p*-type silicon forms the substrate in which the two embedded heavily doped n^+-type regions form the low-resistance connections to the drain and source terminals. Between the two n^+-type regions is the lightly doped *n*-type region that forms the conducting channel between the drain and source terminals. By oxidizing the silicon, a thin layer of silicon dioxide is formed on top of the *p*-type substrate and the *n*-type region. Ohmic contacts are formed by leaving two windows in the silicon dioxide on top of the two n^+-type regions, which are filled with a layer of metal (aluminum) to provide connections to the source and drain terminals. A layer of aluminum is also deposited on top of the silicon dioxide opposite the *n*-type region that provides connection to the gate terminal. The substrate *B* is usually connected to the source terminal.

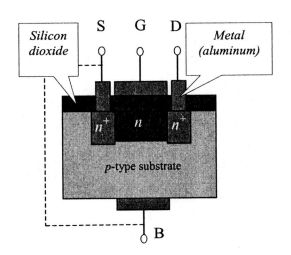

Figure 7-34: Structure of an *n*-channel *depletion* MOSFET

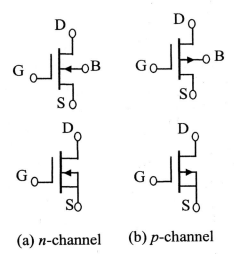

Figure 7-35: Symbols of *n*-channel and *p*-channel *depletion* MOSFETs

Operation of Depletion-Type MOSFET

The application of a small positive voltage V_{DS}, as shown in Figure 7-36(a), causes the flow of current I_D through the channel from drain to source terminal (conventional flow). The resistance of the channel limits the amount of current I_D. Thus, an increase in V_{DS} causes a proportional increase in the drain current according to Ohm's law. However, this direct relationship between V_{DS} and I_D ends at the pinch-off voltage V_P, beyond which a further increase in V_{DS} does not cause significant increase in I_D. That is, the drain current starts leveling off and stays fairly constant at and beyond the pinch-off voltage. The drain current that flows with $V_{DS} = V_P$ (with $V_{GS} = 0$) is called the *drain-to-source saturation current* I_{DSS}. Apparently, the operation of the depletion-type MOSFET is very similar to that of a JFET and has similar characteristics, with the only difference being that there is no reverse-biased *p-n* junction with depletion MOSFET. Hence, with $V_{GS} \geq 0$, there is no formation of a depletion region and the channel width is controlled by the action of the electric field.

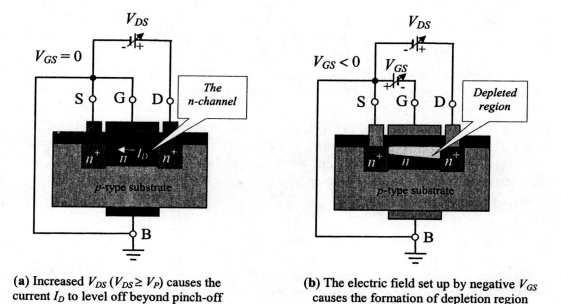

(a) Increased V_{DS} ($V_{DS} \geq V_P$) causes the current I_D to level off beyond pinch-off

(b) The electric field set up by negative V_{GS} causes the formation of depletion region

Figure 7-36: Operation of *n*-channel, *depletion*-type MOSFET

The application of a negative V_{GS}, however, sets up an electric field that repels the charge carriers (electrons) from a portion of the channel beneath the silicon dioxide layer, causing the formation of a non-uniform depletion region, as shown in Figure 7-36(b). The width of the depletion region is directly proportional to the amount of the negative V_{GS} applied at the gate-source terminals. The negative voltage V_{GS} that causes the total depletion of the channel and stops the flow of the drain current completely is called the *gate-to-source cut-off voltage* $V_{GS(off)}$. The voltage $V_{GS(off)}$ is equal to the *pinch-off voltage* V_P in absolute value. That is:

$$|V_{GS(off)}| = V_p$$

Since the gate in depletion MOSFET is insulated from the channel, the gate-to-source voltage V_{GS} can also be positive, in which case more charge carriers are attracted to the channel, thereby enhancing the conductivity of the channel and, as a result, causing an increase in the channel current I_D. This mode of operation of the depletion MOSFET is referred to as the *enhancement mode* operation, whereas the normal mode of operation with negative V_{GS} is called the *depletion mode* operation.

In conclusion, the amount of the drain current I_D that can flow through the channel is controlled by the magnitude of the voltage V_{GS}, which may be positive or negative. The drain and transfer characteristics of an *n*-channel depletion MOSFET, which are similar to the JFET characteristics, are shown in Figure 7-37 below.

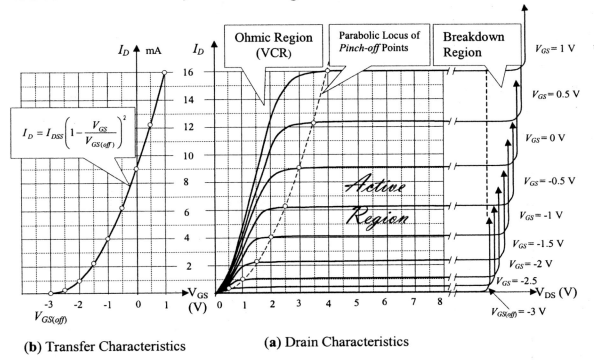

(b) Transfer Characteristics   **(a)** Drain Characteristics

Figure 7-37: Plot of drain and transfer characteristics of *n*-channel *depletion*-type MOSFET

The equation of the *pinch-off* locus parabola and the parabolic transfer characteristic curve of Figure 7-37(b), which is known as *Shockley's equation*, is as follows:

$$I_D = I_{DSS}\left(\frac{V_{GS(off)} - V_{GS}}{V_{GS(off)}}\right)^2 = I_{DSS}\left(1 - \frac{V_{GS}}{V_{GS(off)}}\right)^2 \qquad (7\text{-}27)$$

Like the JFET, the region to the left of the *pinch-off* locus parabola on the *drain characteristics* is called the *ohmic region*, where the resistance between the *drain* and *source* can be controlled by the applied *gate-to-source* voltage V_{GS}. As shown in Figure 7-37(a), the transistor breakdown may occur at excessively large values of V_{DS}, causing a large and sudden increase in the drain current. The region between the *ohmic* and the *breakdown* regions is the *active region*, also known as the *saturation region*; in this region, the device is normally operated to amplify the small signal.

Example 7-9 — depletion MOSFET — ANALYSIS

Let us algebraically determine and verify the drain current at $V_{GS} = \pm 1$ V for the *depletion* MOSFET whose characteristics are shown in Figure 7-37.

Solution:

Refer to Figure 7-37. $I_{DSS} = 9$ mA, and $V_{GS(off)} = -3$ V. The drain current can be determined with Shockley's equation, as follows:

a) At $V_{GS} = -1$ V:

$$I_D = I_{DSS}\left(1 - \frac{V_{GS}}{V_{GS(off)}}\right)^2 = 9\,\text{mA}\left(1 - \frac{-1}{-3}\right)^2 = 4\,\text{mA}$$

b) At $V_{GS} = +1$ V:

$$I_D = I_{DSS}\left(1 - \frac{V_{GS}}{V_{GS(off)}}\right)^2 = 9\,\text{mA}\left(1 - \frac{+1}{-3}\right)^2 = 16\,\text{mA}$$

Returning to Figure 7-37, both of the above answers match the currents on the transfer characteristics that correspond to $V_{GS} = \pm 1$ V. Evidently, Shockley's equation applies equally to the operation of the device in the *depletion* mode (negative V_{GS}) and the *enhancement* mode (positive V_{GS}).

Since the *depletion* MOSFET has characteristics similar to JFET, the DC biasing methods, analysis, and DC bias design of the JFET that we have discussed in detail apply wholly and equally to *depletion* MOSFET. Hence, it would not be worthwhile to repeat the same procedures over again for the *depletion* MOSFET. However, a number of exercises on biasing and DC bias design of *depletion* MOSFET amplifiers is given at the end of the chapter. The approach to solving these problems is the same approach we have used for analyzing or designing the DC bias of JFET amplifiers.

7.3.2 Enhancement MOSFET

The principal difference between the *enhancement*-type MOSFET and the *depletion-type* MOSFET is that with the *depletion* MOSFET a channel is actually built in between the *drain* and the *source*. Recall the somewhat lightly doped *n*-type region stretched between the two heavily doped n^+-type regions that formed the *n*-channel of the *depletion* MOSFET (Figure 7-36). With the *enhancement* MOSFET, however, there is no physical channel built in; in all other ways, the structure is similar to that of *depletion* MOSFET. Like the *depletion* MOSFET, there are two types of *enhancement* MOSFET: *n*-channel and *p*-channel. The structure of an *n*-channel *enhancement* MOSFET is illustrated in Figure 7-38. The substrate is usually connected to the *source*.

Chapter 7

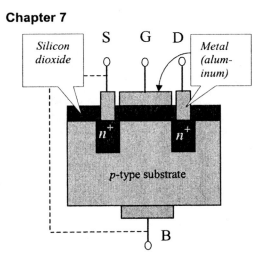

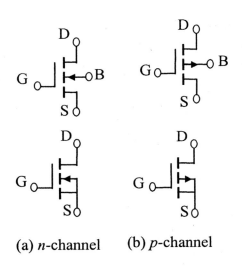

(a) *n*-channel   (b) *p*-channel

Figure 7-38: Structure of an *n*-channel *enhancement* MOSFET

Figure 7-39: Symbols of *n*-channel and *p*-channel *enhancement* MOSFETs

Application of a positive V_{GS}, as shown in Figure 7-40, causes the free holes in the *p*-type substrate to be repelled from the region directly beneath the gate. The absence of the neutralizing holes leaves behind a negatively charged region, which is the first step in the formation of the induced *n*-channel. In addition, the positive gate voltage attracts electrons to the channel region from the heavily doped n^+ regions of *drain* and *source*, where the electrons are in abundance.

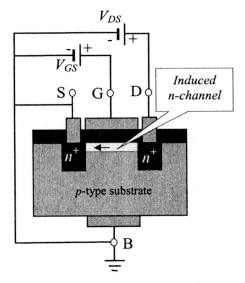

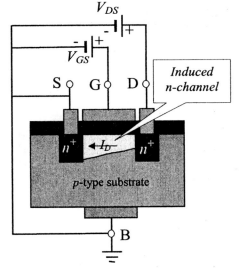

Figure 7-40: *Ohmic* or *VCR* region
An *n*-channel enhancement MOSFET with $V_{GS} > V_t$ and a small V_{DS} applied. A channel is induced and small current I_D flows according to Ohm's law.

Figure 7-41: *Saturation* region
As V_{DS} is increased I_D increases and the induced *n*-channel widens toward the *source*. With a further increase in V_{DS}, channel resistance also increases causing I_D to level off for $V_{DS} \geq V_{GS} - V_t$.

As a result of increased gate voltage, when a sufficient number of electrons are attracted to the channel region beneath the gate, an *n*-channel is in effect established, joining the

Chapter 7

source and *drain* regions (Figure 7-40). The *gate-to-source* voltage V_{GS} that causes the accumulation of a sufficient number of electrons in the channel region to form the conducting channel is called the *threshold voltage* (V_t). Once the channel is established, an increase in V_{DS} corresponds to an increase in I_D up to $V_{DS(sat)} = V_{GS} - V_t$, beyond which a further increase in V_{DS} causes the channel to pinch off and the drain current to level off and remain fairly constant for $V_{DS} \geq V_{GS} - V_t$, as illustrated in Figure 7-41.

An alternate approach to understand the operation of the *enhancement* MOSFET is to view the *gate* electrode and the substrate as the parallel plates of a capacitor separated by the dioxide layer acting as the dielectric. The positive *gate-to-source* voltage V_{GS} causes positive charges to accumulate on the gate electrode, the top plate of the capacitor. The flow of the induced electrons from *drain* and *source* regions to the channel region of the substrate forms the negative charge on the bottom plate of the capacitor. Hence, the MOSFET resembles a charged parallel-plate capacitor with an electric field perpendicular to the gate electrode. Varying the gate voltage V_{GS} causes variations in the electric field. The electric field controls the amount of charge in the channel, and thus the channel conductivity. In other words, assuming that a voltage V_{DS} is applied, the voltage V_{GS} controls the electric field, which controls the channel conductivity, and as a result, the amount of current flow through the channel.

As illustrated in Figure 7-42, with the gate-to-source voltage $V_{GS} > V_t$, initially the current I_D increases linearly with increased V_{DS}. However, like the JFET and *depletion-type* MOSFET, the channel resistance increases and the current starts leveling off with a further increase in V_{DS}.

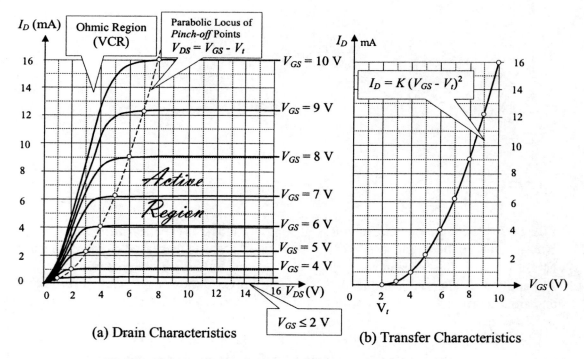

Figure 7-42: Plot of *drain* and *transfer* characteristics of an *n*-channel *enhancement-type* MOSFET

The parabolic transfer characteristics curve is defined by the following equation:

$$I_D = K(V_{GS} - V_t)^2 \quad \text{for } V_{GS} \geq V_t \qquad (7\text{-}28)$$

where K is the conductivity parameter of the device in A/V^2, which depends on the device geometry or *aspect ratio* and the fabrication process constant.
I_D is the drain current, V_t is the threshold voltage, and V_{GS} is the *gate-to-source* voltage.

7.3.3 DC Biasing the Enhancement MOSFET

Since the *n*-channel enhancement MOSFET requires a positive V_{GS}, it may be biased in one of the two methods: voltage-divider bias, and the drain-feedback bias.

Voltage-Divider Bias   ANALYSIS

This bias circuit is similar to the voltage-divider bias circuit of JFET, and the approach to analyze the circuit is about the same. That is, the solution or the operating point is at the intersection of the parabolic transfer characteristics curve and the bias line. Consider the voltage-divider biased circuit of Figure 7-43 below.

$V_t = 2$ V, and
$K = 1$ mA/V^2

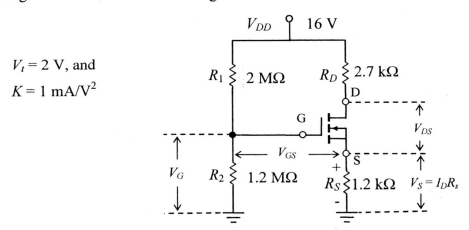

Figure 7-43: Voltage-divider biased *enhancement* MOSFET amplifier circuit

The analysis of the circuit begins with determining the gate voltage V_G by applying the voltage-divider rule at the *gate*, as follows:

$$V_G = V_{DD} \frac{R_2}{R_1 + R_2} \qquad (7\text{-}29)$$

$$V_G = 16\text{ V} \frac{1.2\text{ M}\Omega}{2\text{ M}\Omega + 1.2\text{ M}\Omega} = 6\text{ V}$$

By definition,

$$V_{GS} = V_G - V_S = V_G - I_D R_S$$

Solving for I_D yields the following:

$$I_D = \frac{V_G - V_{GS}}{R_S} \qquad (7\text{-}30)$$

Equating the above equation to Equation 7-28 results in the following:

$$\frac{V_G - V_{GS}}{R_S} = K(V_{GS} - V_t)^2$$

Multiplying both sides by R_S, squaring the right-hand side, and with some more algebraic manipulation we arrive at the following general solution:

$$KR_S V_{GS}^2 + (1 - 2KR_S V_t)V_{GS} + (KR_S V_t^2 - V_G) = 0 \qquad (7\text{-}31)$$

The above equation is in the form of the following general quadratic equation:

$$ax^2 + bx + c = 0$$

which has the following general solution:

$$V_{GS}\Big|_{n-channel} = \frac{-b + \sqrt{b^2 - 4ac}}{2a} \qquad (7\text{-}32)$$

$$V_{GS}\Big|_{p-channel} = \frac{+b - \sqrt{b^2 - 4ac}}{2a} \qquad (7\text{-}33)$$

where,

$$a = KR_S \qquad b = 1 - 2KR_S|V_t| \qquad c = KR_S V_t^2 - |V_G|$$

After the V_{GS} is determined with the above equations, I_D can be determined with Equation 7-30, and V_{DS} can be determined with Equation 7-8 similar to JFET.

$$V_{DS} = V_{DD} - I_D(R_D + R_S)$$

We will now determine the parameters a, b, c, and then V_{GS}, I_D, and V_{DS}.

$a = KR_S = 1\text{ m} \times 1.2\text{ k}\Omega = 1.2$

$b = 1 - 2KR_S|V_t| = 1 - (2 \times 1\text{ m} \times 1.2\text{ k}\Omega \times 2\text{ V}) = -3.8$

$c = KR_S V_t^2 - |V_G| = (1\text{ m} \times 1.2\text{ k}\Omega \times 4) - 6 = -1.2$

$$V_{GS} = \frac{+3.8 + \sqrt{3.8^2 + (4 \times 1.2 \times 1.2)}}{2.4} = 3.456\text{ V}$$

$$I_D = \frac{V_G - V_{GS}}{R_S} = \frac{6\text{ V} - (3.456\text{ V})}{1.2\text{ k}\Omega} = \mathbf{2.12\text{ mA}}$$

$V_{DS} = V_{DD} - I_D(R_D + R_S) = 16\text{ V} - 2.12\text{ mA}(2.7\text{ k}\Omega + 1.2\text{ k}\Omega) = \mathbf{7.732\text{ V}}$

Let us now plot the graphical solution with MATLAB and verify the algebraic solution. To determine two points for plotting the bias line, the equation of the line is:

$$I_D = \frac{V_G - V_{GS}}{R_S}$$

Let $V_{GS} = 0\text{ V} \Rightarrow I_D = 5\text{ mA}$

V_{GS}	I_D (mA)
0 V	5
6 V	0

Chapter 7

Let $V_{GS} = V_G = 6 \text{ V} \Rightarrow I_D = 0 \text{ mA}$

MATLAB commands for plotting the transfer characteristic curve and the bias line:

1. Enter the transfer characteristic equation $I_D = K(V_{GS} - V_t)^2$.
```
» y='1*(x-2)^2';
```
2. Plot the curve in blue.
```
» fplot(y,[2 6 0 16])
```
3. Draw the grid lines and hold the current plot.
```
» grid, hold
```
4. Enter the two points on the bias line (0 V, 5 mA) and (6 V, 0 mA).
```
» x1=[0 6]; y1=[5 0];
```
5. Plot the bias line in red.
```
» plot(x1,y1)
```
6. Define the ranges of both axes.
```
» axis([0 6 0 16]);
```
7. Print x and y labels.
```
» xlabel('Gate-to-Source Voltage (V)')
» ylabel('Drain Current (mA)')
```
8. Print a title for the plot.
```
» title('Graphical Solution Plotted with MATLAB')
```

The graphical solution plotted with MATLAB is presented in Figure 7-44 below.

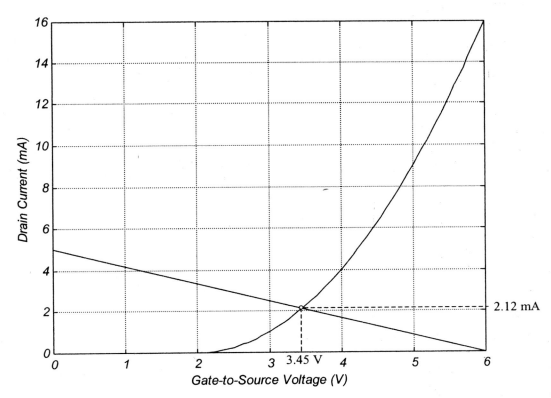

Figure 7-44: MATLAB plot of graphical solution for Figure 7-43

Drain-Feedback Bias — EMOS — ANALYSIS

With the drain-feedback bias, a feedback resistor is connected between the drain and the gate terminals, as shown in Figure 7-45 below.

$V_t = 2$ V, and
$K = 2$ mA/V^2

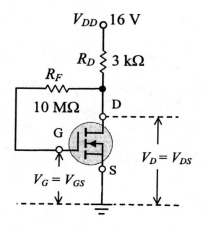

Figure 7-45: Drain-feedback biased *enhancement* MOSFET amplifier circuit

Note that the feedback resistor R_F carries no current, hence, the drain voltage V_D is fed directly to the gate; that is, the drain voltage V_D equals the gate voltage V_G, which equals the gate-to-source voltage V_{GS}.

$$V_{GS} = V_G = V_D = V_{DS} = V_{DD} - I_D R_D \qquad (7\text{-}34)$$

The feedback helps stabilize the Q-point, for instance, if the current I_D increases as a result of variations in the device characteristics, then the voltage V_{GS} decreases. Recall that the current I_D is a function of V_{GS}:

$$I_D = K(V_{GS} - V_t)^2 \qquad (7\text{-}35)$$

Hence, the current I_D will have to decrease because of the decrease in V_{GS}, which was the result of increase in I_D. Now consider the reverse process, that is, assume that I_D decreases as a result of variations in the device characteristics, then the voltage V_{GS} increases, causing the current to increase and stabilize. Evidently, any variation in the Q-point will encounter a counteraction that will immediately reverse the process and maintain a stable operation.

The analysis of the circuit begins with Equation 7-34 above followed by rearranging the equation and solving for I_D, which results in the following:

$$I_D = \frac{V_{DD} - V_{GS}}{R_D} \qquad (7\text{-}36)$$

Equating the two I_D equations (Equations 7-35 and 7-36) yields the following:

Chapter 7

$$\frac{V_{DD} - V_{GS}}{R_D} = K(V_{GS} - V_t)^2$$

By multiplying both sides by R_D, squaring the right-hand side, and with some more algebraic manipulation we arrive at the following general solution:

$$KR_D V_{GS}^2 + (1 - 2KR_D V_t)V_{GS} + (KR_D V_t^2 - V_{DD}) = 0$$

The above equation is in the form of the following general quadratic equation:

$$ax^2 + bx + c = 0$$

which has the following general solution:

$$V_{GS}\Big|_{n-channel} = \frac{-b + \sqrt{b^2 - 4ac}}{2a} \qquad (7\text{-}37)$$

$$V_{GS}\Big|_{p-channel} = \frac{+b - \sqrt{b^2 - 4ac}}{2a} \qquad (7\text{-}38)$$

where,

$$a = KR_D \qquad b = 1 - 2KR_D|V_t| \qquad c = KR_D V_t^2 - |V_{DD}|$$

After V_{GS} is determined with the above equations, I_D can be determined with Equation 7-35 or Equation 7-36, and V_{DS} equals V_{GS}.

$$V_{DS} = V_{GS} \qquad (7\text{-}39)$$

We will now determine the parameters a, b, c, and then V_{GS}, I_D, and V_{DS}.

$a = KR_D = 2\,m \times 3\,k\Omega = 6$

$b = 1 - 2KR_D|V_t| = 1 - (2 \times 2\,m \times 3\,k\Omega \times 2\,V) = -23$

$c = KR_D V_t^2 - V_{DD} = (2\,m \times 3\,k\Omega \times 4) - 16 = 8$

$$V_{GS} = \frac{23 + \sqrt{23^2 - (4 \times 6 \times 8)}}{2 \times 6} = 3.446\,V \cong 3.45\,V$$

$$I_D = \frac{V_{DD} - V_{GS}}{R_D} = \frac{16V - (3.446V)}{3\,k\Omega} = 4.1845\,mA \cong 4.2\,mA$$

$V_{DS} = V_{GS} = 3.446\,V \cong 3.45\,V$

Let us now plot the graphical solution with MATLAB and verify the algebraic solution. To determine two points for plotting the bias line, the equation of the line is:

$$I_D = \frac{V_{DD} - V_{GS}}{R_D} = \frac{16 - V_{GS}}{3\,k\Omega}$$

Let $V_{GS} = 0\,V \Rightarrow I_D = 5.33\,mA$

Let $V_{GS} = 4\,V \Rightarrow I_D = 4\,mA$

V_{GS}	I_D (mA)
0 V	5.33
4 V	4

Chapter 7

MATLAB commands for plotting the transfer characteristics curve and the bias line:

1. Enter the transfer characteristic equation $I_D = K(V_{GS} - V_t)^2$.
```
» y ='2*(x-2)^2';
```
2. Plot the curve in blue.
```
» fplot(y,[2 5  0 18],'b')
```
3. Draw the grid lines and hold the current plot.
```
» grid, hold
```
4. Enter the two points on the bias line (0 V, 5.3 mA) and (4 V, 4 mA)
```
» x1=[0  4]; y1=[5.33   4];
```
5. Plot the bias line in red.
```
» plot(x1,y1)
```
6. Define the ranges of both axes.
```
» axis([0 4  0 8]);
```
7. Print x and y labels.
```
» xlabel('Gate-to-Source Voltage (V)')
» ylabel('Drain Current (mA)')
»
```
The graphical solution plotted with MATLAB is presented in Figure 7-46 below.

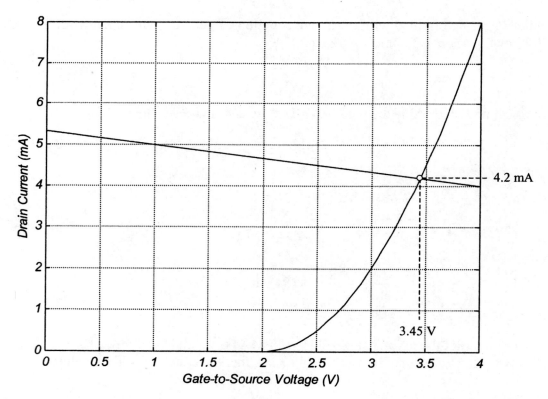

Figure 7-46: MATLAB plot of the graphical solution for the circuit of Figure 7-45

As you can see, the graphical solution is identical to the algebraic solution and the slope of the bias line is almost horizontal, which ensures the stability of the circuit operation.

Practice Problem 7-8

ANALYSIS

Determine V_{GS}, I_D, and V_{DS} for the circuit of Figure 7-47:
a) Algebraically
b) Plot the graphical solution with MATLAB and verify the algebraic solution.

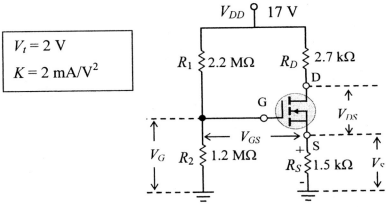

Figure 7-47

Answers:

$I_D = 2$ mA, $V_{GS} = 3$ V, $V_{DS} = 8.6$ V

Practice Problem 7-9

ANALYSIS

Determine V_{GS}, I_D, and V_{DS} for the circuit of Figure 7-48:
a) Algebraically
b) Plot the graphical solution with MATLAB and verify your algebraic solution.

$V_t = 2$ V
$K = 10$ mA/V^2

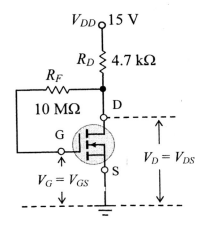

Figure 7-48

Answers:

$V_{GS} = V_{DS} = 2.515$ V, $I_D = 2.869$ mA

Chapter 7

Section Summary — MOSFET DC-Bias — ANALYSIS

Summary of Equations for the Analysis of the MOSFET Amplifier

Depletion MOSFET (Voltage-Divider Bias):

$$V_{GS}\Big|_{n-channel} = \frac{-b + \sqrt{b^2 - 4ac}}{2a} \qquad V_{GS}\Big|_{p-channel} = \frac{+b - \sqrt{b^2 - 4ac}}{2a}$$

where,

$$a = \frac{I_{DSS} R_S}{V_P^2} \qquad b = \frac{2 I_{DSS} R_S}{|V_P|} + 1 \qquad c = I_{DSS} R_S - |V_G|$$

$$I_D = \frac{V_G - V_{GS}}{R_S} \qquad V_{DS} = V_{DD} - I_D(R_D + R_S)$$

Enhancement MOSFET (Voltage-Divider Bias):

$$V_G = V_{DD} \frac{R_2}{R_1 + R_2}$$

$$V_{GS}\Big|_{n-channel} = \frac{-b + \sqrt{b^2 - 4ac}}{2a} \qquad V_{GS}\Big|_{p-channel} = \frac{+b - \sqrt{b^2 - 4ac}}{2a}$$

where,

$$a = K R_S \qquad b = 1 - 2 K R_S |V_t| \qquad c = K R_S V_t^2 - |V_G|$$

$$I_D = \frac{V_G - V_{GS}}{R_S} \qquad V_{DS} = V_{DD} - I_D(R_D + R_S)$$

Enhancement MOSFET (Drain-Feedback Bias):

$$V_{GS}\Big|_{n-channel} = \frac{-b + \sqrt{b^2 - 4ac}}{2a} \qquad V_{GS}\Big|_{p-channel} = \frac{+b - \sqrt{b^2 - 4ac}}{2a}$$

where,

$$a = K R_D \qquad b = 1 - 2 K R_D |V_t| \qquad c = K R_D V_t^2 - |V_{DD}|$$

$$V_{DS} = V_{GS} \qquad I_D = \frac{V_{DD} - V_{GS}}{R_D}$$

7.3.4 DC Bias Design of the MOSFET Amplifier

The approach to designing the DC bias of the enhancement MOSFET amplifier is similar to that of the depletion MOSFET and JFET amplifiers; that is, the slope of the bias line is a major factor in maintaining the amplifier stability. The only difference will be the parameters that define the transistor characteristics, such as K and V_t or $V_{GS(th)}$, instead of I_{DSS} and $V_{GS(off)}$. The following design example demonstrates the whole design process.

Example 7-10 — DESIGN

Let us design the DC bias circuit of the *n*-channel enhancement MOSFET amplifier of Figure 7-49 to meet the specifications given below:

$V_{DD} = 21$ V
$I_{DQ} = 3$ mA
$V_{DSQ} = 6$ V

MOSFET characteristics:

$V_{GS(th)} = V_t = 3$ V
$K = 110$ mA/V^2

Stability considerations:
As the MOSFET characteristics vary by ± 20%, the I_{DQ} is not to shift more than ±10% from its original position.

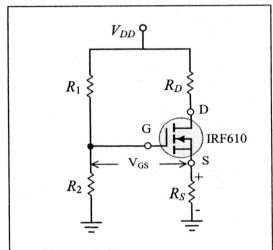

Figure 7-49:
A voltage-divider biased enhancement MOSFET amplifier

Solution:

We will begin the process with Equation 7-35, which defines the drain current.
$$I_D = K(V_{GS} - V_t)^2$$

Solving for V_{GS} in the above equation results in the following:

$$V_{GS} = V_t + \sqrt{\frac{I_D}{K}} \qquad (7\text{-}40)$$

Substituting for all variables in the above equation yields V_{GSQ}.

Nominal characteristics and the Q-point are as follows:
$\quad K = 110$ mA/V$^2 \qquad V_{GS(th)} = V_t = 3$ V $\qquad I_{DQ} = 3$ mA

Upward deviation of the characteristics and the limit of the corresponding shift in I_D:
$\quad K_2 = 132$ mA/V$^2 \qquad V_{t2} = 3.6$ V $\qquad I_{D2} = 3.3$ mA

Downward deviation of the characteristics and the limit of the corresponding shift in I_D:
$\quad K_1 = 88$ mA/V$^2 \qquad V_{t1} = 2.4$ V $\qquad I_{D1} = 2.7$ mA

With Equation 7-40 we can now determine the V_{GSQ}, V_{GS1}, and V_{GS2} that correspond to I_{DQ}, I_{D1}, and I_{D2}, respectively.

$$V_{GSQ} = V_t + \sqrt{\frac{I_D}{K}} = 3\,\text{V} + \sqrt{\frac{3\,\text{mA}}{110\,\text{mA/V}^2}} = 3.165\,\text{V}$$

$$V_{GS2} = V_{t_2} + \sqrt{\frac{I_{D2}}{K_2}} = 3.6\,\text{V} + \sqrt{\frac{3.3\,\text{mA}}{132\,\text{mA/V}^2}} = 3.758\,\text{V}$$

$$V_{GS1} = V_{t_1} + \sqrt{\frac{I_{D1}}{K_1}} = 2.4\,\text{V} + \sqrt{\frac{2.7\,\text{mA}}{88\,\text{mA/V}^2}} = 2.575\,\text{V}$$

The slope of the bias line can be determined with Equation 7-20.

$$|m| = \frac{\Delta I_D}{\Delta V_{GS}} = \frac{I_{D2} - I_{D1}}{V_{GS2} - V_{GS1}}$$

$$|m| = \frac{\Delta I_D}{\Delta V_{GS}} = \frac{3.3\,\text{mA} - 2.7\,\text{mA}}{3.758\,\text{V} - 2.575\,\text{V}} = \frac{0.6\,\text{mA}}{1.183\,\text{V}} = 0.507\,\frac{\text{mA}}{\text{V}}$$

A plot of the transfer characteristics and the bias line of a voltage-divider biased E-MOS circuit is shown below from which the slope of the bias line can be determined, as follows:

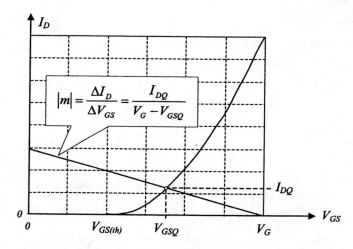

Figure 7-50: Determining the slope of the bias line

With reference to Figure 7-50, the slope of the bias line can also be expressed as follows:

$$|m| = \frac{I_{DQ}}{V_G - V_{GSQ}} \quad (7\text{-}41)$$

Solving for V_G in the above equation yields the following:

$$|V_G| = \frac{I_{DQ}}{|m|} + |V_{GSQ}| \quad (7\text{-}42)$$

$$V_G = \frac{3\,\text{mA}}{0.507\,\text{mA/V}} + 3.165\,\text{V} = 9.08\,\text{V}$$

Chapter 7

By definition,

$$V_{GS} = V_G - V_S$$
$$V_S = V_G - V_{GS}$$

$V_S = 9.08 \text{ V} - 3.165 \text{ V} = 5.915 \text{ V}$

R_S can be determined with Equation 7-24.

$$R_S = \frac{V_S}{I_D}$$

$$R_S = \frac{5.915 \text{ V}}{3 \text{ mA}} = 1.97 \text{ k}\Omega \quad \text{use } R_S = 2 \text{ k}\Omega$$

Applying KVL around the output path and solving for the voltage drop across R_D results in the following:

$$|V_{RD}| = |V_{DD}| - |V_S + V_{DS}|$$

$V_{RD} = 21 - (6 \text{ V} + 6 \text{ V}) = 9 \text{ V}$

R_D can be determined with Equation 7-25, as follows:

$$R_D = \frac{V_{RD}}{I_D}$$

$$R_D = \frac{9 \text{ V}}{3 \text{ mA}} = 3 \text{ k}\Omega$$

Assuming $R_2 = 1 \text{ M}\Omega$, R_1 can be determined with Equation 7-26, as follows:

$$R_1 = \left(\frac{V_{DD}}{V_G} - 1\right) \cdot R_2$$

$$R_1 = \left(\frac{21 \text{ V}}{9.08 \text{ V}} - 1\right) \cdot R_2 = 1.31 R_2 = 1.31 \text{ M}\Omega \quad \text{Use } R_1 = 1.3 \text{ M}\Omega$$

Summary of design computations:

$R_1 = 1.3 \text{ M}\Omega$	$R_2 = 1 \text{ M}\Omega$	$R_S = 2 \text{ k}\Omega$	$R_D = 3 \text{ k}\Omega$
$I_D = 3 \text{ mA}$	$V_{DS} = 6 \text{ V}$	$V_G = 9.08 \text{ V}$	$V_{GS} = 3.16 \text{ V}$

Test Run and Design Verification:

We will now test the design by simulating the circuit with Electronics Workbench. The result of the simulation is presented in Figure 7-51.

Chapter 7

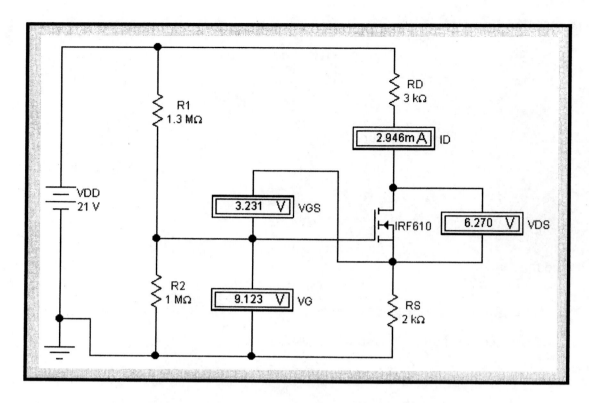

Figure 7-51: Circuit setup with simulation results of Design Example 7-10

Table 7-3: Summary of Results:

Parameter	Specification	Expected	Measured
I_{DQ}	3 mA	3 mA	2.95 mA
V_{DSQ}	6 V	6 V	6.27 V
V_{GSQ}	-	3.16 V	3.23 V
V_G	-	9.08 V	9.12 V

Practice Problem 7-10 — DESIGN

Design the DC bias of a voltage-divider biased MOSFET amplifier circuit to meet the specifications given below:

$$V_{DD} = 20 \text{ V}, \quad I_{DQ} = 3 \text{ mA}, \quad V_{DSQ} = 7.4 \text{ V}$$

MOSFET: IRF610, $V_{GS(th)} = V_t = 3$ V, $K = 110$ mA/V^2

Stability considerations:

As the MOSFET characteristics vary by ± 20%, the I_{DQ} is not to shift more than ±0.5 mA from its original position.

Answers: $R_S = 1.2$ kΩ, $R_D = 3$ kΩ, $R_1 = 910$ kΩ, $R_2 = 470$ kΩ

7.4 SUMMARY

- Like the BJT, the field-effect transistor (FET) is a three-terminal device, but it operates under different principles than the BJT.

- BJT is a bipolar semiconductor device because its current flow depends on both types of charge carriers: majority and minority carriers, or holes and electrons. The FET, however, is a unipolar semiconductor device in which the current flow depends on only one kind of charge carrier: the majority carrier, either holes or electrons.

- The field-effect transistor is so named because its output current depends on the electric field that is set up within the device by an external voltage applied at the input terminal. Hence, the FET is a voltage-controlled device, whereas the BJT is a current-controlled device in which the output current depends on the input current.

- There are two FET types: junction field-effect transistor (JFET), and metal-oxide-semiconductor field-effect transistor (MOSFET). Both types have applications as discrete components and as building blocks of integrated circuits; however, MOSFET is becoming more popular in both areas.

- MOSFET is currently the device of choice in very-large-scale integrated (VLSI) circuits such as microprocessors and memory chips.

- The voltage V_{DS} at which the current levels off (with $V_{GS} = 0$) is referred to as the *pinch-off voltage* V_P. The maximum current I_D that flows through the channel with $V_{DS} \geq V_P$ and with $V_{GS} = 0$ is called the *drain-source saturation current* (I_{DSS}).

- The voltage V_{GS} that causes the total depletion of the channel and stops the current flow is referred to as the *gate-to-source cutoff voltage* ($V_{GS(off)}$), which is equal to the *pinch-off voltage* V_P in absolute value.

- The equation of the *pinch-off* locus parabola and the parabolic transfer characteristic curve [Figure 7-37(b)] is known as *Shockley's equation*.

- The region to the left of the *pinch-off* locus parabola on the drain characteristics is called the *ohmic region*, where the resistance between drain and source can be controlled by the applied voltage V_{GS}.

- The transistor *breakdown* may occur at excessively large values of V_{DS} causing a large and sudden increase in the drain current [Figure 7-37(a)].

- The region between the *ohmic* and the *breakdown* regions is the *active region*, also known as the *saturation region*; in this region, the device is normally operated to amplify the small signal.

Chapter 7

- The principal difference between a metal-oxide-semiconductor FET (MOSFET) and a JFET is that the gate region is physically isolated from the channel region by means of a thin layer of silicon dioxide, eliminating the possibility of a gate current even with a forward biased gate-source junction. There are two types of MOSFETs: the *depletion*-type (D-MOS) and the *enhancement*-type (E-MOS).

- The operation of the *depletion*-type MOSFET is very similar to that of a JFET and has similar characteristics, with the only difference being that there is no reverse-biased *p-n* junction with depletion MOSFET. Hence, with $V_{GS} \geq 0$, there is no formation of a depletion region and the channel width is controlled by the action of the electric field.

- The principal difference between the *enhancement* type MOSFET and the *depletion*-type MOSFET is that with the *depletion* MOSFET a channel is actually built in between the *drain* and the *source*. With the *enhancement*-type MOSFET, however, there is no physical channel built in; in all other ways, the structure is similar to that of *depletion* MOSFET.

- The *gate-to-source* voltage V_{GS} that causes the accumulation of a sufficient number of electrons in the channel region to form the conducting channel is called the *threshold voltage* $V_{GS(th)}$ or V_t.

- The most popular configuration that provides the most stable operation for both the JFET and MOSFET amplifiers is a correctly designed voltage-divider bias, in which the slope of the bias line determines the level of stability.

- Analysis of the JFET and MOSFET amplifiers may be accomplished both algebraically and graphically. For an accurate graphical solution, it is best to use computer software such as MATLAB.

7.5 DEVICE SPECIFICATIONS AND DATA SHEETS

Data sheets of the following devices discussed in this chapter are included for your reference and convenience.

- MPF102 N-Channel JFET
  Copyright of Semiconductor Component Industries, LLC. Used by permission.
- 2N7000 N-Channel Enhancement MOSFET
  Copyright of Semiconductor Components Industries, LLC. Used by permission.

MOTOROLA
SEMICONDUCTOR TECHNICAL DATA

Order this document by MPF102/D

JFET VHF Amplifier
N–Channel — Depletion

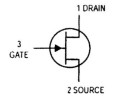

MPF102

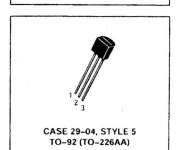

CASE 29–04, STYLE 5
TO–92 (TO–226AA)

MAXIMUM RATINGS

Rating	Symbol	Value	Unit
Drain–Source Voltage	V_{DS}	25	Vdc
Drain–Gate Voltage	V_{DG}	25	Vdc
Gate–Source Voltage	V_{GS}	–25	Vdc
Gate Current	I_G	10	mAdc
Total Device Dissipation @ T_A = 25°C Derate above 25°C	P_D	350 2.8	mW mW/°C
Junction Temperature Range	T_J	125	°C
Storage Temperature Range	T_{stg}	–65 to +150	°C

ELECTRICAL CHARACTERISTICS (T_A = 25°C unless otherwise noted)

Characteristic	Symbol	Min	Max	Unit
OFF CHARACTERISTICS				
Gate–Source Breakdown Voltage (I_G = –10 μAdc, V_{DS} = 0)	$V_{(BR)GSS}$	–25	—	Vdc
Gate Reverse Current (V_{GS} = –15 Vdc, V_{DS} = 0) (V_{GS} = –15 Vdc, V_{DS} = 0, T_A = 100°C)	I_{GSS}	— —	–2.0 –2.0	nAdc μAdc
Gate–Source Cutoff Voltage (V_{DS} = 15 Vdc, I_D = 2.0 nAdc)	$V_{GS(off)}$	—	–8.0	Vdc
Gate–Source Voltage (V_{DS} = 15 Vdc, I_D = 0.2 mAdc)	V_{GS}	–0.5	–7.5	Vdc
ON CHARACTERISTICS				
Zero–Gate–Voltage Drain Current[1] (V_{DS} = 15 Vdc, V_{GS} = 0 Vdc)	I_{DSS}	2.0	20	mAdc
SMALL–SIGNAL CHARACTERISTICS				
Forward Transfer Admittance[1] (V_{DS} = 15 Vdc, V_{GS} = 0, f = 1.0 kHz) (V_{DS} = 15 Vdc, V_{GS} = 0, f = 100 MHz)	$|y_{fs}|$	2000 1600	7500 —	μmhos
Input Admittance (V_{DS} = 15 Vdc, V_{GS} = 0, f = 100 MHz)	$Re(y_{is})$	—	800	μmhos
Output Conductance (V_{DS} = 15 Vdc, V_{GS} = 0, f = 100 MHz)	$Re(y_{os})$	—	200	μmhos
Input Capacitance (V_{DS} = 15 Vdc, V_{GS} = 0, f = 1.0 MHz)	C_{iss}	—	7.0	pF
Reverse Transfer Capacitance (V_{DS} = 15 Vdc, V_{GS} = 0, f = 1.0 MHz)	C_{rss}	—	3.0	pF

1. Pulse Test: Pulse Width ≤ 630 ms, Duty Cycle ≤ 10%.

© Motorola, Inc. 1997

MOTOROLA
SEMICONDUCTOR TECHNICAL DATA

Order this document
by 2N7000/D

TMOS FET Transistor
N–Channel — Enhancement

2N7000
Motorola Preferred Device

CASE 29–04, STYLE 22
TO–92 (TO–226AA)

MAXIMUM RATINGS

Rating	Symbol	Value	Unit
Drain Source Voltage	V_{DSS}	60	Vdc
Drain–Gate Voltage (R_{GS} = 1.0 MΩ)	V_{DGR}	60	Vdc
Gate–Source Voltage — Continuous — Non–repetitive ($t_p \leq 50$ μs)	V_{GS} V_{GSM}	±20 ±40	Vdc Vpk
Drain Current Continuous Pulsed	I_D I_{DM}	200 500	mAdc
Total Power Dissipation @ T_C = 25°C Derate above 25°C	P_D	350 2.8	mW mW/°C
Operating and Storage Temperature Range	T_J, T_{stg}	–55 to +150	°C

THERMAL CHARACTERISTICS

Characteristic	Symbol	Max	Unit
Thermal Resistance, Junction to Ambient	$R_{\theta JA}$	357	°C/W
Maximum Lead Temperature for Soldering Purposes, 1/16″ from case for 10 seconds	T_L	300	°C

ELECTRICAL CHARACTERISTICS (T_C = 25°C unless otherwise noted)

Characteristic	Symbol	Min	Max	Unit
OFF CHARACTERISTICS				
Drain–Source Breakdown Voltage (V_{GS} = 0, I_D = 10 μAdc)	$V_{(BR)DSS}$	60	—	Vdc
Zero Gate Voltage Drain Current (V_{DS} = 48 Vdc, V_{GS} = 0) (V_{DS} = 48 Vdc, V_{GS} = 0, T_J = 125°C)	I_{DSS}	— —	1.0 1.0	μAdc mAdc
Gate–Body Leakage Current, Forward (V_{GSF} = 15 Vdc, V_{DS} = 0)	I_{GSSF}	—	–10	nAdc
ON CHARACTERISTICS[1]				
Gate Threshold Voltage (V_{DS} = V_{GS}, I_D = 1.0 mAdc)	$V_{GS(th)}$	0.8	3.0	Vdc
Static Drain–Source On–Resistance (V_{GS} = 10 Vdc, I_D = 0.5 Adc) (V_{GS} = 4.5 Vdc, I_D = 75 mAdc)	$r_{DS(on)}$	— —	5.0 6.0	Ohm
Drain–Source On–Voltage (V_{GS} = 10 Vdc, I_D = 0.5 Adc) (V_{GS} = 4.5 Vdc, I_D = 75 mAdc)	$V_{DS(on)}$	— —	2.5 0.45	Vdc

1. Pulse Test: Pulse Width ≤ 300 μs, Duty Cycle ≤ 2.0%.

Preferred devices are Motorola recommended choices for future use and best overall value.

REV 3

 MOTOROLA

© Motorola, Inc. 1997

2N7000

ELECTRICAL CHARACTERISTICS (T_C = 25°C unless otherwise noted) (Continued)

Characteristic		Symbol	Min	Max	Unit
ON CHARACTERISTICS[1] (continued)					
On–State Drain Current (V_{GS} = 4.5 Vdc, V_{DS} = 10 Vdc)		$I_{d(on)}$	75	—	mAdc
Forward Transconductance (V_{DS} = 10 Vdc, I_D = 200 mAdc)		g_{fs}	100	—	µmhos
DYNAMIC CHARACTERISTICS					
Input Capacitance	(V_{DS} = 25 V, V_{GS} = 0, f = 1.0 MHz)	C_{iss}	—	60	pF
Output Capacitance		C_{oss}	—	25	
Reverse Transfer Capacitance		C_{rss}	—	5.0	
SWITCHING CHARACTERISTICS[1]					
Turn–On Delay Time	(V_{DD} = 15 V, I_D = 500 mA, R_{gen} = 25 ohms, R_L = 25 ohms)	t_{on}	—	10	ns
Turn–Off Delay Time		t_{off}	—	10	

1. Pulse Test: Pulse Width ≤ 300 µs, Duty Cycle ≤ 2.0%.

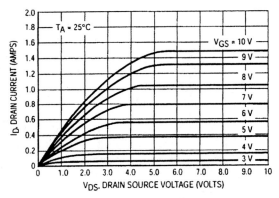

Figure 1. Ohmic Region

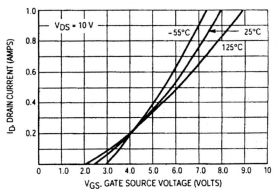

Figure 2. Transfer Characteristics

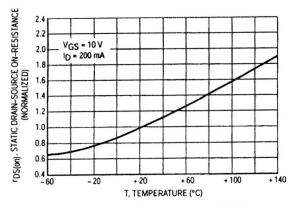

Figure 3. Temperature versus Static Drain–Source On–Resistance

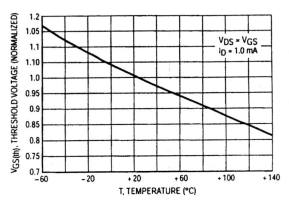

Figure 4. Temperature versus Gate Threshold Voltage

Chapter 7

PROBLEMS

Junction FET — ANALYSIS

7.1 A JFET is known to have a $V_{GS(off)} = -3$ V, and $I_{DSS} = 12$ mA. Determine the drain current I_D with $V_{GS} = -2$ V.

7.2 Carry out Practice Problem 7-1.

7.3 Using computer software such as MATLAB or EXCEL, plot the transfer characteristics of the JFET given in problem 7.1.

7.4 Carry out Practice Problem 7-2.

7.5 Determine the V_{GS}, I_D, and V_{DS} for the circuit of Figure 7-1P.

7.6 Determine the V_{GS}, I_D, and V_{DS} for the circuit of Figure 7-2P.

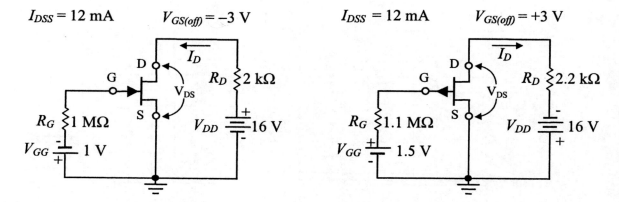

Figure 7-1P: A fixed bias circuit   **Figure 7-2P**: A fixed bias circuit

7.7 Algebraically determine the V_{GS}, I_D, and V_{DS} for the circuit of Figure 7-3P.

7.8 Graphically determine the V_{GS}, I_D, and V_{DS} for the circuit of Figure 7-3P.

7.9 Algebraically determine the V_{GS}, I_D, and V_{DS} for the circuit of Figure 7-4P.

7.10 Graphically determine the V_{GS}, I_D, and V_{DS} for the circuit of Figure 7-4P.

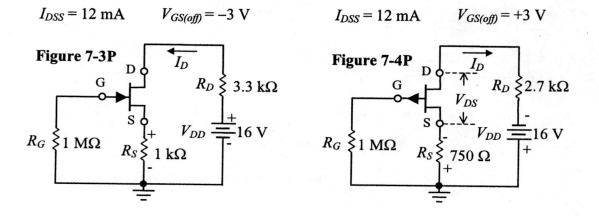

7.11 Carry out Practice Problem 7-3.

7.12 Carry out Practice Problem 7-4.

7.13 Algebraically determine the V_{GS}, I_D, and V_{DS} for the circuit of Figure 7-5P.

7.14 Graphically determine the V_{GS}, I_D, and V_{DS} for the circuit of Figure 7-5P.

7.15 Determine the V_{GS}, I_D, and V_{DS} for the circuit of Figure 7-5P by simulation.

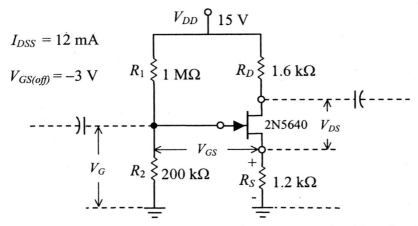

Figure 7-5P: A voltage-divider biased JFET amplifier circuit

7.16 Carry out Practice Problem 7-5.

7.17 Algebraically determine the V_{GS}, I_D, and V_{DS} for the circuit of Figure 7-6P.

7.18 Graphically determine the V_{GS}, I_D, and V_{DS} for the circuit of Figure 7-6P.

7.19 Determine the V_{GS}, I_D, and V_{DS} for the circuit of Figure 7-6P by simulation.

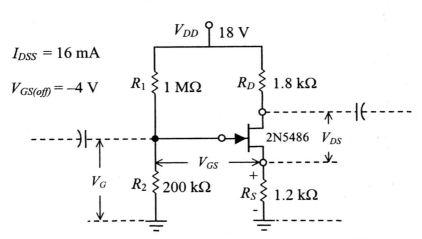

Figure 7-6P: A voltage-divider biased JFET amplifier circuit

7.20 Algebraically determine the V_{GS}, I_D, and V_{DS} for the circuit of Figure 7-7P.

7.21 Graphically determine the V_{GS}, I_D, and V_{DS} for the circuit of Figure 7-7P.

7.22 Determine the V_{GS}, I_D, and V_{DS} for the circuit of Figure 7-7P by simulation.

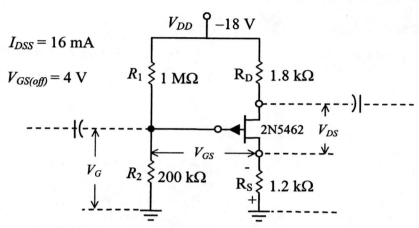

Figure 7-7P: A voltage-divider biased JFET amplifier circuit

Junction FET — DESIGN

7.23 Carry out Practice Problem 7-6.

7.24 Carry out Practice Problem 7-7.

7.25 Design the DC bias of the JFET amplifier circuit of Figure 7-8P for the specifications as given below:

V_{DD} = 18 V

V_{DS} = 8 V, I_{DQ} = 3 mA

I_{DSS} = 12 mA

$V_{GS(off)}$ = −3 V

Stability considerations:

The I_{DQ} not to shift more than ±0.5 mA from its original position as the JFET characteristics change by ±25%.

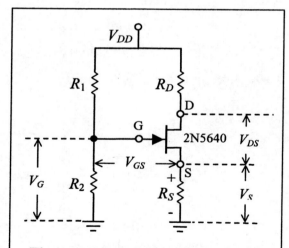

Figure 7-8P: Voltage-divider biased JFET amplifier circuit

Depletion MOSFET — ANALYSIS

7.26 Algebraically determine the V_{GS} and I_D, and V_{DS} for the circuit of Figure 7-9P.

7.27 Graphically determine V_{GS}, I_D, and V_{DS} for the circuit of Figure 7-9P.

7.28 Algebraically determine V_{GS}, I_D, and V_{DS} for the circuit of Figure 7-10P.

7.29 Graphically determine V_{GS}, I_D, and V_{DS} for the circuit of Figure 7-10P.

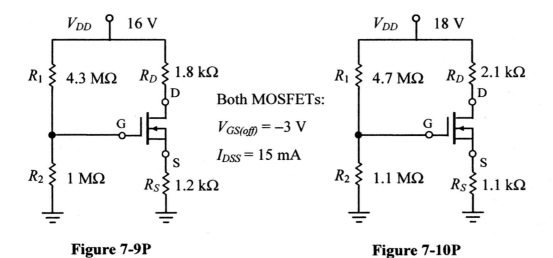

Figure 7-9P   Figure 7-10P

Section 7.3.2 Enhancement MOSFET — ANALYSIS

7.30 Carry out Practice Problem 7-8.

7.31 Algebraically determine V_{GS}, I_D, and V_{DS} for the circuit of Figure 7-11P.

7.32 Graphically determine V_{GS}, I_D, and V_{DS} for the circuit of Figure 7-11P.

7.33 Algebraically determine V_{GS}, I_D, and V_{DS} for the circuit of Figure 7-12P.

7.34 Graphically determine V_{GS}, I_D, and V_{DS} for the circuit of Figure 7-12P.

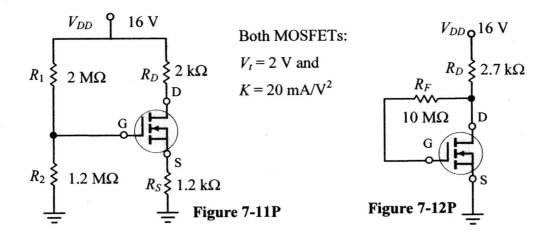

Figure 7-11P   Figure 7-12P

7.35 Carry out Practice Problem 7-9.

Chapter 7

Enhancement MOSFET — DESIGN

7.36 Design the DC bias of a voltage-divider biased EMOS amplifier for the specifications given below:

$V_{DD} = 20$ V

$V_{DS} = 8$ V, $I_{DQ} = 2.5$ mA

$K = 110$ mA/V^2

$V_t = 3$ V

Stability considerations:

The I_{DQ} is not to shift more than ±0.5 mA from its original position as the MOSFET characteristics change by ±25%.

7.37 Carry out Practice Problem 7-10.

Chapter 8

Chapter

SMALL-SIGNAL OPERATION OF THE FET AMPLIFIER

Analysis & Design

8.1 INTRODUCTION

Having discussed the basic theory of the field-effect transistors, followed by an in-depth and comprehensive analysis and design of the DC bias of JFET and MOSFET amplifiers in the previous chapter, we now wish to explore the small-signal operation of JFET and MOSFET amplifiers. The general amplifier fundamentals that we have discussed in Chapter 5 (section 5.2) with BJT amplifiers are wholly and equally applicable to the FET amplifiers. Because field-effect transistors are voltage-controlled devices and thus require no input current in order to operate, they offer extremely high input resistance; ideally, infinite. Consequently, JFET and MOSFET amplifiers can offer a very high input resistance; however, they do not offer the same high voltage gain that is expected from BJT amplifiers. The lower voltage gain seems to be the tradeoff for the high input resistance with FET amplifiers, although a relatively high voltage gain can be achieved with some MOSFETs.

8.2 SMALL-SIGNAL CIRCUIT MODEL OF FET

As mentioned previously, the FET is a voltage-controlled device. However, to be more specific, the FET may be viewed as a voltage-controlled current source. That is, the application of a small-signal v_{gs} across the gate-source terminals provides a current $g_m v_{gs}$ at the drain terminal, where g_m is the transconductance of the transistor as defined with Equation 8-1 and realized with Equation 8-2. Because the FET requires no input current in order to operate as a current source, the input resistance of this voltage-controlled current source is very high, nearly infinite. Hence, the small-signal equivalent circuit of the FET may be modeled as the current source shown in Figure 8-1 below. Note that this model is equally applicable to JFET and MOSFET.

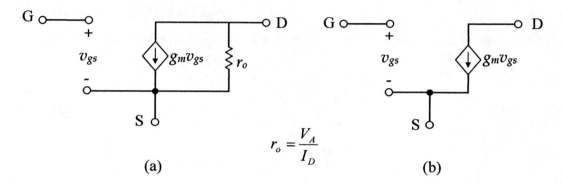

Figure 8-1: Small-signal low-frequency models of FET
(a) Exact, r_o considered. (b) Approximate, r_o neglected.

The Transconductance g_m

As we have seen in the previous chapter, the drain current I_D is a function of V_{GS}, and changes from a minimum of zero to a maximum of I_{DSS} as V_{GS} changes from $V_{GS(off)}$ to zero, respectively. One important small-signal parameter of JFET and MOSFET is the transconductance g_m, which is the slope of the transfer characteristics curve at the operating point. Mathematically, g_m is defined as follows:

$$g_m = \left.\frac{dI_D}{dV_{GS}}\right|_{V_{DS} = constant} \tag{8-1}$$

where I_D is defined by Shockley's equation for both the JFET and *depletion*-type MOSFET, as follows:

$$I_D = I_{DSS}\left(1 - \frac{V_{GS}}{V_{GS(off)}}\right)^2$$

Differentiating the above equation with respect to V_{GS} results in the following:

$$g_m = \frac{dI_D}{dV_{GS}} = 2I_{DSS}\left(1 - \frac{V_{GS}}{V_{GS(off)}}\right)\left(\frac{-1}{V_{GS(off)}}\right) = \frac{-2I_{DSS}}{V_{GS(off)}}\left(1 - \frac{V_{GS}}{V_{GS(off)}}\right)$$

Chapter 8

Since $V_{GS(off)}$ will be negative for an *n*-channel FET, it will cancel out the negative sign in the numerator. Hence, the equation of g_m for JFET and *depletion*-type MOSFET can be expressed as follows:

$$g_m = \frac{2I_{DSS}}{|V_{GS(off)}|}\left(1 - \frac{V_{GS}}{V_{GS(off)}}\right) \quad (8\text{-}2)$$

The maximum value of g_m, which corresponds to the maximum slope of the transfer characteristics curve, occurs at $V_{GS} = 0$, and is referred to as g_{mo}.

$$g_{mo} = \frac{2I_{DSS}}{|V_{GS(off)}|} \quad (8\text{-}3)$$

The drain current I_D for the *enhancement*-type MOSFET is defined by Equation 7-35, from which the transconductance g_m can be derived, as follows:

$$I_D = K(V_{GS} - V_t)^2$$
$$g_m = \frac{dI_D}{dV_{GS}} = 2K(V_{GS} - V_t) \quad (8\text{-}4)$$

Hence, the equation of the transconductance g_m for the *enhancement*-type MOSFET is the following:

$$g_m = 2K(V_{GS} - V_t) \quad (8\text{-}5)$$

The Output Resistance r_o

Since the drain characteristics of the FET are similar to the collector characteristics of the BJT, the output resistance of the FET may be determined in a way similar to that of the BJT, from the drain characteristic curves.

The Early voltage V_A, as presented in Chapter 3 and revisited in Chapter 4, is equally applicable to field-effect transistors, as illustrated in Figure 8-2 below.

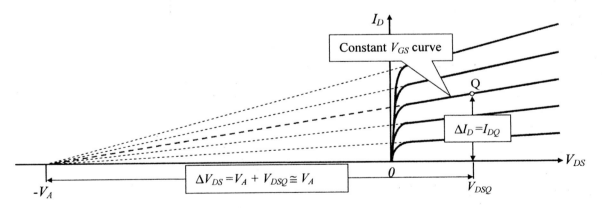

Figure 8-2: Determining the r_o, the common-source output resistance of the FET

Figure 8-2 above exhibits the FET drain characteristics, from which the output conductance of the device may be defined and derived as follows:

Chapter 8

$$\frac{1}{r_o} = \frac{di_D}{dv_{DS}}\bigg|_{v_{GS}=constant} = \frac{\Delta I_D}{\Delta V_{DS}}\bigg|_{V_{GS}=constant} \qquad (8\text{-}6)$$

The above expression for the output conductance ($1/r_o$) is basically the slope of the constant V_{GS} curve upon which the Q-point is residing. Since the extensions of all V_{GS} curves intercept the V_{DS} axis at the negative Early voltage ($-V_A$), thus the r_o can be defined and determined as follows:

$$r_o = \frac{\Delta V_{DS}}{\Delta I_D}\bigg|_{V_{GS}=constant} = \frac{V_A + V_{DSQ}}{I_{DQ}} \cong \frac{V_A}{I_{DQ}} \qquad (8\text{-}7)$$

Therefore, the r_o can be determined simply by knowing the I_{DQ} and the Early voltage.

$$r_o = \frac{V_A}{I_{DQ}} \text{ ohms} \qquad (8\text{-}8)$$

The typical value of the Early voltage V_A for FET may range from 40 V to 200 V.

Example 8-1 — n-channel JFET — SS MODEL

Let us determine the transconductance g_m and draw the small-signal equivalent circuit of the JFET amplifier of Figure 8-3.

Solution:
Recall that the analysis of the voltage-divider bias begins with determining the gate voltage V_G, with Equation 7-14.

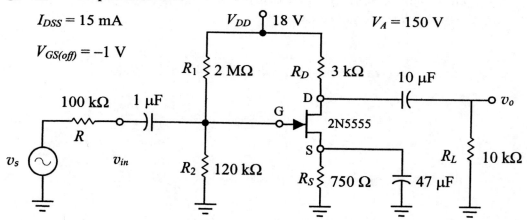

Figure 8-3: Common-source JFET amplifier circuit

$$V_G = V_{DD}\frac{R_2}{R_1+R_2} \qquad V_G = 18\text{ V}\frac{120\text{ k}\Omega}{2\text{ M}\Omega + 120\text{ k}\Omega} = 1.02\text{ V}$$

V_{GS} is determined with Equation 7-13.

$$V_{GS}\bigg|_{n-channel} = \frac{-b+\sqrt{b^2-4ac}}{2a}$$

Chapter 8

where,

$$a = \frac{I_{DSS} R_S}{V_P^2} \qquad a = \frac{15\,mA \times 0.75\,k\Omega}{(1\,V)^2} = 11.25 \qquad V_P = |V_{GS(off)}|$$

$$b = \frac{2 I_{DSS} R_S}{|V_P|} + 1 \qquad b = \frac{2 \times 15\,mA \times 0.75\,k\Omega}{1\,V} + 1 = 23.5$$

$$c = I_{DSS} R_S - |V_G| \qquad c = (15\,mA \times 0.75\,k\Omega) - 1.02 = 10.23$$

$$V_{GS} = \frac{-23.5 + \sqrt{(23.5)^2 - 4 \times 11.25 \times 10.23}}{2 \times 11.25} = -0.618\,V$$

Having determined V_{GS}, I_D can be determined with Equation 7-15, as follows:

$$I_D = \frac{V_G - V_{GS}}{R_S} \qquad I_D = \frac{1.02\,V + 0.618\,V}{0.75\,k\Omega} = 2.184\,mA$$

We can now determine the transconductance g_m with Equation 8-2 and r_o with Equation 8-8, as follows:

$$g_m = \frac{2 I_{DSS}}{|V_{GS(off)}|}\left(1 - \frac{V_{GS}}{V_{GS(off)}}\right) \qquad g_m = \frac{2 \times 15\,mA}{1\,V}\left(1 - \frac{0.618\,V}{1.V}\right) = 11.46\,mA/V$$

$$r_o = \frac{V_A}{I_{DQ}} \qquad r_o = \frac{150\,V}{2.18\,mA} \cong 69\,k\Omega$$

Having determined the FET small-signal model transconductance g_m and output resistance r_o, we now draw the small-signal equivalent circuit as shown in Figure 8-4.

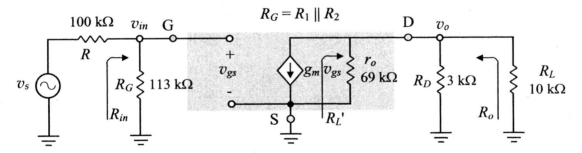

Figure 8-4: Small-signal equivalent circuit of Figure 8-3

For convenience and simplicity, we may use the approximate model of the FET without the r_o, especially when the V_A is not known.

Practice Problem 8-1 — SS MODEL

Determine the g_m and draw the small-signal equivalent circuit for the common-source JFET amplifier of Figure 8-5.

Chapter 8

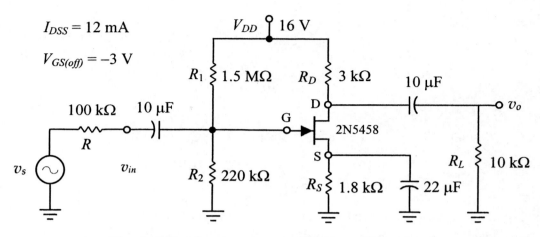

Figure 8-5: Common-source JFET amplifier circuit

Answer: $g_m = 3.33$ mA/V

8.3 ANALYSIS OF THE COMMON-SOURCE JFET AMPLIFIER

The approach to analyzing the small-signal operation of the FET amplifier is similar to that of the BJT amplifier, and it involves determining the amplifier parameters such as voltage gain, current gain, input resistance, and output resistance of the amplifier. Let us begin the analysis by analyzing the amplifier circuit of Figure 8-3 of Example 8-1, for which the transconductance g_m and small-signal equivalent circuit has already been determined, and is re-exhibited below without the r_o.

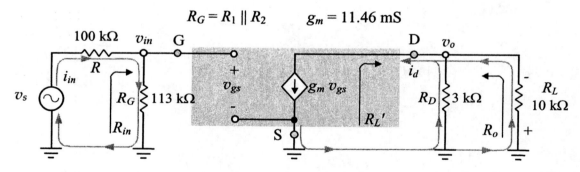

Figure 8-6: Small-signal equivalent circuit of Figure 8-3, using the approximate model

We begin the next phase of the analysis by determining the resistances R_{in}, R_o, and R_L'.

Input resistance:

The input resistance R_{in} is simply the resistance R_G, which is the parallel combination of R_1 and R_2.

$$R_{in} = R_G = R_1 \| R_2 \qquad (8-9)$$

$R_{in} = 2$ M$\Omega \|$ 120 kΩ = 113 kΩ

Chapter 8

Output resistance:

The output resistance R_o is simply the parallel combination of r_o and R_D

$$R_o = R_D \qquad (8\text{-}10)$$

$R_o = 3 \text{ k}\Omega$

Total load resistance:

The total AC load resistance is the parallel combination of r_o, R_D, and R_L

$$R_L' = r_o \| R_D \| R_L \approx R_D \| R_L \qquad (8\text{-}11)$$

$R_L' = 3 \text{ k}\Omega \| 10 \text{ k}\Omega = 2.3 \text{ k}\Omega$

Voltage gain with load:

The voltage gain with load is defined as the ratio of the output voltage v_o with load to the input voltage v_{in}. However, the input voltage v_{in} is v_{gs} and the output voltage is the product of the total output current $-g_m v_{gs}$ and the total AC load R_L'

$$v_{in} = v_{gs} \quad \text{and} \quad v_o = -g_m v_{gs} R_L'$$

$$A_{v(WL)} = \frac{v_o}{v_{in}} = \frac{-g_m v_{gs} R_L'}{v_{gs}} = -g_m R_L'$$

$$A_{v(WL)} = \frac{v_o}{v_{in}} = -g_m R_L' \qquad (8\text{-}12)$$

$$A_{v(WL)} = \frac{v_o}{v_{in}} = -11.46 \text{ mA/V} \times 2.3 \text{ k}\Omega = -26.358$$

Voltage gain without load:

The no-load voltage gain is defined as the ratio of the output voltage v_o without load to the input voltage v_{in}. The internal load of the amplifier without an external load will be defined as R_L'', which is simply R_D in parallel with r_o

$$R_L'' = r_o \| R_D \qquad (8\text{-}13)$$

$R_L'' \approx R_D = 3 \text{ k}\Omega$

$$A_{v(WL)} = \frac{v_o}{v_{in}} = \frac{-g_m v_{gs} R_L''}{v_{gs}} = -g_m R_L''$$

$$A_{v(NL)} = \frac{v_o}{v_{in}} = -g_m R_L'' \qquad (8\text{-}14)$$

$$A_{v(NL)} = \frac{v_o}{v_{in}} = -11.46 \text{ mA/V} \times 3 \text{ k}\Omega = -34.4$$

The current gain:

Although there is no input current to the FET, there is, however, a low current i_{in} through the input resistance R_{in} of the amplifier. Hence, the current gain can be determined as follows:

$$i_{in} = \frac{v_{in}}{R_{in}} \quad \text{and} \quad i_o = \frac{v_o}{R_L}$$

Chapter 8

$$A_i = \frac{i_o}{i_{in}} = \frac{v_o/R_L}{v_{in}/R_{in}} = A_{v(WL)}\frac{R_{in}}{R_L}$$

$$A_i = \frac{i_o}{i_{in}} = A_{v(WL)}\frac{R_{in}}{R_L} \qquad (8\text{-}15)$$

$$A_i = \frac{i_o}{i_{in}} = -26.358 \times \frac{113\,k\Omega}{10\,k\Omega} = -297.845$$

The power gain:

$$A_p = \frac{P_o}{P_{in}} = A_{v(WL)}A_i \qquad (8\text{-}16)$$

$$A_p = (-26.358) \times (-297.845) = 7850.6$$

Let us now verify the results of our analysis by simulation. The results of simulation with Electronics Workbench are presented in Figures 8-7 through 8-11. Some of the amplifier parameters, such as the input and output resistances, will be calculated from the simulation results depicted on Figures 8-7 and 8-9. DC bias voltages and currents, however, can be viewed directly on the DMMs of the Electronics Workbench.

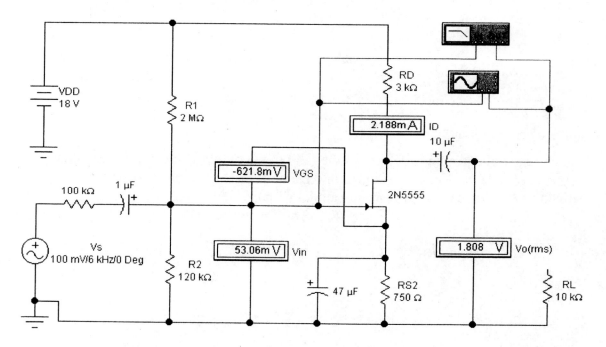

Figure 8-7: Circuit setup with simulation results of Example 8-1 with **no load** created in Electronics Workbench

The results of the above simulation show that the DC bias voltage $V_{GS} = -0.621$ V and the current $I_D = 2.188$ mA. These readings are almost identical to the algebraically determined $V_{GS} = -0.6218$ V and $I_D = 2.18$ mA. The Bode plotter of Figure 8-8 shows the no-load voltage gain of 34.3, which is very close to the algebraically determined voltage gain of 34.35.

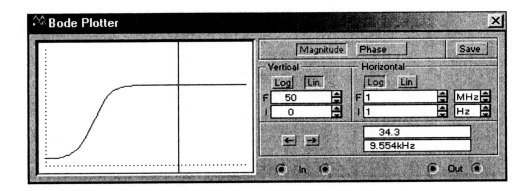

Figure 8-8: Electronics Workbench Bode plotter showing the voltage gain with **no load**

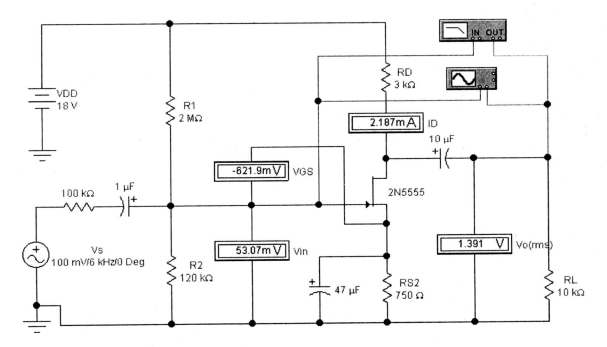

Figure 8-9: Circuit setup with simulation results of Example 8-1 **with load** created in Electronics Workbench

The input resistance:

The input resistance R_{in} is determined from DMM recordings of Figure 8-7 and with Equation 5-8, as follows:

$$R_{in} = v_{in}\frac{R}{v_S - v_{in}} = 53.06\,\text{mV}\frac{100\,\text{k}\Omega}{100\,\text{mV} - 53.06\,\text{mV}} = 113\,\text{k}\Omega$$

The output resistance:

The output resistance R_o is determined from DMM recordings of Figures 8-7, 8-9, and with Equation 5-13, as follows:

$$R_o = \frac{v_o}{v_{in}} = \frac{v_{o(NL)} - v_{o(WL)}}{v_{o(WL)}} R_L = \frac{1.808\text{ V} - 1.391\text{ V}}{1.391\text{ V}} 10\text{ k}\Omega = 2.99\text{ k}\Omega$$

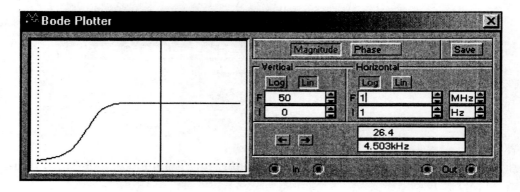

Figure 8-10: Electronics Workbench Bode plotter showing the voltage gain **with load**

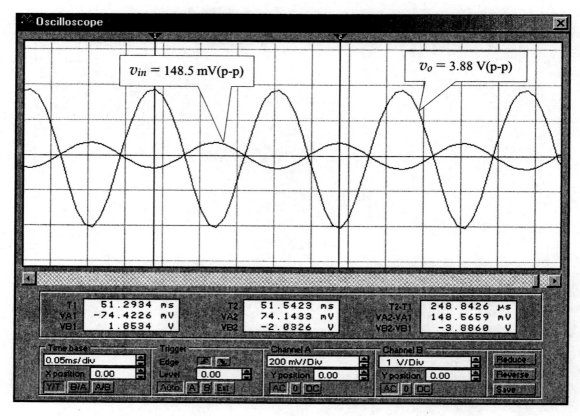

Figure 8-11: Electronics Workbench oscilloscope showing the input and output signals of Example 8-1 with load

The output waveform being 180° out-of-phase with the input waveform as depicted in Figure 8-11 above verifies the inverting characteristic of the common-source amplifier. A summary of the analytically expected results of the JFET amplifier of Example 8-1 versus the results obtained by simulation are exhibited in Table 8-1 below:

Table 8-1: Results summary of Example 8-1

Parameter	Result of analysis	Result of simulation
I_{DQ}	2.18 mA	2.188 mA
V_{GSQ}	−0.618 V	−0.62 V
$|A_{v(NL)}|$	34.35	34.3
$|A_{v(WL)}|$	26.4	26.4
R_{in}	113 kΩ	113 kΩ
R_o	3 kΩ	2.99 kΩ
Phase	180°	180°

As you can see, all of the results of our analysis of the common-source JFET amplifier are either identical or very close to the results obtained by simulation.

Hints on looking up the FET characteristics:

The nominal FET characteristics may be looked up in the Electronics Workbench library by clicking on the FET symbol, then proceeding as follows:
1. Click on *n-channel* or *p-channel* FETs.
2. Select the desired FET by highlighting it and click on the *EDIT* button.
3. A menu opens that contains many FET parameters. The parameters that we are interested in are V_p (in volts) and K or β (in mA/V^2). For a JFET, however, we need the I_{DSS}, which can be determined as follows:

$$I_{DSS} = K V_p^2$$

where V_p is the $|V_{GS(off)}|$

Example 8-2   *n-channel JFET*   ANALYSIS

Determine the R_{in}, R_o, $A_{v(WL)}$, $A_{v(NL)}$, A_i, A_p, and v_o for the JFET amplifier of Figure 8-12.

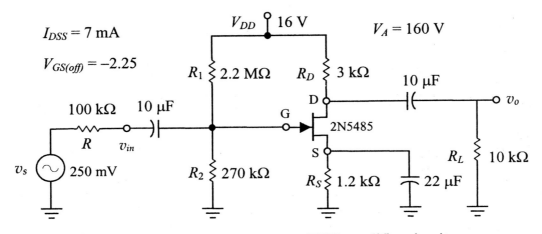

Figure 8-12: Common-source JFET amplifier circuit

Chapter 8

Solution:

The dc voltage V_G can be calculated from Equation 7-14

$$V_G = V_{DD}\frac{R_2}{R_1+R_2} \qquad V_G = 16\,\text{V}\frac{270\,\text{k}\Omega}{2.2\,\text{M}\Omega + 270\,\text{k}\Omega} = 1.75\,\text{V}$$

The voltage V_{GS} can be computed from Equation 7-12

$$V_{GS}\Big|_{n-channel} = \frac{-b+\sqrt{b^2-4ac}}{2a}$$

where

$$a = \frac{I_{DSS}R_S}{V_P^2} \qquad a = \frac{7\,\text{mA}\times 1.2\,\text{k}\Omega}{(2.25\,\text{V})^2} = 1.66\ 1/\text{V} \qquad V_P = |V_{GS(off)}|$$

$$b = \frac{2I_{DSS}R_S}{|V_P|}+1 \qquad b = \frac{2\times 7\,\text{mA}\times 1.2\,\text{k}\Omega}{2.25\,\text{V}}+1 = 8.467$$

$$c = I_{DSS}R_S - |V_G| \qquad c = (7\,\text{mA}\times 1.2\,\text{k}\Omega) - 1.75\,\text{V} = 6.65\,\text{V}$$

$$V_{GS} = \frac{-8.46+\sqrt{(8.46)^2-4\times 1.66\times 6.65}}{2\times 1.66} \cong -1\,\text{V}$$

Having determined voltage V_{GS}, current I_D can be determined from Equation 7-15

$$I_D = \frac{V_G-V_{GS}}{R_S} \qquad I_D = \frac{1.75\,\text{V}+1\,\text{V}}{1.2\,\text{k}\Omega} = 2.29\,\text{mA}$$

We can now find the transconductance g_m from Equation 8-2 and r_o from Equation 8-8

$$g_m = \frac{2I_{DSS}}{|V_{GS(off)}|}\left(1-\frac{V_{GS}}{V_{GS(off)}}\right) \qquad g_m = \frac{2\times 7\,\text{mA}}{2.25\,\text{V}}\left(1-\frac{1\,\text{V}}{2.25\,\text{V}}\right) \cong 3.5\,\text{mA/V}$$

$$r_o = \frac{V_A}{I_{DQ}} \qquad r_o = \frac{160\,\text{V}}{2.29\,\text{mA}} = 70\,\text{k}\Omega$$

Having determined the transconductance g_m and r_o, we will now draw the small-signal equivalent circuit as shown in Figure 8-13.

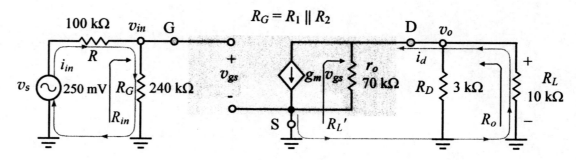

Figure 8-13: Small-signal equivalent circuit of Figure 8-12, using the exact model

Chapter 8

Input resistance:

The input resistance R_{in} can be determined from Equation 8-9

$R_{in} = R_G = R_1 \| R_2$ $R_{in} = 2.2\ \text{M}\Omega \| 270\ \text{k}\Omega = \mathbf{240\ k\Omega}$

Output resistance:

The output resistance R_o is calculated from Equation 8-10

$R_o = r_o \| R_D$ $R_o = 70\ \text{k}\Omega \| 3\ \text{k}\Omega = \mathbf{2.88\ k\Omega}$

Total AC load resistance:

$R_L' = r_o \| R_D \| R_L$ $R_L' = 70\ \text{k}\Omega \| 3\ \text{k}\Omega \| 10\ \text{k}\Omega = 2.23\ \text{k}\Omega$

Voltage gain with load:

The voltage gain with external load is determined from Equation 8-12

$A_{v(WL)} = \dfrac{v_o}{v_{in}} = -g_m R_L'$ $A_{v(WL)} = \dfrac{v_o}{v_{in}} = -3.5\ \text{mA/V} \times 2.23\ \text{k}\Omega = \mathbf{-7.805}$

Voltage gain without load:

The internal load of the amplifier without the external load is defined as R_L'' and can be calculated from Equation 8-13, and the voltage gain without the load is determined from Equation 8-14

$R_L'' = r_o \| R_D$ $R_L'' = 70\ \text{k}\Omega \| 3\ \text{k}\Omega = 2.88\ \text{k}\Omega$

$A_{v(NL)} = \dfrac{v_o}{v_{in}} = -g_m R_L''$ $A_{v(NL)} = \dfrac{v_o}{v_{in}} = -3.46\ \text{mA/V} \times 2.9\ \text{k}\Omega = \mathbf{-10}$

The current gain:

The current gain A_i can be determined from Equation 8-15

$A_i = \dfrac{i_o}{i_{in}} = A_{v(WL)} \dfrac{R_{in}}{R_L}$ $A_i = \dfrac{i_o}{i_{in}} = -7.805 \times \dfrac{240\ \text{k}\Omega}{10\ \text{k}\Omega} = \mathbf{-187.32}$

The power gain:

The power gain A_p can be calculated from Equation 8-16

$A_p = \dfrac{P_o}{P_{in}} = A_{v(WL)} A_i$ $A_p = (-7.805) \times (-187.32) = \mathbf{1462}$

Output voltage with load:

To determine the output voltage, we need to determine the input voltage first. Thus,

$v_{in} = v_s \dfrac{R_{in}}{R + R_{in}}$ $v_{in} = 250\ \text{mV} \times \dfrac{240\ \text{k}\Omega}{100\ \text{k}\Omega + 240\ \text{k}\Omega} = 176.5\ \text{mV}$

$v_{o(WL)} = A_{v(WL)} v_{in}$ $v_{o(WL)} = -7.805 \times 176.5\ \text{mV} = \mathbf{-1.378\ V}$

Practice Problem 8-2

n-channel JFET — ANALYSIS

Determine the R_{in}, R_o, $A_{v(WL)}$, $A_{v(NL)}$, A_i, A_p, and v_o for the JFET amplifier of Figure 8-14. Verify your results by simulation.

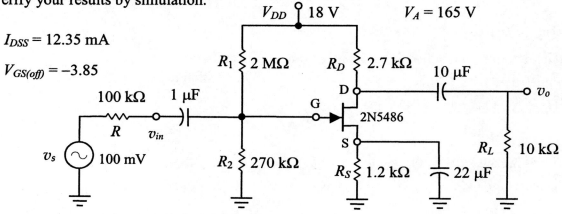

I_{DSS} = 12.35 mA

$V_{GS(off)}$ = −3.85

Figure 8-14: Common-source JFET amplifier circuit

Section Summary

Junction FET — *Depletion MOSFET* — ANALYSIS

Summary of Equations for the Analysis of CS Amplifier utilizing JFET or DMOS

$$V_G = V_{DD} \frac{R_2}{R_1 + R_2} \qquad V_{GS}\bigg|_{n-ch} = \frac{-b + \sqrt{b^2 - 4ac}}{2a} \qquad V_{GS}\bigg|_{p-ch} = \frac{+b - \sqrt{b^2 - 4ac}}{2a}$$

$$a = \frac{I_{DSS} R_S}{V_P^2} \qquad b = \frac{2 I_{DSS} R_S}{|V_P|} + 1 \qquad c = I_{DSS} R_S - |V_G|$$

$$I_D = \frac{V_G - V_{GS}}{R_S} \qquad g_m = \frac{2 I_{DSS}}{|V_{GS(off)}|}\left(1 - \frac{V_{GS}}{V_{GS(off)}}\right) \qquad V_P = |V_{GS(off)}|$$

$$R_{in} = R_G = R_1 \| R_2 \qquad R_o = R_D \qquad r_o = \frac{V_A}{I_{DQ}}$$

$$R_L' = R_D \| R_L \qquad A_{v(WL)} = \frac{v_o}{v_{in}} = -g_m R_L' \qquad A_i = \frac{i_o}{i_{in}} = A_{v(WL)} \frac{R_{in}}{R_L}$$

$$R_L'' = R_D \qquad A_{v(NL)} = \frac{v_o}{v_{in}} = -g_m R_L'' \qquad A_p = \frac{P_o}{P_{in}} = A_{v(WL)} A_i$$

$$v_{in} = v_s \frac{R_{in}}{R + R_{in}} \qquad v_{o(WL)} = A_{v(WL)} \cdot v_{in} \qquad v_{o(NL)} = A_{v(NL)} \cdot v_{in}$$

8.4 ANALYSIS OF THE COMMON-DRAIN JFET AMPLIFIER

The common-drain amplifier, also referred to as a source follower, shown in Figure 8-15 is the FET version of the common-collector amplifier or the emitter follower. Like the emitter follower, the source follower yields some current gain, but offers no voltage gain. Hence, its main application is to be used as a buffer.

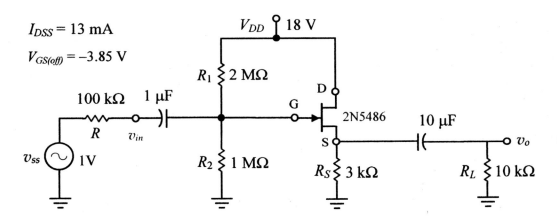

Figure 8-15: JFET source follower

In order to evaluate the transconductance g_m, we need to determine V_G and then V_{GS}

$$V_G = V_{DD}\frac{R_2}{R_1+R_2} \qquad V_G = 18\text{ V}\frac{1\text{ M}\Omega}{2\text{ M}\Omega+1\text{ M}\Omega} = 6\text{ V}$$

The voltage V_{GS} can be determined from Equation 7-12

$$V_{GS}\bigg|_{n-channel} = \frac{-b+\sqrt{b^2-4ac}}{2a}$$

where

$$a = \frac{I_{DSS}R_S}{V_P^2} \qquad a = \frac{13\text{ mA}\times 3\text{ k}\Omega}{(3.85\text{ V})^2} = 2.63$$

$$b = \frac{2I_{DSS}R_S}{|V_P|}+1 \qquad b = \frac{2\times 13\text{ mA}\times 3\text{ k}\Omega}{3.85\text{ V}}+1 = 21.26$$

$$c = I_{DSS}R_S - |V_G| \qquad c = (13\text{ mA}\times 3\text{ k}\Omega) - 6 = 33$$

$$V_{GS} = \frac{-21.26+\sqrt{(21.26)^2-4\times 2.63\times 33}}{2\times 2.63} \cong -2.1\text{ V}$$

Having determined V_{GS}, I_D can be determined from Equation 7-15

$$I_D = \frac{V_G - V_{GS}}{R_S} \qquad I_D = \frac{6\text{ V}+2.1\text{ V}}{3\text{ k}\Omega} = 2.7\text{ mA}$$

We can now find the transconductance g_m from Equation 8-2

Chapter 8

$$g_m = \frac{2I_{DSS}}{|V_{GS(off)}|}\left(1 - \frac{V_{GS}}{V_{GS(off)}}\right) \qquad g_m = \frac{2 \times 13 \text{ mA}}{3.85 \text{ V}}\left(1 - \frac{2.1 \text{ V}}{3.85 \text{ V}}\right) \cong 3.07 \text{ mA/V}$$

Having determined the transconductance g_m, we will now draw the small-signal equivalent circuit shown in Figure 8-16.

$$R_G = R_1 \parallel R_2$$

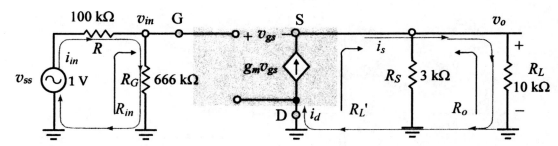

Figure 8-16: The small-signal equivalent circuit of Figure 8-15

For a precise analysis in order to determine the input resistance R_{in} and the output resistance R_o, we must consider the amplifier circuit as a two-port network and apply the two-port network theory as discussed in Chapter 4 and previously applied to the BJT amplifier in Examples 5-3 and 5-4.

Input resistance:

According to the two-port network theory, the input resistance is determined from Equation 5-45, which requires setting $v_o = 0$ by grounding the output

$$R_{in} = \frac{v_{in}}{i_{in}}\bigg|_{v_o = 0}$$

Since the output is isolated from the input, grounding the output will have no effect on the input, and the input resistance R_{in} as defined by the above equation is

$$R_{in} = \frac{v_{in}}{i_{in}}\bigg|_{v_o = 0} = \frac{v_{in}}{v_{in}/R_G} = R_G$$

Hence, the input resistance R_{in} is simply the parallel combination of R_1 and R_2

$$R_{in} = R_G = R_1 \parallel R_2$$

R_{in} = 2 MΩ $\parallel$ 1 MΩ = 666.67 kΩ

Output resistance:

According to the two-port network theory, the output resistance is determined from Equation 5-47, which requires the application of a test signal v_t at the output while reducing the voltage source v_s to zero.

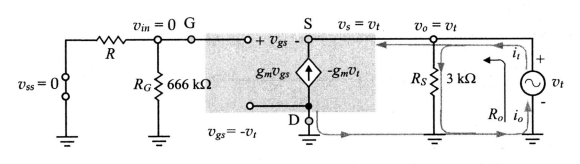

Figure 8-17: Applying the two-port network theory to the source-follower circuit of Figure 8-15 in order to determine the R_o

$$R_o = \left.\frac{v_o}{i_o}\right|_{v_{in}=0} = \frac{v_t}{i_t}$$

$v_{ss} = v_{in} = v_g = 0 \qquad v_s = v_t \qquad v_{gs} = v_g - v_s = 0 - v_s = -v_t \qquad g_m v_{gs} = -g_m v_t$

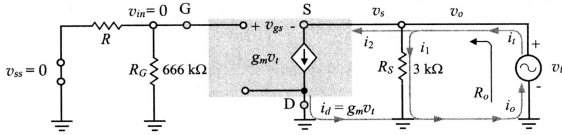

Figure 8-18: Figure 8-17 redrawn and simplified

The application of the test signal v_t at the output causes the flow of a test current i_t, which branches out into i_1 and i_2 at the source terminal, where i_1 is the current through R_S, and i_2 is the current through the current source $g_m v_t$.

$$i_t = i_1 + i_2 \tag{8-17}$$

$$i_1 = \frac{v_t}{R_S} \tag{8-18}$$

$$i_2 = g_m v_t \tag{8-19}$$

$$i_t = \frac{v_t}{R_S} + g_m v_t \tag{8-20}$$

$$R_o = \left.\frac{v_o}{i_o}\right|_{v_{in}=0} = \frac{v_t}{i_t} = \frac{v_t}{v_t\left[\dfrac{1}{R_S}+g_m\right]} = \frac{1}{\dfrac{1}{R_S}+g_m} \tag{8-21}$$

$$R_o = \frac{1}{\dfrac{1}{R_S}+g_m} = \frac{1}{\dfrac{1}{R_S}+\dfrac{1}{r_m}} = \frac{R_S \times r_m}{R_S + r_m} = R_S \parallel r_m \tag{8-22}$$

$$R_o = R_S \parallel r_m \tag{8-23}$$

Chapter 8

where

$$r_m = \frac{1}{g_m} \qquad (8\text{-}24)$$

$$r_m = \frac{1}{g_m} = \frac{1}{3.07 \text{ mA/V}} = 325.732 \text{ V/A}$$

Therefore, the output resistance R_o is

$R_o = r_m \| R_S = 325.732 \, \Omega \| 3 \text{ k}\Omega = \mathbf{294 \, \Omega}$

Total AC load resistance:

Returning to Figure 8-16, the total AC load resistance R_L' seen by the small-signal current dependent current source is the parallel combination of R_S and R_L

$$R_L' = R_S \| R_L \qquad (8\text{-}25)$$

$R_L' = 3 \text{ k}\Omega \| 10 \text{ k}\Omega = 2.3 \text{ k}\Omega$

Voltage gain with load:

The voltage gain with an external load is defined as v_o/v_{in} and is determined by

$$v_o = g_m v_{gs} R_L' \qquad (8\text{-}26)$$

$$v_{in} = v_g = v_{gs} + v_o = v_{gs} + g_m v_{gs} R_L' = v_{gs}(1 + g_m R_L') \qquad (8\text{-}27)$$

$$A_{v(WL)} = \frac{v_o}{v_{in}} = \frac{g_m v_{gs} R_L'}{v_{gs}(1+g_m R_L')} = \frac{g_m R_L'}{1+g_m R_L'} \qquad (8\text{-}28)$$

$$A_{v(WL)} = \frac{v_o}{v_{in}} = \frac{g_m R_L'}{1+g_m R_L'} \qquad (8\text{-}29)$$

$$A_{v(WL)} = \frac{v_o}{v_{in}} = \frac{3.07 \text{ mA/V} \times 2.3 \text{ k}\Omega}{1 + 3.07 \text{ mA/V} \times 2.3 \text{ k}\Omega} = 0.876$$

Voltage gain without load:

The internal load of the amplifier without an external load is defined as R_L'', and the voltage gain without an load is determined from Equation 8-29 above except that R_L' is replaced with R_L''

$$R_L'' = R_S \qquad (8\text{-}30)$$

$R_L'' = 3 \text{ k}\Omega$

$$A_{v(NL)} = \frac{v_o}{v_{in}} = \frac{g_m R_L''}{1+g_m R_L''} \qquad (8\text{-}31)$$

$$A_{v(NL)} = \frac{v_o}{v_{in}} = \frac{3.07 \text{ mA/V} \times 3 \text{ k}\Omega}{1 + 3.07 \text{ mA/V} \times 3 \text{ k}\Omega} = 0.9$$

The current gain:

The current gain A_i can be calculated from Equation 8-15

$$A_i = \frac{i_o}{i_{in}} = A_{v(WL)} \frac{R_{in}}{R_L}$$

$$A_i = \frac{i_o}{i_{in}} = 0.876 \times \frac{666.67\,k\Omega}{10\,k\Omega} = 58.4$$

<u>The power gain:</u>

The power gain A_p can be determined from Equation 8-16

$$A_p = \frac{P_o}{P_{in}} = A_{v(WL)} A_i$$

$$A_p = 0.876 \times 58.2 = 51.158$$

<u>Output voltage with load:</u>

To determine the output voltage, we need to determine the input voltage as

$$v_{in} = v_s \frac{R_{in}}{R + R_{in}} \qquad v_{in} = 1\,V \times \frac{666.67\,k\Omega}{100\,k\Omega + 666.67\,k\Omega} = 0.87\,V$$

$$v_{o(WL)} = A_{v(WL)} v_{in} \qquad v_{o(WL)} = 0.876 \times 0.87\,V = 0.762\,V$$

We will now test the result of our analysis by simulation using Electronics Workbench.

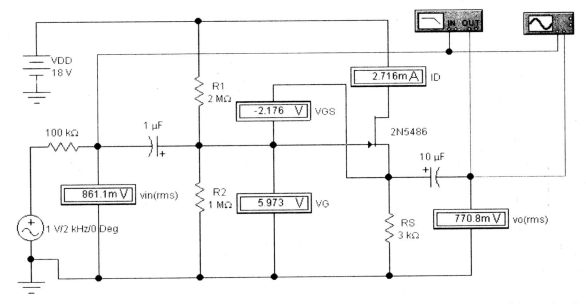

Figure 8-19: Circuit setup with simulation results of the source-follower without load

The results of simulation as depicted in Figures 8-19 and 8-20 are the following:

<u>No-load voltage gain:</u>

As recorded on the DMMs of Figure 8-19, the input signal is 861.1 mV$_{(rms)}$ and the output signal without load is 770.8 mV$_{(rms)}$. Hence, the no-load voltage gain is

$$A_{v(NL)} = \frac{v_o}{v_{in}} = \frac{770.8 \text{ mV}}{861.1 \text{ mV}} = 0.895$$

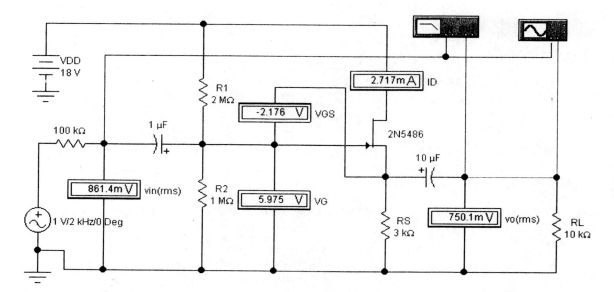

Figure 8-20: Circuit setup with simulation results of the source-follower with load

Voltage gain with load:

As recorded on the DMMs of Figure 8-20, the input signal is 861.4 mV$_{(rms)}$ and the output signal without load is 750.1 V$_{(rms)}$. Hence, the voltage gain with load is

$$A_{v(WL)} = \frac{v_o}{v_{in}} = \frac{750.1 \text{ mV}}{861.4 \text{ mV}} = 0.87$$

The input resistance:

The input resistance R_{in} is determined from the DMM recordings of Figure 8-19 and from Equation 5-10

$$R_{in} = v_{in} \frac{R}{v_s - v_{in}} = 0.861 \text{ V} \times \frac{100 \text{ k}\Omega}{1 \text{ V} - 0.861 \text{ V}} = 619 \text{ k}\Omega$$

The output resistance:

The output resistance R_o is determined from the DMM recordings of Figures 8-19, 8-20, and from Equation 5-13

$$R_o = \frac{v_o}{v_{in}} = \frac{v_{o(NL)} - v_{o(WL)}}{v_{o(WL)}} R_L = \frac{770.8 \text{ mV} - 750.1 \text{ mV}}{750.1 \text{ mV}} \times 10 \text{ k}\Omega = 276 \text{ }\Omega$$

A summary of the analytically expected results of the JFET source-follower of Example 8-1 versus the results obtained by simulation are exhibited in Table 8-2.

Table 8-2: Results Summary for the Source Follower

Parameter	Analysis Results	Simulation Results
I_{DQ}	2.7 mA	2.71 mA
V_{GSQ}	−2.1 V	−2.17 V
$A_{v(NL)}$	0.9	0.895
$A_{v(WL)}$	0.875	0.87
R_{in}	666.67 kΩ	619 kΩ
R_o	294 Ω	276 Ω
$v_{o(WL)}$	0.76 V	0.75 V
Phase	0°	0°

As you can see, the results of our analysis of the source-follower are either identical or very close to the results obtained by simulation.

Section Summary — Junction FET / Depletion MOSFET — ANALYSIS

Summary of Equations for the Analysis of CD Amplifier (Source-Follower)

$$V_G = V_{DD} \frac{R_2}{R_1 + R_2} \qquad V_{GS}\bigg|_{n-ch} = \frac{-b + \sqrt{b^2 - 4ac}}{2a} \qquad V_{GS}\bigg|_{p-ch} = \frac{+b - \sqrt{b^2 - 4ac}}{2a}$$

$$a = \frac{I_{DSS} R_S}{V_P^2} \qquad b = \frac{2 I_{DSS} R_S}{|V_P|} + 1 \qquad c = I_{DSS} R_S - |V_G|$$

$$I_D = \frac{V_G - V_{GS}}{R_S} \qquad g_m = \frac{2 I_{DSS}}{|V_{GS(off)}|}\left(1 - \frac{V_{GS}}{V_{GS(off)}}\right)$$

$$R_{in} = R_G = R_1 \| R_2 \qquad R_o = R_S \| r_m \qquad r_m = \frac{1}{g_m}$$

$$R_L' = R_S \| R_L \qquad A_{v(WL)} = \frac{v_o}{v_{in}} = \frac{g_m R_L'}{1 + g_m R_L'} \qquad A_i = \frac{i_o}{i_{in}} = A_{v(WL)} \frac{R_{in}}{R_L}$$

$$R_L'' = R_S \qquad A_{v(NL)} = \frac{v_o}{v_{in}} = \frac{g_m R_L''}{1 + g_m R_L''} \qquad A_p = \frac{P_o}{P_{in}} = A_{v(WL)} A_i$$

$$v_{in} = v_S \frac{R_{in}}{R + R_{in}} \qquad v_{o(WL)} = A_{v(WL)} v_{in} \qquad v_{o(NL)} = A_{v(NL)} v_{in}$$

Practice Problem 8-3　　　　　　　　　　　　　　　　　　　　　ANALYSIS

Determine R_{in}, R_o, $A_{v(WL)}$, $A_{v(NL)}$, and v_o for the source-follower circuit of Figure 8-21. Verify your answers by simulation.

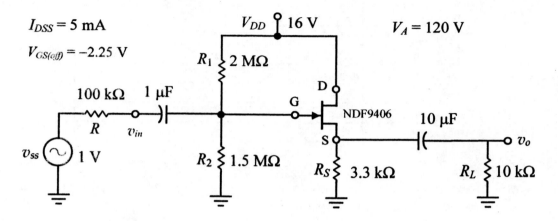

Figure 8-21: JFET source follower

8.5 ANALYSIS OF THE CS ENHANCEMENT MOSFET AMPLIFIER

The approach to analyzing the *enhancement* MOSFET amplifier will be similar to that of the JFET or DMOS amplifier. We will start the process with the DC analysis in order to determine the transconductance g_m, and then continue with the small-signal equivalent circuit.

Example 8-3　　　　　　　　　Enhancement MOSFET　　　　　　　　　ANALYSIS

Let us now analyze the common-source *enhancement* MOSFET amplifier of Figure 8-22 and determine R_{in}, R_o, $A_{v(WL)}$, $A_{v(NL)}$, A_i, A_p, and v_o.

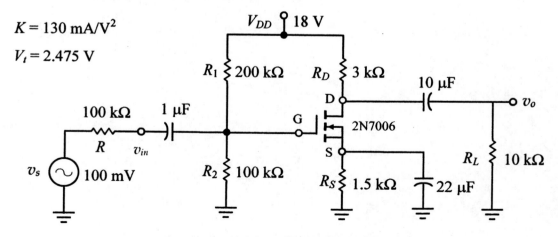

Figure 8-22: Common-source MOSFET amplifier

Solution:

The first item to be calculated is V_G, which is determined from Equation 7-29

$$V_G = V_{DD}\frac{R_2}{R_1+R_2} \qquad V_G = 18\text{ V}\frac{100\text{ k}\Omega}{200\text{ k}\Omega+100\text{ k}\Omega} = 6\text{ V}$$

The voltage V_{GS} is determined from Equation 7-32

$$V_{GS}\bigg|_{n-channel} = \frac{-b+\sqrt{b^2-4ac}}{2a}$$

where

$a = KR_S$ $\qquad a = 0.130 \times 1500 = 195$

$b = 1 - 2KR_S|V_t|$ $\qquad b = 1 - (2 \times 0.130 \times 1500 \times 2.475) = -964$

$c = KR_S V_t^2 - |V_G|$ $\qquad c = (0.130 \times 1500 \times 2.475^2) - 6.75 = 1188$

$$V_{GS} = \frac{964+\sqrt{(964)^2 - 4\times 195 \times 1188}}{2\times 195} = 2.606\text{ V}$$

Having determined V_{GS}, I_D can be determined from Equation 7-15

$$I_D = \frac{V_G - V_{GS}}{R_S} \qquad I_D = \frac{6\text{ V} - 2.606\text{ V}}{1.5\text{ k}\Omega} = 2.263\text{ mA}$$

We can now determine the transconductance g_m from Equation 8-5

$$g_m = 2K(V_{GS}-V_t) = 2\sqrt{KI_D} \qquad g_m = 2 \times 0.13\,(2.606 - 2.475) = \mathbf{34.06\text{ mA/V}}$$

We will now draw the small-signal equivalent circuit using the approximate model of the MOSFET, as shown in Figure 8-23.

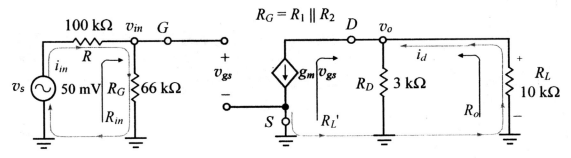

Figure 8-23: The small-signal equivalent circuit of Figure 8-22

Input resistance:

The input resistance R_{in} can be determined from Equation 8-9

$R_{in} = R_G = R_1 \parallel R_2$ $\qquad R_{in} = 200\text{ k}\Omega \parallel 100\text{ k}\Omega = \mathbf{66.67\text{ k}\Omega}$

Output resistance:

The output resistance R_o can be calculated from Equation 8-10

$R_o = R_D$ $\qquad R_o = 3\text{ k}\Omega$

Chapter 8

Total load resistance:

$R_L' = R_D \| R_L$ $R_L' = 3\text{ k}\Omega \| 10\text{ k}\Omega = 2.3\text{ k}\Omega$

Voltage gain with load:

The voltage gain with external load is determined from Equation 8-12

$A_{v(WL)} = \dfrac{v_o}{v_{in}} = -g_m R_L'$ $A_{v(WL)} = \dfrac{v_o}{v_{in}} = -34.06\text{ mA/V} \times 2.3\text{ k}\Omega = \mathbf{-78.338}$

Voltage gain without load:

The internal load of the amplifier without an external load is defined as R_L'', and the voltage gain without an external load is determined from Equation 8-14

$R_L'' = R_D$ $R_L'' = 3\text{ k}\Omega$

$A_{v(NL)} = \dfrac{v_o}{v_{in}} = -g_m R_L''$ $A_{v(NL)} = \dfrac{v_o}{v_{in}} = -34.06\text{ mA/V} \times 3\text{ k}\Omega = \mathbf{-102.18}$

The current gain:

The current gain can be determined from Equation 8-15

$A_i = \dfrac{i_o}{i_{in}} = A_{v(WL)} \dfrac{R_{in}}{R_L}$ $A_i = \dfrac{i_o}{i_{in}} = -78.338 \times \dfrac{66.67\text{ k}\Omega}{10\text{ k}\Omega} = -522.23$

The power gain:

$A_p = \dfrac{P_o}{P_{in}} = A_{v(WL)} A_i$ $A_p = (-78.338) \times (-522.23) = 40{,}914$

Output voltage with load:

$v_{in} = v_S \dfrac{R_{in}}{R + R_{in}}$ $v_{in} = 100\text{ mV} \times \dfrac{66.67\text{ k}\Omega}{100\text{ k}\Omega + 66.67\text{ k}\Omega} = 40\text{ mV}$

$v_{o(WL)} = A_{v(WL)} v_{in}$ $v_{o(WL)} = -78.338 \times 40\text{ mV} = -3.133\text{ V}$

Practice Problem 8-4   **ANALYSIS**

Analyze the MOSFET amplifier circuit of Figure 8-24 and determine R_{in}, R_o, $A_{v(WL)}$, $A_{v(NL)}$, A_i, A_p, and v_o.

$K = 110\text{ mA/V}^2$

$V_t = 3\text{ V}$

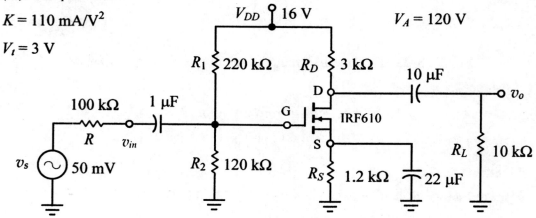

Figure 8-24: Common-source MOSFET amplifier circuit

The EMOS Common-Source Amplifier with a Partially Unbypassed R_S

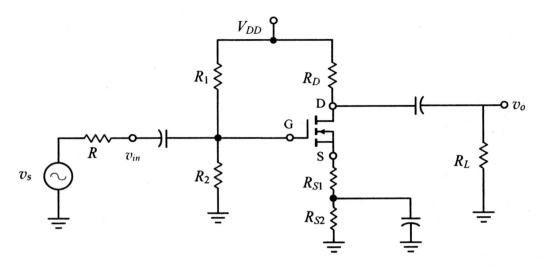

Figure 8-25: Common-source MOSFET amplifier circuit

This is the amplifier circuit of Example 8-3 with the addition of an unbypassed source resistor R_{S1}, which will have a very small value compared to R_{S2} (i.e., 33 Ω versus 1.5 kΩ). Being so small in value, the addition of R_{S1} will have very little or no effect on the DC bias; however, it can have significant effect on the small-signal operation, including gain control and more stable operation. Hence, we will begin the small-signal analysis by presenting the small-signal equivalent circuit. However, to determine the effect of r_o and evaluate its significance in the circuit operation, we will initially use the exact model of the MOSFET in the small-signal equivalent.

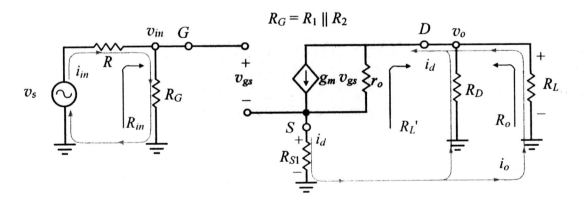

Figure 8-26: The small-signal equivalent circuit of Figure 8-25

The total AC load R_L' at the output terminal is the parallel combination of R_D and R_L.

$$R_L' = R_D \parallel R_L \qquad (8\text{-}32)$$

Without the external load R_L, the internal load of the amplifier will be the parallel combination of r_o and R_D, which will be designated as R_L''.

$$R_L'' = r_o \| R_D \tag{8-33}$$

Voltage gain with load:

$$v_{in} = v_g = v_{gs} + i_d R_{S1} \tag{8-34}$$

The current source $g_m v_{gs}$ branches out into i_d and i_{ro} at the source terminal, where i_d flows out of the source terminal, through the parallel combination of R_D and R_L, back to the drain terminal. The i_{ro} is the fraction of $g_m v_{gs}$ that loops through r_o and is quite small compared to i_d. The exact value of i_d can be evaluated by applying the *KCL* at the source node, as follows:

Let $R_P = R_D \| R_L$

$$i_d = g_m v_{gs} \frac{r_o}{r_o + R_P} = g_m v_{gs} \frac{1}{1 + \frac{R_P}{r_o}} \cong g_m v_{gs} \tag{8-35}$$

The resistance R_P (parallel combination of R_D and R_L) is usually much smaller than the r_o; hence, i_d may be approximated as the total $g_m v_{gs}$.

$$i_d \cong g_m v_{gs}$$

$$v_{in} = v_g = v_{gs} + i_d R_{S1} \qquad v_{in} = v_{gs} + g_m v_{gs} R_{S1} \qquad v_o = -g_m v_{gs} R_L' \tag{8-36}$$

$$A_{v(WL)} = \frac{v_o}{v_{in}} = \frac{-g_m v_{gs} \cdot R_L'}{v_{gs} + g_m v_{gs} R_{S1}} = \frac{-g_m v_{gs} \cdot R_L'}{v_{gs}(1 + g_m R_{S1})} = \frac{-g_m \cdot R_L'}{(1 + g_m R_{S1})} \tag{8-37}$$

$$A_{v(WL)} = \frac{v_o}{v_{in}} = \frac{-g_m \cdot R_L'}{(1 + g_m R_{S1})} \tag{8-38}$$

Voltage gain without load:

To determine the voltage gain without external load, R_L' in the above expression for the voltage gain with load will be replaced by R_L''.

$$R_L'' = r_o \| R_D$$

$$A_{v(NL)} = \frac{v_o}{v_{in}} = \frac{-g_m \cdot R_L''}{(1 + g_m R_{S1})} \tag{8-39}$$

The input resistance:

Since the gate is isolated from the source, the addition of R_{S1} at the source terminal has no impact on the input resistance, which is the parallel combination of R_1 and R_2.

$$R_{in} = R_G = R_1 \| R_2 \tag{8-40}$$

The output resistance:

For a precise analysis in order to determine the output resistance R_o, we will consider the above MOSFET amplifier as a two-port network and apply the two-port network theory, as we did previously with the source-follower of Figure 8-15. According to the two-port network theory, the output resistance is determined with Equation 5-47, which requires the application of a test signal v_t at the output while cutting off the input current i_{in} by shorting out the voltage source v_{ss}.

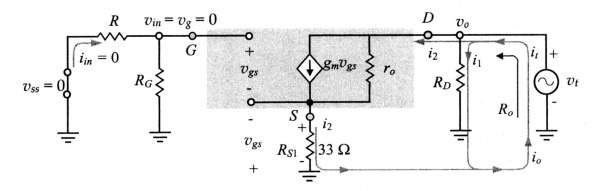

Figure 8-27: Determining the output resistance of Figure 8-25

The application of the test signal v_t at the output will cause the flow of a test current i_t, which will branch out into i_1 and i_2 at the drain terminal, where i_1 is simply the current through R_D, and i_2 is the drain current. Also note that since $v_{in} = 0$, the gate terminal is at the ground potential ($v_g = 0$). Hence, the voltage at the source terminal $v_s = -v_{gs}$. By definition, we can also write the following for v_{gs}:

$$v_{gs} = v_g - v_s = 0 - i_2 R_{S1} \tag{8-41}$$

Solving for v_s and i_2 in the above equation yields the following:

$$v_s = -v_{gs} \tag{8-42}$$

$$i_2 = -\frac{v_{gs}}{R_{S1}} \tag{8-43}$$

For the sake of convenience, we will convert the current source with its shunt resistance to an equivalent voltage source and a series resistance, as shown in Figure 8-28.

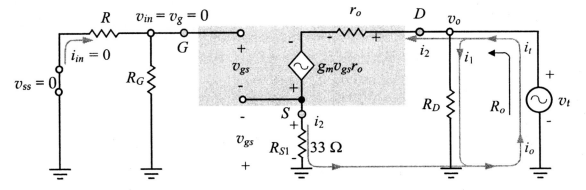

Figure 8-28: Converted equivalent circuit of Figure 8-27

Applying *KVL* around the loop containing the R_D, r_o, and R_{S1} yields the following:

$$v_o = v_t = i_2 r_o - g_m v_{gs} r_o + i_2 R_{S1} \tag{8-44}$$

Substituting Equation 8-32 for v_{gs} in the above equation, we obtain the following:

$$v_o = i_2 r_o - g_m(-i_2 R_{S1}) r_o + i_2 R_{S1} \tag{8-45}$$

Chapter 8

$$v_o = i_2 [r_o + g_m R_{S1} r_o + R_{S1}] = i_2 [r_o + (1 + g_m r_o) R_{S1}]$$

Solving for i_2 in the equation yields

$$i_2 = \frac{v_o}{r_o + (1 + g_m r_o) R_{S1}} \tag{8-46}$$

Applying the KCL at the drain terminal results in

$$i_t = i_1 + i_2 \tag{8-47}$$

$$i_1 = \frac{v_t}{R_D} = \frac{v_o}{R_D} \tag{8-48}$$

$$i_t = \frac{v_o}{R_D} + \frac{v_o}{r_o + (1 + g_m r_o) R_{S1}} = v_o \frac{R_D + [r_o + (1 + g_m r_o) R_{S1}]}{R_D [r_o + (1 + g_m r_o) R_{S1}]} = i_o \tag{8-49}$$

Having determined that $i_o = i_t$, we can now determine R_o from Equation 5-47

$$R_o = \frac{v_o}{i_o}\bigg|_{i_{in}=0} = \frac{R_D [r_o + (1 + g_m r_o) R_{S1}]}{R_D + [r_o + (1 + g_m r_o) R_{S1}]} \tag{8-50}$$

The above expression, being the product over the sum of two terms, is the parallel equivalent of the two terms (resistances). Hence, the output resistance is

$$R_o = R_D \parallel [r_o + (1 + g_m r_o) R_{S1}] \tag{8-51}$$

which is going to be approximately equal to R_D. Hence, we can still use the approximate model of the FET without r_o. In this case,

$$R_o \cong R_D$$
$$R_L' \cong R_D \parallel R_L$$
$$R_L'' \cong R_D$$

Example 8-4 — Enhancement MOSFET — ANALYSIS

We will now analyze a common-source MOSFET amplifier circuit with a partially un-bypassed source capacitor, as shown in Figure 8-29. We will, however, use the approximate model and verify the result of our analysis by simulation.

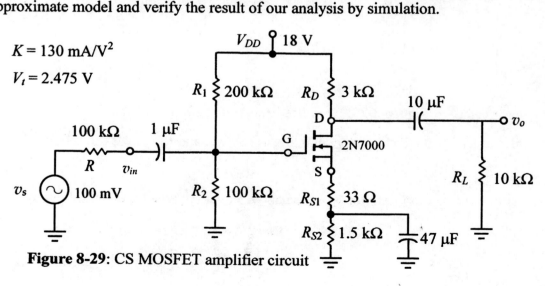

Figure 8-29: CS MOSFET amplifier circuit

Solution:

Note that this is the amplifier circuit of Example 8-3 with the addition of the un-bypassed source resistor $R_{S1} = 33\ \Omega$. Being so small in value, the addition of R_{S1} will have very little or no effect on the DC bias. We will, however, recalculate the DC bias voltages and currents, just to see how significant the impact of R_{S1} on the DC bias is.

$$V_G = V_{DD}\frac{R_2}{R_1+R_2} \qquad V_G = 18\text{ V}\times\frac{100\text{ k}\Omega}{200\text{ k}\Omega + 100\text{ k}\Omega} = 6\text{ V}$$

The voltage V_{GS} is determined from Equation 7-32

$$V_{GS}\Big|_{n-channel} = \frac{-b+\sqrt{b^2-4ac}}{2a}$$

where

$a = KR_S \qquad\qquad a = 0.130 \times 1533 = 199$

$b = 1 - 2KR_S|V_t| \qquad b = 1 - (2 \times 0.130 \times 1533 \times 2.475) = -985$

$c = KR_S V_t^2 - |V_G| \qquad c = (0.130 \times 1533 \times 2.475^2) - 6 = 1215$

$$V_{GS} = \frac{985 + \sqrt{(985)^2 - 4 \times 199 \times 1215}}{2 \times 199} = \mathbf{2.61\text{ V}}$$

The dc drain current I_D can be determined from Equation 7-15

$I_D = \dfrac{V_G - V_{GS}}{R_S} \qquad I_D = \dfrac{6\text{ V} - 2.61\text{ V}}{1.533\text{ k}\Omega} = 2.21\text{ mA}$

$g_m = 2K(V_{GS} - V_t) \qquad g_m = 2 \times 130\text{ mA/V}^2\,(2.61 - 2.475) \cong 35\text{ mA/V}$

Having determined the transconductance g_m, we will now draw the small-signal equivalent circuit using the approximate FET model, as shown in Figure 8-30.

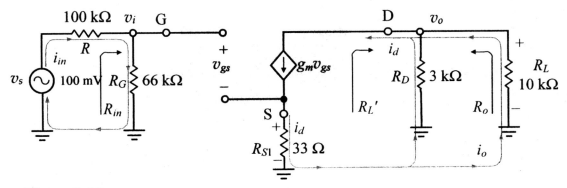

Figure 8-30:
Small-signal equivalent circuit of Figure 8-29, using approximate FET model

Input resistance:

The input resistance is determined from Equation 8-40, as follows:

$R_{in} = R_G = R_1 \parallel R_2 \qquad\qquad R_{in} = 200\text{ k}\Omega \parallel 100\text{ k}\Omega = 66.67\text{ k}\Omega$

Output resistance:

The output resistance can be determined from Equation 8-51

$R_o \cong R_D$ $\qquad\qquad R_o = 3\ k\Omega$

Total load resistance:

The total load resistance seen by the small-signal is determined by

$R_L' = R_D \parallel R_L$ $\qquad\qquad R_L' = 3\ k\Omega \parallel 10\ k\Omega = 2.3\ k\Omega$

Voltage gain with load:

The voltage gain with load can be computed from Equation 8-38

$$A_{v(WL)} = \frac{v_o}{v_{in}} = \frac{-g_m R_L'}{1 + g_m R_{S1}} \qquad A_{v(WL)} = \frac{v_o}{v_{in}} = \frac{-35\ mA/V \times 2.3\ k\Omega}{1 + (35\ mA/V \times 33\ \Omega)} = -37.35$$

Voltage gain without load:

The internal load of the amplifier without an external load is defined as R_L'', and the voltage gain without an external load is determined from Equation 8-39

$R_L'' = R_D$ $\qquad\qquad R_L'' = 3\ k\Omega$

$$A_{v(NL)} = \frac{v_o}{v_{in}} = \frac{-g_m R_L''}{1 + g_m R_{S1}} \qquad A_{v(NL)} = \frac{v_o}{v_{in}} = \frac{-35\ mA/V \times 3\ k\Omega}{1 + (35\ mA/V \times 33\ \Omega)} = -48.7$$

Output voltage with load:

First, we need to determine the input voltage, and then the output voltage

$$v_{in} = v_s \frac{R_{in}}{R + R_{in}} \qquad\qquad v_{in} = 100\ mV \times \frac{66.67\ k\Omega}{100\ k\Omega + 66.67\ k\Omega} = 40\ mV$$

$$v_{o(WL)} = A_{v(WL)} v_{in} \qquad\qquad v_{o(WL)} = -40\ mV \times 37.35 = -1.5\ V$$

Let us now verify the results of our analysis by simulation. The results of simulation are presented in Figures 8-31 through 8-35.

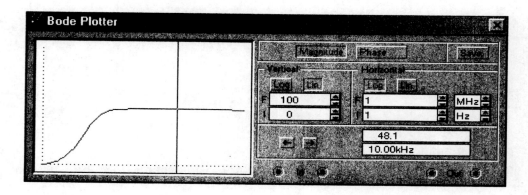

Figure 8-31: Electronics Workbench Bode plotter showing the voltage gain of the MOSFET amplifier of Example 8-4 without load.

Chapter 8

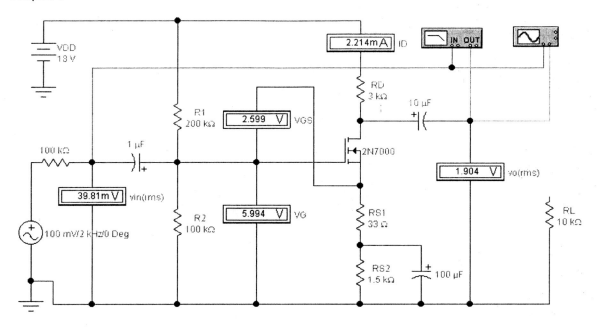

Figure 8-32: Circuit setup with simulation results of Example 8-4 without load created in Electronics Workbench

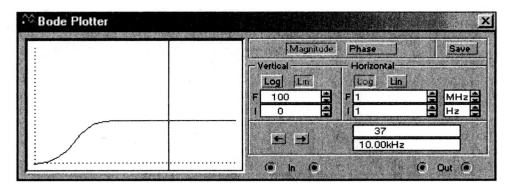

Figure 8-33: Electronics Workbench Bode plotter showing the voltage gain of the MOSFET amplifier of Example 8-4 with load.

The DC bias voltage V_{GS} = 2.599 V and current I_D = 2.214 mA, as recorded in the DMMs of the Electronics Workbench in Figure 8-33, are practically identical to the algebraically determined V_{GS} of 2.61 V and I_D of 2.21 mA.

The input resistance:

The input resistance R_{in} is determined from DMM recordings of Figure 8-32 and with Equation 5-10, as follows:

$$R_{in} = v_{in}\frac{R}{v_S - v_{in}} = 39.81\,\text{mV}\frac{100\,\text{k}\Omega}{100\,\text{mV} - 39.81\,\text{mV}} = 66.14\,\text{k}\Omega$$

319

Chapter 8

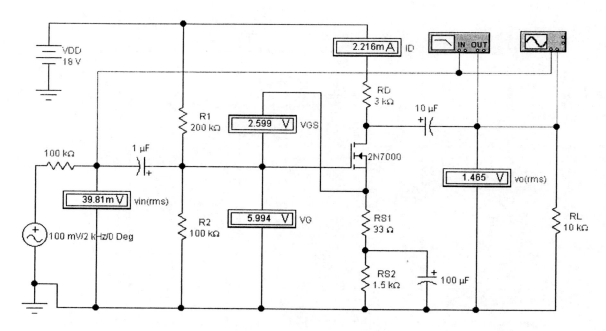

Figure 8-34: Circuit setup with simulation results of Example 8-4 with load created in Electronics Workbench

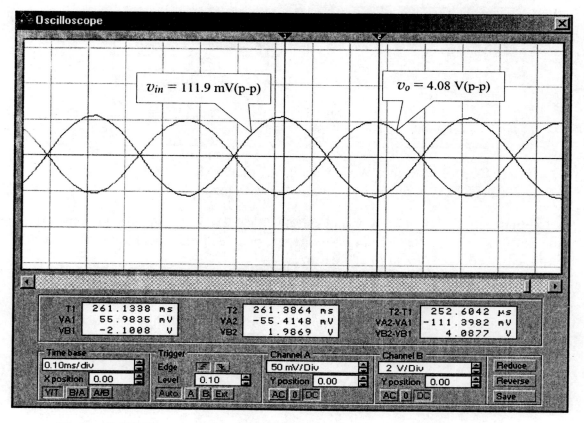

Figure 8-35: Electronics Workbench oscilloscope showing the input and output waveforms of Example 8-4 with load.

The output resistance:

The output resistance R_o is determined from the DMM recordings of Figures 8-32 and 8-34 and from Equation 5-13

$$R_o = \frac{v_o}{v_{in}} = \frac{v_{o(NL)} - v_{o(WL)}}{v_{o(WL)}} R_L = \frac{1.808\text{ V} - 1.391\text{ V}}{1.391\text{ V}} \times 10\text{ k}\Omega = 3\text{ k}\Omega$$

A summary of the analytically expected results of the MOSFET amplifier of Example 8-4 versus the results obtained from simulation are exhibited in Table 8-3.

Table 8-3: Results Summary for Example 8-4

Parameter	Analysis Results	Simulation Results
I_{DQ}	2.21 mA	2.21 mA
V_{GSQ}	2.61 V	2.6 V
$A_{v(NL)}$	−48.7	48.1
$A_{v(WL)}$	−37.35	37
R_{in}	66.67 kΩ	66.1 kΩ
R_o	3 kΩ	3 kΩ
$v_{o(WL)}$	1.5 V(rms)	1.465 V(rms)
Phase	180°	180°

All of the results obtained from analysis are either identical or agree very closely with the results obtained from simulation.

Practice Problem 8-5 — Enhancement MOSFET — ANALYSIS

Analyze the common-source MOSFET amplifier circuit of Figure 8-36 and verify the result of your analysis by simulation.

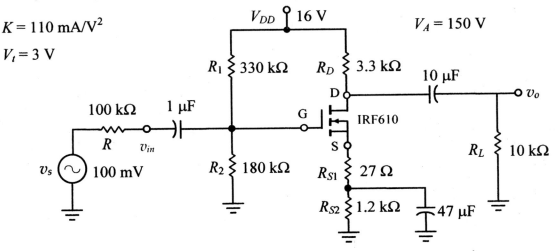

Figure 8-36: A common-source MOSFET amplifier

Section Summary — Enhancement MOSFET — ANALYSIS

Summary of Equations for the Analysis of CS Amplifier with an Unbypassed R_S

$$V_G = V_{DD} \frac{R_2}{R_1 + R_2} \qquad V_{GS}\Big|_{n-ch} = \frac{-b + \sqrt{b^2 - 4ac}}{2a} \qquad V_{GS}\Big|_{p-ch} = \frac{+b - \sqrt{b^2 - 4ac}}{2a}$$

$$a = KR_S \qquad b = 1 - 2KR_S|V_t| \qquad c = KR_S V_t^2 - |V_G|$$

$$I_D = \frac{V_G - V_{GS}}{R_S} \qquad g_m = 2K(V_{GS} - V_t) \qquad r_o = \frac{V_A}{I_{DQ}}$$

$$R_{in} = R_G = R_1 \| R_2 \qquad R_o = R_D \| [r_o + (1 + g_m r_o)R_{S1}] \cong R_D$$

$$R_L'' = r_o \| R_D \cong R_D \qquad A_{v(NL)} = \frac{v_o}{v_{in}} = \frac{-g_m \cdot R_L''}{(1 + g_m R_{S1})} \qquad A_i = \frac{i_o}{i_{in}} = A_{v(WL)} \frac{R_{in}}{R_L}$$

$$R_L' = R_D \| R_L \qquad A_{v(WL)} = \frac{v_o}{v_{in}} = \frac{-g_m \cdot R_L'}{(1 + g_m R_{S1})} \qquad A_p = \frac{P_o}{P_{in}} = A_{v(WL)} A_i$$

$$v_{in} = v_S \frac{R_{in}}{R + R_{in}} \qquad v_{o(WL)} = A_{v(WL)} \cdot v_{in} \text{ (inverted)} \qquad v_{o(NL)} = A_{v(NL)} \cdot v_{in}$$

Practice Problem 8-6 — Enhancement MOSFET — ANALYSIS

Analyze the common-source MOSFET amplifier circuit of Figure 8-37.

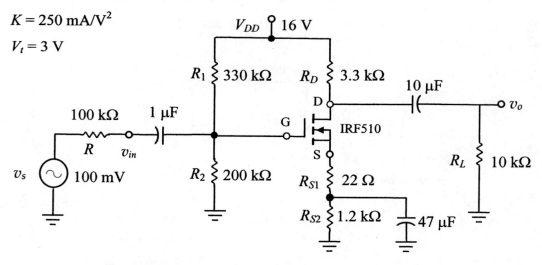

Figure 8-37: Common-source MOSFET amplifier circuit

8.6 ANALYSIS OF THE COMMON-GATE MOSFET AMPLIFIER

So far, we have discussed the small-signal analysis of the common-source amplifier with JFET and MOSFET, and the common-drain or source-follower with JFET only. The small-signal analysis technique that we have developed for the source-follower with JFET applies equally to the source-follower circuit utilizing MOSFET. We now wish to present the small-signal analysis of the common-gate MOSFET amplifier of Figure 8-38, as shown below. Note that the method of analysis presented here for the MOSFET common-gate amplifier applies equally to the JFET common-gate amplifier.

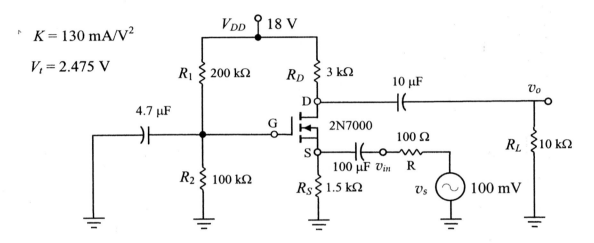

Figure 8-38: Common-gate MOSFET amplifier circuit

As you may have noticed, the common-gate amplifier is analogous to the common-base configuration in the BJT family of amplifiers and has similar characteristics. Recall that the common-base amplifier was a non-inverter and suffered from a very low input resistance. Likewise, we can expect non-inversion and a relatively low input resistance from the common-gate amplifier.

The above common-gate circuit of Figure 8-38 is DC equivalent to the common-source amplifier of Figure 8-22 presented in Example 8-3. Hence, we will refrain from repeating the DC analysis here again; and instead, we will use the results of analysis that we have already determined in Example 8-3, as follows:

$$V_{GS} = 2.6 \text{ V}, \qquad I_D = 2.266 \text{ mA}, \qquad g_m = 32.5 \text{ mA/V}$$

Next, we will draw the small-signal equivalent circuit using the approximate FET model, as shown in Figure 8-39.

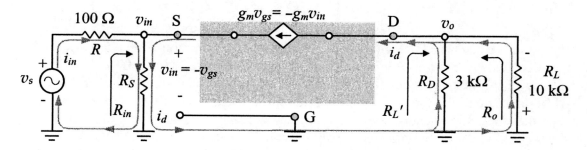

Figure 8-39: The small-signal equivalent circuit of Figure 8-38

Voltage gain with load:

The output voltage with load will be the following:

$$v_o = -i_d R_L' = -g_m v_{gs} R_L' \tag{8-52}$$

where,

$$R_L' = R_D || R_L \tag{8-53}$$

$$A_{v(WL)} = \frac{v_o}{v_{in}} = \frac{v_o}{-v_{gs}} = \frac{-g_m v_{gs} R_L'}{-v_{gs}} = g_m R_L' \tag{8-54}$$

$$A_{v(WL)} = \frac{v_o}{v_{in}} = g_m R_L' \tag{8-55}$$

$A_{v(WL)}$ = 32.5 mA/V (3 kΩ || 10 kΩ) = 75

No-load voltage gain:

Since the internal load of the amplifier without external load is simply the R_D, R_L' in Equation 8-55 above will be replaced with R_D, as follows:

$$A_{v(NL)} = \frac{v_o}{v_{in}} = g_m R_D \tag{8-56}$$

$A_{v(NL)}$ = 32.5 mA/V × 3 kΩ = 97.5

Note that voltage gains with and without load as determined above for this common-gate amplifier, which is DC equivalent to the common-source amplifier of Example 8-3, are identical to voltage gains of the common-source amplifier of Example 8-3, except that the output is in-phase with the input signal.

Input resistance:

The input resistance R_{in} will have to be determined by considering the amplifier as a two-port network and applying the two-port network theory as we have discussed in Chapter 4, and have used it previously in determining the expressions for the input and output resistances when it became necessary.

According to the two-port network theory, the input resistance is determined with Equation 5-45, which requires setting the $v_o = 0$ by grounding the output.

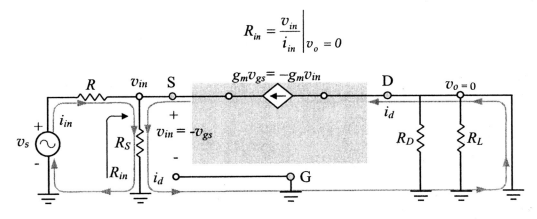

Figure 8-40: Determining the input resistance of Figure 8-37

With the output grounded, both the gate and drain are at the ground level, shorting out the R_D and R_L to ground.

The current through the resistor R_S is the sum of two currents i_{in} and i_d.

$$i_{R_S} = i_{in} + i_d = \frac{v_{in}}{R_S} \tag{8-57}$$

and,

$$i_{in} = \frac{v_{in}}{R_S} - i_d \tag{8-58}$$

$$i_{in} = \frac{v_{in}}{R_S} - i_d = \frac{v_{in}}{R_S} - g_m v_{gs} = \frac{v_{in}}{R_S} - (-g_m v_{in}) \tag{8-59}$$

$$i_{in} = \frac{v_{in}}{R_S} - i_d = \frac{v_{in}}{R_S} + g_m v_{in} = v_{in}\left(\frac{1}{R_S} + g_m\right) = v_{in}\frac{(1 + g_m R_S)}{R_S} \tag{8-60}$$

Substituting for i_{in} in Equation 5-45 results in the following for R_{in}:

$$R_{in} = \frac{v_{in}}{i_{in}} = \frac{R_S}{1 + g_m R_S}$$

$$R_{in} = \frac{R_S}{1 + g_m R_S} = r_m \| R_S \tag{8-61}$$

$$R_{in} = \frac{1.5\,k\Omega}{1 + (33\,mA/V \times 1.5\,k\Omega)} \cong 30\,\Omega$$

Output resistance:

The output for both the common-source and the common-gate are taken at the drain terminal. Hence, The output resistance of the common-gate would be the same as that of the common-source amplifier.

$$R_o = r_o \| R_D \cong R_D \tag{8-62}$$

$R_o = 3\,k\Omega$

The current gain:

The current gain is determined with Equation 8-15, as follows:

$$A_i = \frac{i_o}{i_{in}} = A_{v(WL)}\frac{R_{in}}{R_L} \qquad A_i = \frac{i_o}{i_{in}} = 75\frac{30\,\Omega}{10\,k\Omega} = 0.225$$

Note that like the common-base, the common-gate has no current gain.

The power gain:

$$A_p = \frac{P_o}{P_{in}} = A_{v(WL)}A_i \qquad A_p = 75 \times 0.225 = 16.875$$

Output voltage with load:

To determine the output voltage, we need to determine the input voltage v_{in} first.

$$v_{in} = v_S \frac{R_{in}}{R + R_{in}} \qquad v_{in} = 100\,mV\frac{30\,\Omega}{100\,\Omega + 30\,\Omega} = 23\,mV$$

$$v_{o(WL)} = A_{v(WL)} \times v_{in} \qquad v_{o(WL)} = 75 \times 23\,mV = 1.725\,V$$

Conclusion:

A MOSFET amplifier designed as common-source with a certain gain will have the same gain and output resistance if operated as common-gate. However, the output will be in-phase with the input and it will have a very low input resistance.

Section Summary **ANALYSIS**

Summary of Equations for the Analysis of the EMOS Common-Gate Amplifier

$$V_G = V_{DD}\frac{R_2}{R_1 + R_2} \qquad V_{GS}\bigg|_{n-ch} = \frac{-b + \sqrt{b^2 - 4ac}}{2a} \qquad V_{GS}\bigg|_{p-ch} = \frac{+b - \sqrt{b^2 - 4ac}}{2a}$$

$$a = KR_S \qquad b = 1 - 2KR_S|V_t| \qquad c = KR_S V_t^2 - |V_G|$$

$$I_D = \frac{V_G - V_{GS}}{R_S} \qquad g_m = 2K(V_{GS} - V_t) \qquad V_t = V_{GS(th)}$$

$$R_{in} = r_m \| R_S = \frac{R_S}{1 + g_m R_S} \qquad R_o = R_D \| r_o \qquad r_o = \frac{V_A}{I_{DQ}} \qquad r_m = \frac{1}{g_m}$$

$$R_L'' = r_o \| R_D \qquad A_{v(NL)} = \frac{v_o}{v_{in}} = g_m \cdot R_L'' \qquad A_i = \frac{i_o}{i_{in}} = A_{v(WL)}\frac{R_{in}}{R_L}$$

$$R_L' = r_o \| R_D \| R_L \qquad A_{v(WL)} = \frac{v_o}{v_{in}} = g_m \cdot R_L' \qquad A_p = \frac{P_o}{P_{in}} = A_{v(WL)}A_i$$

$$v_{in} = v_S \frac{R_{in}}{R + R_{in}} \qquad v_{o(WL)} = A_{v(WL)} \cdot v_{in} \qquad v_{o(NL)} = A_{v(NL)} \cdot v_{in}$$

Chapter 8

8.7 ANALYSIS OF THE FET *CASCODE* AMPLIFIER

Similar to the BJT *cascode* amplifier, which is a cascade of common-emitter and common-base configurations, the FET *cascode* amplifier is a cascade of common-source and common-gate configuratios. Recall that the common-source amplifier can provide a very high input resistance, and the common-gate offers a much desirable high-frequency response, but it suffers from a very low input resistance. The FET *cascode* configuration utilizes the superior characteristics of both configurations, while avoiding their undesirable characteristics. The circuit of a JFET cascode amplifier is shown in Figure 8-41 below.

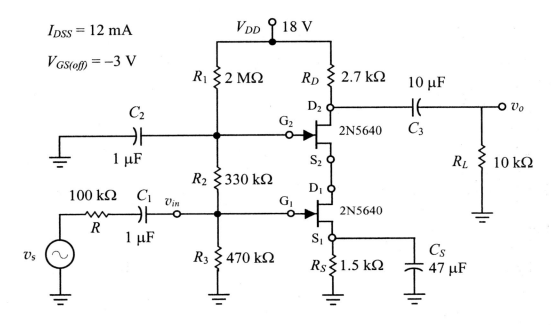

Figure 8-41: A JFET *cascode* amplifier circuit

DC analysis:

Applying the voltage divider rule at G_1, we can solve for the voltage V_{G1}, as follows:

$$V_{G1} = V_{DD} \frac{R_3}{R_1 + R_2 + R_3}$$

$$V_{G1} = 18\text{ V} \frac{470\text{ k}\Omega}{2\text{ M}\Omega + 330\text{ k}\Omega + 470\text{ k}\Omega} \cong 3\text{ V}$$

V_{GS1} is determined with Equation 7-32, as follows:

$$V_{GS}\bigg|_{n-channel} = \frac{-b + \sqrt{b^2 - 4ac}}{2a}$$

where,

$$a = \frac{I_{DSS} R_S}{V_P^2} \qquad a = \frac{12\text{ mA} \times 1.5\text{ k}\Omega}{(3\text{ V})^2} = 2 \qquad V_p = |V_{GS(off)}|$$

327

Chapter 8

$$b = \frac{2I_{DSS}R_S}{|V_P|} + 1 \qquad b = \frac{2 \times 12\,\text{mA} \times 1.5\,\text{k}\Omega}{3\,\text{V}} + 1 = 13$$

$$c = I_{DSS}R_S - |V_G| \qquad c = (12\,\text{mA} \times 1.5\,\text{k}\Omega) - 3 = 15$$

$$V_{GS1} = \frac{-13 + \sqrt{(13)^2 - 4 \times 2 \times 15}}{2 \times 2} \cong -1.5\,\text{V}$$

Having determined V_{GS}, I_D can be determined with Equations 7-15, as follows:

$$I_D = \frac{V_G - V_{GS}}{R_S} \qquad I_D = I_{D1} = I_{D2} = \frac{3\,\text{V} + 1.5\,\text{V}}{1.5\,\text{k}\Omega} = 3\,\text{mA}$$

$$g_m = \frac{2I_{DSS}}{|V_{GS(off)}|}\left(1 - \frac{V_{GS}}{V_{GS(off)}}\right) \qquad g_m = \frac{2 \times 12\,\text{mA}}{3\,\text{V}}\left(1 - \frac{1.5\,\text{V}}{3\,\text{V}}\right) = 4\,\text{mA/V}$$

Applying the voltage divider rule at G_2, we can solve for the voltage V_{G2}, as follows:

$$V_{G2} = V_{DD}\frac{R_2 + R_3}{R_1 + R_2 + R_3}$$

$$V_{G2} = 18\,\text{V}\frac{330\,\text{k}\Omega + 470\,\text{k}\Omega}{2\,\text{M}\Omega + 330\,\text{k}\Omega + 470\,\text{k}\Omega} = 5.14\,\text{V}$$

Both transistors being identical, and drain currents being equal, the gate-source voltages V_{GS1} and V_{GS2} must be equal. This can also be verified easily with Shockley's equation.

$$V_{GS1} = V_{GS2} = -1.5\,\text{V}$$

$$V_{GS2} = V_{G2} - V_{S2} = -1.5\,\text{V}$$

$$V_{S2} = V_{G2} + 1.5\,\text{V}$$

$$V_{S2} = V_{D1} = 5.14\,\text{V} + 1.5\,\text{V} = 6.64\,\text{V}$$

$$V_{S1} = I_D R_S = 3\,\text{mA} \times 1.5\,\text{k}\Omega = 4.5\,\text{V}$$

$$V_{DS1} = V_{D1} - V_{S1} = 6.64\,\text{V} - 4.5\,\text{V} = 2.14\,\text{V}$$

$$V_{D2} = V_{DD} - I_D R_D = 18\,\text{V} - (3\,\text{mA} \times 2.7\,\text{k}\Omega) = 9.9\,\text{V}$$

$$V_{DS2} = V_{D2} - V_{S2} = 9.9\,\text{V} - 6.64\,\text{V} = 3.26\,\text{V}$$

Small-signal analysis:

The small-signal equivalent circuit of the cascode amplifier of Figure 8-41 is shown in Figure 8-42 below.

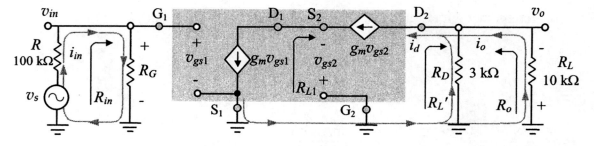

Figure 8-42: Small-signal equivalent circuit of Figure 8-41

Chapter 8

Input resistance:

The input resistance R_{in} is the parallel combination of R_3 and R_2:
$R_{in} = R_G = R_3 \| R_2$ $\quad\quad R_{in} = 470 \text{ k}\Omega \| 330 \text{ k}\Omega = 194 \text{ k}\Omega$

Output resistance:

The output resistance R_o is simply R_D
$R_o = R_D$ $\quad\quad R_o = 2.7 \text{ k}\Omega$

First stage load resistance:

The load of the first stage (R_{L1}) is the input resistance (R_{in2}) of the second stage, which is a common-gate configuration, and its input resistance is simply given by
$R_{L1} = R_{in2} = r_m = 1/g_m = 1/4 \text{ mA/V} = 250 \text{ A/V}$

Second stage load resistance:

$R_{L2} = R_L' = R_D \| R_L$ $\quad\quad R_L' = 2.7 \text{ k}\Omega \| 10 \text{ k}\Omega = 2.126 \text{ k}\Omega$

First stage voltage gain:

$$A_{v1} = \frac{v_{o1}}{v_{in}} = \frac{-g_m v_{gs} r_m}{v_{gs}} \quad\quad A_{v1} = \frac{v_{o1}}{v_{in}} = -g_m r_m \quad\quad (8\text{-}63)$$

Second stage voltage gain:

$$A_{v2} = \frac{v_o}{v_{o1}} = \frac{-g_m v_{gs} R_L'}{-g_m v_{gs} r_m} \quad\quad A_{v2} = \frac{v_o}{v_{o1}} = \frac{R_L'}{r_m} \quad\quad (8\text{-}64)$$

Overall voltage gain:

$$A_v = A_{v1} A_{v2} = -g_m r_m \frac{R_L'}{r_m} = -g_m R_L' \quad\quad (8\text{-}65)$$

$$A_v = -g_m R_L' \quad\quad (8\text{-}66)$$

$A_v = -g_m R_L' = -4 \text{ mA/V} \times 2.125 \text{ k}\Omega = -8.5$

Note that the effect of the transistor r_o was neglected in the process for two reasons: (1) to make the analysis simpler, (2) its effect is negligible with the cascode configuration. However, if r_o is to be accounted for, then R_o and R_L' would be as follows:

$$R_o = 2r_o \| R_D \cong R_D$$
$$R_L' = 2r_o \| R_D \| R_L \cong R_D \| R_L$$

Chapter 8

Practice Problem 8-7 — JFET Cascode — ANALYSIS

Analyze the JFET *cascode* amplifier of Figure 8-43, and determine the following:
$V_{GS}, I_D, V_{DS1}, V_{DS2}, R_{in}, R_o, A_v$.

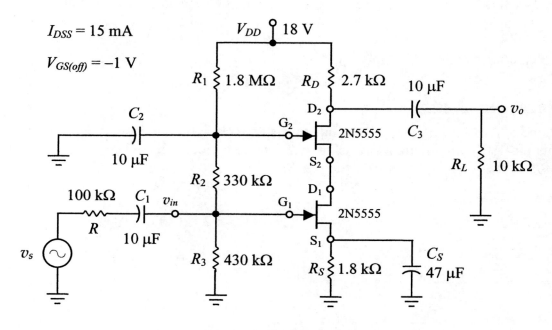

Figure 8-43: A JFET *cascode* amplifier circuit

Answers:

$V_{GS} = -0.633$ V, $I_D = 2$ mA, $V_{DS1} = 2.37$ V, $V_{DS2} = 6.6$ V, $R_{in} = 187$ kΩ, $R_o = 2.7$ kΩ, $A_v = -23.37$

8.8 DESIGN OF SMALL-SIGNAL JFET AND MOSFET AMPLIFIERS

Although the approach to designing the JFET or MOSFET amplifiers for a set of given specifications is somewhat different from that of the BJT amplifiers; however, the same basic design strategies apply equally to FET amplifiers. That is, we will first design the DC bias for an optimum and stable operating point that allows maximum signal swing at the output. In addition, we will employ a configuration that permits designing for a specific gain without significantly altering the established Q-point.

Example 8-5 — n-channel JFET — DESIGN

Let us design a common-source JFET amplifier for the following set of specifications:
  Supply voltage: $V_{DD} = 16$ V   Drain current: $I_{DQ} = 2$ mA
  No load voltage gain: $|A_{v(NL)}| = 26$ Output voltage swing: $v_{o(max)} \geq 6$ V(p-p)
  Input Resistance $R_{in} \geq 330$ kΩ Output Resistance $R_o \leq 3.3$ kΩ
  Stability considerations:

As the JFET characteristics vary by ±20%, the Q-point I_D is not to vary more than ±5%.

Solution:
In order to achieve the specified stability and the voltage gain, we will employ the voltage-divider bias with a partially unbypassed source resistor, as shown in Figure 8-44.

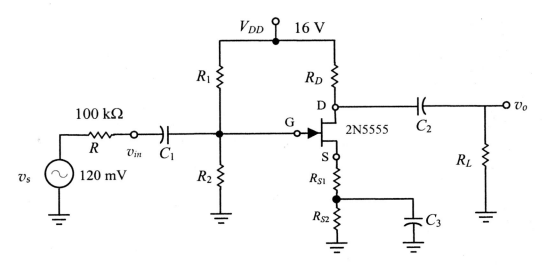

Figure 8-44: Common-source JFET amplifier circuit

Next, we will begin the DC bias design according to the procedures established in the previous chapter by clarifying the stability specifications, as follows:
Nominal characteristics and the Q-point:

$$I_{DSS} = 15 \text{ mA}, \quad V_{GS(off)} = -1 \text{ V}, \quad I_{DQ} = 2 \text{ mA}$$

Upward deviation of the characteristics and the limit of the corresponding shift in I_D:

$$I_{DSS2} = 18 \text{ mA}, \quad V_{GS(off)2} = -1.2 \text{ V}, \quad I_{D2} = 2.1 \text{ mA}$$

Downward deviation of the characteristics and the limit of the corresponding shift in I_D:

$$I_{DSS1} = 12 \text{ mA}, \quad V_{GS(off)1} = -0.8 \text{ V}, \quad I_{D1} = 1.9 \text{ mA}$$

The next step in the design process is to determine the slope of the bias line in order to meet the specified stability. With Equation 7-19 we can determine the V_{GSQ}, V_{GS1}, and V_{GS2} that correspond to I_{DQ}, I_{D1}, and I_{D2}, respectively.

$$V_{GS} = \left(1 - \sqrt{\frac{I_D}{I_{DSS}}}\right) \cdot V_{GS(off)}$$

$$V_{GSQ} = \left(1 - \sqrt{\frac{I_{DQ}}{I_{DSS}}}\right) \cdot V_{GS(off)} = \left(1 - \sqrt{\frac{2 \text{ mA}}{15 \text{ mA}}}\right) \times (-1 \text{ V}) = -0.635 \text{ V}$$

Chapter 8

$$V_{GS2} = \left(1 - \sqrt{\frac{I_{D2}}{I_{DSS2}}}\right)V_{GS(off)2} = \left(1 - \sqrt{\frac{2.1\,\text{mA}}{18\,\text{mA}}}\right) \times (-1.2\,\text{V}) = -0.79\,\text{V}$$

The slope of the bias line can be determined with Equation 7-20

$$m = \frac{\Delta I_D}{\Delta V_{GS}} = \frac{I_{D2} - I_{D1}}{V_{GS2} - V_{GS1}} \qquad m = \frac{2.1\,\text{mA} - 1.9\,\text{mA}}{-0.79\,\text{V} - (-0.48\,\text{V})} = \frac{0.2\,\text{mA}}{-0.31\,\text{V}} = -0.645\,\frac{\text{mA}}{\text{V}}$$

Using Equation 7-22, we can solve for V_G to obtain

$$V_G = \frac{I_{DQ}}{|m|} - |V_{GSQ}| \qquad V_G = \frac{2\,\text{mA}}{0.645\,\text{mA/V}} - 0.635\,\text{V} = 2.466\,\text{V}$$

Next, V_S is determined from Equation 7-23

$$V_S = V_G - V_{GS} \qquad V_S = 2.466\,\text{V} - (-0.635\,\text{V}) = 3.1\,\text{V}$$

Having determined V_S, R_S can be determined from Equation 7-24

$$R_S = \frac{V_S}{I_D} \qquad R_S = \frac{3.1\,\text{V}}{2\,\text{mA}} = 1.55\,\text{k}\Omega$$

From $V_{DD} = 16$ V, 3.1 V drops across R_S and we will allocate 6.6 V across R_D, leaving $V_{DS} = 6.3$ V. Then, R_D can be determined from Equation 7-25

$$R_D = \frac{V_{RD}}{I_D} \qquad R_D = \frac{6.6\,\text{V}}{2\,\text{mA}} = 3.3\,\text{k}\Omega$$

Assuming $R_2 = 400$ kΩ, R_1 can be determined from Equation 7-26

$$R_1 = \left(\frac{V_{DD}}{V_G} - 1\right)R_2 \qquad R_1 = \left(\frac{16\,\text{V}}{2.466\,\text{V}} - 1\right)R_2 = 5.49 R_2 = 2.2\,\text{M}\Omega$$

The transconductance g_m is determined from Equation 8-2

$$g_m = \frac{2 I_{DSS}}{|V_{GS(off)}|}\left(1 - \frac{V_{GS}}{V_{GS(off)}}\right) \qquad g_m = \frac{2 \times 15\,\text{mA}}{1\,\text{V}}\left(1 - \frac{0.635\,\text{V}}{1\,\text{V}}\right) = 11\,\text{mA/V}$$

The internal load of the amplifier without an external load is simply R_D, and the voltage gain without an external load is determined from Equation 8-39

$$R_L'' = R_D = 3.3\,\text{k}\Omega$$

$$A_{v(NL)} = \frac{v_o}{v_{in}} = \frac{-g_m R_D}{(1 + g_m R_{S1})} = -26$$

$$|A_v| = \frac{g_m R_D}{1 + g_m R_{S1}} = \frac{R_D}{r_m + R_{S1}}$$

$$r_m + R_{S1} = \frac{R_D}{|A_v|}$$

Solving for R_{S1} in the above equation results in

$$R_{S1} = \frac{R_D}{|A_v|} - r_m \qquad (8\text{-}67)$$

$$R_{S1} = \frac{R_D}{|A_v|} - r_m \qquad (8\text{-}67)$$

$$r_m = \frac{1}{g_m} = \frac{1}{0.011} = 91\ \Omega \qquad\qquad R_{S1} = \frac{3300}{26} - 91 = 36\ \Omega$$

$$R_S = R_{S1} + R_{S2} = 1.55\ k\Omega \qquad\qquad R_{S2} = R_S - R_{S1} \cong 1.5\ k\Omega$$

Summary of design computations:

$I_D = 2$ mA, $V_{DS} = 6.3$ V, $V_{DD} = 16$ V, $V_{GS} = -0.635$ V, $V_S = 3.1$ V

$R_1 = 2.2$ MΩ, $R_2 = 400$ kΩ, $R_{S2} = 1.5$ kΩ, $R_{S1} = 36$ Ω, $R_D = 3.3$ kΩ

Test Run and Design Verification:

We will now test the design by simulating the circuit with Electronics Workbench. The results of the simulation are presented in Figures 8-45 through 8-48.

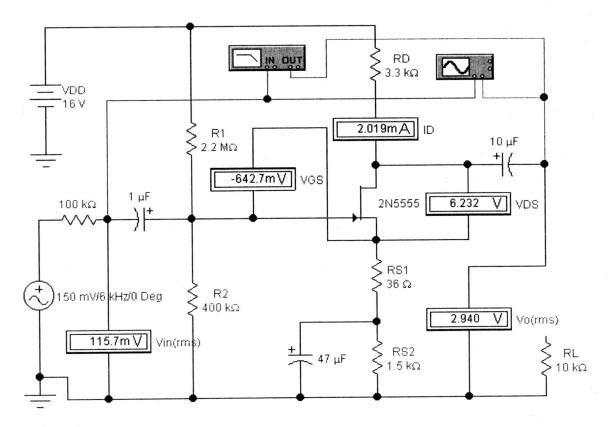

Figure 8-45: Circuit setup with simulation results of Design Example 8-5 without load created in Electronics Workbench

The input resistance:

The input resistance R_{in} is determined from the DMM recordings of Figure 8-44, as follows:

$$R_{in} = v_{in} \frac{R}{v_S - v_{in}} = 115.7 \text{ mV} \frac{100 \text{ k}\Omega}{150 \text{ mV} - 115.7 \text{ mV}} = 337 \text{ k}\Omega$$

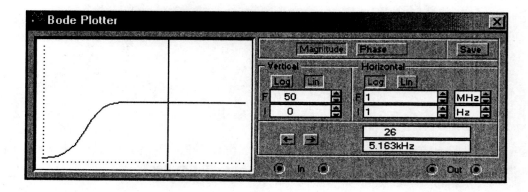

Figure 8-46: Electronics Workbench Bode plotter showing the voltage gain $A_{v(NL)} = 26$

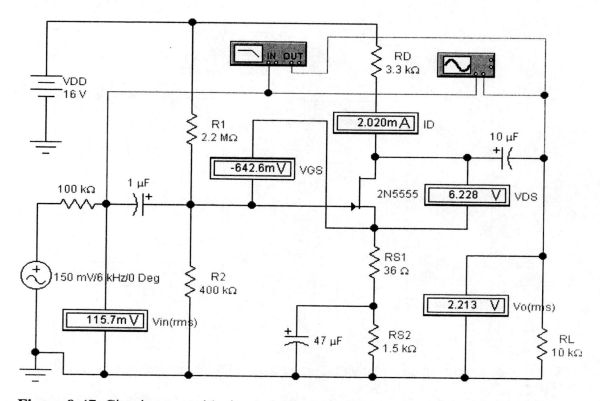

Figure 8-47: Circuit setup with simulation results of Design Example 8-5 with load created in Electronics Workbench

The output resistance:

The output resistance R_o is determined from the DMM recordings of Figures 8-44, 8-47, and with Equation 5-13, as follows:

$$R_o = \frac{v_o}{v_{in}} = \frac{v_{o(NL)} - v_{o(WL)}}{v_{o(WL)}} R_L = \frac{2.37 \text{ V} - 1.783 \text{ V}}{1.783 \text{ V}} 10 \text{ k}\Omega = 3.3 \text{ k}\Omega$$

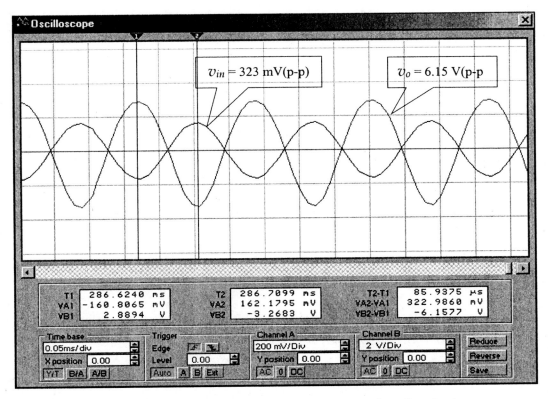

Figure 8-48: Electronics Workbench oscilloscope showing the input and output signals of design Example 8-5

The results of simulation versus design specifications are exhibited in Table 8-4 below:

Table 8-4: Results summary of Example 8-5

Parameter	Spec. / Expected	Result of simulation
I_{DQ}	2 mA	2.02 mA
V_{GSQ}	−0.635 V	−0.64 V
V_{DSQ}	6.3 V	6.23 V
$|A_{v(NL)}|$	26	26
R_{in}	≥ 330 kΩ	337 kΩ
R_o	≤ 3.3 kΩ	3.3 kΩ
$v_{o(WL)}$	≥ 6 V(p-p)	6.15 V(p-p)
Phase	180°	180°

Almost all of the results obtained by simulation are identical to or agree very closely with the design specifications.

Practice Problem 8-8 DESIGN

Design a common-source JFET amplifier for a set of specifications as given below:

Supply voltage: $V_{DD} = 18$ V
No-load voltage gain: $|A_{v(NL)}| = 20$
Input resistance $R_{in} \geq 220$ kΩ

Drain current: $I_{DQ} = 2.2$ mA
Output voltage swing: $v_{o(max)} \geq 6$ V(p-p)
Output resistance $R_o \leq 3$ kΩ

Stability considerations:

As the JFET characteristics vary by ±20%, the Q-point I_D is not to vary more than ±5%. Verify the result of your design by simulation. Use JFET 2N5555.

Example 8-7 DESIGN

Let us now design a common-source amplifier utilizing enhancement-type MOSFET that meets the following set of specifications:

Supply voltage: $V_{DD} = 20$ V
No-load voltage gain: $|A_{v(NL)}| = 50$
Input resistance $R_{in} \geq 400$ kΩ

Drain current: $I_{DQ} = 2$ mA
Output voltage swing: $v_{o(max)} \geq 6$ V(p-p)
Output resistance $R_o \leq 3$ kΩ

Stability considerations:

As the MOSFET characteristics vary by ±15%, I_{DQ} is not to vary more than ±5%.

Solution:

In order to achieve the specified stability and the voltage gain, again we will utilize the voltage-divider bias with a partially unbypassed source resistor as shown in Figure 8-49. We will use the 2N7000 MOSFET, which has the following typical characteristics:

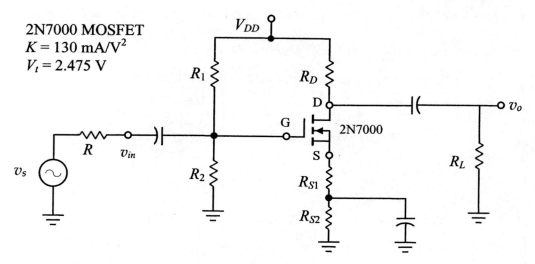

Figure 8-49: Common-source MOSFET amplifier circuit

Chapter 8

We will begin the design process from Equation 7-40 and clarify the stability requirements, as follows:

$$V_{GS} = V_t + \sqrt{\frac{I_D}{K}}$$

Nominal characteristics and the Q-point are as follows:

$$K = 130 \text{ mA/V}^2 \qquad V_{GS(th)} = V_t = 2.475 \text{ V} \qquad I_{DQ} = 2 \text{ mA}$$

Upward deviation of the characteristics and the limit of the corresponding shift in I_D:

$$K_2 = 150 \text{ mA/V}^2 \qquad V_{t2} = 2.846 \text{ V} \qquad I_{D2} = 2.1 \text{ mA}$$

Downward deviation of the characteristics and the limit of the corresponding shift in I_D:

$$K_1 = 110 \text{ mA/V}^2 \qquad V_{t1} = 2.1 \text{ V} \qquad I_{D1} = 1.9 \text{ mA}$$

Using Equation 7-40 we can now determine V_{GSQ}, V_{GS1}, and V_{GS2} that correspond to I_{DQ}, I_{D1}, and I_{D2}, respectively

$$V_{GSQ} = V_t + \sqrt{\frac{I_D}{K}} = 2.475 \text{ V} + \sqrt{\frac{2 \text{ mA}}{130 \text{ mA/V}^2}} = 2.6 \text{ V}$$

$$V_{GS2} = V_{t2} + \sqrt{\frac{I_{D2}}{K_2}} = 2.846 \text{ V} + \sqrt{\frac{2.1 \text{ mA}}{150 \text{ mA/V}^2}} = 2.964 \text{ V}$$

$$V_{GS1} = V_{t1} + \sqrt{\frac{I_{D1}}{K_1}} = 2.1 \text{ V} + \sqrt{\frac{1.9 \text{ mA}}{110 \text{ mA/V}^2}} = 2.23 \text{ V}$$

The slope of the bias line can be determined from Equation 7-20

$$|m| = \frac{\Delta I_D}{\Delta V_{GS}} = \frac{I_{D2} - I_{D1}}{V_{GS2} - V_{GS1}} \qquad |m| = \frac{2.1 \text{ mA} - 1.9 \text{ mA}}{2.964 \text{ V} - 2.23 \text{ V}} = 0.2725 \frac{\text{mA}}{\text{V}}$$

Next, V_G is determined from Equation 7-42

$$|V_G| = \frac{I_{DQ}}{|m|} + |V_{GSQ}| \qquad V_G = \frac{2 \text{ mA}}{0.2725 \text{ mA/V}} + 2.475 \text{ V} = 9.81 \text{ V}$$

$$V_{GS} = V_G - V_S \qquad V_S = V_G - V_{GS} \qquad V_S = 9.81 \text{ V} - 2.6 \text{ V} = 7.21 \text{ V}$$

R_S can be determined from Equation 7-24

$$R_S = \frac{V_S}{I_D} \qquad R_S = \frac{7.21 \text{ V}}{2 \text{ mA}} = 3.61 \text{ k}\Omega = R_{S2} + R_{S1}$$

Let $V_{DS} = 7.3 \text{ V}$, then V_{RD} is determined as follows:

$$V_{RD} = V_{DD} - V_S - V_{DS} = 20 \text{ V} - 7.21 \text{ V} - 7.3 \text{ V} \cong 5.49 \text{ V}$$

R_D can be determined from Equation 7-25

$$R_D = \frac{V_{RD}}{I_D} \qquad R_D = \frac{5.4 \text{ V}}{2 \text{ mA}} = 2.7 \text{ k}\Omega$$

Assuming $R_2 = 1 \text{ M}\Omega$, R_1 can be determined from Equation 7-26

$$R_1 = \left(\frac{V_{DD}}{V_G} - 1\right)R_2 \qquad R_1 = \left(\frac{20 \text{ V}}{9.81 \text{ V}} - 1\right)R_2 \cong R_2 = 1.039 \text{ M}\Omega \qquad R_1 = 1 \text{ M}\Omega$$

Chapter 8

The transconductance g_m is determined from Equation 8-5

$$g_m = 2K(V_{GS} - V_t) \qquad g_m = 2 \times 130\, m(2.6\,V - 2.475\,V) = 32.5\,mA/V$$

Next, the resistor R_{S1} is determined from Equation 8-67

$$R_{S1} = \frac{R_D}{|A_{v(NL)}|} - r_m$$

$$r_m = \frac{1}{g_m} = \frac{1}{0.0325\,A/V} = 31\,V/A \qquad R_{S1} = \frac{2700\,\Omega}{50} - 30\,\Omega = 24\,\Omega$$

$R_S = R_{S2} + R_{S1} = 3.67\,k\Omega$ Let $R_{S2} = 3.6\,k\Omega$, and $R_{S1} = 24\,\Omega$.

A summary of the design computations:

$R_1 = 1\,M\Omega$, $R_2 = 1\,M\Omega$, $R_{S2} = 3.6\,k\Omega$, $R_{S1} = 24\,\Omega$, $R_D = 2.7\,k\Omega$,
$I_D = 2\,mA$, $V_{DD} = 20\,V$, $V_{DS} = 7.3\,V$, $V_G = 9.81\,V$, $V_{GS} = 2.6\,V$.

Test Run and Design Verification:

We will now test the design by simulating the circuit with Electronics Workbench. The results of the simulation are presented in Figures 8-50 through 8-52.

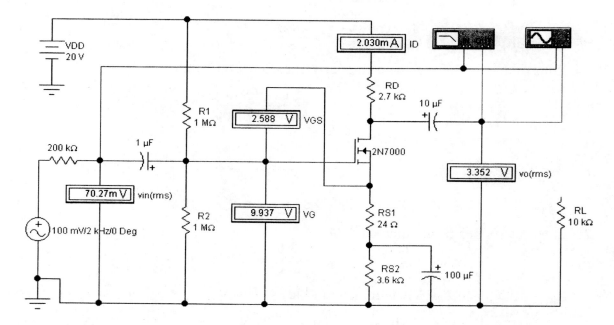

Figure 8-50: Circuit setup with simulation results of Design Example 8-6 without load created in Electronics Workbench

The input resistance:

The input resistance R_{in} is determined from the DMM recordings of Figure 8-50, as follows:

$$R_{in} = v_{in}\frac{R}{v_S - v_{in}} = 70.27\text{ mV}\frac{200\text{ k}\Omega}{100\text{ mV} - 70.27\text{ mV}} = 473\text{ k}\Omega$$

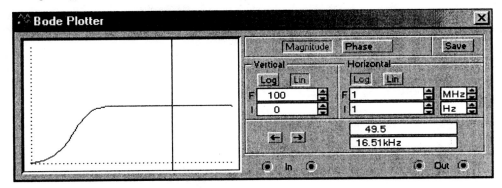

Figure 8-51: Electronics Workbench Bode plotter showing the voltage gain of Design Example 8-6

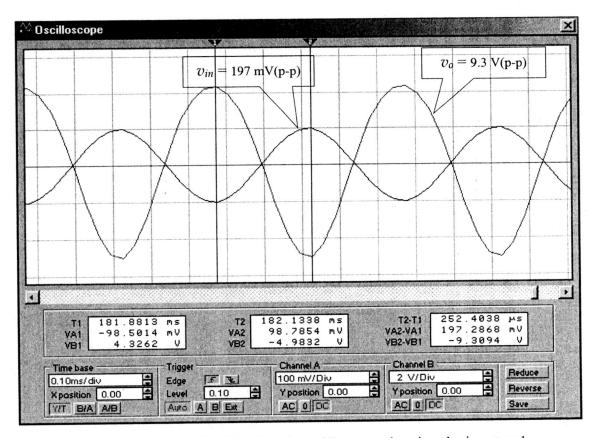

Figure 8-52: Electronics Workbench oscilloscope showing the input and output signals of Design Example 8-6

One would expect the R_{in} to be the parallel combination of R_1 and R_2, which would be 500 kΩ. However, the simulation result of 473 kΩ is due to meter (DMM) loading at the gate terminal.

The output resistance:
The output resistance R_o is approximately equal to $R_D = 2.7$ kΩ.

The results of simulation versus design specifications are exhibited in Table 8-5 below:

Table 8-5: Results summary of Example 8-6

Parameter	Spec. / Expected	Result of simulation
I_{DQ}	2 mA	2.03 mA
V_{GSQ}	2.6V	2.588 V
$|A_{v(NL)}|$	50	49.5
R_{in}	≥ 400 kΩ	473 kΩ
R_o	≤ 3 kΩ	2.7 kΩ
$v_{o(NL)}$	≥ 6 V(p-p)	9.3 V(p-p)
Phase	180°	180°

All the results obtained by simulation agree very closely with the design specifications. Also note that output signal swing of 9.3 V without distortion, as displayed in Figure 8-52, is indicative of a properly placed Q-point.

Practice Problem 8-9 — Enhancement MOSFET — DESIGN

Design a common-source MOSFET amplifier for a set of specifications as given below:

Supply voltage: $V_{DD} = 20$ V
No-load voltage gain: $|A_{v(NL)}| = 40$
Input resistance $R_{in} \geq 360$ kΩ

Drain current: $I_{DQ} = 2.2$ mA
Output voltage swing: $v_{o(max)} \geq 6$ V(p-p)
Output resistance $R_o \leq 3.6$ kΩ

Stability considerations:
As the MOSFET characteristics vary by ±20%, the Q-point I_D not to vary more than ±10%. Verify the result of your design by simulation. Use MOSFET IRF610.

8.9 SUMMARY

- The general amplifier fundamentals that were discussed in Chapter 6 (section 6.2) with BJT amplifiers are wholly and equally applicable to the FET amplifiers.

- FET is a voltage-controlled device. However, to be more specific, FET may be viewed as a voltage-controlled current source. That is, the application of a small-signal v_{gs} across the gate-source terminals provides a current $g_m v_{gs}$ at the drain terminal, where g_m is the transconductance of the transistor as defined with Equation 8-1 and realized with Equation 8-2.

Chapter 8

- Because FET requires no input current in order to operate as a current source, the input resistance of this voltage-controlled current source is very high, nearly infinite. Hence, JFET and MOSFET amplifiers can offer a very high input resistance.

- In general, FET amplifiers do not offer the same high voltage gain that is expected from BJT amplifiers. The lower voltage gain seems to be the tradeoff for the high input resistance with FET amplifiers, although a relatively high voltage gain can be achieved with some MOSFETs.

- The FET amplifier may be configured in one of the three basic configurations: common-source, common-gate, and common-drain or the source-follower.

- The common-source, which is the most popular of the three configurations, is an inverting amplifier and provides moderately high voltage gain, reasonably high current gain, very high input resistance, and a moderate output resistance.

- The common-gate is a non-inverting amplifier and provides the same voltage gain and output resistance expected from the CS amplifier. However, it has no current gain and suffers from a very low input resistance.

- The common-drain or the source-follower, which is the FET version of the emitter-follower, is also a non-inverter, has moderately high current gain, high input resistance, low output resistance, and no voltage gain. Its main application is as a buffer.

- A comprehensive analysis of a JFET or MOSFET amplifier begins with the DC analysis in order to determine the V_{GSQ} and I_{DQ} upon which depend some of the small-signal parameters such as g_m and r_o. Next, we begin the AC analysis by drawing the small-signal equivalent circuit with the FET model substituted for the transistor. From the equivalent circuit, we can determine the amplifier parameters such as voltage gain, current gain, power gain, input resistance, output resistance, and the output voltage.

- For a summary of formulas and equations that were derived for each FET amplifier configuration and were used in the examples that followed, refer to the corresponding *section summary* throughout the chapter.

- Most stable operation is achieved by a properly designed voltage-divider bias in which the slope of the bias line determines the level of stability; that is, the smaller the slope of the bias line, the more stable will be the amplifier operation.

- The FET *cascode* amplifier is a cascade of common-source and common-gate configurations, which utilizes the superior characteristics of both configurations, while avoiding their undesirable characteristics.

Chapter 8

- When designing an FET amplifier to operate at a certain Q-point, we assume small deviations in the Q-point for larger deviations in the FET characteristics, from which the maximum slope of the bias line is determined.

- Splitting the source resistor R_S into R_{S1} and R_{S2}, where R_{S1} is usually very small compared to R_{S2} and remains unbypassed for the small signal, makes it possible to design the FET amplifier for a specific voltage gain. In addition, R_{S1} contributes to system stability by providing a voltage-series feedback for the small signal operation.

- For a complete design procedure refer to Examples 8-5 and 8-6.

Chapter 8

PROBLEMS

Junction FET   ANALYSIS

8.1 Determine the following for the JFET amplifier circuit of Figure 8-1P.
   a) I_{DQ}, V_{GSQ}, and g_m.
   b) Draw the small-signal equivalent and determine R_{in}, R_o, $A_{v(NL)}$, $A_{v(WL)}$, A_i, and A_p.

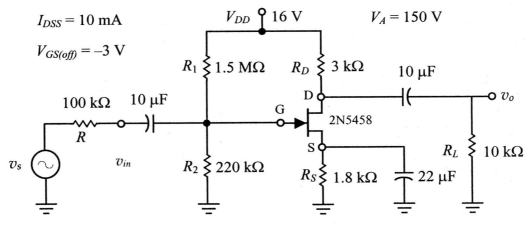

Figure 8-1P: Common-source JFET amplifier circuit

8.2 Simulate the amplifier circuit of Figure 8-1P and verify the theoretically obtained results in problem 8.1.

8.3 Determine the following for the JFET amplifier circuit of Figure 8-2P.
   a) I_{DQ}, V_{GSQ}, and g_m.
   b) Draw the small-signal equivalent and determine R_{in}, R_o, $A_{v(NL)}$, $A_{v(WL)}$, A_i, and A_p.

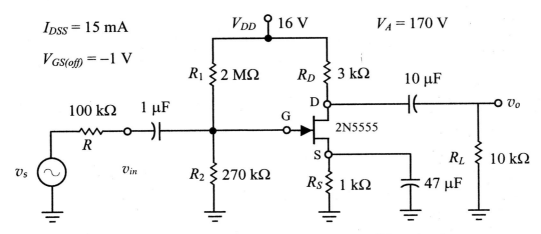

Figure 8-2P: Common-source JFET amplifier circuit

Chapter 8

8.4 Simulate the amplifier circuit of Figure 8-2P and verify the theoretically obtained results in problem 8.3.

8.5 Carry out Practice Problem 8-2.

8.6 Analyze the JFET amplifier circuit of Figure 8-3P and determine the following:
   a) I_{DQ}, V_{GSQ}, and g_m.
   b) Draw the small-signal equivalent and determine R_{in}, R_o, $A_{v(NL)}$, $A_{v(WL)}$, A_i, and A_p.

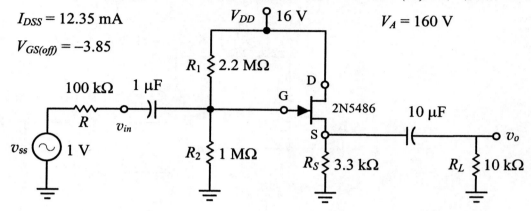

Figure 8-3P: JFET Source-Follower

8.7 Simulate the source-follower circuit of Figure 8-3P and verify the theoretically obtained results in problem 8.6.

8.8 Carry out Practice Problem 8-3.

Depletion MOSFET — ANALYSIS

8.9 Analyze the DMOS amplifier of Figure 8-4P and determine the following:
   a) V_{GSQ}, I_{DQ}, and g_m.
   b) R_{in}, R_o, $A_{v(WL)}$, $A_{v(NL)}$, A_i, A_p, and v_o.

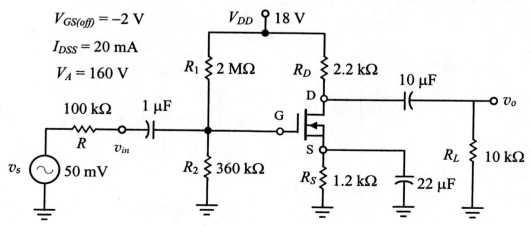

Figure 8-4P: Common-source DMOS amplifier circuit

Chapter 8

8.10 Analyze the DMOS amplifier of Figure 8-5P and determine the following:

   a) V_{GSQ}, I_{DQ}, and g_m.
   b) R_{in}, R_o, $A_{v(WL)}$, $A_{v(NL)}$, A_i, A_p, and v_o.

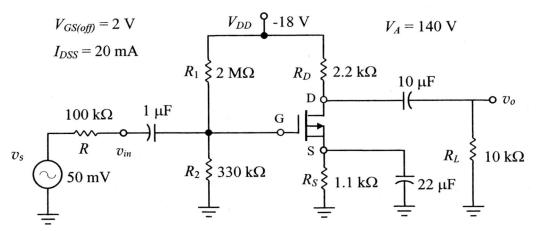

Figure 8-5P: Common-source MOSFET amplifier circuit

Enhancement MOSFET   ANALYSIS

8.11 Analyze the MOSFET amplifier of Figure 8-6P and determine the following:

   a) V_{GSQ}, I_{DQ}, and g_m.
   b) R_{in}, R_o, $A_{v(WL)}$, $A_{v(NL)}$, A_i, A_p, and v_o.

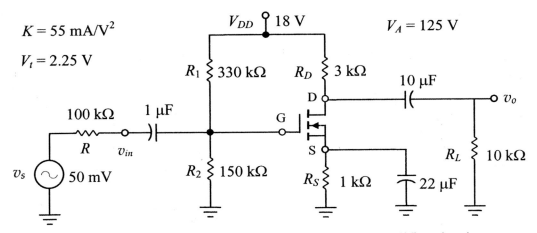

Figure 8-6P: Common-source MOSFET amplifier circuit

8.12 Carry out Practice Problem 8-4.

8.13 Analyze the MOSFET amplifier of Figure 8-7P and determine the following:

a) V_{GSQ}, I_{DQ}, and g_m.
b) R_{in}, R_o, $A_{v(WL)}$, $A_{v(NL)}$, A_i, A_p, and v_o.

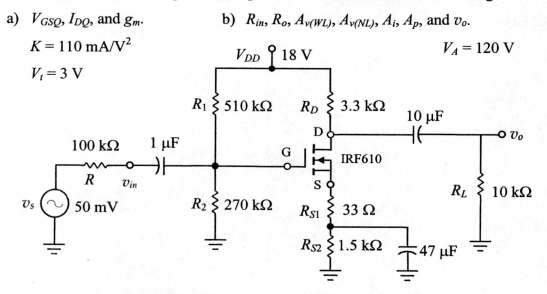

Figure 8-7P: Common-source MOSFET amplifier circuit

8.14 Carry out Practice Problem 8-5.

8.15 Carry out Practice Problem 8-6.

8.16 Analyze the common-gate MOSFET amplifier circuit of Figure 8-8P and determine the following:

a) V_{GSQ}, I_{DQ}, and g_m.
b) R_{in}, R_o, $A_{v(WL)}$, $A_{v(NL)}$, A_i, A_p, and v_o.

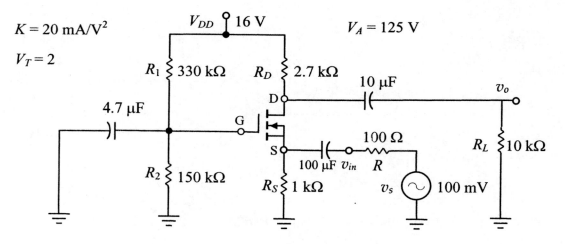

Figure 8-8P: Common-gate MOSFET amplifier circuit

8.17 Analyze the cascode amplifier circuit of Figure 8-9P and determine the following:

a) V_{GSQ}, I_{DQ}, and g_m
b) R_{in}, R_o, $A_{v(WL)}$

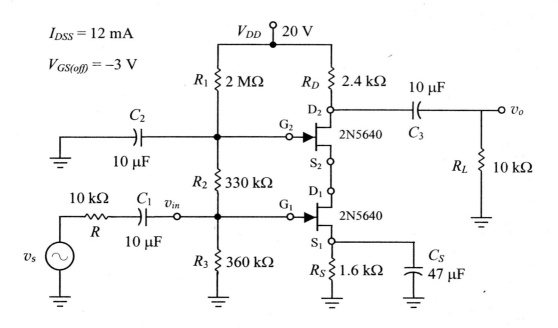

Figure 8-9P: Circuit schematic for problem 8.17

8.18 Carry out Practice Problem 8-7.

Junction FET — DESIGN

8.19 Design a common-source JFET amplifier for the following set of specifications:

Supply voltage: $V_{DD} = 18$ V   Drain current: $I_{DQ} = 2$ mA
No-load voltage gain: $|A_{v(NL)}| = 30$ Output voltage swing: $v_{o(max)} \geq 6$ V(p-p)
Input resistance $R_{in} \geq 300$ kΩ   Output resistance $R_o \leq 3$ kΩ

Stability considerations:
As the JFET characteristics vary by ±20%, the Q-point I_D is not to vary more than ±5%.
Verify the result of your design by simulation.   Use JFET 2N5555.

8.20 Carry out Practice Problem 8-8.

8.21 Design a common-source JFET amplifier for the following set of specifications:

Supply voltage: $V_{DD} = 20$ V   Drain current: $I_{DQ} = 2.5$ mA
No-load voltage gain: $|A_{v(NL)}| = 25$ Output voltage swing: $v_{o(max)} \geq 6$ V(p-p)
Input resistance $R_{in} \geq 360$ kΩ   Output resistance $R_o \leq 3.6$ kΩ

Stability considerations:
As the JFET characteristics vary by ±25%, the Q-point I_D is not to vary more than ±5%.   Verify the result of your design by simulation.   Use JFET 2N5555.

Chapter 8

MOSFET — EMOS — **DESIGN**

8.22 Design a common-source MOSFET amplifier for the following set of specifications:

Supply voltage: $V_{DD} = 18$ V
No-load voltage gain: $|A_{v(NL)}| = 60$
Input resistance $R_{in} \geq 500$ kΩ

Drain current: $I_{DQ} = 2$ mA
Output voltage swing: $v_{o(max)} \geq 6$ V(p-p)
Output resistance $R_o \leq 4$ kΩ

Stability considerations:

As the FET characteristics vary by $\pm 20\%$, the Q-point, I_D is not to vary more than $\pm 10\%$.

Verify the result of your design by simulation.

8.23 Carry out Practice Problem 8-9.

8.24 Design a common-source MOSFET amplifier for the following set of specifications:

Supply voltage: $V_{DD} = 20$ V
No-load voltage gain: $|A_{v(NL)}| = 50$
Input resistance $R_{in} \geq 500$ kΩ

Drain current: $I_{DQ} = 2.5$ mA
Output voltage swing: $v_{o(max)} \geq 6$ V(p-p)
Output resistance $R_o \leq 3.6$ kΩ

Stability considerations:

As the FET characteristics vary by $\pm 25\%$, the Q-point, I_D is not to vary more than $\pm 10\%$.

Verify the result of your design by simulation.

Chapter 9

Chapter

FREQUENCY RESPONSE OF BJT AND FET AMPLIFIERS

9.1 INTRODUCTION

In our analysis of the BJT and FET amplifiers, we have assumed a range of operating frequency in which the coupling and bypass capacitors were replaced by a short circuit; that is, the operating frequency was considered high enough so that the impedance of the capacitors would be very low, nearly zero. The upper range of the operating frequency was considered low enough so that the internal transistor capacitances would not have significant effect on the amplifier operation; that is, their impedances were considered to be very high, nearly infinite. We wish now to investigate the effect of these capacitors, external and internal, and the limitations they impose on the amplifier operation and its frequency response. We will also introduce the decibel (dB) gain and logarithmic scale plot, which is known as the Bode plot.

9.2 FREQUENCY RESPONSE FUNDAMENTALS

We will begin our study of the frequency response by analyzing the response of a simple first-order low-pass and a first-order high-pass *RC* network.

Chapter 9

9.2.1 First-Order Low-Pass RC Network:

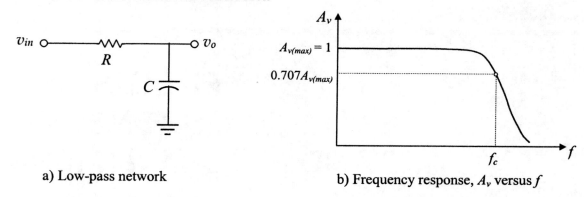

a) Low-pass network

b) Frequency response, A_v versus f

Figure 9-1: A first-order low-pass RC network and its general frequency response plot

The frequency response of this simple RC network is the plot of voltage gain (A_v) versus frequency (f). Notice that at zero frequency (DC), the capacitor is an open circuit and the output voltage v_o equals the input voltage v_{in}; hence, the voltage gain A_v is 1. As the frequency is increased, the reactance of the capacitor decreases gradually; however, at lower frequencies, where the capacitive reactance is much larger than the resistance of the resistor R, the gain A_v stays fairly constant. Then the gain begins dropping for further increases in frequency. The frequency, at which the gain A_v drops from 1 to 0.707 is called the *corner frequency* or the *cutoff frequency f_c*, at which the magnitude of the capacitive reactance (X_C) of the capacitor equals the resistance of the resistor R. Expressing this definition in mathematical terms, we can determine the expression for the cutoff frequency in terms of the circuit components R and C, as follows:

At the cutoff frequency, the magnitude of the capacitive reactance X_C equals R, that is:

$$\frac{1}{\omega_c \cdot C} = R = \frac{1}{2\pi f_c \cdot C}$$

Solving for ω_c or f_c, yields the following:

$$\omega_c = \frac{1}{RC} \qquad (9\text{-}1)$$

$$f_c = \frac{1}{2\pi RC} \qquad (9\text{-}2)$$

Derivation of the gain expression A_v as a function of frequency (f):

Applying the voltage-divider rule at the output of Figure 9-1(a) results in the following:

$$v_o = v_{in} \frac{X_C}{R + X_C} = v_{in} \frac{-j\frac{1}{\omega \cdot C}}{R - j\frac{1}{\omega \cdot C}} \qquad (9\text{-}3)$$

Dividing both sides by v_{in} and multiplying by ($j\omega C$) yields the following expression for the voltage gain A_v.

Chapter 9

$$A_v = \frac{v_o}{v_{in}} = \frac{1}{1+j\omega RC} \qquad (9\text{-}4)$$

Substituting for $RC = 1/\omega_c$ results in the following equation:

$$A_v = \frac{v_o}{v_{in}} = \frac{1}{1+j\dfrac{\omega}{\omega_c}} = \frac{1}{1+j\dfrac{f}{f_c}} \qquad (9\text{-}5)$$

Hence, the expression for the voltage gain as a function of frequency is the following:

$$A_v(f) = \frac{v_o}{v_{in}} = \frac{1}{1+j(f/f_c)} \qquad (9\text{-}6)$$

The above expression, which is a complex quantity, can be expressed in magnitude and phase as follows:

$$A_v(f) = \frac{1}{\sqrt{1+(f/f_c)^2}} \angle\left(-\tan^{-1}(f/f_c)\right) \qquad (9\text{-}7)$$

where the magnitude of the gain is:

$$|A_v| = \frac{1}{\sqrt{1+(f/f_c)^2}} \qquad (9\text{-}8)$$

and the phase is:

$$\phi = -\tan^{-1}(f/f_c) \qquad (9\text{-}9)$$

Recall that the cutoff frequency $f_c = 1/(2\pi RC)$ is a constant and is determined by the value of the resistor and the capacitor. At the cutoff frequency, that is, when $f = f_c$, Equations 9-8 and 9-9 yield the following values:

$$|A_v| = \frac{1}{\sqrt{1+(1)^2}} = \frac{1}{\sqrt{2}} = 0.707 \quad \text{and} \quad \phi = -\tan^{-1}(1) = -45°$$

9.2.2 Decibel and Logarithmic Plots (Bode Plots)

Figure 9-1(b) exhibits the general frequency response plot of a low-pass RC network, with magnitude gain plotted versus frequency in a linear scale. However, frequency response data are often presented with the gain plotted in decibel and the frequency in logarithmic scale, which is usually referred to as the Bode plot. The decibel unit (dB) is used to compare two power levels P_1 and P_2 and is defined as follows:

$$dB = 10\log\left(\frac{P_2}{P_1}\right) \qquad (9\text{-}10)$$

Substituting for P_1 and P_2 in terms of v^2/R results in the following:

$$dB = 10\log\left(\frac{(v_2)^2/R_2}{(v_1)^2/R_1}\right) \qquad (9\text{-}11)$$

If resistances R_1 and R_2 are equal in value, that is, $R_1 = R_2 = R$, then we obtain the following equation for dB in terms of voltages v_1 and v_2.

Chapter 9

$$dB = 10\log\left(\frac{(v_2)^2/R}{(v_1)^2/R}\right) = 10\log\left(\frac{v_2}{v_1}\right)^2 = 20\log\left(\frac{v_2}{v_1}\right) \qquad (9\text{-}12)$$

Hence, the expression for decibel (dB) in terms of two power levels is given by Equation 9-10, and in terms of two voltage levels is given by Equation 9-13, as follows:

$$dB = 20\log\left|\frac{v_2}{v_1}\right| \qquad (9\text{-}13)$$

The voltage gain of an amplifier is often represented in dB units, where v_2 and v_1 in the above equation are replaced with v_o and v_{in}, respectively.

Example 9-1 — DECIBEL

Determine the voltage gain in dB of an amplifier that produces 4 V(p-p) output for an input of 40 mV(p-p).

Solution:
The dB gain (dBG) of the amplifier is determined with Equation 9-13, as follows:

$$dBG = 20\log\left(\frac{v_o}{v_{in}}\right) = 20\log\left(\frac{4\text{ V}}{40\text{ mV}}\right) = 20\log(100) = 40\text{ dB}$$

The expression for the magnitude gain of the simple low-pass RC network defined by Equation 9-8 can be converted to dB with Equation 9-13, as follows:

$$dBG = 20\log|A_v| = 20\log\frac{1}{\sqrt{1+(f/f_c)^2}} = 20\log\left[1+(f/f_c)^2\right]^{-1/2} \qquad (9\text{-}14)$$

Applying the logarithmic properties to the above equation results in the following:

$$dBG = -10\log\left[1+(f/f_c)^2\right] \qquad (9\text{-}15)$$

Equation 9-15 above expresses the magnitude of the gain in dB as a function of frequency f. Note that the cutoff frequency f_c is a constant and its value depends on R and C. Let us now consider a few data points at different frequencies and plot the above expression as the magnitude gain in dB versus frequency f in logarithmic scale. We will begin data points at a frequency two decades below the *cutoff* ($f = 0.01 f_c$) and continue up to two decades above the cutoff ($f = 100 f_c$).

$f = 0.01 f_c$ $f/f_c = 0.01$ $dBG = -10\log[1+(0.01)^2] \cong -10\log(1) \cong 0$ dB

$f = 0.1 f_c$ $f/f_c = 0.1$ $dBG = -10\log[1+(0.1)^2] \cong -10\log(1) \cong 0$ dB

$f = f_c$ $f/f_c = 1$ $dBG = -10\log[1+(1)^2] = -10\log(2) \cong -3$ dB

$f = 10 f_c$ $f/f_c = 10$ $dBG = -10\log[1+(10)^2] = -10\log(101) \cong -20$ dB

$f = 100 f_c$ $f/f_c = 100$ $dBG = -10\log[1+(100)^2] \cong -10\log(10^4) \cong -40$ dB

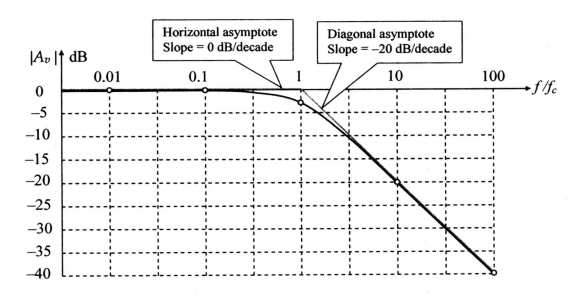

Figure 9-2: Bode plot of a first-order low-pass filter $|A_v|$ versus f in log scale

Given the transfer function in the *s-domain*, the Bode diagram can be plotted with MATLAB. Starting from Equation 9-4, which is the transfer function of the first-order low-pass filter, we will first apply some algebraic manipulation and then replace $j\omega$ with s so that it is in the standard *s-domain* format acceptable to MATLAB.

$$A_v = \frac{v_o}{v_{in}} = \frac{1}{1+j\omega RC} = \frac{\frac{1}{RC}}{\frac{1}{RC}+j\omega} \tag{9-16}$$

$$A_v = \frac{v_o}{v_{in}} = \frac{\frac{1}{RC}}{s+\frac{1}{RC}} = \frac{\omega_c}{s+\omega_c} = \frac{1}{1+\frac{s}{\omega_c}} \tag{9-17}$$

Let $R = 1\ \text{k}\Omega$ and $C = 1\ \mu\text{F}$, then $RC = 10^{-3}$ s and $\omega_c = \frac{1}{RC} = 10^3$ rad/s.

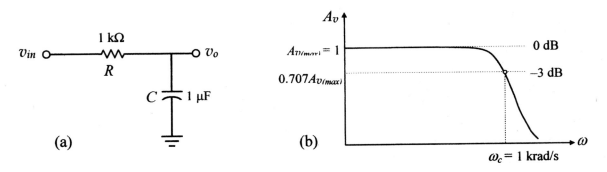

Figure 9-3: (a) First-order low-pass network. (b) Frequency response A_v versus ω (rad/s)

Substituting for ω_c in Equation 9-17, we obtain the transfer function in the *s-domain*

$$T(s) = \frac{1000}{s+1000}$$

Now we will ask MATLAB to plot both the magnitude (in dB) and the phase (in degrees) versus frequency (in rad/sec). In the MATLAB command window, at the MATLAB prompt, we enter the numerator polynomial *n*, then the denominator polynomial *d*, and then ask for the Bode plot, as follows:

```
» n=[0 1000];
» d=[1 1000];
» bode(n,d)
```

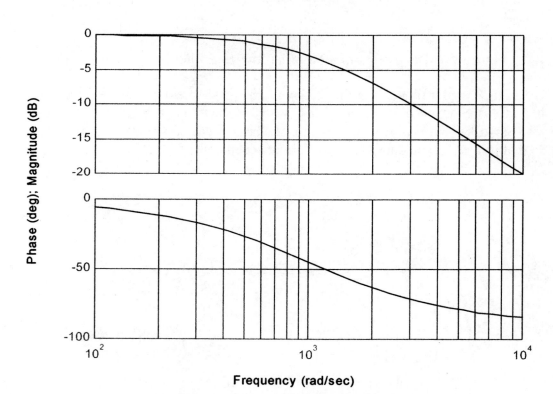

Figure 9-4: Frequency response plot (Bode plot) of the first-order low-pass *RC* network of Figure 9-3 with MATLAB

As you can see, $A_{v(max)} = 0$ dB and $\omega_c = 1000$ rad/sec corresponds to −3 dB gain. In addition, the phase plot extends from 0° at the very low frequency to −90° at the very high frequency, and corresponds to −45° at $\omega_c = 1000$ rad/sec. Also, notice the slope of the magnitude plot at the higher frequency region, which is −20 dB/decade.

Chapter 9

Practice Problem 9-1 — Low-Pass — BODE PLOT

Determine ω_c and f_c for the circuit of Figure 9-5, and plot the Bode diagram

a) by hand, use straight-edge to draw asymptotes.
b) with MATLAB

Answer: $\omega_c = 100$ rad/sec, $f_c = 15.9$ Hz

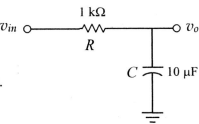

Figure 9-5:
First-order low-pass RC network

9.2.3 First-Order High-Pass RC Network

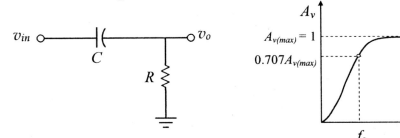

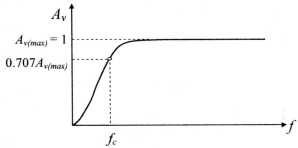

(a) High-pass network

(b) Frequency response, A_v vs. f

Figure 9-6: Circuit diagram and frequency response of simple high-pass RC network.

Derivation of the gain expression A_v as a function of frequency (f):

Applying the voltage-divider rule at the output of Figure 9-6(a) results in the following:

$$v_o = v_{in}\frac{R}{R+X_C} = v_{in}\frac{R}{R-j\dfrac{1}{\omega \cdot C}} \qquad (9\text{-}18)$$

Dividing both sides by v_{in} and R yields the following expression for the voltage gain A_v:

$$A_v = \frac{v_o}{v_{in}} = \frac{1}{1-j\dfrac{1}{\omega RC}} \qquad (9\text{-}19)$$

Substituting for $1/RC = \omega_c$ results in the following equation:

$$A_v = \frac{v_o}{v_{in}} = \frac{1}{1-j\dfrac{\omega_c}{\omega}} = \frac{1}{1-j\dfrac{f_c}{f}} \qquad (9\text{-}20)$$

Hence, the expression for the voltage gain as a function of frequency is the following:

$$A_v(f) = \frac{v_o}{v_{in}} = \frac{1}{1-j(f_c/f)} \qquad (9\text{-}21)$$

where the magnitude of the gain is

$$|A_v| = \frac{1}{\sqrt{1+(f_c/f)^2}} \quad (9\text{-}22)$$

and the phase is

$$\phi = +\tan^{-1}(f_c/f) \quad (9\text{-}23)$$

At the cutoff frequency, that is, when $f = f_c$, Equations 9-22 and 9-23 yield the following values:

$$|A_v| = \frac{1}{\sqrt{1+(1)^2}} = \frac{1}{\sqrt{2}} = 0.707 \quad \text{and} \quad \phi = \tan^{-1}(1) = +45°$$

The expression for the magnitude gain of the simple high-pass RC network defined by Equation 9-22, can be converted to dB with Equation 9-13, as follows:

$$dBG = 20\log|A_v| = 20\log\frac{1}{\sqrt{1+(f_c/f)^2}} = 20\log\left[1+(f_c/f)^2\right]^{-1/2} \quad (9\text{-}24)$$

Applying the logarithmic properties to the above equation results in the following:

$$dBG = -10\log[1+(f_c/f)^2] \quad (9\text{-}25)$$

Let us now consider a few data points at different frequencies and plot the above expression as the magnitude gain in dB versus frequency f in logarithmic scale. We will begin data points at a frequency two decades below the *cutoff* ($f = 0.01 f_c$) and continue up to two decades above the cutoff ($f = 100 f_c$).

$f = 0.01 f_c$ $f_c/f = 100$ $dBG = -10\log[1+(100)^2] \cong -10\log(10^4) \cong -40$ dB
$f = 0.1 f_c$ $f_c/f = 10$ $dBG = -10\log[1+(10)^2] \cong -10\log(10)^2 \cong -20$ dB
$f = f_c$ $f_c/f = 1$ $dBG = -10\log[1+(1)^2] = -10\log(2) \cong -3$ dB
$f = 10 f_c$ $f_c/f = 0.1$ $dBG = -10\log[1+(0.1)^2] = -10\log(1.01) \cong 0$ dB
$f = 100 f_c$ $f_c/f = 0.01$ $dBG = -10\log[1+(0.01)^2] \cong -10\log(1.0001) \cong 0$ dB

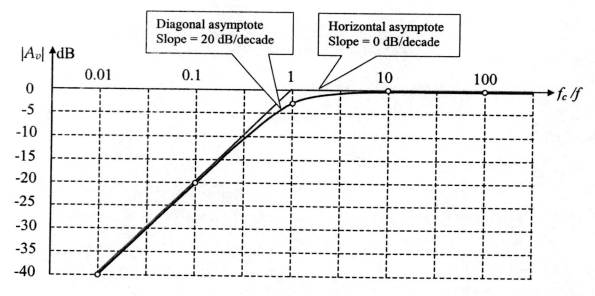

Figure 9-7: Bode plot of a first-order high-pass network, dBG versus f (*log scale*)

Again, we can plot the above Bode diagram with MATLAB by converting the transfer function to *s-domain*. Starting from Equation 9-19, which is the transfer function of the first-order high-pass filter, we will first apply some algebraic manipulation and then replace $j\omega$ with s, so that, it is in the standard *s-domain* format acceptable to MATLAB.

$$A_v = \frac{v_o}{v_{in}} = \frac{1}{1-j\dfrac{1}{\omega RC}} = \frac{j\omega}{j\omega + \dfrac{1}{RC}} = \frac{j\omega}{j\omega + \omega_c} \quad (9\text{-}26)$$

$$A_v = \frac{v_o}{v_{in}} = \frac{s}{s+\omega_c} = \frac{1}{1+\dfrac{\omega_c}{s}} \quad (9\text{-}27)$$

Let $R = 1\ k\Omega$ and $C = 1\ \mu F$, then $RC = 10^{-3}$ s and $\omega_c = \dfrac{1}{RC} = 10^3$ rad/s.

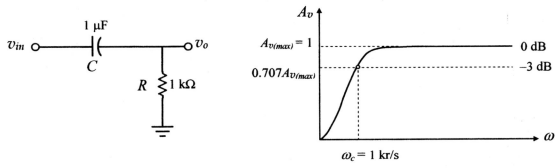

a) High-pass network

b) Frequency response A_v vs. ω

Figure 9-8: (a) First-order high-pass network, (b) frequency response, A_v vs. ω (r/s)

Practice Problem 9-2   High-Pass   BODE PLOT

Determine ω_c and f_c for the circuit of Figure 9-9, and plot the Bode diagram.

a) by hand, use straight-edge to draw asymptotes.

b) with MATLAB.

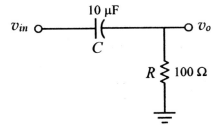

Figure 9-9: First-order high-pass RC filter

Answer: $\omega_c = 1000$ rad/sec, $f_c = 159$ Hz

Substituting for ω_c in Equation 9-27, we obtain the following *s-domain* transfer function

$$T(s) = \frac{s}{s+1000} \quad (9\text{-}36)$$

Having determined the transfer function in the *s-domain*, we will now use MATLAB to plot both the magnitude (in dB) and the phase (in degrees) versus frequency (in rad/sec).

Chapter 9

In the MATLAB command window, we enter the numerator polynomial (*n*), then the denominator polynomial (*d*), and then ask for the Bode plot, as follows:

```
» n=[1  0];
» d=[1 1000];
» bode(n,d)
```

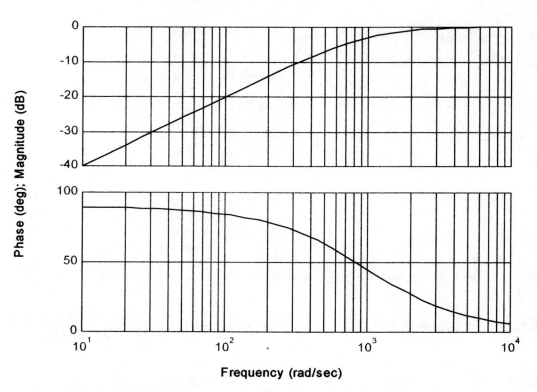

Figure 9-10: Frequency response plot (Bode plot) of the first-order high-pass *RC* filter of Figure 9-9 with MATLAB

As expected, $A_{v(max)} = 0$ dB and $\omega_c = 1000$ rad/sec corresponds to -3 dB gain. In addition, the phase plot extends from 90° at very low frequency to 0° at very high frequency, and corresponds to 45° at $\omega_c = 1000$ rad/sec. Also, notice the slope of the magnitude plot at the lower frequency region, which is 20 dB/decade.

Chapter 9

9.3 FREQUENCY RESPONSE OF THE CE BJT AMPLIFIER

Consider the common-emitter amplifier of Figure 9-11 as shown below.

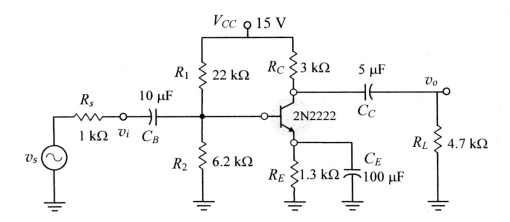

Figure 9-11: A common-emitter amplifier

The frequency response of this amplifier, that is, the plot of voltage gain A_v versus frequency, will be similar to the response illustrated in Figure 9-12. Notice that at zero frequency (DC), the voltage gain A_v is zero, because capacitors are an open circuit for DC and as a result the output v_o is zero. Then the gain rises as the frequency increases and levels off for further increases in frequency, and starts dropping again at higher frequencies. The range of frequency over which the gain is nearly constant is called the midband gain $A_{v(mid)}$. The two frequencies at which the gain drops to $0.707 A_{v(mid)}$ are called the *lower* and *upper* cutoff frequencies, f_L and f_H, respectively. The band of frequency between the two cutoff frequencies is called the *bandwidth* of the amplifier.

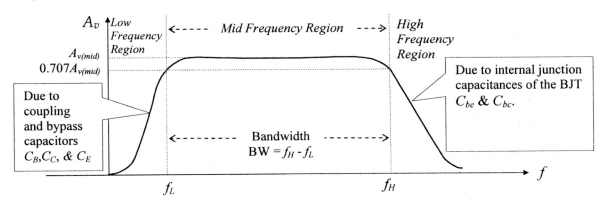

Figure 9-12: Typical frequency response diagram of a BJT amplifier

The upper cutoff frequency f_H is introduced by the two internal junction capacitances C_{be} and C_{bc} of the transistor, and also due to frequency dependence of transistor β at higher frequencies. Coupling capacitors C_B and C_C and the bypass capacitor C_E introduce the lower cutoff frequency f_L. In fact, each of these three capacitors along with their associated resistance introduces a cutoff frequency. The largest of the three cutoffs is the dominant and the lower cutoff frequency f_L of the amplifier. Because of the very low resistance associated with the emitter bypass capacitor, generally the cutoff frequency introduced by C_E is the dominant low cutoff f_L.

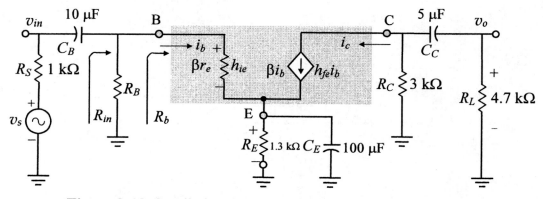

Figure 9-13: Small-signal low-frequency model of Figure 9-11

The results of the DC analysis are given below. However, students are encouraged to verify these results.

$I_E = 2$ mA. Referring to 2N2222 data sheets, I_C of 2 mA corresponds to a $\beta \cong 180$.

$R_B = R_1 \| R_2 = 22 \text{ k}\Omega \| 6.2 \text{ k}\Omega = 4.8 \text{ k}\Omega$, $\qquad r_e = 13 \text{ }\Omega$

$R_b = \beta r_e = 180 \times 13 \text{ }\Omega = 2.34 \text{ k}\Omega$, $\qquad R_{in} = R_B \| R_b = 4.8 \text{ k}\Omega \| 2.34 \text{ k}\Omega = 1.57 \text{ k}\Omega$

9.3.1 The Lower Cutoff Frequency f_L

The cutoff frequency introduced by each capacitor is defined as the frequency at which the magnitude of the capacitive reactance equals Thevenin's equivalent resistance external to the capacitor terminals. Let us express this definition in mathematical notation. At the cutoff frequency f_c,

$$X_C = R_{th}$$

Substituting for X_C results in the following:

$$\frac{1}{2\pi f_c C} = R_{th}$$

Solving for f_c results in the following expression of the cutoff frequency

$$f_c = \frac{1}{2\pi R_{th} C} \qquad (9\text{-}29)$$

where:
f_c is the individual cutoff frequency introduced by each capacitor, and
R_{th} is the Thevenin's equivalent resistance external to the capacitor terminals.

When determining the cutoff frequency due to one capacitor, the general rule is as follows. The other capacitors are considered short circuit, and the cutoff frequency introduced by each capacitor is determined according to Equation 9-29.

Cutoff frequency introduced by the coupling capacitor C_B:

The low-frequency small-signal equivalent circuit of the amplifier for determining the cutoff frequency due to coupling capacitor C_B is exhibited in Figure 9-14 below.

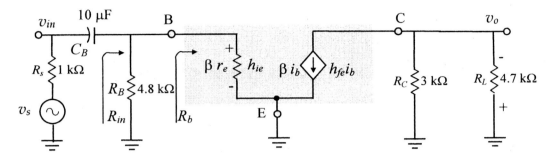

Figure 9-14: Low-frequency small-signal model of Figure 9-11 for determining the low cutoff frequency f_{LB} due to C_B.

First, we need to determine the Thevenin's equivalent resistance external to C_B. The total resistance appearing across the terminals of C_B is the series combination of R_s on one side and R_{in} on the other side, where R_{in} is the parallel combination of R_B and R_b.

$$R_{th(CB)} = R_s + R_{in}$$

$$f_{LB} = \frac{1}{2\pi(R_S + R_{in})C_B} \qquad (9\text{-}30)$$

$$f_{LB} = \frac{1}{2\pi(1\,k\Omega + 1.57\,k\Omega)(10\,\mu F)} = 6.2\,\text{Hz}$$

Cutoff frequency introduced by the coupling capacitor C_C:

The low-frequency small-signal equivalent circuit of the amplifier for determining the cutoff frequency due to coupling capacitor C_C is exhibited in Figure 9-15. The total equivalent resistance appearing across the terminals of C_C is the series combination of R_C on one side and R_L on the other side.

$$R_{th(Cc)} = R_C + R_L$$

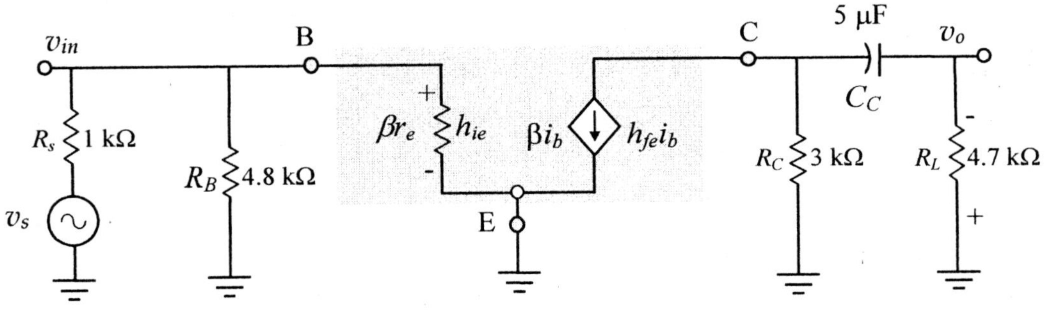

Figure 9-15: Low-frequency small-signal model of Figure 9-11 for determining the low cutoff frequency f_{LC} due to C_C.

The cutoff frequency is determined by substituting for R_{th} in Equation 9-29, as follows:

$$f_{LC} = \frac{1}{2\pi(R_C + R_L)C_C} \tag{9-31}$$

$$f_{LC} = \frac{1}{2\pi(3\text{ k}\Omega + 4.7\text{ k}\Omega)(5\text{ μF})} = 4\text{ Hz}$$

<u>Cutoff frequency introduced by the bypass capacitor C_E:</u>

The low-frequency small-signal model of the amplifier for determining the cutoff frequency due to the emitter bypass capacitor C_E is exhibited in Figure 9-16 below.

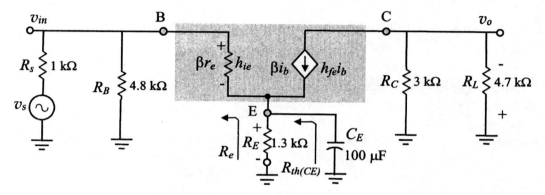

Figure 9-16: Low-frequency small-signal model of Figure 9-11 for determining the cutoff frequency f_{LE} due to C_E.

Thevenin's equivalent resistance appearing across C_E is the parallel combination of R_E and R_e, which is similar to the R_o of the common-collector amplifier, and is determined the same way as follows:

$$R_{th(C_E)} = R_E \| R_e \cong R_e$$

$$R_e = r_e + \frac{R_S \| R_B}{\beta}$$

Therefore,

$$R_{th(C_E)} = r_e + \frac{R_S \| R_B}{\beta} \tag{9-32}$$

$$R_{th(C_E)} = 13\,\Omega + \frac{1\text{ k}\Omega \| 4.8\text{ k}\Omega}{180} = 17.6\,\Omega$$

Having determined the R_{th}, cutoff frequency is determined with Equation 9-29, as follows:

$$f_{LE} = \frac{1}{2\pi(R_{th(C_E)})C_E} \tag{9-33}$$

$$f_{LE} = \frac{1}{2\pi(17.6\,\Omega)(100\text{ μF})} = 90\text{ Hz}$$

The largest of the three cutoff frequencies is the dominant and lower cutoff frequency of the amplifier. As expected, the dominant cutoff frequency is the one introduced by the emitter bypass capacitor C_E, which is 90 Hz. If this lower cutoff frequency happens to be too high for a certain application, the capacitor value (C_E) can be increased to obtain the desired cutoff frequency. For example, if the lower cutoff frequency is specified to be a maximum of 50 Hz, the size of C_E can be increased from 100 µF to 200 µF, yielding a lower cutoff frequency equal to 45 Hz.

Practice Problem 9-3   ANALYSIS

Determine the low cutoff frequencies f_{LB}, f_{LC}, f_{LE}, the dominant cutoff frequency f_L, and $A_{v(mid)}$ in dB for the amplifier circuit of Figure 9-17. Assume $\beta = 200$.

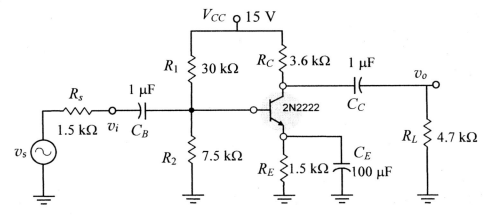

Figure 9-17: A common-emitter amplifier

Answers: $f_{LB} = 43$ Hz, $f_{LC} = 20$ Hz, $f_{LE} = 69$ Hz, $f_L = 85$ Hz, $A_{v(mid)} = 41.41$ dB

Example 9-2   ANALYSIS

Let us now analyze and determine the low cutoff frequency f_L for the common-emitter amplifier circuit of Figure 9-18, as shown below.

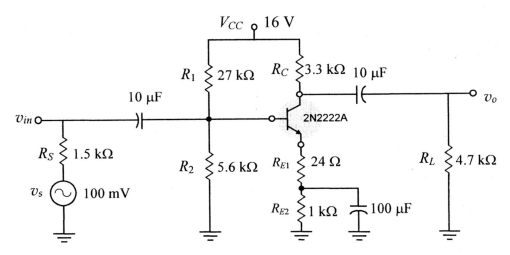

Figure 9-18: A CE amplifier circuit with partially unbypassed R_E

Chapter 9

Solution:
Details of basic calculations are left to the student as an exercise, and only the results are presented for such operations.
Referring to the data sheets of 2N2222A, I_C of 2 mA corresponds to $\beta = 160$.

$$I_E = 2 \text{ mA}, \quad r_e = 13 \text{ }\Omega, \quad \beta r_e \cong 2 \text{ k}\Omega, \quad R_B = R_1 \| R_2 \cong 4.64 \text{ k}\Omega$$
$$R_b = \beta(r_e + R_{E1}) = 5.92 \text{ k}\Omega, \quad R_{in} = R_B \| R_b = 4.64 \text{ k}\Omega \| 5.92 \text{ k}\Omega = 2.6 \text{ k}\Omega$$

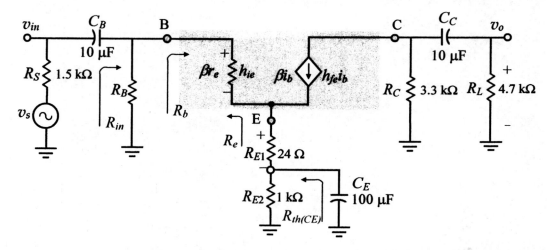

Figure 9-19: Small-signal low-frequency equivalent circuit of Figure 9-18

The cutoff frequency f_{LB} introduced by coupling capacitor C_B is determined from Equation 9-30, as follows:

$$f_{LB} = \frac{1}{2\pi(R_S + R_{in})C_B} = \frac{1}{2\pi(1.5 \text{ k}\Omega + 2.6 \text{ k}\Omega) \times 10 \text{ }\mu\text{F}} \cong 4 \text{ Hz}$$

The cutoff frequency introduced by coupling capacitor C_C is determined from Equation 9-31, as follows:

$$f_{LC} = \frac{1}{2\pi(R_C + R_L)C_C} = \frac{1}{2\pi(3.3 \text{ k}\Omega + 4.7 \text{ k}\Omega)(10 \text{ }\mu\text{F})} = 2 \text{ Hz}$$

<u>Cutoff frequency introduced by the bypass capacitor C_E:</u>

Thevenin's equivalent resistance seen by the capacitor C_E is the following:

$$R_{th(CE)} = R_{E2} \| (R_{E1} + R_e) \quad \quad (9\text{-}34)$$

where

$$R_e = \frac{\beta r_e + (R_B \| R_S)}{\beta} \quad \quad (9\text{-}35)$$

$$R_e = \frac{2 \text{ k}\Omega + (4.6 \text{ k}\Omega \| 1.5 \text{ k}\Omega)}{160} = 20 \text{ }\Omega$$

$$R_{th(CE)} = 1 \text{ k}\Omega \| (24 \text{ }\Omega + 20 \text{ }\Omega) = 42 \text{ }\Omega$$

The cutoff frequency f_{LE} is determined from Equation 9-29, as follows:

$$f_{LE} = \frac{1}{2\pi R_{th}C} = \frac{1}{2\pi \times 42\,\Omega \times 100\,\mu F} = 38\,\text{Hz}$$

As expected, the highest of all three cutoffs is f_{LE}, and thus, is the dominant low-cutoff frequency f_L of the amplifier.

Practice Problem 9-4 — ANALYSIS

Analyze and determine the low-cutoff frequency f_L for the amplifier circuit of Figure 9-20. Also, determine the midband gain, the output voltage at midband and at the cutoff frequency. Assume $\beta = 160$.

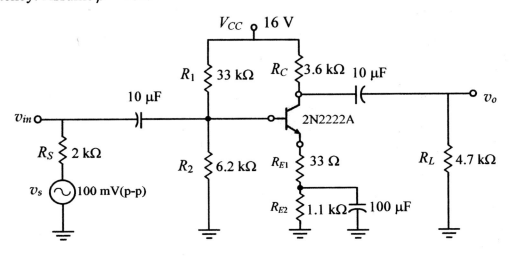

Figure 9-20: A CE amplifier circuit with partially unbypassed R_E

Answers:

$f_{LB} \cong 3\,\text{Hz}$, $f_{LC} \cong 1.9\,\text{Hz}$, $f_{LE} \cong 29.48\,\text{Hz}$, $v_{o(mid)} = 2.538\,\text{V}_{(p-p)}$, $v_o\,@\,f_L = 1.794\,\text{V}_{(p-p)}$

9.3.2 The Upper Cutoff Frequency f_H

The drop in the gain of an amplifier in the higher frequency range is caused by the transistor internal junction capacitance C_{bc} and the difffusion capacitance C_{be}, as shown in Figure 9-21. These capacitances are very small in value, generally a few pF, and show no significant effect at low or midband frequencies. At higher frequencies, however, their reactances become noticeably low and can practically short out the transistor terminals as the frequency is further increased. The capacitance C_{be} of the forward-biased junction is generally higher than C_{bc}, which is the capacitance of the reverse-biased junction. Typically, C_{be} can range from a few pF to a few tens of pF, whereas C_{bc} can range from a fraction of pF to a few pF.

In data sheets, the internal capacitance C_{bc} is usually referred to as the *output capacitance* C_{obo}. The actual capacitance of C_{bc} for the 2N2222A transistor is plotted versus the reverse voltage V_{CB} in Figure 9 of the 2N2222A data sheets and is labeled as such.* Device specifications and data sheets for the 2N2222A and 2N3904 transistors are included at the end of Chapter 3 for your reference. The capacitance C_{ibo} in data sheets represents the junction capacitance of the reverse-biased base-emitter junction. This junction in amplifier applications is usually forward-biased and its capacitance is dominated by the diffusion capacitance. The emitter-base capacitance is determined by

$$C_{be} = \frac{g_m}{2\pi f_T} - C_{bc} = \frac{\frac{I_C}{V_T}}{2\pi f_T} - C_{bc}$$

where f_T is the unity current gain frequecy given in data sheets.

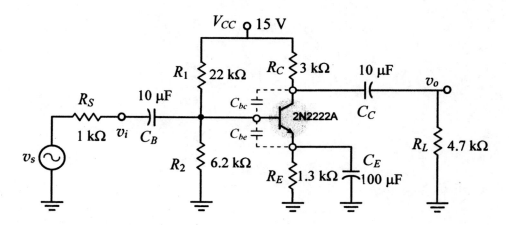

Figure 9-21: A common-emitter amplifier with coupling and bypass capacitors C_B, C_C, and C_E, and the two internal capacitances C_{be} and C_{bc} of the transistor.

At the higher frequency range, the coupling capacitors C_B and C_C and the emitter bypass capacitor C_E are practically short-circuited. Hence, the internal junction capacitance C_{be} will be shunted to ground at the input terminal. The other junction capacitor C_{bc}, which is across the input and output terminals, will behave like a feedback capacitor. According to Miller's theorem, the feedback element, in this case, a capacitor, can be converted to two shunt elements, one to appear at the input side and the other at the output side. In addition to simplifying the circuit analysis, this conversion also demonstrates the significant effect that the feedback capacitance C_{bc} has on the high-frequency response of the amplifier.

Miller's Theorem

Miller's theorem states that a feedback component Y connected across the input and output terminals of an amplifier, as illustrated in Figure 9-22, may be replaced by two equivalent shunt components, Y_1 at the input terminal and Y_2 at the output terminal, as shown in Figure 9-23.

---

* Some textbooks currently on the market show three internal capacitances: C_{be}, C_{bc}, and C_{ce}. However, as you can see on the data sheets of two different BJTs from two different manufacturers, there are only two internal junction capacitances C_{be} and C_{bc}, or C_{ibo} and C_{obo}, as listed in the data sheets.

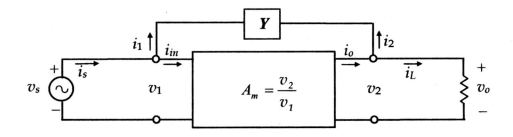

Figure 9-22: An amplifier with a feedback element Y

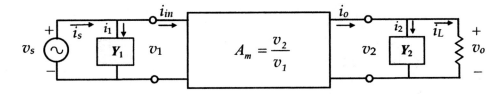

Figure 9-23: Miller's equivalent circuit of Figure 9-22

The Input Circuit:

The component Y connected across the input and output terminals of Figure 9-22 is assumed to be a susceptance, thus, components Y_1 and Y_2 are also susceptances. Applying Ohm's law across the feedback component Y in Figure 9-22, we can write the following equation for i_1:

$$i_1 = Y(v_1 - v_2) = Y(v_1 - A_m v_1) = Y v_1 (1 - A_m) = Y(1 - A_m) v_1 \qquad (9\text{-}36)$$

We can also write the following equation for I_1, by applying Ohm's law across the component Y_1 in Figure 9-23

$$i_1 = Y_1 v_1 \qquad (9\text{-}37)$$

Comparing two Equations 9-36 and 9-37, we obtain the following equation for Y_1

$$Y_1 = Y(1 - A_m) \qquad (9\text{-}38)$$

Note that the Equation (9-38) is in the general form, which applies equally to reactive or resistive feedback elements. In the case of the common-emitter BJT amplifier or common-source FET amplifier, the feedback component is a capacitance $C = C_{bc}$ or $C = C_{gd}$ and its susceptance is as follows:

$$Y = j\omega C \qquad (9\text{-}39)$$

Hence,

$$Y_1 = j\omega C_B = j\omega C(1 - A_m) \qquad (9\text{-}40)$$

Dividing both sides by $j\omega$ results in the following expression for C_1, which is the Miller's equivalent capacitance at the input terminal

$$C_1 = C(1 - A_m) = C_{Mi} \qquad (9\text{-}41)$$

The Output Circuit:

Again, applying Ohm's law across the feedback component y in Figure 9-22, we can write the following equation for i_2:

$$i_2 = Y(v_2 - v_1) = Y(v_2 - v_2/A_m) = Yv_2(1 - 1/A_m) = Y(1 - 1/A_m)v_2 \quad (9\text{-}42)$$

We can also write the following equation for i_2, by applying Ohm's law across the component Y_2 in Figure 9-23:

$$i_2 = Y_2 v_2 \quad (9\text{-}43)$$

Comparing Equations 9-42 and 9-43, we obtain

$$Y_2 = Y(1 - 1/A_m) \quad (9\text{-}44)$$

Again, since the susceptance of the feedback element $Y = j\omega C$ and $Y_2 = j\omega C_2$, Miller's equivalent capacitance C_2 or C_{Mo} at the output terminal is given by

$$C_2 = C(1 - 1/A_m) = C_{Mo} \quad (9\text{-}45)$$

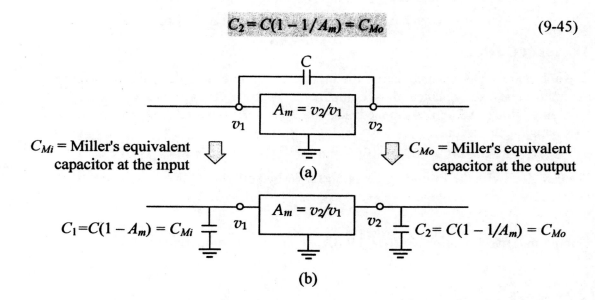

Figure 9-24: (a) Feedback capacitance   (b) Miller's equivalent capacitances

Example 9-3   ANALYSIS

We will now determine the upper cutoff frequencies f_{H1}, f_{H2}, and the dominant upper cutoff frequency f_H for the amplifier circuit of Figure 9-21, for which the low-frequency response analysis has already been carried out. Let $C_{be} = 8$ pF and $C_{bc} = 4$ pF. We will first draw the high-frequency model of the amplifier circuit.

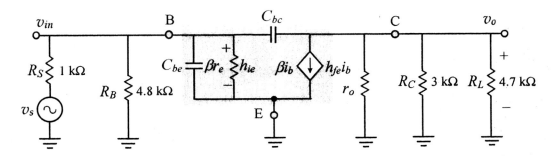

Figure 9-25: High-frequency small-signal model of Figure 9-21

According to Miller's theorem, the feedback capacitance C_{bc} can be converted to two equivalent shunt capacitances C_{Mi} and C_{Mo}, as shown in Figure 9-26.

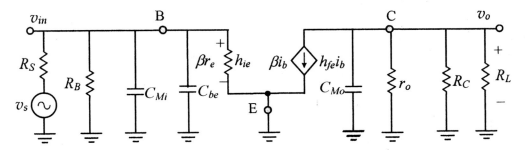

Figure 9-26: High-frequency small-signal model of Figure 9-21 with Miller's equivalent capacitances

In order to evaluate Miller capacitances C_{Mi} and C_{Mo}, we need to determine the amplifier voltage gain $A_m \approx A_{v(mid)}$. As an exercise, the student is asked to verify the following DC and midband frequency computations:

$I_E = 2$ mA, $r_e = 13\,\Omega$, $r_b = \beta r_e = 2.34$ kΩ, $R_B = R_1 \| R_2 = 4.8$ kΩ, $R_{in} = 1.57$ kΩ, $R_L' = r_o\|R_C\|R_L = 1.77$ kΩ, $A_m = -g_m R_L' = -R_L'/r_e = -136$ $|A_v|$ (dB) $= 20\log|A_v| = 42.7$ dB

We can now determine the Miller's equivalent capacitances C_{Mi} at the input and C_{Mo} at the output from Equations 9-41 and 9-45, respectively:

$$C_{Mi} = C_{bc}(1 - A_m) = 4 \text{ pF } [1 - (-136)] = 4 \text{ pF }(137) = 548 \text{ pF}$$

$$C_{Mo} = C_{bc}\left(1 - \frac{1}{A_m}\right) = 4 \text{ pF}\left(1 - \frac{1}{-136}\right) \cong 4 \text{ pF}$$

Referring to Figure 9-26, there are two shunt capacitors C_{be} and C_{Mi} at the input side of the amplifier, which can be combined into one capacitor C_i at the input. There is only one shunt capacitor C_{Mo} at the output and we will label it as C_o at the output, as illustrated in Figure 9-27.

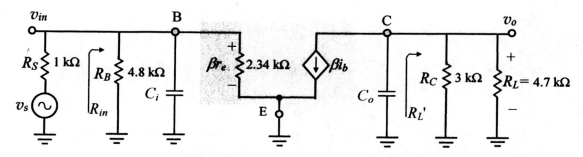

Figure 9-27: Simplified high frequency small-signal model of Figure 9-21

where
$$C_i = C_{be} + C_{Mi} \tag{9-46}$$
and
$$C_o = C_{Mo} \tag{9-47}$$

Having determined C_i and C_o, the upper cutoff frequencies introduced by these two capacitors can then be determined from Equation 9-29, as follows:

$C_i = C_{be} + C_{Mi} = 8\ \text{pF} + 564\ \text{pF} = 572\ \text{pF}$

$C_o = C_{Mo} = 4\ \text{pF}$

$$f_{Hi} = \frac{1}{2\pi R_{th(Ci)} C_i} \tag{9-48}$$

$$f_{Ho} = \frac{1}{2\pi R_{th(Co)} C_o} \tag{9-49}$$

Thevenin's equivalent resistance, R_{th}, in each case is the parallel combination of all resistances across the corresponding capacitor.

$R_{th(Ci)} = R_S \| R_B \| R_b = R_S \| R_{in} = 1\ \text{k}\Omega \| 1.57\ \text{k}\Omega = 611\ \Omega$

$R_{th(Co)} = R_L' = R_C \| R_L = 3\ \text{k}\Omega \| 4.7\ \text{k}\Omega = 1.83\ \text{k}\Omega$

$f_H \approx f_{Hi} = \dfrac{1}{2\pi R_{th(Ci)} C_i} = \dfrac{1}{2\pi \times 611\ \Omega \times 572\ \text{pF}} = 455\ \text{kHz}$

$f_{Ho} = \dfrac{1}{2\pi R_{th(Co)} C_o} = \dfrac{1}{2\pi \times 1.83\ \text{k}\Omega \times 4\ \text{pF}} = 21.75\ \text{MHz}$

Frequencies f_β and f_T

Important high-frequency parameters of BJTs are frequencies f_β and f_T. To derive an expression for the short-circuit current-gain cutoff frequency f_β, consider the transistor test circuit shown in Figure 9-28.

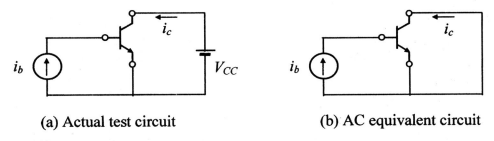

(a) Actual test circuit (b) AC equivalent circuit

Figure 9-28: BJT test circuit for measuring the current gain h_{fe} or β_{ac}

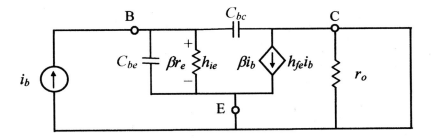

Figure 9-29: High-frequency small-signal model of Figure 9-28

The short-circuit across the collector-emitter terminals shorts out r_o while shunting C_{bc} from base to emitter in parallel with C_{be}. The resultant equivalent circuit is shown in Figure 9-29.

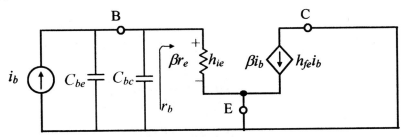

Figure 9-30: Equivalent circuit of Figure 9-29

The short-circuit current-gain cutoff frequency f_β is determined by the input circuit time constant

$$f_\beta = \frac{1}{2\pi r_b (C_{be} + C_{bc})} \quad (9\text{-}50)$$

$r_b = \beta r_e = 2.34 \text{ k}\Omega$

$f_\beta = \dfrac{1}{2\pi \times 2.34 \text{ k}\Omega(8 \text{ pF} + 4 \text{ pF})} = 5.67 \text{ MHz}$

The smaller of the three upper cutoff frequencies is the dominant and the upper cutoff frequency f_H of the amplifier. Hence, the upper cutoff frequency of this amplifier is 456

Chapter 9

kHz, where the lower cutoff frequency was found to be 90 Hz, making the bandwidth of the amplifier approximately 456 kHz.

Note that on transistor data sheets, usually a parameter f_T is given, which varies with the bias current I_C, and is defined as the transistor's *current-gain bandwidth product*. Its relation to f_β is as follows:

$$f_T = \beta f_\beta \qquad (9\text{-}51)$$

Example 9-4 — Frequency Response — ANALYSIS

Let us now analyze and determine the upper cutoff frequency f_H for the amplifier circuit of Figure 9-31, for which we have already determined the lower cutoff frequency f_L. Assume: $\beta = 160$, $C_{bc} = 3$ pF, and $C_{be} = 9$ pF.

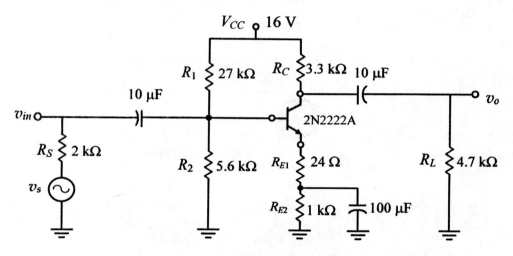

Figure 9-31: A CE amplifier circuit with partially unbypassed R_E

First, we will draw the high-frequency model with Miller's equivalent capacitances C_{Mi} and C_{Mo} already accounted for in C_i and C_o, as illustrated in Figure 9-32.

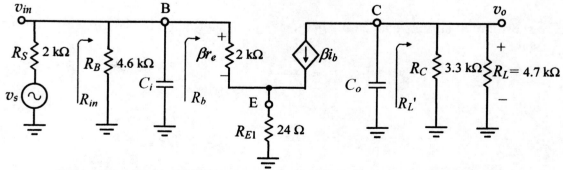

Figure 9-32: Simplified high-frequency small-signal model of Figure 9-31

$$A_m = -\frac{R_L'}{r_e + R_{E1}} = -\frac{1.94 \text{ k}\Omega}{13 \Omega + 24 \Omega} = -52.4$$

$$C_{Mi} = C_{bc}(1 - A_m) = 3 \text{ pF}[1-(-52.4)] = 3 \text{ pF}(53.4) = 160 \text{ pF}$$

$$C_{Mo} = C_{bc}\left(1 - \frac{1}{A_m}\right) = 3 \text{ pF}\left(1 - \frac{1}{-52.4}\right) = 3.057 \text{ pF}$$

$C_i = C_{be} + C_{Mi} = 9 \text{ pF} + 160 \text{ pF} = 169 \text{ pF}$
$C_o = C_{Mo} = 3.057 \text{ pF}$
$R_b = \beta(r_e + R_{E1}) = 160(13 \Omega + 24 \Omega) = 5.92 \text{ k}\Omega$

$R_{th(Ci)} = R_s \| R_B \| R_b = 2 \text{ k}\Omega \| 4.6 \text{ k}\Omega \| 5.92 \text{ k}\Omega = 1.13 \text{ k}\Omega$

$$f_{Hi} = \frac{1}{2\pi R_{th(Ci)} C_i} = \frac{1}{6.28 \times 1.13 \text{ k}\Omega \times 169 \text{ pF}} = 834 \text{ kHz}$$

$R_{th(Co)} = R_C \| R_L = R_L' = 1.94 \text{ k}\Omega$

$$f_{Ho} = \frac{1}{2\pi R_{th(Co)} C_o} = \frac{1}{6.28 \times 1.94 \text{ k}\Omega \times 3.057 \text{ pF}} = 26.85 \text{ MHz}$$

Thus, the upper cutoff frequency f_H of this amplifier is 834 kHz.

Note that the cutoff frequency f_β determined from Equation 9-50 is

$$f_\beta = \frac{1}{2\pi R_b (C_{be} + C_{bc})} = \frac{1}{6.28 \times 5.92 \text{ k}\Omega \times 12 \text{ pF}} = 2.24 \text{ MHz}$$

Practice Problem 9-5

Determine the midband gain $A_{v(mid)}$, lower and upper cutoff frequencies f_L and f_H, and the output voltage $v_{o(p-p)}$ at the cutoff frequency for the amplifier circuit of Figure 9-33. Assume: $\beta = 170$, $C_{be} = 10 \text{ pF}$, and $C_{bc} = 5 \text{ pF}$.

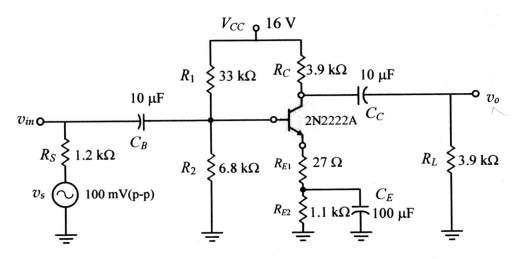

Figure 9-33: CE amplifier circuit with partially unbypassed R_E

Answers: $f_L = 35.6 \text{ Hz}$, $f_H = 733.8 \text{ kHz}$

Section Summary — Frequency Response

ANALYSIS

Summary of Equations for the Frequency Response Analysis of CE Amplifier

Low-Frequency Analysis

Cutoff frequency introduced by C_B:

$$f_{LB} = \frac{1}{2\pi(R_S + R_{in})C_B}$$

Cutoff frequency introduced by C_C:

$$f_{LC} = \frac{1}{2\pi(R_C + R_L)C_C}$$

Cutoff frequency introduced by C_E:

$$f_{LE} = \frac{1}{2\pi(R_{th(C_E)})C_E}$$

With a single unbypassed R_E:

$$R_{th(C_E)} = r_e + \frac{R_S \parallel R_B}{\beta}$$

With a partially bypassed R_E:

$$R_{th(CE)} = R_{E2} \parallel (R_{E1} + R_e)$$

$$R_e = \frac{\beta r_e + (R_B \parallel R_S)}{\beta}$$

The dominant low-cutoff frequency f_L is the highest of the above three cutoff frequencies, which is usually f_{LE} because of the very low resistance seen by C_E.

High-Frequency Analysis:

Cutoff frequency introduced by C_i:

$$f_{Hi} = \frac{1}{2\pi R_{th(Ci)} C_i}$$

Cutoff frequency introduced by C_o:

$$f_{Ho} = \frac{1}{2\pi R_{th(Co)} C_o}$$

$R_{th(Ci)} = R_S \parallel R_{in}$

$R_{th(Co)} = R_C \parallel R_L = R_L'$

$C_i = C_{be} + C_{Mi}$

$C_o = C_{Mo}$

Miller capacitances:

$C_{Mi} = C_{bc}(1 - A_m)$

$$C_{Mo} = C_{bc}\left(1 - \frac{1}{A_m}\right)$$

where A_m is the midband voltage gain between the base and collector.

The dominant high-cutoff frequency f_H is the lowest of the above two cutoff frequencies.

9.4 OVERALL FREQUENCY RESPONSE OF THE CE BJT AMPLIFIER

Having discussed the general frequency response of an amplifier, the inherent lower and upper cutoff frequencies, and the rolloff rate (dB/decade) beyond the cutoff frequency of a simple low-pass or high-pass network, we wish now to plot the overall frequency response for the amplifier of Figure 9-34. Assume: $\beta = 200$, $V_A = 170$ V, $C_{bc} = 5$ pF, and $C_{be} = 10$ pF.

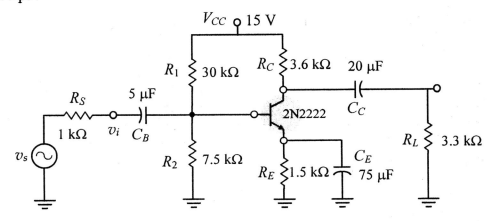

Figure 9-34: A common-emitter amplifier

DC and midband analysis:

Details of some basic computations are not shown. Students, however, are asked to verify the given results as an exercise.

$I_E = 1.533$ mA, $r_e = 17\ \Omega$, $r_o = 111$ kΩ, $r_b = \beta r_e \cong 3.4$ kΩ, $R_B = R_1 \| R_2 = 6$ kΩ, $R_{in} = R_B \| R_b \cong 2.17$ kΩ, $R_L' = R_C \| R_L \| r_o = 1.7$ kΩ, and $A_m \approx A_{v(mid)} = -100$.

Lower cutoff frequencies:

$$f_{LB} = \frac{1}{2\pi(R_S + R_{in})C_B} = \frac{1}{2\pi(1\text{ k}\Omega + 2.17\text{ k}\Omega)(5\ \mu\text{F})} = 10 \text{ Hz}$$

$$f_{LC} = \frac{1}{2\pi(R_C + R_L)C_C} = \frac{1}{2\pi(13.6\text{ k}\Omega + 4.7\text{ k}\Omega)(20\ \mu\text{F})} = 1 \text{ Hz}$$

$$R_{th(C_E)} = r_e + \frac{R_S \| R_B}{\beta} = 17\ \Omega + \frac{1\text{ k}\Omega \| 6\text{ k}\Omega}{200} = 21\ \Omega$$

$$f_{LE} = \frac{1}{2\pi(R_{th(C_E)})C_E} = \frac{1}{2\pi(21\ \Omega)(75\ \mu\text{F})} = \mathbf{101 \text{ Hz}}$$

$$f_z = \frac{1}{2\pi R_E C_E} = \frac{1}{2\pi(1.5\text{ k}\Omega)(75\ \mu\text{F})} = 1.415 \text{ Hz}$$

Upper cutoff frequencies:

$C_{Mi} = C_{bc}(1 - A_m) = 5 \text{ pF}[1 - (-100)] = 5 \text{ pF}(121) = 505 \text{ pF}$

$$C_{Mo} = C_{bc}\left(1 - \frac{1}{A_m}\right) = 5 \text{ pF}\left(1 - \frac{1}{-100}\right) \cong 5 \text{ pF}$$

$C_i = C_{be} + C_{Mi} = 10 \text{ pF} + 505 \text{ pF} = 515 \text{ pF}$ $C_o = C_{Mo} = 5 \text{ pF}$

$R_{th(Ci)} = R_S \| R_B \| R_b = 1 \text{ k}\Omega \| 6 \text{ k}\Omega \| 3.4 \text{ k}\Omega = 685 \text{ }\Omega$

$$f_{Hi} = \frac{1}{2\pi R_{th(Ci)} C_i} = \frac{1}{6.28 \times 685 \text{ }\Omega \times 515 \text{ pF}} = \mathbf{451 \text{ kHz}}$$

$R_{th(Co)} = R_C \| R_L \| r_o = 3.6 \text{ k}\Omega \| 3.3 \text{ k}\Omega \| 111 \text{ k}\Omega = 1.7 \text{ k}\Omega$

$$f_{Ho} = \frac{1}{2\pi R_{th(Co)} C_o} = \frac{1}{6.28 \times 1.7 \text{ k}\Omega \times 5 \text{ pF}} = 18.7 \text{ MHz}$$

As you have noticed, the dominant lower cutoff frequency is 100 Hz, and the dominant upper cutoff frequency is 451 kHz. We shall now plot the full frequency response, the gain in dB versus frequency in Hz. The dB scale will be linear, but the frequency scale will be logarithmic and this is generally referred to as a *semilog scale* plot. The overall Bode plot of the CE amplifier of Figure 9-34 is exhibited in Figure 9-35, covering a range of frequency from 0.1 Hz up to 1 GHz. Note that the straight lines are the asymptotes, each with different slope. At low frequencies, the slope changes from 0 dB at the midband to 20 dB/decade at $f_{LE} = 100$ Hz, then to 40 dB/decade at $f_{LB} = 10$ Hz, to 20 dB/decade at $f_z = 1.415$ Hz, and then 40 dB/decade at $f_{LC} = 1$ Hz. At high frequencies, the slope changes from 0 dB at the midband to -20 dB/decade at $f_{Hi} = 451$ kHz and then to -40 dB/decade at $f_{Ho} = 18.7$ MHz. The actual Bode plot, the blue curve, follows the asymptotic rolloff rate, passing through -3dB at both the lower and upper dominant cutoff frequencies. The band of frequency confined between the two dominant cutoff frequencies is the amplifier *bandwidth* $BW = f_H - f_L$.

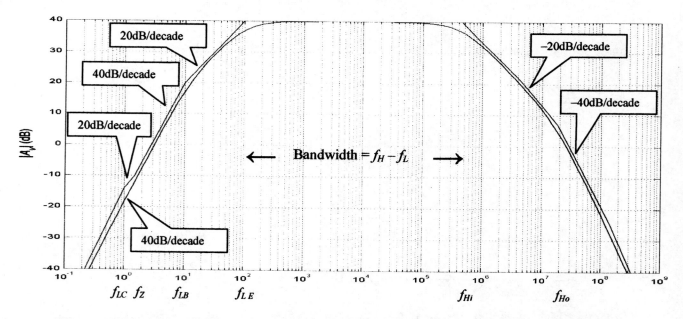

Figure 9-35: Overall frequency response (Bode plot) for the CE amplifier of Figure 9-34

Practice Problem 9-6

Determine the midband gain $A_{v\ (mid)}$, lower and upper cutoff frequencies f_L and f_H, and plot the overall frequency response for the amplifier of Figure 9-36.
Assume: $\beta = 175$, $C_{be} = 9$ pF, and $C_{bc} = 3$ pF.

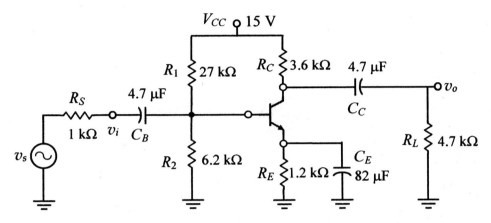

Figure 9-36: A common-emitter amplifier

Answers:

$f_{LB} = 12.5$ Hz, $f_{LC} = 4$ Hz, $f_{LE} = 100$ Hz, $f_{Hi} = 596$ kHz, $f_{Ho} = 26$ MHz, $f_\beta = 5.1$ MHz

9.5 FREQUENCY RESPONSE OF THE CB BJT AMPLIFIER

Consider the common-base amplifier circuit of Figure 9-37, which is DC equivalent to the CE amplifier of Figure 9-21. Assume: $C_{eb} = 9$ pF and $C_{cb} = 3$ pF.

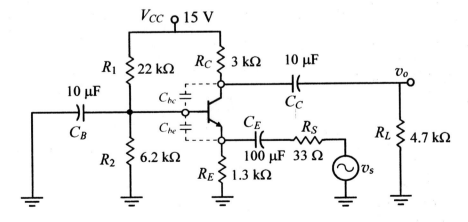

Figure 9-37: A common-base amplifier with coupling and bypass capacitors (C_B, C_C, C_E), and the two internal junction capacitances (C_{be}, C_{bc}) of the BJT

Chapter 9

9.5.1 Low-Frequency Analysis

Analysis of the low-frequency response of the common-base is fundamentally the same as with the common-emitter amplifier; that is, each of the capacitors C_B, C_C, and C_E, in conjunction with its Thevenin's equivalent resistance, will introduce a cutoff frequency, and the largest of the three is the dominant low cutoff frequency f_L. Figure 9-38 below shows the small-signal low-frequency model of Figure 9-37.

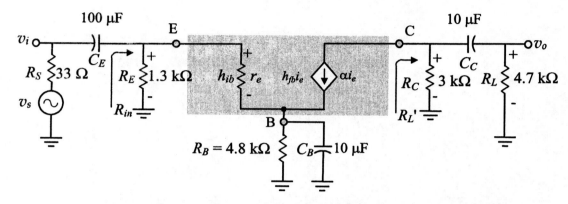

Figure 9-38: Small-signal low-frequency model of Figure 9-37

$$I_E = 2\text{ mA}, \quad r_e = 13\text{ }\Omega, \quad R_L' = 1.83\text{ k}\Omega, \quad R_{in} \cong 13\Omega, \quad A_{v(mid)} = 140$$

Lower cutoff frequencies:

Recall that when determining the cutoff due to one of the three capacitors, the other two capacitors are assumed short-circuited.

$$f_{LE} = \frac{1}{2\pi(R_S + R_{in})C_E} \tag{9-52}$$

$$f_{LE} = \frac{1}{2\pi(33\text{ }\Omega + 13\text{ }\Omega)(100\text{ }\mu\text{F})} = 35\text{ Hz}$$

$$f_{LC} = \frac{1}{2\pi(R_C + R_L)C_C} \tag{9-53}$$

$$f_{LC} = \frac{1}{2\pi(3\text{ k}\Omega + 4.7\text{ k}\Omega)(10\text{ }\mu\text{F})} = 2\text{ Hz}$$

$$R_{th(C_B)} = R_B \parallel \beta[r_e + (R_S \parallel R_E)] \tag{9-54}$$

$$R_{th(C_B)} = 4.8\text{k }\Omega \parallel 175[13\text{ }\Omega + (33\text{ }\Omega \parallel 1.3\text{ k}\Omega)] = 3\text{ k}\Omega$$

$$f_{LB} = \frac{1}{2\pi(R_{th(C_B)})C_B} \tag{9-55}$$

$$f_{LB} = \frac{1}{2\pi(3\text{ k}\Omega)(10\text{ }\mu\text{F})} = 5\text{ Hz}$$

Hence, the dominant low-cutoff frequency f_L is again introduced by C_E and equals 35 Hz.

9.5.2 High-Frequency Analysis

The small-signal high-frequency model of the common-base amplifier of Figure 9-37 is shown in Figure 9-39. Unlike the common-emitter circuit, there is no feedback capacitor between the input and output terminals, and thus, the high-frequency response of the common-base in not limited by the Miller effect. Each of the two junction capacitances, C_{eb} and C_{cb}, is shunted to ground at the input and output, respectively.

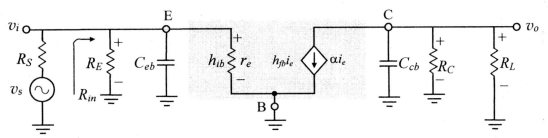

Figure 9-39: Small-signal high-frequency model of Figure 9-37

Upper cutoff frequencies:

$$C_i = C_{eb} = 9 \text{ pF}, \quad C_o = C_{cb} = 3 \text{ pF}$$

$R_{th(Ci)} = R_S \| R_{in} = 33 \text{ }\Omega \| 13 \text{ }\Omega = 9.3 \text{ }\Omega$

$$f_{Hi} = \frac{1}{2\pi R_{th(Ci)} C_i} = \frac{1}{6.28 \times 9.3 \text{ }\Omega \times 9 \text{ pF}} = 1.49 \text{ GHz}$$

$R_{th(Co)} = R_C \| R_L = 3 \text{ k}\Omega \| 4.7 \text{ k}\Omega = 1.83 \text{ k}\Omega$

$$f_{Ho} = \frac{1}{2\pi R_{th(Co)} C_o} = \frac{1}{6.28 \times 1.83 \text{ k}\Omega \times 3 \text{ pF}} = \mathbf{29 \text{ MHz}}$$

Hence, the upper cutoff frequency f_H of this common-base amplifier is 29 MHz.

The absence of the Miller effect in the common-base makes it a superior amplifier in high-frequency applications. However, as you may recall, the common-base configuration suffers from a very low input resistance, and this severely limits its application. At microwave frequencies, a 50-Ω standard is widely used; i.e., the input and load resistances are 50 Ω.

To take advantage of the superior high-frequency response of the common-base amplifier and avoid its inferior input resistance, it is usually cascaded with a common-emitter amplifier, and the whole configuration is called a *cascode* amplifier.

Practice Problem 9-7 — Frequency Response — ANALYSIS

Determine the midband gain $A_{v(mid)}$, and lower and upper cutoff frequencies f_L and f_H for the amplifier circuit of Figure 9-40. Assume: $\beta = 180$, $C_{bc} = 4$ pF, $C_{be} = 12$ pF

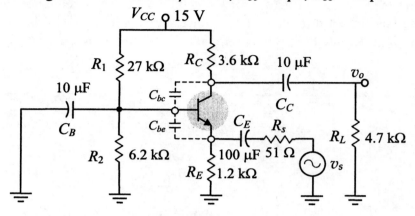

Figure 9-40: A common-base amplifier

Answers: $I_E = 1.7$ mA, $A_{V(mid)} = 137$, $f_L = 24$ Hz, $f_H = 19.9$ MHz

9.6 FREQUENCY RESPONSE OF THE BJT *CASCODE* AMPLIFIER

Consider the cascode amplifier circuit of Figure 9-41 as shown below.
Assume: $\beta = 180$, $C_{eb} = 9$ pF, and $C_{cb} = 3$ pF, for both transistors.

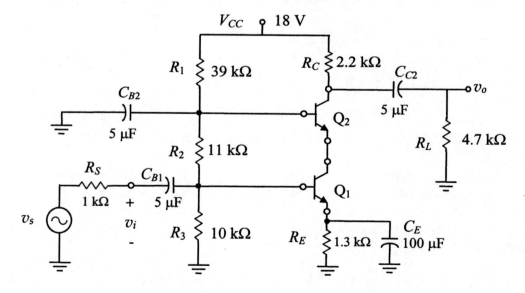

Figure 9-41: A BJT *cascode* amplifier

9.6.1 DC and Midband Analysis

Details of basic calculations for the mid-frequency analysis are not shown; students, however, are asked to verify the following results.

$$V_{B1} = 3 \text{ V}, \quad V_{E1} = 2.3 \text{ V}, \quad I_E = 1.77 \text{ mA}, \quad r_e = 14.7 \text{ }\Omega, \quad V_{B2} = 6.33 \text{ V},$$
$$V_{CE1} = 3.3 \text{ V}, \quad V_{CE2} = 8.5 \text{ V}, \quad R_B = R_2 \| R_3 = 5.238 \text{ k}\Omega, \quad R_b = \beta r_e = 2.65 \text{ k}\Omega,$$
$$R_{in} = R_B \| R_b = 1.76 \text{ k}\Omega, \quad R_o = R_C = 2.2 \text{ k}\Omega, \text{ and } A_v = -102$$

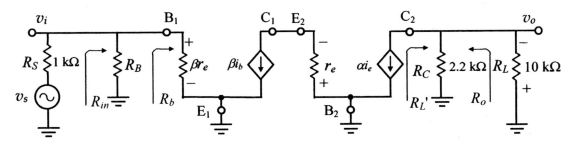

Figure 9-42: Small-signal mid-frequency model of Figure 9-41

9.6.2 Low-Frequency Analysis

The low-frequency model of the cascode amplifier is shown in Figure 9-43. There are a total of four capacitors and each, along with its associated resistance, introduces a cutoff frequency. The highest of all four cutoffs is the dominant low cutoff frequency f_L of the amplifier.

$$R_B = R_2 \| R_3 = 5.238 \text{ k}\Omega, \quad R_b = \beta r_e = 2.65 \text{ k}\Omega, \quad R_{in} = R_B \| R_b = 1.76 \text{ k}\Omega$$

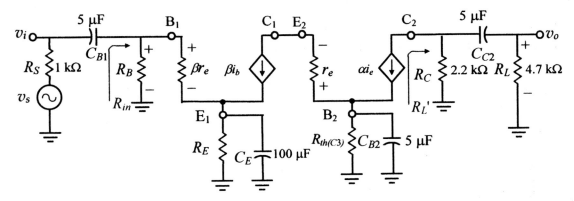

Figure 9-43: Small-signal low-frequency model of Figure 9-41

<u>Cutoff frequency introduced by the coupling capacitor C_{B1}:</u>

$$f_{L1} = \frac{1}{2\pi(R_S + R_{in})C_{B1}} \tag{9-56}$$

$$f_{L1} = \frac{1}{2\pi(1\text{ k}\Omega + 1.76\text{ k}\Omega)(5\text{ }\mu\text{F})} = 11.5 \text{ Hz}$$

Chapter 9

Cutoff frequency introduced by the coupling capacitor C_{C2}:

$$f_{L2} = \frac{1}{2\pi(R_C + R_L)C_{C2}} \tag{9-57}$$

$$f_{L2} = \frac{1}{2\pi(2.2\text{ k}\Omega + 4.7\text{ k}\Omega)(5\text{ }\mu\text{F})} = 4.6\text{ Hz}$$

Cutoff frequency introduced by the base bypass capacitor C_{B2}:

$$f_{L3} = \frac{1}{2\pi(R_{th(C3)})C_{B2}} \tag{9-58}$$

$R_{th(CB)} = R_1 \| [R_2 + (R_S \| R_{in})] = 8.9\text{ k}\Omega$

$$f_{L3} = \frac{1}{2\pi \times 8.9\text{ k}\Omega \times 5\text{ }\mu\text{F}} = 3.6\text{ Hz}$$

Cutoff frequency introduced by the emitter bypass capacitor C_E:

$$f_{LE} = \frac{1}{2\pi(R_{th(CE)})C_E} \tag{9-59}$$

$$R_{th(C_E)} = R_E \left\| \left(r_e + \frac{R_S \| R_B}{\beta} \right) \right. = 1.3\text{ k}\Omega \left\| \left(14.7\text{ }\Omega + \frac{1\text{ k}\Omega \| 5.24\text{ k}\Omega}{180} \right) \right. = 19\text{ }\Omega$$

$$f_{LE} = \frac{1}{2\pi(R_{th(C_E)})C_E} = \frac{1}{2\pi(19\text{ }\Omega)(100\text{ }\mu\text{F})} = \mathbf{84\text{ Hz}}$$

As expected, the dominant low cutoff frequency is the one introduced by the emitter bypass capacitor, which is 84 Hz.

9.6.3 High-Frequency Analysis

The high-frequency model of the cascode amplifier is shown in Figure 9-44.

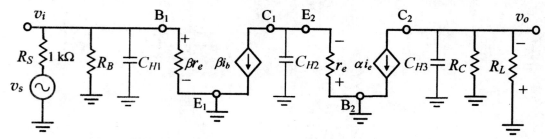

Figure 9-44: Small-signal high-frequency model of Figure 9-41

Note that the first stage, which is a common-emitter, has an inherent Miller effect, but not as severe as in a single-stage common-eitter amplifier. The second stage, which is a common-base amplifier, has no Miller effect. Capacitors denoted as C_{H1}, C_{H2}, and C_{H3} are equivalent capacitances defined as follows:

$$C_{H1} = C_{Mi} + C_{be}, \quad C_{H2} = C_{Mo} + C_{eb}, \quad C_{H3} = C_{bc}$$
$$C_{Mi} = C_{bc}(1 - A_{m1}) \quad\quad\quad\quad C_{Mo} = C_{bc}(1 - 1/A_{m1})$$

$$A_{m1} = -\frac{R_{L1}}{r_e} = -\frac{r_e}{r_e} = -1 \qquad (9\text{-}60)$$

$C_{Mi} = C_{bc}(1 - A_{m1}) = 5\text{ pF}(1 + 1) = 10\text{ pF}$

$C_{Mo} = C_{bc}(1 - 1/A_{m1}) = 5\text{ pF}(1 + 1) = 10\text{ pF}$

$C_{H1} = C_{Mi} + C_{be} = 10\text{ pF} + 10\text{ pF} = 20\text{ pF}$

$C_{H2} = C_{Mo} + C_{be} = 10\text{ pF} + 10\text{ pF} = 20\text{ pF}$

$C_{H3} = C_{bc} = 5\text{ pF}$

$$f_{H1} = \frac{1}{2\pi(R_S\|R_{in})C_{H1}} = \frac{1}{2\pi \times 637\text{ }\Omega \times 20\text{ pF}} = 12.5\text{ MHz}$$

$$f_{H2} = \frac{1}{2\pi r_e C_{H2}} = \frac{1}{2\pi \times 14.7\text{ }\Omega \times 20\text{ pF}} = 541\text{ MHz}$$

$$f_{H3} = \frac{1}{2\pi(R_C\|R_L)C_{H3}} = \frac{1}{2\pi \times 1.5\text{ k}\Omega \times 50\text{ pF}} = 21.2\text{ MHz}$$

The smallest of the three high cutoffs is the upper cutoff frequency f_H, which is 12.5 MHz. In conclusion, the cascode amplifier has all the characteristics of a common-emitter amplifier, but offers the superior high-frequency response of the common-base.

Practice Problem 9-8 — Cascode Amplifier — ANALYSIS

Determine the low and high cutoff frequencies for the cascode amplifier of Figure 9-54. Also, determine the midband voltage gain in dB.
Assume $\beta = 200$, $C_{bc} = 3\text{ pF}$, and $C_{be} = 9\text{ pF}$ for both transistors.

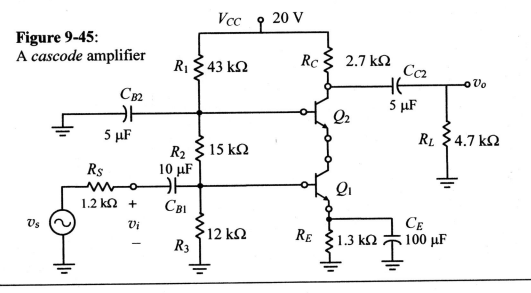

Figure 9-45: A *cascode* amplifier

Answers: $f_{L1} = 5.3$ Hz, $f_{L2} = 4.3$ Hz, $f_{L3} = 2.77$ Hz, $f_{LE} = 26.5$ Hz, $f_{H1} = 14.7$ MHz, $f_{H2} = 856$ MHz, $f_{H3} = 31$ MHz, and $A_{v(mid)} = 42.8$ dB.

Chapter 9

9.7 FREQUENCY RESPONSE OF THE CS FET AMPLIFIER

Frequency response of the FET amplifier is fundamentally the same as the BJT amplifier. All the concepts and theoretical analyses that were developed for the BJT amplifier apply wholly and equally to the FET amplifier.

9.7.1 Frequency Response of the CS JFET Amplifier

Let us determine the lower cutoff and upper cutoff frequencies for the common-source JFET amplifier of Figure 9-46.

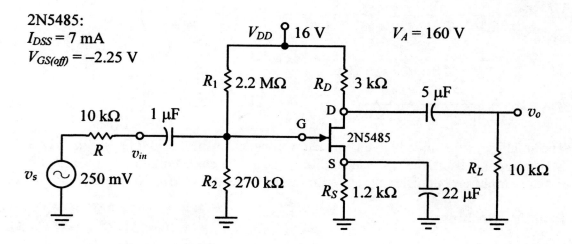

Figure 9-46: A common-source JFET amplifier circuit

DC and Midband Analysis:

Applying the analysis procedure developed in Chapters 7 and 8, we obtain the following results. Details of basic computations will be omitted; however, students may want to refresh their memories by verifying the answers given below:

$$V_{GS} = -1 \text{ V}, \quad I_D = 2.29 \text{ mA}, \quad g_m = 3.5 \text{ mA/V}$$

$$r_o = 70 \text{ k}\Omega, \quad r_m = 1/g_m = 286 \text{ V/A}$$

$R_{in} = R_G = R_1 \| R_2 = 2.2 \text{ M}\Omega \| 270 \text{ k}\Omega = 240 \text{ k}\Omega$

$R_o = r_o \| R_D = 70 \text{ k}\Omega \| 3 \text{ k}\Omega = 2.88 \text{ k}\Omega$

$R_L' = r_o \| R_D \| R_L = 70 \text{ k}\Omega \| 3 \text{ k}\Omega \| 10 \text{ k}\Omega = 2.25 \text{ k}\Omega$

$A_m = \dfrac{v_o}{v_{gs}} = \dfrac{v_o}{v_{in}} = -g_m R_L' = -3.5 \text{ mA/V} \times 2.25 \text{ k}\Omega = -7.875$

$A_{v(mid)} = \dfrac{v_o}{v_s} = -g_m R_L' \dfrac{R_{in}}{R_{in} + R_S} = -7.875 \times \dfrac{240}{240 + 10} = -7875 \times 0.96 = -7.56$

Low-Frequency Analysis:

The general low-frequency model of the amplifier is presented in Figure 9-47.

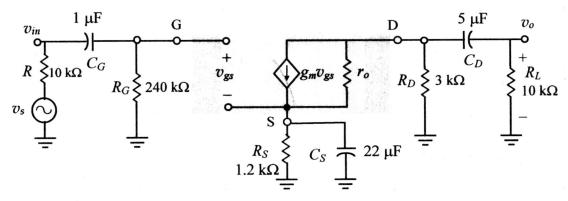

Figure 9-47: Small-signal low-frequency model of Figure 9-46

Each of the three capacitors, along with its associated resistance, will introduce a cutoff frequency and the largest of the three will be the dominant low cutoff frequency f_L of the amplifier.

<u>Cutoff frequency introduced by the coupling capacitor C_G:</u>

$$f_{LG} = \frac{1}{2\pi(R+R_G)C_G} \tag{9-61}$$

$$f_{LG} = \frac{1}{2\pi(10\,\text{k}\Omega + 240\,\text{k}\Omega)(1\,\mu\text{F})} = 0.6\,\text{Hz}$$

<u>Cutoff frequency introduced by the coupling capacitor C_D:</u>

$$f_{LD} = \frac{1}{2\pi[(R_D \| r_o) + R_L]C_D} \tag{9-62}$$

$$f_{LD} = \frac{1}{2\pi(2.87\,\text{k}\Omega + 10\,\text{k}\Omega)(5\,\mu\text{F})} = 2.5\,\text{Hz}$$

<u>Cutoff frequency introduced by the source bypass capacitor C_S:</u>

$$f_{LS} = \frac{1}{2\pi(R_{th(Cs)})C_S} \tag{9-63}$$

$R_{th(Cs)} = r_o \| R_S \| r_m = 70\,\text{k}\Omega \| 1.2\,\text{k}\Omega \| 286\,\Omega = 230\,\Omega$

$$f_{LS} = \frac{1}{2\pi \times 230\,\Omega \times 22\,\mu\text{F}} = 31\,\text{Hz}$$

The lower cutoff frequency f_L is the largest of all three cutoffs and equals 31 Hz, which is introduced by the source bypass capacitor C_S.

Chapter 9

High-Frequency Analysis:

The data sheets for this transistor list three parasitic capacitances, as follows:

Input capacitance $C_{iss\,(max)} = 5$ pF.
Output capacitance $C_{oss\,(max)} = 2$ pF.
Reverse transfer capacitance $C_{rss\,(max)} = 1$ pF.

Assuming the worst case and using the maximum parasitic capacitance values, the interelectrode capacitances of the transistor may be determined from the above data, as follows:

$$C_{gs} = C_{iss} - C_{rss} = 5\text{ pF} - 1\text{ pF} = 4\text{ pF}$$

$$C_{gd} = C_{rss} = 1\text{ pF}$$

$$C_{ds} = C_{oss} - C_{rss} = 2\text{ pF} - 1\text{ pF} = 1\text{ pF}$$

A high-frequency model of the amplifier, including the inter-electrode capacitances is shown in Figure 9-48.

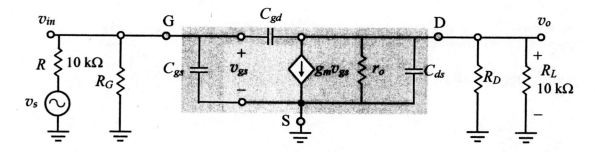

Figure 9-48: Small-signal high-frequency model of Figure 9-46

The result of Miller's theorem developed for the common-emitter amplifier applies equally to the common-source amplifier. Hence, Miller's equivalent capacitances of the feedback capacitance C_{gd}, denoted as C_{Mi} and C_{Mo}, are determined as follows:

$$C_{Mi} = C_{gd}(1 - A_m) \quad (9\text{-}64)$$

$C_{Mi} = 1\text{ pF}(1 + 7.875) = 8.875\text{ pF}$

$$C_{Mo} = C_{gd}(1 - 1/A_m) \quad (9\text{-}65)$$

$C_{Mo} = 1\text{ pF}(1 + 1/7.875) = 1.127\text{ pF}$

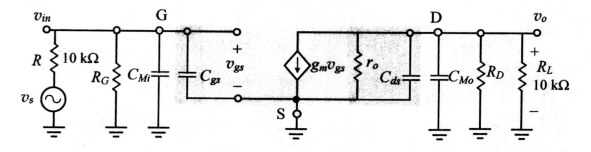

Figure 9-49: Miller's equivalent high-frequency model of Figure 9-46

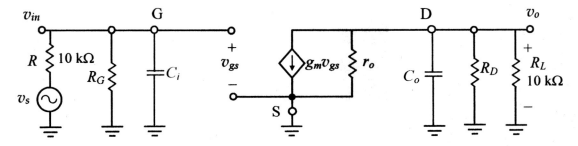

Figure 9-50: Simplified Miller's equivalent high-frequency model

where

$$C_i = C_{Mi} + C_{gs} \qquad (9\text{-}66)$$

$C_i = 8.875 \text{ pF} + 4 \text{ pF} = 12.875 \text{ pF}$

$$C_o = C_{Mo} + C_{ds} \qquad (9\text{-}67)$$

$C_o = 1.127 \text{ pF} + 1 \text{ pF} = 2.127 \text{ pF}$

<u>Cutoff frequency introduced by the input capacitor C_i:</u>

$$f_{Hi} = \frac{1}{2\pi (R_G \| R) C_i} \qquad (9\text{-}68)$$

$$f_{Hi} = \frac{1}{2\pi \times 9.6 \text{ k}\Omega \times 12.875 \text{ pF}} = 1.288 \text{ MHz}$$

<u>Cutoff frequency introduced by the output capacitor C_o:</u>

$$f_{Ho} = \frac{1}{2\pi (r_o \| R_D \| R_L) C_o} \qquad (9\text{-}69)$$

$$f_{Ho} = \frac{1}{2\pi \times 2.25 \text{ k}\Omega \times 2.127 \text{ pF}} = 33.3 \text{ MHz}$$

Therefore, the upper cutoff frequency f_H is approximately 1.3 MHz.

Practice Problem 9-9 — JFET Amplifier — ANALYSIS

Determine the low and high cutoff frequencies for the common-source amplifier of Figure 9-51. Assume: $C_{iss} = 6$ pF, $C_{oss} = 4$ pF, and $C_{rss} = 2$ pF.

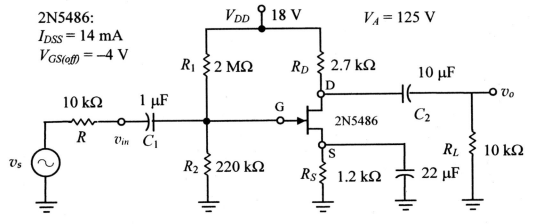

Figure 9-51: Common-source JFET amplifier circuit

Answers:

$f_{L1} = 0.76$ Hz, $f_{L2} = 1.27$ Hz, $f_{LS} = 30$ Hz, $f_{Hi} = 857$ kHz, $f_{Ho} = 18.5$ MHz

9.7.2 Frequency Response of the CS MOSFET Amplifier

Let us now analyze and determine the lower cutoff and upper cutoff frequencies for the common-source EMOS amplifier of Figure 9-52.

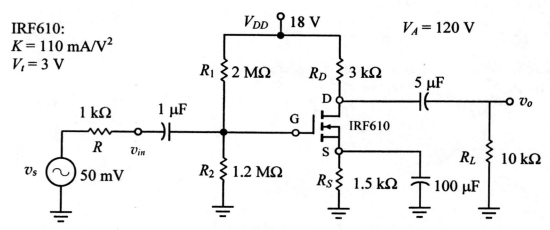

Figure 9-52: Common-source MOSFET amplifier circuit

DC and Midband Analysis:

Applying the analysis procedure of Chapters 8 and 9, we obtain the following results:

$V_{GS} = 3.15$ V, $I_D = 2.4$ mA, $g_m = 33$ mA/V, $r_o = 50$ kΩ, $r_m = 1/g_m = 30$ V/A

$R_{in} = R_G = R_1 \| R_2 = 2$ MΩ $\| 1.2$ MΩ $= 750$ kΩ

$R_o = r_o \| R_D = 50$ kΩ $\| 3$ kΩ $= \mathbf{2.8}$ **kΩ**

$R_L' = r_o \| R_D \| R_L = 50$ kΩ $\| 3$ kΩ $\| 10$ kΩ $= 2.2$ kΩ

$$A_m = \frac{v_o}{v_{in}} = -g_m R_L' = -33 \text{ mA/V} \times 2.2 \text{ kΩ} = -72.6$$

$$A_{v(mid)} = \frac{v_o}{v_s} = -g_m R_L' \cdot \frac{R_{in}}{R_{in} + R_S} = -72.6 \times \frac{750}{750 + 1} = -72.6 \times 0.9868 = -71.64$$

Low- and high-frequency analyses for the EMOS and DMOS amplifiers are identical to those we just applied to JFET amplifier in the previous example.

Low-Frequency Analysis:

The general low-frequency model of the amplifier is presented in Figure 9-53.

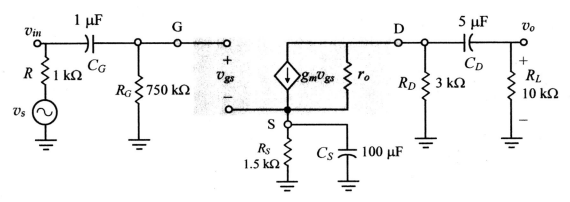

Figure 9-53: Small-signal low-frequency model of Figure 9-52

Cutoff frequency introduced by the coupling capacitor C_G:
$$f_{LG} = \frac{1}{2\pi(R+R_G)C_G} \qquad f_{LG} = \frac{1}{2\pi(1\,k\Omega + 750\,k\Omega)(1\,\mu F)} = 0.2\,Hz$$

Cutoff frequency introduced by the coupling capacitor C_D:
$$f_{LD} = \frac{1}{2\pi(R_D \| r_o + R_L)C_D} \qquad f_{LD} = \frac{1}{2\pi(2.8\,k\Omega + 10\,k\Omega)(5\,\mu F)} = 2.5\,Hz$$

Cutoff frequency introduced by the source bypass capacitor C_S:
$$R_{th(Cs)} = R_S \| r_m = 1.5\,k\Omega \| 30\,\Omega = 29.41\,\Omega$$
$$f_{LS} = \frac{1}{2\pi(R_{th(Cs)})C_S} \qquad f_{LS} = \frac{1}{2\pi \times 29.41\,\Omega \times 100\,\mu F} = 54.1\,Hz$$

$f_z = 1/2\pi R_S C_S = 1/2\pi(1.5\,k\Omega)(100\,\mu F) = 1.061\,Hz$. As expected, the dominant low-cutoff frequency f_L is introduced by the source bypass capacitor, and has a value of 54.1 Hz.

High-Frequency Analysis:

Data sheets of the IR610 MOSFET lists the following as the typical internal capacitances:
$$C_{iss} = 140\,pF, \quad C_{oss} = 53\,pF, \text{ and } C_{rss} = 15\,pF$$
The inter-electrode capacitances are determined from the above data, as follows:
$C_{gs} = C_{iss} - C_{rss} = 140\,pF - 15\,pF = 125\,pF$
$C_{gd} = C_{rss} = 15\,pF$
$C_{ds} = C_{oss} - C_{rss} = 53\,pF - 15\,pF = 38\,pF$
A simplified high-frequency model of the amplifier, including Miller's equivalent capacitors, is shown in Figure 9-54.

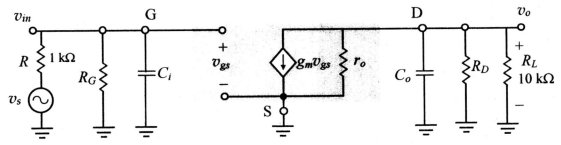

Figure 9-54: Simplified Miller's equivalent high-frequency model

Miller's equivalent capacitors C_{Mi} and C_{Mo} and the overall input and output capacitances C_i and C_o are determined from Equations 9-64 through 9-67, as follows:

$C_{Mi} = C_{gd}(1 - A_m) = 15 \text{ pF}(1 + 72.6) = 1104 \text{ pF}$

$C_{Mo} = C_{gd}(1 - 1/A_m) = 15 \text{ pF}(1 + 1/72.6) = 15 \text{ pF}$

$C_i = C_{Mi} + C_{gs} = 1104 \text{ pF} + 125 = 1229 \text{ pF}$

$C_o = C_{Mo} + C_{ds} = 15 \text{ pF} + 38 \text{ pF} = 53 \text{ pF}$

<u>Cutoff frequency introduced by the input capacitor C_i:</u>

$$f_{Hi} = \frac{1}{2\pi(R_G \| R)C_i} \qquad f_{Hi} = \frac{1}{2\pi \times 1\text{k}\Omega \times 1229 \text{ pF}} = 12.5 \text{ kHz}$$

<u>Cutoff frequency introduced by the output capacitor C_o:</u>

$$f_{Ho} = \frac{1}{2\pi(r_o \| R_D \| R_L)C_o} \qquad f_{Ho} = \frac{1}{2\pi \times 2.2 \text{k}\Omega \times 53 \text{ pF}} = 1.365 \text{ MHz}$$

Section Summary — Frequency Response — ANALYSIS

Summary of Equations for the Frequency Response Analysis of the CS Amplifier

Low-Frequency Analysis:

Cutoff frequency introduced by C_G:

$$f_{Lg} = \frac{1}{2\pi(R + R_G)C_G}$$

Cutoff frequency introduced by C_D:

$$f_{LD} = \frac{1}{2\pi(R_C + R_L)C_D}$$

Cutoff frequency introduced by C_S:

$$f_{LS} = \frac{1}{2\pi(R_{th(Cs)})C_S}$$

where

$R_{th(CS)} = R_S \| r_m$

$r_m = \dfrac{1}{g_m}$

The dominant low-cutoff frequency f_L is the largest of the above three cutoff frequencies, which is usually f_{LS} because of the very low resistance involved with C_S.

Midband voltage gain:

$A_v = -g_m R_L' R_G/(R_G + R)$

High-Frequency Analysis:

$C_{gs} = C_{iss} - C_{rss}$
$C_{gd} = C_{rss}$
$C_{ds} = C_{oss} - C_{rss}$

Cutoff frequency introduced by C_i:

$$f_{Hi} = \frac{1}{2\pi R_{th(Ci)}C_i}$$

Cutoff frequency introduced by C_o:

$$f_{Ho} = \frac{1}{2\pi R_{th(Co)}C_o}$$

where $R_{th(Ci)} = R_G \| R$
$R_{th(Co)} = r_o \| R_D \| R_L = R_L'$
$C_i = C_{gs} + C_{Mi}$
$C_o = C_{ds} + C_{Mo}$

Miller capacitances:

$C_{Mi} = C_{gd}(1 - A_m)$

$C_{Mo} = C_{gd}\left(1 - \dfrac{1}{A_m}\right)$

The dominant high-cutoff frequency f_H is the lowest of the above two cutoff frequencies.

9.8 FREQUENCY RESPONSE OF THE FET CASCODE AMPLIFIER

We will now analyze and determine the lower cutoff and upper cutoff frequencies for the JFET *cascode* amplifier of Figure 9-55.
Assume: $C_{iss} = 18$ pF, $C_{oss} = 8$ pF, and $C_{rss} = 3$ pF.

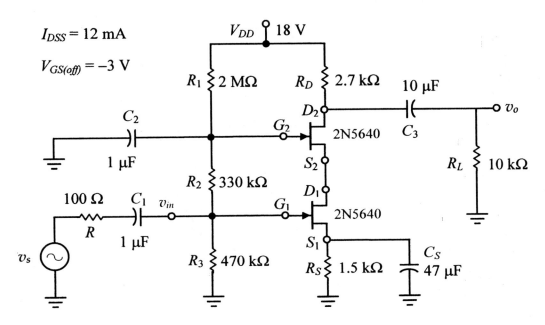

Figure 9-55: A JFET *cascode* amplifier circuit

DC and Mid-Frequency Analysis:

For this cascode amplifier, whose DC and midband analysis have already been carried out, the results are presented below:

$$I_D = 3 \text{ mA}, \quad V_{GS} = -1.5 \text{ V}, \quad g_m = 4 \text{ mA/V}, \quad r_m = 1/g_m = 1/0.004 = 250 \text{ V/A}$$

$$R_{in} = R_2 || R_3 = 330 \text{ k}\Omega || 470 \text{ k}\Omega = 283 \text{ k}\Omega$$

$$R_o = 2.7 \text{ k}\Omega$$

$$R_L' = 2.125 \text{ k}\Omega$$

$$A_v = -8.5$$

Low-Frequency Analysis:

The small-signal low-frequency model of the cascode amplifier of Figure 9-55 is shown in Figure 9-56. There are a total of four capacitors and each, along with its associated resistance, will introduce a low cutoff frequency.

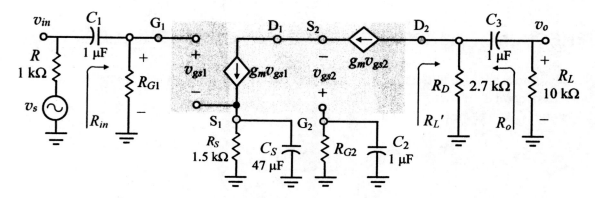

Figure 9-56: Small-signal low-frequency model of Figure 9-55

$$R_{G1} = R_2 \| R_3 = 194 \text{ k}\Omega, \quad R_{G2} = R_1 \| (R_2 + R\|R_3) \cong R_1 \| R_2 = 283 \text{ k}\Omega$$

Cutoff frequency introduced by the coupling capacitor C_1:

$$f_{L1} = \frac{1}{2\pi(R + R_{G1})C_1} \qquad f_{L1} = \frac{1}{2\pi(100\,\Omega + 194 \text{ k}\Omega)(1\,\mu\text{F})} = 0.8 \text{ Hz}$$

Cutoff frequency introduced by the gate bypass capacitor C_2:

$$f_{L2} = \frac{1}{2\pi(R_{G2})C_2} \qquad f_{L2} = \frac{1}{2\pi(283 \text{ k}\Omega)(1\,\mu\text{F})} = 0.56 \text{ Hz}$$

Cutoff frequency introduced by the coupling capacitor C_3:

$$f_{L3} = \frac{1}{2\pi(R_D + R_L)C_3} \qquad f_{L3} = \frac{1}{2\pi(2.7 \text{ k}\Omega + 10 \text{ k}\Omega)(10\,\mu\text{F})} = 1.25 \text{ Hz}$$

Cutoff frequency introduced by the source coupling capacitor C_S:

$$f_{LS} = \frac{1}{2\pi(r_m\|R_S)C_S} \qquad f_{LS} = \frac{1}{2\times\pi\times(250\,\Omega\|1.5 \text{ k}\Omega)47\,\mu\text{F}} = 16 \text{ Hz}$$

The lower cutoff frequency f_L, which is the largest of the above four cutoff frequencies, is introduced by the source coupling capacitor, and has a value of 16 Hz.

High-Frequency Analysis:

$C_{iss} = 18$ pF, $C_{oss} = 8$ pF, $C_{rss} = 3$ pF

$C_{gs} = C_{iss} - C_{rss} = 18$ pF $- 3$ pF $= 15$ pF

$C_{gd} = C_{rss} = 3$ pF

$C_{ds} = C_{oss} - C_{rss} = 8$ pF $- 3$ pF $= 5$ pF

The high-frequency model of the amplifier, including interelectrode capacitances, is shown in Figure 9-57.

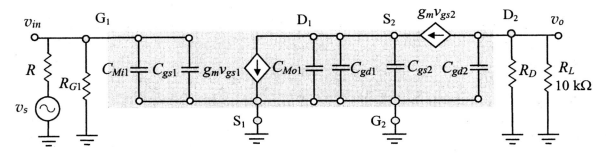

Figure 9-57: High-frequency model of the cascode amplifier of Figure 9-55

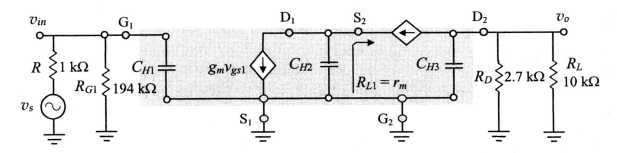

Figure 9-58: Simplified high-frequency model of Figure 9-55

where,

$$C_{H1} = C_{Mi} + C_{gs1}, \qquad C_{H2} = C_{Mo} + C_{gd1} + C_{gs2}, \qquad C_{H3} = C_{gd2}$$

$$C_{Mi} = C_{bc}(1 - A_{v1}) \qquad\qquad C_{Mo} = C_{bc}(1 - 1/A_{v1})$$

$$A_{v1} = -g_m R_{L1} = -\frac{R_{L1}}{r_m} = -\frac{r_m}{r_m} = -1 \tag{9-82}$$

$C_{Mi} = C_{gd}(1 - A_{v1}) = 3\text{ pF }(1+1) = 6\text{ pF}$

$C_{Mo} = C_{gd}(1 - 1/A_{v1}) = 3\text{pF }(1+1) = 6\text{ pF}$

$C_{H1} = C_{Mi} + C_{gs1} = 6\text{ pF} + 15\text{ pF} = 21\text{ pF}$

$C_{H2} = C_{Mo} + C_{gd1} + C_{gs2} = 6\text{ pF} + 3\text{ pF} + 15\text{ pF} = 24\text{ pF}$

$C_{H3} = C_{gd2} = 3\text{ pF}$

$$f_{H1} = \frac{1}{2\pi(R\|R_{G1})C_{H1}} = \frac{1}{6.28 \times (100\,\Omega \| 194\,\text{k}\Omega) \times 21\,\text{pF}} = 75.8\text{ MHz}$$

$$f_{H2} = \frac{1}{2\pi r_m C_{H2}} = \frac{1}{6.28 \times 250\,\Omega \times 24\,\text{pF}} = 26.5\text{ MHz}$$

$$f_{H3} = \frac{1}{2\pi(R_D\|R_L)C_{H3}} = \frac{1}{6.28 \times 2.125\,\text{k}\Omega \times 3\,\text{pF}} = 25\text{ MHz}$$

Chapter 9

The upper cutoff frequency f_H is the smallest of the above three cutoffs and is 25 MHz.

Practice Problem 9-10 — Cascode Amplifier — ANALYSIS

Analyze and determine the lower cutoff and upper cutoff frequencies for the JFET *cascode* amplifier of Figure 9-59, as shown below.
Assume: $C_{iss} = 18$ pF, $C_{oss} = 8$ pF, $C_{rss} = 3$ pF.

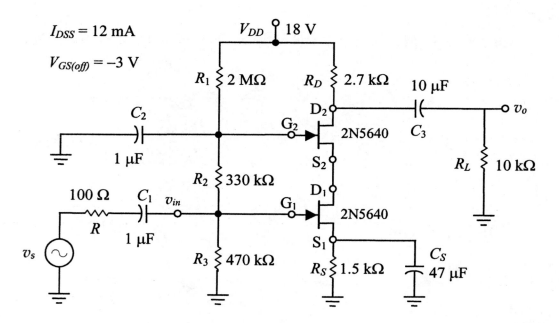

Figure 9-59: A JFET *cascode* amplifier circuit

9.9 SUMMARY

- The midband is a range of operating frequency in which the coupling and bypass capacitors of BJT and FET amplifiers are replaced by a short circuit; that is, the operating frequency is considered high enough so that the impedance of the capacitors are very low, nearly zero. The upper range of the operating frequency is considered low enough, at midband, so that the internal transistor capacitances would not have significant effect on the amplifier operation; that is their impedances are considered to be very high, nearly infinite.

- At the low-frequency range, impedance of the coupling and bypass capacitors tends to increase, and thus, impose limitations on the amplifier operation and its frequency response.

- At the high-frequency range, impedance of the internal capacitances of the transistor tends to decrease. It can even short out transistor terminals with further increase in

Chapter 9

- frequency. Hence, the higher frequencies impose limitations on the amplifier operation and its frequency response.

- The dB gain versus logarithmic scale frequency plot, which is known as the Bode plot, is generally used when plotting the frequency response, instead of the magnitude gain versus linear scale frequency plot.

- The lower and upper cutoff frequencies f_L and f_H occur at 0.707 $A_{v(mid)}$ in the linear scale, and at 3 dB below the midband gain in the dB scale.

- The cutoff frequency f_β, which is caused by dependence of transistor β on frequency, is another high-frequency limitation imposed on the amplifier's performance in the common-emitter and common-collector configurations.

- In addition to the limitations imposed by the internal junction capacitances of the transistor, high-frequency response of the common-emitter amplifier is also limited by the Miller effect and transistor β.

- The independence of the high-frequency response on β, and the absence of the Miller effect in the common-base, makes it a superior amplifier in high-frequency applications. However, the common-base configuration suffers from a very low input resistance, and this severely limits its application.

- To take advantage of the superior high-frequency response of the common-base and avoid its inferior input resitance, common-base is cascaded with a common-emitter, and is called *cascode* amplifier, which is used in high-frequency applications such as *RF* (radio frequency) amplifiers.

- Like the common-emitter BJT amplifier, the common-source FET amplifier's high-frequency response is limited by the Miller effect.

- Similar to the emitter-follower (common-collector BJT amplifier), high-frequency response of the source-follower (common-drain FET amplifier) is not limited by the Miller effect, and thus, has a better high-frequency response compared to the common-source amplifier, but it offers no voltage gain.

- Like the common-base BJT amplifier, high-frequency response of the common-gate FET amplifier is not limited by the Miller effect, and thus, has a superior high-frequency response compared to the common-source amplifier, but it suffers from a very low input resistance.

- Like the BJT *cascode* amplifier, which is a cascade of common-emitter and common-base amplifiers, the FET *cascode* amplifier is a cascade of common-source and common-gate amplifiers, which utilizes the superior characteristics of both, while avoiding their inferior characteristics.

Chapter 9

PROBLEMS

Section 9.2 Frequency Response Fundamentals

9.1 Determine the voltage gain in dB of an amplifier, that produces 6 V(p-p) output for an input of 45 mV(p-p).

9.2 Determine the voltage gain in dB of an amplifier, that produces 6 V(p-p) output for an input of 25 mV(p-p).

9.3 Determine the ω_c and f_c for a low-pass RC network with $R = 1.6$ kΩ, and $C = 0.10$ µF and plot the Bode diagram with MATLAB.

9.4 Determine the ω_c and f_c for a low-pass RC network with $R = 3.3$ kΩ, and $C = 0.02$ µF and plot the Bode diagram with MATLAB.

9.5 Carry out Practice Problems 9-1 and 9-2.

9.6 Determine the ω_c and f_c for a high-pass RC network with $R = 220$ Ω, and $C = 0.47$ µF and plot the Bode diagram with MATLAB.

9.7 Determine the ω_c and f_c for a high-pass RC network with $R = 3$ kΩ, and $C = 0.01$ µF and plot the Bode diagram with MATLAB.

Section 9.3 Frequency Response of CE BJT Amplifier

9.8 Carry out Practice Problem 9-3.

9.9 Determine the low cutoff frequencies f_{L1}, f_{L2}, f_{LE}, and the lower cutoff frequency f_L for the amplifier circuit of Figure 9-1P.

9.10 Determine the lower cutoff frequency f_L for the amplifier circuit of Figure 9-1P, with the following changes: $R_1 = 36$ kΩ, $R_E = 1.2$ kΩ, $R_C = 3.6$ kΩ

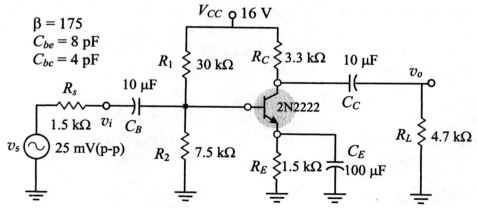

Figure 9-1P: Circuit diagram of a CE amplifier

Chapter 9

9.11 Determine the low cutoff frequencies f_{LB}, f_{LC}, f_{LE}, and the lower cutoff frequency f_L for the amplifier circuit of Figure 9-2P.

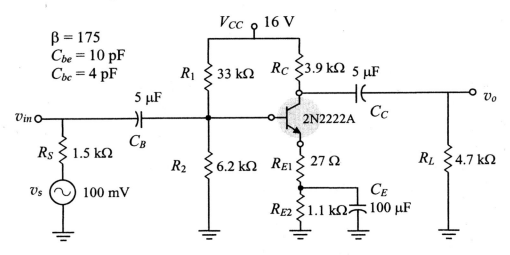

Figure 9-2P: Common-emitter amplifier with partially bypassed R_E

9.12 Carry out Practice Problem 9-4.

9.13 Determine the low cutoff frequencies f_{LB}, f_{LC}, f_{LE}, and the lower cutoff frequency f_L for the amplifier circuit of Figure 9-2P, with the following changes:
$R_1 = 36$ kΩ, $R_E = 1.2$ kΩ, $R_C = 3.6$ kΩ.

9.14 Carry out Practice Problem 9-5.

9.15 Determine the high cutoff frequencies f_{Hi}, f_{Ho}, f_β, and the upper cutoff frequency f_H for the amplifier circuit of Figure 9-1P. Also determine the midband gain in dB, and the output voltage at midband and at the upper cutoff frequency.

9.16 Determine the high cutoff frequencies f_{Hi}, f_{Ho}, f_β, and the upper cutoff frequency f_H for the amplifier circuit of Figure 9-2P. Also determine the midband gain in dB, and the output voltage at midband and at the upper cutoff frequency.

Section 9.4 Frequency Response Plot of the CE BJT Amplifier

9.17 Carry out Practice Problem 9-6.

Section 9.5 Frequency Response of the CB BJT Amplifier

9.18 Determine the lower and upper cutoff frequencies for the amplifier circuit of Figure 9-3P. Also, determine the midband gain in dB, and the output voltage at midband and at the cutoff frequency.

Chapter 9

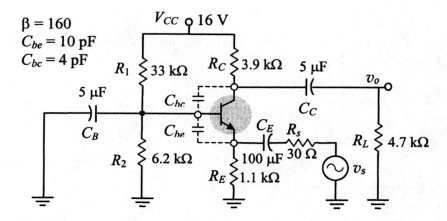

Figure 9-3P: Circuit diagram for problems 9.17 and 9.18

9.19 Determine the lower and upper cutoff frequencies for the amplifier circuit of Figure 9-3P, with the following changes: $R_1 = 36$ kΩ, $R_E = 1.2$ kΩ.

9.20 Carry out Practice Problem 9-7.

9.21 Determine the lower and upper cutoff frequencies for the amplifier circuit of Figure 9-4P.

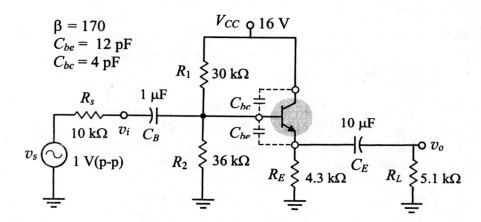

Figure 9-4P: Circuit diagram for problems 9.21 and 9.22

9.22 Determine the lower and upper cutoff frequencies for the amplifier circuit of Figure 9-4P, with the following changes: $R_1 = 33$ kΩ, $R_2 = 39$ kΩ, $R_E = 5.1$ kΩ.

Section 9.6 Frequency Response of the BJT Cascode Amplifier

9.23 Carry out Practice Problem 9-8.

9.24 Determine the lower and upper cutoff frequencies for the amplifier circuit of Figure 9-5P. Also, determine the midband gain in dB.

Chapter 9

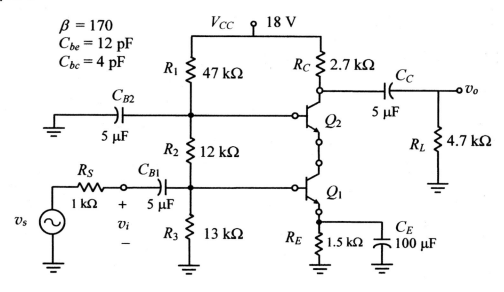

Figure 9-5P: Circuit diagram for Problems 9.24 and 9.25

9.25 Determine the lower and upper cutoff frequencies for the amplifier circuit of Figure 9-5P with the following changes: $R_1 = 51$ kΩ, $R_2 = 15$ kΩ, and $R_3 = 12$ kΩ.

Section 9.7 Frequency Response of the FET Amplifier

9.26 Carry out Practice Problem 9-9.

9.27 Determine the lower and upper cutoff frequencies for the amplifier circuit of Figure 9-6P. Also, determine the midband gain in dB. Assume $C_{iss} = 6$ pF, $C_{oss} = 4$ pF, and $C_{rss} = 2$ pF.

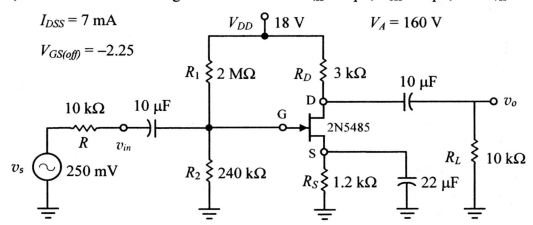

Figure 9-6P: Circuit diagram for Problems 9.27 and 9.28

9.28 Determine the lower and upper cutoff frequencies for the amplifier circuit of Figure 9-6P with the following changes: $R_1 = 1.6$ MΩ and $R_2 = 200$ kΩ.

9.29 Determine the lower and upper cutoff frequencies for the amplifier circuit of Figure 9-7P. Also, determine the midband gain in dB.

Chapter 9

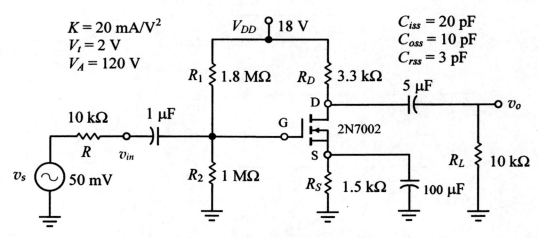

Figure 9-7P: Circuit diagram for Problems 9.29 and 9.30

9.30 Determine the lower and upper cutoff frequencies for the amplifier circuit of Figure 9-7P with the following changes: $R_1 = 1.8$ MΩ and $R_2 = 1$ MΩ.

Section 9.8 Frequency Response of the FET Cascode Amplifier

9.31 Carry out Practice Problem 9-10.

9.32 Determine the lower and upper cutoff frequencies for the cascode amplifier circuit of Figure 9-8P. Assume: $C_{iss} = 15$ pF, $C_{oss} = 5$ pF, and $C_{rss} = 3$ pF.

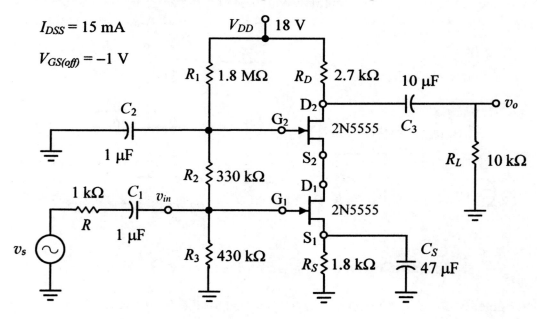

Figure 9-8P: Circuit schematic for Problem 9.32

Chapter 10

CURRENT-MIRROR CURRENT SOURCES AND DIFFERENTIAL AMPLIFIERS

Analysis & Design

10.1 INTRODUCTION

The differential amplifier, which comprises two transistors and a current source, is somewhat different from the amplifiers we have studied in the previous chapters. A differential amplifier amplifies the intended signal, while attenuating the undesirable and/or unwanted signals like noise. This chapter is dedicated to the analysis and design of current-mirror current sources and differential amplifiers. This comprehensive coverage will prepare the student for the study of operational amplifiers and their applications in the following chapter.

10.2 CURRENT-MIRROR CURRENT SOURCES

An ideal current source supplies a steady amount of current and has infinite output resistance. Several types of current sources can be configured with BJTs and FETs; however, the most popular configurations that are widely used in integrated circuits are the *current-mirror* type current sources. We will discuss two types of current-mirrors, starting with the *basic* current-mirror and then, later in the chapter, the much-improved version of it, the *Wilson* current-mirror.

10.2.1 The Basic BJT Current-Mirror

As shown in Figure 10-1 below, the basic *current-mirror* current source comprises a resistor and two transistors.

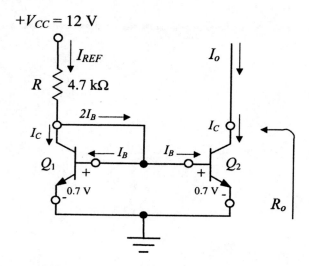

Figure 10-1: Circuit diagram of a basic *current-mirror* current source

The two transistors Q_1 and Q_2 are tied together at the base and the emitter, and both are biased through the same resistor and from the same V_{CC}. In other words, the two transistors are biased in parallel. Hence, assuming that the two transistors are identical, the base currents in both transistors will be equal, and thus, the collector currents will also be equal. The reference current through the resistor R, which is labeled I_{REF}, and its relationship to the output current I_o can be determined as follows:

According to *KVL*, the supply voltage V_{CC} must equal the sum of the voltage drops around the current path; that is,

$$V_{CC} = (I_{REF} \times R) + V_{BE}$$

Solving for I_{REF} results in the following:

$$I_{REF} = \frac{V_{CC} - V_{BE}}{R} \tag{10-1}$$

Chapter 10

$$I_{REF} = \frac{12 - 0.7}{4.7\ k\Omega} = 2.4\ mA$$

To determine the relationship between the I_o and I_{REF}, we will apply *KCL* at the collector node of Q_1, as follows:

$$I_{REF} = 2I_B + I_C = 2I_B + \beta I_B = (\beta + 2)\ I_B$$

The output current I_o, which is the collector current of Q_2, is the following:

$$I_o = I_C = \beta I_B$$

Dividing the I_o by I_{REF} results in the following ratio:

$$\frac{I_o}{I_{REF}} = \frac{\beta \cdot I_B}{(\beta + 2)I_B} = \frac{\beta}{\beta + 2} = \frac{1}{1 + \frac{2}{\beta}} \qquad (10\text{-}2)$$

Assuming that $\beta = 200$, the above ratio will be 0.9900. That is, the I_o, which is the output current of the current source, will be the following:

$$I_o = 0.99 I_{REF}$$

$I_o = 0.99(2.4\ mA) = 2.376\ mA$

The output resistance of the current-mirror R_o is simply the output resistance of the transistor r_o, which can be determined with the Early voltage V_A of the transistor and the output current $I_o = I_C$ as follows:

$$R_o = r_o = \frac{V_A}{I_o} \qquad (10\text{-}3)$$

Assuming $V_A = 120$ V, R_o will be approximately 50 kΩ. According to the above equation, the output resistance is inversely related to the output current; that is, the larger the output current, the smaller the output resistance will be, and vice versa.

The current-mirror of Figure 10-1, which is redrawn in Figure 10-2(a) below, may also be drawn with a negative supply, as shown in Figure 10-2(b).

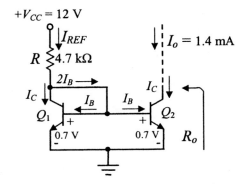

Figure 10-2(a): Circuit of a basic current-mirror with positive supply

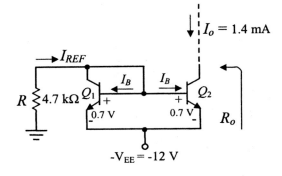

Figure 10-2(b): Circuit of a basic current-mirror with negative supply

Chapter 10

Practice Problem 10-1 — Current Mirror — ANALYSIS

Determine the following for the circuits of Figure 10-3(a) and 10-3(b) below:
a) Output current and output resistance of the current-mirror shown in Figure 10-3(a).
b) The resistance R and output resistance R_o of the current-mirror shown in Figure 10-3(b). Assume $V_A = 125$ V.

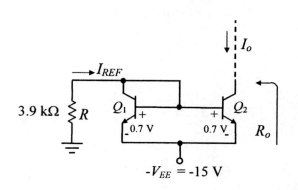

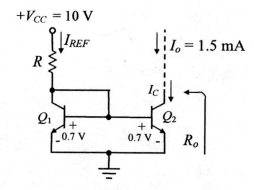

Figure 10-3(a): Basic current-mirror

Figure 10-3(b): Basic current-mirror

Answers: a) $I_o = 3.66$ mA, $R_o = 34$ kΩ b) $R = 6.2$ kΩ, $R_o = 83.3$ kΩ

10.2.2 Wilson Current-Mirror

A circuit diagram of a Wilson current-mirror and its equivalent with negative supply is shown in Figures 10-4(a) and (b) below:

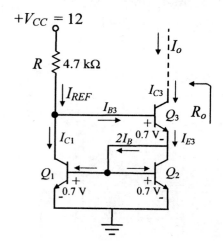

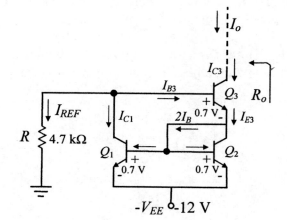

(a) Wilson *current-mirror* with positive supply

(b) Wilson *current-mirror* with negative supply

Figure 10-4: Circuit diagram of a Wilson *current-mirror*

Applying the *KVL* around the path from V_{CC} to ground results in the following:

$$V_{CC} = I_{REF} \times R + V_{BE3} + V_{BE2}$$

Rearranging the variables and solving for I_{REF}, we obtain the following:

$$I_{REF} = \frac{V_{CC} - 2V_{BE}}{R} \tag{10-4}$$

Assuming $Q_1=Q_2=Q_3$ and applying the *KCL* at the collector node of Q_2 results in the following equation:

$$I_{E3} = I_{C2} + 2I_B \tag{10-5}$$

Since Q_1 and Q_2 are biased in parallel by having their base and emitter terminals tied together, let $I_{E1} = I_{E2} = I_E$, $I_{C1} = I_{C2} = I_C$, and $I_{B1} = I_{B2} = I_B$.

Substituting for I_C and I_B in terms of I_E in Equation 10-5 results in the following:

$$I_{E3} = \frac{\beta}{\beta+1}I_E + 2\frac{I_E}{\beta+1} = I_E\left(\frac{\beta}{\beta+1} + \frac{2}{\beta+1}\right) = I_E\left(\frac{\beta+2}{\beta+1}\right) \tag{10-6}$$

Substituting for I_{B3} in terms of I_{E3}, we obtain the following:

$$I_{B3} = \frac{I_{E3}}{\beta+1} = I_E\left(\frac{\beta+2}{(\beta+1)^2}\right) \tag{10-7}$$

Applying the *KCL* at the collector node of Q_1 and substituting for I_C in terms of I_E results in the following equation:

$$I_{REF} = I_C + I_{B3} = \frac{\beta}{\beta+1}I_E + I_E\left(\frac{\beta+2}{(\beta+1)^2}\right) \tag{10-8}$$

Factoring out I_E and taking the common denominator, we obtain the following:

$$I_{REF} = I_E\left(\frac{\beta(\beta+1) + \beta + 2}{(\beta+1)^2}\right) \tag{10-9}$$

Substituting for I_{C3} in terms of I_{E3} results in the following:

$$I_o = I_{C3} = \frac{\beta}{\beta+1}I_{E3} = \frac{\beta}{\beta+1}\left(\frac{\beta+2}{\beta+1}\cdot I_E\right) = \frac{\beta(\beta+2)}{(\beta+1)^2}I_E \tag{10-10}$$

To obtain the ratio of I_o to I_{REF}, we can simply multiply Equation 10-10 with the inverse of Equation 10-9, as follows:

$$\frac{I_o}{I_{REF}} = \frac{\beta(\beta+2)I_E}{(\beta+1)^2} \cdot \frac{(\beta+1)^2}{I_E\beta(\beta+1) + \beta + 2} \tag{10-11}$$

Canceling out the common terms in the denominator and the numerator and dividing both by β results in the following:

$$\frac{I_o}{I_{REF}} = \frac{(\beta+2)}{(\beta+1)+1+\frac{2}{\beta}} = \frac{\beta+2}{(\beta+2)+\frac{2}{\beta}} \tag{10-12}$$

Dividing both the numerator and the denominator by $(\beta + 2)$ results in the following:

$$\frac{I_o}{I_{REF}} = \frac{1}{1+\frac{2}{\beta(\beta+2)}} = \frac{1}{1+\frac{2}{\beta^2+2\beta}} \tag{10-13}$$

$$\frac{I_o}{I_{REF}} \cong \frac{1}{1+\frac{2}{\beta^2}} \tag{10-14}$$

$$I_o \cong \frac{I_{REF}}{1+\frac{2}{\beta^2}} \tag{10-15}$$

For example, if $\beta = 100$, $I_o = 0.9998 I_{REF}$, and if $\beta = 200$, $I_o = 0.99995 I_{REF}$.

The output resistance R_o of the Wilson current mirror can be shown to be approximately $\frac{\beta}{2} r_o$, which is $\frac{\beta}{2}$ times higher than the output resistance of the basic current mirror.

$$R_o = \frac{\beta}{2} r_o \tag{10-16}$$

where

$$r_o = \frac{V_A}{I_{E(Q3)}} = \frac{V_A}{I_{REF}} \tag{10-17}$$

Let us now determine the bias current and the output resistance of the Wilson current mirror of Figure 10-4. Assume $V_A = 135$ V and $\beta = 200$. Then,

$$I_{REF} = \frac{V_{CC} - 2V_{BE}}{R} = \frac{(12 - 1.4) \text{ V}}{4.7 \text{ k}\Omega} = 2.255 \text{ mA} \qquad R_o = \frac{\beta}{2} r_o = 100 \times 60 \text{ k}\Omega = 6 \text{ M}\Omega$$

Practice Problem 10-2 — Current Mirror — ANALYSIS

Determine the reference current I_{REF}, the output current I_o, and the output resistance for the Wilson current-mirror current source of Figure 10-5.
Assume $\beta = 160$ and $V_A = 160$ V.

Figure 10-5:
Wilson current-mirror current source

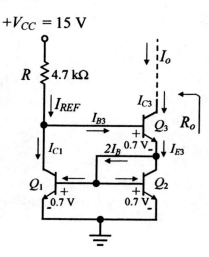

Answers: $I_{REF} = 2.89$ mA,
$I_o = 2.89$ mA, $R_o = 4.4$ MΩ

Section Summary   ANALYSIS

Summary of Equations for the Analysis of Basic and Wilson Current-Mirrors

Basic current-mirror:

$$I_{REF} = \frac{V_{CC} - V_{BE}}{R}$$

$$\frac{I_o}{I_{REF}} = \frac{\beta}{\beta + 2} = \frac{1}{1 + \frac{2}{\beta}}$$

$$I_o = (1 + 2/\beta)I_{REF} \cong I_{REF}$$

$$R_o = r_o \qquad r_o = \frac{V_A}{I_o}$$

Wilson current-mirror:

$$I_{REF} = \frac{V_{CC} - 2V_{BE}}{R}$$

$$\frac{I_o}{I_{REF}} = \frac{\beta}{(\beta + 2)^2} \cong \frac{1}{1 + \frac{2}{\beta^2}}$$

$$I_o = (1 + 2/\beta^2)I_{REF} = I_{REF}$$

$$R_o = \frac{\beta}{2} r_o \qquad r_o = \frac{V_A}{I_o}$$

Practice Problem 10-3   DESIGN

Design the following current sources for an output current of 4 mA, and determine its output resistance if $V_{CC} = 16$ V, $\beta = 160$, and $V_A = 160$ V.
  a) Basic current-mirror current source
  b) Wilson current-mirror current source

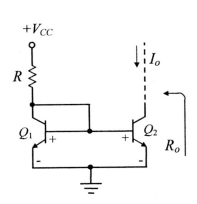
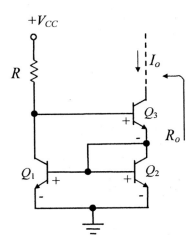

Figure 10-6(a): Basic current-mirror   **Figure 10-6(b):** Wilson current-mirror

Answers: a) $R = 3.9$ kΩ, $R_o = 40$ kΩ,   b) $R = 3.6$ kΩ, $R_o = 3.2$ MΩ

10.2.3 MOSFET Current-Mirror Current Sources

Basic MOSFET Current-Mirror

The basic MOSFET current-mirror can be achieved with two MOSFETs connected in a similar fashion as the basic BJT current-mirror, as shown in Figure 10-7 below:

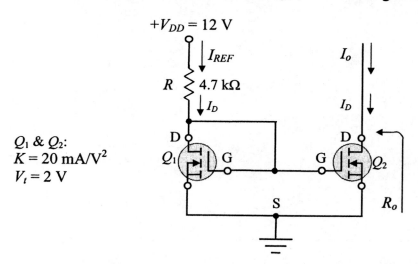

Figure 10-7: Circuit diagram of a basic MOSFET *current-mirror* current source

Since V_{GS1} equals V_{GS2}, I_{D1} equals I_{D2}, and because there is no gate current, I_{RFE} equals I_o.

$$V_{GS1} = V_{GS2} = V_{GS}$$

$$I_{D1} = I_{D2} = I_D$$

$$I_{RFE} = I_D = I_o$$

According to *KVL*, the supply voltage V_{DD} must equal the sum of voltage drops along the path to ground; that is,

$$V_{DD} = I_D \times R + V_{GS}$$

Solving for I_D in the above equation results in the following: (10-18)

$$I_D = \frac{V_{DD} - V_{GS}}{R} \qquad (10\text{-}19)$$

I_D can also be determined as follows:

$$I_D = K(V_{GS} - V_t)^2$$

Equating the above two equations for I_D we obtain the following:

$$I_D = \frac{V_{DD} - V_{GS}}{R} = K(V_{GS} - V_t)^2$$

$$V_{DD} - V_{GS} = KR(V_{GS} - V_t)^2$$

$$V_{DD} - V_{GS} = KR(V_{GS}^2 - 2V_{GS}V_t + V_t^2) \qquad (10\text{-}20)$$

Further algebraic manipulation and separation of variables yields the following second order equation in terms of the variable V_{GS} and other known circuit parameters:

$$KRV_{GS}^2 + (1 - 2KRV_t)V_{GS} + (KRV_t^2 - V_{DD}) = 0 \qquad (10\text{-}21)$$

Chapter 10

The above equation is in the form of the following general quadratic equation:

$$ax^2 + bx + c = 0 \qquad (10\text{-}22)$$

which has the following general solution:

$$V_{GS}\Big|_{n-channel} = \frac{-b + \sqrt{b^2 - 4ac}}{2a} \qquad (10\text{-}23)$$

$$V_{GS}\Big|_{p-channel} = \frac{+b - \sqrt{b^2 - 4ac}}{2a} \qquad (10\text{-}24)$$

where,

$$a = KR \qquad b = 1 - 2KR|V_t| \qquad c = KRV_t^2 - |V_{DD}|$$

After V_{GS} is determined with the above equations, I_D can be determined with Equation 10-19, but first we need to determine the parameters a, b, c, and then V_{GS} and I_D.

$a = KR = 20\text{ mA/V}^2 \times 4.7\text{ k}\Omega = 94$

$b = 1 - 2KR|V_t| = 1 - (2 \times 20\text{ mA/V}^2 \times 4.7\text{ k}\Omega \times 2\text{ V}) = -375$

$c = KRV_t^2 - |V_{DD}| = 20\text{ mA/V}^2 \times 4.7\text{ k}\Omega \times 4\text{ V}^2 - 12 = 364$

$$V_{GS} = \frac{+375 + \sqrt{375^2 - (4 \times 94 \times 364)}}{2 \times 94} = 2.32\text{ V}$$

$$I_D = \frac{V_{DD} - V_{GS}}{R} = \frac{12\text{V} - 2.32\text{V}}{4.7\text{ k}\Omega} = 2.06\text{ mA} \cong 2\text{ mA}$$

$$I_{RFE} = I_D = I_o = 2\text{ mA}$$

Assuming $V_A = 120$ V, the output resistance is determined as follows:

$$R_o = r_o = \frac{V_A}{I_o} = \frac{120\text{ V}}{2\text{ mA}} = 60\text{ k}\Omega$$

Practice Problem 10-4 — Current Mirror — ANALYSIS

Determine the I_o and R_o for the basic current-mirror as shown below:

Q_1 & Q_2:
$K = 20\text{ mA/V}^2$
$V_t = 2\text{ V}$
$V_A = 120\text{ V}$

Answers:

$I_o = 2.48\text{ mA}, \quad R_o = 48\text{ k}\Omega$

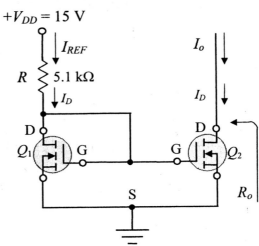

Cascode Current-Mirror Current Source

A substantially higher output resistance can be achieved with another MOSFET current source called *cascode current mirror*, shown in Figure 10-8.

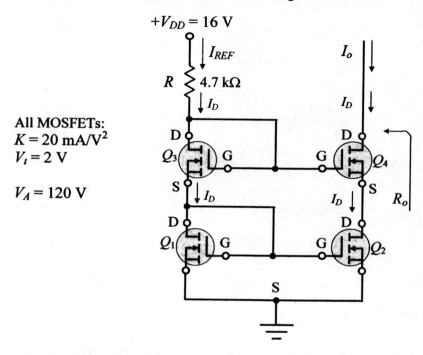

Figure 10-8: Circuit diagram of a *cascode current-mirror* current source

Each MOSFET pair in the above diagram represents a basic current mirror. Hence, all drain currents are equal to I_{REF}, and thus all gate-to-source voltages will be the same

$$V_{GS1} = V_{GS2} = V_{GS3} = V_{GS4} = V_{GS}$$
$$I_{D1} = I_{D2} = I_{D3} = I_{D4} = I_{REF} = I_o$$

According to KVL, the supply voltage V_{DD} must equal the sum of voltage drops along the path to ground

$$V_{DD} = I_D R + 2V_{GS} \tag{10-25}$$

Solving for I_D in the above equation leads to the drain current

$$I_D = \frac{V_{DD} - 2V_{GS}}{R} \tag{10-26}$$

The dain current I_D can also be determined from Equation 7-25

$$I_D = K(V_{GS} - V_t)^2$$

Equating the above two equations for I_D and further algebraic manipulation and separation of variables yields the following second-order equation in terms of the variable V_{GS} and other known circuit parameters:

$$KRV_{GS}^2 + 2(1 - KRV_t)V_{GS} + (KRV_t^2 - V_{DD}) = 0 \tag{10-27}$$

The above equation is in the form of the following general quadratic equation:

Chapter 10

$$ax^2 + bx + c = 0 \quad (10\text{-}28)$$

which has the following general solution:

$$V_{GS}\Big|_{n-channel} = \frac{-b + \sqrt{b^2 - 4ac}}{2a} \quad (10\text{-}29)$$

where,

$$a = KR \qquad b = 2(1 - KR|V_t|) \qquad c = KRV_t^2 - |V_{DD}|$$

After the V_{GS} is determined with the above equations, I_D can be determined with Equation 10-26, but first we need to determine the parameters a, b, and c, and then V_{GS} and I_D.

$a = KR = 20 \text{ mA/V}^2 \times 4.7 \text{ k}\Omega = 94$

$b = 2(1 - KR|V_t|) = 2[1 - (20 \text{ mA/V}^2 \times 4.7 \text{ k}\Omega \times 2 \text{ V})] = -374$

$c = KRV_t^2 - |V_{DD}| = (20 \text{ mA/V}^2 \times 4.7 \text{ k}\Omega \times 4 \text{ V}^2) - 16 = 360$

$$V_{GS} = \frac{+374 + \sqrt{374^2 - (4 \times 94 \times 360)}}{2 \times 94} = 2.35 \text{ V}$$

$$I_D = \frac{V_{DD} - 2V_{GS}}{R} = \frac{16 \text{ V} - 4.7 \text{ V}}{4.7 \text{ k}\Omega} = 2.4 \text{ mA}$$

$$I_{RFE} = I_D = I_o = 2.4 \text{ mA}$$

The output resistance of the cascode current-mirror can be shown to be as follows:

$$R_o = g_m r_o^2 \quad (10\text{-}30)$$

where,

$g_m = 2K(V_{GS} - V_t) = 2 \times 20 \text{ mA/V}^2 (2.35 \text{ V} - 2 \text{ V}) = 14 \text{ mS}$

Assuming $V_A = 120 \text{ V}$, the output resistance is determined as follows:

$$r_o = \frac{V_A}{I_o} = \frac{120 \text{ V}}{2.4 \text{ mA}} = 50 \text{ k}\Omega$$

$R_o = g_m r_o^2 = 14 \text{ mA/V} \times (50 \text{ k}\Omega)^2 = 35 \text{ M}\Omega$

Summary of Equations for the Analysis of MOSFET Current-Mirrors

MOSFET basic current-mirror:	MOSFET cascode current-mirror:
$V_{GS}\Big\|_{n-channel} = \dfrac{-b + \sqrt{b^2 - 4ac}}{2a}$	$V_{GS}\Big\|_{n-channel} = \dfrac{-b + \sqrt{b^2 - 4ac}}{2a}$
$a = KR$, $b = 1 - 2KR\|V_t\|$	$a = KR$, $b = 2(1 - KR\|V_t\|)$
$c = KRV_t^2 - \|V_{DD}\|$, $I_{REF} = I_D = I_o$	$c = KRV_t^2 - \|V_{DD}\|$, $I_{REF} = I_D = I_o$
$I_D = \dfrac{V_{DD} - V_{GS}}{R}$, $R_o = r_o = \dfrac{V_A}{I_o}$	$I_D = \dfrac{V_{DD} - 2V_{GS}}{R}$, $R_o = g_m r_o^2$
	$g_m = 2K(V_{GS} - V_t)$

Practice Problem 10-5 — Current Mirror — ANALYSIS

Determine the I_o and R_o for the cascode current-mirror of Figure 10-9 below:

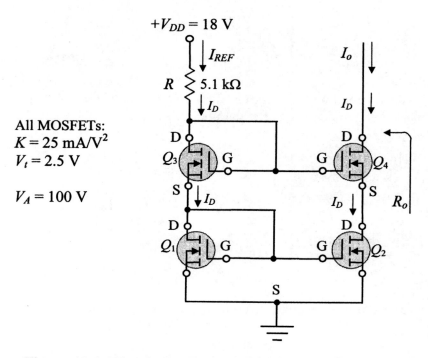

All MOSFETs:
$K = 25$ mA/V^2
$V_t = 2.5$ V

$V_A = 100$ V

Figure 10-9: Circuit diagram of a *cascode current-mirror* current source

Answers: $V_{GS} = 2.8115$ V, $I_D = 2.426$ mA, $R_o = 26.46$ MΩ

10.3 DIFFERENTIAL AMPLIFIER

The differential amplifier (diff-amp) is an essential component of the operational amplifier; in fact, the first two stages of an operational amplifier (op-amp) are differential amplifiers. Thus, before we begin the study of the operational amplifier and its applications, it would be useful to explore the characteristics, capabilities, and limitations of the differential amplifier. A differential amplifier is an amplifier with two input terminals and two output terminals. Its block diagram is shown below:

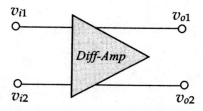

Figure 10-10: Differential amplifier block diagram

A differential amplifier amplifies the difference between the two input signals while suppressing the signal that is common to both inputs. To clarify the *difference* and the *common* signals, consider Figures 10-10(a) and 10-10(b) as shown below:

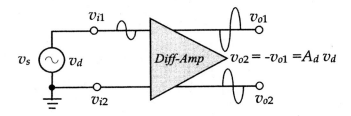

$$v_d = v_{i1} - v_{i2} = v_s \quad (10\text{-}31)$$

where v_s is the applied signal and v_d is the *difference* signal.
A_d is the differential voltage gain.
v_{o1} and v_{o2} are of equal amplitude but out-of-phase by 180°.

Figure 10-10(a): *Differential mode* connection

$$v_c = v_{i1} = v_{i2} = v_s \quad (10\text{-}32)$$

where v_s is the applied signal and v_c is the *common* signal.

v_c is also defined as:

$$v_c = \tfrac{1}{2}(v_{i1} + v_{i2}) \quad (10\text{-}33)$$

Figure 10-10(b): *Common mode* connection

Single-Ended and Double-Ended Outputs with Differential Mode Connection

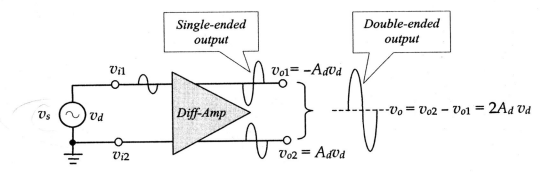

Figure 10-11: Differential mode connection with *single-ended* and *double-ended* outputs

As shown in Figure 10-11 above, when the output is taken from either v_{o1} or v_{o2} terminal with respect to ground, it is referred to as the *single-ended output*. However, when the output is taken across the v_{o1} and v_{o2} terminals, it is called the *double-ended output*.

Single-ended output:

$$v_{o2} = A_d v_d \text{ and } v_{o1} = -A_d v_d \quad (10\text{-}34)$$

Double-ended output:

$$v_o = v_{o2} - v_{o1} = |v_{o2}| + |v_{o1}| = A_d v_d + A_d v_d = 2 A_d v_d \quad (10\text{-}35)$$

Chapter 10

Now consider the following cascaded diff-amps:

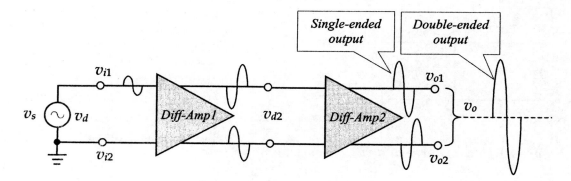

Figure 10-12: Cascaded differential amplifiers

Single-ended output:

$$v_{o1} = A_{d2}v_{d2} = A_{d2}(2A_{d1}v_d) = 2A_{d2}A_{d1}v_d \qquad (10\text{-}36)$$

Double-ended output:

$$v_o = 2A_{d2}v_{d2} = 2A_{d2}(2A_{d1}v_d) = 4A_{d2}A_{d1}v_d \qquad (10\text{-}37)$$

Practice Problem 10-6 (Cascaded Diff-amps) ANALYSIS

Determine the following for the cascaded diff-amp stage of Figure 10-13 below, given that $v_s = 1$ mV(p-p), $A_{d1} = 50$, and $A_{d2} = 40$.

a) v_{d2}
b) v_{o1}
c) v_{o2}
d) v_o

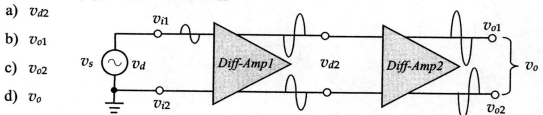

Figure 10-13: Cascaded differential amplifiers

Answers:

a) $v_{d2} = 0.1$ V(p-p) b) $v_{o1} = 4$ V(p-p) c) $v_{o2} = -4$ V(p-p) d) $v_o = 8$ V(p-p)

10.4 THE BASIC DIFFERENTIAL AMPLIFIER CIRCUIT

The circuit diagram of a basic BJT diff-amp and its DC equivalent circuit are shown in Figure 10-14(a) and 10-14(b) below. The bases of the two transistors are the two inputs and the collectors are the two outputs.

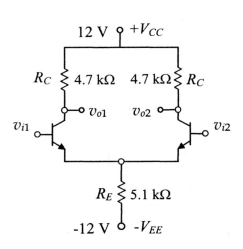

Figure 10-14(a):
Circuit diagram of a basic diff-amp

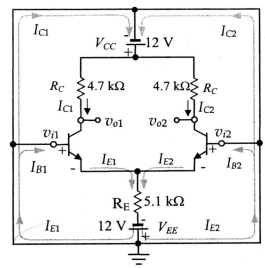

Figure 10-14(b):
DC equivalent circuit of a basic diff-amp

DC Analysis:

Applying *KVL* around either of the loops with the *base-emitter* junction in Figure 10-14(b) results in the following equation:

$$V_{EE} = V_{BE} + 2I_E R_E$$
$$V_{EE} - V_{BE} = 2I_E R_E$$

Solving for I_E in the above equation yields the following:

$$I_E = \frac{V_{EE} - V_{BE}}{2R_E} \cong I_C \tag{10-38}$$

$$I_E = \frac{12\text{ V} - 0.7\text{ V}}{2 \times 5.1\text{ k}\Omega} = 1.1\text{ mA}$$

Since the bases of both transistors are at the ground, the voltage $V_E = V_{EB} = -0.7$ V. The voltage at the collector of either transistor is the difference between the V_{CC} and the voltage drop across the R_C, as follows:

$$V_C = V_{CC} - I_C R_C \tag{10-39}$$

$V_C = 12$ V $-$ (1.1 mA $\times$ 4.7 kΩ) $= 12$ V $- 5.17$ V $= 6.83$ V

The voltage across the collector-emitter terminals is the difference between the two terminal voltages, as follows:

$$V_{CE} = V_C - V_E \tag{10-40}$$

$V_{CE} = 6.83$ V $- (-0.7$ V$) = 6.83$ V $+ 0.7$ V $= 7.53$ V

Chapter 10

Small-Signal Operation of the Basic Diff-Amp (AC Analysis)

As mentioned earlier, there are two modes of operation with regard to AC analysis: *differential mode* and *common mode*. With the differential mode, we are interested in finding the amplitude of the output signal due to the differential input $v_d = v_{i1} - v_{i2}$. With the common mode, we are interested in finding the amplitude of the output signal due to the common input $v_c = \frac{1}{2}(v_{i1} + v_{i2})$. The ratio of the output signal to the input signal in the differential mode is called *the differential mode gain A_d*, and the same ratio in the common mode is the *common mode gain A_c*.

Differential Mode Gain:

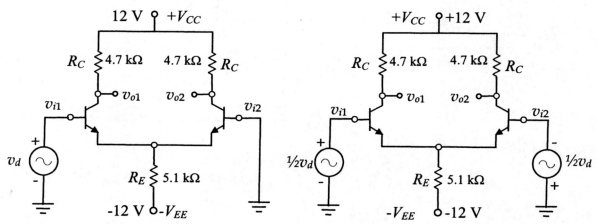

Figure 10-15(a): Practical differential mode connection

Figure 10-15(b): Ideal differential mode connection

The differential input and the common input voltages in Figure 10-15(a) are as follows:
$$v_{i1} - v_{i2} = v_d \qquad v_c = \frac{1}{2}(v_{i1} + v_{i2}) = \frac{1}{2} v_d$$

The differential input and the common input voltages in Figure 10-15(b) are as follows:
$$v_{i1} - v_{i2} = \frac{1}{2} v_d - (-\frac{1}{2} v_d) = v_d \qquad v_c = \frac{1}{2}(v_{i1} + v_{i2}) = 0$$

Hence, Figure 10-15(b) is the ideal differential mode connection and is better suited for differential mode analysis.

The small-signal equivalent circuit of the basic diff-amp in an ideal differential mode is illustrated in Figure 10-16 below.

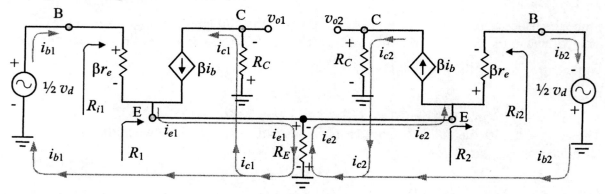

Figure 10-16: Differential mode small-signal equivalent circuit of the basic diff-amp

Chapter 10

The resistance R_2 is the resistance seen from the emitter to base terminal and by definition this resistance is referred to as the r_e.

$$R_2 = r_e = \frac{26\,mV}{I_E} = \frac{26\,mV}{1.1\,mA} = 23.5\,\Omega$$

$$R_1 = R_E \| r_e \cong r_e \tag{10-41}$$

the input resistance seen from either input terminal is as follows:

$$R_i = \beta r_e + (\beta + 1) R_1 = \beta r_e + (\beta + 1) r_e \cong 2\beta r_e \tag{10-42}$$

The output voltage at v_{o1} and v_{o2} are the following:

$$v_{o2} = -v_{o1} = i_c R_C = \beta i_b R_C \tag{10-43}$$

The voltage at the input terminals v_{i1} and v_{i2} is as follows:

$$v_{i1} = -v_{i2} = \tfrac{1}{2} v_d = (i_b \beta r_e) + (i_{e1} - i_{e2})R_E = i_b \beta r_e \tag{10-44}$$

$$v_d = 2(i_b \beta r_e) = 2\, i_c r_e \tag{10-45}$$

The differential gain is the ratio of the output signal to the input signal as follows:

$$A_d = \frac{v_{o2}}{v_d} = \left|\frac{-v_{o1}}{v_d}\right| = \frac{i_c R_C}{2 i_c r_e} = \frac{R_C}{2 \times r_e}$$

$$|A_d| = \frac{R_C}{2 \times r_e} \tag{10-46}$$

$$|A_d| = \frac{4.7\,k\Omega}{2 \times 23.5\,\Omega} = 100$$

Note that the output signal v_{o2} is in-phase but v_{o1} is out-of-phase with the differential input signal. Hence, the two outputs are out-of-phase with each other.

Common Mode Gain:

The small-signal equivalent circuit of the basic diff-amp in the common mode is illustrated in Figure 10-17 below, in which the same input signal v_s is common to both inputs and thus, it is referred to as the common input v_c.

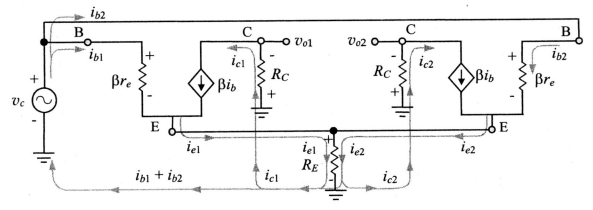

Figure 10-17: Common mode small-signal equivalent circuit of the basic diff-amp

Chapter 10

The output voltages v_{o1} and v_{o2} are as follows:
$$v_{o1} = v_{o2} = \beta i_b R_C \tag{10-47}$$

The voltage at the input terminals v_{i1} and v_{i2} are the following:
$$v_c = v_{i1} = v_{i2} = i_b \beta r_e + 2i_e R_E = \beta i_b(r_e + 2R_E) \cong \beta i_b(2R_E) \tag{10-48}$$
$$v_c = 2 i_c R_E \tag{10-49}$$

The common mode gain is the ratio of the output signal to the input signal, as follows:
$$A_c = \frac{-v_{o1}}{v_c} = \frac{-v_{o2}}{v_c} = -\frac{\beta \cdot i_b \cdot R_C}{2(\beta \cdot i_b)R_E} = -\frac{R_C}{2R_E} \tag{10-50}$$

$$|A_c| = \frac{R_C}{2R_E} \tag{10-51}$$

$$|A_c| = \frac{4.7 \text{ k}\Omega}{2 \times 5.1 \text{ k}\Omega} = 0.46$$

Note that the two outputs v_{o1} and v_{o2} are in-phase with each other but out-of-phase with the common input signal.

Common Mode Rejection Ratio:

The common mode rejection ratio (CMRR) is a measure of how well the diff-amp rejects the common mode signals and is defined as the ratio of the differential mode gain to the common mode gain. For a well-designed diff-amp, the CMRR would be very large in magnitude; hence, the CMRR is usually stated in dB.

$$CMRR = \frac{A_d}{A_c} \tag{10-52}$$

$$CMRR(dB) = 20 \log\left(\frac{A_d}{A_c}\right) \tag{10-53}$$

$$CMRR = \frac{A_d}{A_c} = \frac{100}{0.46} = 217 \qquad CMRR(dB) = 20\log(217) = 46.7 \text{ dB}$$

Example 10-1 — Basic Diff-Amp — ANALYSIS

Determine the DC voltages and currents for the basic diff-amp of Figure 10-18. Also determine the differential gain, common mode gain, and the CMRR.

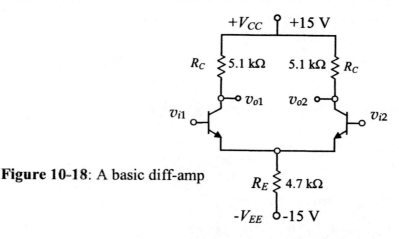

Figure 10-18: A basic diff-amp

Chapter 10

Solution:

$$I_E = \frac{V_{EE} - V_{BE}}{2R_E} = \frac{15V - 0.7V}{2 \times 4.7 \text{ k}\Omega} = 1.52 \text{ mA} \qquad r_e = \frac{26 \text{ mV}}{I_E} = 17.1 \Omega$$

$$V_C = V_{CC} - I_C R_C = 15 \text{ V} - (1.52 \text{ mA} \times 5.1 \text{ k}\Omega) = 7.25 \text{ V}$$

$$V_{CE} = V_C - V_E = 7.25 \text{ V} - (-0.7 \text{ V}) = 7.95 \text{ V}$$

$$|A_d| = \frac{R_C}{2 \times r_e} = \frac{5.1 \text{ k}\Omega}{2 \times 17.1 \Omega} = 149$$

$$|A_c| = \frac{R_C}{2 \times R_E} = \frac{5.1 \text{ k}\Omega}{2 \times 4.7 \text{ k}\Omega} = 0.54$$

$$CMRR = \frac{A_d}{A_c} = \frac{149}{0.54} = 276 = 48.8 \text{ dB}$$

Design of the Basic Differential Amplifier

We will try to center the Q-point in order to achieve a maximum signal swing at the output. For a diff-amp without external load, the AC load line is the same as the DC load line. Hence, we will center the Q-point on the DC load line by letting $V_{CE} = \frac{1}{2} V_{CC}$.

$$V_{CE} = \frac{1}{2} V_{CC} = V_C - V_E = V_{CC} - I_C R_C + 0.7 \text{ V}$$

Rearranging the equation and solving for V_{CC} results in the following:

$$V_{CC} = 2 I_C R_C - 1.4 \text{ V} \qquad (10\text{-}54)$$

The expression for the differential gain is

$$|A_d| = \frac{R_C}{2 \times r_e} \qquad (10\text{-}55)$$

Solving for R_C results in the following:

$$R_C = A_d (2 \times r_e) = A_d \left(\frac{2 \times 26 \text{ mV}}{I_E} \right) = \frac{A_d \times 52 \text{ mV}}{I_E}$$

$$R_C = \frac{A_d \times 52 \text{ mV}}{I_E} \qquad (10\text{-}56)$$

Let us design a basic diff-amp with a differential gain of 135 and $I_{CQ} = 2$ mA.

Note that in our analysis we have diregarded the effect of the output resistance r_o of Q_1 and Q_2, which would be in parallel with R_C and thus cause some reduction in the differential gain A_d. To compensate for this loss, we will design for a gain approximately 10% higher than the specified gain. Hence, we will design for a differential gain of 150, in order to achieve the specified gain of 135.

$$R_C = \frac{A_d \times 52 \text{ mV}}{I_E} = \frac{150 \times 52 \text{ mV}}{2 \text{ mA}} = 3.9 \text{ k}\Omega$$

Having determined R_C, we can now determine V_{CC}, as follows:

$$V_{CC} = 2 I_C R_C - 1.4 \text{ V} \qquad (10\text{-}57)$$

Chapter 10

$V_{CC} = 2 \, (2 \text{ mA} \times 3.9 \text{ k}\Omega) - 1.4 \text{ V} = 14.2 \text{ V}$ Let $V_{CC} = V_{EE} = 15 \text{ V}$

To determine the R_E, the following equation can be written around the loop containing the R_E:

$$V_{EE} = 2I_E R_E + V_{BE} \tag{10-58}$$

Solving for R_E results in the following:

$$R_E = \frac{V_{EE} - V_{BE}}{2 \times I_E} \tag{10-59}$$

$R_E = \dfrac{15 \text{ V} - 0.7 \text{ V}}{2 \times 2 \text{ mA}} = 3.575 \text{ k}\Omega$ use $R_E = 3.6 \text{ k}\Omega$

Summary of design specifications and computations:

$A_d = 135$, $I_{EQ} = 2 \text{ mA}$, $V_{CC} = 15 \text{ V}$, $R_C = 3.9 \text{ k}\Omega$, $R_E = 3.6 \text{ k}\Omega$

$A_c = \dfrac{R_C}{2R_E} = \dfrac{3.9 \text{ k}\Omega}{2 \times 3.6 \text{ k}\Omega} = 0.54$ $CMRR = \dfrac{A_d}{A_c} = \dfrac{135}{0.54} = 250 = 48 \text{ dB}$

Test Run and Design Verification:

Having completed the design computations, let us now simulate the circuit with Electronics Workbench in order to verify the accuracy of the design. The results of simulation are presented in Figures 10-19 through 10-24.

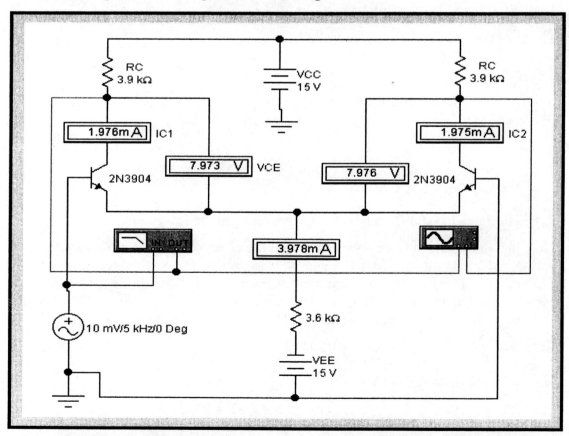

Figure 10-19: Differential mode circuit setup with simulation results (DC) created in Electronics Workbench

Figure 10-19 above displays the circuit setup for the differential mode with simulation results of DC voltages and currents depicted in DMMs of Electronics Workbench. The registered voltages and currents on DMMs agree very closely with the design specifications. The Bode plotter of Figure 10-20 below displays a differential gain of 135, which is the exact gain of the diff-amp that was designed for.

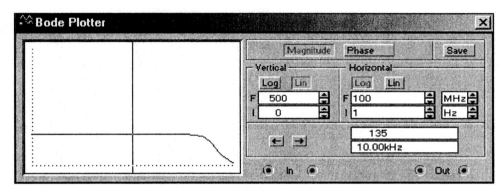

Figure 10-20: Electronics Workbench Bode plotter displaying the *differential mode* gain

The output signals v_{o1} and v_{o2} are depicted on the oscilloscope of Figure 10-21 below. As expected, both signals are of the same amplitude but out-of-phase by 180°.

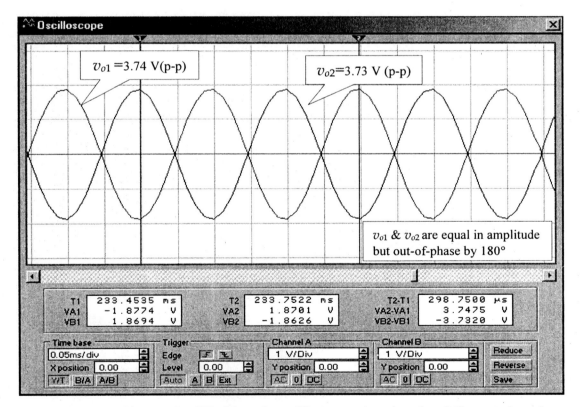

Figure 10-21: Electronics Workbench oscilloscope showing the *differential mode* outputs v_{o1} and v_{o2}

Figure 10-22 below exhibits the circuit setup and the DC simulation results for the common mode, which, as expected, are not different from the results of the differential mode connection. The Bode plotter of Figure 10-23 depicts the common mode gain of 0.59, which agrees fully with the theoretically expected common mode gain.

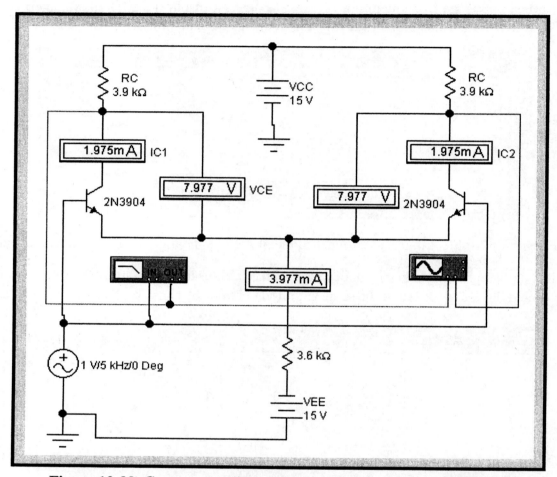

Figure 10-22: Common mode circuit setup with simulation results (DC) created in Electronics Workbench

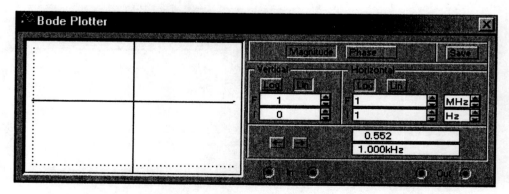

Figure 10-23: Electronics Workbench Bode plotter displaying the *common mode* gain $A_c = 0.55$

Chapter 10

Figure 10-24 below depicts the common mode output signals v_{o1} and v_{o2}, and as expected, they are equal in amplitude and phase.

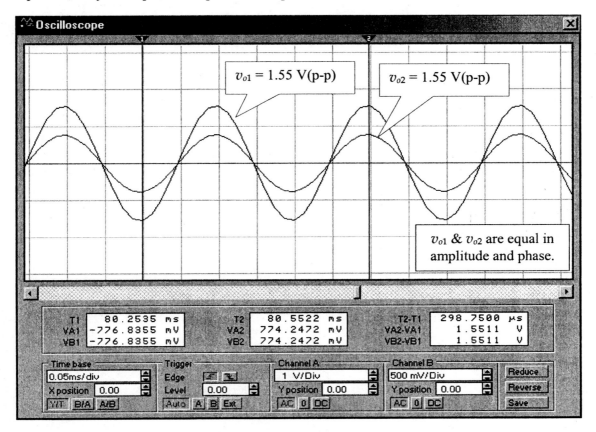

Figure 10-24: Electronics Workbench oscilloscope showing the *common mode* outputs v_{o1} and v_{o2}

Practice Problem 10-7

Basic Diff-Amp — DESIGN

Design a basic diff-amp for a differential gain of 127, with an $I_{CQ} = 1.3$ mA.

Verify your design by simulation.

Answer:

$V_{CC} = 13$ V, $R_C = 5.6$ kΩ, $R_E = 4.7$ kΩ

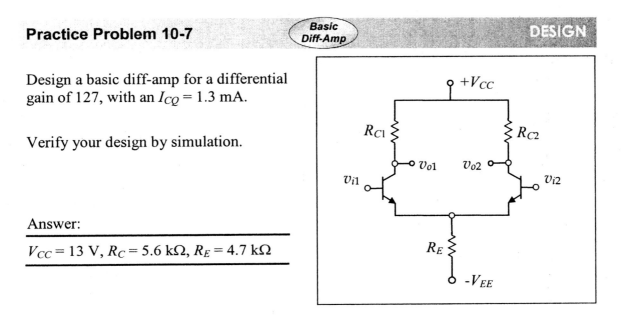

10.5 DIFF-AMP WITH CURRENT-MIRROR CURRENT SOURCE

Recall that equations for the *CMRR*, common mode gain, and differential mode gain are as follows:

$$CMRR = \frac{A_d}{A_c} \qquad |A_c| = \frac{R_C}{2R_E} \qquad |A_d| = \frac{R_C}{2r_e}$$

In order to increase the *CMRR* of the basic diff-amp to a respectable level, the common mode gain A_c will have to be very small, which seems to be achievable with a very large R_E. However, a large R_E will drastically limit the bias current I_E. Note that the R_E appearing in the denominator of the equation for A_c is an AC resistance and has no effect on the differential gain A_d. Hence, a smaller A_c can be achieved if the R_E is replaced with a constant current source with a high output resistance as shown in Figure 10-25 below.

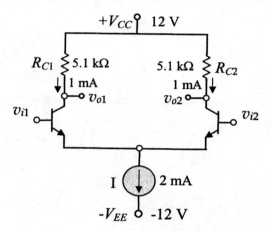

Figure 10-25: Diff-amp with a constant current source

In the above diagram, the current source has a value of 2 mA, which is equal to $2I_E$. Hence, a 1 mA current will have to flow down the emitter of each transistor in order to make up a total of 2 mA. Assuming that the output resistance of the current source is 100 kΩ, let us perform a complete analysis of this circuit.

$$V_C = V_{CC} - I_C R_C = 12 \text{ V} - (1 \text{ mA} \times 5.1 \text{ k}\Omega) = 12 - 5.1 \text{ V} = 6.9 \text{ V}$$
$$V_{CE} = V_C - V_E = 6.9 \text{ V} - (-0.7 \text{ V}) = 7.6 \text{ V}$$

$$r_e = 26 \text{ mV} / 1 \text{ mA} = 26 \text{ }\Omega$$

$$A_d = \frac{R_C}{2r_e} = \frac{5100 \text{ }\Omega}{2 \times 26 \text{ }\Omega} = 98$$

$$A_c = \frac{R_C}{2R_o} \tag{10-60}$$

$$A_c = \frac{R_C}{2R_o} = \frac{5.1 \text{ k}\Omega}{2 \times 100 \text{ k}\Omega} = 0.0255$$

$$CMRR = \frac{A_d}{A_c} = \frac{98}{0.0255} = 3843 = 71.7 \text{ dB}$$

10.5.1 Diff-Amp with Basic Current-Mirror

The R_E of the basic diff-amp is replaced with a current-mirror current source, as shown in Figure 10-26 below. We can expect a much higher *CMRR* from this configuration compared to the basic diff-amp. Let us carry out a complete analysis of this diff-amp and determine the level of improvement in *CMRR*.

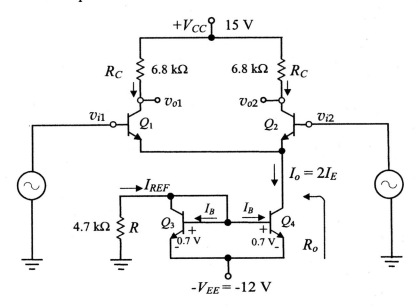

Figure 10-26: Circuit diagram of a diff-amp with basic current-mirror

The first step in the analysis is to determine the I_{REF} of the current source with Equation 10-1, as follows:

$$I_{REF} = \frac{V_{EE} - V_{BE}}{R}$$

$$I_{REF} = \frac{12\,V - 0.7\,V}{4.7\,k\Omega} = 2.4\,mA$$

$$I_C \cong I_E = \tfrac{1}{2} I_o \cong \tfrac{1}{2} I_{REF}$$

$$I_C \cong I_E = \tfrac{1}{2} I_{REF} = 1.2\,mA$$

Assuming $V_A = 120\,V$, we can now determine the r_e and the output resistance of the current source r_o, as follows:

$$r_e = \frac{26\,mV}{I_{E(Q_1)}} = \frac{26\,mV}{1.2\,mA} = 21.66\,\Omega \qquad R_o = r_o = \frac{V_A}{I_{C(Q4)}} = \frac{120\,V}{2.4\,mA} = 50\,k\Omega$$

The DC voltages V_C and V_{CE} are determined as follows:

$$V_{C1} = V_{C2} = V_{CC} - I_C R_C$$

$$V_{C1} = V_{C2} = 15\,V - (1.2\,mA \times 6.8\,k\Omega) = 15\,V - 8.16\,V = 6.84\,V$$

$$V_{CE} = V_C - V_E$$

$$V_{CE} = V_C - V_E = 6.84\,V - (-0.7\,V) = 7.54\,V$$

Chapter 10

The differential gain, common mode gain, and CMRR are determined with Equations 10-46, 10-51, and 10-52, as follows:

$$A_d = \frac{R_C}{2r_e}$$

$$A_d = \frac{R_C}{2r_e} = \frac{6.8\,k\Omega}{2 \times 21.66\,\Omega} = 157$$

$$A_c = \frac{R_C}{2R_o}$$

$$A_c = \frac{R_C}{2R_o} = \frac{6.8\,k\Omega}{2 \times 50\,k\Omega} = 0.068$$

$$CMRR = \frac{A_d}{A_c}$$

$$CMRR = \frac{A_d}{A_c} = \frac{157}{0.068} = 2308.8 = 67.26\,dB$$

Practice Problem 10-8 — ANALYSIS

Carry out a complete analysis of the diff-amp shown in Figure 10-27.
Assume $V_A = 140$ V

Answers:

$I_{REF} = 1.9$ mA, $V_{CE} = 7.2$ V

$A_d = 182$, $A_c = 0.0678$

$CMRR = 2684 = 68.575$ dB

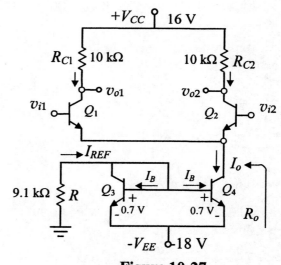

Figure 10-27

10.5.2 Design of the Diff-Amp with Basic Current-Mirror

The design procedure for the diff-amp with basic current-mirror is similar to that of the basic diff-amp. To demonstrate the process, let us design one for the same specifications given for the design of the basic diff-amp.

Example 10-2 **DESIGN**

Design a diff-amp with basic current mirror for a differential gain of 135 and $I_{CQ} = 2$ mA.

Solution: Again, we will design for a gain of 150 (approximately 10% higher than the specified gain) for the same reasons discussed in the design of the basic diff-amp. The equation for the differential voltage gain is given by

$$A_d = \frac{v_{o2}}{v_d} = \frac{R_C}{2r_e}$$

Solving for R_C gives

$$R_C = 2A_d r_e = \frac{2A_d \times 26 \text{ mV}}{I_E} = \frac{A_d \times 52 \text{ mV}}{I_E} \quad (10\text{-}61)$$

$$R_C = \frac{A_d \times 52 \text{ mV}}{I_E} = \frac{150 \times 52 \text{ mV}}{2 \text{ mA}} = 3.9 \text{ k}\Omega$$

We can now determine V_{CC} from Equation 10-54, as follows:

$$V_{CC} = 2I_C R_C - 1.4 \text{ V}$$

$V_{CC} = 2 \, (2 \text{ mA} \times 3.9 \text{ k}\Omega) - 1.4 \text{ V} = 14.2 \text{ V}$

Let $V_{EE} = V_{CC} = 15$ V, and determine the value of R, as follows:

$$V_{EE} = I_{REF} R + V_{BE} = 2I_E R + V_{BE}$$

Solving for R yields

$$R = \frac{V_{EE} - V_{BE}}{I_{REF}} = \frac{V_{EE} - V_{BE}}{2I_E} \quad (10\text{-}62)$$

$$R = \frac{15 \text{ V} - 0.7 \text{ V}}{2 \times 2 \text{ mA}} = 3.575 \text{ k}\Omega \quad \text{Use } R = 3.6 \text{ k}\Omega.$$

Assuming $V_A = 120$ V, we can determine the output resistance of the current mirror, as follows:

$$R_o = r_o = \frac{V_A}{I_{REF}} = \frac{120}{2 \times 2 \text{ mA}} = 30 \text{ k}\Omega$$

Summary of design specifications and computations:

$$A_d = 135, \quad I_{EQ} = 2 \text{ mA}, \quad V_{CC} = 15 \text{ V}, \quad R_C = 3.9 \text{ k}\Omega, \quad R = 3.6 \text{ k}\Omega$$

$$A_c = -\frac{R_C}{2R_o} = -\frac{3.9 \text{ k}\Omega}{2 \times 30 \text{ k}\Omega} = -0.065 \quad \text{CMRR} = \left|\frac{A_d}{A_c}\right| = \frac{135}{0.065} = 2077 = 66.35 \text{ dB}$$

Practice Problem 10-9 **DESIGN**

Design the diff-amp of Figure 10-28 for a differential voltage gain of 125 and $I_{CQ} = 1.25$ mA. Assume $V_A = 125$ V.

Chapter 10

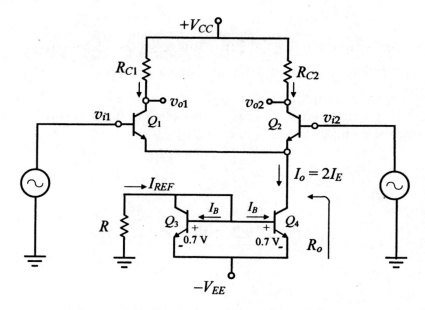

Figure 10-28: Circuit diagram of a diff-amp with basic current mirror

Answers:

$V_{CC} = 11.6$ V, $R_C = 5.6$ kΩ, $R = 4.3$ kΩ, $A_c = -0.056$, $CMRR = 67$ dB

10.5.3 Diff-Amp with Wilson Current Mirror

The complete circuit diagram of a diff-amp biased with Wilson current mirror is shown in Figure 10-29.

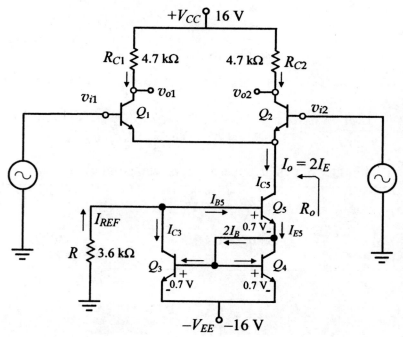

Figure 10-29: Circuit diagram of a diff-amp with Wilson current mirror

428

Let us carry out a complete analysis of this differential amplifier and compare the level of improvement in *CMRR* to that of the diff-amp with basic current mirror. The reference current I_{REF} is determined from Equation 10-4

$$I_{REF} = \frac{V_{EE} - 2V_{BE}}{R} = \frac{16\,V - 1.4\,V}{3.6\,k\Omega} = 4\,mA$$

$$I_{E1} = I_{E2} = \tfrac{1}{2} I_{REF} = 2\,mA$$

$$V_{C1} = V_{C2} = V_{CC} - I_C R_C = 16 - (2\,mA \times 4.7\,k\Omega) = 6.6\,V$$

$$V_{CE1} = V_{CE2} = V_C - V_E = 6.6\,V - (-0.7\,V) = 7.3\,V$$

$$r_e = \frac{V_T}{I_{E(Q1)}} = \frac{26\,mV}{I_{E(Q1)}}$$

$$r_e = \frac{V_T}{I_{E(Q1)}} = \frac{26\,mV}{2\,mA} = 13\,\Omega$$

Assuming $V_A = 100$ V and $\beta = 200$, the output resistance of the current source is determined from Equation 10-16

$$r_{o(Q5)} = \frac{V_A}{I_{E(Q5)}} = \frac{V_A}{I_{REF}} = \frac{100\,V}{4\,mA} = 25\,k\Omega$$

$$R_o = \frac{\beta}{2} r_{o(Q5)}$$

$$R_o = \frac{\beta}{2} r_{o(Q5)} = \frac{200}{2} \times 25\,k\Omega = 2.5\,M\Omega$$

The differential-voltage gain A_d is identical to that of the basic diff-amp and the diff-amp with basic current mirror and given by

$$A_d = \frac{v_{o2}}{v_d} = \frac{R_C}{2r_e} = \frac{g_m R_C}{2} \tag{10-63}$$

$$A_d = \frac{R_C}{2r_e} = \frac{4700\,\Omega}{26\,\Omega} = 181$$

The expression for the common-mode voltage gain is the same, but R_o of the current mirror is much higher in value, resulting in a very low A_c. Thus,

$$A_c = \frac{v_{o1}}{v_c} = \frac{v_{o2}}{v_c} = -\frac{R_C}{2R_o} \tag{10-64}$$

$$A_c = -\frac{R_C}{2R_o} = -\frac{4.7\,k\Omega}{2 \times 2.5\,M\Omega} = -9.4 \times 10^{-4}$$

$$CMRR = \left|\frac{A_d}{A_c}\right| = \frac{R_o}{r_e} = g_m R_o = \frac{\beta g_m r_{o(Q5)}}{2} = \frac{\beta V_A}{2V_T} \tag{10-65}$$

$$CMRR = \left|\frac{A_d}{A_c}\right| = \frac{180}{9.4 \times 10^{-4}} = 191489$$

$$CMRR(dB) = 20 \log(191489) = 105.64\,dB$$

Practice Problem 10-10 ANALYSIS

Carry out a complete analysis of the Wilson current mirror of Figure 10-29 with $V_{CC} = V_{EE} = 18$ V, $R = 5$ kΩ, and $R_C = 5.6$ kΩ. Assume $V_A = 100$ V and $\beta = 150$.

Answers: $I_{REF} = 3.32$ mA, $I_{CQ} = 1.66$ mA, $R_o = 2.25$ MΩ,
$A_d = 178.8$, $A_c = -1.244 \times 10^{-3}$, CMRR $= 103.15$ dB

10.5.4 Design of the Diff-Amp with Wilson Current Mirror

The design procedure for the diff-amp with Wilson current mirror is similar to that of the diff-amp with basic current mirror. To demonstrate the process, let us design one for the same specifications given for the design of the basic diff-amp and the diff-amp with basic current mirror.

Example 10-3 DESIGN

Design a diff-amp with Wilson current mirror for a differential voltage gain of 135 and $I_C = 2$ mA.

Solution: Again, we will design for a differential voltage gain of 150 (approximately 10% higher than the specified gain) for the same reasons discussed in the design of the basic diff-amp. The equation for the differential voltage gain is

$$A_d = \frac{v_{o2}}{v_d} = \frac{R_C}{2r_e} = \frac{g_m R_C}{2}$$

Solving for the resistor R_C yields

$$R_C = 2A_d r_e = \frac{|A_d| \times 52 \text{ mV}}{I_E}$$

$$R_C = \frac{|A_d| \times 52 \text{ mV}}{I_E} = \frac{150 \times 52 \text{ mV}}{2 \text{ mA}} = 3.9 \text{ k}\Omega$$

We can now determine V_{CC} from Equation 10-54

$$V_{CC} = 2I_C R_C - 1.4 \text{ V}$$

$V_{CC} = 2(2 \text{ mA} \times 3.9 \text{ k}\Omega) - 1.4 \text{ V} = 14.2$ V.

Let $V_{EE} = V_{CC} = 15$ V, and determine the resistance of R

$$V_{EE} = I_{REF}R + 2V_{BE} = 2I_E R + 2V_{BE} \qquad (10\text{-}66)$$

Solving for the resistor R gives

$$R = \frac{V_{EE} - 2V_{BE}}{2I_E} \qquad (10\text{-}67)$$

$$R = \frac{15\text{ V} - 1.4\text{ V}}{2 \times 2\text{ mA}} = 3.4\text{ k}\Omega \qquad \text{Use } R = 3.3\text{ k}\Omega.$$

Assuming $V_A = 120$ V, we can determine the output resistance of the current mirror from Equation 10-16

$$r_{o(Q5)} = \frac{V_A}{I_{REF}} = \frac{120\text{ V}}{2 \times 2\text{ mA}} = 30\text{ k}\Omega$$

$$R_o = \frac{\beta}{2} r_o = \frac{200}{2} \times 30\text{ k}\Omega = 3\text{ M}\Omega$$

Summary of design specifications and computations:

$$A_d = 135, \quad I_{EQ} = 2\text{ mA}, \quad V_{CC} = 15\text{ V}, \quad R_C = 3.9\text{ k}\Omega, \quad R = 3.3\text{ k}\Omega$$

$$A_c = -\frac{R_C}{2R_o} = -\frac{3.9\text{ k}\Omega}{2 \times 3\text{ M}\Omega} = -0.00065, \quad CMRR = \left|\frac{A_d}{A_c}\right| = \frac{135}{0.00065} = 207692 = 106.35\text{ dB}$$

Table 10-1: Comparison of the Three Designs

Diff-amp type	Basic	Basic current mirror	Wilson current mirror
I_{CQ}	2 mA	2 mA	2 mA
A_d	135	135	135
A_c	−0.55	−0.065	−0.00065
CMRR(dB)	48 dB	66.35 dB	106.35 dB

As you can see in Table 10-1, the improvement in CMRR of the Wilson current mirror over the basic current mirror is 40 dB and 5.35 dB over the basic diff-amp. In other words, the diff-amp designed with Wilson current mirror can reject the *common* signals 100 times better than the diff-amp designed with basic current mirror, and approximately 1000 times better than the basic diff-amp.

Practice Problem 10-11 **DESIGN**

Design a diff-amp with Wilson current mirror for a gain of $A_d = 120$ and $I_{CQ} = 1.2$ mA. Assume $V_A = 120$ V and $\beta = 150$. Upon completion of the design, determine the common-mode voltage gain and CMRR of the diff-amp.

Answers:

If designed for $A_d = 120$: $R_C = 5.6$ kΩ, $V_{CC} = 12$ V, $R = 4.3$ kΩ, $A_c = -7.467 \times 10^{-4}$

If designed for 10% more than the given A_d to recover the losses; that is, for $A_d = 132$:

$R_C = 5.27$ kΩ, $V_{CC} = 11.5$ V, $R = 4.6$ kΩ, $A_c = -7.6 \times 10^{-4}$, CMRR = 105 dB.

Let us now test our design of the diff-amp with Wilson current-mirror by a simulation. The results of the simulation are presented in Figures 10-30 through 10-33.

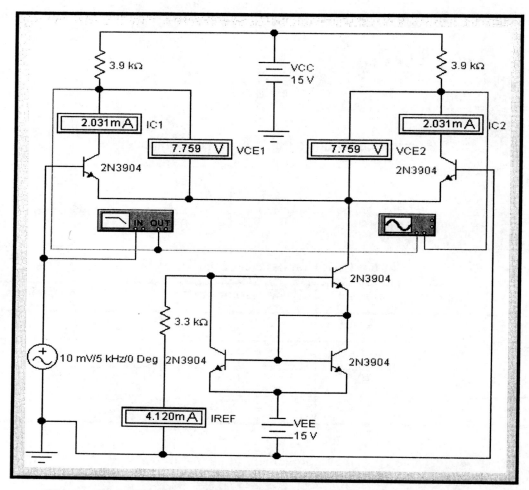

Figure 10-30: Differential mode circuit setup and simulation results of Design Example 10-3 created in Electronics Workbench

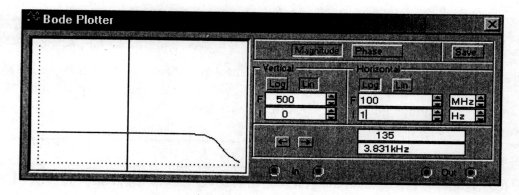

Figure 10-31: Electronics Workbench Bode plotter showing the differential gain $A_d = 135$ for Design Example 10-3

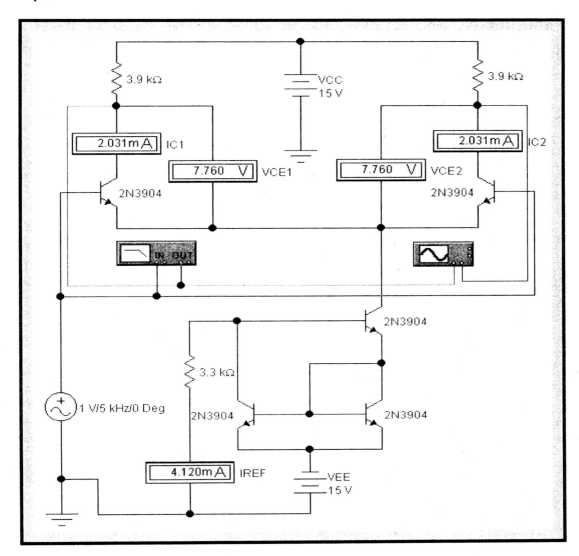

Figure 10-32: Common mode circuit setup with simulation results of Design Example 10-3 created in Electronics Workbench

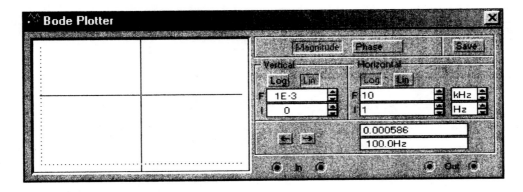

Figure 10-33: Electronics Workbench Bode plotter showing the common-mode gain ($A_c = -0.000586$) of Design Example 10-3

The *CMRR* can now be determined from the Bode plotter readings of Figures 10-31 and 10-33

$$CMRR = \left|\frac{A_d}{A_c}\right| = \frac{135}{5.86 \times 10^{-4}} = 230375 = 107.25 \text{ dB}$$

Table 10-2: Summary of Calculated and Measured Results

Parameter	Specification	Expected Results	Measured Results
I_{REF}	–	4 mA	4.12 mA
I_{CQ}	2 mA	2 mA	2.03 mA
V_{CEQ}	–	7.9 V	7.76 V
A_d	135	135	135
A_c	–	-6.5×10^{-4}	-5.86×10^{-4}
CMRR	–	106.35 dB	107.25 dB

As you can see in Table 10-2, the measured values agree very closely with the expected results.

Practice Problem 10-12 — Diff-Amp — DESIGN

Design the following diff-amps for a differential gain of 100 and I_{CQ} of 1.5 mA. After the completion of the design, determine the output resistance and *CMRR* in dB. Assume $\beta = 150$ and $V_A = 150$ V.
a) Basic diff-amp.
b) Diff-amp with basic current mirror.
c) Diff-amp with Wilson current mirror.

Answers:

$R_C = 3.6$ kΩ for all three designs.
a) $R_o = R_E = 3$ kΩ, $R_E = 3$ kΩ, $V_{CC} = 9$ V, $CMRR = 44.437$ dB
b) $R_o = r_o = 50$ kΩ, $V_{CC} = 10$ V, $CMRR = 68.87$ dB
c) $R_o = \dfrac{\beta}{2} r_o = 3.75$ MΩ, $V_{CC} = 10$ V, $CMRR = 106.375$ dB

Section Summary — Diff-Amp — ANALYSIS

Summary of Equations for the Analysis of the CM biased Differential Amplifier

Diff-amp with basic current-mirror

DC Analysis:

$$I_{REF} = \frac{V_{EE} - V_{BE}}{R} \cong I_o$$

$$I_{C1} = I_{C2} = I_C = \tfrac{1}{2} I_o$$

$$V_{C1} = V_{C2} = V_C = V_{CC} - I_C R_C$$

$$V_{CE} = V_C - V_E$$

AC Analysis:

$$r_e = \frac{26\,\text{mV}}{I_E}$$

$$R_o = r_o, \quad r_o = \frac{V_A}{I_o}$$

$$A_d = \frac{R_C}{2 r_e}, \quad A_c = \frac{R_C}{2 R_o}$$

$$CMRR = \frac{A_d}{A_c}$$

$$CMRR(dB) = 20 \log \frac{A_d}{A_c}$$

Diff-amp with Wilson current-mirror

DC Analysis:

$$I_{REF} = \frac{V_{EE} - 2V_{BE}}{R} = I_o$$

$$I_{C1} = I_{C2} = I_C = \tfrac{1}{2} I_o$$

$$V_{C1} = V_{C2} = V_C = V_{CC} - I_C R_C$$

$$V_{CE} = V_C - V_E$$

AC Analysis:

$$r_e = \frac{26\,\text{mV}}{I_E}$$

$$R_o = \frac{\beta}{2} r_o, \quad r_o = \frac{V_A}{I_o}$$

$$A_d = \frac{R_C}{2 r_e}, \quad A_c = \frac{R_C}{2 R_o}$$

$$CMRR = \frac{A_d}{A_c}$$

$$CMRR(dB) = 20 \log \frac{A_d}{A_c}$$

Practice Problem 10-13 — Diff-Amp — DESIGN

A differential amplifier is to be designed to amplify the signal detected in a biomedical engineering experiment. The magnitude of the detected signal ranges from 0 to ±5 mV. The laboratory environment in which the experiment is conducted induces a 60 Hz undesirable noise signal up to 100 mV to each of the electrodes from the room lights and electrical wiring. The design specifications are such that the diff-amp must produce a single-ended output of 0 to ±1 V. In addition, the diff-amp must attenuate the 60 Hz noise signal such that the noise component of the output signal is no more than 1 mV. Design a suitable differential amplifier that meets the given specifications.

Work in groups of two; show complete work in an orderly fashion. Let $I_C = 1.15$ mA. Use commercially available standard value resistors.

Chapter 10

Use 2N222A transistors: $\beta = 160$, $V_A = 115$ V. Verify your design by simulation.

Answers:

$V_{CC} = V_{EE} = 22$ V,	$R_C = 10$ kΩ,	$R = 8.95$ kΩ $= 8.2$ kΩ $+ 750$ Ω,	$A_d = 200$
$A_c = 1.25 \times 10^{-3}$,	CMRR $= 104$ dB	Current source is Wilson current-mirror.	

Example 10-4 — Cascaded Diff-Amps — ANALYSIS

Figure 10-34 below exhibits a two-stage cascaded differential amplifier representing partial circuitry of an operational amplifier. Analysis of this and the following circuit will help us understand the operation and significance of the operational amplifier.

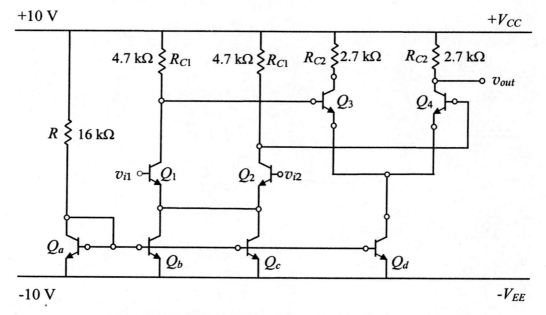

Figure 10-34: A two-stage cascaded differential amplifier

DC Analysis:

The objective of DC analysis is to determine all the DC voltages and currents such as: I_R, I_{C1}, I_{C3}, V_{C1}, V_{C3}, and V_{CE4}. Note that transistor Q_a is the reference current source, whose current is mirrored into transistors Q_b, Q_c, and Q_d, which all have their base-emitter terminals shunted with that of Q_a. Hence, the first step is to determine the reference current I_{REF}, as follows:

$$I_{REF} = \frac{V_{CC} - V_{BE} - (-V_{EE})}{R} \qquad (10\text{-}68)$$

$$I_{REF} = \frac{10\text{ V} - 0.7\text{ V} + 10\text{ V}}{16\text{ k}\Omega} = 1.2\text{ mA}$$

$$I_{Qa} = I_{Qb} = I_{Qc} = I_{Qd} = I_{REF} = 1.2\text{ mA}$$

Collector currents I_{C1} and I_{C2} are equal to I_{Qa}, but I_{C3} and I_{C4} are equal to one-half of I_{Qa}.

$I_{C1} = I_{C2} = I_{Qa} = 1.2 \text{ mA}$

$I_{C3} = I_{C4} = 0.5 I_{Qa} = 0.6 \text{ mA}$

$V_{C1} = V_{C2} = V_{CC} - I_{C1}R_{C1} = 10 \text{ V} - (1.2 \text{ mA} \times 4.7 \text{ k}\Omega) = 4.36 \text{ V}$

$V_{C3} = V_{C4} = V_{CC} - I_{C3}R_{C2} = 10 \text{ V} - (0.6 \text{ mA} \times 2.7 \text{ k}\Omega) = 8.38 \text{ V}$

$V_{E4} = V_{E3} = V_{B3} - 0.7 \text{ V} = V_{C1} - 0.7 \text{ V} = 4.36 \text{ V} - 0.7 \text{ V} = 3.66 \text{ V}$

$V_{CE4} = V_{C4} - V_{E4} = 8.38 \text{ V} - 3.66 \text{ V} = 4.72 \text{ V}$

Small-Signal (AC) Analysis:

Assuming that $v_{i1} = 1$ mV(p-p), v_{i2} = GND, let us determine the single-ended output v_{out}. The single-ended output of a cascaded two-stage diff-amp is determined by Equation 10-36, as follows:

$$v_{out} = A_{d2}v_{d2} = A_{d2}(A_{dm1}v_d) = A_{d2}(-2A_{d1}v_d) = -2A_{d2}A_{d1}v_d = -2A_{d2}A_{d1}v_{i1}$$

where

$$A_{dm1} = \frac{v_{d2}}{v_{d1}} = \frac{v_{c1} - v_{c2}}{v_{i1} - v_{i2}} = -\frac{R_{C1}}{r_{e1}} \qquad A_{d1} = \frac{v_{c2}}{v_{i1}} = \frac{R_{C1}}{2r_{e1}} = -\frac{A_{dm1}}{2}$$

$$A_{d1} = \frac{4.7 \text{ k}\Omega}{2 \times 21.66 \text{ }\Omega} = 108.5$$

$$A_{d2} = \frac{v_{out}}{v_{d2}} = \frac{v_{c4}}{v_{c1} - v_{c2}} = \frac{R_{C3}}{2r_{e3}}$$

$$A_{d2} = \frac{2.7 \text{ k}\Omega}{2 \times 43.33 \text{ }\Omega} = 31.15$$

$v_{out} = -2A_{d2}A_{d1}v_d = -2 \times 108.6 \times 31.15 \times 1 \text{ mV} = -6.765 \text{ V(p-p)}$

Practice Problem 10-14 **ANALYSIS**

Determine the following for the cascaded dif-amps of Figure 10-35, if $\pm V_{CC} = \pm 11.35$ V, $R = 10 \text{ k}\Omega$, $R_{C1} = 3.3 \text{ k}\Omega$, $R_{C2} = 2.7 \text{ k}\Omega$, $v_{i1} = 500$ μV, and v_{i2} = GND.

a) I_{REF}, I_{C2}, I_{C4}, V_{C2}, V_{C4}, V_{CE4}

b) A_{d1}, A_{d2}, A_{dm} in dB

c) v_{out}

Answers:

$I_{REF} = 2.2 \text{ mA}$, $I_{C2} = 2.2 \text{ mA}$, $I_{C4} = 1.1 \text{ mA}$, $V_{C2} = 4.09 \text{ V}$, $V_{C4} = 7.97 \text{ V}$, $V_{CE4} = 4.99 \text{ V}$

$A_{d1} = 139.6$, $A_{d2} = 57.1$, $A_{dm} = 84$ dB, $v_{out} = 7.97$ V.

Example 10-5 — Cascaded Diff-Amps — ANALYSIS

Figure 10-35 below is the cascaded diff-amp circuit of Figure 10-34 with the addition of a third stage, which is basically an emitter-follower serving as a DC level shifter. Our objective is to determine the value of the resistor R_E so that $V_{out(dc)} = 0$.

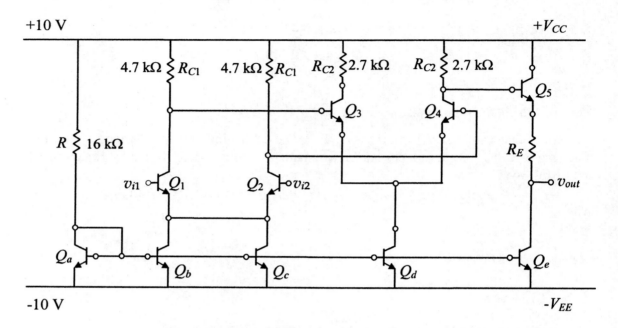

Figure 10-35: A two-stage cascaded differential amplifier with DC level shifter

With the base-emitter of Q_e shunted with Q_a, the reference current I_{REF} is mirrored into Q_e. Thus,

$$I_{Qe} = I_{Q5} = I_{REF} = 1.2 \text{ mA}$$

$$V_{E5} = V_{B5} - 0.7 \text{ V} = V_{C4} - 0.7 \text{ V} = 8.38 \text{ V} - 0.7 \text{ V} = 7.68 \text{ V}$$

The output voltage $V_{o(dc)}$ is determined by subtracting the voltage drop across R_E from the emitter voltage V_{E5}.

$$V_{o(dc)} = V_{E5} - I_{E5}R_E = 0$$

$$V_{E5} = I_{E5}R_E$$

Having already determined V_{E5}, we can now solve for R_E, as follows:

$$R_E = \frac{V_{E5}}{I_{E5}}$$

$$R_E = \frac{7.68 \text{ V}}{1.2 \text{ mA}} = 6.4 \text{ k}\Omega$$

Note that without the third stage the DC output would have been $V_{C4} = 8.38$ V. With the addition of the DC level shifter, however, the DC at the output has now dropped to zero.

The DC level shifter, being an emitter-follower, has a gain of approximately 1, and has no significant effect on the small-signal operation. Hence, the amplified signal at the output will have no DC shift if the input signal has no DC component. This feature and the absence of coupling or bypass capacitors make this configuration suitable for amplifying AC as well as DC voltages.

Practice Problem 10-15 **ANALYSIS/DESIGN**

Add a DC level shifter to the circuit of Figure 10-36 as shown below, and determine the value of R_E. Then determine the output signal (DC and AC) for the following inputs with v_{i2} = GND and
   a) $v_{i1} = 400\ \mu V$(p-p) with no DC content.
   b) $v_{i1} = 400\ \mu V$(p-p) with 200 μV DC content.

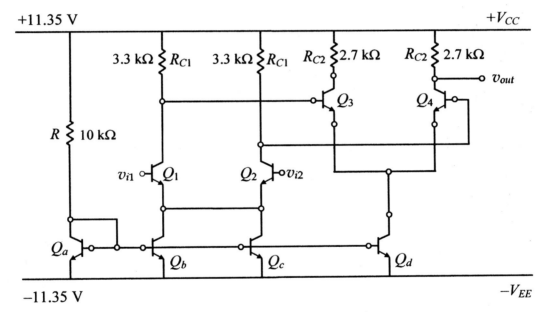

Figure 10-36: A two-stage cascaded differential amplifier

Answers:

R_E = 3.5 kΩ, (a) v_{out} = 6.377 V(p-p), no DC (b) v_{out} = 6.377 V(p-p) + 3.18 V (DC)

10.6 MOSFET DIFFERENTIAL AMPLIFIER

As we have demonstrated with BJT diff-amps, a diff-amp with a reasonably high *CMRR* can only be achieved when the diff-amp transistors are biased with a current source having a high output resistance. The same principle applies equally to the FET diff-amps.

10.6.1 Basic MOSFET Differential Amplifier

Let us now analyze the MOSFET diff-amp of Figure 10-37, which is biased wth a basic current-mirror current source of Figure 10-7, for which the output current and the output resistance have already been determined in Section 10.2.3. Thus,

$$I_{REF} = I_o = I_{D4} = I_{D1} + I_{D2} = 2I_D = 2 \text{ mA} \qquad R_o = r_{o4} = \frac{V_A}{I_o} = \frac{V_A}{I_{D4}} = \frac{120 \text{ V}}{2 \text{ mA}} = 60 \text{ k}\Omega$$

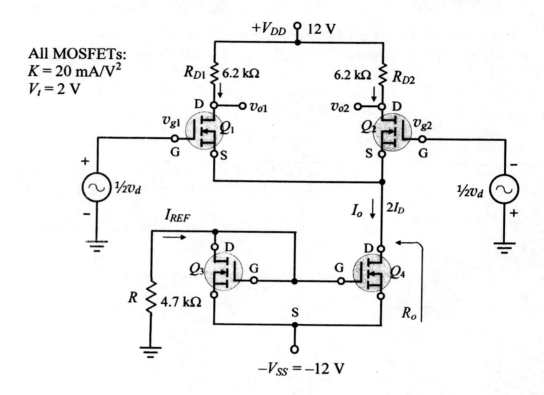

Figure 10-37: Current-mirror biased MOSFET diff-amp in ideal differential-mode connection

The analysis begins with computation of V_{GS} and I_o of the current mirror. The output current I_o of the current mirror, which has already been determined to be 2 mA, splits evenly between the two transistors Q_1 and Q_2

$$I_{D1} = I_{D2} = I_D = \frac{I_o}{2} \tag{10-69}$$

Hence,

$I_{D1} = I_{D2} = I_D = I_o/2 = 2/2 = 1 \text{ mA}$

$V_{D1} = V_{D2} = V_{DD} - I_{D1}R_{D1} = 12 \text{ V} - 1 \text{ mA} \times 6.2 \text{ k}\Omega = 5.8 \text{ V}$

$V_{S1} = V_{S2} = V_{SG1} = -V_{GS1}$

Having determined I_{D1}, V_{GS1} can be determined from Equation 7-40, as follows:

$$V_{GS} = V_t + \sqrt{\frac{I_D}{K}}$$

$$V_{GS1} = V_{GS2} = V_t + \sqrt{\frac{I_{D1}}{K}} = 2 + \sqrt{\frac{1\,\text{mA}}{20\,\text{mA/V}^2}} = 2.22\,\text{V}$$

$V_{S1} = V_{S2} = V_{SG1} = -V_{GS1} = -2.22\,\text{V}$

$V_{DS1} = V_{DS2} = V_{D1} - V_{S1} = V_{D1} + V_{GS1} = 5.8\,\text{V} - (-2.22\,\text{V}) \cong 8\,\text{V}$.

The diff-amp circuit of Figure 10-37 is redrawn in Figure 10-38, where the current mirror is repalced by an ideal dc current source.

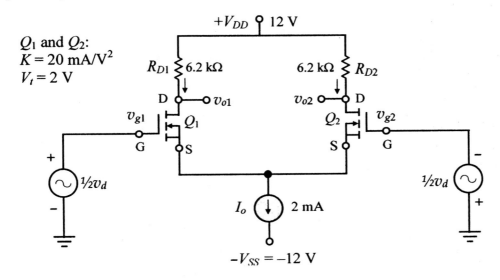

Figure 10-38: MOSFET diff-amp of Figure 10-37 in differential mode with the current mirror replaced by a *current source*

The differential input voltage v_d, which is the difference between the two input voltages v_{g1} and v_{g2}, is given by

$$v_d = v_{g1} - v_{g2} \tag{10-70}$$

where

$$v_{g1} = v_{gs1} + v_{s1} = v_{gs1} + v_s \tag{10-71}$$

and

$$v_{g2} = v_{gs2} + v_{s2} = v_{gs2} + v_s \tag{10-72}$$

Hence, (10-73)

$$v_d = v_{g1} - v_{g2} = v_{gs1} + v_s - (v_{gs2} + v_s) = v_{gs1} - v_{gs2}$$

$$v_d = v_{g1} - v_{g2} = v_{gs1} - v_{gs2} \tag{10-74}$$

Therefore,

$$v_{gs1} = \tfrac{1}{2} v_d \tag{10-75}$$

$$v_{gs2} = -\tfrac{1}{2} v_d \tag{10-76}$$

Chapter 10

The small-signal equivalent circuit of the output section of the diff-amp of Figure 10-39 is shown in Figure 10-39 below.

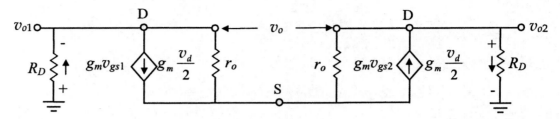

Figure 10-39: Small-signal equivalent circuit of the output section in differential mode

Converting two current sources and their shunt output resistances to voltage sources with a series resistance results in the following simple series circuit.

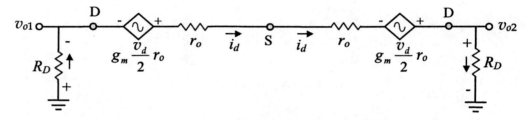

Figure 10-40: Converted equivalent circuit of Figure 10-39

Applying *KVL* around the current path and solving for i_d results in the following:

$$i_d = \frac{g_m v_d r_o}{2(R_D + r_o)} \tag{10-77}$$

Having determined i_d, output voltages v_{o1} and v_{o2} are determined as follows:

$$|v_{o1}| = |v_{o2}| = i_d R_D = \frac{g_m v_d r_o R_D}{2(r_o + R_D)} = \frac{g_m v_d}{2} \frac{r_o R_D}{r_o + R_D} \tag{10-78}$$

$$|v_{o1}| = |v_{o2}| = \frac{g_m v_d}{2}(r_o \| R_D) \tag{10-79}$$

$$|A_{d1}| = \left|\frac{v_{o1}}{v_d}\right| = \frac{g_m}{2}(r_o \| R_D) \tag{10-80}$$

$$|A_{d2}| = \left|\frac{v_{o2}}{v_d}\right| = \frac{g_m}{2}(r_o \| R_D) \tag{10-81}$$

$$|A_{dm}| = \left|\frac{v_o}{v_d}\right| = g_m(r_o \| R_D) \tag{10-82}$$

$g_m = 2K(V_{GS} - V_T) = 2 \times 20 \text{ mA/V}^2 \, (2.22 \text{ V} - 2 \text{ V}) = 8.8 \text{ mS}$

Chapter 10

Assuming $V_A = 120$ V, $\quad r_{o(Q1)} = \dfrac{V_A}{I_{D1}} = \dfrac{120 \text{ V}}{1 \text{ mA}} = 120 \text{ k}\Omega$

$|A_{d1}| = |A_{d2}| = 4.4 \text{ mS } (120 \text{ k}\Omega \| 6.2 \text{ k}\Omega) \cong 26$

$|A_{dm}| = 2A_{d1} = 52$

$R_o = r_{o(Q4)} = \dfrac{V_A}{I_o} = \dfrac{120 \text{ V}}{2 \text{ mA}} = 60 \text{ k}\Omega$

The common mode gain can be determined with Equation 10-51, as follows:

$$|A_c| = \left|\dfrac{v_o}{v_c}\right| = \dfrac{R_D}{2R_o} \quad (10\text{-}83)$$

$|A_c| = \dfrac{R_D}{2R_o} = \dfrac{6.2 \text{ k}\Omega}{120 \text{ k}\Omega} = 0.05166$

The common mode rejection ratio is determined with Equations 10-52 and 10-53.

$CMRR = \left|\dfrac{A_d}{A_c}\right| = \dfrac{26}{0.05166} = 503 = 54 \text{ dB}$

Practice Problem 10-16 *(MOSFET Diff-Amp)* — ANALYSIS

Determine the following for the MOSFET diff-amp of Figure 10-41. Assume $V_A = 150$ V.
  a) I_{REF}, I_o, $I_{D(Q1)}$, and $V_{DS(Q2)}$.
  b) A_{d1}, A_{d2}, and A_{dm}
  c) A_c, and CMRR

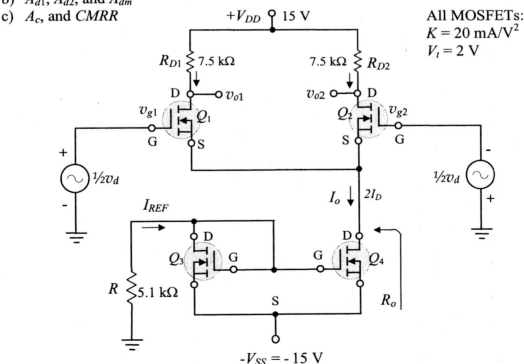

Figure 10-41: Current-mirror biased MOSFET diff-amp

Answers:

$$I_o = I_{REF} = I_{D(Q4)} = 2.48 \text{ mA}, \quad I_{D(Q1)} = 1.24 \text{ mA}, \quad V_{DS(Q2)} = 7.95 \text{ V}, \quad g_{m1} = 10 \text{ mA/V}$$

$$A_{d1} = -A_{d2} = -35, \quad A_{dm} = -70, \quad A_c = 0.0625, \quad CMRR = 55 \text{ dB}.$$

10.6.2 MOSFET Diff-Amp with Cascode Current-Mirror

The circuit of a MOSFET diff-amp with cascode current-mirror is shown in Figure 10-42 below.

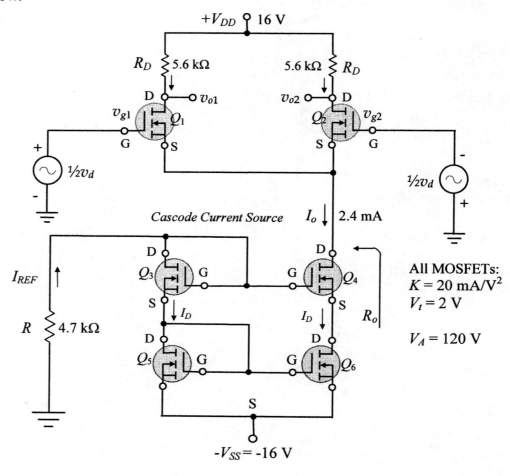

Figure 10-42: MOSFET diff-amp with cascode current-mirror current-source shown in ideal differential mode connection.

The above cascode current-mirror is the same current-mirror shown in Figure 10-8, for which the output current I_o and output resistance R_o have already been determined, and have the following values:

$$I_o = 2.4 \text{ mA} \qquad R_o = 35 \text{ M}\Omega$$

This diff-amp will have a very low A_c and very high $CMRR$ because of the very high R_o of the cascode current mirror. The output current I_o of 2.4 mA will split evenly between the two MOSFETs Q_1 and Q_2 of the diff-amp; hence,

$$I_{DQ1} = I_{DQ2} = 1.2 \text{ mA}$$

V_{GS} of the diff-amp MOSFETs Q_1 and Q_2 are from Equation 7-40

$$V_{GS} = V_t + \sqrt{\frac{I_D}{K}}$$

$$V_{GS1} = V_{GS2} = V_t + \sqrt{\frac{I_D}{K}} = 2 \text{ V} + \sqrt{\frac{1.2 \text{ mA}}{20 \text{ mA/V}^2}} = 2.25 \text{ V}$$

Having determined V_{GS}, the transconductance g_m for Q_1 and Q_2 is determined as follows:

$$g_m = 2K(V_{GS} - V_t) = 40 \text{ mA/V}^2 (2.25 \text{ V} - 2 \text{ V}) = 10 \text{ mA/V}$$

The differential-mode gain is determined from Equation 10-80, as follows:

$$A_{d1} = \frac{v_{o1}}{v_d} = -\frac{g_m}{2}(r_{o1} \| R_D)$$

$$A_{d1} = -\frac{10 \text{ mA/V}}{2}(100 \text{ k}\Omega \| 5.6 \text{ k}\Omega) = -26.59$$

where

$$r_{o1} = \frac{V_A}{I_{D1}} = \frac{120 \text{ V}}{1.2 \text{ mA}} = 100 \text{ k}\Omega$$

The common mode gain can be determined from Equation 10-83, as follows:

$$A_c = -\frac{R_D}{2R_o} = -\frac{5.6 \text{ k}\Omega}{2 \times 35 \text{ M}\Omega} = -8 \times 10^{-5}$$

The common-mode rejection ratio is determined from Equations 10-52 and 10-53.

$$CMRR = \left|\frac{A_d}{A_c}\right| = \frac{g_m r_o R_o}{R_D + r_o} = \frac{26.59}{8 \times 10^{-5}} = 3.32375 \times 10^5 = 110.44 \text{ dB}$$

Practice Problem 10-17 ANALYSIS

If $\pm V_{DD} = \pm 15$ V, $R_D = 4.7$ kΩ, and $R = 5.1$ kΩ, determine the following for the MOSFET diff-amp of Figure 10-42:

a) I_{REF}, I_o, $I_{D(Q1)}$, and $V_{DS(Q2)}$.

b) A_{d1}, A_{d2}, and A_{dm}

c) A_c and $CMRR$

Assume MOSFETs are 2N7002:
$K = 20$ mA/V^2
$V_t = 2$ V
$V_A = 120$ V

Answers:

$I_o = I_{REF} \cong 2$ mA, $R_o = 45$ GΩ, $g_{m(Q4)} = 11.8$ mA/V, $I_{D(Q1)} = 1$ mA, $V_{DS2} = 12.52$ V, $g_{m(Q1)} = 8.8$ mA/V, $A_{d1} = -A_{d2} \cong -20$, $A_c = -1.044 \times 10^{-6}$, $CMRR = 145$ dB

10.7 SUMMARY

- An ideal current source supplies a steady amount of current and has infinite output resistance. The *Wilson current-mirror* current source has nearly ideal characteristics.

- The *basic current-mirror* current source, which can be realized with two transistors, BJT or FET, provides fairly steady current and has moderately high output resistance.

- The *cascode current-mirror* current source, which is realized with four MOSFETs and is an improved version of the basic current-mirror, provides substantially high output resistance.

- A differential amplifier (diff-amp) amplifies the difference between the two input signals while suppressing the signal that is common to both inputs. The differential amplifier is an essential component of the operational amplifier; in fact, the first two stages of an operational amplifier (op-amp) are differential amplifiers.

- There are two modes of operation regarding the AC analysis of a diff-amp: *differential mode* and *common mode*. With the differential mode, we are interested in finding the amplitude of the output signal due to the differential input $v_d = v_{i1} - v_{i2}$. With the common mode, we are interested in finding the amplitude of the output signal due to the common input $v_c = \frac{1}{2}(v_{i1} + v_{i2})$. The ratio of the output signal to the input signal in the differential mode is called the *differential mode gain* A_d, and the same ratio in the common mode is the *common mode gain* A_c.

- The common mode rejection ratio (*CMRR*) is a measure of how well the diff-amp rejects or attenuates the *common mode* signals and is defined as the ratio of the differential mode gain to the common mode gain.

- A nearly ideal BJT differential amplifier with superior characteristics can be achieved if biased with a Wilson current-mirror current source.

- The MOSFET diff-amp biased with a basic current-mirror provides the same moderate characteristics as the BJT diff-amp biased with a basic current-mirror. However, the MOSFET diff-amp can provide a much higher input resistance, and as a result, imposes almost no loading on the differential input signals.

- The MOSFET diff-amp biased with a cascode current-mirror provides much more desirable characteristics compared to the MOSFET diff-amp biased with a basic current-mirror, which is due to the extremely high output resistance of the cascode current-mirror.

Chapter 10

- The BJT *Wilson current-mirror* may be utilized to source the MOSFET diff-amp (see problem 10.26), which provides a very high input resistance, nearly infinite.

PROBLEMS

Section 10.2 Current-Mirror Current Sources

10.1 Carry out Practice Problem 10-1.

10.2 Determine the following for the circuits of Figure 10-1P(a) and 10-1P(b) below:
   a) Output current and output resistance of the current-mirror of Figure 10-1P(a).
   b) Resistance R and output resistance of the current-mirror of Figure 10-1P(b). Assume $V_A = 140$ V.

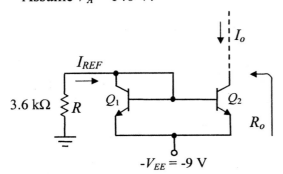

Figure 10-1P(a): Basic current-mirror

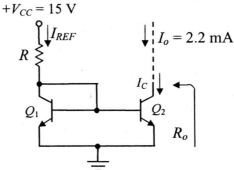

Figure 10-1P(b): Basic current-mirror

10.3 Carry out Practice Problem 10-2.

10.4 Determine the following for the circuits of Figure 10-2P(a) and 10-2P(b) below:
   a) Output current and output resistance of the current-mirror of Figure 10-2P(a).
   b) Resistance R and output resistance of the current-mirror of Figure 10-2P(b). Assume $\beta = 150$ and $V_A = 150$ V.

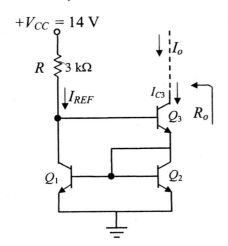

Figure 10-2P(a):
Wilson current-mirror

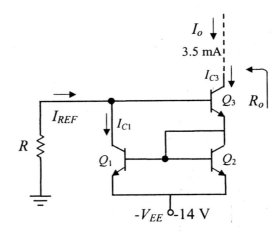

Figure 10-2P(b):
Wilson current-mirror

10.5 Carry out Practice Problem 10-3.

10.6 Carry out Practice Problem 10-5.

10.7 Determine the output current I_o and output resistance R_o of the basic current-mirror of Figure 10-3P.

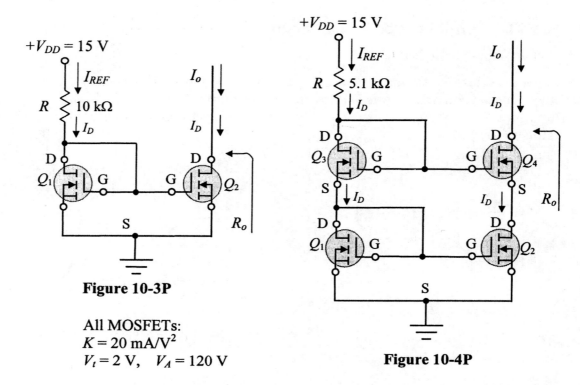

Figure 10-3P

All MOSFETs:
$K = 20$ mA/V^2
$V_t = 2$ V, $V_A = 120$ V

Figure 10-4P

10.8 Determine the output current I_o and the output resistance R_o of the cascode current-mirror of Figure 10-4P.

Section 10.3 Differential Amplifier

10.9 Carry out Practice Problem 10-6.

10.10 Determine the DC voltages and currents for the basic diff-amp of Figure 10-5P. Also determine the differential gain, common mode gain, and CMRR.

10.11 Carry out Practice Problem 10-7.

Section 10.5.1 Diff-Amp with Basic Current-Mirror

10.12 Carry out Practice Problem 10-8.

10.13 Determine the DC voltages and currents for the diff-amp of Figure 10-6P. Also determine the differential gain, common mode gain, and the CMRR. Assume $V_A = 150$ V.

Chapter 10

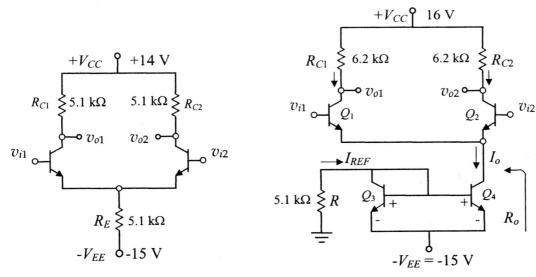

Figure 10-5P: A basic diff-amp

Figure 10-6P: Diff-amp with basic current-mirror

10.14 Carry out Practice Problem 10-9.

Section 10.5.2 Diff-Amp with Wilson Current-Mirror

10.15 Carry out Practice Problem 10-10.

10.16 Determine the DC voltages and currents for the diff-amp of Figure 10-7P. Also determine the differential gain, common mode gain, and CMRR. Assume $\beta = 160$ and $V_A = 140$ V.

10.17 Carry out Practice Problem 10-11.

10.18 Carry out Practice Problem 10-12.

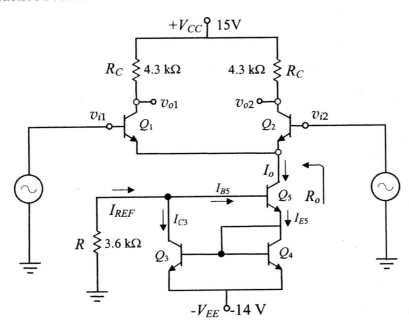

Figure 10-7P: Diff-amp with Wilson current-mirror

449

Chapter 10

10.19 Carry out Practice Problem 10-13. Test your design by simulation.

10.20 Design the following diff-amps for a differential gain of 200 and I_{CQ} of 2 mA. After the completion of the design work, determine the output resistance and *CMRR* in dB. Assume $\beta = 160$ and $V_A = 140$ V. Test your designs by simulation.
   a) Basic diff-amp.
   b) Diff-amp with basic current-mirror.
   a) Diff-amp with Wilson current-mirror.

Cascaded Diff-Amp Systems

10.21 Carry out Practice Problem 10-14.

10.22 Carry out Practice Problem 10-15.

10.23 Determine the following for the cascaded diff-amp system of Figure 10-8P as shown below:
   a) DC currents and voltage: I_R, I_{C1}, I_{C2}, I_{C3}, I_{C4}, I_{C5}, and V_{CE4}.
   b) Determine the value of R_E, so that $V_{o(DC)} = 0$ V.
   c) Determine the output signal (DC and AC) for the following inputs:
   b) $v_{i1} = 500$ μV(p-p) with no DC content, and v_{i2} = GND.
   e) $v_{i1} = 400$ μV(p-p) with 20 μV DC content, and v_{i2} = GND.

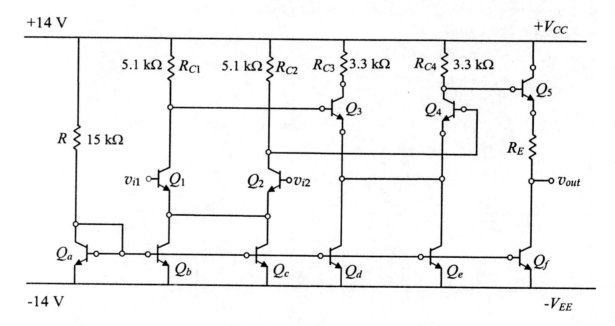

Figure 10-8P: A two-stage cascaded differential amplifier with DC level shifter

Section 10.6 MOSFET Differential Amplifier

10.24 Carry out Practice Problem 10-16.

Chapter 10

10.25 Determine I_o, I_{D1}, V_{DS1}, A_d, A_c, and *CMRR* in dB for the MOSFET diff-amp of Figure 10-9P.

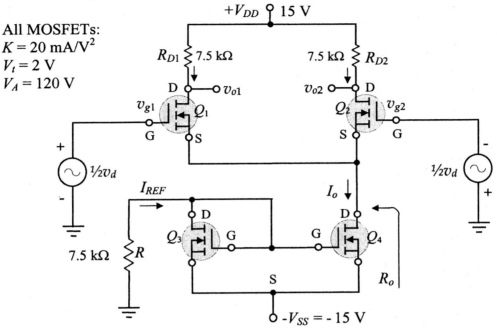

Figure 10-9P: Current-mirror biased MOSFET diff-amp

10.26 Carry out Practice Problem 10-17.

10.27 Determine I_o, I_{D1}, V_{DS1}, A_d, A_c, and *CMRR* in dB for the diff-amp of Figure 10-10P.

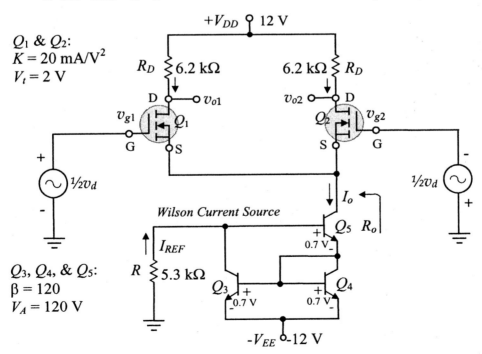

Figure 10-10P: MOSFET diff-amp with Wilson current-mirror current-source shown in ideal differential mode connection

451

10.28 Design the Wilson current-mirror biased differential amplifier of Figure 10-10P for an I_D of 2 mA and a differential gain of 25. Upon completion of the design determine the common-mode gain and the $CMRR$. Test and verify your design by simulation. Let $+V_{DD} = +16$ V and $-V_{EE} = -16$ V.

Chapter 11

OPERATIONAL AMPLIFIERS
Analog Integrated Circuits

11.1 INTRODUCTION

Having studied the concept and the underlying theory of cascaded stages, current sources, and differential amplifiers in the previous two chapters, we now turn our attention to one of the earlier wonders in electronics engineering, the analog integrated circuit known as the *operational amplifier* (op-amp). As mentioned previously, the op-amp is a cascade of four main stages: The first two stages of an op-amp are differential amplifiers; the third stage is a DC level shifter; and the last or the output stage is a current driver. Although the op-amp is a precision-engineered complex circuitry, its low cost and ease of use allow it to be treated as a simple three-terminal electronic device. Hence, we will avoid the detailed analysis of the internal circuitry and instead focus our attention on the characteristics, limitations, and applications of the op-amp. As we will see in this and the following chapters, the operational amplifier is an extremely versatile device with a broad range of practical applications. For example, we can use the op-amp as a comparator and LED driver. With the addition of two external resistors we can design an amplifier for a specified gain with nearly infinite input resistance (in $G\Omega$) and extremely low output resistance (in $m\Omega$). Also, with the addition of a few external resistors and capacitors we can design active circuits such as oscillators, filters, integrators, differentiators, etc.

Chapter 11

11.2 OPERATIONAL AMPLIFIER

The internal stages and circuitry of a simple operational amplifier are illustrated in Figure 11-1 below.

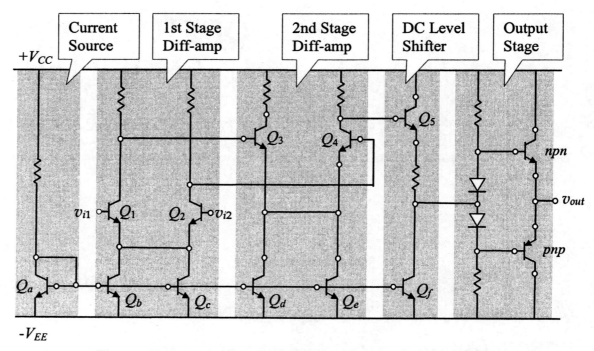

Figure 11-1: Internal circuitry of a simple operational amplifier

As we have seen in the previous chapter, the current source in the above diagram is used to bias the transistors of the following three stages with a steady current. The gain of the operational amplifier is achieved from the two cascaded diff-amps, and the level shifter shifts the DC voltage at the collector of Q_4 down to zero, which is then fed to the output stage. The output stage is generally a class AB complementary symmetry push-pull configuration or current driver, which will be discussed in chapter 15. In addition to supplying current and thus power to the load, the output stage also provides low output resistance. The absence of coupling and bypass capacitors enables the op-amp to amplify AC as well as DC signals. In addition to the two power supply terminals $+V_{CC}$ and $-V_{EE}$ (usually referred to as $\pm V_{CC}$), the external terminals of the above simple op-amp are the two input terminals v_{i1} with v_{i2} and the output terminal v_{out}. Hence, no matter how complex the internal circuitry of the op-amp happens to be, we can treat it as a simple three-terminal device, as illustrated in Figure 11-2 below:

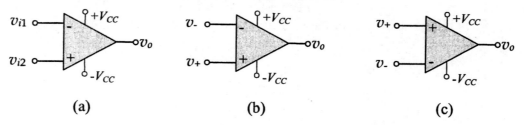

Figure 11-2: Op-amp block diagrams

According to the circuit of Figure 11-1, the input terminal v_{i1} is the inverting terminal and v_{i2} is the non-inverting terminal. That is, a signal applied at the input v_{i1}, with v_{i2} grounded, will have an output that is out-of-phase with the input signal v_{i1}. Similarly, a signal applied at the input v_{i2}, with v_{i1} grounded, will have an output that is in-phase with the input signal v_{i2}. Hence, as shown in Figure 11-2(a), v_{i1} is labeled with a minus sign, which stands for inverting, and v_{i2} is labeled with a plus sign, which stands for non-inverting. To make it even simpler, the inverting terminal may be designated as v_- and the non-inverting terminal may be designated as v_+, as illustrated in Figures 11-2(b) and (c) above.

11.2.1 Op-Amp Characteristics

Some of the typical characteristics of three all-purpose operational amplifiers versus the expected characteristics of an ideal op-amp are listed in Table 11-1.

Table 11-1: Typical versus ideal op-amp characteristics

Parameter	LM741	LF347	LM318	Ideal
Open-Loop Gain (A_{OL})	2×10^5	1×10^5	2×10^5	∞
Input Resistance (R_{in})	2 MΩ	10^{12} Ω	3 MΩ	∞ Ω
Output Resistance (R_o)	75 Ω	75 Ω	75 Ω	0 Ω
Gain Bandwidth Product	1 MHz	4 MHz	15 MHz	∞ Hz
CMRR	90 dB	100 dB	100 dB	∞

Note that the 741 op-amp may not be the latest and the most up-to-date op-amp; however, it is the industry's standard reference op-amp to which all other op-amps are compared. Hence, one should be thoroughly familiar with its characteristics and applications. Assuming that $\pm V_{CC} = \pm 12$ V, let us apply different levels of input voltage as shown in Table 11-2 to the op-amp of Figure 11-3 and plot the corresponding transfer characteristic curve.

Figure 11-3:
Circuit setup for plotting the transfer characteristics of the op-amp

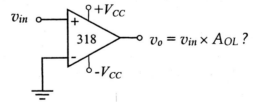

The output voltage can never exceed V_{CC}; in fact, the output will reach the saturation level V_{sat} at 1 to 2 V less than V_{CC}. Assuming the worst case, $\pm V_{sat}$ is defined as follows:

$$+V_{sat} = +V_{CC} - 2 \text{ V} \qquad (11\text{-}1)$$

$$-V_{sat} = -V_{CC} + 2 \text{ V} \qquad (11\text{-}2)$$

Hence, the saturation levels for the op-amp of Figure 11-4 are the following:

$$+V_{sat} = +12\text{ V} - 2\text{ V} = +10\text{ V}$$

$$-V_{sat} = -12\text{ V} + 2\text{ V} = -10\text{ V}$$

Table 11-2: Input and Output Voltages of Open-Loop Op-amp

v_{in}	–1 V	–1 mV	–50 µV	–10 µV	0	+10 µV	+50 µV	+1 mV	+1 V
v_o	–10 V	–10 V	–10 V	–2 V	0	+2 V	+10 V	10 V	+10 V

Plotting the above data yields the transfer characteristic as shown in Figure 11-4.

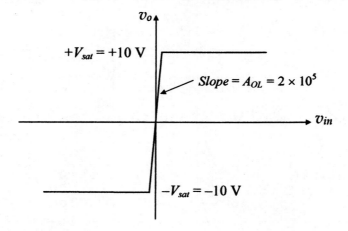

Figure 11-4: Transfer characteristic of the LM318 op-amp

The transition from $-V_{sat}$ to $+V_{sat}$, which has the slope of $A_{OL} = 2 \times 10^5$, is nearly vertical. The output will be at saturation level with an input v_{in} as low as one-tenth of a mV.

11.3 OPEN-LOOP OPERATION

The op-amp may be used in the open-loop configuration as a comparator. Because of the very large open-loop gain A_{OL} and according to the characteristics shown in Figure 11-4, the output voltage of the op-amp will reach $+V_{sat}$ if the voltage at the non-inverting input v^+ is slightly higher than the voltage at the inverting input v^-. Conversely, the output voltage of the op-amp will reach $-V_{sat}$ if the voltage at the non-inverting input v^+ is slightly lower than the voltage at the inverting input v^-. In a short form, the above statement may be expressed algebraically as follows:

$$\text{if } v^+ > v^- \Rightarrow v_o = +V_{sat} \tag{11-3}$$

$$\text{if } v^+ < v^- \Rightarrow v_o = -V_{sat} \tag{11-4}$$

Example 11-1 — Op-Amp Comparator — ANALYSIS

Let us now analyze the op-amp comparator of Figure 11-5 with the input as shown in Figure 11-6.

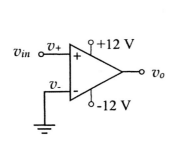

Figure 11-5: Op-amp comparator

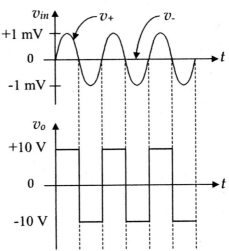

Figure 11-6: Input and output waveforms of the op-amp

The inverting input of the op-amp is at the ground level ($v_- = 0$) and the sinusoidal input signal is applied to the non-inverting input ($v_+ = v_{in}$). Hence, when the input signal is above the zero line ($v_+ > v_-$) output reaches $+V_{sat}$, and when the input signal is below the zero line ($v_- > v_+$) output drops to $-V_{sat}$, as shown in Figure 11-6.

Practice Problem 11-1 — Op-Amp Comparator — ANALYSIS

Determine and plot the output for the op-amp comparator of Figure 11-7 as shown below:

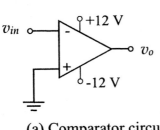

(a) Comparator circuit

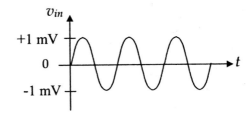

(b) Input signal v_{in}

Figure 11-7: Op-amp comparator

Example 11-2 — Op-amp Comparator — ANALYSIS

In the previous example, the reference was chosen to be the ground level. However, the reference level can be a non-zero voltage level, as shown in the following example.

Chapter 11

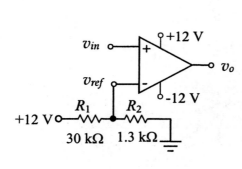

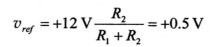

$$v_{ref} = +12\,V \frac{R_2}{R_1 + R_2} = +0.5\,V$$

Figure 11-8: Op-amp comparator

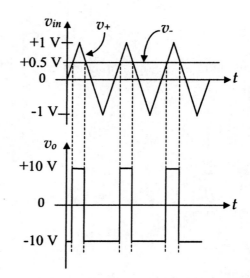

Figure 11-9: Input and output waveforms

The inverting input of the op-amp is at the reference voltage ($v_- = v_{ref} = +0.5$ V), and the triangular input signal is applied to the non-inverting input ($v_+ = v_{in}$). Hence, when the input signal is above the $v_{ref} = 0.5$ V line ($v_+ > v_-$), the output reaches $+V_{sat}$, and when the input signal is below the v_{ref} ($v_- > v_+$), the output drops to $-V_{sat}$, as shown in Figure 11-9.

Practice Problem 11-2 — Op-Amp Comparator — ANALYSIS

Plot the output waveform for the op-amp comparator of Figure 11-10 for the given input as shown below.

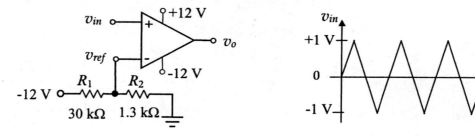

(a) Comparator Circuit (b) Input to the comparator

Figure 11-10: Op-amp comparator

11.3.1 Op-Amp as LED Driver

Most LEDs require 10 to 20 mA of current with a voltage drop of 1 to 2 V for emitting a reasonably bright light. The 741 op-amp, for instance, has a maximum output current of 25 mA, and thus it may be used to drive an LED. Consider the comparator and LED driver circuit of Figure 11-11 as shown below:

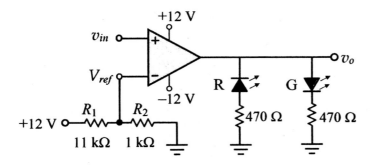

Figure 11-11: Op-amp as LED driver

Applying the voltage-divider rule, $v^- = v_{ref}$ can be found to be exactly 1 V. For the input $v_{in} = v^+ = 0.5$ V, $v^+ < v^-$ and thus, $v_o = -V_{sat} = -10$ V. The output having dropped to -10 V causes the red LED, R, to be forward-biased (on) and the green LED, G, reverse-biased (off). Conversely, for the input $v_{in} = v^+ = 1.5$ V, $v^+ > v^-$ and as a result $v_o = +V_{sat} = +10$ V, which turns the red LED off and the green LED on. The current through the LED will be about 17 mA, when on.

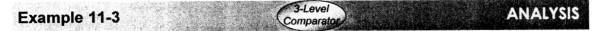

Example 11-3 — 3-Level Comparator — ANALYSIS

Now let us analyze the three-level comparator or window comparator of Figure 11-12.

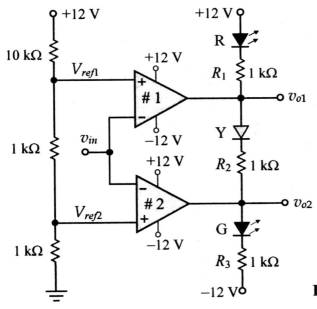

v_{in}	R	Y	G
0.5 V	OFF	OFF	ON
1.5 V	OFF	ON	OFF
2.5 V	ON	OFF	OFF

Figure 11-12: A three-level comparator

The non-inverting inputs of both op-amps are the reference voltages V_{ref1} and V_{ref2}, and thus, according to the voltage-divider rule:

$$V_{ref1} = V_1^+ = 2 \text{ V}$$
$$V_{ref2} = V_2^+ = 1 \text{ V}$$

The input signal is applied to the inverting inputs of both op-amps; that is,

$$v_1^- = v_2^- = v_{in}$$

For $v_{in} = 0.5$ V,

$$v_1^+ > v_1^- \Rightarrow v_{o1} = +V_{sat} = +10 \text{ V, and } v_2^+ > v_2^- \Rightarrow v_{o2} = +V_{sat} = +10 \text{ V}$$

and as a result, the green LED is forward-biased (G is on), the yellow LED is unbiased (Y is off), and the red LED is unbiased (R is off). Note that one can argue that the red LED is forward-biased because of the 2-V potential difference between $+V_{CC} = +12$ V at the anode of the LED and $+V_{sat} = +10$ V at v_{o1}, which is a correct statement. However, assuming a 2 V drop across the LED, the voltage drop across the resistor R_1 is zero; hence, no current flows and the LED remains unbiased or at the verge of being biased. For $v_{in} = 1.5$ V,

$$v_1^+ > v_1^- \Rightarrow v_{o1} = +V_{sat} = +10 \text{ V, and } v_2^+ < v_2^- \Rightarrow v_{o2} = -V_{sat} = -10 \text{ V}$$

and as a result, the green LED is unbiased (G is off), the yellow LED is forward biased (Y is on), and the red LED is unbiased (R is off). For $v_{in} = 2.5$ V,

$$v_1^- > v_1^+ \Rightarrow v_{o1} = -V_{sat} = -10 \text{ V, and } v_2^+ < v_2^- \Rightarrow v_{o2} = -V_{sat} = -10 \text{ V}$$

and as a result, the green LED is unbiased (G is on), the yellow LED is unbiased (Y is off), and the red LED is forward biased (R is on). When on, the amount of current through each LED will be as follows:

$$I_{RED} = I_{GREEN} = \frac{V_{CC} - (-V_{sat}) - V_{LED}}{R_1} = \frac{12 \text{ V} + 10 \text{ V} - 2 \text{ V}}{1 \text{ k}\Omega} = 20 \text{ mA}$$

$$I_{YEL} = \frac{+V_{sat} - (-V_{sat}) - V_{LED}}{R_2} = \frac{10 \text{ V} + 10 \text{ V} - 2 \text{ V}}{1 \text{ k}\Omega} = 18 \text{ mA}$$

Practice Problem 11-3   **ANALYSIS**

Analyze the three-level comparator of Figure 11-13 and determine the ON and OFF conditions of each LED according to the given input levels. Also, determine the amount of current through each LED, when on.

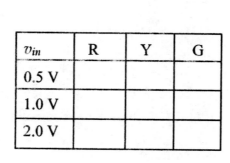

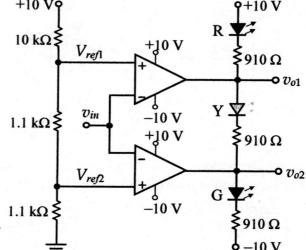

Figure 11-13: A three-level comparator

11.3.2 Schmitt Trigger

The Schmitt trigger is a bi-reference level comparator with positive feedback. Its advantage over the simple comparator with a single reference level is demonstrated in Example 11-4, shortly. The circuit diagram of the Schmitt trigger, its response to the same analog input of Example 11-2, and its transfer characteristics are exhibited in Figures 11-14(a) through 11-14(c). As shown in Figure 11-14(a), the output is fed back through the resistor R_1 to the non-inverting terminal of the op-amp, causing a positive feedback. The addition of the resistor R_2 establishes a simple voltage-divider network from the output to ground through R_1 and R_2.

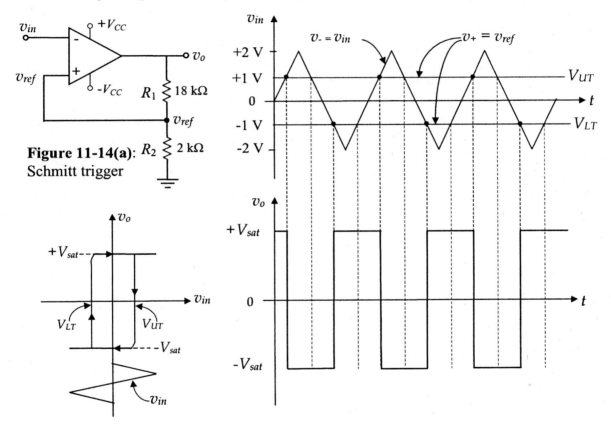

Figure 11-14(a): Schmitt trigger

Figure 11-14(c): Transfer characteristics exhibiting the *Hysteresis* loop

Figure 11-14(b): Input and output waveforms v_{in} and v_o

Applying the voltage-divider rule, the reference voltage v_{ref} at the non-inverting terminal is determined in terms of R_1 and R_2, as follows:

$$v_{ref} = v_o \frac{R_2}{R_1 + R_2} \qquad (11\text{-}5)$$

However, since the output will be at either the positive or the negative saturation $\pm V_{sat}$, the reference voltage v_{ref} will change accordingly, as the output v_o switches from $+V_{sat}$ to $-V_{sat}$ and vice versa. Hence,

Chapter 11

$$V_{ref} = \pm V_{sat} \frac{R_2}{R_1 + R_2} = v^+ \qquad (11\text{-}6)$$

When the output is at $+V_{sat}$, the reference level is positive and is referred to as the *upper threshold* voltage V_{UT}

$$V_{UT} = +V_{sat} \frac{R_2}{R_1 + R_2} \qquad (11\text{-}7)$$

When the output is at $-V_{sat}$, the reference level is negative and is referred to as the *lower threshold* voltage V_{LT}

$$V_{LT} = -V_{sat} \frac{R_2}{R_1 + R_2} \qquad (11\text{-}8)$$

If the supply voltage $\pm V_{CC} = \pm 12$ V, then $\pm V_{sat} = \pm 10$ V, $V_{UT} = +1$ V, and $V_{LT} = -1$ V. Assuming that the output is initially at $+V_{sat}$, then the reference level V_{ref} is $V_{UT} = +1$V, and the output will switch to $-V_{sat}$ only when the input v_{in} rises above the V_{UT}; that is, when $v^- > v^+$ then $v_o = -V_{sat}$. When the output switches to $-V_{sat}$, then the reference level v_{ref} is no longer V_{UT}, but changes to $V_{LT} = -1$V. Hence, the output will switch to $+V_{sat}$ only when the input v_{in} drops below the V_{LT}; that is, when $v^- < v^+$, then $v_o = +V_{sat}$, as illustrated in Figure 11-14(b). As a result, the output v_o changes the state only when the input signal v_{in} rises above the *upper threshold* or drops below the *lower threshold*, and does not respond to changes of v_{in} within the boundaries of the two threshold levels. Because of this important feature, the Schmitt trigger has some advantages over the simple op-amp comparator and is widely used in certain applications, where the input signal is likely to be affected by spurious noise, which is to be ignored by the comparator. The transfer characteristics of the Schmitt trigger exhibiting the *hysteresis* loop, as illustrated in Figure 11-14(c). Let us now compare the responses of the Schmitt trigger to that of a simple op-amp comparator in the following example (Example 11-4), where the signal is affected by some spurious noise.

As you can see in Figure 11-15(b), the Schmitt trigger ignores the noise and responds only to the changes in signal levels outside the boundaries set by the upper and lower thresholds. The simple comparator of Figure 11-16(a), however, whose response is illustrated in Figure 11-16(b), responds to all variations in the signal each time the corrupted signal crosses the zero reference level.

One major application of the Schmitt trigger comparator is in voice and data communications, in which the analog signal is converted to stream of pulses, called the pulse-code modulation (PCM), or the pulses may be the original data representing 0s and 1s, which are to be transmitted over the communication channel. When they are received at the receiver or at a repeater, the pulses are very likely to be corrupted by noise and the first step is to generate fresh pulses or 0s and 1s by means of a comparator. A properly designed Schmitt trigger is generally the comparator of choice in such applications, which produces fresh pulses based on the transitions of the original signal while ignoring the fluctuations and transitions due to the spurious noise.

Example 11-4 — Schmitt Trigger — ANALYSIS

In the following example, we will analyze the response of the Schmitt trigger circuit of Figure 11-14 with the same analog input signal but affected by some spurious noise. We will also analyze the response of a simple op-amp comparator to the same noisy signal and compare the results, as illustrated in Figures 11-15(b) and 11-16(b). Again, we will assume that output is initially at $+V_{sat}$.

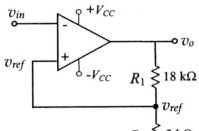

Figure 11-15(a): Schmitt trigger

The Schmitt trigger does not respond to the spurious noise and thus avoids false triggering if the noise level is limited to within the boundaries of the upper and lower thresholds V_{UT} and V_{LT}.

The simple comparator responds to the spurious noise and causes false triggering every time the noise crosses the single threshold level v_{ref}.

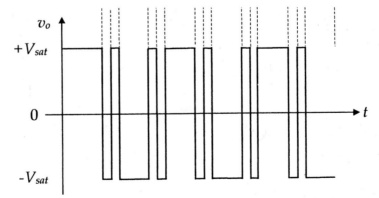

Figure 11-15(b): Schmitt trigger response to the signal corrupted with spurious noise

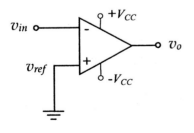

Figure 11-16(a): Simple op-amp comparator

Figure 11-16(b): Response of the simple comparator to a signal corrupted with noise

Practice Problem 11-4 — Schmitt Trigger — DESIGN

Design a Schmitt trigger with upper and lower threshold voltages at ±0.5 V. Assume $\pm V_{CC} = \pm 10$ V. Let $R_1 = 30$ kΩ. Show complete work in an orderly fashion. Assuming that the output is initially at $+V_{sat}$, plot the output for the given input signal as shown in Figure 11-17(b). Also, plot the *hysteresis* loop and label all levels accurately.

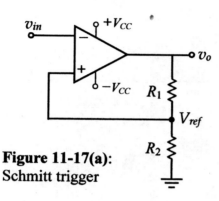

Figure 11-17(a): Schmitt trigger

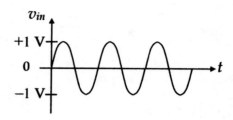

Figure 11-17(b): Input signal v_{in}

Answers: $\pm V_{CC} = \pm 10$ V, $V_{UT} = +0.5$ V, $V_{LT} = -0.5$ V, $R_1 = 30$ kΩ, $R_2 = 2$ kΩ

11.4 CLOSED-LOOP OPERATION

The closed-loop operation of the op-amp involves a feedback from the output to the input. There are two kinds of feedback: positive feedback and negative feedback. Positive feedback is used exclusively with oscillator circuits, and all other op-amp configurations mainly use negative feedback with the output fed back to the inverting input through a feedback resistor or capacitor. The negative feedback theory will be discussed in detail in the following chapter. For the time being, consider the closed-loop circuit of Figure 11-18 with negative feedback.

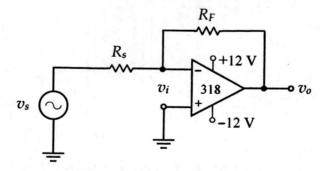

Figure 11-18: Op-amp with negative feedback

Because of the very high voltage gain of the op-amp, it takes an extremely small v_i to produce a reasonably large output voltage v_o. Assuming that the op-amp is a LM318 with $A_v = 2 \times 10^5$ and $R_i = 3$ MΩ, let us determine what level of v_i is needed to produce an output of about 10 V(p-p). Thus,

$$v_i = \frac{v_o}{A_v} = \frac{10 \text{ V(p-p)}}{2 \times 10^5} = 0.05 \text{ mV(p-p)}$$

The input voltage v_i of 0.05 mV is so small compared to the other voltage levels that it can be considered to be approximately zero

$$v_i \cong 0 \tag{11-9}$$

Because of the very high input resistance of the op-amp, there will be no significant current flow from one terminal of the op-amp to the other, especially with such a small input voltage v_i. Thus,

$$i_i \cong 0 \tag{11-10}$$

11.4.1 Virtual Short Circuit and Virtual Ground

Combining the above two equations ($v_i \cong 0$ and $i_i \cong 0$) leads to the conclusion that a *virtual short circuit* exists across the input terminals of the op-amp of Figure 11-18, in which both the input voltage and the input current are approximately zero. In addition, because of the virtual short circuit between the two terminals, the inverting terminal is at *virtual ground*. Note that even if the non-inverting terminal is not directly at the ground and is connected to ground through another resistor, there would still be no current flow through the resistor ($i_i \cong 0$), resulting in zero voltage drop; hence, the inverting terminal would still be at virtual ground.

11.5 BASIC OP-AMP CONFIGURATIONS

11.5.1 Inverting Amplifier

The schematic diagram of an inverting amplifier configuration is shown in Figure 11-19.

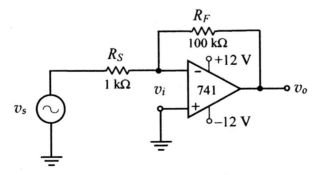

Figure 11-19: Inverting amplifier

In order to analyze the above op-amp circuit and determine the output in terms of the applied source voltage v_s and other external components, we will draw a virtual equivalent circuit, in which only the input and output terminals of the op-amp and the external circuitry will be shown.

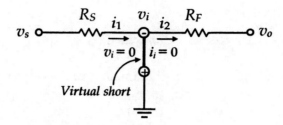

Figure 11-20: Virtual equivalent circuit

Because $i_i = 0$, then $i_1 = i_2$, where

$$i_1 = \frac{v_s - v_i}{R_S} = \frac{v_s}{R_S} \tag{11-11}$$

$$i_2 = \frac{v_i - v_o}{R_F} = \frac{-v_o}{R_F} \tag{11-12}$$

Equating the above two equations and solving for v_o, we obtain the following:

$$v_o = -\frac{R_F}{R_S} v_s \tag{11-13}$$

For $R_S = 1$ kΩ and $R_F = 100$ kΩ, the voltage gain is

$$A_v = \frac{v_o}{v_s} = -\frac{R_F}{R_S} = -100$$

Input resistance:

Since the inverting terminal is at virtual ground, the input resistance seen by the signal source is simply R_S. Thus,

$$R_{in} = R_S \tag{11-14}$$

Output resistance:

Although the output resistance of the 741 op-amp is 75 Ω, because of the negative feedback that will be demonstrated in the following chapter, the output resistance of the circuit seen by an external load is a few mΩ, approximately zero

$$R_o \cong 0 \tag{11-15}$$

Bandwidth:

The *Gain-Bandwidth Product* (*GBW*) of any op-amp, which is also referred to as the *unity gain bandwidth*, is given in the data sheets. For 741 op-amps, it is 1 MHz, from which the bandwidth of the closed-loop circuit can be determined as follows:

$$BW = \frac{GBW}{|A_v|} \tag{11-16}$$

Chapter 11

Hence, the bandwidth of the above inverting amplifier is 10 kHz.
Note that because the op-amp can amplify both DC and AC signals, there is no lower cutoff frequency; hence, the bandwidth extends from zero Hz to the upper cutoff frequency f_H.

Example 11-5  ANALYSIS

Determine the gain, input resistance, and output resistance of the inverting amplifier of Figure 11-21, as shown below:

Figure 11-21: Inverting amplifier

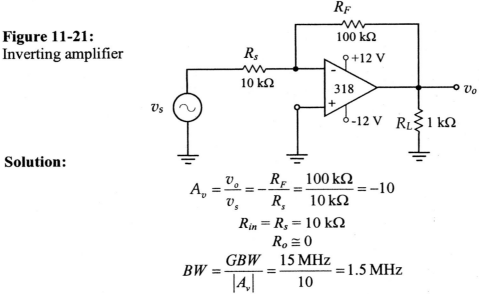

Solution:

$$A_v = \frac{v_o}{v_s} = -\frac{R_F}{R_s} = -\frac{100\,k\Omega}{10\,k\Omega} = -10$$

$$R_{in} = R_s = 10\,k\Omega$$

$$R_o \cong 0$$

$$BW = \frac{GBW}{|A_v|} = \frac{15\,MHz}{10} = 1.5\,MHz$$

Let us now test and verify the above solution with simulation. The results of simulation with Electronics Workbench are exhibited in Figures 11-22 through 11-25 below:

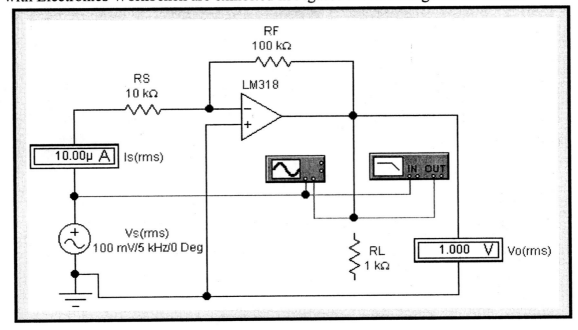

Figure 11-22: Circuit setup with simulation results of Example 11-5 with **no load** created in Electronics Workbench

467

Chapter 11

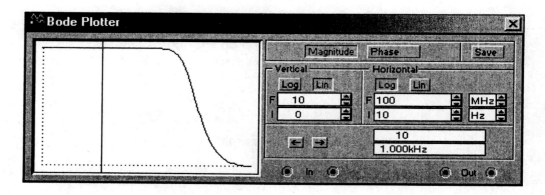

Figure 11-23: Electronics Workbench Bode plotter showing frequency response plot and midband voltage gain for the inverting amplifier of Example 11-5

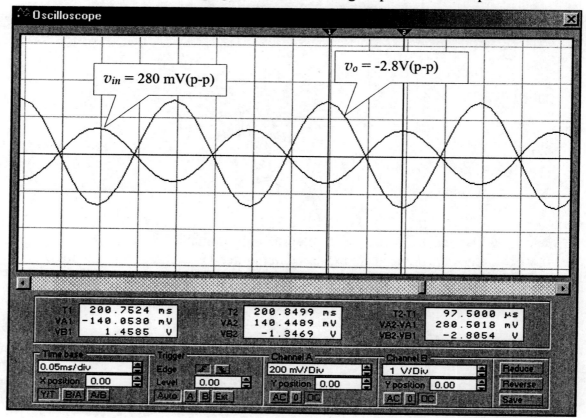

Figure 11-24: Electronics Workbench oscilloscope showing the input and output signals and phase relationship of the inverting amplifier of Example 11-5

As exhibited on the Electronics Workbench oscilloscope of Figure 11-24 above, the output is an amplified but inverted version of the input signal. The magnitude voltage gain of 10 can be viewed directly on the Bode plotter of Figure 11-23.

Chapter 11

Input resistance:

The input resistance may be determined from the DMM readings of Figure 11-22 or Figure 11-25, as follows:

$$R_i = \frac{v_s}{i_s} = \frac{100 \text{ mV}}{10 \text{ μA}} = 10 \text{ kΩ}$$

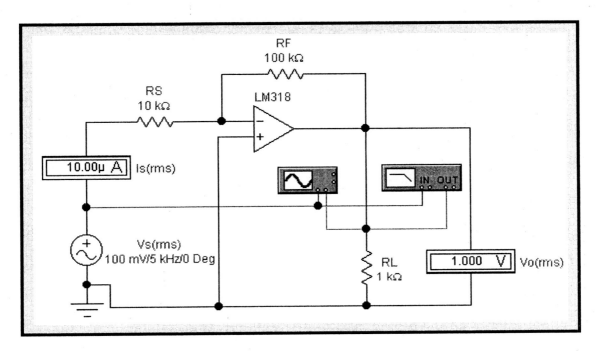

Figure 11-25: Circuit setup with the simulation results of Example 11-5 **with load** created in Electronics Workbench

Output resistance:

The output resistance is determined from the DMM readings of Figures 11-22 and 11-25 as follows:

$$R_o = \frac{v_{o(NL)} - v_{o(WL)}}{v_{o(WL)}} R_L = \frac{1 \text{ V} - 1 \text{ V}}{1 \text{ V}} 1 \text{ kΩ} = 0 \text{ Ω}$$

As you can see, the simulation results for the gain, input resistance, and output resistance are identical to the theoretically expected results of −10, 10 kΩ, and 0 Ω, respectively.

Practice Problem 11-5 *Inverting Amplifier* **DESIGN**

Design an inverting amplifier with a gain of $A_v = -50$ and input resistance $R_i \geq 1$ kΩ. Test and verify the voltage gain and input and output resistances by simulation.

The inverting amplifier may also be viewed as a *current-to-voltage converter*. Let us reconsider the circuit of the inverting amplifier and its output equation.

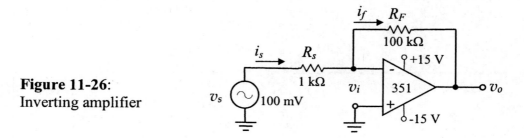

Figure 11-26: Inverting amplifier

The output voltage can be determined with Equation 11-13, as follows:

$$v_o = -\frac{R_F}{R_S} v_S = -\frac{100\,\text{k}\Omega}{1\,\text{k}\Omega}(100\,\text{mV}) = -10\,\text{V}$$

The above equation may also be rewritten as follows, which yields the same output voltage.

$$v_o = -\frac{v_S}{R_S} R_F = -i_S R_F \tag{11-17}$$

$$v_o = -i_S R_F = -(0.1\,\text{mA})(100\,\text{k}\Omega) = -10\,\text{V}$$

Note that because of the virtual short across the two input terminals of the op-amp, $v_i = 0$, and $i_i = 0$. Hence, $i_s = i_f$.

Now, instead of a voltage source v_s and a series resistor R_s, a current source may be used to supply the input current i_s, as shown below:

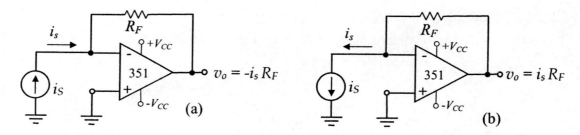

Figure 11-27: Op-amp utilized as *current-to-voltage converter*

With an input current $i_S = 10$ mA and $R_F = 100$ kΩ, the output voltage $v_{o(a)} = -1$ V, and $v_{o(b)} = +1$ V. Since the output resistance of the op-amp with negative feedback is approximately zero, the output voltage with and without load will be the same.

Because the output voltage v_o is controlled by the input current i_s, this configuration is also referred to as *current-controlled voltage source*.

11.5.2 Non-Inverting Amplifier

A schematic diagram of a non-inverting amplifier configuration is shown in Figure 11-28. Note that in closed-loop configurations with negative feedback (inverting or non-inverting) both R_F and R_S are always tied to the inverting terminal.

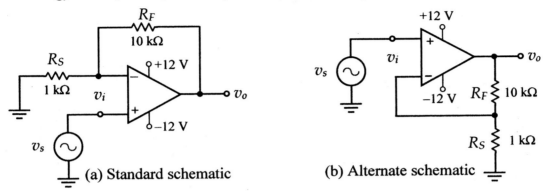

(a) Standard schematic (b) Alternate schematic

Figure 11-28: Non-inverting amplifier

Again, to analyze the above circuit, we will draw a virtual equivalent circuit as follows:

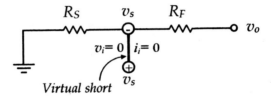

Figure 11-29: Virtual equivalent circuit

Applying the voltage-divider rule from the output to ground through R_F and R_S and solving for v_s results in the following:

$$v_s = v_o \frac{R_S}{R_F + R_S} \tag{11-18}$$

$$v_o = v_s \frac{R_F + R_S}{R_S} = v_s \left(\frac{R_F}{R_S} + 1 \right) \tag{11-19}$$

$$A_v = \frac{v_o}{v_s} = \left(\frac{R_F}{R_S} + 1 \right) \tag{11-20}$$

Input resistance:

Because of the negative feedback, as will be demonstrated in the next chapter, the input resistance of the non-inverting amplifier is in the GΩ range, which is practically infinite.

$$R_{in} \cong \infty \tag{11-21}$$

Output resistance:

The output resistance of the non-inverting amplifier is similar to that of the inverting amplifier; that is,

$$R_o \cong 0$$

Bandwidth:

The bandwidth of the non-inverting amplifier is also determined by Equation 11-16, as follows:

$$BW = \frac{GBW}{|A_v|}$$

11.5.3 Unity-Gain Buffer

The unity-gain buffer is a version of the non-inverter in which R_F is replaced by a short circuit, eliminating the need for R_s.

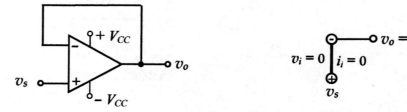

Figure 11-30(a): Unity-gain buffer

Figure 11-30(b): Virtual equivalent circuit

$$A_v = \frac{v_o}{v_s} = +1 \tag{11-22}$$

Input and output resistances of the unity-gain buffer are the same as those of the non-inverting amplifier

$$R_{in} \cong \infty$$
$$R_o \cong 0$$
$$BW = GBW$$

Because of the almost infinite input resistance and nearly zero output resistance, the unity-gain buffer configured with op-amp is nearly an ideal buffer, which can be utilized to isolate signal sources and loads from the main circuits. An ideal buffer behaves like a short circuit for voltage and like an open-circuit for current.

Like the inverting amplifier, the non-inverting amplifier and the unity gain buffer may also be utilized as *current-to-voltage converter or transresistance amplifier* with the addition of a resistor R, as shown in Figure 11.31.

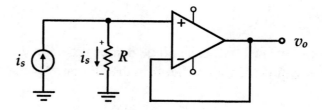

Figure 11-31:
Unity-gain buffer utilized as *current-to-voltage converter*

$$v_o = v^+ = Ri_s \qquad (11\text{-}23)$$

Example 11-6 — Non-Inv. Amplifier — ANALYSIS

Determine the gain in dB, input resistance, output resistance, and the bandwidth for the non-inverting amplifier of Figure 11-32.

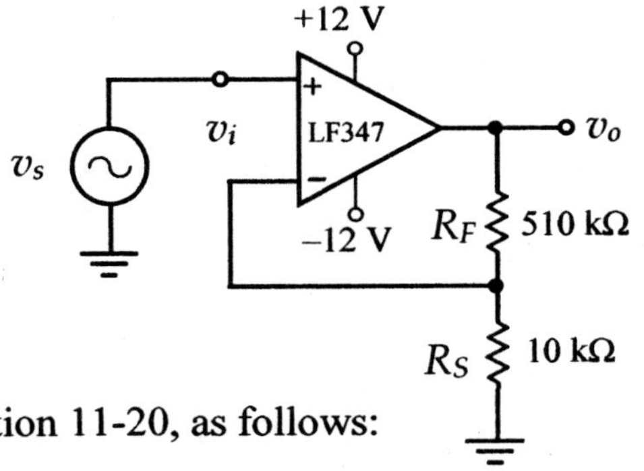

Figure 11-32:
Non-inverting amplifier

Solution:

The voltage gain is determined from Equation 11-20, as follows:

$$A_v = \frac{v_o}{v_s} = \frac{R_F}{R_S} + 1 = \frac{510\ \text{k}\Omega}{10\ \text{k}\Omega} + 1 = 52$$

$$A_v = 20 \log (52) = 34.32\ \text{dB}$$

$$R_{in} \cong \infty$$
$$R_o \cong 0$$

$$BW = \frac{GBW}{A_v} = \frac{4\ \text{MHz}}{52} = 77\ \text{kHz}$$

Example 11-7 — Non-Inv. Amplifier — ANALYSIS

Determine the gain, input resistance, output resistance, and bandwidth for the non-inverting amplifier of Figure 11-33.

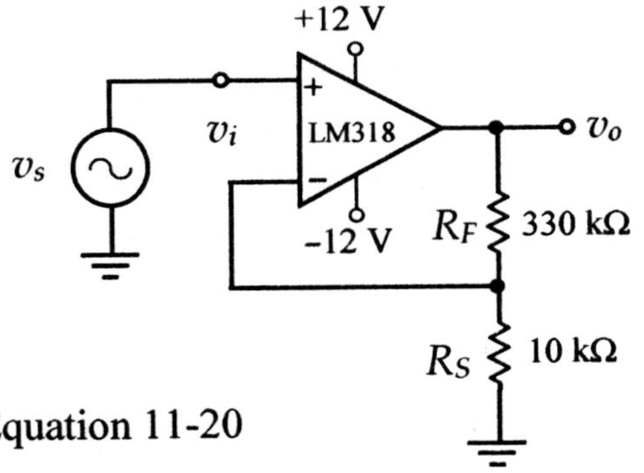

Figure 11-33:
Non-inverting amplifier

Solution:

Again, the voltage gain is determined by Equation 11-20

$$A_v = \frac{v_o}{v_s} = \frac{R_F}{R_S} + 1 = \frac{330\ \text{k}\Omega}{10\ \text{k}\Omega} + 1 = 34$$

$$A_v = 20 \log (34) = 30.63\ \text{dB}$$

$$R_{in} \cong \infty$$
$$R_o \cong 0$$

$$BW = \frac{GBW}{A_v} = \frac{15\,\text{MHz}}{34} = 441\text{ kHz}$$

Example 11-8 — Non-Inv. Amplifier — DESIGN

Let us now design a non-inverting amplifier for a gain of 11, then test and verify the gain, input resistance, and output resistance by simulation.

Solution:

Let $R_F = 100\text{ k}\Omega$ and $R_S = 10\text{ k}\Omega$. Hence, $A_v = R_F/R_S + 1 = 11$
The expected input and output resistances are $R_i \cong \infty\ \Omega$, and $R_o \cong 0\ \Omega$.
The results of simulation are exhibited in Figures 11-34 through 11-36 below:

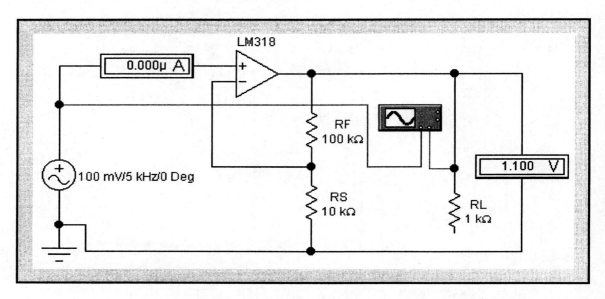

Figure 11-34: Circuit setup and simulation result of non-inverting amplifier with **no load**

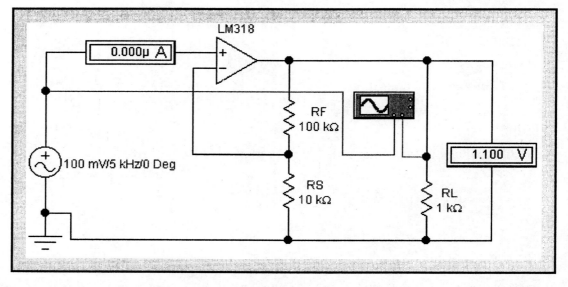

Figure 11-35: Circuit setup and simulation result of non-inverting amplifier **with load** created in Electronics Workbench

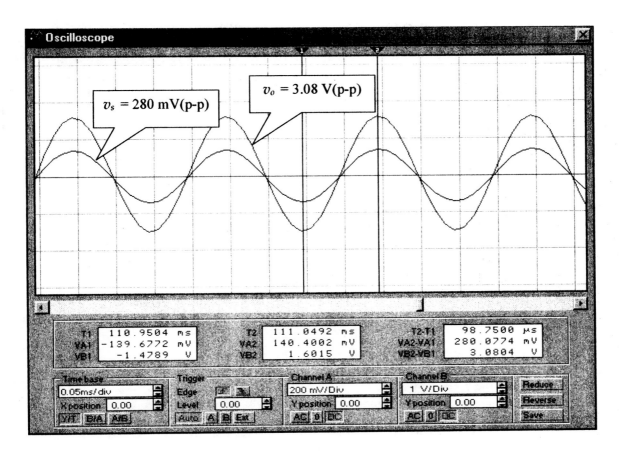

Figure 11-36: Electronics Workbench oscilloscope showing the input and output waveforms of the non-inverting amplifier of Example 11-8

As exhibited on the EWB oscilloscope of Figure 11-36, the output is an amplified version of, and in-phase with, the input signal. The voltage gain of 11 can be verified from the oscilloscope display and from the DMM readings of Figures 11-34 and 11-35.

Input resistance:

The input resistance may be determined from the DMM readings of Figure 11-34 or Figure 11-35, as follows:

$$R_i = \frac{v_s}{i_s} = \frac{100\,\text{mV}}{0\,\mu\text{A}} = \infty$$

Output resistance:

The output resistance is determined from the DMM readings of Figures 11-34 and 11-35 and from Equation 6-13, as follows:

$$R_o = \frac{v_{o(NL)} - v_{o(WL)}}{v_{o(WL)}} R_L = \frac{1.1\,\text{V} - 1.1\,\text{V}}{1.1\,\text{V}} \times 1\,\text{k}\Omega = 0$$

11.5.4 The Compensation Resistor

Recall that the two inputs of an op-amp are the two inputs of the first stage diff-amp. To help visualize the balanced operation required from an ideal diff-amp, the schematic of a basic diff-amp is shown below:

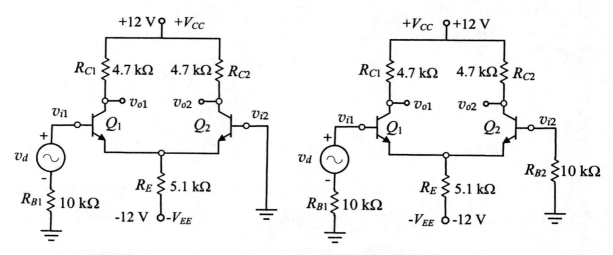

Figure 11-37(a):
A basic diff-amp (unbalanced)

Figure 11-37(b):
A compensated (balanced) basic diff-amp

If a resistance R_{B1} is introduced at the base side of Q_1, a resistance R_{B2} equal to R_{B1} must be introduced at the base of Q_2 in order to maintain the balance and symmetry for DC biasing of both transistors, as shown in Figure 11-37(b). Consider the inverting amplifier of Figure 11-19, which is repeated below in Figure 11-38 for convenience.

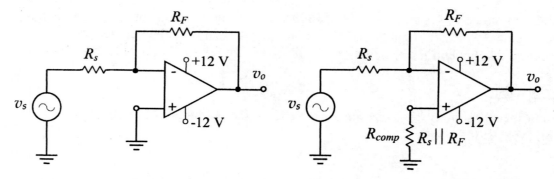

Figure 11-38(a): Inverting amplifier

Figure 11-38(b): Inverting amplifier with compensation resistor (R_{comp})

Note that the addition of the compensation resistor will only help minimize the DC offset errors and will not affect the AC operation at all. The compensation resistor may be utilized in applications where the DC offset at the output is of great importance.

Example 11-9 — Inverting Amplifier (ANALYSIS)

Determine the minimum and maximum gain, bandwidth, and input and output resistances for the amplifier of Figure 11-39.

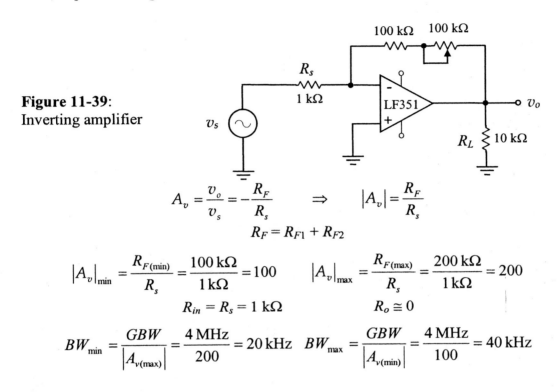

Figure 11-39: Inverting amplifier

$$A_v = \frac{v_o}{v_s} = -\frac{R_F}{R_s} \quad \Rightarrow \quad |A_v| = \frac{R_F}{R_s}$$

$$R_F = R_{F1} + R_{F2}$$

$$|A_v|_{min} = \frac{R_{F(min)}}{R_s} = \frac{100\ k\Omega}{1\ k\Omega} = 100 \qquad |A_v|_{max} = \frac{R_{F(max)}}{R_s} = \frac{200\ k\Omega}{1\ k\Omega} = 200$$

$$R_{in} = R_s = 1\ k\Omega \qquad\qquad R_o \cong 0$$

$$BW_{min} = \frac{GBW}{|A_{v(max)}|} = \frac{4\ MHz}{200} = 20\ kHz \qquad BW_{max} = \frac{GBW}{|A_{v(min)}|} = \frac{4\ MHz}{100} = 40\ kHz$$

Example 11-10 — Inverting Amplifier (DESIGN)

Design an inverting amplifier with an adjustable gain from 34 dB to 48 dB and with an input resistance $R_{in} \geq 1\ k\Omega$.

Solution:
$34\ dB = 20\ \log(A_{v(min)}) \quad \Rightarrow \quad 1.7 = \log(A_{v(min)}) \quad \Rightarrow \quad A_{v(min)} = 10^{1.7} = 50.1$
$48\ dB = 20\ \log(A_{v(max)}) \quad \Rightarrow \quad 2.4 = \log(A_{v(max)}) \quad \Rightarrow \quad A_{v(max)} = 10^{2.4} = 251$

Let $R_s = 1\ k\Omega$
$R_{F(min)} = A_{v(min)}\ R_s = 50.1 \times 1\ k\Omega = 50.1\ k\Omega$
$R_{F(max)} = A_{v(max)}\ R_s = 251 \times 1\ k\Omega = 251\ k\Omega$
$R_F = R_{F1} + R_{F2} = 51\ k\Omega + 200\ k\Omega$ (pot.)

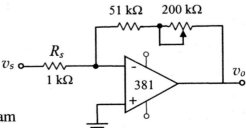

Figure 11-40: Circuit diagram

Example 11-11 — Non-Inv. Amplifier — DESIGN

Design an amplifier with an adjustable gain from 0 dB to 40 dB, and with an input resistance $R_{in} \geq 1$ MΩ.

Solution:
The specified input resistance $R_{in} \geq 1$ MΩ can only be achieved with a non-inverting amplifier. Hence, our design will be a non-inverter.

$0 \text{ dB} = 20 \log(A_{v(min)}) \quad \Rightarrow \quad 0 = \log(A_{v(min)}) \quad \Rightarrow \quad A_{v(min)} = 10^0 = 1$

$40 \text{ dB} = 20 \log(A_{v(max)}) \quad \Rightarrow \quad 2 = \log(A_{v(max)}) \quad \Rightarrow \quad A_{v(max)} = 10^2 = 100$

Let $R_S = 1$ kΩ $\quad R_F = A_{v(max)} R_S = 100 \times 1 \text{ k}\Omega = 100$ kΩ (pot)

$R_{in} \cong \infty$
$R_o \cong 0$

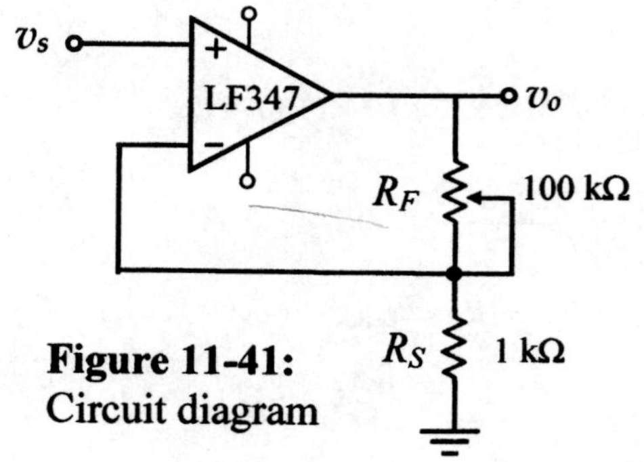

Figure 11-41: Circuit diagram

Practice Problem 11-6 — Non-Inv. Amplifier — ANALYSIS

Determine the output expression for v_o, the voltage gain, input resistance, and output resistance for the amplifier of Figure 11-42.

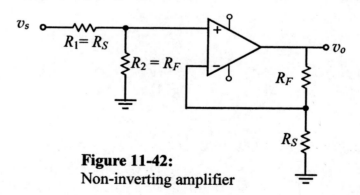

Figure 11-42: Non-inverting amplifier

Answers: $A_v = \dfrac{v_o}{v_s} = \dfrac{R_F}{R_S} \qquad v_o = v_s \dfrac{R_F}{R_S} \qquad R_{in} = R_1 + R_2 = R_S + R_F \qquad R_o \cong 0$

11.6 SUMMING AMPLIFIERS

A summing amplifier, which may be inverting or non-inverting, is an amplifier with more than one input whose output voltage is the scaled or weighted sum of the input voltages.

11.6.1 Inverting Summer

The circuit diagram of a two-input inverting summer and its virtual equivalent are shown in Figures 11-43(a) and (b).

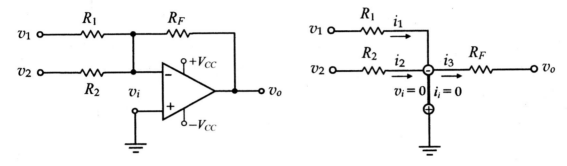

Figure 11-43(a): Inverting summer   **Figure 11-43(b):** Virtual equivalent circuit

Applying *KCL* at the inverting terminal node of Figure 11-43(b), we obtain

$$i_1 + i_2 = i_3 \tag{11-24}$$

$$i_1 = \frac{v_1 - v_i}{R_1} = \frac{v_1}{R_1} \tag{11-25}$$

$$i_2 = \frac{v_2 - v_i}{R_2} = \frac{v_2}{R_2} \tag{11-26}$$

$$i_3 = \frac{v_i - v_o}{R_F} = \frac{-v_o}{R_F} \tag{11-27}$$

$$\frac{-v_o}{R_F} = \frac{v_1}{R_1} + \frac{v_2}{R_2} \tag{11-28}$$

Solving the above equation for v_o, we obtain

$$v_o = -\left(\frac{R_F}{R_1}v_1 + \frac{R_F}{R_2}v_2\right) \tag{11-29}$$

For *n* inputs, it can be shown that the output equation has the following form:

$$v_o = -\left(\frac{R_F}{R_1}v_1 + \frac{R_F}{R_2}v_2 + \cdots + \frac{R_F}{R_n}v_n\right) \tag{11-30}$$

For equal channel resistances ($R_1 = R_n = R$), the above equation takes the following form:

$$v_o = -\frac{R_F}{R}(v_1 + v_2 + \cdots + v_n) \tag{11-31}$$

Example 11-12 — Inverting Summer — ANALYSIS

Determine and plot the output signal for the summer of Figure 11-44 with the given inputs v_1 and v_2 below:

$v_1 = -0.5$ V DC
$v_2 = 0.5$ V(p-p) sinusoid

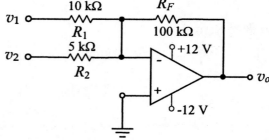

Figure 11-44: Inverting summer

Solution:

$$v_o = -\left[\frac{R_F}{R_1}v_1 + \frac{R_F}{R_2}v_2\right]$$

$$v_o = -\left[\frac{100\,k\Omega}{10\,k\Omega}(-0.5\,V_{DC}) + \frac{100\,k\Omega}{5\,k\Omega}0.5\,V_{(p-p)}\right] = 5\,V_{DC} - 10\,V_{(p-p)}$$

The negative sign at the output signifies inversion with respect to the input

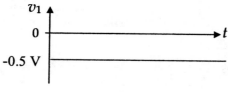

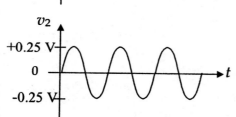

Figure 11-45: Input signals

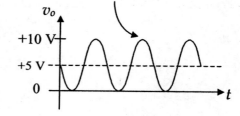

Figure 11-46: Output signal

Practice Problem 11-7 — Inverting Summer — ANALYSIS

Determine and plot the output signal for the summer of Figure 11-47 with the given inputs v_1 and v_2 below:

$v_1 = 0.25$ V DC
$v_2 = 0.25$ V(p-p) sinusoid

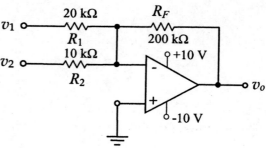

Answer:

$v_o = -(2.5\,V_{DC} + 5\,V_{(p-p)})$, plot not shown

Figure 11-47: Inverting summer

11.6.2 Non-Inverting Summer

The circuit diagram of a two-input non-inverting summer is shown in Figure 11-48.

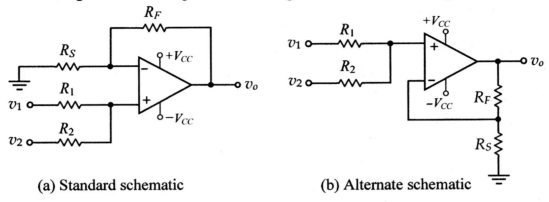

(a) Standard schematic   (b) Alternate schematic

Figure 11-48: Non-inverting summer

To determine the output expression for the above two-input non-inverting summer, we will simply use the superposition theorem. That is, we will apply one input at a time and determine the corresponding output, then combine the results.

<u>Output due to v_1 only:</u>

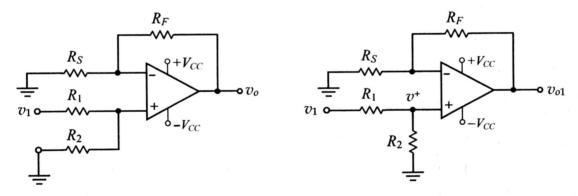

Figure 11-49(a): Non-inverting summer with input v_2 grounded

Figure 11-49(b): Figure 11-49(a) redrawn

The amount of voltage reaching the non-inverting terminal v^+ can be determined by the voltage-divider rule, as follows:

$$v^+ = v_1 \frac{R_2}{R_1 + R_2} \qquad (11\text{-}32)$$

Having determined the voltage v^+ at the non-inverting terminal, the output voltage v_{o1} can now be determined as follows:

$$v_{o1} = v^+\left(1 + \frac{R_F}{R_S}\right) = v_1 \frac{R_2}{R_1 + R_2}\left(1 + \frac{R_F}{R_S}\right) \qquad (11\text{-}33)$$

Chapter 11

Output due to v_2 only:

In a similar manner, the output voltage v_{o2} due to v_2 only can be determined to be

$$v_{o2} = v_2 \frac{R_1}{R_1 + R_2}\left(1 + \frac{R_F}{R_S}\right) \tag{11-34}$$

Applying the superposition theorem and combining the above two individual results, we obtain

$$v_o = v_{o1} + v_{o1} = v_1 \frac{R_2}{R_1 + R_2}\left(1 + \frac{R_F}{R_S}\right) + v_2 \frac{R_1}{R_1 + R_2}\left(1 + \frac{R_F}{R_S}\right) \tag{11-35}$$

$$v_o = \left(1 + \frac{R_F}{R_S}\right)\left(v_1 \frac{R_2}{R_1 + R_2} + v_2 \frac{R_1}{R_1 + R_2}\right) \tag{11-36}$$

For n inputs and with $R_1 = R_2 = R_n = R$, the above equation takes the following form:

$$v_o = \left(1 + \frac{R_F}{R_S}\right)\left(\frac{v_1 + v_2 + \ldots + v_n}{n}\right) \tag{11-37}$$

Example 11-13 (Non-inv. Summer) ANALYSIS

Determine the output voltage v_o of the summing amplifier of Figure 11-50 with the input v_1 and v_2 given below:

$v_1 = 0.1$ V DC
$v_2 = 0.05\sin\omega t$ (V)

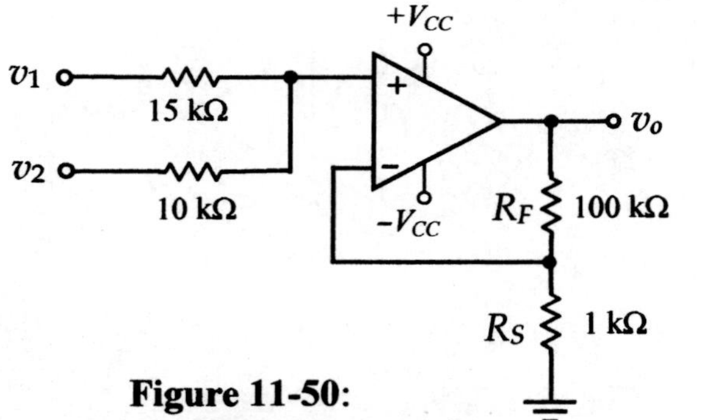

Solution:

$$v_o = \left(1 + \frac{R_F}{R_s}\right)\left(v_1 \frac{R_2}{R_1 + R_2} + v_2 \frac{R_1}{R_1 + R_2}\right)$$

$$v_o = \left(1 + \frac{100}{1}\right)\left(0.1 \frac{10\,k\Omega}{25\,k\Omega} + 0.05\sin\omega t \frac{15\,k\Omega}{25\,k\Omega}\right)$$

$$v_o = 4 + 3\sin\omega t \quad (V)$$

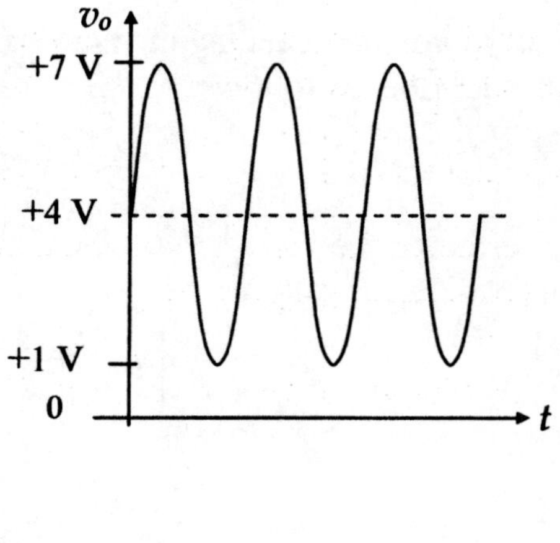

Figure 11-50: Non-inverting summer

Figure 11-51: Output waveform

482

11.6.3 Applications of the Op-Amp Buffer and Summing Amplifier in Digital Communication Circuits

Nowadays, the transmission, reception, and processing of information is mainly digital. That is, although the information might originally be analog such as audio or video, it is digitized before being processed by a computer and/or transmitted over a local area network or the public telephone network.

The first step in digitizing some analog signal is to sample it and then convert each sample to an *n*-bit code. As shown in Figure 11-52, the sampling circuit comprises an EMOS, a capacitor, and two op-amp buffers. The role of the EMOS is to sample the analog signal by turning on momentarily. The capacitor is to charge up immediately and hold on to the sample value when the EMOS turns off. The two op-amp buffers, in addition to isolating the input and output signals from the EMOS and from each other, provide almost zero resistance across the capacitor through the output of the input buffer when the MOSFET turns on, and as a result, the capacitor charges up immediately ($\tau = RC = R_o C \cong 0$). When the EMOS turns off, the capacitor holds on to the charge because the only resistance appearing across the capacitor is the input resistance of the output buffer, which is approximately infinite ($\tau = RC = R_{in} C \cong \infty$).

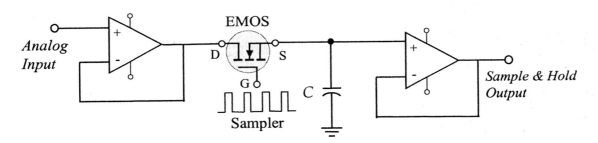

Figure 11-52: Sample-and-hold circuit

EMOS acts like an open switch when the gate voltage is zero, and like a closed switch when the gate voltage exceeds a certain *threshold voltage* (V_t). Hence, the sampler pulse must have amplitude larger than V_t.

Another advantage of sampling is that several analog signals can be sampled in sequence and then multiplexed over a single transmission medium. For example, the T1 line of the telephone network, which is simply a twisted wire pair, is a multiplex of 24 telephone channels. That is, each of the 24 lines is sampled sequentially and outputted onto T1 line, one sample at a time. This kind of multiplexing is generally referred to as *time-division multiplexing* (TDM), because the frame time is divided into 24 time slots to allow transmission of one sample from each channel during each frame. The circuit of a 4-channel TDM system is shown in Figure 11-53, where the outputs of sample-and-hold circuits are fed to a summing amplifier. Note that at any given time, only one sample appears at the output of the summing amplifier from the corresponding channel.

Chapter 11

By connecting the FET gate terminals (G_1, G_2, G_3, G_4) to a suitable counter type circuit (ring counter), MOSFETs can be turned on in sequence in accordance with a master clock.

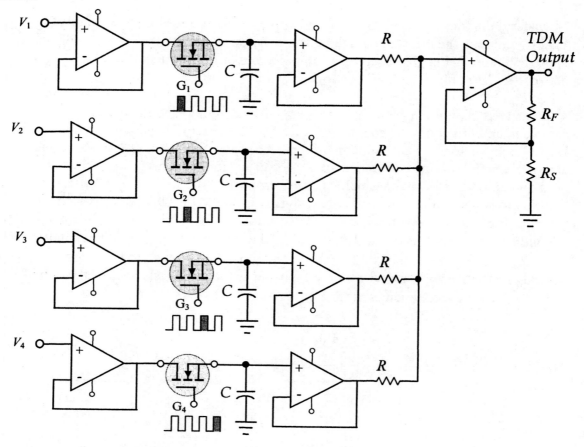

Figure 11-53: A 4-channel TDM circuit with sample-and-hold

$$v_o = \left(1 + \frac{R_F}{R_s}\right)\left(\frac{v_1 + v_2 + ...v_n}{n}\right)$$

Since only one of the n signals appears at the output at a time, gain of the summing amplifier ($1 + R_F/R_S$) can be adjusted to be larger than or equal to n in the denominator.

To help demonstrate the sample-and-hold signal and verify the underlying theory, let us simulate the circuit of Figure 11-52 and observe its output. Simulation results are displayed in Figures 11-54 and 11-55, as follows:

Chapter 11

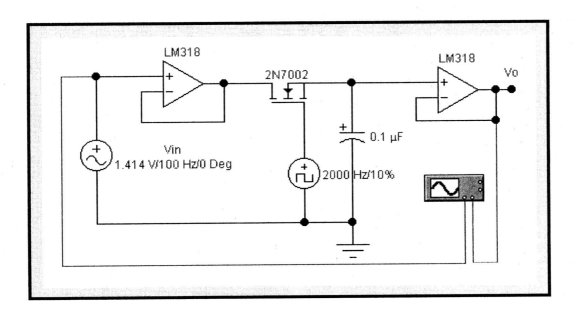

Figure 11-54: Sample-and-hold circuit simulation created in Electronics Workbench

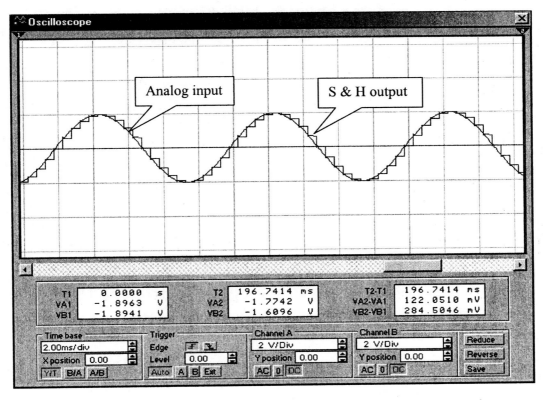

Figure 11-55: Electronics Workbench oscilloscope showing the analog input and sample-and-hold output signals

11.7 DIFFERENCE AMPLIFIER

The circuit diagram of a difference amplifier is shown in Figure 11-56 below:
For a balanced operation and to minimize the DC offset errors, the DC resistance external to both inputs must be equal. Hence, let $R_1 = R_2 = R$ and $R_3 = R_F$.

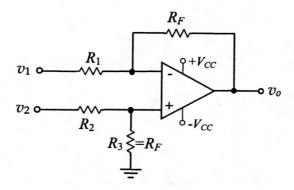

Figure 11-56: Difference amplifier

To analyze the above circuit and determine the output in terms of the two inputs and other circuit components, we will again make use of the superposition theorem.

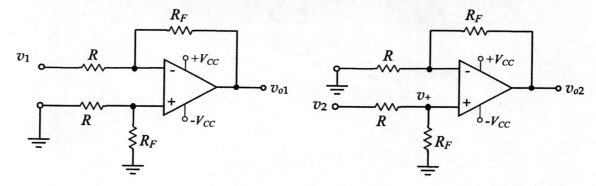

Figure 11-57(a): Difference amplifier with the non-inv. input grounded

Figure 11-57(b): Difference amplifier with the inverting input grounded

Grounding the non-inverting input changes the diff-amp circuit of Figure 11-56 into an inverting amplifier with the two resistors at the non-inverting terminal playing the role of the compensation resistor.

$$v_{o1} = -\frac{R_F}{R}v_1 \tag{11-38}$$

Grounding the inverting input changes the diff-amp circuit of Figure 11-56 into a non-inverting amplifier. However, only a portion of v_2 reaching the non-inverting terminal of the op-amp (v_+) will be amplified, as follows:

$$v_+ = v_2 \frac{R_F}{R_F + R} \tag{11-39}$$

$$v_{o2} = v_+ \left(1 + \frac{R_F}{R}\right) \tag{11-40}$$

$$v_{o2} = v_2 \frac{R_F}{R_F + R}\left(1 + \frac{R_F}{R}\right) = v_2 \frac{R_F}{R_F + R} \cdot \frac{R + R_F}{R} \tag{11-41}$$

$$v_{o2} = v_2 \frac{R_F}{R} \tag{11-42}$$

$$v_o = v_{o2} + v_{o1} = v_2 \frac{R_F}{R} - v_1 \frac{R_F}{R} \tag{11-43}$$

$$v_o = (v_2 - v_1)\frac{R_F}{R} = -(v_1 - v_2)\frac{R_F}{R} = -\frac{R_F}{R} v_d \tag{11-44}$$

Therefore, the difference of the two inputs will be amplified by the differential gain A_d of the difference amplifier, which is defined and determined as follows:

$$A_d = \frac{v_o}{v_d} = -\frac{R_F}{R} \tag{11-45}$$

Example 11-14 — Difference Amplifier — ANALYSIS

Determine the v_o at the output of the difference amplifier of Figure 11-58 below for an input signal $v_s = 25$ mV(p-p).

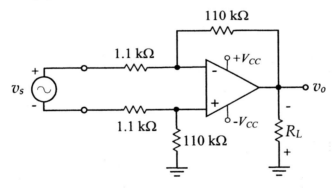

Figure 11-58: Difference amplifier

Solution:

$$v_o = (v_2 - v_1)\frac{R_F}{R} = -(v_1 - v_2)\frac{R_F}{R} = -v_s \frac{R_F}{R}$$

$$v_o = -(25\text{ mV})\frac{110\text{ k}\Omega}{1.1\text{ k}\Omega} = -2.5\text{ V(p-p)}$$

Note that the negative sign at the output signifies inversion with respect to the input.

11.7.1 Noise Suppression with Diff-Amp

Recall that an ideal diff-amp amplifies the difference signal v_d while suppressing the common signal v_c. With a practical diff-amp, however, the amount of suppression of the common signal is directly proportional to the *CMRR*. An op-amp utilized in diff-amp configuration makes use of the op-amp's *CMRR*, which is usually quite high. Consider the following differential mode and common mode connections:

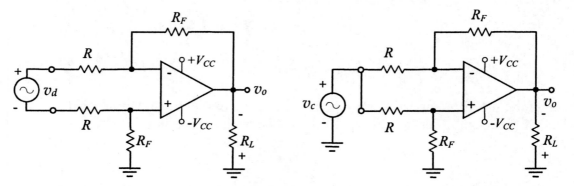

Figure 11-59(a): Differential mode   **Figure 11-59(b)**: Common mode

The differential mode gain (A_d) and common mode gain (A_c) are defined as follows:

$$A_d = \frac{v_o}{v_d} = -\frac{R_F}{R} \qquad v_d = v_1 - v_2 \qquad A_c = \frac{v_o}{v_c}$$

$$CMRR = \frac{A_d}{A_c} = -\frac{R_F}{R} \cdot \frac{v_c}{v_o} \qquad (11\text{-}46)$$

Solving for v_o in the above equation results in the following:

$$v_o = -\frac{R_F}{R} \cdot \frac{v_c}{CMRR} \qquad (11\text{-}47)$$

Hence, the common-mode signal v_c is first amplified by the gain of the diff-amp and then attenuated by the *CMRR* of the op-amp.

Let us now compare the noise suppression ability of an inverting amplifier to that of a difference amplifier with equal gain, as in the following example.

Chapter 11

Example 11-15 — Difference Amplifier — ANALYSIS

Determine the output signal plus noise for the following two configurations, assuming that both circuits are operating in a noisy environment with a noise voltage $v_n = 10$ mV.

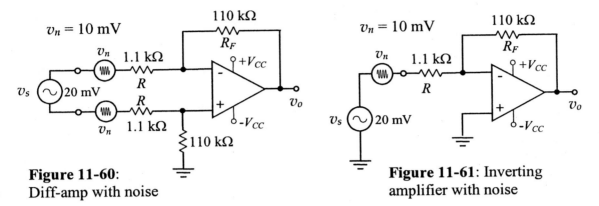

Figure 11-60: Diff-amp with noise

Figure 11-61: Inverting amplifier with noise

Difference amplifier:

The noise voltage v_n that is present in the operating environment couples equally to both inputs of the diff-amp as illustrated in Figure 11-60. Hence, v_n is the common signal to both inputs and will be attenuated by the *CMRR* of the op-amp according to Equation 11-47. The input signal v_s, however, which is the differential input, will be amplified by the differential gain of the diff-amp according to Equation 11-45.

$$v_o = -\frac{R_F}{R}v_s - \frac{R_F}{R} \cdot \frac{v_n}{CMRR} \qquad (11\text{-}48)$$

$$v_o = -\frac{R_F}{R}\left(v_s + \frac{v_n}{CMRR}\right) \qquad (11\text{-}49)$$

Assuming an average *CMRR* of 80 dB for the 741 op-amp, which equals 10^4 when converted to linear scale, which is the *CMRR* value to be used in the above equation:

$$|v_o| = \frac{110\,k\Omega}{1.1\,k\Omega}\left(20\,mV(\text{signal}) + \frac{10\,mV(\text{noise})}{10^4}\right) = 2\,V(\text{signal}) + 0.1\,mV(\text{noise})$$

Inverting amplifier:

The noise voltage present in the operating environment is coupled to the only input of the amplifier, which appears in series with the signal source as illustrated in Figure 11-61. Hence, the noise will be amplified equally with the signal, without attenuation.

$$v_o = -\frac{R_F}{R}(v_s + v_n) \qquad (11\text{-}50)$$

$$|v_o| = \frac{110\,k\Omega}{1.1\,k\Omega}[20\,mV(\text{signal}) + 10\,mV(\text{noise})] = 2\,V(\text{signal}) + 1\,V(\text{noise})$$

It is, therefore, advantageous to employ the diff-amp configuration when operating in a noisy environment, which amplifies the signal according to the gain of the diff-amp, but attenuates the noise proportional to the *CMRR* of the op-amp.

11.7.2 Buffered Diff-Amp and the Instrumentation Amplifier

The difference amplifier of Figure 11-60 provides neither equal nor adequately high input channel resistances. Nearly infinite and equal input resistances can be achieved with buffered inputs, as shown in Figure 11-62 below.

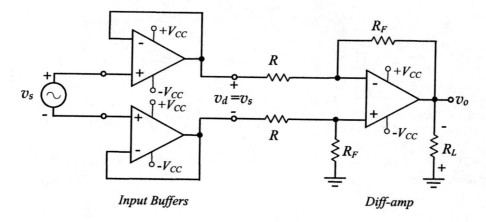

Figure 11-62: Buffered diff-amp

The signal source is completely isolated from the diff-amp and the resistances seen by the source at both input terminals are nearly infinite. However, the total differential input signal v_s is present across the input terminals of the diff-amp, which is amplified by the differential gain A_d of the diff-amp.

$$v_o = -\frac{R_F}{R} v_s \qquad (11\text{-}51)$$

Note that the above equation solves for the output due to the differential input v_s only. The possible presence of noise is not accounted for in this equation, in which case a complete solution can be obtained with Equation 11-48.

Practice Problem 11-8 — Buffered Diff-amp — ANALYSIS

Determine the output signal plus noise for the amplifier of Figure 11-62 assuming that $v_s = 50$ mV$_{rms}$ and a noise voltage $v_n = 25$ mV$_{rms}$ is also present. Also determine the input resistance seen by the signal source and the output resistance seen by the load. Other circuit components are as follows: $R_F = 1$ MΩ, $R = 10$ kΩ, and $CMRR = 74$ dB.

Answers: $|v_o| = 5$ V$_{signal}$ + 0.5 mV$_{noise}$ $R_i = \infty$, $R_o = 0$.

The modified version of the buffered diff-amp with six equal resistors R and one variable resistor R_A, which yields an adjustable gain, is referred to as the *instrumentation amplifier*, as depicted in Figure 11-63 below:

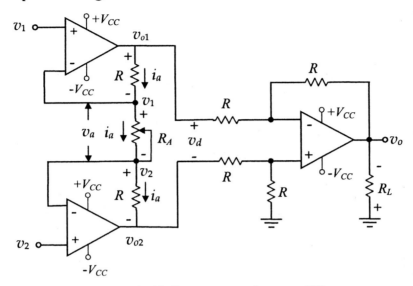

Figure 11-63: Instrumentation amplifier

Recall that because of the virtual short across the two input terminals of the op-amp with feedback, the voltage at the upper end of the resistor R_A equals v_1 and at the lower end equals v_2. Hence, assuming that $v_1 > v_2$, the voltage across R_A equals the difference between v_1 and v_2,

$$v_a = v_1 - v_2 \tag{11-52}$$

$$i_a = \frac{v_a}{R_A} = \frac{v_1 - v_2}{R_A} \tag{11-53}$$

Since no current flows into the input terminals of the op-amp, the current flows from the output terminal v_{o1} to the output terminal v_{o2}. Hence, the current through all three resistors equals i_a, and as a result, we obtain the following equations for v_{o1} and v_{o2}.

$$v_{o2} = v_2 - i_a R \tag{11-54}$$

$$v_{o1} = v_1 + i_a R \tag{11-55}$$

$$v_d = v_{o1} - v_{o2} = v_1 + i_a R - (v_2 - i_a R) \tag{11-56}$$

$$v_d = v_1 - v_2 + 2 i_a R = v_1 - v_2 + 2 \frac{v_1 - v_2}{R_A} R \tag{11-57}$$

Factoring out $(v_1 - v_2)$ yields the following:

$$v_d = (v_1 - v_2)\left(1 + 2\frac{R}{R_A}\right) \tag{11-58}$$

Since R_F and R of the diff-amp section are equal, and the diff-amp is an inverter, the gain of the diff-amp section equals -1, and the output v_o is the inverted version of v_d, as follows:

$$v_0 = -v_d = (v_2 - v_1)\left(1 + 2\frac{R}{R_A}\right) \tag{11-59}$$

$$A_d = \frac{v_0}{v_2 - v_1} = 1 + \frac{2R}{R_A} \tag{11-60}$$

Example 11-16 — Instrumentation Amplifier — ANALYSIS

Let us determine the minimum and maximum output signal if $R = 10\ k\Omega$ and R_A is a 200 Ω potentiometer in series with a 200 Ω resistor, and $v_2 - v_1 = 20\ mV(rms)$.

Solution:

$$v_{o(min)} = (20\ mV)\left(1 + 2\frac{10\ k\Omega}{400\ \Omega}\right) = 1.02\ V(rms)$$

$$v_{o(max)} = (20\ mV)\left(1 + 2\frac{10\ k\Omega}{200\ \Omega}\right) = 2.02\ V(rms)$$

Practice Problem 11-9 — Instrumentation Amplifier — ANALYSIS

Determine the output v_o for the instrumentation amplifier of Figure 11-64 below:

$v_1 = 10\ mV$
$v_2 = 50\ mV$
$R = 5.1\ k\Omega$

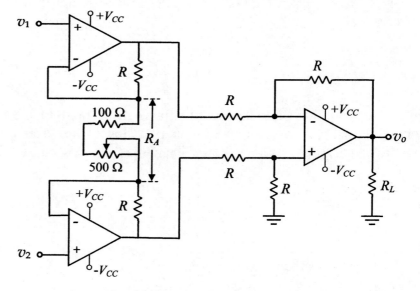

Figure 11-64: Instrumentation amplifier

Answers: $v_{o(min)} = 0.72\ V$, $v_{o(max)} = 4.12\ V$

Chapter 11

Section Summary — Op-Amp Analysis

Summary of Equations for the Analysis of the Basic Op-Amp Circuits

Open-Loop Operation

In open-loop operation, the op-amp is generally used as a comparator

if $v_+ > v_-$ ⇒ $v_o = +V_{sat}$ $+V_{sat} = +V_{CC} - 2\text{ V}$

if $v_- > v_+$ ⇒ $v_o = -V_{sat}$ $-V_{sat} = -V_{CC} + 2\text{ V}$

Closed-Loop Operation

The closed-loop operation of the op-amp involves negative feedback

Inverting Amplifier

$$A_v = \frac{v_o}{v_s} = -\frac{R_F}{R_S} \qquad v_o = -\frac{R_F}{R_S} v_s$$

$$R_i = R_s \qquad R_o \cong 0\ \Omega$$

Non-inverting Amplifier

$$A_v = \frac{v_o}{v_s} = \frac{R_F}{R_S} + 1 \qquad v_o = \left(\frac{R_F}{R_S} + 1\right) v_s$$

$$R_i \cong \infty\ \Omega \qquad R_o \cong 0\ \Omega$$

Inverting Summer

$$v_o = -\left(\frac{R_F}{R_1} v_1 + \frac{R_F}{R_2} v_2\right)$$

Non-inverting Summer

$$v_o = \left(\frac{R_F}{R_S} + 1\right)\left(\frac{R_2}{R_1 + R_2} v_1 + \frac{R_1}{R_1 + R_2} v_2\right)$$

For n inputs and with

$R_1 = R_2 = R_n = R$

$$v_o = -\frac{R_F}{R_S}(v_1 + v_2 + \dots + v_n)$$

For n inputs and with

$R_1 = R_2 = R_n = R$

$$v_o = \left(\frac{R_F}{R_S} + 1\right)\left(\frac{v_1 + v_2 + \dots + v_n}{n}\right)$$

Difference Amplifier

$$A_d = \frac{v_o}{v_d} = -\frac{R_F}{R} \qquad v_o = -\frac{R_F}{R} v_d$$

Instrumentation Amplifier

$$A_d = \frac{v_o}{v_d} = \frac{v_o}{v_2 - v_1} = \left(1 + \frac{2R}{R_A}\right)$$

Difference Amplifier with Noise

$$v_o = -\frac{R_F}{R}\left(v_s + \frac{v_n}{CMRR}\right)$$

Inverting Amplifier with Noise

$$v_o = -\frac{R_F}{R}(v_s + v_n)$$

Chapter 11

11.8 INTEGRATOR AND DIFFERENTIATOR

11.8.1 The Integrator

The circuit diagram of a basic op-amp integrator is shown in Figure 11-65 below:

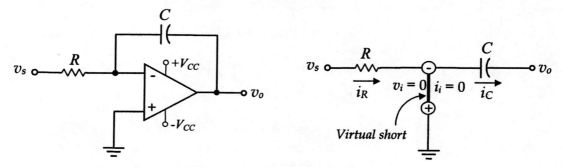

Figure 11-65: Basic integrator circuit   **Figure 11-66**: Virtual equivalent circuit

From the virtual equivalent circuit of Figure 11-66 we observe that i_R equals i_C, and since $v_i = 0$, i_R and i_C can be determined as follows:

$$i_R = \frac{v_s - v_i}{R} = \frac{v_s}{R} \qquad (11\text{-}61)$$

$$i_C = C\frac{dv_C}{dt} = C\frac{d(v_i - v_o)}{dt} = C\frac{d(-v_o)}{dt} = -C\frac{dv_o}{dt} \qquad (11\text{-}62)$$

Equating the above two equations and solving for v_o, we obtain the following:

$$\frac{v_s}{R} = -C\frac{dv_o}{dt} \qquad (11\text{-}63)$$

$$-\frac{v_s}{RC} = \frac{dv_o}{dt} \qquad (11\text{-}64)$$

$$v_o = -\frac{1}{RC}\int_0^t v_s\, dt + v_o(0) \qquad (11\text{-}65)$$

For a more accurate integration, the time-constant $\tau = RC$ of the integrator must be small compared to the period $T = 1/f_o$ of the input signal, at least ten times smaller.

$$RC \leq 0.1T = \frac{0.1}{f_o} = \frac{1}{10 f_o} \qquad (11\text{-}66)$$

$$R \leq \frac{0.1}{f_o C} = \frac{1}{10 f_o C} \qquad (11\text{-}67)$$

where f_o is the operating frequency.

As the frequency decreases, the resistance R can take a larger value. For a given operating frequency range, the maximum R can then be determined, as follows:

$$R_{max} = \frac{1}{10 f_{min} C} \qquad (11\text{-}68)$$

The magnitude gain of the basic integrator of Figure 11-65 is the ratio of the capacitive reactance X_C to the resistance R, that is:

$$A_v = -\frac{X_C}{R} = -\frac{1}{\omega_o CR} = -\frac{1}{2\pi f_o CR} \qquad (11\text{-}69)$$

The gain varies with frequency and will be very high at lower frequencies, approaching the open-loop gain at DC. To control the low-frequency response and to limit the DC gain to a modest level, a shunt resistor R_F may be added in parallel with the capacitor as shown in Figure 11-67 below:

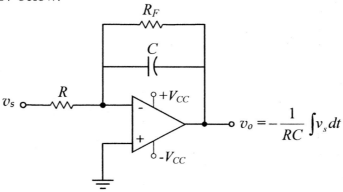

Figure 11-67: A practical integrator

$$\left|A_{v(max)}\right| = \frac{R_F}{R} \qquad (11\text{-}70)$$

As a trade-off to limiting the DC gain, the feedback resistor R_F introduces a low cutoff frequency f_L, below which integration does not take place.

$$f_L = \frac{1}{2\pi R_F C} \qquad (11\text{-}71)$$

For a more accurate integration, the operating frequency f_o is kept at least one decade above the f_L.

$$f_o \geq 10 f_L \qquad (11\text{-}72)$$

Combining Equations 11-67 and 11-72 results in the following range of operating frequency f_o for the input signal v_s:

$$10 f_L \leq f_o \leq \frac{1}{10 RC} \qquad (11\text{-}73)$$

Example 11-17 *Integrator* DESIGN

Let us design a practical integrator with a minimum operating frequency of about 5 kHz and with a maximum gain of about 30.

Solution:
We will begin the design process by selecting a non-electrolytic capacitor less than or equal to 0.1 µF. Let $C = 0.01$ µF.
Let $f_{o(min)} = 10 f_L \Rightarrow f_L = 0.1 f_{o(min)} = 500$ Hz

$$f_L = \frac{1}{2\pi R_F C} \Rightarrow R_F \geq \frac{1}{2\pi f_L C} = \frac{1}{2\pi \times 500 \text{ Hz} \times 0.01\,\mu\text{F}} = 32\text{ k}\Omega \quad \text{Use } R_F = 33\text{ k}\Omega$$

$$R_{max} = \frac{1}{2\pi f_{o(min)} C} = \frac{1}{2\pi \times 5\text{ kHz} \times 0.01\,\mu\text{F}} = 2\text{ k}\Omega$$

$$|A_{v(max)}| = \frac{R_F}{R} \Rightarrow R = \frac{R_F}{A_{v(max)}} = \frac{33\text{ k}\Omega}{30} = 1.1\text{ k}\Omega, \quad R = 1.1\text{ k}\Omega$$

$$RC = 11\,\mu\text{s} \ll T = 200\,\mu\text{s}$$

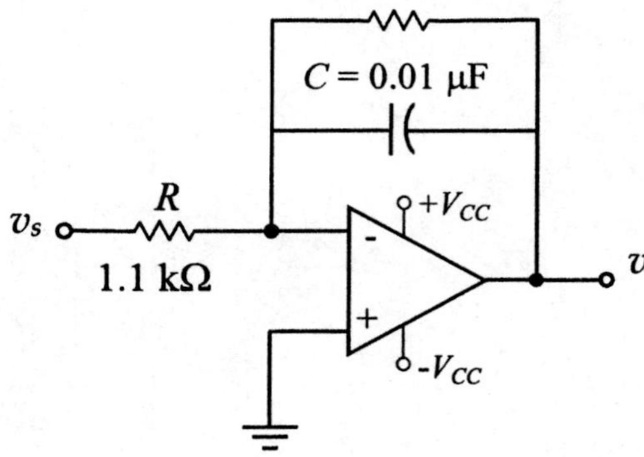

Figure 11-68:
The integrator circuit diagram with component values

Test Run and Design Verification:

The circuit setup for simulation and the results of simulation with Electronics Workbench are exhibited in Figures 11-69 through 11-71. A simple method for testing the accuracy of an integrator is to observe the response of the circuit to a sinusoidal waveform. Since the integral of a sine is negative cosine and the circuit is an inverter, the negative cosine becomes positive cosine; and thus, a sinusoidal waveform with 90° lead is expected at the output. If necessary, minor adjustments can then be made to the circuit components to achieve the 90° lead at the output. As exhibited on the EWB oscilloscope of Figure 11-70, the time difference between the two signals ($t_2 - t_1$) is 50 μs, which is a quarter of the 200 μs period T of the input signal. A quarter of the period T in time domain corresponds to 90° in the phase domain.

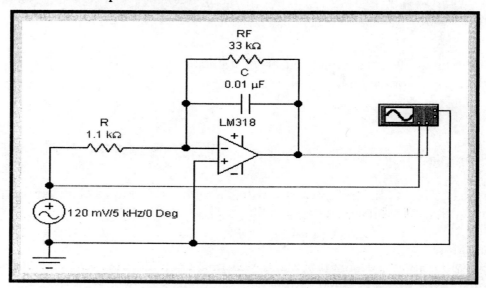

Figure 11-69: Circuit setup for simulation of the integrator circuit of Example 11-17 created in Electronics Workbench

Chapter 11

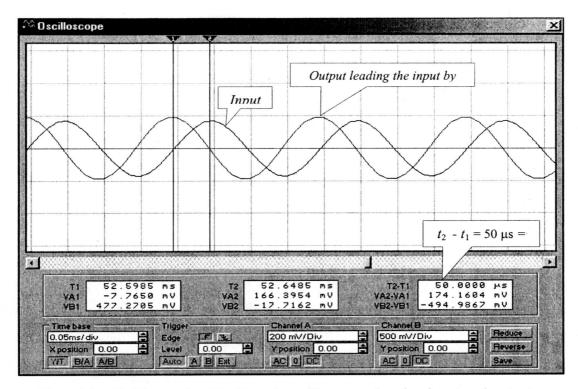

Figure 11-70: Electronics Workbench oscilloscope showing input and output waveforms for Example 11-17

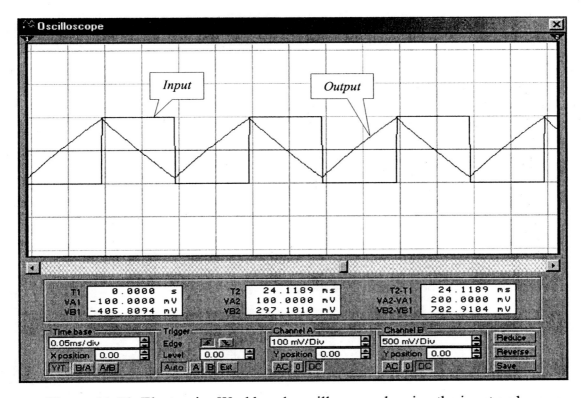

Figure 11-71: Electronics Workbench oscilloscope showing the input and output waveforms for the design Example 11-17

497

Another simple test for accurate integration of an integrator circuit is to change the input signal to a bipolar square-wave and observe the output. The expected output signal is a triangular waveform with no phase shift, as exhibited on the EWB oscilloscope of Figure 11-71 above displaying the simulation result of our design with a square-wave input.

Since the output is the integral or anti-derivative of the input signal, the input signal is the derivative or slope of the output signal. This relationship between the two waveforms can be observed clearly in the EWB oscilloscope display of Figure 11-71. For example, during the positive half-cycle of the square-wave input, the output is a ramp with a fixed negative slope. However, since the circuit is inverting, the negative slope corresponds to the +100 mV level of the input signal. Conversely, the positive slope of the output during the negative half-cycle of the input corresponds to the −100 mV level.

11.8.2 The Differentiator

The circuit diagram of a basic op-amp differentiator is shown in Figure 11-72 below:

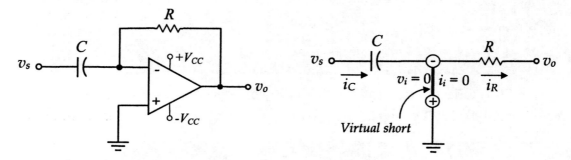

Figure 11-72: Basic differentiator circuit **Figure 11-73**: Virtual equivalent circuit

From the virtual equivalent circuit of Figure 11-72 we observe that i_R equals i_C, and since $v_i = 0$, i_R and i_C can be determined as follows:

$$i_C = C\frac{dv_C}{dt} = C\frac{d(v_s - v_i)}{dt} = C\frac{dv_s}{dt} \quad (11\text{-}74)$$

$$i_R = \frac{v_i - v_o}{R} = \frac{-v_o}{R} \quad (11\text{-}75)$$

Equating the above two equations and solving for v_o we obtain the following:

$$\frac{-v_o}{R} = C\frac{dv_s}{dt} \quad (11\text{-}76)$$

$$v_o = -RC\frac{dv_s}{dt} \quad (11\text{-}77)$$

Like the integrator circuit, the time-constant RC of the differentiator must also be small, at least ten times smaller, compared to the period T of the input signal.

$$RC \leq 0.1T = \frac{0.1}{f_o} \quad (11\text{-}78)$$

Chapter 11

$$R \leq \frac{0.1}{f_o C} = \frac{1}{10 f_o C} \qquad (11\text{-}79)$$

For a given operating frequency range, the maximum value of R can then be determined from the above equation, as follows:

$$R_{max} = \frac{1}{10 f_{min} C} \qquad (11\text{-}80)$$

The magnitude gain of the basic differentiator of Figure 11-71 is the ratio of the resistance R to the capacitive reactance X_C, that is:

$$A_v = -\frac{R}{X_C} = -\omega_o CR = -2\pi f_o CR \qquad (11\text{-}81)$$

where f_o is the operating frequency.

According to the above equation, the gain varies with frequency and will be very high at higher frequencies and very low at lower frequencies. To improve and stabilize the operation of the basic differentiator over the range of some operating frequency, a shunt capacitor C_F and a series resistor R_s may be added as shown in Figure 11-74. The size of the shunt capacitor C_F is generally about one-tenth of the capacitor C or smaller, and the size of the series resistor R_s is chosen to be about one-tenth of the resistor R or smaller.

$$C_F \leq 0.1\, C$$
$$R_s \leq 0.1\, R$$

The addition of the series resistor R_s can also help establish a minimum input impedance at higher frequencies.

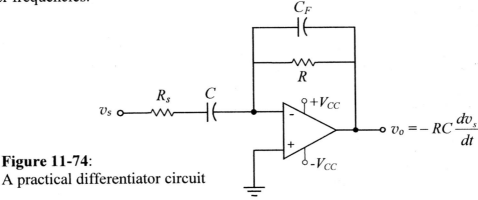

Figure 11-74:
A practical differentiator circuit

Example 11-18 DESIGN

Let us design a practical differentiator with minimum operating frequency 1 kHz and with minimum input impedance of about 1 kΩ.

Solution:

We will begin the design by selecting a non-electrolytic capacitor less than or equal to 0.1 µF. Let $C = 0.005$ µF. Using Equation 11-80, the resistance R is determined as follows:

Chapter 11

$$R_{max} = \frac{1}{10 f_{min} C} = \frac{1}{10 \times 1000 \text{ Hz} \times 0.005 \text{ μF}} = 20 \text{ kΩ} \qquad \text{Let } R = 15 \text{ kΩ}$$

$$C_F \leq 0.1 \, C \quad \text{and} \quad R_s \leq 0.1 \, R$$

Let $C_F = 0.0001$ μF $= 100$ pF, and $R_s = 1$ kΩ.

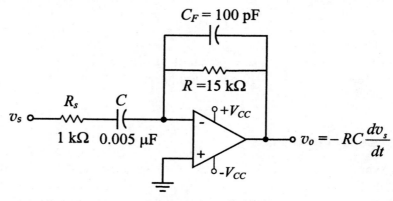

Figure 11-75: The differentiator circuit diagram with component values

Test Run and Design Verification:

The circuit setup for simulation and the results of simulation are exhibited in Figures 11-76 through 11-78. Again, a simple method for testing the accuracy of a differentiator is to observe the response of the circuit to a sinusoidal waveform. Since the derivative of a sine is cosine and the circuit is an inverter, the output is a negative cosine; and thus, a sinusoidal waveform with 90° lag is expected at the output. If necessary, minor adjustments can be made to circuit components to achieve the 90° lag at the output. As exhibited on the EWB oscilloscope of Figure 11-77, the time difference between the two signals $(t_1 - t_2)$ is 50 μs, which is a quarter of the 200 μs period T of the input signal. A quarter of the period T in the time domain corresponds to exactly 90° in the phase domain.

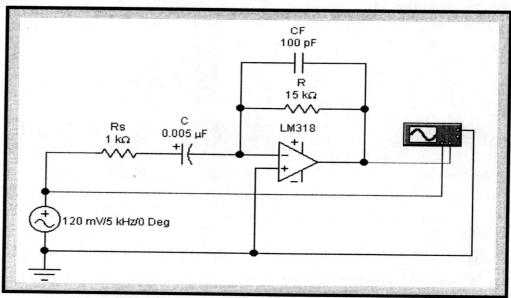

Figure 11-76: Circuit setup for simulation of the design Example 11-18 created in Electronics Workbench

Chapter 11

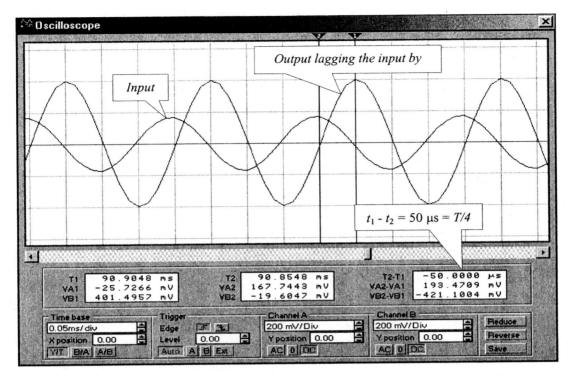

Figure 11-77: Electronics Workbench oscilloscope showing the input and output waveforms for the differentiator circuit of the design Example 11-18

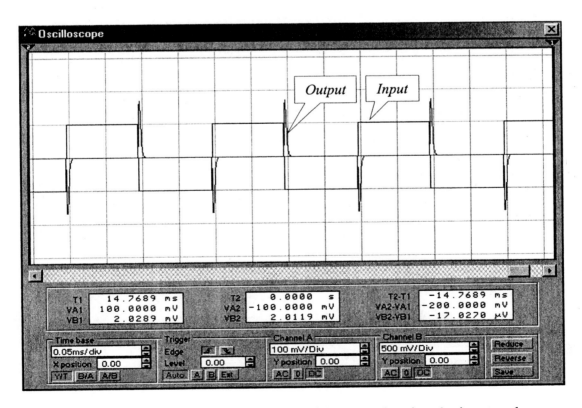

Figure 11-78: Electronics Workbench oscilloscope showing the input and output waveforms for the differentiator circuit of the design Example 11-18

As with the integrator, another simple test for accurate differentiation of a differentiator circuit is to change the input signal to a bipolar square-wave and observe the output. The expected output is simply a train of impulses occurring at each transition, alternately positive and negative, as exhibited on the EWB oscilloscope of Figure 11-78 above displaying the simulation result of the design example with a square-wave input.

With a square-wave input, the derivative is zero for both the positive and negative half-cycles except for the transitions, at which the derivative approaches infinity. Since the circuit is inverting, the output exhibits a negative impulse at positive transitions and a positive impulse at negative transitions, as shown on the Electronics Workbench oscilloscope display of Figure 11-78.

Practice Problem 11-10 — Integrator — DESIGN

Design a practical integrator for an operating frequency ranging from 1 kHz to 10 kHz. Let $C = 0.01 \; \mu F$. Test your design by simulation with sinusoidal and square-wave inputs.

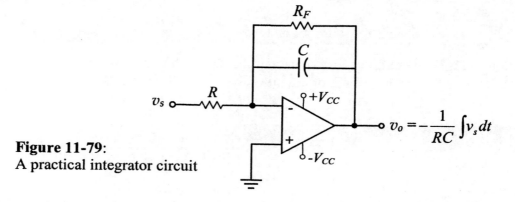

Figure 11-79:
A practical integrator circuit

$$v_o = -\frac{1}{RC}\int v_s \, dt$$

Answers: $C = 0.01 \; \mu F$, $R_F = 160 \; k\Omega$, $R = 10 \; k\Omega$

Practice Problem 11-11 — Differentiator — DESIGN

Design a practical differentiator for an operating frequency ranging from 1 kHz to 10 kHz. Let $C = 0.0047 \; \mu F$. Test your design by simulation with sinusoidal and square-wave inputs.

Answers:

$C = 0.0047 \; \mu F$, $R = 20 \; k\Omega$,
$R_s = 2 \; k\Omega$, $C_F = 200 \; pF$

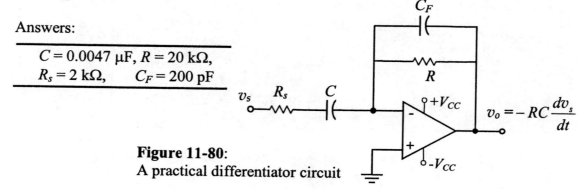

Figure 11-80:
A practical differentiator circuit

$$v_o = -RC\frac{dv_s}{dt}$$

Chapter 11

Section Summary — Integrator/Differentiator — ANALYSIS/DESIGN

Summary of Equations for the Analysis/Design of Integrator/Differentiator

The Integrator:

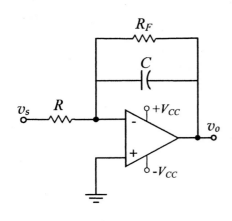

$$v_o = -\frac{1}{RC}\int_0^t v_s\, dt + v_o(0)$$

$$C \leq 0.1\ \mu F$$

$$R_{max} = \frac{1}{10 f_{min} C}$$

$$|A_{v(max)}| = \frac{R_F}{R}$$

$$f_L = \frac{1}{2\pi R_F C}$$

$$10 f_L \leq f_o \leq \frac{1}{10 RC}$$

The Differentiator:

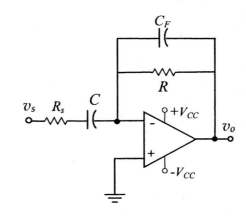

$$v_o = -RC \frac{dv_s}{dt}$$

$$C \leq 0.1\ \mu F$$

$$R_{max} = \frac{1}{10 f_{min} C}$$

$$|A_v| = \frac{R}{X_C} = 2\pi f_o CR$$

$$C_F \leq 0.1\ C$$

$$R_s \leq 0.1\ R$$

$$f_o \leq \frac{0.1}{RC}$$

11.9 SUMMARY

- The operational amplifier is a cascade of four main stages. The first two stages of an op-amp are differential amplifiers, the third stage is a DC level shifter, and the last or output stage is a current driver. The absence of coupling and bypass capacitors enables the op-amp to amplify AC as well as DC signals.

- Although the op-amp is a precision-engineered complex circuit, its low cost and ease of use allow it to be treated as a simple three-terminal electronic device. Two of the three terminals are the inverting and non-inverting inputs, and the third is the output terminal.

- The operational amplifier is an extremely versatile device with a broad range of practical applications such as: comparator, LED driver, inverting and non-inverting amplifier, nearly ideal buffer, summing amplifier, difference amplifier, integrator, differentiator, active filter, oscillator, etc.

- When the op-amp is used a comparator, the output of the op-amp is either at the $+V_{sat}$ or $-V_{sat}$. That is, if the voltage level at the non-inverting input (v_+) is larger the voltage level at the inverting input (v_-), then the output goes to $+V_{sat}$. Conversely, if v_- is larger than v_+, then the output drops to $-V_{sat}$.

- The Schmitt trigger is a type of comparator with positive feedback that ignores the spurious noise on a corrupted signal, and thus avoids false triggering if the noise level is limited to within the boundaries of the upper and lower thresholds V_{UT} and V_{LT}.

- There exists a *virtual short* across the input terminals of the op-amp in which both the input voltage and the input current are approximately zero.

- The non-inverting amplifier has nearly infinite input resistance and almost zero output resistance. Its gain can be designed for a desired finite value with two external resistors.

- The inverting amplifier shares the almost ideal low output resistance characteristics of the non-inverting amplifier, but not its ideal input resistance characteristics.

- An ideal buffer behaves like short-circuit for voltage and like open-circuit for current. The unity gain buffer configured with an op-amp has nearly such characteristics.

- The inverting amplifier, non-inverting amplifier, and op-amp buffer may all be utilized as a *current-to-voltage converter*.

- A summing amplifier amplifies the sum of n inputs. However, each input signal may be amplified by a different gain, which depends on the channel resistance.

Chapter 11

- A *difference amplifier* amplifies the difference of two input signals while suppressing the signal common to those input terminals. The diff-amp designed with op-amp uses the *CMRR* of the op-amp to suppress the common signals like noise.

- The basic difference amplifier provides neither equal nor adequately high input channel resistances. Nearly infinite and equal input resistances can be achieved with buffered inputs, which is referred to as the *instrumentation amplifier*.

- A basic *integrator* circuit can be achieved with a resistor R, a capacitor C, and an op-amp. However, to control the low frequency response and to limit the DC gain to a modest level, a shunt resistor R_F is placed in parallel with the capacitor. The addition of the resistor R_F introduces a low cutoff frequency f_L, below which integration does not take place.

- For a more accurate integration, the operating frequency f_o is kept at least one decade above the f_L. In addition, the time-constant RC of the integrator must be small, at least ten times smaller, compared to the period $T = 1/f_o$ of the input signal.

- An *integrator* circuit may be tested for accuracy by applying a sinusoidal input signal at the operating frequency and by observing the output. The sinusoidal output must lead the input by 90°. If necessary, minor adjustments can be made to the design to achieve the 90° lead.

- A basic *differentiator* circuit can also be achieved with a resistor R, a capacitor C, and an op-amp. To improve and stabilize the operation of the basic differentiator, a shunt capacitor C_F and a series resistor R_s may be added. The size of the shunt capacitor C_F is generally about one-tenth of the capacitor C or smaller, and the size of the series resistor R_s is chosen to be about one-tenth of the resistor R or smaller.

- As with the integrator, the time-constant RC of the differentiator must be small, at least ten times smaller, compared to the period $T = 1/f_o$ of the input signal.

- A *differentiator* may also be tested for accuracy by applying a sinusoidal input signal at the operating frequency and by observing the output. The sinusoidal output must lag the input by 90°. If necessary, minor adjustments can be made to the design to achieve the 90° lag.

Chapter 11

11.10 DEVICE SPECIFICATIONS AND DATA SHEETS

Data sheets of the following devices discussed in this chapter are included for your reference and convenience:

- LM741 Operational Amplifier
  Courtesy of National Semiconductor

- LF347, LF351, and LF353 Operational Amplifiers
  Copyright of Semiconductor Component Industries, LLC. Used by permission.

- LM118, LM218, and LM318 Operational Amplifiers
  Courtesy of National Semiconductor

Chapter 11

National Semiconductor

November 1994

LM741 Operational Amplifier

General Description

The LM741 series are general purpose operational amplifiers which feature improved performance over industry standards like the LM709. They are direct, plug-in replacements for the 709C, LM201, MC1439 and 748 in most applications.

The amplifiers offer many features which make their application nearly foolproof: overload protection on the input and output, no latch-up when the common mode range is exceeded, as well as freedom from oscillations.

The LM741C/LM741E are identical to the LM741/LM741A except that the LM741C/LM741E have their performance guaranteed over a 0°C to +70°C temperature range, instead of −55°C to +125°C.

Schematic Diagram

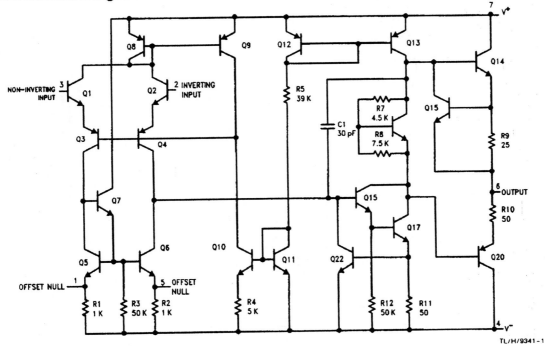

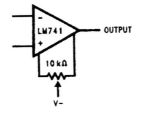

Offset Nulling Circuit

507

Absolute Maximum Ratings

If Military/Aerospace specified devices are required, please contact the National Semiconductor Sales Office/Distributors for availability and specifications.
(Note 5)

	LM741A	LM741E	LM741	LM741C
Supply Voltage	±22V	±22V	±22V	±18V
Power Dissipation (Note 1)	500 mW	500 mW	500 mW	500 mW
Differential Input Voltage	±30V	±30V	±30V	±30V
Input Voltage (Note 2)	±15V	±15V	±15V	±15V
Output Short Circuit Duration	Continuous	Continuous	Continuous	Continuous
Operating Temperature Range	−55°C to +125°C	0°C to +70°C	−55°C to +125°C	0°C to +70°C
Storage Temperature Range	−65°C to +150°C	−65°C to +150°C	−65°C to +150°C	−65°C to +150°C
Junction Temperature	150°C	100°C	150°C	100°C
Soldering Information				
N-Package (10 seconds)	260°C	260°C	260°C	260°C
J- or H-Package (10 seconds)	300°C	300°C	300°C	300°C
M-Package				
Vapor Phase (60 seconds)	215°C	215°C	215°C	215°C
Infrared (15 seconds)	215°C	215°C	215°C	215°C

See AN-450 "Surface Mounting Methods and Their Effect on Product Reliability" for other methods of soldering surface mount devices.

	LM741A	LM741E	LM741	LM741C
ESD Tolerance (Note 6)	400V	400V	400V	400V

Electrical Characteristics (Note 3)

Parameter	Conditions	LM741A/LM741E Min	Typ	Max	LM741 Min	Typ	Max	LM741C Min	Typ	Max	Units
Input Offset Voltage	$T_A = 25°C$ $R_S \leq 10\ k\Omega$ $R_S \leq 50\Omega$		0.8	3.0		1.0	5.0		2.0	6.0	mV mV
	$T_{AMIN} \leq T_A \leq T_{AMAX}$ $R_S \leq 50\Omega$ $R_S \leq 10\ k\Omega$			4.0			6.0			7.5	mV mV
Average Input Offset Voltage Drift				15							μV/°C
Input Offset Voltage Adjustment Range	$T_A = 25°C$, $V_S = ±20V$		±10			±15			±15		mV
Input Offset Current	$T_A = 25°C$		3.0	30		20	200		20	200	nA
	$T_{AMIN} \leq T_A \leq T_{AMAX}$			70		85	500			300	nA
Average Input Offset Current Drift				0.5							nA/°C
Input Bias Current	$T_A = 25°C$		30	80		80	500		80	500	nA
	$T_{AMIN} \leq T_A \leq T_{AMAX}$			0.210			1.5			0.8	μA
Input Resistance	$T_A = 25°C$, $V_S = ±20V$	1.0	6.0		0.3	2.0		0.3	2.0		MΩ
	$T_{AMIN} \leq T_A \leq T_{AMAX}$, $V_S = ±20V$	0.5									MΩ
Input Voltage Range	$T_A = 25°C$							±12	±13		V
	$T_{AMIN} \leq T_A \leq T_{AMAX}$				±12	±13					V
Large Signal Voltage Gain	$T_A = 25°C$, $R_L \geq 2\ k\Omega$ $V_S = ±20V, V_O = ±15V$ $V_S = ±15V, V_O = ±10V$	50				50	200	20	200		V/mV V/mV
	$T_{AMIN} \leq T_A \leq T_{AMAX}$, $R_L \geq 2\ k\Omega$, $V_S = ±20V, V_O = ±15V$ $V_S = ±15V, V_O = ±10V$ $V_S = ±5V, V_O = ±2V$	32 10				25			15		V/mV V/mV V/mV

Chapter 11

Electrical Characteristics (Note 3) (Continued)

Parameter	Conditions	LM741A/LM741E Min	LM741A/LM741E Typ	LM741A/LM741E Max	LM741 Min	LM741 Typ	LM741 Max	LM741C Min	LM741C Typ	LM741C Max	Units
Output Voltage Swing	$V_S = \pm 20V$ $R_L \geq 10 k\Omega$ $R_L \geq 2 k\Omega$	±16 ±15									V V
	$V_S = \pm 15V$ $R_L \geq 10 k\Omega$ $R_L \geq 2 k\Omega$				±12 ±10	±14 ±13		±12 ±10	±14 ±13		V V
Output Short Circuit Current	$T_A = 25°C$ $T_{AMIN} \leq T_A \leq T_{AMAX}$	10 10	25	35 40		25			25		mA mA
Common-Mode Rejection Ratio	$T_{AMIN} \leq T_A \leq T_{AMAX}$ $R_S \leq 10 k\Omega, V_{CM} = \pm 12V$ $R_S \leq 50\Omega, V_{CM} = \pm 12V$	80	95		70	90		70	90		dB dB
Supply Voltage Rejection Ratio	$T_{AMIN} \leq T_A \leq T_{AMAX}$, $V_S = \pm 20V$ to $V_S = \pm 5V$ $R_S \leq 50\Omega$ $R_S \leq 10 k\Omega$	86	96		77	96		77	96		dB dB
Transient Response Rise Time Overshoot	$T_A = 25°C$, Unity Gain		0.25 6.0	0.8 20		0.3 5			0.3 5		μs %
Bandwidth (Note 4)	$T_A = 25°C$	0.437	1.5								MHz
Slew Rate	$T_A = 25°C$, Unity Gain	0.3	0.7			0.5			0.5		V/μs
Supply Current	$T_A = 25°C$					1.7	2.8		1.7	2.8	mA
Power Consumption	$T_A = 25°C$ $V_S = \pm 20V$ $V_S = \pm 15V$		80	150		50	85		50	85	mW mW
LM741A	$V_S = \pm 20V$ $T_A = T_{AMIN}$ $T_A = T_{AMAX}$			165 135							mW mW
LM741E	$V_S = \pm 20V$ $T_A = T_{AMIN}$ $T_A = T_{AMAX}$			150 150							mW mW
LM741	$V_S = \pm 15V$ $T_A = T_{AMIN}$ $T_A = T_{AMAX}$					60 45	100 75				mW mW

Note 1: For operation at elevated temperatures, these devices must be derated based on thermal resistance, and T_j max. (listed under "Absolute Maximum Ratings"). $T_j = T_A + (\theta_{jA} P_D)$.

Thermal Resistance	Cerdip (J)	DIP (N)	HO8 (H)	SO-8 (M)
θ_{jA} (Junction to Ambient)	100°C/W	100°C/W	170°C/W	195°C/W
θ_{jC} (Junction to Case)	N/A	N/A	25°C/W	N/A

Note 2: For supply voltages less than ±15V, the absolute maximum input voltage is equal to the supply voltage.

Note 3: Unless otherwise specified, these specifications apply for $V_S = \pm 15V$, $-55°C \leq T_A \leq +125°C$ (LM741/LM741A). For the LM741C/LM741E, these specifications are limited to $0°C \leq T_A \leq +70°C$.

Note 4: Calculated value from: BW (MHz) = 0.35/Rise Time(μs).

Note 5: For military specifications see RETS741X for LM741 and RETS741AX for LM741A.

Note 6: Human body model, 1.5 kΩ in series with 100 pF.

Chapter 11

Connection Diagrams

Metal Can Package

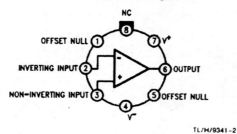

Order Number LM741H, LM741H/883*,
LM741AH/883 or LM741CH
See NS Package Number H08C

Ceramic Dual-In-Line Package

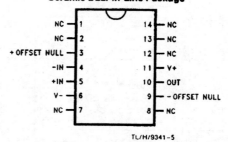

Order Number LM741J-14/883*, LM741AJ-14/883**
See NS Package Number J14A

*also available per JM38510/10101
**also available per JM38510/10102

Dual-In-Line or S.O. Package

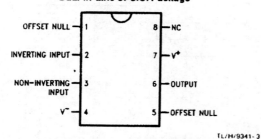

Order Number LM741J, LM741J/883,
LM741CM, LM741CN or LM741EN
See NS Package Number J08A, M08A or N08E

Ceramic Flatpak

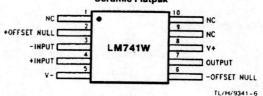

Order Number LM741W/883
See NS Package Number W10A

*LM741H is available per JM38510/10101

Chapter 11

Order this document by LF347/D

LF347, B
LF351
LF353

JFET Input Operational Amplifiers

These low cost JFET input operational amplifiers combine two state-of-the-art analog technologies on a single monolithic integrated circuit. Each internally compensated operational amplifier has well matched high voltage JFET input devices for low input offset voltage. The JFET technology provides wide bandwidths and fast slew rates with low input bias currents, input offset currents, and supply currents.

These devices are available in single, dual and quad operational amplifiers which are pin-compatible with the industry standard MC1741, MC1458, and the MC3403/LM324 bipolar devices.

- Input Offset Voltage of 5.0 mV Max (LF347B)
- Low Input Bias Current: 50 pA
- Low Input Noise Voltage: 16 nV/$\sqrt{Hz}$
- Wide Gain Bandwidth: 4.0 MHz
- High Slew Rate: 13V/μs
- Low Supply Current: 1.8 mA per Amplifier
- High Input Impedance: 10^{12} Ω
- High Common Mode and Supply Voltage Rejection Ratios: 100 dB

FAMILY OF JFET OPERATIONAL AMPLIFIERS

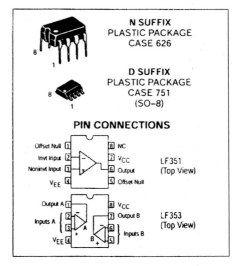

N SUFFIX
PLASTIC PACKAGE
CASE 626

D SUFFIX
PLASTIC PACKAGE
CASE 751
(SO-8)

PIN CONNECTIONS

LF351 (Top View)

LF353 (Top View)

MAXIMUM RATINGS

Rating	Symbol	Value	Unit
Supply Voltage	V_{CC}	+18	V
	V_{EE}	−18	
Differential Input Voltage	V_{ID}	±30	V
Input Voltage Range (Note 1)	V_{IDR}	±15	V
Output Short Circuit Duration (Note 2)	t_{SC}	Continuous	
Power Dissipation at T_A = +25°C	P_D	900	mW
Derate above T_A = +25°C	$1/\theta_{JA}$	10	mW/°C
Operating Ambient Temperature Range	T_A	0 to +70	°C
Operating Junction Temperature Range	T_J	115	°C
Storage Temperature Range	T_{stg}	−65 to +150	°C

NOTES: 1. Unless otherwise specified, the absolute maximum negative input voltage is limited to the negative power supply.
2. Any amplifier output can be shorted to ground indefinitely. However, if more than one amplifier output is shorted simultaneously, maximum junction temperature rating may be exceeded.

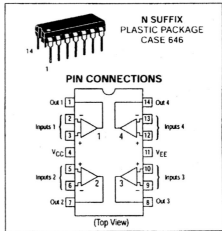

N SUFFIX
PLASTIC PACKAGE
CASE 646

PIN CONNECTIONS

(Top View)

ORDERING INFORMATION

Device	Function	Operating Temperature Range	Package
LF351D	Single		SO-8
LF351N	Single		Plastic DIP
LF353D	Dual	T_A = 0° to +70°C	SO-8
LF353N	Dual		Plastic DIP
LF347BN	Quad		Plastic DIP
LF347N	Quad		Plastic DIP

© Motorola, Inc. 1996   Rev 0

Chapter 11

LF347, B LF351 LF353

ELECTRICAL CHARACTERISTICS (V_{CC} = +15 V_{EE} = –15 V, T_A = 25°C, unless otherwise noted.)

Characteristic	Symbol	LF347B Min	LF347B Typ	LF347B Max	LF347, LF351, LF353 Min	LF347, LF351, LF353 Typ	LF347, LF351, LF353 Max	Unit
Input Offset Voltage ($R_S \leq 10$ k, V_{CM} = 0) T_A = +25°C 0°C ≤ T_A ≤ +70°C	V_{IO}	– –	1.0 –	5.0 8.0	– –	5.0 –	10 13	mV
Avg. Temperature Coefficient of Input Offset Voltage $R_S \leq 10$ k, 0°C ≤ T_A ≤ +70°C	$\Delta V_{IO}/\Delta T$	–	10	–	–	10	–	µV/°C
Input Offset Current (V_{CM} = 0, Note 3) T_A = +25°C 0°C ≤ T_A ≤ +70°C	I_{IO}	– –	25 –	100 4.0	– –	25 –	100 4.0	pA nA
Input Bias Current (V_{CM} = 0, Note 3) T_A = +25°C 0°C ≤ T_A ≤ +70°C	I_{IB}	– –	50 –	200 8.0	– –	50 –	200 8.0	pA nA
Input Resistance	r_i	–	10^{12}	–	–	10^{12}	–	Ω
Common Mode Input Voltage Range	V_{ICR}	±11	+15 –12	–	±11	+15 –12	–	V
Large–Signal Voltage Gain (V_O = ±10 V, R_L = 2.0 k) T_A = +25°C 0°C ≤ T_A ≤ +70°C	A_{VOL}	50 25	100 –	– –	25 15	100 –	– –	V/mV
Output Voltage Swing (R_L = 10 k)	V_O	±12	±14	–	±12	±14	–	V
Common Mode Rejection ($R_S \leq 10$ k)	CMR	80	100	–	70	100	–	dB
Supply Voltage Rejection ($R_S \leq 10$ k)	PSRR	80	100	–	70	100	–	dB
Supply Current LF347 LF351 LF353	I_D	– – –	7.2 – –	11 – –	– – –	7.2 1.8 3.6	11 3.4 6.5	mA
Short Circuit Current	I_{SC}	–	25	–	–	25	–	mA
Slew Rate (A_V = +1)	SR	–	13	–	–	13	–	V/µs
Gain–Bandwidth Product	BWp	–	4.0	–	–	4.0	–	MHz
Equivalent Input Noise Voltage (R_S = 100 Ω, f = 1000 Hz)	e_n	–	24	–	–	24	–	nV/√Hz
Equivalent Input Noise Current (f = 1000 Hz)	i_n	–	0.01	–	–	0.01	–	pA/√Hz
Channel Separation (LF347, LF353) 1.0 Hz ≤ f ≤ 20 kHz (Input Referred)	–	–	–120	–	–	–120	–	dB

For Typical Characteristic Performance Curves, refer to MC34001, 34002, 34004 data sheet.

NOTE: 3. Input bias currents of JFET input op amps approximately double for every 10°C rise in junction temperature. To maintain junction temperatures as close to ambient as is possible, pulse techniques are utilized during test.

Chapter 11

National Semiconductor

August 2000

LM118/LM218/LM318
Operational Amplifiers

General Description

The LM118 series are precision high speed operational amplifiers designed for applications requiring wide bandwidth and high slew rate. They feature a factor of ten increase in speed over general purpose devices without sacrificing DC performance.

The LM118 series has internal unity gain frequency compensation. This considerably simplifies its application since no external components are necessary for operation. However, unlike most internally compensated amplifiers, external frequency compensation may be added for optimum performance. For inverting applications, feedforward compensation will boost the slew rate to over 150V/μs and almost double the bandwidth. Overcompensation can be used with the amplifier for greater stability when maximum bandwidth is not needed. Further, a single capacitor can be added to reduce the 0.1% settling time to under 1 μs.

The high speed and fast settling time of these op amps make them useful in A/D converters, oscillators, active filters, sample and hold circuits, or general purpose amplifiers. These devices are easy to apply and offer an order of magnitude better AC performance than industry standards such as the LM709.

The LM218 is identical to the LM118 except that the LM218 has its performance specified over a −25°C to +85°C temperature range. The LM318 is specified from 0°C to +70°C.

Features

- 15 MHz small signal bandwidth
- Guaranteed 50V/μs slew rate
- Maximum bias current of 250 nA
- Operates from supplies of ±5V to ±20V
- Internal frequency compensation
- Input and output overload protected
- Pin compatible with general purpose op amps

Fast Voltage Follower

(Note 1)

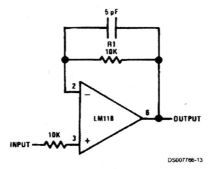

Note 1: Do not hard-wire as voltage follower (R1 ≥ 5 kΩ)

© 2000 National Semiconductor Corporation DS007766

www.national.com

Absolute Maximum Ratings (Note 7)

If Military/Aerospace specified devices are required, please contact the National Semiconductor Sales Office/Distributors for availability and specifications.

Supply Voltage	±20V
Power Dissipation (Note 2)	500 mW
Differential Input Current (Note 3)	±10 mA
Input Voltage (Note 4)	±15V
Output Short-Circuit Duration	Continuous
Operating Temperature Range	
LM118	−55°C to +125°C
LM218	−25°C to +85°C
LM318	0°C to +70°C
Storage Temperature Range	−65°C to +150°C
Lead Temperature (Soldering, 10 sec.)	
Hermetic Package	300°C
Plastic Package	260°C
Soldering Information	
Dual-In-Line Package	
Soldering (10 sec.)	260°C
Small Outline Package	
Vapor Phase (60 sec.)	215°C
Infrared (15 sec.)	220°C

See AN-450 "Surface Mounting Methods and Their Effect on Product Reliability" for other methods of soldering surface mount devices.

ESD Tolerance (Note 8)	2000V

Electrical Characteristics (Note 5)

Parameter	Conditions	LM118/LM218 Min	LM118/LM218 Typ	LM118/LM218 Max	LM318 Min	LM318 Typ	LM318 Max	Units
Input Offset Voltage	T_A = 25°C		2	4		4	10	mV
Input Offset Current	T_A = 25°C		6	50		30	200	nA
Input Bias Current	T_A = 25°C		120	250		150	500	nA
Input Resistance	T_A = 25°C	1	3		0.5	3		MΩ
Supply Current	T_A = 25°C		5	8		5	10	mA
Large Signal Voltage Gain	T_A = 25°C, V_S = ±15V, V_{OUT} = ±10V, R_L ≥ 2 kΩ	50	200		25	200		V/mV
Slew Rate	T_A = 25°C, V_S = ±15V, A_V = 1 (Note 6)	50	70		50	70		V/μs
Small Signal Bandwidth	T_A = 25°C, V_S = ±15V		15			15		MHz
Input Offset Voltage				6			15	mV
Input Offset Current				100			300	nA
Input Bias Current				500			750	nA
Supply Current	T_A = 125°C		4.5	7				mA
Large Signal Voltage Gain	V_S = ±15V, V_{OUT} = ±10V, R_L ≥ 2 kΩ	25			20			V/mV
Output Voltage Swing	V_S = ±15V, R_L = 2 kΩ	±12	±13		±12	±13		V
Input Voltage Range	V_S = ±15V	±11.5			±11.5			V
Common-Mode Rejection Ratio		80	100		70	100		dB
Supply Voltage Rejection Ratio		70	80		65	80		dB

Note 2: The maximum junction temperature of the LM118 is 150°C, the LM218 is 110°C, and the LM318 is 110°C. For operating at elevated temperatures, devices in the H08 package must be derated based on a thermal resistance of 160°C/W, junction to ambient, or 20°C/W, junction to case. The thermal resistance of the dual-in-line package is 100°C/W, junction to ambient.

Note 3: The inputs are shunted with back-to-back diodes for overvoltage protection. Therefore, excessive current will flow if a differential input voltage in excess of 1V is applied between the inputs unless some limiting resistance is used.

Note 4: For supply voltages less than ±15V, the absolute maximum input voltage is equal to the supply voltage.

Note 5: These specifications apply for ±5V ≤ V_S ≤ ±20V and −55°C ≤ T_A ≤ +125°C (LM118), −25°C ≤ T_A ≤ +85°C (LM218), and 0°C ≤ T_A ≤ +70°C (LM318). Also, power supplies must be bypassed with 0.1 μF disc capacitors.

Note 6: Slew rate is tested with V_S = ±15V. The LM118 is in a unity-gain non-inverting configuration. V_{IN} is stepped from −7.5V to +7.5V and vice versa. The slew rates between −5.0V and +5.0V and vice versa are tested and guaranteed to exceed 50V/μs.

Note 7: Refer to RETS118X for LM118H and LM118J military specifications.

Note 8: Human body model, 1.5 kΩ in series with 100 pF.

Absolute Maximum Ratings (Note 7)

If Military/Aerospace specified devices are required, please contact the National Semiconductor Sales Office/Distributors for availability and specifications.

Supply Voltage	±20V
Power Dissipation (Note 2)	500 mW
Differential Input Current (Note 3)	±10 mA
Input Voltage (Note 4)	±15V
Output Short-Circuit Duration	Continuous
Operating Temperature Range	
LM118	−55°C to +125°C
LM218	−25°C to +85°C
LM318	0°C to +70°C
Storage Temperature Range	−65°C to +150°C
Lead Temperature (Soldering, 10 sec.)	
Hermetic Package	300°C
Plastic Package	260°C
Soldering Information	
Dual-In-Line Package	
Soldering (10 sec.)	260°C
Small Outline Package	
Vapor Phase (60 sec.)	215°C
Infrared (15 sec.)	220°C

See AN-450 "Surface Mounting Methods and Their Effect on Product Reliability" for other methods of soldering surface mount devices.

ESD Tolerance (Note 8)	2000V

Electrical Characteristics (Note 5)

Parameter	Conditions	LM118/LM218 Min	LM118/LM218 Typ	LM118/LM218 Max	LM318 Min	LM318 Typ	LM318 Max	Units
Input Offset Voltage	$T_A = 25°C$		2	4		4	10	mV
Input Offset Current	$T_A = 25°C$		6	50		30	200	nA
Input Bias Current	$T_A = 25°C$		120	250		150	500	nA
Input Resistance	$T_A = 25°C$	1	3		0.5	3		MΩ
Supply Current	$T_A = 25°C$		5	8		5	10	mA
Large Signal Voltage Gain	$T_A = 25°C$, $V_S = ±15V$ $V_{OUT} = ±10V$, $R_L \geq 2$ kΩ	50	200		25	200		V/mV
Slew Rate	$T_A = 25°C$, $V_S = ±15V$, $A_V = 1$ (Note 6)	50	70		50	70		V/μs
Small Signal Bandwidth	$T_A = 25°C$, $V_S = ±15V$		15			15		MHz
Input Offset Voltage				6			15	mV
Input Offset Current				100			300	nA
Input Bias Current				500			750	nA
Supply Current	$T_A = 125°C$		4.5	7				mA
Large Signal Voltage Gain	$V_S = ±15V$, $V_{OUT} = ±10V$ $R_L \geq 2$ kΩ	25			20			V/mV
Output Voltage Swing	$V_S = ±15V$, $R_L = 2$ kΩ	±12	±13		±12	±13		V
Input Voltage Range	$V_S = ±15V$	±11.5			±11.5			V
Common-Mode Rejection Ratio		80	100		70	100		dB
Supply Voltage Rejection Ratio		70	80		65	80		dB

Note 2: The maximum junction temperature of the LM118 is 150°C, the LM218 is 110°C, and the LM318 is 110°C. For operating at elevated temperatures, devices in the H08 package must be derated based on a thermal resistance of 160°C/W, junction to ambient, or 20°C/W, junction to case. The thermal resistance of the dual-in-line package is 100°C/W, junction to ambient.

Note 3: The inputs are shunted with back-to-back diodes for overvoltage protection. Therefore, excessive current will flow if a differential input voltage in excess of 1V is applied between the inputs unless some limiting resistance is used.

Note 4: For supply voltages less than ±15V, the absolute maximum input voltage is equal to the supply voltage.

Note 5: These specifications apply for ±5V ≤ V_S ≤ ±20V and −55°C ≤ T_A ≤ +125°C (LM118), −25°C ≤ T_A ≤ +85°C (LM218), and 0°C ≤ T_A ≤ +70°C (LM318). Also, power supplies must be bypassed with 0.1 μF disc capacitors.

Note 6: Slew rate is tested with V_S = ±15V. The LM118 is in a unity-gain non-inverting configuration. V_{IN} is stepped from −7.5V to +7.5V and vice versa. The slew rates between −5.0V and +5.0V and vice versa are tested and guaranteed to exceed 50V/μs.

Note 7: Refer to RETS118X for LM118H and LM118J military specifications.

Note 8: Human body model, 1.5 kΩ in series with 100 pF.

Chapter 11

PROBLEMS

11.1 Draw the four stages of an operational amplifier in block diagram and briefly describe the function of each stage.

Section 11.3 Open-Loop Operation

11.2 Carry out Practice Problem 11-1.

11.3 Determine and plot the output of the comparator of Figure 11-1P(a) for the given input signal shown in Figure 11-1P(b).

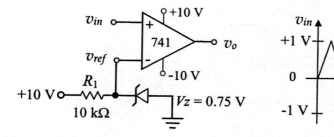
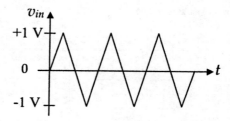

Figure 11-1P(a): Op-amp   **Figure 11-1P(b):** Input signal

11.4 Carry out Practice Problem 11-2.

11.5 Determine and plot the output of the comparator of Figure 11-2P(a) for the given input signals shown in Figure 11-2P(b).

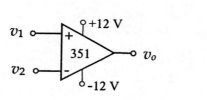
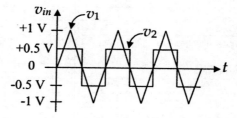

Figure 11-2P(a): Comparator   **Figure 11-2P(b):** Input signals

11.6 Determine the conditions of the LEDs (ON/OFF) of Figure 11-3P for the given input levels. Also determine the amount of current through each LED, when on.

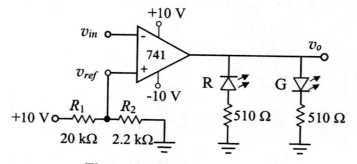

v_{in}	R	G
0.75 V		
1.5 V		

Figure 11-3P: LED driver

Chapter 11

11.7 Determine the conditions of the LEDs (ON/OFF) of Figure 11-4P for the given input levels. Also determine the amount of current through each LED, when on.

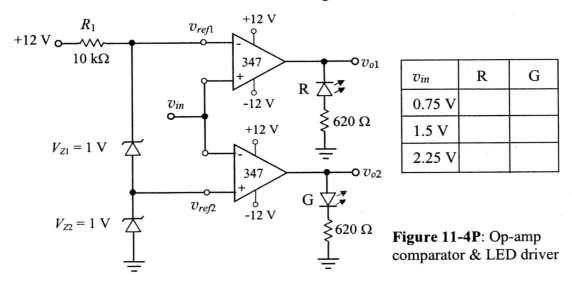

Figure 11-4P: Op-amp comparator & LED driver

11.8 Carry out Practice Problem 11-3.

11.9 Plot the output of the Schmitt trigger of Figure 11-5P(a) for the given input as shown in Figure 11-5P(b). Assume that the output is initially at $-V_{sat}$.

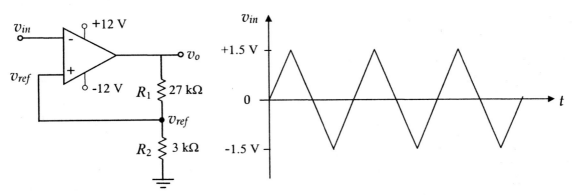

Figure 11-5P(a): Schmitt trigger   Figure 11-5P(b): Input signal

11.10 Carry out Practice Problem 11-4.

Section 11.5 Basic Op-Amp Configurations

11.11 Describe the concept of *virtual short* and *virtual ground* in an op-amp.

11.12 Carry out Practice Problem 11-5.

11.13 Determine the gain, input resistance, output resistance, and bandwidth for the amplifier of Figure 11-6P.

Chapter 11

11.14 Determine the minimum and maximum gains, input resistance, output resistance, and bandwidth for the amplifier of Figure 11-7P.

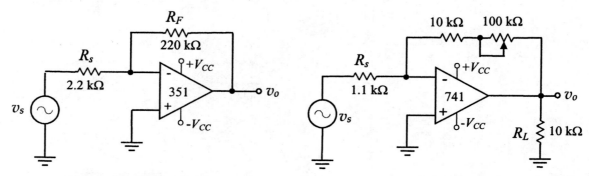

Figure 11-6P: Problem 11.13

Figure 11-7P: Problem 11.14

11.15 Carry out Practice Problem 11-6.

11.16 Determine the voltage gain, input resistance, output resistance, and bandwidth for the amplifier of Figure 11-8P.

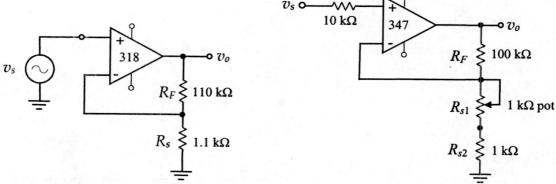

Figure 11-8P: Problem 11.16

Figure 11-9P: Problem 11.17

11.17 Determine the minimum and maximum gain, input resistance, output resistance, and the minimum bandwidth for the amplifier of Figure 11-9P.

11.18 Design an amplifier with $A_{v(min)} = 0$ dB and $A_{v(max)} \cong 40$ dB. Let $R_s = 1$ kΩ. Show your complete work. Draw the circuit schematic with component values. Upon completion of the design determine the input resistance and output resistance of the amplifier.

11.19 Design an amplifier with an input resistance about 10 kΩ and with an adjustable gain from $|A_{v(min)}| = 0$ to $|A_{v(max)}| = 100$. Show your complete work. Draw the circuit schematic with component values. Upon completion of the design determine the output resistance of the amplifier.

Section 11.6 Summing Amplifiers

11.20 Determine and plot the output signal for the summing amplifier of Figure 11-10P with the given inputs v_1 and v_2 below:

$v_1 = 0.5$ V(p-p) sinusoid

$v_2 = -0.5$ V DC

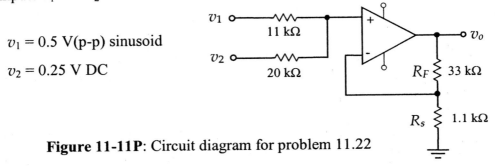

Figure 11-10P: Circuit diagram for problem 11.20

11.21 Carry out Practice Problem 11-7.

11.22 Determine and plot the output signal for the summing amplifier of Figure 11-11P with the given inputs v_1 and v_2 below:

$v_1 = 0.5$ V(p-p) sinusoid

$v_2 = 0.25$ V DC

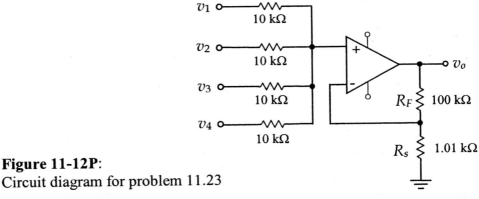

Figure 11-11P: Circuit diagram for problem 11.22

11.23 Determine the output v_o for the summing amplifier of Figure 11-12P, in terms of v_1, v_2, v_3, and v_4.

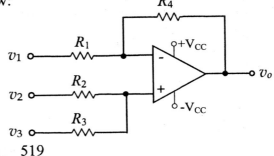

Figure 11-12P: Circuit diagram for problem 11.23

11.24 Utilizing the results of derivations of the appropriate closed-loop configurations and making use of the superposition theorem, determine the output expression v_o for the circuit of Figure 11-13P, as shown below:

Figure 11-13P: Circuit diagram for problem 11.24

Chapter 11

11.25 Utilizing the results of derivations of the appropriate closed-loop configurations and making use of the superposition theorem, draw the circuit schematic of an op-amp configuration that produces the output expression v_o as given below. Label the feedback resistor as R_F, and the other resistors as R_1 through R_4, accordingly.

$$v_o = v_1 A_1 + v_2 A_2 - v_3 A_3 - v_4 A_4$$

Section 11.7 Difference Amplifier

11.26 Determine the peak-to-peak amplitude and the phase of the output v_o for the amplifier of Figure 11-14P, as shown below:

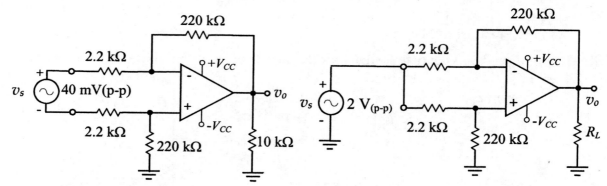

Figure 11-14P: Circuit diagram for problem 11.26

Figure 11-15P: Circuit diagram for problem 11.27

11.27 Determine the peak-to-peak amplitude and the phase of the output v_o for the amplifier of Figure 11-15P, as shown below. Assume a typical *CMRR* of 86 dB for the op-amp utilized in the circuit. Show your complete work.

11.28 Determine the output signal plus noise for the amplifiers whose circuit diagrams are shown in Figures 11-16P(a) and 11-16P(b). Assume a *CMRR* of 86 dB for the op-amp. There is also 20 mV of spurious noise present in the operating environment.

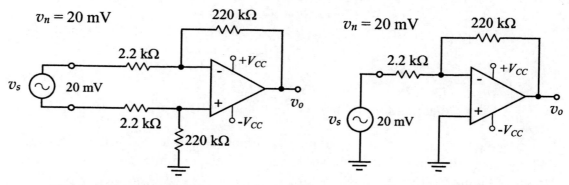

Figure 11-16P(a): Difference amplifier

Figure 11-16P(b): Inv. amplifier

11.29 Carry out Practice Problem 11-8.

11.30 Carry out Practice Problem 11-9.

Chapter 11

11.31 Determine the $A_{v(min)}$ and $A_{v(max)}$ for the instrumentation amplifier of Figure 11-17P as shown below. Furthermore, assuming that there exists 10 mV of noise in the operating environment, determine the output v_o (signal plus noise) if $R = 20$ kΩ and R_A is set at 500 Ω. The op-amp CMRR equals 90 dB.

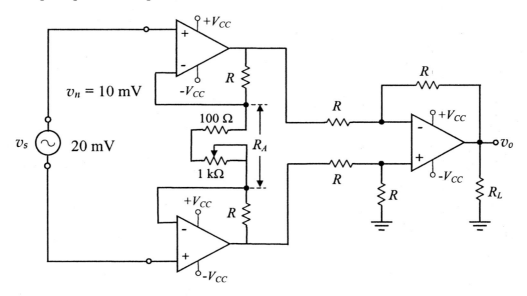

Figure 11-17P: Instrumentation amplifier

Section 11.8 Integrator and Differentiator

11.32 Design a practical integrator with a minimum operating frequency of about 2 kHz, and with a maximum gain of about 20. Test your design by simulation with sinusoidal and square-wave inputs.

11.33 Carry out Practice Problem 11-10.

11.34 Design a practical differentiator for a minimum operating frequency of 2 kHz, and with minimum input impedance of about 2 kΩ. Test your design by simulation with sinusoidal and square-wave inputs.

11.35 Carry out Practice Problem 11-11.

Chapter 12

NEGATIVE FEEDBACK, OP-AMP CHARACTERISTICS, AND SINGLE-SUPPLY OPERATION

12.1 INTRODUCTION

As you have seen in the previous chapter, we have been able, through negative feedback, to control and set the gain, input resistance, and output resistance of the op-amp to a desired level. Furthermore, a more stable operation and wider bandwidth have also been achieved, which will be discussed shortly. These are actually the advantages and improvements achievable through negative feedback: *controlled gain, improved input and output resistances, wider bandwidth,* and *more stable operation.* By presenting the basic theory of negative feedback in this chapter, we intend to demonstrate, for example, why and how the input and output resistances of the non-inverting amplifier are in the GΩ range and mΩ range, respectively. We also intend to demonstrate how the system's stability and bandwidth increase considerably with negative feedback. Additionally, we will define and discuss some of the more important characteristics and/or parameters of the op-amp in this chapter. Furthermore, we will also discuss the operation of basic op-amp configurations with a single supply.

Chapter 12

12.2 NEGATIVE FEEDBACK

Consider the two amplifier systems whose block diagrams with and without feedback are shown below. Figure 12-1 shows the block diagram of a basic amplifier without feedback, whose output v_o is the product of the input v_{in} and the amplifier voltage gain. Figure 12-2 shows the same amplifier with a negative feedback, where the output is connected to the input through a feedback network, which is generally referred to as the *feedback network* or *β-network*. The output of the system v_o is still the product of the input v_{in} and the gain; however, the input v_{in} is the difference of the source voltage v_s and the feedback voltage v_f. That is, if the output somehow changes in amplitude due to change in temperature, etc., the feedback will cause it to change in the opposite direction and maintain a fairly constant and stable output.

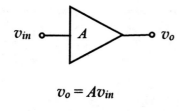

$v_o = Av_{in}$

Figure 12-1:
Block diagram of an amplifier without feedback

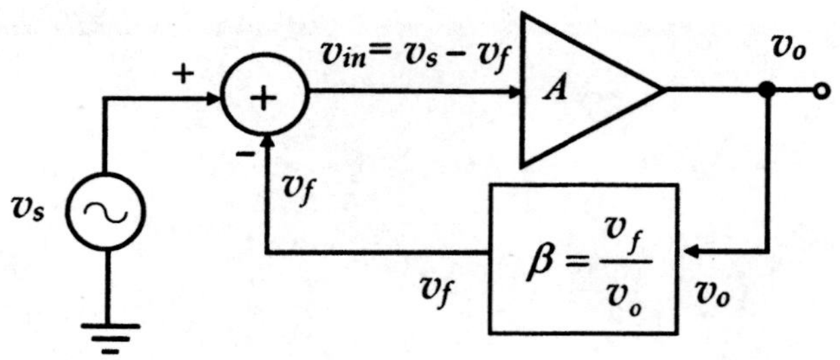

Figure 12-2: Block diagram of an amplifier with negative feedback

From the above two block diagrams, we conclude that

$$v_o = Av_{in} = A(v_s - v_f) = A(v_s - \beta v_o) = Av_s - \beta Av_o \qquad (12\text{-}1)$$

Rearranging the variables, we obtain

$$v_o + \beta Av_o = Av_s \qquad (12\text{-}2)$$

$$v_o(1 + \beta A) = Av_s \qquad (12\text{-}3)$$

$$A_f = \frac{v_o}{v_s} = \frac{A}{1 + \beta A} = \frac{A}{1 + T} \qquad (12\text{-}4)$$

where A is the gain without feedback, A_f is the gain with feedback, and $T = \beta A$ is the loop gain. In order to establish pure negative feedback βA must be positive ($\beta A > 0$). If $\beta A = 0$, there is no feedback, and if βA is negative ($\beta A < 0$), then the feedback is positive.

12.2.1 Gain Desensitivity

The gain of a more predictable system must be less sensitive or *desensitive* to changes in the system's internal component characteristics and/or changes in the operating environment. One major achievement with negative feedback is the increased *gain desensitivity* to the component values in the forward path A. In this section, we intend to determine the fractional change in gain and the percentile improvement in system stability with negative feedback. For $\beta A \gg 1$, $A_f \approx 1/\beta$ and is almost independent on A.

The gain with single-loop feedback is given by

$$A_f = \frac{v_o}{v_s} = \frac{A}{1+\beta A} \qquad (12\text{-}5)$$

Assuming that β is constant and differentiating the gain with feedback A_f with respect to gain without feedback A, we obtain

$$\frac{dA_f}{dA} = \frac{(1+\beta A) - \beta A}{(1+\beta A)^2} = \frac{1}{(1+\beta A)^2} \qquad (12\text{-}6)$$

$$dA_f = \frac{dA}{(1+\beta A)^2} \qquad (12\text{-}7)$$

Then the *fractional change* (% change) in gain with feedback is

$$\frac{dA_f}{A_f} = \frac{dA}{(1+\beta A)^2}\frac{(1+\beta A)}{A} = \frac{1}{1+\beta A}\frac{dA}{A} \qquad (12\text{-}8)$$

$$\frac{dA_f}{A_f} = \left(\frac{1}{1+\beta A}\right)\frac{dA}{A} \qquad (12\text{-}9)$$

The percentile *improvement in system stability* (*ISS*) is determined by

$$ISS = \frac{dA/A - dA_f/A_f}{dA/A} \qquad (12\text{-}10)$$

Similarly,

$$\frac{dA_f}{A_f} = -\left(\frac{\beta A}{1+\beta A}\right)\frac{d\beta}{\beta} \approx -\frac{d\beta}{\beta}$$

ANALYSIS

Example 12-1

Let the gain of an amplifier without feedback $A = -4950$ and $\beta = v_f/v_o = -0.02$. Determine the fractional change in the gain of the amplifier with feedback and the percentile improvement in system stability if the change in gain without feedback is 10%.

Solution:

$$\frac{dA}{A} = 0.1 = 10\%$$

Then the fractional change with feedback is determined from Equation 12-9

$$\frac{dA_f}{A_f} = \left(\frac{1}{1+\beta A}\right)\frac{dA}{A} = \frac{0.1}{1+(-0.02)(-4950)} = 0.001 = 0.1\%$$

The 10% change in the gain without feedback corresponds to only 0.1% change in the gain with feedback. Therefore, the system with feedback is 100 times more stable than it is without feedback.

The percentile improvement in system stability is determined from Equation 12-10 as

$$ISS = \frac{dA/A - dA_f/A_f}{dA/A} = \frac{10\% - 0.1\%}{10\%} = 0.99 = 99\%$$

Practice Problem 12-1   *Desensitivity*   **ANALYSIS**

Assuming that the gain without feedback $A = -10,000$ and $\beta = -0.05$ determine the fractional change in gain with feedback if the change in gain without feedback is 11%. Also, determine the percentile improvement in system stability.

Answer: $dA_f/A_f = 0.022\%$, $ISS = 99.8\%$

12.3 BASIC FEEDBACK TOPOLOGIES

In classical feedback theory, there are four basic feedback topologies:

- Series-shunt feedback (also referred to as *voltage-series feedback*)
- Shunt-shunt feedback (also referred to as *voltage-shunt feedback*)
- Shunt-series feedback (also referred to as *current-shunt feedback*)
- Series-series feedback (also referred to as *current-series feedback*)

Since we do not intend to do a full classical analysis of all the above feedback topologies, we will only discuss the most popular and practical of the four topologies used in the op-amp applications, the series-shunt feedback.

12.3.1 Series-Shunt Feedback

A block diagram of a series-shunt feedback topology is shown in Figure 12-3.

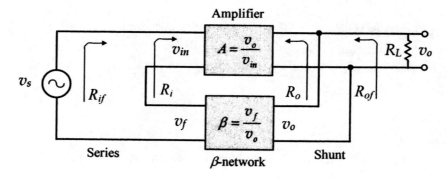

Figure 12-3: Block diagram of a series-shunt feedback topology

The reason this feedback topology is referred to as *series-shunt* is now apparent from the above block diagram. The connection to the *β-network* at the input side is a series connection, while it is a shunt connection at the output side; hence, *series-shunt*.

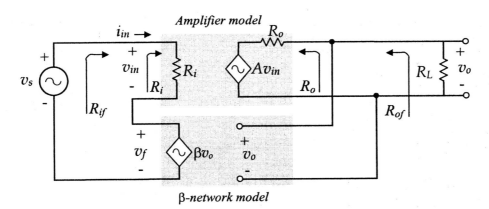

Figure 12-4: Equivalent model of the *series-shunt* feedback

Assuming that the gain, input resistance, and output resistance of the basic amplifier without feedback are A, R_i, and R_o, respectively, we would like to determine the same parameters with feedback A_f, R_{if}, and R_{of} for the *series-shunt* connection as shown in Figure 12-4 above. The gain with feedback is determined with Equation 12-4 as follows:

$$A_f = \frac{v_o}{v_s} = \frac{A}{1+\beta A}$$

Applying *KVL* around the input loop results in the following:

$$v_s = v_{in} + v_f = v_{in} + \beta v_o \qquad (12\text{-}11)$$

$$v_s = v_{in} + \beta A v_{in} = v_{in}(1+\beta A) \qquad (12\text{-}12)$$

Input resistance R_{if}:

The input resistance of the amplifier without feedback R_i and the input resistance with feedback of the system R_{if} are defined as follows:

$$R_i = \frac{v_{in}}{i_{in}} \qquad (12\text{-}13)$$

$$R_{if} = \frac{v_s}{i_s} = \frac{v_s}{i_{in}} \qquad (12\text{-}14)$$

$$R_{if} = \frac{v_s}{i_s} = \frac{v_{in}(1+\beta A)}{i_{in}} = R_i(1+\beta A) \qquad (12\text{-}15)$$

$$R_{if} = R_i(1+\beta A) \qquad (12\text{-}16)$$

Output resistance R_{of}:

To determine the output resistance of the system, we will apply the two-port network theory developed in Chapter 4, which requires the application of a test voltage at the output with the input current set to zero.

Chapter 12

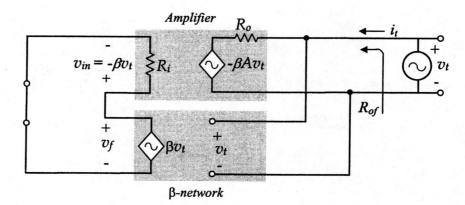

Figure 12-5: Series-shunt feedback circuit with input source set to zero, and a test voltage applied at the output

The output resistance with feedback can now be defined and determined as follows:

$$R_{of} = \left.\frac{v_t}{i_t}\right|_{v_s = 0} \quad (12\text{-}17)$$

The current i_t is the current through the resistance R_o and is determined as follows:

$$i_t = \frac{v_t - (-\beta A v_t)}{R_o} = \frac{v_t(1 + \beta A)}{R_o} \quad (12\text{-}18)$$

Substituting for i_t in Equation 12-17 yields the following equation for R_{of}:

$$R_{of} = \frac{v_t}{i_t} = \frac{v_t}{\frac{v_t(1+\beta A)}{R_o}} = \frac{R_o}{1+\beta A} \quad (12\text{-}19)$$

$$R_{of} = \frac{R_o}{1+\beta A} \quad (12\text{-}20)$$

Therefore, in *series-shunt* feedback, the input resistance increases with a factor of $(1 + \beta A)$ while the output resistance decreases by the same factor.

Note that the negative feedback effectiveness can be maintained within a finite frequency range, because the gain A drops at higher frequencies.

Modeling the β-network:

The two-port network model and the underlying theory have been discussed thoroughly in Chapter 4. Since the *β-network* is actually a two-port network, we will model it as such, as illustrated in Figure 12-6 below:

Figure 12-6: Two-port network model of the β-network

In the above model, v_1 and i_2 are considered to be functions of i_1 and v_2, where i_1 and v_2 are the test signals. The *hybrid* equivalent circuit or the *h-parameter model* of the above two-port network is illustrated in Figure 12-7 below:

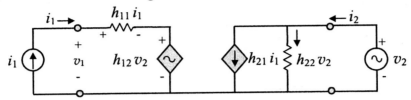

Figure 12-7: The *hybrid* or *h-parameter* model of the β-network

where,

$$v_1 = h_{11}i_1 + h_{12}v_2 \qquad (12\text{-}21)$$
$$i_2 = h_{21}i_1 + h_{22}v_2 \qquad (12\text{-}22)$$

The *h-parameters* utilized in the above model were defined and evaluated in Chapter 4 and are repeated below for convenience:

$$h_{11} = \left.\frac{v_1}{i_1}\right|_{v_2=0} \qquad\qquad h_{12} = \left.\frac{v_1}{v_2}\right|_{i_1=0}$$

$$h_{21} = \left.\frac{i_2}{i_1}\right|_{v_2=0} \qquad\qquad h_{22} = \left.\frac{i_2}{v_2}\right|_{i_1=0}$$

The input current i_i is extremely small; that is, $i_s = i_i = i_1 \cong 0$, and the parameter h_{21} of the *β-network* is also very small in value. Hence, the current source $h_{21}i_1$, whose amplitude is approximately zero, can be replaced with an open-circuit.

The equivalent *series-shunt* feedback circuit, with the *β-network* replaced with its hybrid model, is shown in Figure 12-8 below:

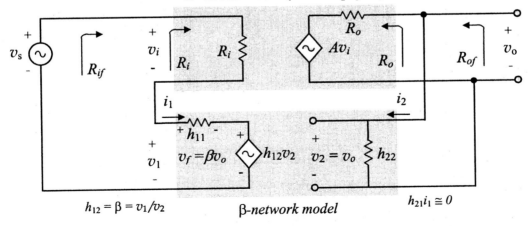

Figure 12-8: Equivalent circuit of the *series-shunt* feedback with the *hybrid* or *h-parameter* model of the *β-network*

Example 12-2 — Series-Shunt Feedback — ANALYSIS

Having developed the *hybrid* model for the *series-shunt* feedback network, let us now apply it to the analysis of a practical and familiar op-amp configuration with *series-shunt* feedback, the non-inverting amplifier.

741 op-amp:

$A = A_{OL} = 2 \times 10^5$

$R_i = R_{i(OL)} = 2 \text{ M}\Omega$

$R_o = R_{o(OL)} = 75 \text{ }\Omega$

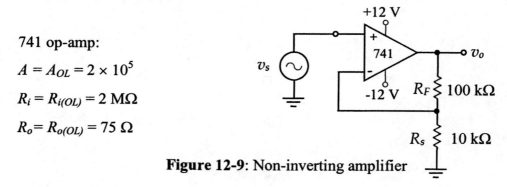

Figure 12-9: Non-inverting amplifier

The open-loop model of the op-amp, derived from the open-loop characteristics, is illustrated in Figures 12-10 and 12-11 below:

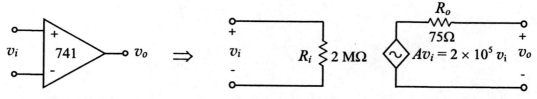

Figure 12-10: 741 op-amp

Figure 12-11: Open-loop model of 741 op-amp

We will now illustrate the circuit diagram of the non-inverting amplifier utilizing the open-loop model of the op-amp and its feedback loop, known as the *β-network*, as shown in Figure 12-12 below:

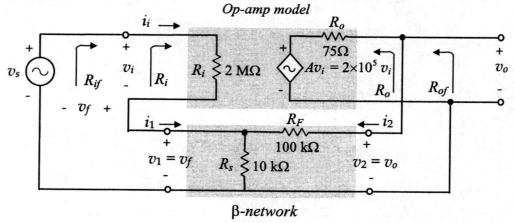

Figure 12-12: *Series-shunt* feedback illustration of the non-inverting amplifier

From the above *series-shunt* feedback circuit, we intend to determine the gain with feedback A_f, input resistance with feedback R_{if}, and output resistance with feedback R_{of}. But first, we will isolate the *β-network* and determine its *h-parameters*, as follows:

Figure 12-13: The β-network

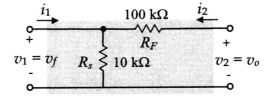

Determining the *h-parameter* h_{11}:

The *h-parameter* h_{11} is defined by the following equation and is realized as shown in Figure 12-14 below:

$$h_{11} = \left.\frac{v_1}{i_1}\right|_{v_2=0}$$

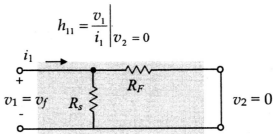

Figure 12-14: β-network with v_2 set to zero

$$i_1 = \frac{v_1}{R_s \| R_F} \tag{12-23}$$

$$h_{11} = \frac{v_1}{i_1} = \frac{v_1}{v_1} R_s \| R_F = R_s \| R_F \tag{12-24}$$

$$h_{11} = R_s \| R_F \tag{12-25}$$

$h_{11} = 100\ \text{k}\Omega \| 10\ \text{k}\Omega \cong 9.1\ \text{k}\Omega$

Determining the *h-parameter* h_{12}:

The *h-parameter* h_{12} is defined by the following equation and is realized as shown in Figure 12-15 below:

$$h_{12} = \left.\frac{v_1}{v_2}\right|_{i_1=0}$$

Figure 12-15: β-network with i_1 set to zero

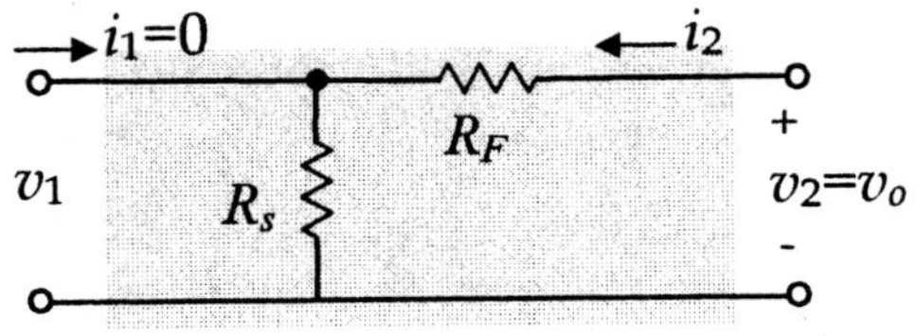

$$v_1 = v_2 \frac{R_s}{R_s + R_F} \tag{12-26}$$

Chapter 12

$$h_{12} = \frac{v_1}{v_2} = \frac{R_s}{R_s + R_F} = \beta \quad (12\text{-}27)$$

$$h_{12} = \beta = \frac{R_s}{R_s + R_F} = \frac{10\,k\Omega}{10\,k\Omega + 100\,k\Omega} \cong 0.091\text{ mS}$$

Determining the *h-parameter* h_{22}:

The *h-parameter* h_{22} is defined by the following equation, which can also be realized with Figure 12-15 above:

$$h_{22} = \left.\frac{i_2}{v_2}\right|_{i_1 = 0} \quad (12\text{-}28)$$

where,

$$i_2 = \frac{v_2}{R_s + R_F} \quad (12\text{-}29)$$

Hence,

$$h_{22} = \frac{i_2}{v_2} = \frac{1}{R_s + R_F} \quad (12\text{-}30)$$

or

$$\frac{1}{h_{22}} = R_s + R_F \quad (12\text{-}31)$$

$$h_{22} = \frac{1}{R_s + R_F} = \frac{1}{10\,k\Omega + 100\,k\Omega} \cong 0.0091\text{mS} \qquad \frac{1}{h_{22}} = R_s + R_F = 110\,k\Omega$$

Having determined the *h-parameters* of the *β-network*, let us now redraw the equivalent circuit of the non-inverting amplifier with the *β-network* replaced with its *hybrid* model.

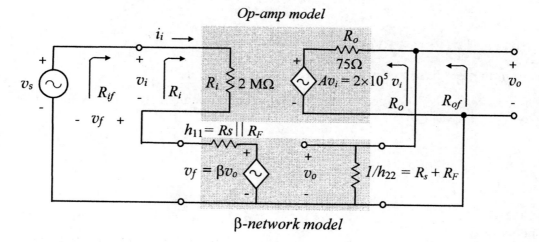

Figure 12-16: Equivalent circuit model of the non-inverting amplifier

Chapter 12

Voltage gain without feedback (A_v):

To determine the gain without feedback ($A_v = v_o/v_i$), we need to analyze the basic amplifier circuit without feedback ($v_f = 0$), but loaded with the β-network, as shown in Figure 12-17 below:

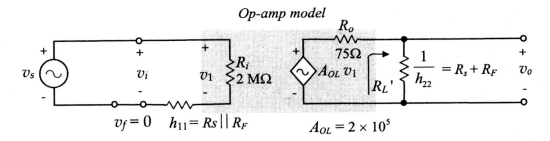

Figure 12-17: Equivalent circuit model of the op-amp without feedback but loaded with the β-network.

The output voltage v_o may be determined by applying the voltage-divider rule at the output, as follows:

$$v_o = A_{OL}\, v_1 \frac{R_L'}{R_o + R_L'} = A_{OL}\, v_1 \frac{R_s + R_F}{R_o + (R_s + R_F)} \tag{12-32}$$

$$\frac{v_o}{v_1} = A_{OL} \frac{R_L'}{R_o + R_L'} = A_{OL} \frac{R_s + R_F}{R_o + (R_s + R_F)} \tag{12-33}$$

$$v_1 = v_i \frac{R_i}{R_i + h_{11}} \tag{12-34}$$

$$\frac{v_1}{v_i} = \frac{R_i}{R_i + h_{11}} \tag{12-35}$$

$$A_v = \frac{v_o}{v_i} = \frac{v_o}{v_1} \cdot \frac{v_1}{v_i} \tag{12-36}$$

$$A_v = \frac{v_o}{v_i} = A_{OL} \frac{R_L'}{R_o + R_L'} \times \frac{R_i}{R_i + h_{11}} \tag{12-37}$$

$$A_v = \frac{v_o}{v_i} = A_{OL} \frac{R_s + R_F}{R_o + (R_s + R_F)} \times \frac{R_i}{R_i + h_{11}} \cong A_{OL} \tag{12-38}$$

$$A_v = \frac{v_o}{v_i} = 2\times 10^5 \frac{110\,\text{k}\Omega}{75\,\Omega + 110\,\text{k}\Omega} \times \frac{2\,\text{M}\Omega}{2\,\text{M}\Omega + 9.1\,\text{k}\Omega} = 1.99 \times 10^5 \cong A_{OL}$$

Voltage gain with feedback (A_{vf}):

Having determined the gain without feedback (A_v), we will now determine the gain with feedback (A_{vf}) with Equation 12-4, as follows:

$$A_{vf} = \frac{v_{of}}{v_{if}} = \frac{A_v}{1 + \beta A_v} \tag{12-39}$$

$$A_{vf} = \frac{1.99 \times 10^5}{1 + (0.091 \times 1.99 \times 10^5)} = 10.988$$

With a large open-loop gain, the product βA_v is much larger than 1 and thus,

$$(1 + \beta A_v) \cong \beta A_v \tag{12-40}$$

Hence,

$$A_{vf} = \frac{v_{of}}{v_{if}} \cong \frac{A_v}{\beta \cdot A_v} = \frac{1}{\beta} = \frac{R_F + R_S}{R_s} \tag{12-41}$$

$$A_{vf} \cong \frac{R_F}{R_s} + 1 \tag{12-42}$$

$$A_{vf} \cong \frac{100 \text{ k}\Omega}{10 \text{ k}\Omega} + 1 = 11$$

The above equation, which is the result of a detailed analysis with negative feedback utilizing the *series-shunt* feedback topology, is identical to Equation 11-20, which was obtained by analyzing the virtual equivalent circuit of the non-inverting amplifier.

Input resistance with feedback (R_{if}):

The input resistance is determined with Equation 12-16, as follows:

$$R_{if} = R_i (1 + \beta A_v)$$

$$R_{if} = 2 \text{ M}\Omega (1 + 0.091 \times 1.99 \times 10^5) = \mathbf{36.22 \text{ G}\Omega}$$

Recall that it was mentioned in the previous chapter that the input resistance of the non-inverter (non-inverting amplifier and unity-gain buffer) is in the GΩ range, and thus it can be considered approximately infinite. Although you have had to accept this possibility as a fact, you were probably never completely satisfied and could not wait to find out how and why. Now that we have demonstrated how large the input resistance can really be, it must be clear that the input resistance of a non-inverter is equivalent to an open-circuit and can certainly be approximated by infinity.

Output resistance with feedback (R_{of}):

The output resistance is determined with Equation 12-20, as follows:

$$R_{of} = \frac{R_o}{1 + \beta A_v}$$

$$R_{of} = \frac{75 \, \Omega}{(1 + 0.091 \times 1.99 \times 10^5)} = \mathbf{4.1 \text{ m}\Omega}$$

Again, in the previous chapter, it was mentioned that the output resistance of the op-amp with negative feedback is a few mΩ and thus can be considered to be approximately zero. Having evaluated the exact output resistance of a typical non-inverting amplifier, now we have an idea of how small it can really be, as a result of negative feedback.

Chapter 12

Bandwidth with feedback:

Since the gain of the op-amp, with or without feedback, is a maximum at DC and drops at higher frequencies, it can be modeled as a first order low-pass system, as illustrated in Figure 12-18 below:

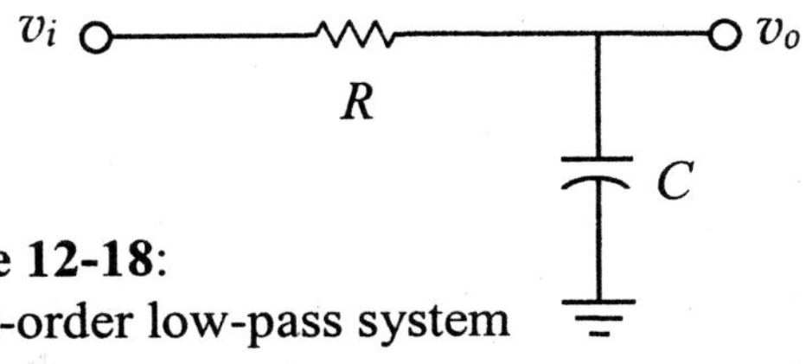

$$X_C = \frac{1}{j\omega C}$$

$$|X_C| = \frac{1}{\omega C} = \frac{1}{2\pi f C}$$

Figure 12-18: A first-order low-pass system

Applying the voltage-divider rule, the output v_o is determined as follows:

$$v_o = v_i \frac{X_C}{R + X_C} \tag{12-43}$$

$$A_v = \frac{v_o}{v_i} = \frac{X_C}{R + X_C} \tag{12-44}$$

Dividing the numerator and the denominator by X_C and substituting for X_C in terms of $j\omega$ yields the following for the gain A_v:

$$A_v = \frac{v_o}{v_i} = \frac{1}{1 + \frac{R}{X_C}} = \frac{1}{1 + j\omega CR} \tag{12-45}$$

At DC, the magnitude gain is 1 and the output is a maximum. As the frequency is increased, the gain decreases, approaching zero eventually. However, at some frequency, referred to as the *high cutoff* frequency (ω_H), the magnitude of the capacitive reactance $|X_C|$ equals the resistance R; that is, at ω_H:

$$R = |X_C| = \frac{1}{\omega_H C} \tag{12-46}$$

Solving for ω_H yields the following for the cutoff frequency:

$$\omega_H = \frac{1}{RC} \tag{12-47}$$

Substituting for $RC = 1/\omega_H$ in Equation 12-45, we obtain the following for the gain:

$$A_v = \frac{v_o}{v_i} = \frac{1}{1 + j\omega CR} = \frac{1}{1 + \frac{j\omega}{\omega_H}} \tag{12-48}$$

For the sake of simplicity and convenience, let $s = j\omega$ in the above equation; then the equation for the gain as a function of s or $j\omega$ is the following:

$$A_v(s) = \frac{v_o}{v_i} = \frac{1}{1 + \frac{j\omega}{\omega_H}} = \frac{1}{1 + \frac{s}{\omega_H}} \tag{12-49}$$

535

For the simple *RC* circuit of Figure 12-18, the DC gain or A_o is 1. However, the DC gain for the non-inverting amplifier $A_o \geq 1$. Hence, Equation 12-49 is slightly modified to reflect the DC gain of the op-amp configuration, as follows:

$$A_v(s) = \frac{v_o}{v_i} = \frac{A_{vo}}{1 + \frac{s}{\omega_H}} \tag{12-50}$$

Having determined the gain without feedback, the gain with feedback is determined from Equation 12-4

$$A_{vf}(s) = \frac{A_v(s)}{1 + \beta A_v(s)} \tag{12-51}$$

$$A_{vf}(s) = \frac{\frac{A_{vo}}{1 + (s/\omega_H)}}{1 + \beta \frac{A_{vo}}{1 + (s/\omega_H)}} \tag{12-52}$$

Multiplying the numerator and the denominator by $(1 + s/\omega_H)$ results in the following:

$$A_{vf}(s) = \frac{A_{vo}}{1 + \frac{s}{\omega_H} + \beta A_o} \tag{12-53}$$

$$A_{vf}(s) = \frac{A_{vo}}{1 + \frac{s}{\omega_H} + \beta A_{vo}} = \frac{A_{vo}}{\frac{s}{\omega_H} + (1 + \beta A_{vo})} \tag{12-54}$$

Dividing the numerator and the denominator by $(1 + \beta A_o)$, we obtain

$$A_{vf}(s) = \frac{\frac{A_{vo}}{1 + \beta A_o}}{1 + \frac{s}{\omega_H(1 + \beta A_o)}} = \frac{\left(\frac{A_o}{1 + \beta A_o}\right)}{1 + \left(\frac{s}{\omega_H(1 + \beta A_o)}\right)} = \frac{A_{vfo}}{1 + \frac{s}{\omega_{Hf}}} \tag{12-55}$$

Inspecting the above equation and comparing its numerator to Equation 12-50, the numerator is the DC gain with feedback given by

$$A_{vfo} = \frac{A_{vo}}{1 + \beta A_{vo}} \tag{12-56}$$

Again, inspecting the denominator and comparing it to Equation 12-50, we can conclude that the cutoff frequency with feedback is expressed by

$$\omega_{Hf} = \omega_H (1 + \beta A_{vo}) \tag{12-57}$$

Therefore, as the gain with feedback A_{of} drops by a factor of $(1+\beta A_{vo})$, the bandwidth with feedback $BW_f = \omega_{Hf}$ increases by the same factor, as illustrated in Figure 12-19.

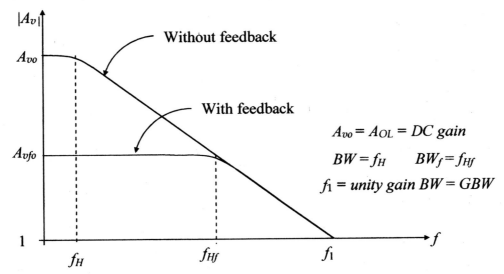

Figure 12-19: The op-amp gain with and without feedback versus frequency.

Example 12-3 — Feedback — ANALYSIS

Let us plot the gain with and without feedback versus frequency for the non-inverting amplifier of Example 12-2.

Solution:

For the 741 op-amp, $A_{OL} = 2 \times 10^5$ and the bandwidth without feedback is

$$f_H = \frac{f_1}{A_{OL}} = \frac{1\,\text{MHz}}{2 \times 10^5} = 5\,\text{Hz}$$

For $A_{vf} = 11$, the bandwidth with feedback for small-signal operation ($V_{om} < 1$ V) is

$$BW_f = f_{Hf} = \frac{f_1}{A_{vfo}} = \frac{1\,\text{MHz}}{11} = 90.9\,\text{kHz}$$

Let us also check and verify the bandwidth with feedback f_{Hf} from Equation 12-57.

$$\omega_{Hf} = \omega_H (1 + \beta A_{vo})$$

or

$$BW_f = f_{Hf} = f_H (1 + \beta A_{vo}) = 5\,\text{Hz}\,(1 + 0.091 \times 1.99 \times 10^5) = 90.55\,\text{kHz} \text{ for } V_{om} < 1\text{ V}$$

The plot of the gain with and without feedback versus frequency is exhibited in Figure 12-20.

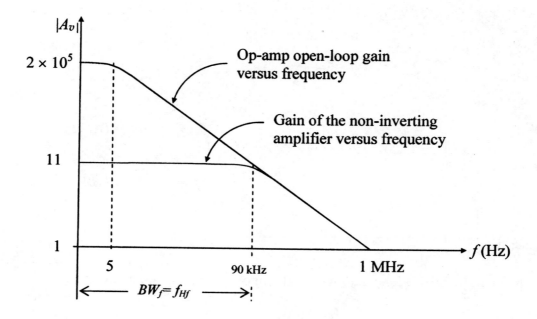

Figure 12-20: The non-inverting amplifier voltage gain versus frequency

Section Summary — Series-shunt Feedback — ANALYSIS

Summary of Equations for the Analysis of Series-shunt Feedback Amplifiers

The non-inverting amplifier is a practical example of the *series-shunt* feedback.

$A_{vo} \cong A_{OL}$ = open-loop gain

$\beta = \dfrac{R_S}{R_F + R_S}$ = gain of the β-network

$A_{vf} = \dfrac{A_v}{1 + \beta A_v}$

$\omega_{Hf} = \omega_H (1 + \beta A_v)$

$f_{Hf} = BW_f = \dfrac{f_1}{A_{vfo}}$

$f_H = BW = \dfrac{f_1}{A_o} = \dfrac{GBW}{A_{OL}}$

$R_{if} = R_i (1 + \beta A_v)$

$R_{of} = \dfrac{R_o}{1 + \beta A_v}$

R_i and R_o are the open-loop input and output resistances of the op-amp, respectively.

Practice Problem 12-2 — Series-shunt Feedback — ANALYSIS

Determine the gain, bandwidth, and the exact input and output resistances for the non-inverting amplifier of Figure 12-21, as shown below:

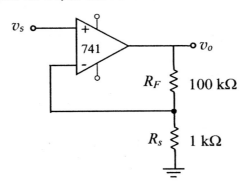

Figure 12-21: A non-inverting amplifier

Answers: $A_{vf} = 100.95$, $R_{if} = 3.96$ GΩ, $R_{of} = 37.8$ mΩ, BW = 9.9 kHz.

12.4 OP-AMP CHARACTERISTICS

We have introduced and utilized some of the typical characteristics and/or parameters of some general-purpose operational amplifiers such as: the *open-loop gain, input resistance, output resistance, gain-bandwidth product,* and *common-mode rejection ratio,* as listed in manufacturers' data sheets. In this section, we would like to introduce several more important parameters/characteristics of op-amps as listed in the attached data sheets.

12.4.1 DC Characteristics

Input bias current I_{IB}:
The two input DC bias currents to the op-amp are the base currents I_{B1} and I_{B2} of the first stage diff-amp, which are equal for an ideal diff-amp and/or op-amp, and are nearly equal for a practical op-amp. The typical value of the I_{IB} for a general-purpose op-amp such as the 741 is about 80 nA.

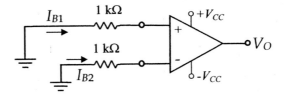

Figure 12-22: Input bias currents

Input offset current I_{IO}:
The *input offset current* is the difference between the two input DC bias currents I_{B1} and I_{B2}, and thus, it equals zero for an ideal op-amp. However, in the case of a practical op-

amp, the input offset current may not be exactly zero. The typical value of the I_{IO} for a general-purpose op-amp such as the 741 is 20 nA.

$$I_{IO} = |I_{B2} - I_{B1}| \qquad (12\text{-}58)$$

Input offset voltage V_{IO}:

Recall that the two input terminals of the op-amp are the two inputs of the first stage diff-amp, and if both are tied to ground as shown in Figure 12-23(a) through a resistor R, the DC voltages V_1 and V_2 must be equal for an ideal op-amp. For the practical op-amp, however, V_2 might be somewhat different from V_1. The difference between the two DC voltages at the two input terminals of the op-amp is the *input offset voltage V_{IO}*. The typical value of the V_{IO} for a general-purpose op-amp such as the 741 is 1 mV.

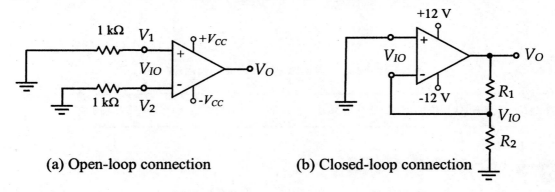

(a) Open-loop connection (b) Closed-loop connection

Figure 12-23: Measuring the V_{IO}

With the closed-loop connection as shown in Figure 12-23(b), the V_{IO} is amplified, which makes it easier to measure such a small offset voltage. Applying the voltage-divider rule at the output results in the following equation for the V_{IO}:

$$V_{IO} = V_O \frac{R_2}{R_1 + R_2} \qquad (12\text{-}59)$$

or

$$V_O = V_{IO} \frac{R_2 + R_1}{R_2} = V_{IO}\left(1 + \frac{R_1}{R_2}\right) \qquad (12\text{-}60)$$

Supply voltage rejection ratio (PSRR):

Ideally all ripple, hum, and noise from the power supply must be prevented from reaching the output of the op-amp. The supply voltage rejection ratio or power supply rejection ratio (*PSRR*) is a measure of how well the op-amp rejects these undesirable effects and parameters from the power supply and prevent them from reaching the output. The *PSRR* is generally expressed in dB and is determined as follows:

$$PSRR = \frac{\Delta V_{CC}}{\Delta V_{IO}} \qquad (12\text{-}61)$$

That is, V_{IO} is measured with the supply voltage $\pm V_{CC}$ at a certain level such as ± 15 V, and then the same measurement is repeated with $\pm V_{CC}$ at a different level such as ± 10 V. Then the *PSRR* is determined as the ratio of ($\Delta V_{CC} = V_{CC1} - V_{CC2}$) to ($\Delta V_{IO} = V_{IO1} - V_{IO2}$).

$$PSRR(dB) = 20 \log(PSRR) \tag{12-62}$$

The typical value of the PSRR for a general-purpose op-amp such as the 741 is 96 dB.

Output short circuit current I_{SC}:
All op-amps are usually protected against overload and short-circuit. That is, if the output is overloaded or short-circuited, the op-amp will only supply the maximum output current it has been designed for. The maximum output current or the output short-circuit current for the 741 op-amp is 25 mA.

12.4.2 AC Characteristics

Large-signal voltage gain A_{VOL}:
The large-signal voltage gain is actually the open-loop gain of the op-amp, which is listed as such in most data sheets. The typical value of the A_{VOL} for a general-purpose op-amp such as the 741 is 200 V/mV as listed in the data sheets, which is another way of saying that the open-loop gain $A_{VOL} = 2 \times 10^5$.

Common mode rejection ratio CMRR:
When the op-amp is utilized in the diff-amp configuration, ideally any signal common to both inputs will be prevented from reaching the output of the op-amp. For a practical op-amp, however, a fraction of the common signal may reach the output. The CMRR, which is generally expressed in dB, is a measure of how well the op-amp can reject or attenuate the common signals such as noise, which would be the common signal to both inputs, if the op-amp is used in diff-amp configuration. With the common mode connection, as shown in Figure 12-24, the output is determined as follows:

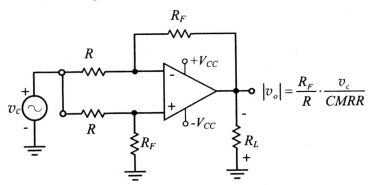

$$|v_o| = \frac{R_F}{R} \cdot \frac{v_c}{CMRR}$$

Figure 12-24: Common mode connection for measuring the CMRR of the op-amp

Solving for CMRR in the above equation for v_o, results in the following:

$$CMRR = \frac{R_F}{R} \cdot \frac{v_c}{v_o} \tag{12-63}$$

Hence, the CMRR of the op-amp can be measured simply by measuring the output of the above common mode connection of the diff-amp configuration with an applied input signal v_c. The typical value of the CMRR for a general-purpose op-amp such as the 741 is 90 dB, as listed in the data sheets.

Example 12-4 — CMRR — ANALYSIS

Assuming that $R_F = 100\ k\Omega$ and $R_S = 1\ k\Omega$ in the diff-amp of Figure 12-24, let us determine the *CMRR* for the op-amp if $v_o = 1\ mV$ is observed at the output with an application of $v_c = 1\ V$.

Solution:

$$CMRR = \frac{R_F}{R} \cdot \frac{v_c}{v_o} = 100 \times \frac{1\ V}{1\ mV} = 10^5$$

$$20 \log(10^5) = 100\ dB$$

Gain-bandwidth product (GBW):
Gain-bandwidth product is actually the unity-gain bandwidth of the op-amp, which may be referred to as *GBW*, *GBP*, *BW*, *GBWP*, or f_1 in some data sheets. For the 741 op-amp the *GBW* is 1 MHz, and for the 351 op-amp it is 4 MHz. Hence, for an amplifier with a gain of 10, the bandwidth is as follows:

$$BW\bigg|_{741} = \frac{GBW}{|A_v|} = \frac{1\ MHz}{10} = 100\ kHz$$

$$BW\bigg|_{351} = \frac{GBW}{|A_v|} = \frac{4\ MHz}{10} = 400\ kHz$$

Practice Problem 12-3 — CMRR — ANALYSIS

Determine the *CMRR* for the following op-amp circuit:

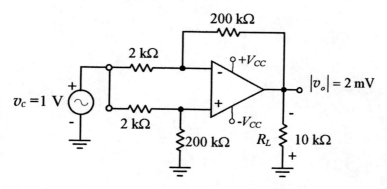

Answer: $CMRR = 94\ dB$

Slew rate (SR):
The slew rate is defined as the maximum rate of change of the output voltage in time and it is usually expressed in V/μs. Algebraically, slew rate is defined as follows:

$$SR = \frac{dv_o}{dt}\bigg|_{max} \tag{12-64}$$

Chapter 12

The slew rates of the 741 and 351 op-amps are 0.5 V/μs and 13 V/μs, respectively. This means that the output of the 741 op-amp cannot change more than 0.5 V/μs.

Example 12-5 *Slew Rate* ANALYSIS

Consider the non-inverting amplifier of Figure 12-25 as shown below with unipolar square-wave input representing a stream of binary data:

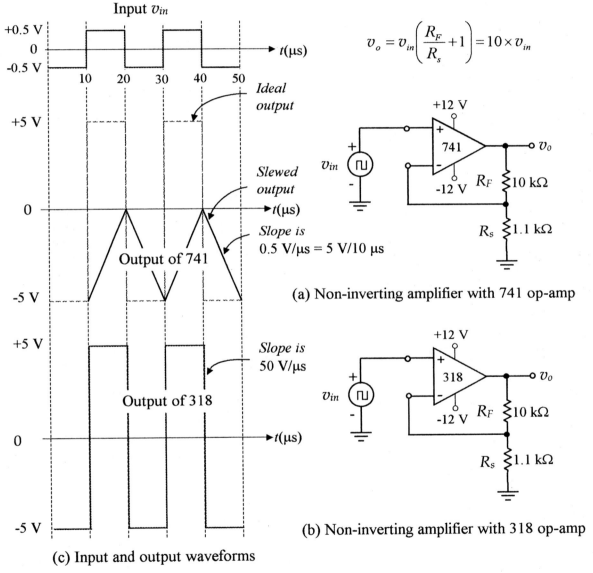

$$v_o = v_{in}\left(\frac{R_F}{R_s}+1\right) = 10 \times v_{in}$$

(a) Non-inverting amplifier with 741 op-amp

(b) Non-inverting amplifier with 318 op-amp

(c) Input and output waveforms

Figure 12-25: Examining the slew rate of two different op-amps

As you can see, because of an inferior slew rate, the output of the 741 op-amp is not even close to the amplified version of the input signal. However, the 318 op-amp, whose slew rate is 100 times better than that of the 741, produces a nearly perfect amplified version of the input signal, which is due to its superior slew rate.

Chapter 12

Effect of slew rate limitation on sinusoidal signals:

Let the input to the non-inverting amplifier of Figure 12-25 have the following standard sinusoidal format:

$$v_{in} = V_m \sin(\omega t) = V_m \sin(2\pi f t)$$

$$v_o = A_v \times v_{in} = A_v [V_m \sin(\omega t)] = A_v [V_m \sin(2\pi f t)]$$

$$SR = \left.\frac{dv_o}{dt}\right|_{max}$$

$$SR = \left. A_v \omega V_m \cos(\omega t) \right|_{max}$$

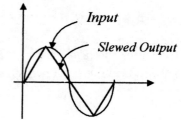

Figure 12-26: Effect of slewing

Since the maximum value of a cosine is 1, the slew rate is the following:

$$SR = A_v \omega V_m = 2\pi f V_m A_v \qquad (12\text{-}66)$$

Therefore, the maximum operating frequency of an op-amp with a given slew rate is the following:

$$f_{max} = \frac{SR}{2\pi V_m A_v} = \frac{SR}{2\pi V_{o(m)}} \qquad (12\text{-}67)$$

Example 12-6 (Slew Rate) ANALYSIS

If the input signal to the amplifier of Figure 12-27 is given as follows:

$$v_{in} = 0.5 \sin(10{,}000\pi t)$$

Determine the minimum slew rate required of the op-amp.

Solution: $A_v = 21$, $f = 5$ kHz

$V_{o(m)} = 0.5 \text{ V} \times 21 = 10.5 \text{ V}$

$$f_{max} = \frac{SR}{2\pi V_{o(m)}}$$

$$SR_{(min)} = 2\pi V_{o(m)} f_{max}$$

$SR_{(min)} = 2\pi \times 10.5 \text{ V} \times 5 \text{ kHz} = 3.3 \times 10^5 \text{ VHz}$

$SR_{(min)} = 3.3 \times 10^5 \text{ V} \dfrac{10^{-6}}{10^{-6}} = 0.33 \dfrac{\text{V}}{\mu\text{s}}$

Hence, $SR \geq 0.33$ V/μs

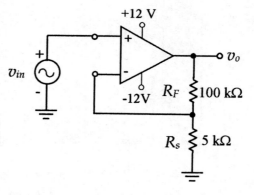

Figure 12-27: Non-inverting amplifier

Chapter 12

12.5 SINGLE-SUPPLY OPERATION

Although most operational amplifiers are designed to operate from a dual power supply ($\pm V_{CC}$), they can also function with a single power supply ($+V_{CC}$). However, certain op-amps such as LM324, TL3472, TLC074, MC33171, and MC33272A are specifically designed to operate from a single-supply.

Consider the following single-supply inverting amplifier:

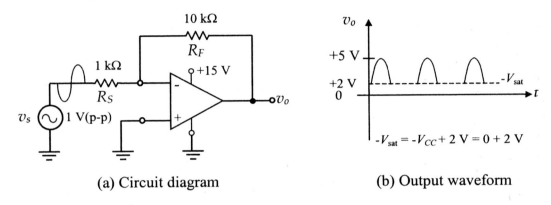

(a) Circuit diagram   (b) Output waveform

Figure 12-28: Inverting amplifier with single-supply connection

If the op-amp were connected to a dual power supply ($\pm V_{CC} = \pm 15$ V), then the expected output would be a 10 V(p-p) full sinusoidal signal. However, with $-V_{CC} = 0$, $-V_{sat} = +2$V and thus the output will be clipped off at +2 V, as shown in Figure 12-28(b) above. This clipped-off signal certainly is not a satisfactory output. In order to avoid clipping and to obtain a full sinusoid at the output, we may shift the signal upward by introducing a DC offset voltage as follows:

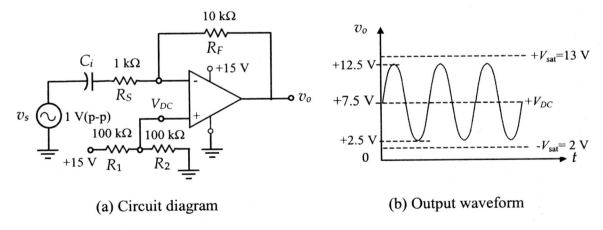

(a) Circuit diagram   (b) Output waveform

Figure 12-29: Single-supply connection with DC offset

Applying the voltage-divider rule at the non-inverting terminal of the op-amp, the DC voltage V_{DC} is the following:

545

$$V_{DC} = V_{CC} \frac{R_2}{R_1 + R_2} \qquad (12\text{-}67)$$

$$V_{DC} = 15 \text{ V} \frac{100 \text{ k}\Omega}{200 \text{ k}\Omega} = 7.5 \text{ V}$$

According to the superposition theorem, we can analyze the above circuit for DC and AC separately and then combine the results.

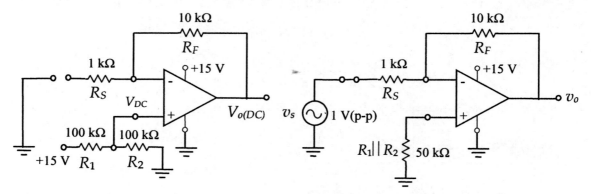

Figure 12-30: DC equivalent circuit of the single-supply amplifier of Figure 12-29

Figure 12-31: AC equivalent circuit of the single-supply amplifier of Figure 12-29

DC Analysis:

The DC equivalent circuit of Figure 12-29 is illustrated in Figure 12-30 above. The capacitor C_i at the input is an open-circuit to DC, and with the offset voltage V_{DC} at the non-inverting terminal the DC voltage at the output is determined as follows:

$$V_{O(DC)} = V_{DC}\left(\frac{R_F}{R_S} + 1\right) = V_{DC}\left(\frac{R_F}{\infty} + 1\right) = V_{DC} \qquad (12\text{-}68)$$

$$V_{O(DC)} = V_{I(DC)} \qquad (12\text{-}69)$$

$V_{O(DC)} = 7.5$ V

AC Analysis:

The AC equivalent circuit of Figure 12-29 is illustrated in Figure 12-31 above. The capacitor C_i at the input is a short-circuit to AC and the signal level at the output is determined as follows:

$$v_o = -v_s \frac{R_F}{R_S}$$

$v_{o(AC)} = -10 \text{ V}_{(p\text{-}p)}$ \hfill (12-70)

Combining Equations 12-68 and 12-70 results in the following equation for the output:

$$v_o = V_{CC} \frac{R_2}{R_1 + R_2} - v_s \frac{R_F}{R_S} \qquad (12\text{-}71)$$

$v_o = 7.5 \text{ V}_{(DC)} - 10 \text{ V}_{(p\text{-}p)}$

For an optimum signal swing at the output, R_1 must equal R_2, so that $V_{DC} = 0.5 V_{CC}$. As shown in Figure 12-29(b), the output of 10 V(p-p) swings symmetrically around the 7.5 V DC offset, producing a full sinusoid at the output. Connecting another coupling capacitor C_o at the output blocks the DC and produces a full sinusoid without DC offset, as shown in Figure 12-32 below:

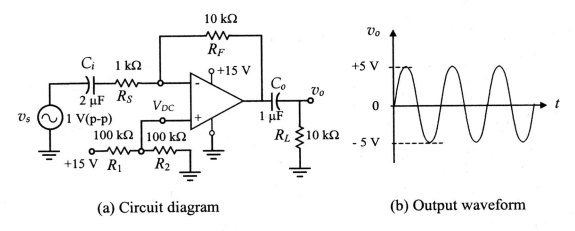

(a) Circuit diagram (b) Output waveform

Figure 12-32: Single-supply output after the output capacitor

Recall that due to the absence of coupling and bypass capacitors with dual power supply, there were no frequency response limitations at the low end of the frequency spectrum. With the single-supply, however, the addition of the two capacitors C_i and C_o introduces two cutoff frequencies f_{Li} and f_{Lo}, where the larger of the two is the dominant low cutoff frequency f_L.

$$f_{Li} = \frac{1}{2\pi R_s C_i} \quad (12\text{-}72)$$

$$f_{Li} = \frac{1}{2\pi \times 1\,k\Omega \times 2\,\mu F} \cong 80\,\text{Hz}$$

$$f_{Lo} = \frac{1}{2\pi R_L C_o} \quad (12\text{-}73)$$

$$f_{Lo} = \frac{1}{2\pi \times 10\,k\Omega \times 1\,\mu F} \cong 16\,\text{Hz}$$

The cutoff frequency f_L corresponds to -3dB gain or $0.707 A_v$. However, to achieve full amplitude at the output and avoid any drop in the gain, the minimum operating frequency $f_{o(min)}$ should be kept approximately one decade above the cutoff frequency f_L. That is,

$$f_{o(min)} \geq 10 f_L \quad (12\text{-}74)$$

Chapter 12

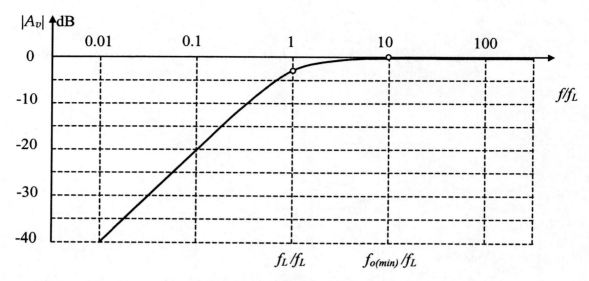

Figure 12-33: Plot of normalized dB gain versus f (*log scale*).

In the above example of Figure 12-32, the dominant cutoff frequency f_L is 80 Hz, and thus the minimum operating frequency $f_{o(min)}$ with full amplitude at the output is 800 Hz.

Table 12-1: Brief table of some single-supply op-amp characteristics

Parameter	LM324	TL3472	MC33171
Supply Voltage (V_{CC})	3 V to 32 V	4 V to 36 V	3 V to 44 V
Gain-bandwidth Product	1 MHz	4 MHz	1.8 MHz
Slew Rate	0.5 V/μs	13 V/μs	2.1 V/μs
Large-signal Voltage Gain	100 dB	-	500 V/mV
Max. Output Current	50 mA	80 mA	5 mA

Practice Problem 12-4   ANALYSIS

Determine the gain A_v, $V_{o(DC)}$, $v_{o(AC)}$, f_L, and $f_{o(min)}$ for the single-supply amplifier of Figure 12-34, as shown below:

Figure 12-34:
Single-supply amplifier

548

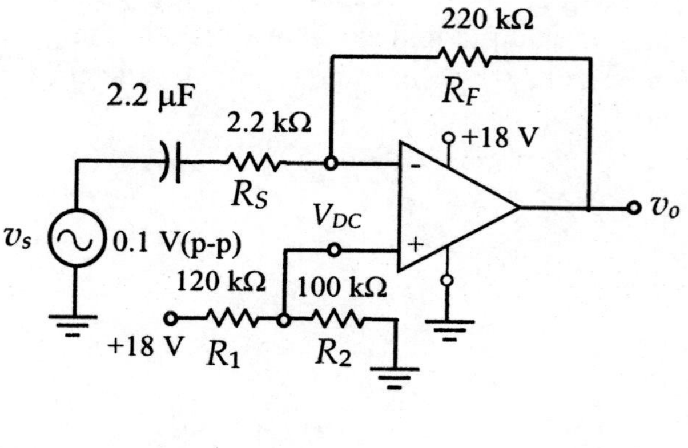

Chapter 12

Answers:

$A_v = -100$, $V_{o(DC)} = 8.18$ V, $v_{o(AC)} = 10$ V(p-p), $f_L = 33$ Hz, and $f_{o(min)} = 330$ Hz

Example 12-7 (Single-Supply) DESIGN

Let us design a single-supply inverting amplifier with a gain of 35 dB, $f_{o(min)} = 33$ Hz, and $f_H \geq 50$ kHz. The expected maximum input signal swing is 100 mV(p-p), and the expected minimum load resistor is 4.7 kΩ.

Solution:

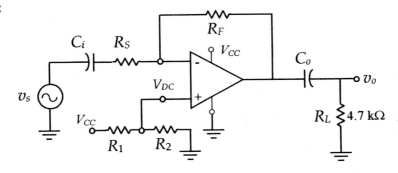

Figure 12-35: Circuit diagram for the design

The 35 dB gain corresponds to a gain of 56 in the linear scale. Hence, let $R_F = 560$ kΩ, and $R_S = 10$ kΩ. The maximum input signal swing of 100 mV(p-p) will produce a maximum output swing of 5.6 V(p-p). The minimum V_{CC} is the maximum output signal swing plus 4 V. Hence, let $V_{CC} = +12$ V.

The minimum operating frequency of 33 Hz requires that the lower cutoff frequency be less than or equal to 3.3 Hz. The coupling capacitors C_i and C_o with their associated resistances R_S and R_L introduce the two lower cutoff frequencies. Therefore, let us determine the minimum capacitor values so that the cutoff frequency $f_L \leq 3.3$ Hz.

$$C_{i(min)} = \frac{1}{2\pi R_s f_L} = \frac{1}{2\pi \times 10 \text{ k}\Omega \times 3.3 \text{ Hz}} = 4.8 \text{ μF}$$

$$C_{o(min)} = \frac{1}{2\pi R_L f_L} = \frac{1}{2\pi \times 4.7 \text{ k}\Omega \times 3.3 \text{ Hz}} = 10.26 \text{ μF}$$

Using standard capacitor values, let $C_i = 4.7$ μF, $C_o = 10$ μF, and $R_1 = R_2 = 100$ kΩ.

$$\text{GBW} = f_H \times A_v = 50 \text{ kHz} \times 56 = 2.8 \text{ MHz}$$

The op-amp must have a GBW ≥ 2.8 MHz. Hence, we will choose the TL3742 op-amp, which has a 4 MHz GBW.

Design summary:

$V_{CC} = 12$ V, $R_F = 560$ kΩ, $R_S = 10$ kΩ, $R_1 = R_2 = 100$ kΩ, $C_i = 4.7$ μF, and $C_o = 10$ μF

Chapter 12

Non-inverting amplifier:
The circuit of a single-supply non-inverting amplifier is shown in Figure 12-36 below.

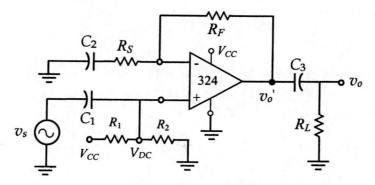

Figure 12-36: Single-supply connection of a non-inverting amplifier

Applying the superposition theorem, as we did with the inverting amplifier, it can be shown that the output is the following:

Before the output capacitor:

$$v_o' = V_{CC} \frac{R_2}{R_1 + R_2} + v_s \left(\frac{R_F}{R_S} + 1 \right) \qquad (12\text{-}75)$$

After the output capacitor:

$$v_o = v_s \left(\frac{R_F}{R_S} + 1 \right) \qquad (12\text{-}76)$$

Each of the coupling capacitors, along with its associated Thevenin's resistance, will introduce a cutoff frequency, as follows:

$$f_{L1} = \frac{1}{2\pi (R_1 \| R_2) C_1} \qquad (12\text{-}77)$$

$$f_{L2} = \frac{1}{2\pi R_S C_2} \qquad (12\text{-}78)$$

$$f_{L3} = \frac{1}{2\pi R_L C_3} \qquad (12\text{-}79)$$

The largest of the above three cutoff frequencies is the dominant cutoff frequency f_L. The upper cutoff frequency is determined with the GBW of the op-amp, as follows:

$$f_H = \frac{GBW}{A_v} \qquad (12\text{-}80)$$

Example 12-8 ANALYSIS

In the non-inverting amplifier of Figure 12-35, let $V_{CC} = 16$ V, $R_1 = 100$ kΩ, $R_2 = 100$ kΩ, $R_S = 2.7$ kΩ, $R_F = 100$ kΩ, $R_L = 4.7$ kΩ, $C_1 = 1$ μF, $C_2 = 4.7$ μF, $C_3 = 4.7$ μF, and $v_s = 0.25$ V(p-p).
  a) Determine the output voltages before and after the output capacitor.
  b) Determine the lower and upper cutoff frequencies.

Solution:
The output before the output capacitor is v_o' and can be determined from Equation 12-75

$$v_o' = V_{CC}\frac{R_2}{R_1+R_2} + v_s\left(\frac{R_F}{R_S}+1\right)$$

$$v_o' = 8 \text{ V(DC)} + 9.5 \text{ V(p-p)}$$

The output after the output capacitor is v_o and is determined from Equation 12-76 to produce

$$v_o = 9.5 \text{ V(p-p)}$$

The lower cutoff frequencies introduced by C_1, C_2, and C_3 are determined from Equations 12-77 through 12-79

$$f_{L1} = \frac{1}{2\pi(R_1\|R_2)C_1}$$

$$f_{L1} = \frac{1}{2\pi \times 50 \text{ k}\Omega \times 1 \text{ μF}} \cong 3.2 \text{ Hz}$$

$$f_{L2} = \frac{1}{2\pi R_S C_2}$$

$$f_{L2} = \frac{1}{2\pi \times 2.7 \text{ k}\Omega \times 4.7 \text{ μF}} \cong 12.5 \text{ Hz}$$

$$f_{L3} = \frac{1}{2\pi R_L C_3}$$

$$f_{L3} = \frac{1}{2\pi \times 4.7 \text{ k}\Omega \times 4.7 \text{ μF}} \cong 7.2 \text{ Hz}$$

Therefore, the dominant lower cutoff frequency f_L is 12.5 Hz. The upper cutoff frequency is determined from Equation 12-80

$$f_H = \frac{GBW}{A_v} = \frac{1 \text{ MHz}}{38} = 26.3 \text{ kHz}$$

Practice Problem 12-5  ANALYSIS

Determine the following for the non-inverting single-supply amplifier of Figure 12-37:

Practice Problem 12-5 *Single-Supply* ANALYSIS

Determine the following for the non-inverting single-supply amplifier of Figure 12-37:
a) The output voltages before and after the output capacitor.

b) The lower and upper cutoff frequencies and $f_{o(min)}$.

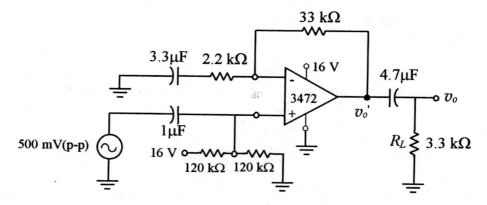

Figure 12-37: A single-supply non-inverting amplifier

Answers:

v_o' = 8 V(DC) + 8 V(p-p), v_o = 8 V(p-p), f_L = 22 Hz, f_H = 250 kHz, $f_{o(min)}$ = 220 Hz

Practice Problem 12-6 *Single-Supply* DESIGN

Design a single-supply non-inverting amplifier with a gain of 26 dB, $f_{o(min)}$ = 100 Hz, and the upper cutoff frequency larger than or equal to 100 kHz. The expected maximum input signal swing is 500 mV(p-p), and the expected minimum load at the output is 2 kΩ. Available op-amps are LM324, MC33171, and TL3472.

Answers:

R_F = 200 kΩ,
R_S = 10 kΩ,
$R_1 = R_2$ = 100 kΩ,
C_1 = 0.33 µF,
C_2 = 2.2 µF,
C_3 = 10 µF
GBW ≥ 2 MHz

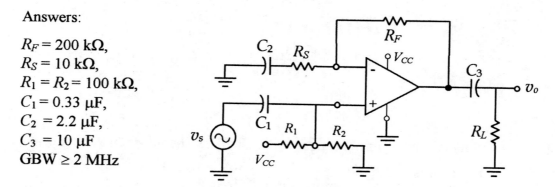

Figure 12-38: Schematic diagram for the design

Analyses of other single-supply configurations such as inverting and non-inverting summing amplifiers are given as homework assignments at the end of the chapter.

Chapter 12

12.6 SUMMARY

- The advantages and improvements achievable through negative feedback are controlled gain, improved input and output resistances, wider bandwidth, and more stable operation.

- In classical feedback theory, there are four basic feedback topologies: *series-shunt* feedback, *shunt-shunt* feedback, *shunt-series* feedback, and *series-series* feedback.

- In this chapter, we have only discussed the most popular and practical of the four topologies used in the op-amp applications, the *series-shunt* feedback.

- The gain of a more stable system is less sensitive or *desensitive* to changes in the system's internal component characteristics and/or changes in the operating environment. One major achievement with negative feedback is the increased gain desensitivity and improved system stability.

- The non-inverting amplifier is a practical example of the *series-shunt* feedback, in which the gain can be set to a desired level through two external resistors, the input resistance rises to several GΩ, and the output resistance drops to a few mΩ. All this is achievable through negative feedback.

- Since the gain-bandwidth product of the op-amp is a fixed value, the bandwidth increases with decreased gain, and vice versa.

- The *input offset current* is the difference between the two input DC bias currents I_{B1} and I_{B2}, and thus, it equals zero for an ideal op-amp. However, in case of a practical op-amp, the input offset current may not be exactly zero.

- The difference between the two DC voltages at the two input terminals of the op-amp is the *input offset voltage* V_{IO}.

- The *CMRR* is a measure of how well the op-amp can reject or attenuate the common signals, such as noise, which would be the common signal to both inputs if the op-amp is used in diff-amp configuration.

- The slew rate is defined as the maximum rate of change of the output voltage in time and it is usually expressed in V/μs.

- The large-signal voltage gain is actually the open-loop gain A_{OL} of the op-amp, which is listed as such in most data sheets.

- Although most operational amplifiers are designed to operate from a dual power supply ($\pm V_{CC}$), they can also function with a single power supply ($+V_{CC}$). However, certain op-amps such as LM324, TL3472, and MC33171 are specifically designed to operate from a single-supply.

- With single-supply operation, full output signal swing is achieved by introducing a DC offset at the output.

- For an optimum signal swing at the output, R_1 must equal R_2, so that $V_{DC} = 0.5 V_{CC}$.

- With the single-supply, each of the coupling capacitors introduces a cutoff frequency, where the largest is the dominant low-cutoff frequency f_L.

- The cutoff frequency f_L corresponds to –3dB gain or $0.707 A_v$. However, to achieve full amplitude at the output and avoid any drop in the gain, the minimum operating frequency $f_{o(min)}$ is kept approximately one decade above the cutoff frequency f_L.

- The upper cutoff frequency is determined with the GBW of the op-amp.

12.7 DEVICE SPECIFICATIONS AND DATA SHEETS

Data sheets of the following devices discussed in this chapter are included for your reference and convenience:

- LM124, LM224, LM324, and LM2902 Single Supply Operational Amplifiers (Courtesy of National Semiconductor).
- MC33171, MC33172, and MC33174 Single Supply Operational Amplifiers. Copyright of Semiconductor Component Industries, LLC. Used by permission.
- MC33272A, and MC33274A Single Supply Operational Amplifiers. Copyright of Semiconductor Components Industries, LLC. Used by permission.

Chapter 12

National Semiconductor

August 2000

LM124/LM224/LM324/LM2902 Low Power Quad Operational Amplifiers

General Description

The LM124 series consists of four independent, high gain, internally frequency compensated operational amplifiers which were designed specifically to operate from a single power supply over a wide range of voltages. Operation from split power supplies is also possible and the low power supply current drain is independent of the magnitude of the power supply voltage.

Application areas include transducer amplifiers, DC gain blocks and all the conventional op amp circuits which now can be more easily implemented in single power supply systems. For example, the LM124 series can be directly operated off of the standard +5V power supply voltage which is used in digital systems and will easily provide the required interface electronics without requiring the additional ±15V power supplies.

Unique Characteristics

- In the linear mode the input common-mode voltage range includes ground and the output voltage can also swing to ground, even though operated from only a single power supply voltage
- The unity gain cross frequency is temperature compensated
- The input bias current is also temperature compensated

Advantages

- Eliminates need for dual supplies
- Four internally compensated op amps in a single package
- Allows directly sensing near GND and V_{OUT} also goes to GND
- Compatible with all forms of logic
- Power drain suitable for battery operation

Features

- Internally frequency compensated for unity gain
- Large DC voltage gain 100 dB
- Wide bandwidth (unity gain) 1 MHz (temperature compensated)
- Wide power supply range:
    Single supply 3V to 32V
    or dual supplies ±1.5V to ±16V
- Very low supply current drain (700 µA)—essentially independent of supply voltage
- Low input biasing current 45 nA (temperature compensated)
- Low input offset voltage 2 mV and offset current: 5 nA
- Input common-mode voltage range includes ground
- Differential input voltage range equal to the power supply voltage
- Large output voltage swing 0V to V^+ − 1.5V

Connection Diagram

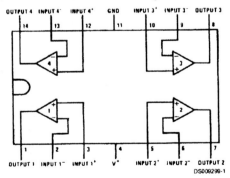

Dual-In-Line Package

Top View

Order Number LM124J, LM124AJ, LM124J/883 (Note 2), LM124AJ/883 (Note 1), LM224J, LM224AJ, LM324J, LM324M, LM324MX, LM324AM, LM324AMX, LM2902M, LM2902MX, LM324N, LM324AN, LM324MT, LM324MTX or LM2902N LM124AJRQML and LM124AJRQMLV(Note 3)
See NS Package Number J14A, M14A or N14A

Note 1: LM124A available per JM38510/11006
Note 2: LM124 available per JM38510/11005

© 2000 National Semiconductor Corporation DS009299

www.national.com

Chapter 12

Absolute Maximum Ratings (Note 12)

If Military/Aerospace specified devices are required, please contact the National Semiconductor Sales Office/Distributors for availability and specifications.

	LM124/LM224/LM324 LM124A/LM224A/LM324A	LM2902
Supply Voltage, V^+	32V	26V
Differential Input Voltage	32V	26V
Input Voltage	−0.3V to +32V	−0.3V to +26V
Input Current ($V_{IN} < -0.3V$) (Note 6)	50 mA	50 mA
Power Dissipation (Note 4)		
Molded DIP	1130 mW	1130 mW
Cavity DIP	1260 mW	1260 mW
Small Outline Package	800 mW	800 mW
Output Short-Circuit to GND (One Amplifier) (Note 5) $V^+ \leq 15V$ and $T_A = 25°C$	Continuous	Continuous
Operating Temperature Range		−40°C to +85°C
LM324/LM324A	0°C to +70°C	
LM224/LM224A	−25°C to +85°C	
LM124/LM124A	−55°C to +125°C	
Storage Temperature Range	−65°C to +150°C	−65°C to +150°C
Lead Temperature (Soldering, 10 seconds)	260°C	260°C
Soldering Information		
Dual-In-Line Package		
Soldering (10 seconds)	260°C	260°C
Small Outline Package		
Vapor Phase (60 seconds)	215°C	215°C
Infrared (15 seconds)	220°C	220°C

See AN-450 "Surface Mounting Methods and Their Effect on Product Reliability" for other methods of soldering surface mount devices.

ESD Tolerance (Note 13)	250V	250V

Electrical Characteristics

$V^+ = +5.0V$, (Note 7), unless otherwise stated

Parameter	Conditions	LM124A			LM224A			LM324A			Units
		Min	Typ	Max	Min	Typ	Max	Min	Typ	Max	
Input Offset Voltage	(Note 8) $T_A = 25°C$		1	2		1	3		2	3	mV
Input Bias Current (Note 9)	$I_{IN(+)}$ or $I_{IN(-)}$, $V_{CM} = 0V$, $T_A = 25°C$		20	50		40	80		45	100	nA
Input Offset Current	$I_{IN(+)}$ or $I_{IN(-)}$, $V_{CM} = 0V$, $T_A = 25°C$		2	10		2	15		5	30	nA
Input Common-Mode Voltage Range (Note 10)	$V^+ = 30V$, (LM2902, $V^+ = 26V$), $T_A = 25°C$	0		V^+−1.5	0		V^+−1.5	0		V^+−1.5	V
Supply Current	Over Full Temperature Range $R_L = \infty$ On All Op Amps										mA
	$V^+ = 30V$ (LM2902 $V^+ = 26V$)		1.5	3		1.5	3		1.5	3	
	$V^+ = 5V$		0.7	1.2		0.7	1.2		0.7	1.2	
Large Signal Voltage Gain	$V^+ = 15V$, $R_L \geq 2k\Omega$, ($V_O = 1V$ to $11V$), $T_A = 25°C$	50	100		50	100		25	100		V/mV
Common-Mode Rejection Ratio	DC, $V_{CM} = 0V$ to $V^+ - 1.5V$, $T_A = 25°C$	70	85		70	85		65	85		dB

Electrical Characteristics (Continued)

$V^+ = +5.0V$, (Note 7), unless otherwise stated

Parameter		Conditions		LM124A			LM224A			LM324A			Units
				Min	Typ	Max	Min	Typ	Max	Min	Typ	Max	
Power Supply Rejection Ratio		$V^+ = 5V$ to $30V$ (LM2902, $V^+ = 5V$ to $26V$), $T_A = 25°C$		65	100		65	100		65	100		dB
Amplifier-to-Amplifier Coupling (Note 11)		$f = 1$ kHz to 20 kHz, $T_A = 25°C$ (Input Referred)			−120			−120			−120		dB
Output Current	Source	$V_{IN}^+ = 1V$, $V_{IN}^- = 0V$, $V^+ = 15V$, $V_O = 2V$, $T_A = 25°C$		20	40		20	40		20	40		mA
	Sink	$V_{IN}^- = 1V$, $V_{IN}^+ = 0V$, $V^+ = 15V$, $V_O = 2V$, $T_A = 25°C$		10	20		10	20		10	20		
		$V_{IN}^- = 1V$, $V_{IN}^+ = 0V$, $V^+ = 15V$, $V_O = 200$ mV, $T_A = 25°C$		12	50		12	50		12	50		µA
Short Circuit to Ground		(Note 5) $V^+ = 15V$, $T_A = 25°C$			40	60		40	60		40	60	mA
Input Offset Voltage		(Note 8)				4			4			5	mV
V_{OS} Drift		$R_S = 0\Omega$			7	20		7	20		7	30	µV/°C
Input Offset Current		$I_{IN(+)} - I_{IN(-)}$, $V_{CM} = 0V$				30			30			75	nA
I_{OS} Drift		$R_S = 0\Omega$			10	200		10	200		10	300	pA/°C
Input Bias Current		$I_{IN(+)}$ or $I_{IN(-)}$			40	100		40	100		40	200	nA
Input Common-Mode Voltage Range (Note 10)		$V^+ = +30V$ (LM2902, $V^+ = 26V$)		0		V^+−2	0		V^+−2	0		V^+−2	V
Large Signal Voltage Gain		$V^+ = +15V$ (V_O Swing = 1V to 11V) $R_L \geq 2$ kΩ		25			25			15			V/mV
Output Voltage Swing	V_{OH}	$V^+ = 30V$ (LM2902, $V^+ = 26V$)	$R_L = 2$ kΩ	26			26			26			V
			$R_L = 10$ kΩ	27	28		27	28		27	28		
	V_{OL}	$V^+ = 5V$, $R_L = 10$ kΩ			5	20		5	20		5	20	mV
Output Current	Source	$V_O = 2V$	$V_{IN}^+ = +1V$, $V_{IN}^- = 0V$, $V^+ = 15V$	10	20		10	20		10	20		mA
	Sink		$V_{IN}^- = +1V$, $V_{IN}^+ = 0V$, $V^+ = 15V$	10	15		5	8		5	8		

Electrical Characteristics

$V^+ = +5.0V$, (Note 7), unless otherwise stated

Parameter	Conditions	LM124/LM224			LM324			LM2902			Units
		Min	Typ	Max	Min	Typ	Max	Min	Typ	Max	
Input Offset Voltage	(Note 8) $T_A = 25°C$		2	5		2	7		2	7	mV
Input Bias Current (Note 9)	$I_{IN(+)}$ or $I_{IN(-)}$, $V_{CM} = 0V$, $T_A = 25°C$		45	150		45	250		45	250	nA
Input Offset Current	$I_{IN(+)}$ or $I_{IN(-)}$, $V_{CM} = 0V$, $T_A = 25°C$		3	30		5	50		5	50	nA
Input Common-Mode Voltage Range (Note 10)	$V^+ = 30V$, (LM2902, $V^+ = 26V$), $T_A = 25°C$	0		V^+−1.5	0		V^+−1.5	0		V^+−1.5	V
Supply Current	Over Full Temperature Range $R_L = \infty$ On All Op Amps										mA
	$V^+ = 30V$ (LM2902 $V^+ = 26V$)		1.5	3		1.5	3		1.5	3	
	$V^+ = 5V$		0.7	1.2		0.7	1.2		0.7	1.2	
Large Signal Voltage Gain	$V^+ = 15V$, $R_L \geq 2$ kΩ, ($V_O = 1V$ to $11V$), $T_A = 25°C$	50	100		25	100		25	100		V/mV
Common-Mode Rejection Ratio	DC, $V_{CM} = 0V$ to $V^+ - 1.5V$, $T_A = 25°C$	70	85		65	85		50	70		dB
Power Supply Rejection Ratio	$V^+ = 5V$ to $30V$ (LM2902, $V^+ = 5V$ to $26V$),	65	100		65	100		50	100		dB

www.national.com

Electrical Characteristics (Continued)

$V^+ = +5.0V$, (Note 7), unless otherwise stated

Parameter		Conditions		LM124/LM224 Min	LM124/LM224 Typ	LM124/LM224 Max	LM324 Min	LM324 Typ	LM324 Max	LM2902 Min	LM2902 Typ	LM2902 Max	Units
		$T_A = 25°C$											
Amplifier-to-Amplifier Coupling (Note 11)		f = 1 kHz to 20 kHz, $T_A = 25°C$ (Input Referred)			−120			−120			−120		dB
Output Current	Source	$V_{IN}^+ = 1V$, $V_{IN}^- = 0V$, $V^+ = 15V$, $V_O = 2V$, $T_A = 25°C$		20	40		20	40		20	40		mA
	Sink	$V_{IN}^- = 1V$, $V_{IN}^+ = 0V$, $V^+ = 15V$, $V_O = 2V$, $T_A = 25°C$		10	20		10	20		10	20		
		$V_{IN}^- = 1V$, $V_{IN}^+ = 0V$, $V^+ = 15V$, $V_O = 200$ mV, $T_A = 25°C$		12	50		12	50		12	50		µA
Short Circuit to Ground		(Note 5) $V^+ = 15V$, $T_A = 25°C$			40	60		40	60		40	60	mA
Input Offset Voltage		(Note 8)				7			9			10	mV
V_{OS} Drift		$R_S = 0Ω$			7			7			7		µV/°C
Input Offset Current		$I_{IN(+)} - I_{IN(-)}$, $V_{CM} = 0V$				100			150		45	200	nA
I_{OS} Drift		$R_S = 0Ω$			10			10			10		pA/°C
Input Bias Current		$I_{IN(+)}$ or $I_{IN(-)}$			40	300		40	500		40	500	nA
Input Common-Mode Voltage Range (Note 10)		$V^+ = +30V$ (LM2902, $V^+ = 26V$)		0		V^+-2	0		V^+-2	0		V^+-2	V
Large Signal Voltage Gain		$V^+ = +15V$ (V_O Swing = 1V to 11V) $R_L \geq 2$ kΩ		25			15			15			V/mV
Output Voltage Swing	V_{OH}	$V^+ = 30V$ (LM2902, $V^+ = 26V$)	$R_L = 2$ kΩ	26			26			22			V
			$R_L = 10$ kΩ	27	28		27	28		23	24		
	V_{OL}	$V^+ = 5V$, $R_L = 10$ kΩ			5	20		5	20		5	100	mV
Output Current	Source	$V_O = 2V$	$V_{IN}^+ = +1V$, $V_{IN}^- = 0V$, $V^+ = 15V$	10	20		10	20		10	20		mA
	Sink		$V_{IN}^- = +1V$, $V_{IN}^+ = 0V$, $V^+ = 15V$	5	8		5	8		5	8		

Note 4: For operating at high temperatures, the LM324/LM324A/LM2902 must be derated based on a +125°C maximum junction temperature and a thermal resistance of 88°C/W which applies for the device soldered in a printed circuit board, operating in a still air ambient. The LM224/LM224A and LM124/LM124A can be derated based on a +150°C maximum junction temperature. The dissipation is the total of all four amplifiers — use external resistors, where possible, to allow the amplifier to saturate or to reduce the power which is dissipated in the integrated circuit.

Note 5: Short circuits from the output to V^+ can cause excessive heating and eventual destruction. When considering short circuits to ground, the maximum output current is approximately 40 mA independent of the magnitude of V^+. At values of supply voltage in excess of +15V, continuous short-circuits can exceed the power dissipation ratings and cause eventual destruction. Destructive dissipation can result from simultaneous shorts on all amplifiers.

Note 6: This input current will only exist when the voltage at any of the input leads is driven negative. It is due to the collector-base junction of the input PNP transistors becoming forward biased and thereby acting as input diode clamps. In addition to this diode action, there is also lateral NPN parasitic transistor action on the IC chip. This transistor action can cause the output voltages of the op amps to go to the V^+ voltage level (or to ground for a large overdrive) for the time duration that an input is driven negative. This is not destructive and normal output states will re-establish when the input voltage, which was negative, again returns to a value greater than −0.3V (at 25°C).

Note 7: These specifications are limited to −55°C ≤ T_A ≤ +125°C for the LM124/LM124A. With the LM224/LM224A, all temperature specifications are limited to −25°C ≤ T_A ≤ +85°C, the LM324/LM324A temperature specifications are limited to 0°C ≤ T_A ≤ +70°C, and the LM2902 specifications are limited to −40°C ≤ T_A ≤ +85°C.

Note 8: $V_O \approx 1.4V$, $R_S = 0Ω$ with V^+ from 5V to 30V; and over the full input common-mode range (0V to $V^+ - 1.5V$) for LM2902, V^+ from 5V to 26V.

Note 9: The direction of the input current is out of the IC due to the PNP input stage. This current is essentially constant, independent of the state of the output so no loading change exists on the input lines.

Note 10: The input common-mode voltage of either input signal voltage should not be allowed to go negative by more than 0.3V (at 25°C). The upper end of the common-mode voltage range is $V^+ - 1.5V$ (at 25°C), but either or both inputs can go to +32V without damage (+26V for LM2902), independent of the magnitude of V^+.

Note 11: Due to proximity of external components, insure that coupling is not originating via stray capacitance between these external parts. This typically can be detected as this type of capacitance increases at higher frequencies.

Note 12: Refer to RETS124AX for LM124A military specifications and refer to RETS124X for LM124 military specifications.

Note 13: Human body model, 1.5 kΩ in series with 100 pF.

Chapter 12

MC33171
MC33172
MC33174

Low Power, Single Supply Operational Amplifiers

Quality bipolar fabrication with innovative design concepts are employed for the MC33171/72/74 series of monolithic operational amplifiers. These devices operate at 180 µA per amplifier and offer 1.8 MHz of gain bandwidth product and 2.1 V/µs slew rate without the use of JFET device technology. Although this series can be operated from split supplies, it is particularly suited for single supply operation, since the common mode input voltage includes ground potential (V_{EE}). With a Darlington input stage, these devices exhibit high input resistance, low input offset voltage and high gain. The all NPN output stage, characterized by no deadband crossover distortion and large output voltage swing, provides high capacitance drive capability, excellent phase and gain margins, low open loop high frequency output impedance and symmetrical source/sink AC frequency response.

The MC33171/72/74 are specified over the industrial/ automotive temperature ranges. The complete series of single, dual and quad operational amplifiers are available in plastic as well as the surface mount packages.

- Low Supply Current: 180 µA (Per Amplifier)
- Wide Supply Operating Range: 3.0 V to 44 V or ±1.5 V to ±22 V
- Wide Input Common Mode Range, Including Ground (V_{EE})
- Wide Bandwidth: 1.8 MHz
- High Slew Rate: 2.1 V/µs
- Low Input Offset Voltage: 2.0 mV
- Large Output Voltage Swing: –14.2 V to +14.2 V (with ±15 V Supplies)
- Large Capacitance Drive Capability: 0 pF to 500 pF
- Low Total Harmonic Distortion: 0.03%
- Excellent Phase Margin: 60°C
- Excellent Gain Margin: 15 dB
- Output Short Circuit Protection
- ESD Diodes Provide Input Protection for Dual and Quad

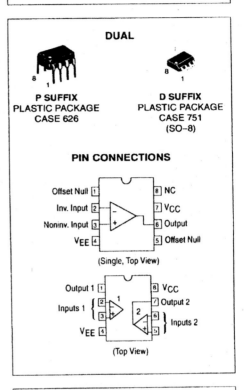

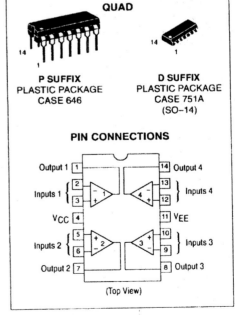

ORDERING INFORMATION

Op Amp Function	Device	Operating Temperature Range	Package
Single	MC33171D MC33171P	T_A = –40° to +85°C T_A = –40° to +85°C	SO–8 Plastic DIP
Dual	MC33172D MC33172P	T_A = –40° to +85°C T_A = –40° to +85°C	SO–8 Plastic DIP
Quad	MC33174D MC33174P	T_A = –40° to +85°C T_A = –40° to +85°C	SO–14 Plastic DIP

© Motorola, Inc. 1996   Rev 0

Chapter 12

MC33272A
MC33274A

Single Supply, High Slew Rate Low Input Offset Voltage, Bipolar Operational Amplifiers

HIGH PERFORMANCE OPERATIONAL AMPLIFIERS

SEMICONDUCTOR TECHNICAL DATA

The MC33272/74 series of monolithic operational amplifiers are quality fabricated with innovative Bipolar design concepts. This dual and quad operational amplifier series incorporates Bipolar inputs along with a patented Zip–R–Trim element for input offset voltage reduction. The MC33272/74 series of operational amplifiers exhibits low input offset voltage and high gain bandwidth product. Dual-doublet frequency compensation is used to increase the slew rate while maintaining low input noise characteristics. Its all NPN output stage exhibits no deadband crossover distortion, large output voltage swing, and an excellent phase and gain margin. It also provides a low open loop high frequency output impedance with symmetrical source and sink AC frequency performance.

The MC33272/74 series is specified over –40° to +85°C and are available in plastic DIP and SOIC surface mount packages.

- Input Offset Voltage Trimmed to 100 μV (Typ)
- Low Input Bias Current: 300 nA
- Low Input Offset Current: 3.0 nA
- High Input Resistance: 16 MΩ
- Low Noise: 18 nV/√Hz @ 1.0 kHz
- High Gain Bandwidth Product: 24 MHz @ 100 kHz
- High Slew Rate: 10 V/μs
- Power Bandwidth: 160 kHz
- Excellent Frequency Stability
- Unity Gain Stable: w/Capacitance Loads to 500 pF
- Large Output Voltage Swing: +14.1 V/ –14.6 V
- Low Total Harmonic Distortion: 0.003%
- Power Supply Drain Current: 2.15 mA per Amplifier
- Single or Split Supply Operation: +3.0 V to +36 V or ±1.5 V to ±18 V
- ESD Diodes Provide Added Protection to the Inputs

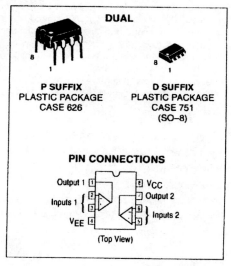

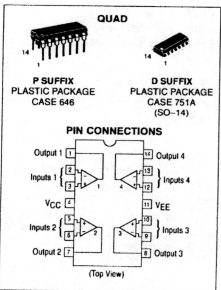

ORDERING INFORMATION

Op Amp Function	Device	Operating Temperature Range	Package
Dual	MC33272AD	$T_A = -40°$ to $+85°C$	SO–8
	MC33272AP		Plastic DIP
Quad	MC33274AD		SO–14
	MC33274AP		Plastic DIP

© Motorola, Inc. 1996   Rev 0

Chapter 12

PROBLEMS

Section 12.2 Negative Feedback

12.1 List the advantages and improvements achievable through negative feedback.

12.2 Carry out Practice Problem 12-1.

12.3 Assuming that the gain without feedback $A = -12{,}500$ and $\beta = -0.025$, determine the fractional change in the gain with feedback if the change in gain without feedback is 12.5%. Also, determine the percentile improvement in the system stability.

12.4 The gain of a system without feedback $A = -20{,}000$ and $\beta = -0.01$. If the change in the gain without feedback is 20%, determine the % change (fractional change) in the gain with feedback and the percentile improvement in system stability.

Section 12.3 Series-Shunt Feedback

12.5 Carry out Practice Problem 12-2.

12.6 Plot the frequency response (gain versus frequency) for Problem 12.5.

12.7 Determine the gain, bandwidth, and the exact input and output resistances for a non-inverting amplifier with $R_S = 7.5$ kΩ and $R_F = 330$ kΩ. The op-amp is 741.

12.8 Determine the gain, bandwidth, and the exact input and output resistances for the op-amp buffer of Figure 12-1P.

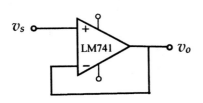

Figure 12-1P: Op-amp buffer

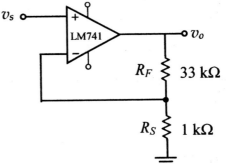

Figure 12-2P: A non-inverting amplifier

12.9 Determine the gain, bandwidth, and the exact input and output resistances for the non-inverting amplifier of Figure 12-2P.

Section 12.4 Op-Amp Characteristics

12.10 Plot the frequency response (gain versus frequency) for Problem 12.9.

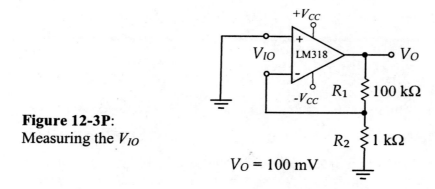

Figure 12-3P:
Measuring the V_{IO}

12.11 Determine the *input offset voltage* V_{IO} for the test circuit of Figure 12-3P.

12.12 Determine the *PSRR* for the circuit of Figure 12-3P, with $\pm V_{CC1} = \pm 10$ V and $\pm V_{CC2} = \pm 15$ V, $V_{O1} = 100$ mV, $V_{O2} = 110$ mV.

12.13 Determine the *CMRR* for the test circuit of Figure 12-4P.

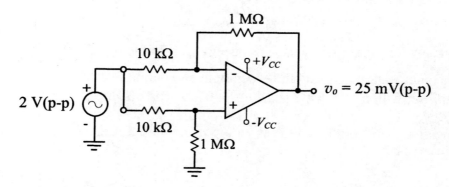

Figure 12-4P: Problem 12.13

12.14 Carry out Practice Problem 12-3.

12.15 Determine and plot the output waveform for the amplifier of Figure 12-5P:

    a) if the op-amp is 741

    b) if the op-amp is 318

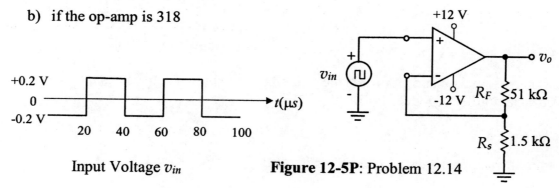

Input Voltage v_{in}

Figure 12-5P: Problem 12.14

Chapter 12

12.16 If the input signal to the amplifier of Figure 12-6P is given as follows:

$$v_{in} = 0.25 \sin(20{,}000\pi)t$$

determine the minimum slew rate in V/μs required of the op-amp. Can the 741 op-amp be used in this application?

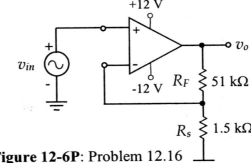

Figure 12-6P: Problem 12.16

Section 12.5 Single-Supply Operation

12.17 Carry out Practice Problem 12-4.

12.18 Determine the output voltage before and after the output capacitor, lower and upper cutoff frequencies, and the minimum operating frequency $f_{o(min)}$ for the single-supply amplifier of Figure 12-7P.

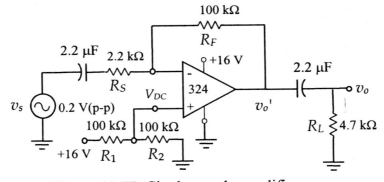

Figure 12-7P: Single-supply amplifier

12.19 Design a single-supply inverting amplifier with a gain of 26 dB, minimum operating frequency $f_{o(min)} = 100$ Hz, and $f_H \geq 100$ kHz. The expected maximum input signal swing is 500 mV(p-p), and the expected minimum load at the output is 2.2 kΩ. Available op-amps are LM324, TL3472, and MC33171.

12.20 Carry out Practice Problem 12-5.

12.21 Determine the output voltage before and after the output capacitor, lower and upper cutoff frequencies, and the minimum operating frequency $f_{o(min)}$ for the single-supply amplifier of Figure 12-8P.

Chapter 12

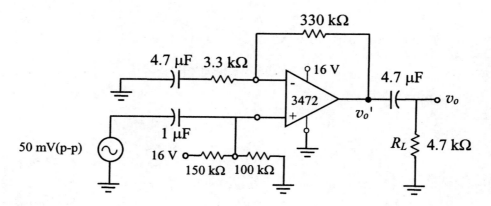

Figure 12-8P: Single-supply amplifier

12.22 Carry out Practice Problem 12-6.

12.23 Design a single-supply non-inverting amplifier with a gain of 40 in the operating frequency range, where minimum operating frequency $f_{o(min)} = 100$ Hz, and $f_H \geq 100$ kHz. The expected maximum input signal swing is 500 mV(p-p), and the expected minimum load at the output is 2.7 kΩ. Available op-amps are LM324, TL3472, and MC33171.

12.24 Determine the outputs v_o' and v_o for the single-supply inverting summer of Figure 12-9P. Also determine the cutoff frequencies introduced by C_i and C_o.

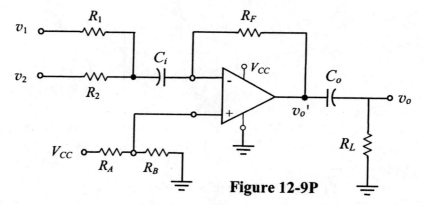

Figure 12-9P

12.25 Determine the outputs v_o' and v_o for the single-supply non-inverting summer of Figure 12-10P. Also determine the cutoff frequencies introduced by C_1 and C_2.

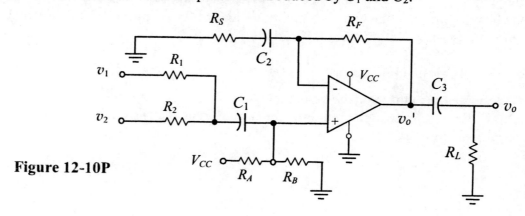

Figure 12-10P

Chapter 13

Chapter 13
POWER AMPLIFIERS AND OUTPUT STAGES
Analysis & Design

13.1 INTRODUCTION

The power amplifiers are usually the output stages in an amplifier system. In addition to being called the output stages, power amplifiers are also referred to as large-signal amplifiers. The input signal to the power amplifier is a large-signal; that is, it has already been sufficiently amplified by the small-signal amplifiers preceding the output stage. The primary function of the output stage is not to amplify the signal, but provide the needed power and deliver it efficiently to the output transducer. Hence, one important feature of the output stage is to have a very low output resistance, so that the power can be delivered to the load efficiently. The transistors used in output stages are power transistors. These power transistors must be capable of handling large amounts of power with minimal dissipation of power in the transistor itself, or else the internal junction temperature of the transistor will rise, causing the destruction of the transistor. In this chapter, we also intend to demonstrate how the utilization of the op-amp with negative feedback, in addition to providing an almost infinite input resistance and nearly zero output resistance, minimizes or practically eliminates the crossover distortion in power amplifiers.

13.2 CLASSES OF OPERATION

There are several classes of amplifier operation: A, B, AB, and C. In this chapter, we will discuss the classes A, B, and AB. Class C operation is mainly utilized in *RF* amplifiers in communications applications. The class of operation is defined by the angle of conduction of the power transistor in one cycle of the input signal. Recall that one cycle of a sinusoidal signal corresponds to a total of 360° in the phase domain. In class A type of operation, the transistor conducts continuously for the whole 360° of each cycle of the input signal, as shown in Figure 13-1(a). In class B, however, the transistor conducts only for 180° of each cycle of the input signal, as shown in Figure 13-1(b). In class AB type of operation, the angle of conduction is slightly more than 180°, as shown in Figure 13-1(c). In class C type of operation, the angle of conduction is less than 180°, as shown in Figure 13-1(d).

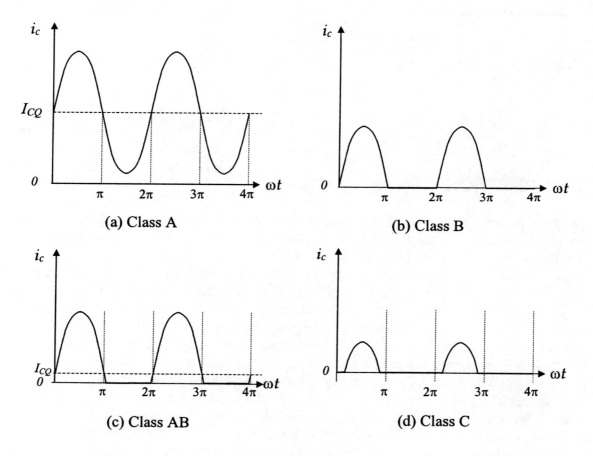

Figure 13-1: Collector current waveforms for a transistor operating in (a) class A, (b) class B, (c) class AB, and (d) class C type of operation.

13.3 CLASS A POWER AMPLIFIER

The low output resistance, which is a primary feature of an output stage, can be achieved with an emitter-follower configuration (CC), as shown in Figure 13-2(a) for a sinusoidal input voltage $v_i = V_{im}\sin\omega t$. To ensure that the transistor is continuously conducting for the full cycle of the input signal, the Q-point must be centered on the load line, as illustrated in Figure 13-2(b). Since $v_o = v_i - 0.7$ V, $A_v = i_o/i_i \approx 1$, $A_i = i_o/i_i = i_e/i_b = \beta + 1$, and $A_p = P_o/P_i = A_i A_v \approx A_i = \beta + 1$. Audio power amplifiers drive loudspeakers. Standard loudspeaker resistances are 2, 4, and 8 Ω.

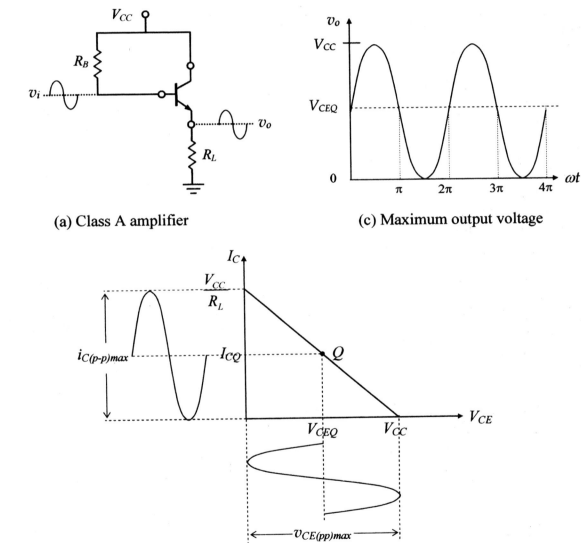

(a) Class A amplifier

(c) Maximum output voltage

(b) Maximum output current and voltage waveforms

Figure 13-2: Class A amplifier and the maximum output current and voltage waveforms

Efficiency of the Class A Amplifier

Assuming $V_{CEsat} = 0$, the following conclusions are drawn from Figure 13-2(b):

$$V_{m(max)} = \frac{V_{CC}}{2} \tag{13-1}$$

$$I_{CQ} = \frac{I_{C(sat)}}{2} = \frac{V_{CC}}{2R_L} \tag{13-2}$$

The output power:

The output power, which is the AC signal power delivered to the load, is determined as

$$P_O = \frac{I_m V_m}{2} = \frac{I_m^2 R_L}{2} = \frac{V_m^2}{2R_L} = \frac{V_{o(rms)}^2}{R_L} = \frac{V_{o(p-p)}^2}{8R_L} \tag{13-3}$$

$$P_{O(max)} = \frac{V_{m(max)}^2}{2R_L} = \frac{V_{o(p-p)max}^2}{8R_L} = \frac{V_{CC}^2}{8R_L} \tag{13-4}$$

The input power:

The input power is the DC power drawn from the power supply and is determined as

$$P_I = V_{CC} I_{CQ} = V_{CC} \times \frac{V_{CC}}{2R_L} = \frac{V_{CC}^2}{2R_L} \tag{13-5}$$

Maximum efficiency of Class A amplifier:

$$\eta_{max} = \frac{P_{O(max)}}{P_I} = \frac{V_{CC}^2}{8R_L} \frac{2R_L}{V_{CC}^2} = \frac{1}{4} = 25\% \tag{13-6}$$

Therefore, the maximum efficiency of the class A amplifier with a resistive load is no more than 25%, which is not very efficient.

Practice Problem 13-1 **ANALYSIS**

Determine the efficiency of a class A amplifier with:

a) $V_{CC} = 12$ V and $V_m = 5$ V.

b) $V_{CC} = 15$ V and $V_m = 5$ V.

Answers: (a) 17.36 % (b) 11.11%

13.4 CLASS B PUSH-PULL POWER AMPLIFIER

For class B operation, the conduction angle of the power transistor is 180°. The transistor conducts only for one half cycle of the input signal and is off for the other half cycle, which produces a half-cycle waveform at the output. To produce the other half cycle, another complementary transistor is employed, as shown in Figure 13-3 for $v_i = V_{im}\sin \omega t$. This forms a Class B complementary-transistor push-pull power amplifier.

Chapter 13

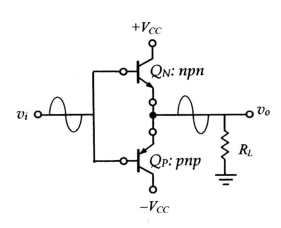

(a) Basic class B push-pull amplifier

Figure 13-3:
Output voltage and current waveforms of the basic class B push-pull amplifier

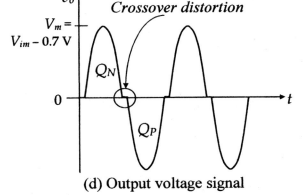

(d) Output voltage signal

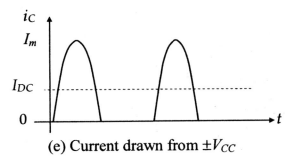

(e) Current drawn from $\pm V_{CC}$

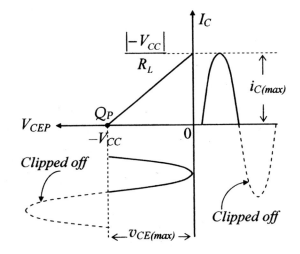

(b) Maximum output current and voltage waveforms of Q_P

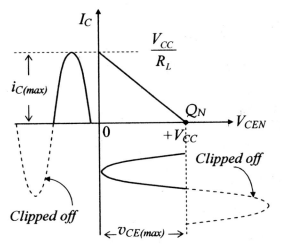

(c) Maximum output current and voltage waveforms of Q_N

In Figure 13-3(a), transistors Q_N and Q_P are both utilized as emitter followers, but none are DC biased from the base; that is, $I_{BQ} = 0$ and thus, $I_{CQ} = 0$. Transistors will be activated with the application of the large-signal voltage at the base, which is common to both transistors. The output is the midpoint between $+V_{CC}$ and $-V_{CC}$; hence, the DC voltage at the output is zero. Furthermore, with both emitters being at zero potential, V_{CEN} equals $+V_{CC}$ and V_{CEP} equals $-V_{CC}$. Thus,

$$V_{O(DC)} = 0 \tag{13-7}$$

569

$$V_{CEQN} = +V_{CC} \tag{13-8}$$

$$V_{CEQP} = -V_{CC} \tag{13-9}$$

Therefore, the Q-point for the *npn* transistor Q_N has the following I_C and V_{CE} values, as illustrated in Figure 13-3(c):

$$I_{CQN} = 0 \text{ and } V_{CEQN} = +V_{CC} \tag{13-10}$$

The Q-point for the *pnp* transistor Q_P has the following I_C and V_{CE} values, as illustrated in Figure 13-3(b):

$$I_{CQP} = 0 \text{ and } V_{CEQP} = -V_{CC} \tag{13-11}$$

With the application of a large sinusoidal input voltage $v_i = V_{im}\sin\omega t$, the *npn* transistor Q_N will conduct for the positive half-cycle, and the *pnp* transistor Q_P will conduct for the negative half-cycle of the input voltage. Of course, in each case, there will be a 0.7 V drop at the base-emitter junction, and as a result, the output voltage will always be approximately 0.7 V less than the input voltage. Furthermore, since the transistors will not conduct until the input voltage reaches beyond ±0.7 V, there will be no conduction for the time interval when Q_P turns on and Q_N turns off, i.e., for $-0.7\text{ V} \leq v_i \leq 0.7\text{ V}$. This period of no conduction is usually referred to as the *dead zone* or *dead band*, which causes a *nonlinear distortion* called a *crossover distortion* of v_o, shown in Figure 13-3(d). The distorted output voltage contains the fundamental component and harmonics.

As illustrated in Figure 13-3(e), the current waveform drawn from $\pm V_{CC}$ is similar to the output current of a half-wave rectifier, which has the following average or DC value:

$$I_{DC} = \frac{1}{2\pi}\int_{-\frac{\pi}{2}}^{\frac{\pi}{2}} I_m \cos\omega t \, d(\omega t) = \frac{I_m}{\pi} \tag{13-12}$$

where

$$I_m = \frac{V_m}{R_L} \tag{13-13}$$

Hence,

$$I_{DC} = \frac{V_m}{\pi R_L} \tag{13-14}$$

Efficiency of the Class B Power Amplifier

The output power P_O:
The output power is determined by

$$P_O = \frac{V_{o(rms)}^2}{R_L} = \frac{V_m^2}{2R_L} \tag{13-15}$$

The input power P_I:
The input power is the DC power drawn from the power supply and is determined as

$$P_I = 2V_{CC}I_{DC} = 2V_{CC} \times \frac{V_m}{\pi R_L} \tag{13-16}$$

Chapter 13

Efficiency η:

$$\eta = \frac{P_O}{P_I} = \frac{V_m^2}{2R_L} \frac{\pi R_L}{2V_{CC} V_m} = \frac{\pi V_m}{4V_{CC}} \qquad (13\text{-}17)$$

Maximum efficiency of class B amplifier:

Inspecting the above equation for efficiency and referring to Figure 13-3, the theoretical maximum efficiency will occur when V_m equals V_{CC}. This efficiency is given by

$$\eta_{max} = \frac{P_{O(max)}}{P_I} = \frac{\pi}{4} = 78.5\% \qquad (13\text{-}18)$$

Hence, the maximum efficiency of the class B amplifier is 78.5%, which proves to be much more efficient compared to the class A amplifier. At $V_{m(max)} = V_{CC} - V_{CEsat}$,

$$\eta = \frac{\pi}{4}\left(1 - \frac{V_{CEsat}}{V_{CC}}\right)$$

The dissipated power in both transistors is given by

$$P_D = P_I - P_O = P_I - \eta P_I = P_I(1 - \eta) \qquad (13\text{-}19)$$

Also,

$$P_D = P_I - P_O = \frac{2V_{CC}V_m}{\pi R_L} - \frac{V_m^2}{2R_L} \qquad (13\text{-}20)$$

Figure 13-4 shows a circuit setup for simulation of a basic class B power amplifier with Electronics Workbench.

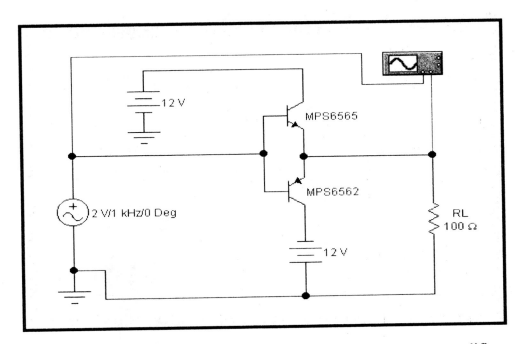

Figure 13-4: Circuit setup for simulation of the basic class B power amplifier created in Electronics Workbench

Chapter 13

Maximum dissipated power:

In Equation 13-20 for P_D, the only variable is V_m and the rest are all constants. Hence, to determine $P_{D(max)}$, we will differentiate the above equation with respect to V_m, set the result equal to zero, and solve for V_m. Thus,

$$\frac{dP_D}{dV_m} = \frac{2V_{CC}}{\pi R_L} - \frac{V_m}{R_L} = 0 \qquad (13\text{-}21)$$

$$V_m = \frac{2V_{CC}}{\pi} \qquad (13\text{-}22)$$

For this case, $\eta = 50\%$. Substituting the above expression for V_m into Equation 13-20 for P_D results in the equation for the maximum power dissipation $P_{D(max)}$ in both transistors Q_N and Q_P:

$$P_{D(max)} = \frac{2V_{CC}(2V_{CC}/\pi)}{\pi R_L} - \frac{(2V_{CC}/\pi)^2}{2R_L} = \frac{4V_{CC}^2}{\pi^2 R_L}\left(1 - \frac{1}{2}\right) = \frac{2V_{CC}^2}{\pi^2 R_L} \qquad (13\text{-}23)$$

$$P_{D(max)} = \frac{2V_{CC}^2}{\pi^2 R_L} \qquad (13\text{-}24)$$

Example 13-1 ANALYSIS

Determine the efficiency and the dissipated power of a class B amplifier with $V_{CC} = 16$ V, $V_m = 12$ V, and $R_L = 20\ \Omega$.

Solution:

The input DC power is determined from Equation 13-16, as follows:

$$P_I = 2V_{CC}I_{DC} = 2V_{CC} \times \frac{V_m}{\pi R_L} = 2 \times 16\text{ V} \times \frac{12\text{ V}}{20\pi} = 6.11\text{ W}$$

The output AC power is determined from Equation 13-15, as follows:

$$P_O = \frac{V_{rms}^2}{R_L} = \frac{V_m^2}{2R_L} = \frac{(12\text{ V})^2}{2 \times 20\ \Omega} = 3.6\text{ W}$$

The dissipated power is the difference between the input power and the output power:

$$P_D = P_I - P_O = 6.11\text{ W} - 3.6\text{ W} = 2.51\text{ W}$$

The efficiency is the ratio of the output power to the input power:

$$\eta = \frac{P_O}{P_I} = \frac{3.6\text{ W}}{6.11\text{ W}} = 58.9\%$$

Practice Problem 13-2 ANALYSIS

Determine the efficiency and the dissipated power of a class B amplifier with:

a) $V_{CC} = 12$ V, $V_m = 5$ V, and $I_{DC} = 500$ mA.
b) $V_{CC} = 15$ V, $V_m = 5$ V, and $R_L = 25\ \Omega$.

Answers: (a) $\eta = 65.45\%$ $P_D = 4.146$ W   (b) $\eta = 52.36\%$ $P_D = 1.82$ W

Reducing the Crossover Distortion with Negative Feedback

The crossover distortion of the basic class B amplifier can be reduced drastically by utilizing negative feedback and an op-amp, as shown in Figure 13-5. In addition to providing an almost infinite input resistance and nearly zero output resistance, the negative feedback and the op-amp reduce the ±0.7 V no conduction band or the dead zone to ±0.7 V/A_{OL}, in which A_{OL} is the open-loop gain of the op-amp. As a result, $v_o \approx v_s$. The slew rate limitation of the op-amp makes the crossover distortion somewhat noticeable at higher frequencies. Hence, an op-amp with higher slew rate yields better results. However, even better result in reducing the crossover distortion can be achieved with the class AB operation, which is the subject of discussion in the next section.

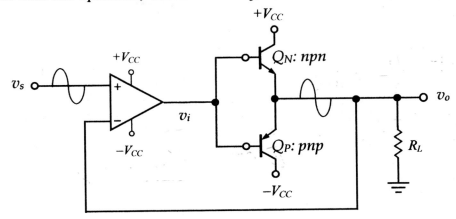

Figure 13-5: Buffered class B push-pull amplifier with negative feedback resulting in reduced crossover distortion

The virtual equivalent circuit of Figure 13-5 is illustrated in Figure 13-6, in which the output v_o equals v_s. Thus, the op-amp must produce at its output a voltage v_i that is 0.7 V higher than the output voltage v_o.

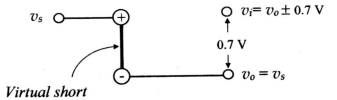

Figure 13-6: Virtual equivalent circuit of Figure 13-5

If necessary, a gain may also be introduced as shown in Figure 13-7, which could be particularly useful in the laboratory for testing a power amplifier, where a large-signal voltage of about 20 V(p-p) or more may not be available from the lab function generator.

Chapter 13

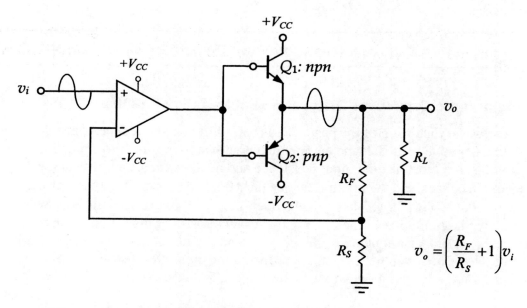

Figure 13-7: Buffered class B push-pull amplifier with gain

Let us now simulate the buffered power amplifier of Figure 13-7 with a gain of 10 and with an input signal of 2 V(p-p) and observe the output. Apparently, we expect an output signal of 20 V(p-p). The results of simulation are exhibited in Figures 13-8 and 13-9, as follows:

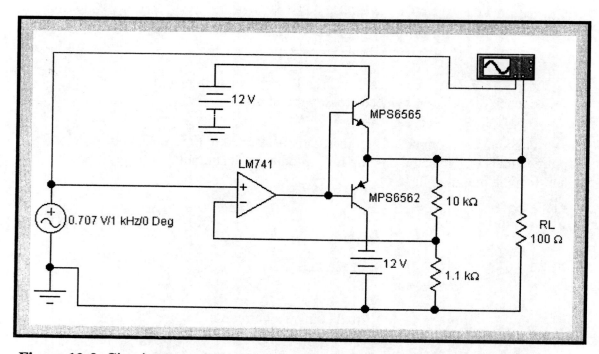

Figure 13-8: Circuit setup of the class B power amplifier with feedback and gain created in Electronics Workbench

574

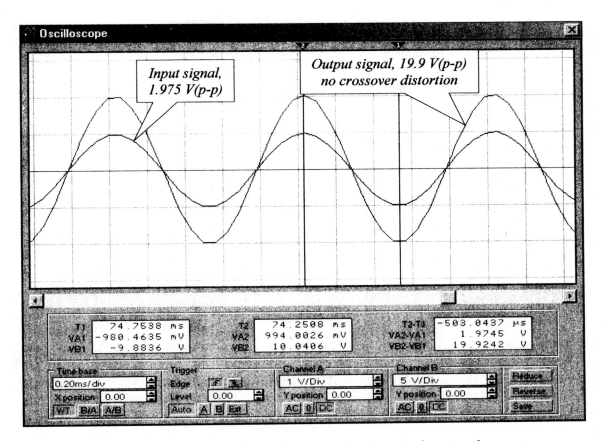

Figure 13-9: Electronics Workbench oscilloscope showing the input and output waveforms of the above class B push-pull amplifier

Example 13-2

Let us design a class B push-pull power amplifier with an output power of 5 W and an output signal swing of 20 V(p-p), given that the input signal swing is 4 V(p-p).

Solution:

Specifications: $P_O = 5$ W, $V_m = 10$ V, $V_{sm} = 2$ V

$$A_v = \frac{V_m}{V_{sm}} = \frac{10\text{ V}}{2\text{ V}} = 5 = \frac{R_F}{R_S} + 1 \qquad \frac{R_F}{R_S} = 5 - 1 = 4$$

Pick $R_F = 12$ kΩ and $R_S = 3$ kΩ. $V_{im(max)} = V_{CC} - 2$ V. $V_{im(max)} - V_{BE} - V_{m(max)} = 0$.

$V_{CC} - 2\text{ V} - V_{BE} - V_{m(max)} = 0$. $V_{CC} = V_{m(max)} + 2\text{ V} + V_{BE} = V_m + 2\text{ V} + 0.7\text{ V} = 10 + 2 + 0.7 = 12.7$ V. Pick $\pm V_{CC} = \pm 13$ V. The output power is

$$P_O = \frac{I_m V_m}{2}$$

$$I_m = \frac{2P_O}{V_m} \tag{13-25}$$

$$I_m = \frac{2P_O}{V_m} = \frac{2 \times 5 \text{ W}}{10 \text{ V}} = 1 \text{ A}$$

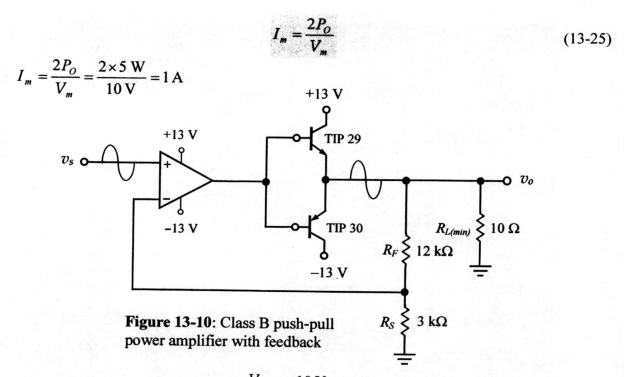

Figure 13-10: Class B push-pull power amplifier with feedback

$$R_{L(min)} = \frac{V_m}{I_{m(max)}} = \frac{10 \text{ V}}{1 \text{ A}} = 10 \text{ }\Omega$$

The average (DC) current drawn from the power supply is

$$I_{DC} = \frac{I_m}{\pi} = \frac{1 \text{ A}}{\pi} = 0.318 \text{ A}$$

The DC supply input power is given by

$$P_I = 2V_{CC}I_{DC} = 2 \times 13 \text{ V} \times 0.318 \text{ A} = 8.268 \text{ W}$$

Let us also verify the output power of 5 W. Thus,

$$P_{O(max)} = \frac{V_m^2}{2R_{L(min)}} = \frac{(10 \text{ V})^2}{2 \times 10 \text{ }\Omega} = 5 \text{ W}$$

The efficiency is

$$\eta = \frac{P_O}{P_I} = \frac{5 \text{ W}}{8.268 \text{ W}} = 60.47\%$$

Practice Problem 13-3   *Class B*   DESIGN

Design a class B push-pull power amplifier that delivers a total of 10 W power to a resistive load of 20 Ω. Also, determine the efficiency of the amplifier and the dissipated power. Assume that the input signal is 4 V(p-p).

Answers: $R_F = 27$ kΩ, $R_S = 3$ kΩ, $\pm V_{CC} = \pm 23$ V, $\eta = 68.3\%$, Q_N = TIP29, Q_P = TIP30

Section Summary — Class B — ANALYSIS/DESIGN

Summary of Equations for the Analysis/Design of Class B Power Amplifier

$$v_o = A_v v_s$$

$$I_m = \frac{V_m}{R_L} = \frac{2P_O}{V_m}$$

$$R_L = \frac{V_m}{I_m} = \frac{V_{m(max)}^2}{2P_O}$$

$$P_O = \frac{V_{m(max)}^2}{2R_L}$$

$$\eta = \frac{P_O}{P_I}$$

$$A_v = \frac{V_m}{V_{sm}} = \frac{R_F}{R_S} + 1$$

$$I_{DC} = \frac{I_m}{\pi} = \frac{V_m}{\pi R_L}$$

$$P_I = 2 V_{CC} I_{DC}$$

$$P_{O(max)} = \frac{V_{m(max)}^2}{2 R_{L(min)}}$$

$$P_D = P_I - P_O = P_I(1 - \eta)$$

13.5 CLASS AB PUSH-PULL POWER AMPLIFIER

In class AB operation, the crossover distortion is minimized or eliminated by having the transistors biased with a small quiescent current I_{CQ} as illustrated in Figure 13-1(c). The circuit of a basic class AB amplifier is shown in Figure 13-11, in which the two diodes are utilized to maintain the voltage $2V_D = 2V_{BE}$ between the base terminals of the two transistors, which is needed to keep the transistors at the verge of conduction. With this small I_{CQ}, the transistors are barely on, and as a result, will not demand a drop of 0.7 V from the input signal in order to turn on. Hence, in an ideal situation, the dead zone or no conduction zone of the class B amplifier is virtually eliminated.

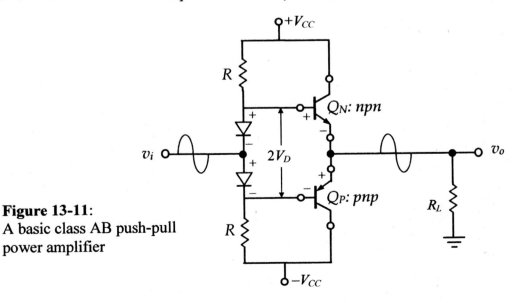

Figure 13-11: A basic class AB push-pull power amplifier

With the diodes continuously on, a small positive voltage v_i will cause Q_1 to conduct, and a small negative voltage v_i will cause Q_2 to conduct. Hence, the crossover distortion is eliminated if there is a perfect match between the diodes and the transistors.

A more practical and more precise biasing method is to use an active current source to bias the diodes and the transistors, as shown in Figure 13-12 below.

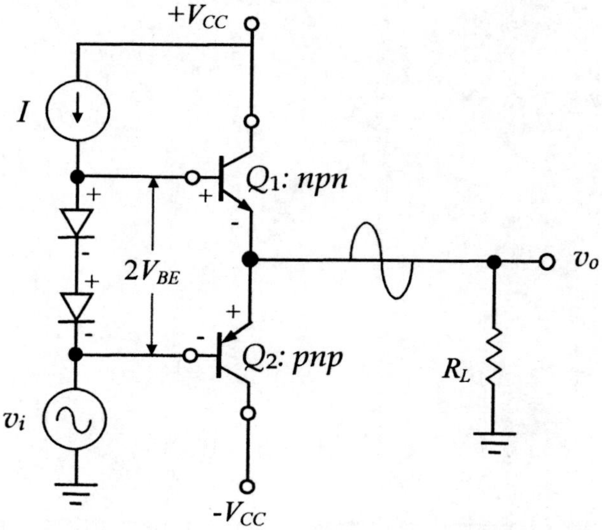

Figure 13-12: A basic class AB push-pull power amplifier biased with an active current source

Following the path from v_i through the base of each transistor, the output v_o is as follows:

$$v_o = v_i + 2\,V_{BE} - V_{BE1} = v_i + 1.4\text{ V} - 0.7\text{ V} = v_i + 0.7\text{ V}$$
$$v_o = v_i + V_{BE2} = v_i + 0.7\text{ V}$$

Hence, no matter which transistor is conducting, there is an offset of 0.7 V at the output.

$$v_o = v_i + 0.7\text{ V}$$

The dead zone and the crossover distortion have been eliminated and the DC offset can be reduced practically to zero with feedback, as shown in Figure 13-13 below. Additionally, the op-amp buffers the input signal by providing an almost infinitely large input resistance and reduces the output resistance seen by the load to nearly zero.

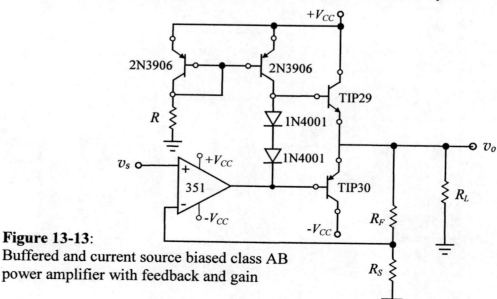

Figure 13-13: Buffered and current source biased class AB power amplifier with feedback and gain

Chapter 13

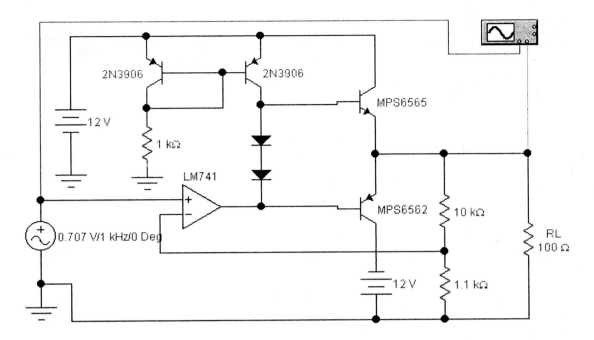

Figure 13-14: Circuit setup of a class AB amplifier with feedback and gain created in Electronics Workbench

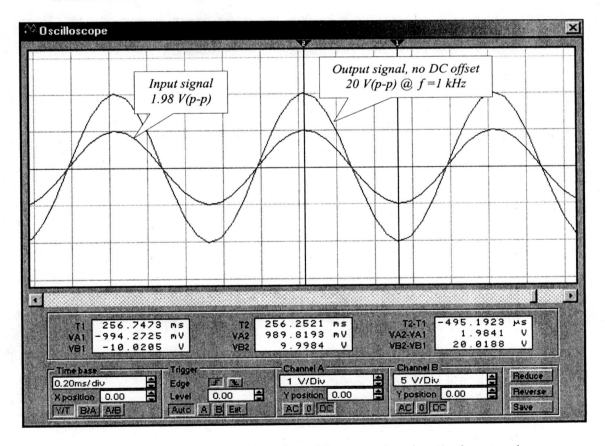

Figure 13-15: Electronics Workbench oscilloscope showing the input and output waveforms of the above class AB amplifier with feedback at $f = 1$ kHz

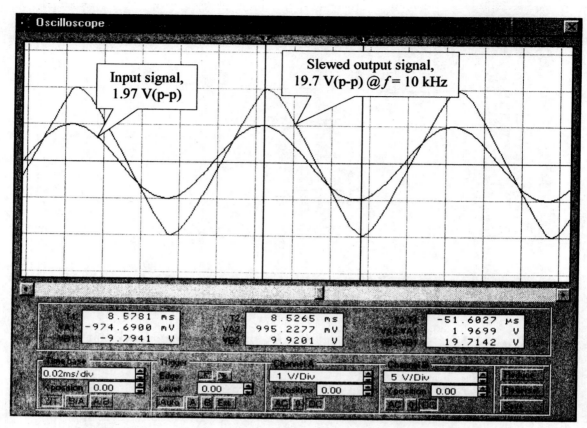

Figure 13-16: Electronics Workbench oscilloscope showing input and output waveforms of class AB power amplifier of Figure 13-11; 741 op-amp at $f = 10$ kHz

The EWB oscilloscope of Figure 13-15 shows the perfect 20 V(p-p) sinusoidal output of the class AB amplifier of Figure 13-14 with absolutely no DC offset at 1 kHz frequency. However, when the frequency is increased to 10 kHz, the output looks more like a triangular waveform as shown on the EWB oscilloscope of Figure 13-16, which is the result of the low slew rate of the 741C op-amp used in the circuit. When the 741C op-amp is replaced with a higher slew rate LM318 op-amp and the frequency is further increasd to 100 kHz, as exhibited in the simulation circuit setup of Figure 13-17, the output is distortion free and shows no sign of slewing, as shown on the EWB oscilloscope of Figure 13-18.

Practice Problem 13-4 — Class AB — ANALYSIS

Determine the output voltage amplitude V_m and the efficiency of the class AB amplifier of Figure 13-13, given the following: $R_F = 12$ kΩ, $R_S = 1.5$ kΩ, $\pm V_{CC} = \pm 16$ V, $V_{sm} = 1.5$ V, and $R_L = 27$ Ω.

Answers: $P_O = 3.375$ W, $P_I = 5.1$ W, $I_{DC} = 0.159$ A, $V_m = 13.5$, $\eta = 66.18$ %

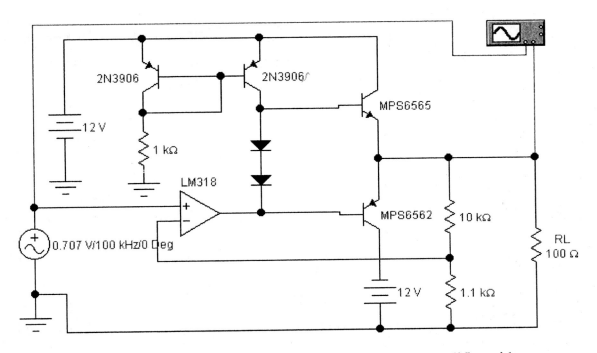

Figure 13-17: Circuit setup for simulation of the class AB amplifier with LM318 op-amp at $f = 100$ kHz, created in Electronics Workbench

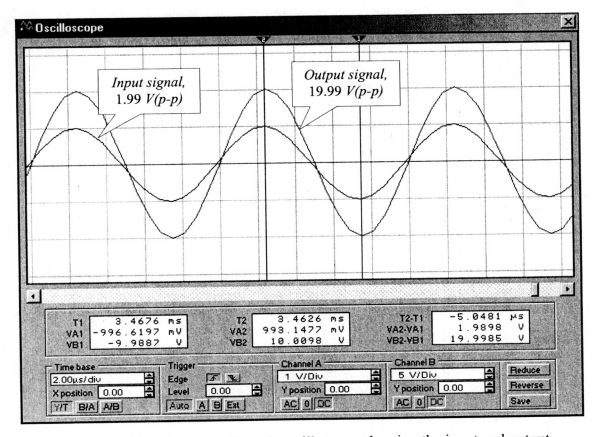

Figure 13-18: Electronics Workbench oscilloscope showing the input and output waveforms of the class AB power amplifier with the **LM318** op-amp at $f = 100$ kHz

Example 13-3 — Class AB — DESIGN

Let us now design a class AB push-pull power amplifier that will deliver a total of 10 W power to a load of 10 Ω, given that the input signal swing is 2.828 V(p-p).

Solution:

Specifications: $P_O = 10$ W, $R_L = 10$ Ω, $V_{sm} = 1.414$ V

$$P_O = \frac{V_m^2}{2R_L} = 10 \text{ W} \qquad V_m = \sqrt{2 P_O R_L} = \sqrt{2 \times 10 \times 10} = 14.14 \text{ V}$$

$V_{CC} \cong V_m + V_{BE} - 2V_D + 2 \text{ V} = V_m + 0.7 - 1.4 + 2 \text{ V} = V_m + 1.3 \text{ V} = 14.14 + 1.3 = 15.44$ V. $\Rightarrow \pm V_{CC} = \pm 16$ V

$$A_v = \frac{V_m}{V_{im}} = \frac{14.14 \text{ V}}{1.414 \text{ V}} = 10 = \frac{R_F}{R_S} + 1$$

Let $R_F = 27$ kΩ and $R_S = 3$ kΩ.

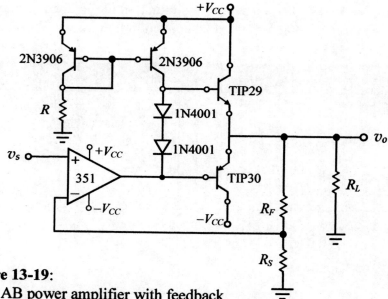

Figure 13-19:
Class AB power amplifier with feedback

$$I_m = \frac{V_m}{R_L} = \frac{14.14 \text{ V}}{10 \text{ Ω}} = 1.414 \text{ A}$$

The average (DC) current drawn from each power supply is

$$I_{DC} = \frac{I_m}{\pi} = \frac{1.414 \text{ A}}{\pi} = 0.45 \text{ A}$$

The power transistors to be utilized in the circuit will be TIP29 and TIP30, which have a maximum DC current rating of 1 A.

The current source:

According to the manufacturer's data sheets, the TIP29 and TIP30 power transistors have $\beta_{max} = 75$. Hence, we will assume an average β of 70. Then the base current of each transistor, which is to be supplied by the current source I, is the following:

$$I = \frac{I_m}{\beta} + i_{D\min} = \frac{1.414 \text{ A}}{70} + 1 \text{ mA} = 21.2 \text{ mA}$$

$$R = \frac{V_{CC} - 0.7 \text{ V}}{I} = \frac{16 \text{ V} - 0.7 \text{ V}}{21.2 \text{ mA}} = 0.72 \text{ k}\Omega \quad \text{Pick } R = 0.75 \text{ k}\Omega.$$

Note that it is important for the current source to provide just the exact amount of needed bias current. Hence, depending on the actual transistor β, the exact value of R may be adjusted for the optimum result when testing your design in the laboratory. The dc input power is

$$P_I = 2V_{CC}I_{DC} = 2 \times 16 \text{ V} \times 0.45 \text{ A} = 14.4 \text{ W}$$

Let us also verify the output power of 10 W

$$P_O = \frac{V_m^2}{2R_L} = \frac{(14.14 \text{ V})^2}{20 \, \Omega} = 10 \text{ W}$$

and the efficiency

$$\eta = \frac{P_O}{P_I} = \frac{10 \text{ W}}{14.4 \text{ W}} = 69.44\%$$

Practice Problem 13-5  ANALYSIS

Determine the output voltage amplitude V_m, efficiency, total dissipated power, and the dissipated power by each transistor for a class AB push-pull power amplifier, given the following: $V_{sm} = 1.25$ V, $R_F = 10$ kΩ, $R_S = 1$ kΩ, $R_L = 25$ Ω, $R = 3$ kΩ, and $\pm V_{CC} = \pm 16$ V. Also, determine β of the power transistors.

Answers: $V_m = 13.75$ V $\eta = 67.5$ %, $P_D = 1.82$ W, $P_{DQN} = P_{DQP} = 0.91$ W, $\beta \cong 135$

Linearity

Power amplifiers are nonlinear circuits and the output voltage contains the fundamental component and harmonics of the input signal, causing *nonlinear distortion*. The linearity of power amplifiers is measured by the *total harmonic distortion*

$$\text{THD} = \sqrt{\frac{P_{O2} + P_{O3} + P_{O4} + \ldots}{P_{O1}}} = \sqrt{\frac{V_{m2}^2 + V_{m3}^2 + V_{m4}^2 + \ldots}{V_{m1}^2}} = \frac{\sqrt{V_{rms2}^2 + V_{rms3}^2 + V_{rms4}^2 + \ldots}}{V_{rms1}} = \frac{V_{rmsh}}{V_{rms1}}$$

Section Summary — Class AB — ANALYSIS/DESIGN

Summary of Equations for the Analysis/Design of Class AB Power Amplifier

The Op-Amp:

$$v_o = A_v v_s \qquad A_v = \frac{V_m}{V_{sm}} = \frac{R_F}{R_S} + 1$$

The Power Amplifier:

$$I_m = \frac{V_m}{R_L} = \frac{2P_O}{V_m} \qquad I_{DC} = \frac{I_m}{\pi} = \frac{V_m}{\pi R_L}$$

$$R_L = \frac{V_m}{I_m} = \frac{V_m^2}{2P_O} \qquad P_I = 2 V_{CC} I_{DC}$$

$$P_O = \frac{V_m^2}{2R_L} \qquad P_{O(max)} = \frac{V_m^2}{2R_{L(min)}}$$

$$\eta = \frac{P_O}{P_I} \qquad P_D = P_I - P_O = P_I(1 - \eta)$$

The Current Source:

$$I = \frac{I_m}{\beta} + i_{D\,min} \qquad R = \frac{V_{CC} - 0.7\ \text{V}}{I}$$

Practice Problem 13-6 — Class AB — DESIGN

Design a class AB push-pull power amplifier with an output power of 7.5 W and V_m of 15 V. Also, determine the efficiency of the amplifier and the dissipated power. Assume that the input signal is 3 V(p-p).

Answers:

$R_F = 27\ \text{k}\Omega$, $R_S = 3\ \text{k}\Omega$, $\pm V_{CC} = \pm 17\ \text{V}$, $\eta = 69.44\%$, $Q_N =$ TIP29, $Q_P =$ TIP30, $R = 1\ \text{k}\Omega$, $P_D = 3.3\ \text{W}$

Chapter 13

13.6 HEAT SINKS AND THERMAL CONSIDERATIONS

As we have noticed in the previous section, the power transistors in power amplifiers are subject to dissipating a large amount of power, which is converted to heat and can cause the transistor's junction temperature to rise. Thus, steps must be taken to keep the junction temperature T_J of the power transistors below the specified maximum junction temperature $T_{J(max)}$ to avoid overheating and destruction of the transistor. A common practice for keeping the junction temperature under control is to mount the power transistor on a properly selected heat sink, which transfers the heat from the transistor to the ambient.

Thermal resistance:

The thermal resistance θ_{JA} is a measure of the heat flow or heat transfer from the transistor junction to the ambient and is defined as follows:

$$\theta_{JA} = \frac{T_J - T_A}{P_D} \text{ (in °C/W)} \qquad (13\text{-}26)$$

where T_J is the junction temperature and T_A is the ambient temperature in °C.

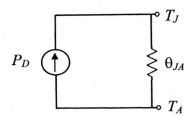

Figure 13-20: Thermal resistance

The above equation (13-26), which represents the heat transfer process from the transistor's junction to the ambient, is analogous to Ohm's law. The thermal resistance model of Figure 13-20 is analogous to a simple electrical circuit, in which the dissipated power P_D corresponds to current I, the thermal resistance θ_{JA} corresponds to resistance R, and the difference in temperature $(T_J - T_A)$ corresponds to the difference in voltage V or the voltage drop across the resistor R.

The thermal equivalent of a transistor mounted on heat sink is illustrated in Figure 13-21, which is analogous to a simple series circuit. Hence, the difference in temperature from the transistor junction T_J to the ambient T_A can be expressed by the following equation:

$$T_J - T_A = P_D(\theta_{JC} + \theta_{CS} + \theta_{SA}) \qquad (13\text{-}27)$$

Similarly, the difference in temperature from the transistor case T_C to the ambient T_A can be expressed by the following equation:

$$(13\text{-}28)$$

T_J is the junction average temperature, in °C,
θ_{JC} is the junction-to-case thermal resistance, in °C/W,
T_C is the case temperature, in °C.
θ_{CS} is the case-to-heat sink thermal resistance, in °C/W,
T_S is the heat sink average temperature, in °C,
θ_{SA} is the heat sink-to-ambient thermal resistance, in °C/W,
T_A is the ambient average temperature, in °C.

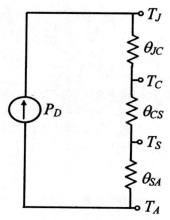

Figure 13-21:
Equivalent thermal resistance model of a transistor on heat sink

The parameters $T_{J(max)}$, $P_{D(max)}$, θ_{JA}, and θ_{JC} are normally specified in the transistor manufacturer's data sheet at 25°C. The manufacturer also provides a derating factor $DF = 1/\theta_{JC}$, which can be used to derate a given $P_{D(max)}$ for temperatures other than 25°C.

Example 13-4 — Power Dissipation — ANALYSIS

Let us determine the maximum dissipated power $P_{D(max)}$ for the TIP29 power transistor, assuming that it is mounted on a heat sink with the following specifications: thermal resistances are $\theta_{CS} = 0.5$°C/W and $\theta_{SA} = 1.5$°C/W. Find T_C.

Looking up the data sheets for TIP29, its specifications are as follows: $P_{D0(max)}$ at $T_{C0} = 25$°C is 30 W to be derated by a derating factor $DF = 0.24$ W/°C and $T_{J(max)} = 150$°C. The ambient temperature is assumed to be 30°C.

Solution:
First, $\theta_{JC(max)} = 1/DF = 1/0.24 = 4.167$°C/W. Assuming that the junction temperature is equal to the maximum temperature $T_{J(max)}$, we determine P_D from Equation 13-27
$$T_J - T_A = P_D (\theta_{JC} + \theta_{CS} + \theta_{SA})$$

$$P_{D(max)} = \frac{T_{J(max)} - T_A}{\theta_{JC} + \theta_{CS} + \theta_{SA}} = \frac{(150-30)°C}{(4.167+0.5+1.5)°C/W} = \frac{120}{6.167} W = 19.458 \text{ W}$$

Second, we determine the case temperature T_C from Equation 13-28

$$T_C = T_A + P_{D(max)} (\theta_{CS} + \theta_{SA}) = 30°C + 19.458 \text{ W} (0.5 + 1.5)°C/W$$
$$= 30°C + 38.916°C = 68.916°C$$

Third, we apply the derating factor $DF = 1/\theta_{JC} = 1/4.167 = 0.24$ W/°C at $T_C = 68.916$°C

$$P_{D(max)}\Big|_{T_C} = P_{D0(max)} - DF \text{ (W/°C)} \times (T_C - T_{C0}) = P_{D0(max)} - \frac{T_C - T_{C0}}{\theta_{JC}} \quad (13\text{-}29)$$

$$P_{D(max)}\Big|_{T_C = 69°C} = 30 \text{ W} - 0.24 \text{ W/°C} \times (68.916 - 25)°C = 30 \text{ W} - 10.54 \text{ W} = 19.46 \text{ W}$$

The result is the same as that in the first step.

Practice Problem 13-7 ANALYSIS

Determine the dissipated power $P_{D(max)}$ for the TIP30 power transistor, assuming that it is mounted on a heat sink with the following specifications: thermal resistances are $\theta_{CS} = 0.7°C/W$ and $\theta_{SA} = 1.8°C/W$. Assume $T_A = 35°C$.

Answer: $P_{D(max)} = 17$ W

Example 13-5 ANALYSIS

Let us determine the thermal specifications of the heat sink for the TIP29 power transistor to achieve $P_{D(max)} = 22$ W. Assume that the ambient temperature is $T_A = 28°C$.

Solution:
First, we will determine the case temperature T_C from Equation 13-29

$$P_{D(max)}\Big|_{T_C} = P_{D0(max)} - [DF\,(W/°C) \times (T_C - T_{C0})]$$

$$= P_{D0(max)} - \frac{T_C - T_{C0}}{\theta_{JC}}$$

$$22\,W = 30\,W - [0.24\,W/°C \times (T_C - 25°C)]$$

$$0.24\,W/°C \times (T_C - 25°C) = 30\,W - 22\,W = 8\,W$$

$$0.24\,W/°C \times T_C = 8\,W + 25°C\,(0.24\,W/°C) = 14\,W$$

$$T_C = \frac{14\,W}{0.24\,W/°C} = 58.33°C$$

Next, we will determine the total required sink thermal resistance from Equation 13-28

$$T_C = T_A + P_{D0(max)}(\theta_{CS} + \theta_{SA})$$

$$\theta_{CS} + \theta_{SA} = \frac{T_C - T_A}{P_{D0(max)}} = \frac{58.33 - 28}{22} = \frac{30.33}{22} = 1.379°C/W$$

Therefore, we need to select a heat sink with a total thermal resistance $\theta_{SA} = 1.379°C/W$.

Practice Problem 13-8 Heat Sink ANALYSIS

Determine the thermal specifications of the heat sink for the power transistor to achieve $P_{D(max)} = 25$ W. Assume $T_A = 33°C$. The power transistor specifications: $P_{D0(max)} = 40$ W at $T_{C0} = 25°C$ and the derating factor is 0.25 W/°C.

Answer: $\theta_{CS} + \theta_{SA} = 2.08°C/W$

Section Summary — Thermal Considerations — ANALYSIS/DESIGN

Summary of Equations for the Analysis/Design of Thermal Considerations

$$T_J - T_A = P_D (\theta_{JA} + \theta_{CS} + \theta_{SA})$$

$$P_D = \frac{T_{J(max)} - T_A}{\theta_{JC} + \theta_{CS} + \theta_{SA}}$$

$$T_C - T_A = P_D (\theta_{CS} + \theta_{SA})$$

$$P_{D(max)}\bigg|_{T_C} = P_{D(max)} - [\text{derating factor } W/°C \times (T_C - T_{C0})\ °C]$$

13.7 SUMMARY

- The power amplifiers are usually the output stages in an amplifier system, which are also referred to as the output stages and large-signal amplifiers.

- The primary function of the output stage is not to amplify the signal, but to provide the needed power and deliver it efficiently to the output transducer. Hence, one important feature of the output stage is to have a very low output resistance, so that the power can be delivered to the load efficiently.

- There are several classes of amplifier operation: A, B, AB, and C. The class of operation is defined by the angle of conduction of the power transistor in one cycle of the input signal.

- In class A type of operation, the transistor conducts continuously for the whole 360° of each cycle of the input signal.

- In class B type of operation, the transistor conducts only for 180° of each cycle of the input signal.

- In class AB type of operation, the angle of conduction is slightly more than 180°.

- In class C type of operation, the angle of conduction is less than 180° and is mainly utilized in *RF* amplifiers in communications applications.

- The maximum efficiency of the class A operation is no more than 25%. However, class B and class AB can yield a maximum efficiency of 78.5%.

Chapter 13

- With class B type of operation, the conduction angle of the power transistor is 180°; that is, the transistor conducts only for the half-cycle of the input signal and is off for the other half-cycle, and, as a result, produces a half-cycle waveform at the output. To produce the other half-cycle, another complementary transistor is employed.

- Since the transistors in class B do not conduct until the input reaches beyond ±0.7 V, there will be no conduction from the time Q_1 turns off and Q_2 turns on. This period of no conduction is usually referred to as the *dead zone*, which causes a *crossover distortion*.

- The crossover distortion of the basic class B amplifier can be reduced substantially by utilizing an op-amp with negative feedback.

- In addition to providing an almost infinite input resistance and nearly zero output resistance, the negative feedback provided by the op-amp reduces the ±0.7 V no conduction band or the dead zone to $\pm 0.7/A_{OL}$ V.

- The slew rate limitation of the op-amp makes the crossover distortion somewhat noticeable at higher frequencies. However, a much better result, in reducing the crossover distortion, can be achieved with the class AB type of operation.

- In class AB type of operation, the crossover distortion is minimized or eliminated by having the transistors biased with a small quiescent current I_{CQ}.

- In a class AB amplifier two diodes are utilized to maintain the voltage $2V_{BE}$ across the base terminals of the two transistors, which is needed to keep the transistors at the verge of conducting. Hence, in an ideal situation, the dead zone or no conduction zone of the class B amplifier is virtually eliminated.

- Power transistors are subject to dissipating a large amount of power, which is converted to heat and can cause the transistor's junction temperature to rise.

- A common practice for keeping the junction temperature under control is to mount the power transistor on a properly selected heat sink, which transfers the heat from the transistor to the ambient.

- The thermal resistance θ_{JA} is a measure of the heat flow or heat transfer from the transistor junction to the ambient.

- The thermal resistance model of a power transistor is analogous to a simple electrical circuit, in which the dissipated power P_D corresponds to current I, the thermal resistance θ_{JA} corresponds to resistance R, and the difference in temperature $(T_J - T_A)$ corresponds to the difference in voltage V or the voltage drop across the resistor R.

- The thermal equivalent of a transistor mounted on a heat sink is analogous to a simple series circuit.

Chapter 13

- The $T_{J(max)}$, $P_{D(max)}$, θ_{JA}, and θ_{JC} are normally specified in the transistor manufacturer's data sheet at 25°C. The manufacturer also provides a derating factor, which can be used to derate the given $P_{D(max)}$ for temperatures other than 25°C.

Data sheets of TIP29B, TIP30B, TIP29C, and TIP30C power transistors are presented in the following two pages for your reference.
Copyright of Semiconductor Component Industries, LLC. Used by permission.

MOTOROLA
SEMICONDUCTOR TECHNICAL DATA

Order this document by TIP29B/D

Complementary Silicon Plastic Power Transistors

... designed for use in general purpose amplifier and switching applications. Compact TO-220 AB package.

NPN TIP29B
TIP29C
PNP TIP30B
TIP30C

1 AMPERE
POWER TRANSISTORS
COMPLEMENTARY
SILICON
80–100 VOLTS
30 WATTS

MAXIMUM RATINGS

Rating	Symbol	TIP29B TIP30B	TIP29C TIP30C	Unit
Collector–Emitter Voltage	V_{CEO}	80	100	Vdc
Collector–Base Voltage	V_{CB}	80	100	Vdc
Emitter–Base Voltage	V_{EB}	5.0		Vdc
Collector Current — Continuous Peak	I_C	1.0 3.0		Adc
Base Current	I_B	0.4		Adc
Total Power Dissipation @ T_C = 25°C Derate above 25°C	P_D	30 0.24		Watts W/°C
Total Power Dissipation @ T_A = 25°C Derate above 25°C	P_D	2.0 0.016		Watts W/°C
Unclamped Inductive Load Energy (See Note 3)	E	32		mJ
Operating and Storage Junction Temperature Range	T_J, T_{stg}	–65 to +150		°C

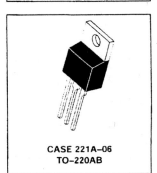

CASE 221A–06
TO–220AB

THERMAL CHARACTERISTICS

Characteristic	Symbol	Max	Unit
Thermal Resistance, Junction to Ambient	$R_{\theta JA}$	62.5	°C/W
Thermal Resistance, Junction to Case	$R_{\theta JC}$	4.167	°C/W

ELECTRICAL CHARACTERISTICS (T_C = 25°C unless otherwise noted)

Characteristic		Symbol	Min	Max	Unit
OFF CHARACTERISTICS					
Collector–Emitter Sustaining Voltage (1) (I_C = 30 mAdc, I_B = 0)	TIP29B, TIP30B TIP29C, TIP30C	$V_{CEO}(sus)$	80 100	— —	Vdc
Collector Cutoff Current (V_{CE} = 60 Vdc, I_B = 0)		I_{CEO}	—	0.3	mAdc
Collector Cutoff Current (V_{CE} = 80 Vdc, V_{EB} = 0) (V_{CE} = 100 Vdc, V_{EB} = 0)	TIP29B, TIP30B TIP29C, TIP30C	I_{CES}	— —	200 200	µAdc
Emitter Cutoff Current (V_{BE} = 5.0 Vdc, I_C = 0)		I_{EBO}	—	1.0	mAdc
ON CHARACTERISTICS (1)					
DC Current Gain (I_C = 0.2 Adc, V_{CE} = 4.0 Vdc) (I_C = 1.0 Adc, V_{CE} = 4.0 Vdc)		h_{FE}	40 15	— 75	—
Collector–Emitter Saturation Voltage (I_C = 1.0 Adc, I_B = 125 mAdc)		$V_{CE}(sat)$	—	0.7	Vdc
Base–Emitter On Voltage (I_C = 1.0 Adc, V_{CE} = 4.0 Vdc)		$V_{BE}(on)$	—	1.3	Vdc
DYNAMIC CHARACTERISTICS					
Current–Gain — Bandwidth Product (2) (I_C = 200 mAdc, V_{CE} = 10 Vdc, f_{test} = 1.0 MHz)		f_T	3.0	—	MHz
Small–Signal Current Gain (I_C = 0.2 Adc, V_{CE} = 10 Vdc, f = 1.0 kHz)		h_{fe}	20	—	—

(1) Pulse Test: Pulse Width ≤ 300 µs, Duty Cycle ≤ 2.0%.
(2) $f_T = |h_{fe}| \cdot f_{test}$.
(3) This rating based on testing with L_C = 20 mH, R_{BE} = 100 Ω, V_{CC} = 10 V, I_C = 1.8 A, P.R.F = 10 Hz.

REV 1

© Motorola, Inc. 1995

Chapter 13

TIP29B TIP29C TIP30B TIP30C

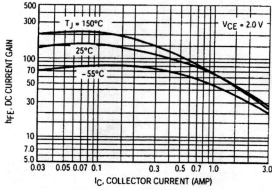

Figure 1. DC Current Gain

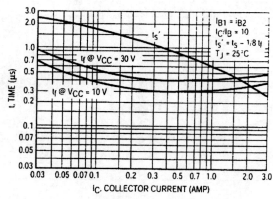

Figure 2. Turn-Off Time

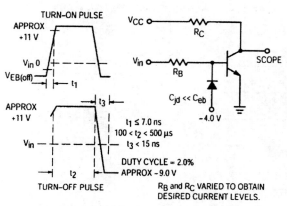

Figure 3. Switching Time Equivalent Circuit

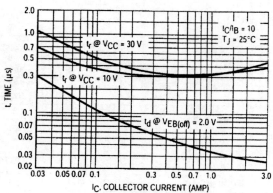

Figure 4. Turn-On Time

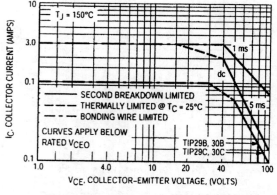

Figure 5. Active Region Safe Operating Area

There are two limitations on the power handling ability of a transistor: average junction temperature and second breakdown. Safe operating area curves indicate $I_C - V_{CE}$ operation; i.e., the transistor must not be subjected to greater dissipation than the curves indicate.

The data of Figure 5 is based on $T_{J(pk)} = 150°C$; T_C is variable depending on conditions. Second breakdown pulse limits are valid for duty cycles to 10% provided $T_{J(pk)} \leq 150°C$. At high case temperatures, thermal limitations will reduce the power that can be handled to values less than the limitations imposed by second breakdown.

Chapter 13

PROBLEMS

Section 13.2 Classes of Operation

13.1 Define and describe classes of power amplifier operation.

Section 13.3 Class A Power Amplifier

13.2 Carry out Practice Problem 13-1.

13.3 Determine the efficiency of a class A amplifier, given the following:
(a) $V_{CC} = 16$ V, $v_{o(p-p)} = 14$ V.
(b) $V_{CC} = 15$ V, $v_{o(p-p)} = 12$ V.

Section 13.4 Class B Power Amplifier

13.4 Describe the crossover distortion in class B and explain how it can be minimized or eliminated.

13.5 What is the maximum efficiency expected from class A and class B power amplifiers?

13.6 Carry out Practice Problem 13-2.

13.7 Determine the efficiency and dissipated power of a class B power amplifier, given the following: $V_{CC} = 15$ V, $v_{o(p-p)} = 25$ V, and $R_L = 25$ Ω.

13.8 Determine the efficiency and dissipated power of a class B power amplifier, given the following: $V_{CC} = 16$ V, $v_{o(p)} = 12$ V, and $I_{DC} = 0.6$ A.

13.9 Determine $v_{o(p-p)}$, efficiency and dissipated power of a class B power amplifier, given the following: $V_{CC} = 15$ V, $v_{i(p-p)} = 2$ V, and $R_F = 75$ kΩ, $R_S = 7.5$ kΩ, $R_L = 25$ Ω.

13.10 Carry out Practice Problem 13-3.

13.11 Design a class B push-pull power amplifier that delivers 7.5 W to a load of 25 Ω, with $v_{i(p-p)} = 1$ V.

13.12 Design a class B push-pull power amplifier with an output power of 12 W, $v_{o(p-p)} = 24$ V, and $v_{i(p-p)} = 2$ V.

Section 13.5 Class AB Power Amplifier

13.13 How is the crossover distortion of class B eliminated in a basic class AB power amplifier?

13.14 What is the function of the current source utilized in a class AB power amplifier?

13.15 What is accomplished by utilizing an op-amp in a class AB power amplifier?

13.16 Explain fully and clearly the advantages of using an LM318 op-amp versus an LM741 op-amp in a class AB power amplifier.

13.17 Carry out Practice Problem 13-4.

13.18 Carry out Practice Problem 13-5.

13.19 Carry out Practice Problem 13-6.

13.20 Design a class AB push-pull power amplifier that delivers a total of 12.5 W to a load of 8 Ω, with $v_{i(p-p)} = 2.828$ V. Assume $\beta = 66$ for the power transistors.

Section 13.6 Heat Sink

13.21 Carry out Practice Problem 13-7.

13.22 Determine the dissipated power $P_{D(max)}$ for a power transistor, assuming that it is mounted on a heat sink with the following specifications:
Transistor: $P_{D(max)}$ (at $T_C = 25°C$) is 50 W, to be derated by 0.4 W/°C.
$\theta_{JC(max)} = 5$ °C/W, $T_{J(max)} = 180°C$.
Heat sink: $\theta_{CS} = 0.6$ °C/W, and $\theta_{SA} = 0.8$ °C/W. Assume $T_A = 35°C$.

13.23 Carry out Practice Problem 13-8.

13.24 Determine the thermal specifications of the heat sink for the power transistor with the following specifications, to achieve a $P_{D(max)}$ of 30 W. Assume $T_A = 35°C$. Power transistor specs: $P_{D(max)} = 50$ W at $T_C = 25°C$, derate by 0.4 W/°C.

Chapter 14

Chapter  ACTIVE FILTERS

Analysis & Design

14.1 INTRODUCTION

Another major application of the operational amplifier is in the area of active filters. With a single op-amp, a couple of small capacitors, and a few resistors, one can design and implement an active low-pass, high-pass, band-pass, or band-stop filter. Active filters have several advantages over passive filters. The op-amp provides the needed gain, very high input resistance, and very low output resistance. In addition, use of inductors, which are bulky and costly, can be avoided with active filters. Furthermore, nearly ideal responses can be achieved with active filters. In this chapter, we will present a thorough analysis and design of all types of active filters including low-pass, high-pass, band-pass, and band-stop.

14.2 ACTIVE FILTERS

Active filters are frequency selective circuits that pass a certain band of frequency while stopping the rest of the frequency spectrum. An ideal low-pass filter will pass a band of frequency from DC up to a cutoff frequency f_H as illustrated in Figure 14-1(a). An ideal high-pass filter will stop the band of frequency from DC up to a cutoff frequency f_L and pass the frequency band above and beyond f_L as illustrated in Figure 14-1(b). Typical frequency response plots of the low-pass, high-pass, band-pass, and band-stop filters are exhibited in Figure 14-1 below:

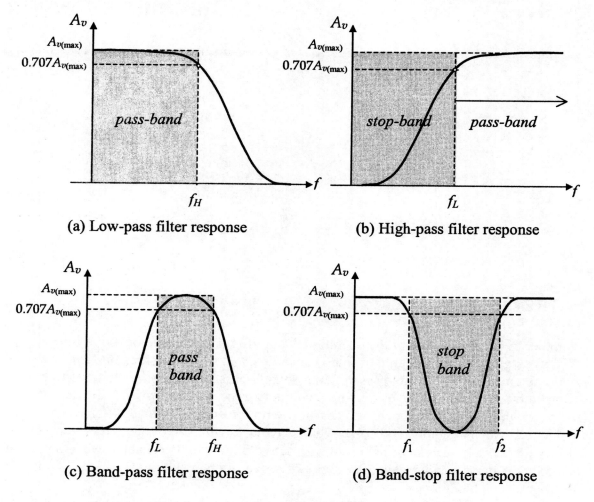

Figure 14-1: Typical frequency response plots of various filter types

14.3 FIRST-ORDER LOW-PASS ACTIVE FILTER

The first-order filter has only one reactive component, as shown in Figure 14-2 below: The capacitor C in conjunction with the resistor R provides the filtering action, while the op-amp with its associated resistors R_F and R_S functions as a non-inverting amplifier and provides the needed gain.

Chapter 14

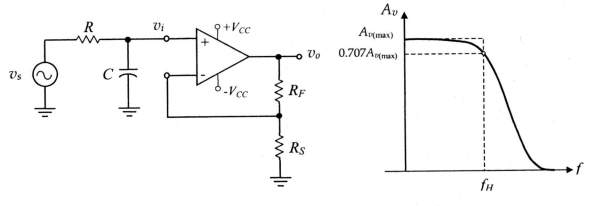

(a) Circuit schematic (b) Typical frequency response

Figure 14-2: First-order low-pass active filter

The input voltage to the op-amp v_i, which is the voltage across the capacitor C, is determined by applying the voltage divider rule to the filter circuit, as follows:

$$v_i = v_s \frac{\mathbf{X_C}}{R + \mathbf{X_C}} = v_s \frac{-jX_C}{R - jX_C} \tag{14-1}$$

Multiplying the numerator and the denominator by j and substituting for X_C results in the following:

$$v_i = v_s \frac{X_C}{X_C + jR} = v_s \frac{1}{1 + j\dfrac{R}{X_C}} = v_s \frac{1}{1 + j\omega RC} = v_s \frac{1}{1 + j2\pi fRC} \tag{14-2}$$

At the cutoff frequency f_H, the magnitude of the capacitive reactance X_C equals the resistance of the resistor R.

$$\frac{1}{2\pi f_H C} = R \tag{14-3}$$

Solving for the cutoff frequency f_H yields the following:

$$f_H = \frac{1}{2\pi RC} \tag{14-4}$$

or

$$\frac{1}{f_H} = 2\pi RC \tag{14-5}$$

Substituting for $2\pi RC$ in Equation 14-2 results in the following equation for v_i:

$$v_i = v_s \frac{1}{1 + j\dfrac{f}{f_H}} \tag{14-6}$$

where f is the operating frequency and f_H is the cutoff frequency.

Chapter 14

The output v_o is the amplified version of v_i.

$$v_o = v_i \left(\frac{R_F}{R_S}+1\right) = v_s \frac{\left(\frac{R_F}{R_S}+1\right)}{1+j\frac{f}{f_H}} = v_s \frac{A_{PB}}{1+j\frac{f}{f_H}} \qquad (14\text{-}7)$$

where A_{PB} is the pass-band voltage gain or the gain of the non-inverting amplifier. Dividing both sides of the equation by v_s yields the gain of the system from v_s to v_o.

$$A_v = \frac{v_o}{v_s} = \frac{A_{PB}}{1+j\frac{f}{f_H}} \qquad (14\text{-}8)$$

$$A_{PB} = \frac{R_F}{R_S}+1 \qquad (14\text{-}9)$$

Since the denominator is complex, it has a magnitude and phase as follows:

$$A_v = \frac{v_o}{v_s} = \frac{A_{PB}}{\left[1+\left(\frac{f}{f_H}\right)^2\right]^{1/2} \angle \tan^{-1}\left(\frac{f}{f_H}\right)} = A_{PB}\left[1+\left(\frac{f}{f_H}\right)^2\right]^{-1/2} \angle -\tan^{-1}\left(\frac{f}{f_H}\right) \qquad (14\text{-}10)$$

Hence, the magnitude and the phase of the system gain are as follows:

$$A_{v(mag)} = A_{PB}\left[1+\left(\frac{f}{f_H}\right)^2\right]^{-1/2} \qquad (14\text{-}11)$$

$$\phi = -\tan^{-1}\left(\frac{f}{f_H}\right) \qquad (14\text{-}12)$$

Magnitude dB gain:

The standard frequency response plot is generally the magnitude gain in dB versus frequency in logarithmic scale. Hence, we will convert the magnitude gain to dB as follows:

$$A_{v(dB)} = 20\log\left\{A_{PB}\left[1+\left(\frac{f}{f_H}\right)^2\right]^{-1/2}\right\} \qquad (14\text{-}13)$$

Applying the log properties to the above equation, we obtain the following:

$$A_{v(dB)} = 20\log A_{PB} + 20\log\left[1+\left(\frac{f}{f_H}\right)^2\right]^{-1/2} \qquad (14\text{-}14)$$

$$A_{v(dB)} = 20\log A_{PB} - 10\log\left[1+\left(\frac{f}{f_H}\right)^2\right] \qquad (14\text{-}15)$$

Chapter 14

Let us now assume that $R_F = 10$ kΩ and $R_S = 1.1$ kΩ so that $A_{PB} = 10$, and then plot the magnitude and phase frequency responses using Equation 14-15.

$$A_{v(dB)} = 20\log 10 - 10\log\left[1+\left(\frac{f}{f_H}\right)^2\right] \quad (14\text{-}16)$$

$$A_{v(dB)} = 20\text{ dB} - 10\log\left[1+\left(\frac{f}{f_H}\right)^2\right] \quad (14\text{-}17)$$

To cover a sufficient range of frequency, we will start the frequency response plot at $0.01f_H$, and continue producing some data points up to $100f_H$. This will cover a frequency range of four decades, two decades below and two decades above the cutoff frequency.

$f = 0.01f_H$: $f/f_H = 0.01$ $A_{v(dB)} = 20\text{ dB} - 10\log[1+(0.01)^2] \cong 20\text{ dB}$ $\phi = -0.57°$

$f = 0.1f_H$: $f/f_H = 0.1$ $A_{v(dB)} = 20\text{ dB} - 10\log[1+(0.1)^2] \cong 20\text{ dB}$ $\phi = -5.7°$

$f = f_H$: $f/f_H = 1$ $A_{v(dB)} = 20\text{ dB} - 10\log[1+(1)^2] = 17\text{ dB}$ $\phi = -45°$

$f = 10f_H$: $f/f_H = 10$ $A_{v(dB)} = 20\text{ dB} - 10\log[1+(10)^2] \cong 0\text{ dB}$ $\phi = -84.2°$

$f = 100f_H$: $f/f_H = 100$ $A_{v(dB)} = 20\text{ dB} - 10\log[1+(100)^2] \cong -20\text{ dB}$ $\phi = -89.4°$

Table 14-1: Data table for plotting the magnitude dB gain and phase for the LPF

We will now plot the above data, dB gain and phase, versus frequency f/f_H as follows:

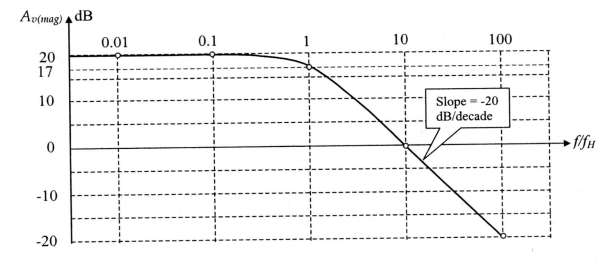

Figure 14-3: Plot of magnitude dB gain versus frequency (*log scale*) for the low-pass active filter with a pass-band gain $A_{PB} = 10 \Rightarrow 20$ dB

As you can see, the frequency response plot starts at a gain of 20 dB at the low frequency range ($f \leq 0.1f_H$), drops 3 dB at the cutoff frequency ($f = f_H$), and rolls off at a rate of -20 dB/decade beyond the cutoff frequency.

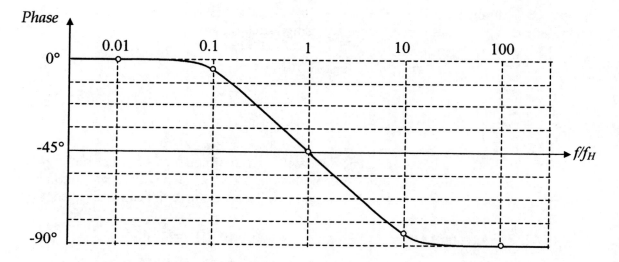

Figure 14-4: Plot of the phase versus frequency (*log scale*) for the first-order LPF

From the data of Table 14-1 and the phase plot of Figure 14-4, it is obvious that the phase starts at 0° in the low frequency range; then, dropping gradually, it reaches –45° at the cutoff frequency, and then approaches –90° at the higher frequency range.

14.4 FIRST-ORDER HIGH-PASS ACTIVE FILTER

The circuit diagram and the typical frequency response plot of the first-order high-pass filter are exhibited in Figure 14-5 below. As you can see, the only difference with the low-pass filter is that the positions of *R* and *C* have been switched.

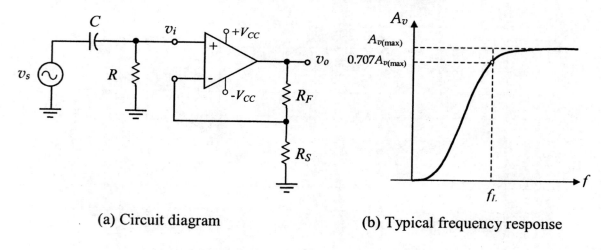

(a) Circuit diagram (b) Typical frequency response

Figure 14-5: First-order high-pass active filter

The derivation process is similar to that of the low-pass and thus it is left to the student as a homework assignment at the end of the chapter. The end result, however, is presented below:

Chapter 14

$$A_{v(mag)} = A_{PB}\left[1+\left(\frac{f_L}{f}\right)^2\right]^{-1/2} \qquad (14\text{-}18)$$

$$\phi = \tan^{-1}\left(\frac{f_L}{f}\right) \qquad (14\text{-}19)$$

$$A_{v(dB)} = 20\log A_{PB} - 10\log\left[1+\left(\frac{f_L}{f}\right)^2\right] \qquad (14\text{-}20)$$

Frequency response plot:

Again, we will assume that $R_F = 10$ kΩ and $R_S = 1.1$ kΩ so that $A_{PB} = 10$, and then plot the magnitude and phase frequency responses using Equations 14-20 and 14-19. We will again cover a frequency range of four decades, two decades below and two decades above the cutoff frequency ($0.01 f_L$ to $100 f_L$).

$f = 0.01 f_L$: $f/f_L = 0.01$ $f_L/f = 100$ $A_{v(dB)} = 20\text{ dB} - 10\log[1+(100)^2] \cong -20\text{ dB}$ $\phi = 89.4°$

$f = 0.1 f_L$: $f/f_L = 0.1$ $f_L/f = 10$ $A_{v(dB)} = 20\text{ dB} - 10\log[1+(10)^2] \cong 0\text{ dB}$ $\phi = 84.2°$

$f = f_L$: $f/f_L = 1$ $f_L/f = 1$ $A_{v(dB)} = 20\text{ dB} - 10\log[1+(1)^2] = 17\text{ dB}$ $\phi = 45°$

$f = 10 f_L$: $f/f_L = 10$ $f_L/f = 0.1$ $A_{v(dB)} = 20\text{ dB} - 10\log[1+(0.1)^2] \cong 20\text{ dB}$ $\phi = 5.7°$

$f = 100 f_L$: $f/f_L = 100$ $f_L/f = 0.01$ $A_{v(dB)} = 20\text{ dB} - 10\log[1+(0.01)^2] \cong 20\text{ dB}$ $\phi = 0.57°$

Table 14-2: Data table for plotting the magnitude dB gain and phase of the HPF

We will now plot the above data, dB gain and phase, versus frequency f/f_L as follows:

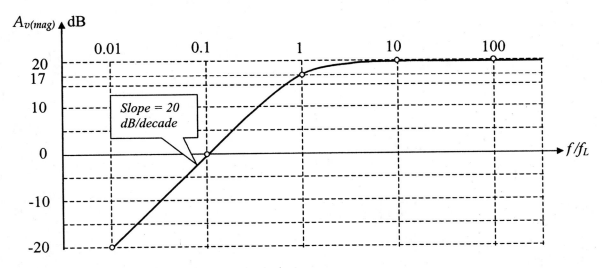

Figure 14-6: Plot of the magnitude dB gain versus frequency (*log scale*) for the high-pass active filter with a pass-band gain $A_{PB} = 10 \Rightarrow 20$ dB

The frequency response plot starts at the pass-band gain of 20 dB at the high-frequency range ($f \geq 10f_L$), drops 3 dB at the cutoff frequency ($f = f_L$), and rolls off at a rate of 20 dB/decade below the cutoff frequency.

From the data of Table 14-2 and the phase plot of Figure 14-7, it is obvious that the phase starts at 90° in the low-frequency range; then, dropping gradually, it reaches 45° at the cutoff frequency, and then approaches 0° at the high-frequency range.

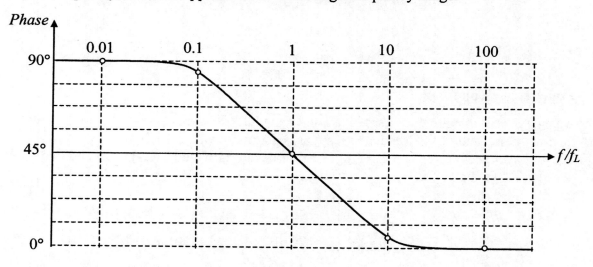

Figure 14-7: Plot of phase versus frequency (*log scale*) for the first-order HPF

Example 14-1 — 1st Order LPF — ANALYSIS

Determine the cutoff frequency, the pass-band gain in dB, and the gain at the cutoff frequency for the active filter of Figure 14-8 below:

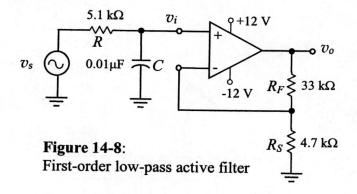

Figure 14-8: First-order low-pass active filter

Solution:

The cutoff frequency f_H is determined with Equation 14-4, as follows:

$$f_H = \frac{1}{2\pi RC} = \frac{1}{2\pi \times 5.1\,\text{k}\Omega \times 0.01\,\mu\text{F}} = 3.12\,\text{kHz}$$

The pass-band gain A_{PB} is determined with Equation 14-9.

$$A_{PB} = \frac{R_F}{R_S} + 1 = \frac{33\,k\Omega}{4.7\,k\Omega} + 1 = 8$$

$$dBG = 20 \log A_{PB} = 20 \log(8) = 18\,dB$$

The gain drops 3 dB at the cutoff frequency; hence, the gain at the cutoff is 15 dB.

Example 14-2 ANALYSIS

Determine the cutoff frequency, the pass-band gain in dB, and the gain at the cutoff frequency for the active filter of Figure 14-9 below:

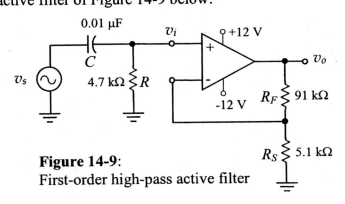

Figure 14-9: First-order high-pass active filter

Solution:

The cutoff frequency f_L is determined with Equation 14-4, as follows:

$$f_L = \frac{1}{2\pi RC} = \frac{1}{2\pi \times 4.7\,k\Omega \times 0.01\,\mu F} = 3.38\,kHz$$

The pass-band gain A_{PB} is determined with Equation 14-9.

$$A_{PB} = \frac{R_F}{R_S} + 1 = \frac{91\,k\Omega}{5.1\,k\Omega} + 1 = 18.8$$

$$dBG = 20 \log A_{PB} = 20 \log(18.8) = 25.5\,dB$$

The gain drops 3 dB at the cutoff frequency; hence, the gain at the cutoff is 22.5 dB.

Practice Problem 14-1 ANALYSIS

Determine the cutoff frequency, the pass-band gain in dB, and the gain at the cutoff frequency for the active filter of Figure 14-8 with $C = 0.022\,\mu F$, $R = 3.3\,k\Omega$, $R_F = 24\,k\Omega$, and $R_S = 2.2\,k\Omega$.

Answers: $f_H = 2.2\,kHz$, $A_{PB} = 21.5\,dB$, $A_{v(at\ cutoff)} = 18.5\,dB$

Practice Problem 14-2

ANALYSIS

Determine the cutoff frequency, the pass-band gain in dB, and the gain at the cutoff frequency for the active filter of Figure 14-9 with $C = 0.02$ μF, $R = 5.1$ kΩ, $R_F = 36$ kΩ, and $R_S = 3.3$ kΩ.

Answers: $f_L = 1.56$ kHz, $A_{PB} = 21.5$ dB, $A_{v(at\ cutoff)} = 18.5$ dB

14.5 DESIGN OF THE FIRST-ORDER ACTIVE FILTER

The following design procedure applies equally to the design of the first-order low-pass and high-pass filters. However, to use a single notation for the cutoff frequency in both cases (f_H and f_L), let us refer to it as f_o; hence,

$$f_o = \frac{1}{2\pi RC} \tag{14-21}$$

1. Pick the capacitor $C \leq 0.1$ μF.
2. Solve for R in Equation 14-21 to achieve the desired cutoff frequency.
3. Determine the values of R_F and R_S for the given gain. However, recall that in order to minimize the offset errors and balance the op-amp operation, the DC resistance at both input terminals of the op-amp must be equal; that is,

$$R = R_F || R_S = \frac{R_F \times R_S}{R_F + R_S} \tag{14-22}$$

Dividing both the numerator and the denominator by R_S and then solving for R_F results in the following:

$$R = \frac{R_F}{\frac{R_F}{R_S}+1} = \frac{R_F}{A_{PB}} \tag{14-23}$$

$$R_F = R \cdot A_{PB} \tag{14-24}$$

The resistor R_S can be determined from Equation 14-9, as follows:

$$A_{PB} = \frac{R_F}{R_S}+1 \tag{14-25}$$

$$\frac{R_F}{R_S} = A_{PB} - 1 \tag{14-26}$$

$$R_S = \frac{R_F}{A_{PB} - 1} \tag{14-27}$$

Example 14-3

DESIGN

Let us design a first-order low-pass active filter for a cutoff frequency at 5 kHz, and with a pass-band gain of 16 dB.

Solution: $16 \text{ dB} = 20 \log A_{PB}$ $\quad 0.8 = \log A_{PB}$ $\quad A_{PB} = 10^{0.8} = 6.31$

Let $C = 0.01 \text{ µF}$.

$R = \dfrac{1}{2\pi f_o C} = \dfrac{1}{2\pi \times 5 \text{ kHz} \times 0.01 \text{ µF}} = 3.18 \text{ k}\Omega$ $\quad$ Use $R = 3 \text{ k}\Omega + 180 \text{ }\Omega$

$R_F = R \cdot A_{PB} = 3.18 \text{ k}\Omega \times 6.31 = 20 \text{ k}\Omega.$ $\quad R_F = 20 \text{ k}\Omega$

$R_S = \dfrac{R_F}{A_{PB} - 1} = \dfrac{20 \text{ k}\Omega}{6.31 - 1} = 3.76 \text{ k}\Omega$ $\quad$ Use $R_S = 3 \text{ k}\Omega + 750 \text{ }\Omega$

Test Run and Design Verification:

Let us now test and verify the above design by simulation. The results of simulation created in Electronics Workbench are presented in Figures 14-10 through 14-11 below:

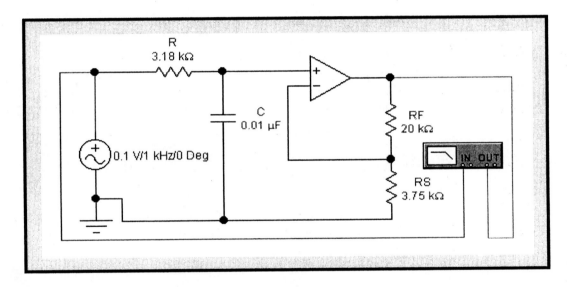

Figure 14-10: Circuit setup for simulation of Design Example 14-3 created in Electronics Workbench

The frequency response plot is depicted on the EWB Bode plotter of Figure 14-11 below, which also exhibits and verifies the pass-band gain of 16 dB at 100 Hz, and the cutoff frequency of 5 kHz, which corresponds to −3 dB gain (13 dB).

Note that the two cursors of the Bode plotter are set such that the first cursor measures the pass-band gain of $y_1 = 16$ dB at the pass-band frequency of $x_1 = 100$ Hz, and the second cursor measures the cutoff frequency of $x_2 = 5$ kHz at −3 dB gain of $y_2 = 13$ dB. Also note that the slope of the diagonal asymptote for one decade of frequency from 30 kHz at 0 dB gain to 130 kHz at −20 dB corresponds to a roll-off rate of −20 dB/decade.

Chapter 14

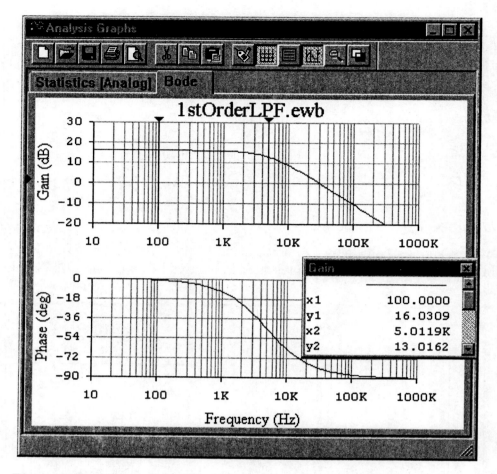

Figure 14-11: Electronics Workbench Bode plotter exhibiting the frequency response of Design Example 14-3 with the pass-band gain of 16 dB at 100 Hz and the cutoff frequency of 5 kHz at 13 dB (−3 dB)

Example 14-4   1st Order HPF   DESIGN

Let us now design a first-order high-pass active filter for a cutoff frequency at 10 kHz, and with a pass-band gain of 26 dB.

Solution:

$$26 \text{ dB} = 20 \log A_{PB}, \quad 1.3 = \log A_{PB}, \quad A_{PB} = 10^{1.3} = 20 \quad \text{Let } C = 0.01 \text{ μF}$$

$$R = \frac{1}{2\pi f_o C} = \frac{1}{2\pi \times 10 \text{ kHz} \times 0.01 \text{ μF}} = 1.59 \text{ k}\Omega \qquad \text{Use } R = 1.6 \text{ k}\Omega$$

$$R_F = R \cdot A_{PB} = 1.6 \text{ k}\Omega \times 20 = 32 \text{ k}\Omega. \qquad \text{Use } R_F = 33 \text{ k}\Omega$$

$$R_S = \frac{R_F}{A_{PB} - 1} = \frac{33 \text{ k}\Omega}{20 - 1} = 1.73 \text{ k}\Omega \qquad \text{Use } R_S = 1.6 \text{ k}\Omega + 130 \text{ }\Omega$$

606

Chapter 14

Test Run and Design Verification:

The results of simulation are presented in Figures 14-12 and 14-13 below.

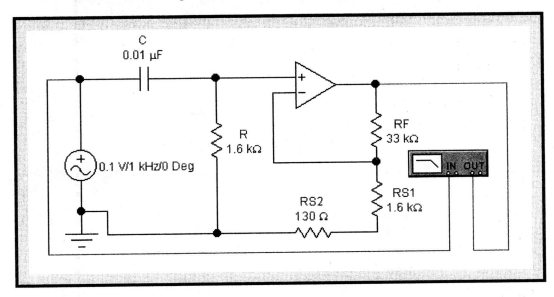

Figure 14-12: Circuit setup for simulation of the Design Example 14-4 created in Electronics Workbench

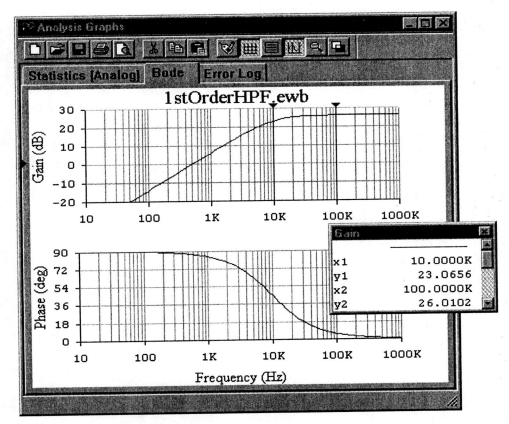

Figure 14-13: Electronics Workbench Bode plotter exhibiting the frequency response plot with the pass-band gain of 26 dB at 100 kHz, and the cutoff frequency of 10 kHz at 23 dB

Also note that the roll-off rate is +20 dB/decade, as expected.

Practice Problem 14-3 — 1st Order LPF — DESIGN

Design a first-order active low-pass filter for a cutoff frequency at 4 kHz, and with a pass-band gain of 22 dB. Verify your design by simulation.

Answers: $C = 0.01\ \mu F$, $R = 3.9\ k\Omega$, $R_F = 49\ k\Omega = 39\ k\Omega + 10\ k\Omega$, $R_S = 4.2\ k\Omega$

Practice Problem 14-4 — 1st Order HPF — DESIGN

Design a first order active high-pass filter for a cutoff frequency at 4 kHz, and with a pass-band gain of 20 dB. Verify your design by simulation.

Answers: $C = 0.01\ \mu F$, $R = 3.9\ k\Omega$, $R_F = 39\ k\Omega$, $R_S = 4.3\ k\Omega$.

14.6 SECOND-ORDER ACTIVE FILTERS

The number of reactive elements in the circuit determines the order of any circuit. For instance, there is only one capacitive element in the first-order filters. Hence, there will be two capacitive elements in the second-order active filters. The higher the order of the filter, the higher the slope of the diagonal asymptote or the roll-off rate. Recall that the roll-off rate for the first order is 20 dB/decade, and it will be 40 dB/decade for the second-order. The general block diagram of a second-order active filter, which applies equally to low-pass and high-pass, is shown in Figure 14-14 below. For a low-pass filter, components z_1 and z_2 are resistive elements (R_1 & R_2) and components z_3 and z_4 are capacitive elements (C_1 & C_2). The opposite is true for a high-pass filter.

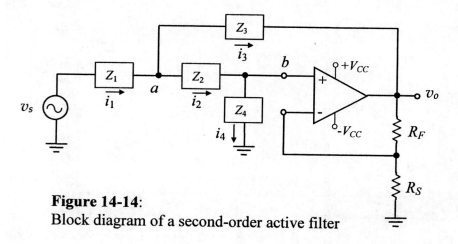

Figure 14-14:
Block diagram of a second-order active filter

Chapter 14

Derivation of the transfer function and other parameters of the second order filter:

Referring to the circuit of Figure 14-14 and applying Ohm's law, currents i_1, i_2, i_3, and i_4 can be determined as follows:

$$i_1 = \frac{v_s - v_a}{z_1} = \frac{v_s}{z_1} - \frac{v_a}{z_1} \tag{14-28}$$

$$i_2 = i_4 = \frac{v_b}{z_4} \tag{14-29}$$

$$v_a = i_2(z_2 + z_4) = \frac{v_b}{z_4}(z_2 + z_4) \tag{14-30}$$

$$i_1 = \frac{v_s}{z_1} - \frac{v_a}{z_1} = \frac{v_s}{z_1} - v_b \frac{z_2 + z_4}{z_1 \cdot z_4} \tag{14-31}$$

$$i_3 = i_1 - i_2 = \frac{v_s}{z_1} - v_b \frac{z_2 + z_4}{z_1 \cdot z_4} - \frac{v_b}{z_4} \tag{14-32}$$

$$v_o = v_a - i_3 z_3 \tag{14-33}$$

Substituting for v_a and i_3 in the above equation results in the following equation for v_o:

$$v_o = \frac{v_b}{z_4}(z_2 + z_4) - \left[\frac{v_s}{z_1} - v_b \frac{z_2 + z_4}{z_1 \cdot z_4} - \frac{v_b}{z_4}\right] z_3 \tag{14-34}$$

The output v_o is also the amplified version of v_b at the non-inverting input.

$$v_o = v_b \cdot A_{PB} \tag{14-35}$$

$$v_b = \frac{v_o}{A_{PB}} \tag{14-36}$$

where,

$$A_{PB} = \frac{R_F}{R_S} + 1 \tag{14-37}$$

Substituting for v_b in Equation 14-34 results in the following equation for v_o:

$$v_o = \frac{v_o}{A_{PB}} \frac{z_2 + z_4}{z_4} - \left[v_s \frac{z_3}{z_1} - \frac{v_o}{A_{PB}} \frac{z_2 + z_4}{z_1 \cdot z_4} z_3 - \frac{v_o}{A_{PB}} \frac{z_3}{z_4}\right] \tag{14-38}$$

Separating the variables and factoring out v_o yields the following:

$$v_o \left[\frac{z_2 + z_4}{A_{PB} z_4} + \frac{(z_2 + z_4) z_3}{A_{PB} z_1 \cdot z_4} + \frac{z_3}{A_{PB} z_4} - 1\right] = v_s \frac{z_3}{z_1} \tag{14-39}$$

Solving for the transfer function or the overall gain v_o/v_s, we obtain the following:

$$\frac{v_o}{v_s} = \frac{z_3}{z_1}\left[\frac{z_2 + z_4}{A_{PB} z_4} + \frac{(z_2 + z_4) z_3}{A_{PB} z_1 \cdot z_4} + \frac{z_3}{A_{PB} z_4} - 1\right]^{-1} \tag{14-40}$$

Taking the common denominator, we arrive at the following equation:

$$\frac{v_o}{v_s} = \frac{z_3}{z_1}\left[\frac{z_1(z_2+z_4)+z_3(z_2+z_4)+z_1z_3-A_{PB}z_1z_4}{A_{PB}z_1z_4}\right]^{-1} \quad (14\text{-}41)$$

$$\frac{v_o}{v_s} = \frac{z_3}{z_1}\left[\frac{A_{PB}z_1z_4}{z_1(z_2+z_4)+z_3(z_2+z_4)+z_1z_3-A_{PB}z_1z_4}\right] \quad (14\text{-}42)$$

$$\frac{v_o}{v_s} = z_3\left[\frac{A_{PB}z_4}{z_1(z_2+z_4)+z_3(z_2+z_4)+z_1z_3-A_{PB}z_1z_4}\right] \quad (14\text{-}43)$$

$$\frac{v_o}{v_s} = \frac{A_{PB}z_3z_4}{z_1z_2+z_1z_4+z_2z_3+z_3z_4+z_1z_3-A_{PB}z_1z_4} \quad (14\text{-}44)$$

Factoring out the common terms (z_1z_4) in the denominator results in the following:

$$\frac{v_o}{v_s} = \frac{A_{PB}z_3z_4}{z_1z_2+z_2z_3+z_3z_4+z_1z_3+z_1z_4(1-A_{PB})} \quad (14\text{-}45)$$

14.6.1 Second-Order Low-Pass Active Filter

As mentioned previously, for a low-pass filter, components z_1 and z_2 are resistive elements (R_1 & R_2) and components z_3 and z_4 are capacitive elements (C_1 & C_2).

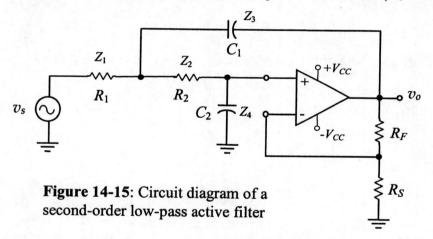

Figure 14-15: Circuit diagram of a second-order low-pass active filter

Hence, in the above transfer function (Equation 14-45), impedances z_1 and z_2 will be replaced with resistances R_1 and R_2, and impedances z_3 and z_4 will be replaced with capacitive reactances X_{C1} and X_{C2}. However, for the sake of convenience, we will carry out the operation in the s-domain (Laplace domain); that is, $j\omega$ will be replaced by s.

$$X_C = \frac{1}{j\omega C} = \frac{1}{sC} \quad (14\text{-}46)$$

$$\frac{v_o}{v_s} = \frac{A_{PB}z_3z_4}{z_1z_2+z_2z_3+z_3z_4+z_1z_3-z_1z_4(1-A_{PB})} \quad (14\text{-}47)$$

Chapter 14

$$\frac{v_o}{v_s} = \frac{A_{PB}\dfrac{1}{sC_1}\cdot\dfrac{1}{sC_2}}{R_1R_2 + R_2\dfrac{1}{sC_1} + \dfrac{1}{sC_1}\cdot\dfrac{1}{sC_2} + R_1\dfrac{1}{sC_1} + R_1\dfrac{1}{sC_2}(1-A_{PB})} \quad (14\text{-}48)$$

Taking a common denominator in the denominator of the above equation yields the following:

$$\frac{v_o}{v_s} = \frac{A_{PB}\dfrac{1}{s^2C_1C_2}}{\dfrac{R_1R_2s^2C_1C_2 + R_2sC_2 + 1 + R_1sC_2 + R_1sC_1(1-A_{PB})}{s^2C_1C_2}}$$

$$\frac{v_o}{v_s} = \frac{A_{PB}}{R_1R_2s^2C_1C_2 + R_2sC_2 + 1 + R_1sC_2 + R_1sC_1(1-A_{PB})}$$

$$\frac{v_o}{v_s} = \frac{A_{PB}}{s^2R_1R_2C_1C_2 + s[R_2C_2 + R_1C_2 + R_1C_1(1-A_{PB})] + 1}$$

Dividing both the numerator and the denominator by $R_1R_2C_1C_2$ results in the following standard transfer function:

$$\frac{v_o}{v_s} = \frac{A_{PB}\dfrac{1}{R_1R_2C_1C_2}}{s^2 + \dfrac{[R_2C_2 + R_1C_2 + R_1C_1(1-A_{PB})]}{R_1R_2C_1C_2}s + \dfrac{1}{R_1R_2C_1C_2}}$$

The above equation is in the form of the following standard transfer function for the second-order low-pass network:

$$\frac{v_o}{v_s} = \frac{A_{PB}\omega_o^2}{s^2 + 2\xi\omega_o s + \omega_o^2} \quad (14\text{-}49)$$

where ω_o is the cutoff frequency or critical frequency, and ξ is the damping factor. Hence,

$$\omega_o^2 = \frac{1}{R_1R_2C_1C_2} \quad (14\text{-}50)$$

and

$$\omega_o = \sqrt{\frac{1}{R_1R_2C_1C_2}} = \frac{1}{\sqrt{R_1R_2C_1C_2}} \quad (14\text{-}51)$$

$$2\xi\omega_o = \frac{R_2C_2 + R_1C_2 + R_1C_1(1-A_{PB})}{R_1R_2C_1C_2} \quad (14\text{-}52)$$

$$2\xi = \frac{2\xi\omega_o}{\omega_o} = \frac{R_2C_2 + R_1C_2 + R_1C_1(1-A_{PB})}{R_1R_2C_1C_2}\sqrt{R_1R_2C_1C_2} \quad (14\text{-}53)$$

$$\xi_{LP} = \frac{R_2C_2 + R_1C_2 + R_1C_1(1-A_{PB})}{2\sqrt{R_1R_2C_1C_2}} \quad (14\text{-}54)$$

14.6.2 Second-Order High-Pass Active Filter

For a high-pass filter, components z_1 and z_2 are capacitive elements (C_1 & C_2) and components z_3 and z_4 are resistive elements (R_1 & R_2).

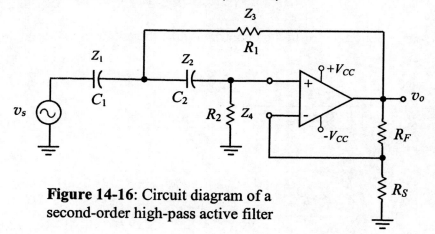

Figure 14-16: Circuit diagram of a second-order high-pass active filter

$$X_C = \frac{1}{j\omega C} = \frac{1}{sC} \tag{14-55}$$

$$\frac{v_o}{v_s} = \frac{A_{PB} z_3 z_4}{z_1 z_2 + z_2 z_3 + z_3 z_4 + z_1 z_3 - z_1 z_4 (1 - A_{PB})} \tag{14-56}$$

$$\frac{v_o}{v_s} = \frac{A_{PB} R_1 R_2}{\dfrac{1}{sC_1} \cdot \dfrac{1}{sC_2} + R_1 \dfrac{1}{sC_2} + R_1 R_2 + R_1 \dfrac{1}{sC_1} + R_2 \dfrac{1}{sC_1}(1 - A_{PB})} \tag{14-57}$$

Taking a common denominator in the denominator of the above equation yields the following:

$$\frac{v_o}{v_s} = \frac{A_{PB} R_1 R_2}{\dfrac{1 + R_1 s C_1 + R_1 R_2 s^2 C_1 C_2 + R_1 s C_2 + R_2 s C_2 (1 - A_{PB})}{s^2 C_1 C_2}} \tag{14-58}$$

$$\frac{v_o}{v_s} = \frac{A_{PB} R_1 R_2 s^2 C_1 C_2}{s^2 R_1 R_2 C_1 C_2 + s[R_1 C_1 + R_1 C_2 + R_2 C_2 (1 - A_{PB})] + 1} \tag{14-59}$$

Dividing both the numerator and the denominator by $R_1 R_2 C_1 C_2$ results in the following standard transfer function:

$$\frac{v_o}{v_s} = \frac{A_{PB} s^2}{s^2 + \dfrac{[R_1 C_1 + R_1 C_2 + R_2 C_2 (1 - A_{PB})]}{R_1 R_2 C_1 C_2} s + \dfrac{1}{R_1 R_2 C_1 C_2}} \tag{14-60}$$

The above equation is in the form of the following standard transfer function for the second order high-pass network:

$$\frac{v_o}{v_s} = \frac{A_{PB} s^2}{s^2 + 2\xi \omega_o s + \omega_o^2} \tag{14-61}$$

where, ω_o is the cutoff frequency or critical frequency, and ξ is the damping factor.

Hence,

$$\omega_o^2 = \frac{1}{R_1 R_2 C_1 C_2} \tag{14-62}$$

and

$$\omega_o = \sqrt{\frac{1}{R_1 R_2 C_1 C_2}} = \frac{1}{\sqrt{R_1 R_2 C_1 C_2}} \tag{14-63}$$

$$2\xi\omega_o = \frac{R_1 C_1 + R_1 C_2 + R_2 C_2 (1 - A_{PB})}{R_1 R_2 C_1 C_2} \tag{14-64}$$

$$2\xi = \frac{2\xi\omega_o}{\omega_o} = \frac{R_1 C_1 + R_1 C_2 + R_2 C_2 (1 - A_{PB})}{R_1 R_2 C_1 C_2} \sqrt{R_1 R_2 C_1 C_2} \tag{14-65}$$

$$\xi_{HP} = \frac{R_1 C_1 + R_1 C_2 + R_2 C_2 (1 - A_{PB})}{2\sqrt{R_1 R_2 C_1 C_2}} \tag{14-66}$$

14.6.3 Sallen-Key Unity-Gain Active Filter

The Sallen-Key unity-gain filter sets the pass-band gain to unity ($A_{PB} = 1$), which simplifies and reduces the equation for the damping factor ξ as follows:

$$\xi_{LP} = \frac{R_2 C_2 + R_1 C_2}{2\sqrt{R_1 R_2 C_1 C_2}} \tag{14-67}$$

$$\xi_{HP} = \frac{R_1 C_1 + R_1 C_2}{2\sqrt{R_1 R_2 C_1 C_2}} \tag{14-68}$$

However, to obtain a desired gain, an amplifier must follow the unity-gain filter. For the sake of simplicity and convenience, generally C_1 is set equal to C_2 ($C_1 = C_2 = C$), which further simplifies the damping factor and the cutoff frequency as follows:

$$\xi_{LP} = \frac{(R_2 + R_1)C}{2C\sqrt{R_1 R_2}} = \frac{R_1 + R_2}{2\sqrt{R_1 R_2}} \tag{14-69}$$

$$\xi_{HP} = \frac{2R_1 C}{2C\sqrt{R_1 R_2}} = \frac{R_1}{\sqrt{R_1 R_2}} = \sqrt{\frac{R_1}{R_2}} \tag{14-70}$$

For example, for the Butterworth response the damping factor $\xi = 0.707$, from which the relationship between R_1 and R_2 can be determined as follows:

$$\xi_{HP} = \sqrt{\frac{R_1}{R_2}} = 0.707 \tag{14-71}$$

Squaring both sides and solving for R_2 yield the following:

$$R_2 = 2R_1 \tag{14-72}$$

However, we cannot arbitrarily choose R_1. It must be determined from the intended cutoff frequency ω_o, as follows:

$$\omega_o = \frac{1}{\sqrt{R_1 R_2 C_1 C_2}} = \frac{1}{C\sqrt{R_1 R_2}} \tag{14-73}$$

Chapter 14

$$\sqrt{R_1 R_2} = \frac{1}{\omega_o C} = \sqrt{2} R_1 \tag{14-74}$$

$$R_1 = \frac{0.707}{\omega_o C} \tag{14-75}$$

With the ω_o given, let $C = 0.1$ μF, and solve for R_1.

The circuit diagram of a second-order unity-gain high-pass Butterworth filter, which has been designed for a cutoff or critical frequency ω_o at 1 krad/sec, and its frequency response plot are displayed in Figures 14-17 and 14-18 below:

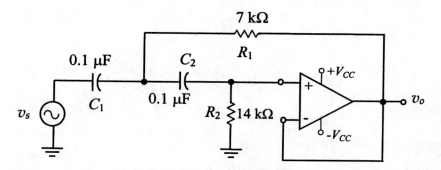

Figure 14-17: Circuit diagram of a Sallen-Key unit-gain second-order Butterworth high-pass active filter

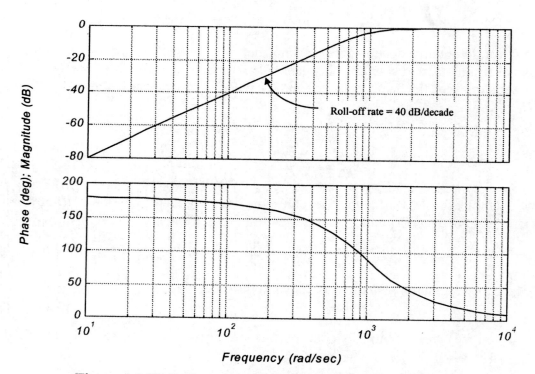

Figure 14-18: Frequency response plot of the second-order unity-gain high-pass Butterworth filter plotted with MATLAB

The frequency response of unity-gain second-order low-pass active filters with different standard damping factors is exhibited in Figure 14-19 below:

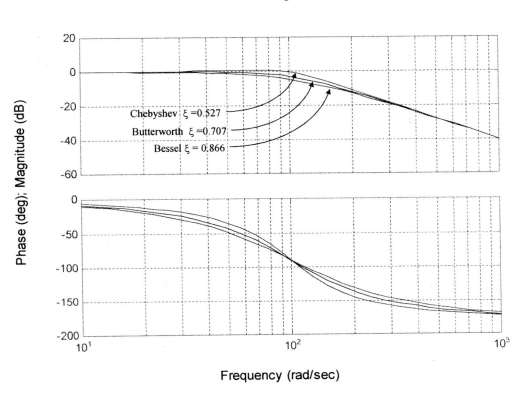

Figure 14-19: Frequency response plot of second-order equal component low-pass unity-gain active filters with different damping factors (ξ)

14.6.4 Sallen-Key Equal Component Active Filter

Another approach that simplifies the analysis and design of active filters is the Sallen-Key method of using equal components; that is, letting $R_1 = R_2 = R$ and $C_1 = C_2 = C$. This approach simplifies and reduces the equations for the damping factor as follows:

$$\xi_{LP} = \frac{R_2 C_2 + R_1 C_2 + R_1 C_1 (1 - A_{PB})}{2\sqrt{R_1 R_2 C_1 C_2}} \tag{14-76}$$

$$\xi_{LP} = \frac{RC + RC + RC(1 - A_{PB})}{2\sqrt{RRCC}} = \frac{RC(3 - A_{PB})}{2RC} \tag{14-77}$$

$$\xi_{LP} = 0.5(3 - A_{PB})$$

$$\xi_{HP} = \frac{R_1 C_1 + R_1 C_2 + R_2 C_2 (1 - A_{PB})}{2\sqrt{R_1 R_2 C_1 C_2}} \tag{14-78}$$

Chapter 14

$$\xi_{HP} = \frac{RC + RC + RC(1 - A_{PB})}{2\sqrt{RRCC}} = \frac{RC(3 - A_{PB})}{2RC} \quad (14\text{-}79)$$

$$\xi_{HP} = 0.5(3 - A_{PB}) \quad (14\text{-}80)$$

Hence, the damping factor for the equal component filter, low-pass or high-pass, is:

$$\xi_{EC} = 0.5(3 - A_{PB})$$

The pass-band gain A_{PB} is a function of the damping factor (ξ), and the critical frequency ω_o is a function of R and C.

$$A_{PB} = 3 - 2\xi \quad (14\text{-}81)$$

$$\omega_o = \frac{1}{RC} \quad (14\text{-}82)$$

The Sallen-Key equal component strategy is the most popular, practical, and convenient approach; thus, we will limit our analysis and design to equal component filters.

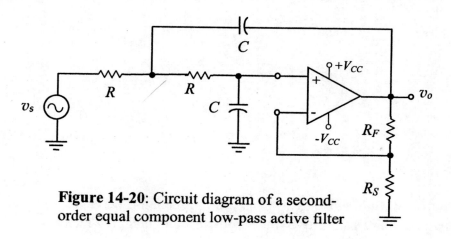

Figure 14-20: Circuit diagram of a second-order equal component low-pass active filter

Example 14-5 — 2nd Order LPF — ANALYSIS

Let us determine the critical frequency f_o, pass-band gain A_{PB}, damping factor ξ, and the roll-off rate for a second-order low-pass active filter with the following circuit components:

$R_1 = R_2 = 4.3$ kΩ, $C_1 = C_2 = 0.05$ µF, $R_F = 7.5$ kΩ, $R_S = 10$ kΩ.

Solution:

$$f_o = \frac{1}{2\pi RC} = \frac{1}{2\pi \times 4.3 \text{ k}\Omega \times 0.05 \text{ µF}} = 740 \text{ Hz}$$

$$A_{PB} = \frac{R_F}{R_S} + 1 = \frac{7.5 \text{ k}\Omega}{10 \text{ k}\Omega} + 1 = 1.75$$

$$\xi = 0.5(3 - A_{PB}) = 0.5(3 - 1.75) = 0.625$$

As exhibited in Figure 14-19, the roll-off rate for the second-order low-pass filter is –40 dB/decade.

Practice Problem 14-5 ANALYSIS

Determine the critical frequency f_o, pass-band gain A_{PB}, and the damping factor ξ for the second-order low-pass active filter whose circuit diagram is shown below:

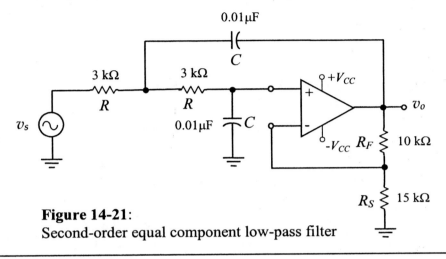

Figure 14-21:
Second-order equal component low-pass filter

Answers: $f_o = 5.3$ kHz, $A_{PB} = 1.66$, $\xi = 0.67$

The frequency response plot of second-order equal component low-pass active filters with different standard damping factors is exhibited in Figure 14-22 below. As you can see, the critical frequency ω_o for all is 100 rad/sec, which corresponds to the $-90°$ phase and they all roll-off at the expected rate of -40 dB/decade. However, each has a different pass-band gain because A_{PB} is a function of the damping factor ($A_{PB} = 3 - 2\xi$).

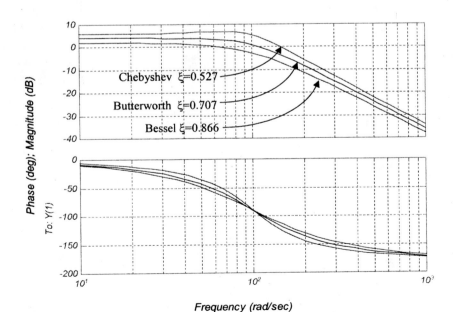

Figure 14-22: Frequency response plot of second-order equal component low-pass active filters with different damping factors (ξ)

Chapter 14

Example 14-6 — 2nd Order HPF — ANALYSIS

Let us now determine the critical frequency f_o, pass-band gain A_{PB}, and the damping factor ξ for the second-order high-pass active filter whose circuit diagram is shown below:

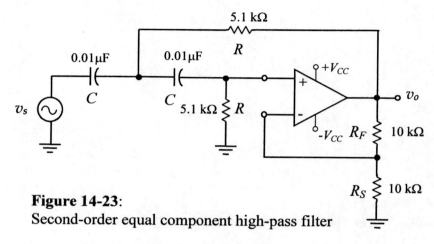

Figure 14-23:
Second-order equal component high-pass filter

Solution:

$$f_o = \frac{1}{2\pi RC} = \frac{1}{2\pi \times 5.1\,\text{k}\Omega \times 0.01\,\mu\text{F}} = 3.12\,\text{kHz}$$

$$A_{PB} = \frac{R_F}{R_S} + 1 = \frac{10\,\text{k}\Omega}{10\,\text{k}\Omega} + 1 = 2$$

$$\text{dBG} = 20 \log 2 = 6\,\text{dB}$$

$$\xi = 0.5(3 - A_{PB}) = 0.5$$

As exhibited in Figure 14-18, the roll-off rate for the second-order high-pass filter is 40 dB/decade.

Practice Problem 14-6 — 2nd Order HPF — ANALYSIS

Determine the critical frequency f_o, pass-band gain A_{PB}, damping factor ξ, and the roll-off rate for a second-order equal component high-pass active filter:
$R_1 = R_2 = 4.7\,\text{k}\Omega$, $C_1 = C_2 = 0.02\,\mu\text{F}$, $R_F = 6.2\,\text{k}\Omega$, $R_S = 10\,\text{k}\Omega$.

Answers: $f_o = 1.69\,\text{kHz}$, $A_{PB} = 4.19\,\text{dB}$, $\xi = 0.69$

For the purpose of comparison and reinforcement, the frequency response plots of unity-gain first- and second-order low-pass active filters are exhibited in Figure 14-24 below. As expected, the first order rolls off at the rate of –20 dB/decade and the second-order rolls off at the rate of –40 dB/decade. In fact, each increase in the order of the filter increases the roll-off rate by 20 dB/decade. As you can see, although each response plot takes a different path, both responses pass through the same –3 dB gain at the cutoff frequency of 100 rad/sec. The phase for the first order extends from 0° to −90° passing through −45° at the cutoff frequency (ω_{-3dB}). The phase for the second-order, however, extends from 0° to −180° passing through −90° at the cutoff frequency.

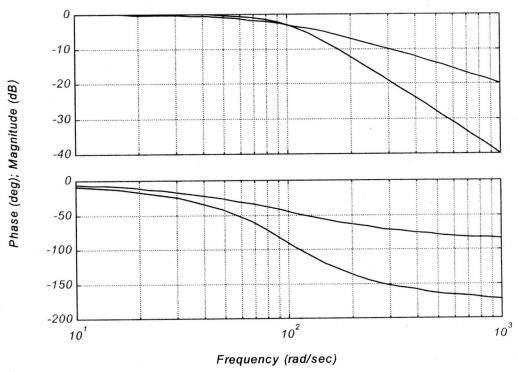

Figure 14-24: Frequency response plot of first- and second-order unity-gain low-pass filters

14.6.5 Design of the Second-Order Active Filter

The design procedure for the second-order active filter is similar to that of the first-order. In general, the same three-step design procedure that we followed with the first-order will have to be followed with the second-order. However, there is a damping factor (ξ) with the second-order that dictates the response type and limits the pass-band gain. With the Butterworth response type the critical frequency f_o equals the f_{-3dB}, which is not the case with the other types. Hence, to keep the design procedure simple and convenient, we will limit our designs to the Butterworth response.

Example 14-7 — 2nd Order LPF — DESIGN

Let us design a second-order equal component low-pass active Butterworth type filter with f_{-3dB} at 5 kHz.

Chapter 14

Solution:

Butterworth: $\xi = 0.707$, $A_{PB} = 3 - 2\xi = 3 - 1.414 = 1.586 = 4$ dB, $f_o = f_{-3dB} = 5$ kHz

Let $C = 0.01$ μF.

$$R = \frac{1}{2\pi f_o C} = \frac{1}{2\pi \times 5 \text{ kHz} \times 0.01 \text{ μF}} = 3.18 \text{ k}\Omega \qquad \text{Use } R = 3 \text{ k}\Omega + 180 \text{ }\Omega$$

Recall that in order to minimize the offset errors and balance the op-amp operation, the DC resistance at both input terminals of the op-amp must be equal; that is,

$$2R = R_F \| R_S = \frac{R_F \times R_S}{R_F + R_S} \qquad (14\text{-}83)$$

Dividing both the numerator and the denominator by R_S, and then solving for R_F results in the following:

$$2R = \frac{R_F}{\frac{R_F}{R_S} + 1} = \frac{R_F}{A_{PB}} \qquad (14\text{-}84)$$

$$R_F = 2RA_{PB} \qquad (14\text{-}85)$$

$R_F = 2RA_{PB} = 2 \times 3.18 \text{ k}\Omega \times 1.586 = 10 \text{ k}\Omega \qquad R_F = 10 \text{ k}\Omega$

The resistor R_S can be determined with Equation 14-27 as follows:

$$R_S = \frac{R_F}{A_{PB} - 1} = \frac{10 \text{ k}\Omega}{1.586 - 1} = 17.2 \text{ k}\Omega \qquad \text{Use } R_S = 15 \text{ k}\Omega + 2.2 \text{ k}\Omega$$

Test Run and Design Verification:

The results of simulation by EWB are presented in Figures 14-25 and 14-26 below. The pass-band gain of 3.98 dB, which is practically equal to the expected gain of 4 dB, and the expected f_{-3dB} of 4.94 kHz ≅ 5 kHz can be viewed on the Bode plotter of Figure 14-26.

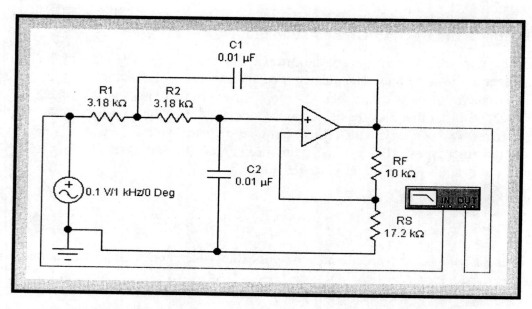

Figure 14-25: Circuit setup for simulation of Design Example 14-7 created in Electronics Workbench

Chapter 14

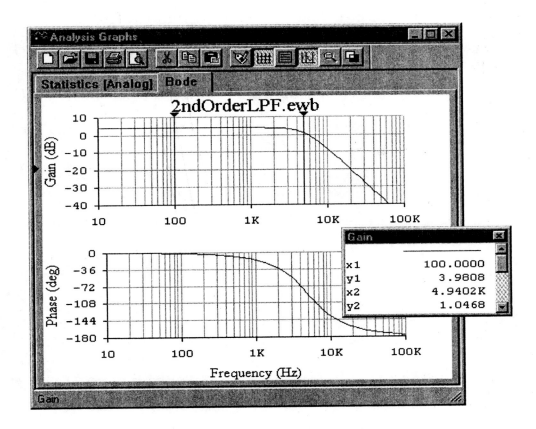

Figure 14-26: Electronics Workbench Bode plotter exhibiting the frequency response plot for the second order low-pass **Butterworth** filter of Design Example 14-7 with the pass-band gain $A_{PB} \cong 4$ dB at 100 Hz, and the $f_{-3dB} \cong 5$ kHz at 1 dB ($A_{PB} - 3$ dB)

Practice Problem 14-7 DESIGN

Design a second-order equal component low-pass active Butterworth filter with the cutoff frequency $f_o = 4$ kHz.

Answers: $R_1 = R_2 = 3.9$ kΩ, $C_1 = C_2 = 0.01$ µF, $R_F = 12$ k$\Omega + 360\Omega$, $R_S = 21$ kΩ

Example 14-8 DESIGN

Let us now design a second-order equal component high-pass active Butterworth filter with f_{-3dB} at 10 kHz.

Solution: $\xi = 0.707$, $A_{PB} = 3 - 2\xi = 1.586 = 4$ dB, $f_o = f_{-3dB} = 10$ kHz

Let $C = 0.001$ µF.

Chapter 14

$$R = \frac{1}{2\pi f_o C} = \frac{1}{2\pi \times 10 \text{ kHz} \times 0.01 \text{ μF}} = 1.59 \text{ k}\Omega \qquad \text{Use } R = 1.6 \text{ k}\Omega$$

Again, in order to minimize the offset errors and balance the op-amp operation, the DC resistance at both input terminals of the op-amp must be equal; that is,

$$R = R_F \| R_S = \frac{R_F \times R_S}{R_F + R_S} \qquad (14\text{-}86)$$

Dividing both the numerator and the denominator by R_S, and then solving for R_F results in the following:

$$R_F = RA_{PB} \qquad (14\text{-}87)$$

$R_F = RA_{PB} = 1.6 \text{ k}\Omega \times 1.586 = 2.54 \text{ k}\Omega$ Use $R_F = 2.4 \text{ k}\Omega + 150 \text{ }\Omega = 2.55 \text{ k}\Omega$

The resistor R_S can be determined with Equation 14-27 as follows:

$$R_S = \frac{R_F}{A_{PB} - 1} = \frac{2.55 \text{ k}\Omega}{1.586 - 1} = 4.35 \text{ k}\Omega \qquad \text{Use } R_S = 4.3 \text{ k}\Omega$$

Test Run and Design Verification:

The results of simulation for the above design are presented in Figures 14-27 and 14-28 below. The frequency response can be viewed on the Bode plotter of Figure 14-28, which exhibits the expected pass-band gain of 4 dB at 100 kHz and the expected f_{-3dB} of 10 kHz. Furthermore, the expected roll-off rate of 40 dB/decade can be verified on the Bode plotter.

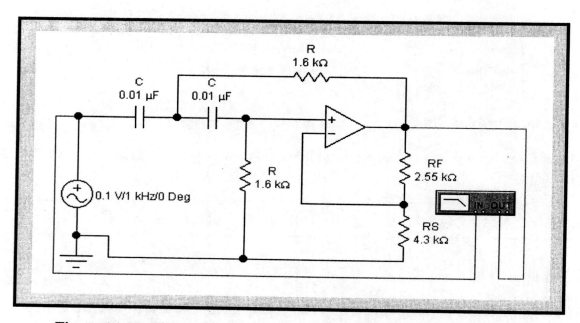

Figure 14-27: Circuit setup for simulation of Design Example 14-8 created in Electronics Workbench

Chapter 14

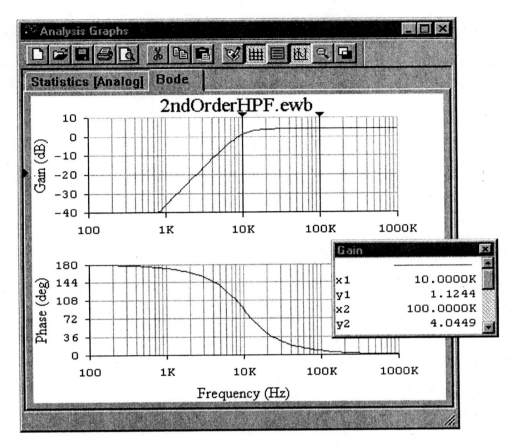

Figure 14-28: Electronics Workbench Bode plotter exhibiting the frequency response plot for the second-order high-pass **Butterworth** filter of the Design Example 14-8 with the expected pass-band gain of 4 dB at 100 kHz, and $f_{-3dB} = 10$ kHz at 1.1 dB ($A_{PB} - 3$ dB)

Practice Problem 14-8 DESIGN

Design a second-order equal component high-pass active Butterworth filter with $f_{-3dB} = 4$ kHz.

Answers: $C_1 = C_2 = 0.01\ \mu F$, $\quad R_1 = R_2 = 3.9\ k\Omega$, $\quad R_F = 6.2\ k\Omega$, $\quad R_S = 10\ k\Omega$

Section Summary — 2nd Order Filters — ANALYSIS/DESIGN

A Summary for the Analysis/Design of Second-Order Equal Component Active Filters

Analysis: $R_1 = R_2 = R$ $C_1 = C_2 = C$

$$f_o = \frac{1}{2\pi RC} \qquad A_{PB} = \frac{R_F}{R_S} + 1 \qquad 2\xi = 3 - A_{PB}$$

Butterworth: $f_{-3dB} = f_o$

Low-Pass: Roll-off rate = -40 dB/decade.

High-Pass: Roll-off rate = $+40$ dB/decade.

Design: Let $R_1 = R_2 = R$ $C_1 = C_2 = C$

Let $C = 0.01\ \mu F$.

Butterworth: $\xi = 0.707$, $f_{-3dB} = f_o$

Low-Pass:

$$R = \frac{1}{2\pi f_o C}$$

$A_{PB} = 3 - 2\xi$

$R_F = 2R A_{PB}$

$$R_S = \frac{R_F}{A_{PB} - 1}$$

High-Pass:

$$R = \frac{1}{2\pi f_o C}$$

$A_{PB} = 3 - 2\xi$

$R_F = R A_{PB}$

$$R_S = \frac{R_F}{A_{PB} - 1}$$

14.7 HIGHER ORDER ACTIVE FILTERS

Thus far, we have observed and studied the responses of first- and second-order filters. We have seen that the first-order rolls off at the rate of 20 dB/decade and the roll-off rate for the second-order is 40 dB/decade. As mentioned previously, each increase in the order of the filter increases the roll-off rate by 20 dB/decade. That is, the roll-off rate for a third-order is 60 dB/decade and for a fourth-order it is 80 dB/decade. As a means of comparison, the frequency response plot of unity-gain low-pass active filters ranging from first-order to fifth-order is exhibited in Figure 14-29 below:

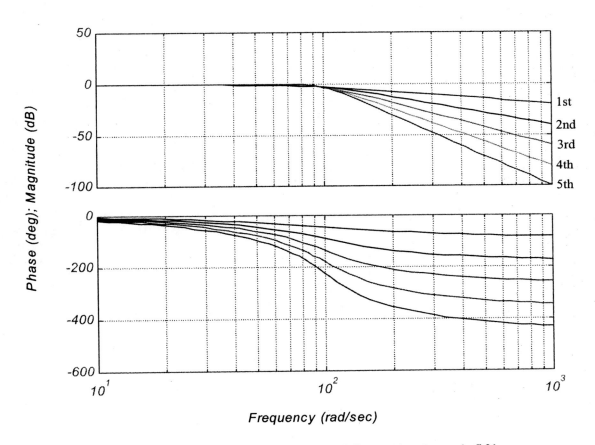

Figure 14-29: Frequency response plot of first-order through fifth-order unity-gain low-pass filters

Higher order filters are generally a cascade of second-order and first-order filters. That is, a third-order is a cascade of a second-order and a first-order, a fourth-order is a cascade of two second-orders, and a fifth-order is a cascade of two second-orders and a first-order. When cascading, however, the damping factor (ξ) and the frequency

Chapter 14

correction factor (k) is usually different from those given for a single-stage second-order filter. Values of ξ and k_{LP} for the higher order cascaded filters are exhibited in Table 14-3 below:

Table 14-3:
Higher Order Filter Damping Factors (ξ) and
Low-Pass Frequency Correction Factors (k_{LP})

Filter Order	Section Order	ξ and k_{LP}	Response Type		
			Bessel	Butterworth	Chebyshev
3	2	ξ	0.7385	0.5	0.248
		k	0.687	1	1.098
	1	ξ	-	-	-
		k	0.753	1	2.212
4	2	ξ	0.958	0.924	0.6375
		k	0.696	1	1.992
	2	ξ	0.620	0.3825	0.1405
		k	0.621	1	1.060
5	2	ξ	0.5455	0.31	0.09
		k	0.549	1	1.040
	2	ξ	0.8875	0.81	0.357
		k	0.619	1	1.577
	1	ξ	-	-	-
		k	0.321	1	3.571
6	2	ξ	1.959	0.966	0.657
		k	0.621	1	2.881
	2	ξ	0.818	0.707	0.2275
		k	0.590	1	1.364
	2	ξ	0.4885	0.259	0.0625
		k	0.523	1	1.023

Second-Order Filter Parameters:

Filter Type	ξ	k_{LP}
Bessel	0.866	0.785
Butterworth	0.707	1
Chebyshev	0.526	1.238

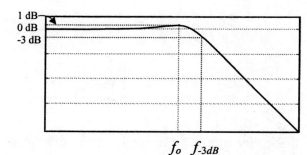

Second-Order Low-Pass
Unity-gain Chebyshev Filter

ξ = Damping factor
k_{LP} = Low-pass frequency correction factor
k_{HP} = High-pass frequency correction factor

Chapter 14

$$k_{HP} = \frac{1}{k_{LP}}, \quad f_o = \frac{1}{2\pi RC}, \quad f_{-3dB} = kf_o$$

Example 14-9   DESIGN

Let us now design a third-order low-pass Chebyshev active filter with f_{-3dB} at 10 kHz and a pass-band gain of 20 dB.

Solution:
A third-order filter is a cascade of a second-order and a first-order. The circuit diagram of a third-order low-pass filter is shown in Figure 14-30 below:

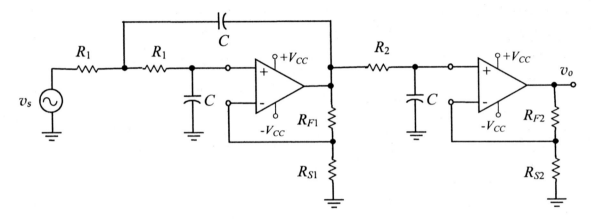

Figure 14-30: Third-order equal component low-pass active filter

$$\text{dBG} = 20 \text{ dB} = 20 \log(A_{PB}), \quad \log(A_{PB}) = 1, \quad A_{PB} = 10^1 = 10 = A_1 \times A_2$$

Recall that the gain of the second-order is limited by the damping factor ξ; however, there is no limitation on the gain of the first-order. Hence, we will first determine the gain of the second-order (A_1) and then we will obtain the rest of the needed gain from the first-order (A_2).

First stage (second-order):

$k_{LP} = 1.098, \quad \xi = 0.248, \quad A_1 = 3 - 2\xi = 2.504 \cong 2.5, \quad f_o = f_{-3dB}/k = 9.1 \text{ kHz}$
Let $C = 0.01 \text{ µF}$ for both stages.

$R_1 = \dfrac{1}{2\pi f_o C} = \dfrac{1}{2\pi \times 9.1 \text{ kHz} \times 0.01 \text{ µF}} = 1.75 \text{ k}\Omega$ 　Use $R_1 = 1 \text{ k}\Omega + 750 \text{ }\Omega$

$R_{F1} = 2 R \times A_1 = 2 \times 1.75 \text{ k}\Omega \times 2.5 = 8.75 \text{ k}\Omega$ 　Use $R_{F1} = 8.6 \text{ k}\Omega$

$R_{S1} = \dfrac{R_{F1}}{A_1 - 1} = \dfrac{8.6 \text{ k}\Omega}{1.504} = 5.7 \text{ k}\Omega$ 　Use $R_{S1} = 4.7 \text{ k}\Omega + 1 \text{ k}\Omega$

Chapter 14

Second stage (first-order): $k_{LP} = 2.212$, $A_2 = \dfrac{A_{PB}}{A_1} = \dfrac{10}{2.5} = 4$ $f_o = f_{-3dB}/k = 4.7$ kHz

$$R_2 = \dfrac{1}{2\pi f_o C} = \dfrac{1}{2\pi \times 4.7 \text{ kHz} \times 0.01 \text{ μF}} = 3.38 \text{ k}\Omega \qquad \text{Use } R_2 = 3.3 \text{ k}\Omega$$

$$R_{F2} = R_2 \times A_2 = 3.3 \text{ k}\Omega \times 4 = 13.2 \text{ k}\Omega \qquad \text{Use } R_{F2} = 12 \text{ k}\Omega + 1.2 \text{ k}\Omega$$

$$R_{S2} = \dfrac{R_{F2}}{A_2 - 1} = \dfrac{13.2 \text{ k}\Omega}{3} = 4.4 \text{ k}\Omega \qquad \text{Use } R_{S2} = 2.2 \text{ k}\Omega + 2.2 \text{ k}\Omega$$

Test Run and Design Verification:

The results of simulation by EWB for the above design are presented in Figures 14-31 and 14-32 below. The frequency response plot with the expected pass-band gain of 20 dB and the expected f_{-3dB} of 10 kHz can be viewed on the Bode plotter of Figure 14-32. In addition, the expected roll-off rate of –60 dB/decade can be viewed and verified for a frequency range of one decade extending from 0 dB to –60 dB.

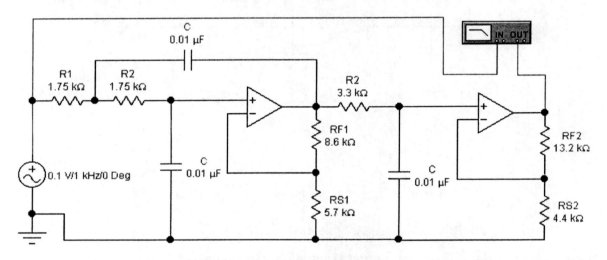

Figure 14-31: Circuit setup for simulation of Design Example 14-9 created in Electronics Workbench

---

Practice Problem 14-9 *3rd Order LPF* **ANALYSIS**

Determine the pass-band gain, f_o, f_{-3dB}, damping factor ξ, and the response type for a third order equal component low-pass active filter with the following circuit components:
Second-order: $C = 0.001$ μF, $R_1 = 10$ kΩ, $R_{F1} = 39$ kΩ, $R_{S1} = 39$ kΩ
First-order: $C = 0.001$ μF, $R_2 = 10$ kΩ, $R_{F2} = 200$ kΩ, $R_{S2} = 22$ kΩ

Answers: $A_{PB} = 26.1$ dB, $f_o = f_{-3dB} = 15.9$ kHz, $\xi = 0.5$, Butterworth

---

Practice Problem 14-10 *3rd Order LPF* **DESIGN**

Chapter 14

Design a third-order equal component low-pass, 1 dB Chebyshev filter with f_{-3dB} = 4 kHz and with a pass-band gain of 32 dB.

Answers: $C = 0.01\ \mu F$, $R_1 = 4.37\ k\Omega$, $R_{F1} = 18.73\ k\Omega$, $R_{S1} = 16.38\ k\Omega$,
                                    $R_2 = 8.8\ k\Omega$, $R_{F2} = 164\ k\Omega$, $R_{S2} = 9.3\ k\Omega$

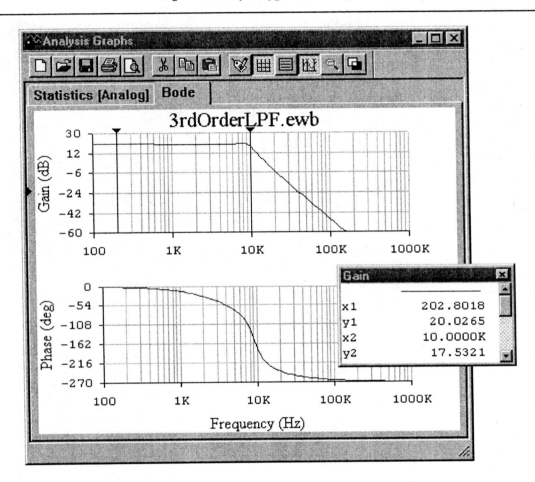

Figure 14-32: Electronics Workbench Bode plotter exhibiting the frequency response for the third-order low-pass **Chebyshev** filter of Design Example 14-9 with the expected A_{PB} of 20 dB at 202 Hz, and f_{-3dB} of 10 kHz at 17.5 d .

The procedure for the analysis and design of the third order high-pass filter is similar to that of the third-order low-pass. However, be aware of the fact that $k_{HP} = 1/k_{LP}$. Two practice problems on the third-order high-pass are given below as an exercise.

Practice Problem 14-11 ANALYSIS

Determine the pass-band gain, f_{-3dB}, damping factor ξ, and the response type for the third-order high-pass active filter of Figure 14-33 below with the following circuit components:

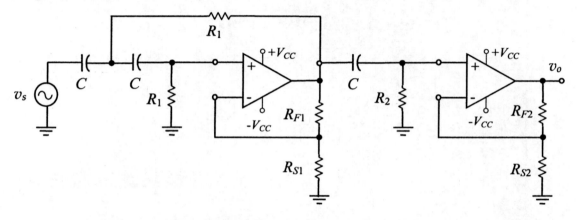

Figure 14-33: Third-order equal component high-pass active filter

Second-order: $C = 0.01\ \mu F$, $R_1 = 1.5\ k\Omega$, $R_{F1} = 3.75\ k\Omega$, $R_{S1} = 2.5\ k\Omega$

First-order: $C = 0.01\ \mu F$, $R_2 = 750\ \Omega$, $R_{F2} = 7.5\ k\Omega$, $R_{S2} = 830\ \Omega$

Answers: $A_{PB} = 25$, $f_{-3dB} = 9.6$ kHz, $\xi \cong 0.25$, Chebyshev

---

Practice Problem 14-12 DESIGN

Design a third-order equal component high-pass Butterworth filter with $f_{-3dB} = 4$ kHz and with a pass-band gain of 22 dB.

Answers: $C = 0.01\ \mu F$, $R_1 = 4\ k\Omega$, $R_{F1} = 8.2\ k\Omega$, $R_{S1} = 8.2\ k\Omega$,
$R_2 = 4\ k\Omega$, $R_{F2} = 25.2\ k\Omega$, $R_{S2} = 4.7\ k\Omega$

Example 14-10  DESIGN

Let us now design a fourth-order low-pass Chebyshev active filter with f_{-3dB} at 10 kHz.

Solution:

A fourth-order filter is a cascade of a two second-order filters, as shown in Figure 14-34.

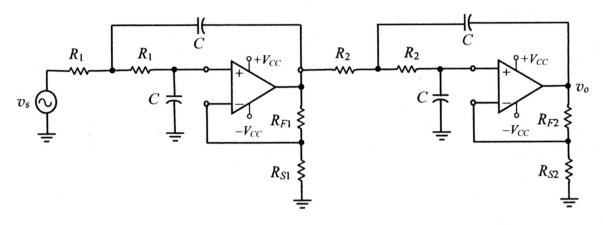

Figure 14-34: Fourth-order equal component low-pass active filter

First stage:

$\xi = 0.6375$, $k_{LP} = 1.992$, $A_1 = 3 - 2\xi = 1.725$, $f_o = f_{-3dB}/k_{LP} = 5.02$ kHz

Let $C = 0.01$ µF.

$R_1 = \dfrac{1}{2\pi f_o C} = \dfrac{1}{2\pi \times 5.02 \text{ kHz} \times 0.01 \text{ µF}} = 3.17$ kΩ   Use $R_1 = 3.3$ kΩ.

$R_{F1} = 2R_1 A_1 = 2 \times 3.3 \text{ kΩ} \times 1.725 = 11.385$ kΩ   Use $R_{F1} = 11$ kΩ.

$R_{S1} = \dfrac{R_{F1}}{A_1 - 1} = \dfrac{11 \text{ kΩ}}{1.725 - 1} = 15.2$ kΩ   Use $R_{S1} = 15$ kΩ.

Second stage:

$\xi = 0.1405$, $k_{LP} = 1.060$, $A_2 = 3 - 2\xi = 2.719$, $f_o = f_{-3dB}/k_{LP} = 9.43$ kHz

Let $C = 0.01$ µF.

$R_2 = \dfrac{1}{2\pi f_o C} = \dfrac{1}{2\pi \times 9.43 \text{ kHz} \times 0.01 \text{ µF}} = 1.68$ kΩ   Use $R_2 = 1.6$ kΩ.

$R_{F2} = 2R_2 A_2 = 2 \times 1.6 \text{ kΩ} \times 2.719 = 8.7$ kΩ   Use $R_{F2} = 9.1$ kΩ.

$R_{S2} = \dfrac{R_{F2}}{A_2 - 1} = \dfrac{9.1 \text{ kΩ}}{1.719} = 5.3$ kΩ   Use $R_{S2} = 5.6$ kΩ.

$A_{PB} = A_1 A_2 = 1.725 \times 2.719 = 4.69$

$|A_{PB}| = 20 \log (A_{PB}) = 13.4$ dB

Chapter 14

Test Run and Design Verification:

The results of simulation by EWB for the above design are presented in Figures 14-35 and 14-36 below. The frequency response plot can be viewed on the Bode plotter of Figure 14-36, which also exhibits the expected pass-band gain of 13.4 dB, and the expected f_{-3dB} of 10 kHz. However, the corresponding gain at 10 kHz is 11.4 dB instead of 10.4 dB. That is, the actual f_{-3dB} is probably at about 10.2 kHz; this small discrepancy could be the result of using standard value resistors.

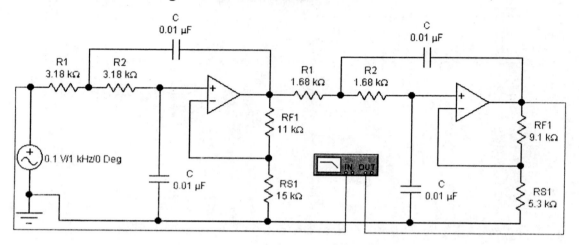

Figure 14-35: Circuit setup for simulation of Design Example 14-10 created in Electronics Workbench

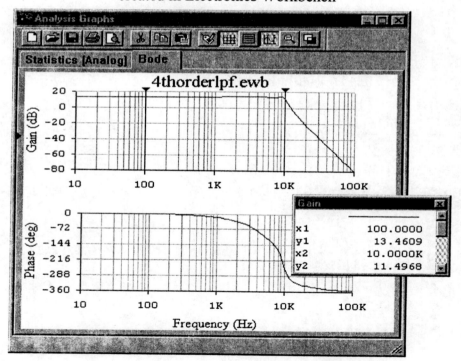

Figure 14-36: Electronics Workbench Bode plotter exhibiting the frequency response plot for the fourth-order low-pass Chebyshev filter of Design Example 14-10 with the expected pass-band gain of 13.46 dB at 100 Hz.

Practice Problem 14-13   DESIGN

Design a fourth-order equal component low-pass Butterworth filter with f_{-3dB} = 10 kHz. Determine the pass-band gain in dB and verify your design by simulation.

Answers: C = 0.01 µF, R_1 = 1.6 kΩ, R_{F1} = 3.6 kΩ, R_{S1} = 24 kΩ,
R_2 = 1.6 kΩ, R_{F2} = 7.15 kΩ, R_{S2} = 5.7 kΩ

Practice Problem 14-14   DESIGN

Design a fourth-order equal component high-pass Butterworth filter with f_{-3dB} = 10 kHz. Determine the pass-band gain in dB and verify your design by simulation.

Answers: C = 0.01 µF, R_1 = 1.6 kΩ, R_{F1} = 1.8 kΩ, R_{S1} = 12.27 kΩ,
R_2 = 1.6 kΩ, R_{F2} = 3.6 kΩ, R_{S2} = 2.9 kΩ

14.8 BAND-PASS ACTIVE FILTER

Band-pass filters pass a certain band of interest while suppressing the bands below and above that pass-band. A wide band-pass filter can be achieved by cascading a low-pass and a high-pass as illustrated in Figure 14-37 below:

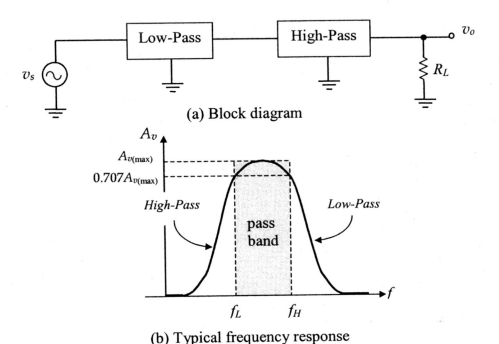

(a) Block diagram

(b) Typical frequency response

Figure 14-37:
Block diagram and typical frequency response of a wide pass-band active filter

Chapter 14

Note that the overall pass-band gain A_{PB} is the product of the two individual low-pass and high-pass gains A_1 and A_2.

$$A_{PB} = A_1 \times A_2$$

The design procedure would be identical to the low-pass and high-pass design procedures discussed earlier. The moderately wide and narrow band-pass filter ($1 \leq Q \leq 20$), however, can be achieved with a single op-amp configuration as shown in Figure 14-38 below:

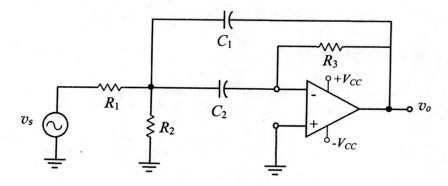

Figure 14-38:
Circuit diagram of a single op-amp band-pass filter

In order to fully analyze and derive the transfer function for the above circuit, we will draw its virtual equivalent circuit as illustrated in Figure 14-39 below:

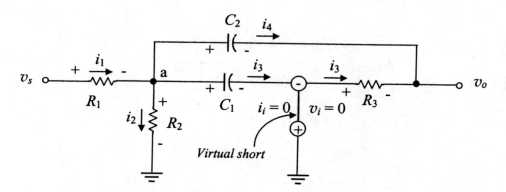

Figure 14-39: Virtual equivalent circuit of the above band-pass filter

Let $C_1 = C_2 = C$

$$X_C = \frac{1}{j\omega C} = \frac{1}{sC} \qquad (14\text{-}88)$$

The voltage v_a, which is the voltage across R_2 and is also equal to the voltage across C_2, is determined by applying Ohm's law, as follows:

$$v_a = v_{R_2} = v_{C_2} = i_3 \frac{1}{sC} \qquad (14\text{-}89)$$

Chapter 14

Currents i_1 through i_4 are determined by applying Ohm's law, as follows:

$$i_3 = -\frac{v_o}{R_3} \tag{14-90}$$

$$v_a = v_{R_2} = v_{C_2} = -\frac{v_o}{R_3 sC} \tag{14-91}$$

$$i_2 = \frac{v_a}{R_2} = -\frac{v_o}{R_3 sCR_2} \tag{14-92}$$

$$i_4 = \frac{v_a - v_o}{1/sC} = (v_a - v_o)sC \tag{14-93}$$

Substituting for v_a in the above equation yields the following:

$$i_4 = \left(\frac{-v_o}{R_3 sC} - v_o\right)sC = -\frac{v_o}{R_3}(1 + R_3 sC) \tag{14-94}$$

$$i_1 = \frac{v_s - v_a}{R_1} = \frac{v_s}{R_1} - \frac{v_a}{R_1} = \frac{v_s}{R_1} + \frac{v_o}{R_3 sCR_1} \tag{14-95}$$

Applying the *KCL* at node *a* results in the following equation:

$$i_1 = i_2 + i_3 + i_4 \tag{14-96}$$

Substituting for currents i_1 through i_4 in the above equation yields the following:

$$\frac{v_s}{R_1} + \frac{v_o}{R_3 sCR_1} = -\frac{v_o}{R_3 sCR_2} - \frac{v_o}{R_3} - \frac{v_o}{R_3}(1 + R_3 sC) \tag{14-97}$$

$$\frac{v_s}{R_1} = -\frac{v_o}{R_3}\left(\frac{1}{R_1 sC} + \frac{1}{R_2 sC} + 1 + (1 + R_3 sC)\right) \tag{14-98}$$

$$\frac{v_s}{R_1} = -\frac{v_o}{R_3}\left(\frac{1}{R_1 sC} + \frac{1}{R_2 sC} + 1 + R_3 sC + 2\right) \tag{14-99}$$

Taking a common denominator and applying some more algebraic manipulation, we arrive at the following transfer function:

$$\frac{v_s}{R_1} = -\frac{v_o}{R_3}\left(\frac{R_2 + R_1 + R_1 R_2 R_3 s^2 C^2 + 2R_1 R_2 sC}{R_1 R_2 sC}\right) \tag{14-100}$$

$$\frac{v_o}{v_s} = -\frac{R_3}{R_1}\left(\frac{R_1 R_2 sC}{R_2 + R_1 + R_1 R_2 R_3 s^2 C^2 + 2R_1 R_2 sC}\right) \tag{14-101}$$

$$\frac{v_o}{v_s} = -\left(\frac{R_2 R_3 sC}{R_2 + R_1 + R_1 R_2 R_3 s^2 C^2 + 2R_1 R_2 sC}\right) \tag{14-102}$$

$$\frac{v_o}{v_s} = -\frac{(R_2 R_3 C)s}{(R_1 R_2 R_3 C^2)s^2 + (2R_1 R_2 C)s + (R_1 + R_2)} \tag{14-103}$$

Chapter 14

Dividing both the numerator and denominator by the coefficient of s^2 term $(R_1R_2R_3C^2)$ yields the following transfer function:

$$\frac{v_o}{v_s} = -\frac{\frac{1}{R_1C}s}{s^2 + \frac{2}{R_3C}s + \frac{R_1+R_2}{R_1R_2R_3C^2}} \quad (14\text{-}104)$$

The above transfer function is in the form of the following standard band-pass network transfer function:

$$\frac{v_o}{v_s} = \frac{A_c(2\xi)\omega_c s}{s^2 + 2\xi\omega_c s + \omega_c^2} \quad (14\text{-}105)$$

where,
ω_c = center frequency
A_c = gain at the center frequency
ξ = damping factor

However, the narrowness and width of the band-pass filter are generally defined by the quality factor Q, which is the reciprocal of 2ξ.

$$Q = \frac{f_c}{BW} = \frac{1}{2\xi} \quad (14\text{-}106)$$

Generally, a band-pass filter with a $Q < 10$ is considered wide-band, and one with a $Q \geq 10$ is considered narrow-band. However, returning to the transfer function, we now can determine the A_c, ω_c, and BW in terms of Q and circuit components.

$$-\frac{1}{R_1C} = A_c(2\xi)\omega_c \quad (14\text{-}107)$$

$$\frac{1}{R_1C} = 2\xi|A_c|\omega_c \quad (14\text{-}108)$$

$$R_1 = \frac{1}{2\xi|A_c|\omega_c C} = \frac{Q}{|A_c|\omega_c C} \quad (14\text{-}109)$$

$$R_1 = \frac{Q}{2\pi|A_c|f_c C} \quad (14\text{-}110)$$

$$2\xi\omega_c = \frac{2}{R_3C} \quad (14\text{-}111)$$

$$\omega_c = \frac{2}{2\xi R_3C} = \frac{2Q}{R_3C} \quad (14\text{-}112)$$

$$R_3 = \frac{2Q}{\omega_c C} = \frac{2Q}{2\pi f_c C} \quad (14\text{-}113)$$

$$R_3 = \frac{Q}{\pi f_c C} \quad (14\text{-}114)$$

$$A_c(2\xi)\omega_c = \frac{-1}{R_1C} \quad (14\text{-}115)$$

Substituting for $2\xi\omega_c$ in the above equation results in the following:

$$A_c\left(\frac{2}{R_3C}\right) = \frac{-1}{R_1C} \quad (14\text{-}116)$$

$$A_c = -\frac{R_3C}{2R_1C} \quad (14\text{-}117)$$

$$A_c = -\frac{R_3}{2R_1}$$

$$\omega_c^2 = \frac{R_1+R_2}{R_1R_2R_3C^2}$$

$$\left(\frac{2Q}{R_3C}\right)^2 = \frac{R_1+R_2}{R_1R_2R_3C^2} \quad (14\text{-}118)$$

Multiplying both sides of the above equation by R_3C^2 results in the following:

$$\frac{4Q^2}{R_3} = \frac{R_1+R_2}{R_1R_2} = \frac{1}{R_1} + \frac{1}{R_2}$$

$$\frac{1}{R_2} = \frac{4Q^2}{R_3} - \frac{1}{R_1} = \frac{4Q^2R_1 - R_3}{R_1R_3}$$

$$R_2 = \frac{R_1R_3}{4Q^2R_1 - R_3} \quad (14\text{-}119)$$

$$BW = \frac{\omega_c}{Q} = \frac{2Q}{R_3CQ} = \frac{2}{R_3C} \text{ (rad/sec)}$$

Example 14-11   BPF   DESIGN

Let us now design a band-pass active filter with $f_c = 5$ kHz, $A_c = 5$, and $Q = 5$.

Solution:

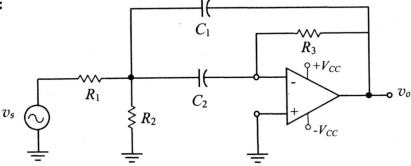

Figure 14-40: Circuit diagram of a single op-amp band-pass filter

Let $C_1 = C_2 = C = 0.01 \, \mu F$. The resistance R_1 is determined from Equation 14-110, as follows:

$$R_1 = \frac{Q}{2\pi |A_c| f_c C} = \frac{5}{2\pi \times 5 \times 5 \, \text{kHz} \times 0.01 \, \mu F} = 3.18 \, k\Omega \qquad \text{Use } R_1 = 3.1 \, k\Omega.$$

The resistance R_3 is determined from Equation 14-114, as follows:

$$R_3 = \frac{Q}{\pi f_c C} = \frac{5}{\pi \times 5 \, \text{kHz} \times 0.01 \, \mu F} = 31.83 \, k\Omega \qquad \text{Use } R_3 = 32 \, k\Omega.$$

The resistance R_2 is determined from Equation 14-119, as follows:

$$R_2 = \frac{R_1 R_3}{4 Q^2 R_1 - R_3} = \frac{3.1 \, k\Omega \times 32 \, k\Omega}{4 \times 5^2 \times 3.1 \, k\Omega - 32 \, k\Omega} = 356.8 \, \Omega \qquad \text{Use } R_4 = 360 \, \Omega.$$

$$BW = \frac{f_c}{Q} = \frac{5 \, \text{kHz}}{5} = 1 \, \text{kHz}$$

Test Run and Design Verification:

The results of simulation by EWB for the above design are presented in Figures 14-41 and 14-42. The frequency response, center frequency $f_c = 5$ kHz, and the expected gain at the center frequency $A_c = 5$ can be viewed on the Bode plotter of Figure 14-42.

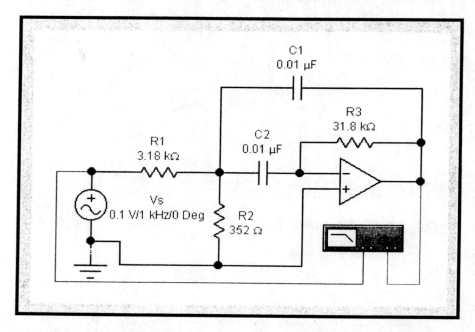

Figure 14-41: Circuit setup for simulation of Design Example 14-11 created in Electronics Workbench

Chapter 14

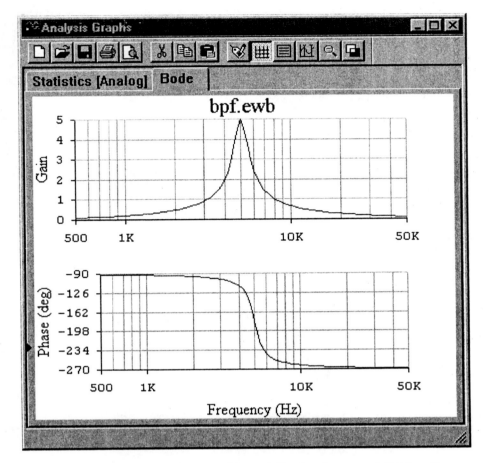

Figure 14-42: Electronics Workbench Bode plotter exhibiting the frequency response of the band-pass filter of Design Example 14-11

14.9 BAND-STOP ACTIVE FILTER

A band-stop or notch filter may be achieved by connecting the output of a band-pass filter and the input signal source to a two-input inverting summer as illustrated in Figure 14-43 below:

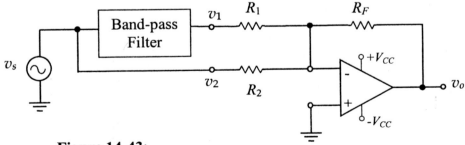

Figure 14-43:
Block diagram of a band-stop (notch) filter

Chapter 14

Recall that the band-pass filter configuration is an inverting one; however, the magnitude frequency response is usually the plot of the absolute value of the gain and does not reflect the inversion. Hence, the input v_1 to the inverting summer is out-of-phase with v_s and v_2 is equal to v_s; that is,

$$v_1 = -A_c v_s \text{ (band-pass)} \tag{14-120}$$

$$v_2 = v_s \text{ (all-pass)} \tag{14-121}$$

In order to provide the same amplitude to both channels at the output of the summing amplifier, we will pick the resistance R_2 to have a value that is A_c times smaller than R_1. Assuming that $A_c = 5$, let $R_1 = 10 \text{ k}\Omega$, $R_F = 10 \text{ k}\Omega$, and $R_2 = 2 \text{ k}\Omega$.

$$v_o = -\left(v_1 \frac{R_F}{R_1} + v_2 \frac{R_F}{R_2} \right) \tag{14-122}$$

$$v_o = -[-A_c v_s \text{ (band-pass)} + 5v_2 \text{ (all-pass)}] \tag{14-123}$$

$$v_o = v_{o1} + v_{o2} = 5v_s \text{ (band-pass)} - 5v_s \text{ (all-pass)} \tag{14-124}$$

The above equation is represented graphically and then summed accordingly, as follows:

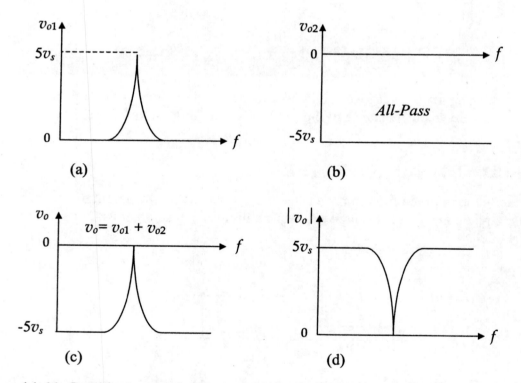

Figure 14-44: Graphical representation of the two individual outputs (v_{o1} & v_{o2}) and the summed output of the summing amplifier

Chapter 14

The response at the output of the summing amplifier is actually Figure 14-44(c); however, as mentioned previously, the magnitude frequency response plot is usually done in absolute value. Hence, the magnitude frequency response plot at the output of the summing amplifier looks like Figure 14-44(d), which is the familiar band-stop filter response.

The addition of the above summing amplifier to the band-pass filter of Design Example 14-9 will produce a band-stop filter with the same specifications: $A_{PB} = 5$, $f_c = 5$ kHz, and $Q = 5$.

Test Run and Design Verification:

Let us now simulate the above proposed design and check the result. The results of simulation by EWB for the above design are presented in Figures 14-45 and 14-46 below. The frequency response plot including the expected pass-band gain at $A_{PB} = 5$ and the center frequency $f_c = 5$ kHz can be viewed on the Bode plotter of Figure 14-46.

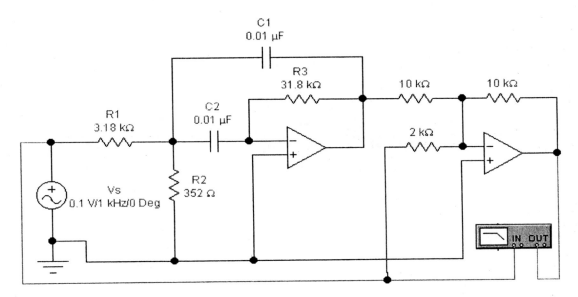

Figure 14-45: Circuit setup for simulation of the band-stop filter created in Electronics Workbench

Chapter 14

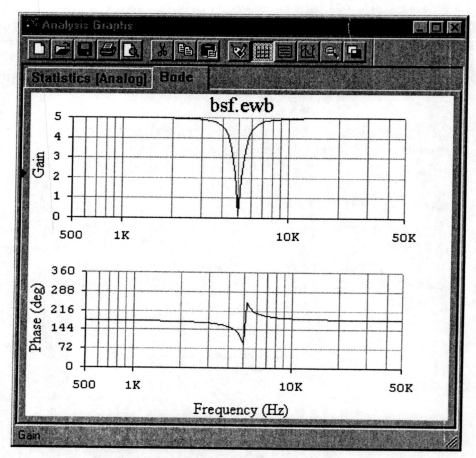

Figure 14-46: Electronics Workbench Bode plotter exhibiting the frequency response with the pass-band gain $A_{PB} = 5$ and $f_c = 5$ kHz for the band-stop filter

Chapter 14

Section Summary — BP/BS Filters — ANALYSIS/DESIGN

Summary of Equations for the Analysis/Design of Band-Pass and Band-Stop Active Filters

Analysis:

$$\omega_c = \sqrt{\frac{R_1 + R_2}{R_1 R_2 R_3 C^2}} \text{ (rad/sec)}$$

$$f_c = \frac{\omega_c}{2\pi} \text{ (Hz)}$$

$$Q = \frac{\omega_c R_3 C}{2}$$

$$|A_c| = \frac{R_3}{2R_1}$$

$$BW = \frac{\omega_c}{Q} = \frac{2}{R_3 C} \text{ (rad/sec)}$$

Design:

Band-Pass: Let $C_1 = C_2 = C \leq 0.1 \, \mu F$

$$R_1 = \frac{Q}{2\pi |A_c| f_c C}$$

$$R_2 = \frac{R_1 R_3}{4Q^2 R_1 - R_3}$$

$$R_3 = \frac{Q}{\pi f_c C}$$

Band-Stop:

Add a two-input inverting summer to the above band-pass filter. Pick the resistance R_2 of the summing amplifier to have a value that is A_c times smaller than R_1 or R_1 is A_c times larger than R_2.

$$R_F = R_1 = A_c R_2$$

643

14.10 SUMMARY

- Active filters are frequency selective circuits that pass a certain band of frequency while stopping the rest of the frequency spectrum.

- An ideal low-pass filter will pass a band of frequency from DC up to a cutoff frequency f_H as illustrated in Figure 14-1(a).

- An ideal high-pass filter will stop the band of frequency from DC up to a cutoff frequency f_L and pass the frequency band above and beyond f_L.

- The number of active elements in the circuit determines the order of any circuit. For instance, there is only one capacitive element in the first-order filters and there are two capacitive elements in the second-order active filters.

- The higher the order of the filter, the higher the slope of the diagonal asymptote or the roll-off rate. The roll-off rate for the first-order is 20 dB/decade, and it is 40 dB/decade for the second-order. In fact, each increase in the order of the filter increases the roll-off rate by 20 dB/decade.

- The Sallen-Key equal component approach simplifies the analysis and design of second-order active filters by letting $R_1 = R_2 = R$ and $C_1 = C_2 = C$.

- The f_{-3dB} is the frequency at which the gain drops 3 dB below the pass-band gain A_{PB}.

- The design procedure for the second-order active filter is similar to that of the first-order. However, there is a damping factor (ξ) with the second-order that dictates the response type and limits the pass-band gain.

- Higher order filters are generally a cascade of second-order and first-order filters. That is, a third-order is a cascade of a second-order and a first-order, a fourth-order is a cascade of two second-orders, and a fifth-order is a cascade of two second-orders and a first-order. When cascading, however, the damping factor (ξ) and the frequency correction factor (k) are usually different from those given for a single-stage second-order filter. Values of ξ and k_{LP} for the higher order cascaded filters are exhibited in Table 14-3.

- Band-pass filters pass a certain band of interest while suppressing the bands below and above that pass-band. A wide band-pass filter can be achieved by cascading a low-pass and a high-pass as illustrated in Figure 14-50.

- The moderately wide and narrow band-pass filter ($1 \leq Q \leq 20$) can be achieved with a single op-amp configuration as shown in Figure 14-46.

Chapter 14

- A band-stop or notch filter may be achieved by connecting the output of a band-pass filter and the input signal source to a two-input inverting summer as illustrated in Figure 14-56.

PROBLEMS

14.1 Determine the cutoff frequency, the pass-band gain in dB, roll-off rate, and the gain at the cutoff frequency for the active filter of Figure 14-1P below:

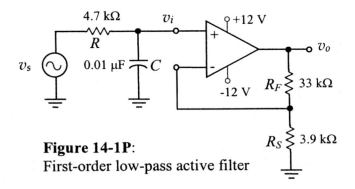

Figure 14-1P:
First-order low-pass active filter

14.2 Determine the cutoff frequency, the pass-band gain in dB, roll-off rate, and the gain at the cutoff frequency for the active filter of Figure 14-2P below:

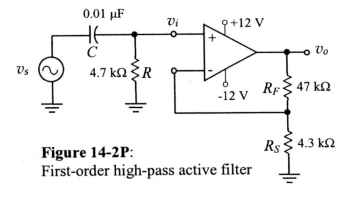

Figure 14-2P:
First-order high-pass active filter

14.3 Carry out Practice Problem 14-1.

14.4 Carry out Practice Problem 14-2.

14.5 Design a first-order low-pass active filter for a cutoff frequency at 20 kHz, and with a pass-band gain of 20 dB. Verify your design by simulation.

14.6 Carry out Practice Problem 14-3.

14.7 Carry out Practice Problem 14-4.

14.8 Design a first-order high-pass active filter for a cutoff frequency at 20 kHz, and with a pass-band gain of 20 dB. Verify your design by simulation.

14.9 Determine the critical frequency f_o, pass-band gain, the damping coefficient, and the roll-off rate for the second-order low-pass active of Figure 14-3P below:

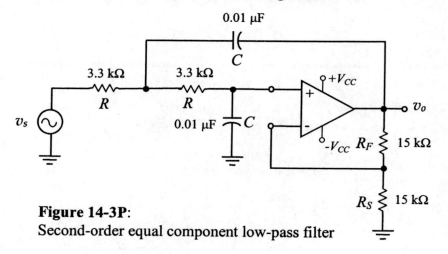

Figure 14-3P:
Second-order equal component low-pass filter

14.10 Determine the critical frequency f_o, pass-band gain in dB, roll-off rate, and damping factor for the second-order high-pass active filter of Figure 14-4P below:

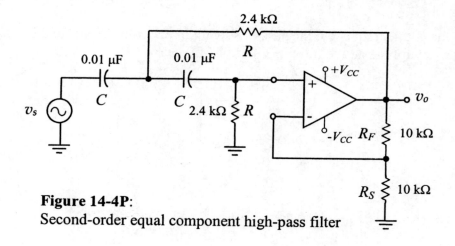

Figure 14-4P:
Second-order equal component high-pass filter

14.11 Carry out Practice Problem 14-5.

14.12 Carry out Practice Problem 14-6.

14.13 Carry out Practice Problem 14-7.

14.14 Carry out Practice Problem 14-8.

14.15 Design a second-order equal component low-pass active Butterworth filter with $f_{-3dB} = 4.7$ kHz.

14.16 Design a second-order equal component high-pass active Butterworth filter with $f_o = 4.7$ kHz.

14.17 Carry out Practice Problem 14-9.

14.18 Carry out Practice Problem 14-10.

14.19 Design a third-order equal component low-pass Chebyshev active filter with $f_{-3dB} = 4.7$ kHz, and $A_{PB} = 26$ dB.

14.20 Carry out Practice Problem 14-11.

14.21 Carry out Practice Problem 14-12.

14.22 Design a third-order equal component high-pass Chebyshev active filter with $f_{-3dB} = 20$ kHz and $A_{PB} = 26$ dB.

14.23 Carry out Practice Problem 14-13.

14.24 Carry out Practice Problem 14-14.

14.25 Design a fourth-order equal component low-pass Chebyshev active filter with $f_{-3dB} = 4.7$ kHz.

14.26 Design a fourth-order equal component high-pass Chebyshev active filter with $f_{-3dB} = 20$ kHz.

14.27 Design a fifth-order equal component low-pass Butterworth active filter with $f_{-3dB} = 4.7$ kHz and $A_{PB} = 26$ dB.

14.28 Design a fifth-order equal component high-pass Butterworth active filter with $f_{-3dB} = 20$ kHz and $A_{PB} = 26$ dB.

14.29 Design a band-pass active filter with $f_c = 20$ kHz, $A_c = 20$, and $Q = 10$.

14.30 Design a band-stop active filter with $f_c = 20$ kHz, $A_{PB} = 20$, and $Q = 10$. Verify your design by simulation.

14.31 Determine the center frequency f_c, A_c in dB, Q, and BW for the band-pass filter of Figure 14-5P.

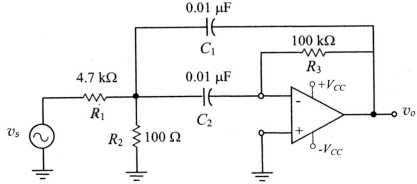

Figure 14-5P:
Circuit diagram of a single op-amp band-pass filter

Chapter 14

14.32 Design a band-stop active filter with $f_c = 10$ kHz, $A_{PB} = 10$, and $Q = 10$. Verify your design by simulation.

Chapter 15

Chapter

IC REGULATORS AND SWITCHING POWER SUPPLIES

Analysis & Design

15.1 INTRODUCTION

The rectification, filtering, and regulation discussed in Chapter 2 are the basic steps in converting the 115 V, 60 Hz utility line voltage to DC, which is generally the supply voltage needed to run most electronic devices and equipment. However, a more efficient regulation with added features such as current boosting, adjustable output, and overload and short-circuit protection can be achieved by utilizing an op-amp as the regulator. Furthermore, a more sophisticated and highly efficient regulation is achievable through the use of switching regulators, generally referred to as switching power supplies or switch-mode power supplies. The primary advantage of the switching regulator is that the physical size and dimension of the power supply can be miniaturized because of the very high switching frequency.

15.2 INTEGRATED CIRCUIT REGULATORS

The operation of the full-wave bridge rectifier and its output voltage regulation by means of a Zener diode is discussed fully in Chapter 2. However, a more efficient way of regulating the output voltage is by means of an *IC* regulator as shown in Figure 15-1 below.

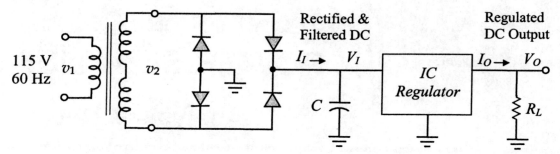

Figure 15-1: Full-wave bridge rectifier with filter capacitor and *IC* regulator

Op-amp as an *IC* regulator:

With the op-amp utilized as a regulator, the Zener diode is still used not as a regulator but as a reference voltage. The primary advantage of using an op-amp is that it provides an almost ideal buffer between the rectified DC voltage and the load.

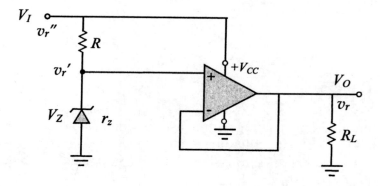

Figure 15-2: A Zener diode and an op-amp buffer utilized as regulator

The rectified and filtered DC voltage V_I, which is the output of the full-wave bridge rectifier, contains a ripple voltage v_r''. The input to the single-supply op-amp is the Zener voltage V_Z with a reduced ripple v_r'. Since the op-amp is utilized as a buffer, the output V_O equals V_Z with a ripple v_r that equals v_r'.

Chapter 15

Adjustable output voltage:

Another advantage of utilizing the op-amp as a voltage regulator is that it can provide gain and thus adjustable output voltage, as shown in Figure 15-3 below:

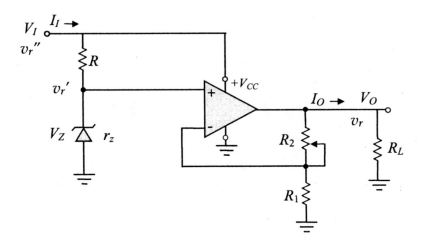

Figure 15-3: Adjustable output voltage regulator

The output DC voltage V_O and the output ripple v_r are the amplified versions of the op-amp input voltages V_Z and v_r', respectively.

$$V_O = V_Z \left(\frac{R_2}{R_1} + 1 \right) \qquad (15\text{-}1)$$

$$v_r = v_r' \left(\frac{R_2}{R_1} + 1 \right) \qquad (15\text{-}2)$$

where,

$$v_r' = v_r'' \left(\frac{r_z}{R + r_z} \right) \qquad (15\text{-}3)$$

and

$$v_r''{}_{(p-p)} = \frac{I_I}{2fC} = \frac{I_I}{120 \times C} \qquad (15\text{-}4)$$

$$I_O = \frac{V_O}{R_L} \leq I_{op-amp(\max)} \qquad (15\text{-}5)$$

$$I_Z = \frac{V_I - V_Z}{R} \qquad (15\text{-}6)$$

Current boosting:

The problem with above circuit of Figure 15-3 is that its output current I_O can be no higher than the maximum output current of the op-amp. This problem can be resolved by utilizing a power transistor, as shown in Figure 15-4.

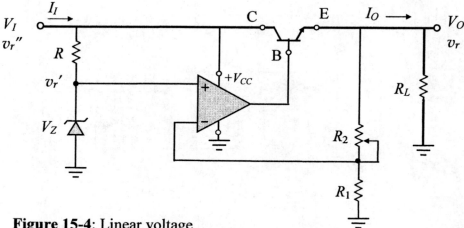

Figure 15-4: Linear voltage regulator with boosted output current and adjustable output voltage

In the above circuit, the input and output currents are approximately equal and their maximum is β times higher than the maximum output current of the op-amp. Thus,

$$I_I \cong I_O = \beta I_{op-amp} = \frac{V_O}{R_L} \tag{15-7}$$

$$I_{O(max)} = \beta I_{op-amp(max)} \tag{15-8}$$

$$P_{Q1(max)} = I_{O(max)} V_{CE} = I_{O(max)} (V_{I(max)} - V_O)$$

Overload and short-circuit protection:

The addition of a general-purpose transistor such as 2N3904 and a current-sense resistor R_S can provide protection against overload and short-circuit, as shown in Figure 15-5.

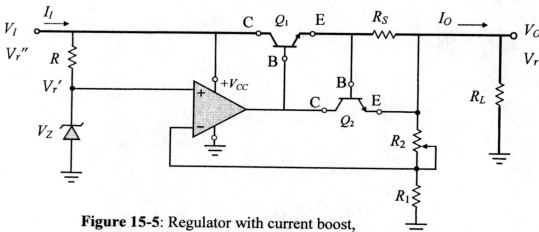

Figure 15-5: Regulator with current boost, adjustable output, and short-circuit protection

Chapter 15

The value of the current-sense resistor R_S is chosen such that with $I_I = I_o \leq I_{o(max)}$, the voltage drop across the $R_S = V_{BE(Q2)}$ is less than 0.7 V, and as a result, Q_2 is off and Q_1 is on. However, when $I_I = I_o > I_{o(max)}$, which could be due to overload or short-circuit, the voltage drop across the $R_S = V_{BE(Q2)}$ reaches or exceeds the nominal 0.7 V, turning the Q_2 on and Q_1 off, cutting off the output current and dropping the output voltage to zero. When the overload or short-circuit is cleared, Q_1 turns back on, Q_2 turns off and normal operation continues.

Example 15-1   ANALYSIS

Let us determine the output voltage V_o, output current I_o and $I_{o(max)}$, and output ripple voltage $v_{r(p-p)}$ for the regulator of Figure 15-6, given the following:
$V_I = 20$ V, $v_r''_{(p-p)} = 3$ V(p-p), R_2 set at 6 kΩ.

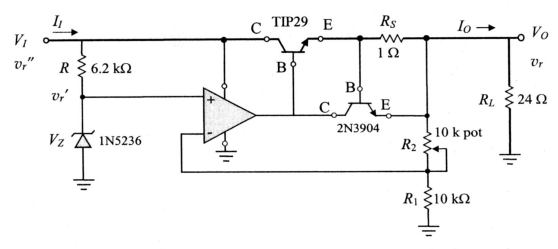

Figure 15-6: Regulator with adjustable output and overload protection

Solution:

From the Zener diode data sheets, the nominal characteristic data for the 1N5236 are as follows:

$V_Z = 7.5$ V, $I_{ZK} = 0.25$ mA, r_z @ $I_{ZT} = 20$ mA is 6 Ω.

$$V_O = V_Z\left(\frac{R_2}{R_1} + 1\right) = 7.5\text{ V}\left(\frac{6\text{ k}\Omega}{10\text{ k}\Omega} + 1\right) = 12\text{ V}$$

$$I_O = \frac{V_O}{R_L} = \frac{12\text{ V}}{24\text{ }\Omega} = 0.5\text{ A}$$

$$I_{O(max)} = \frac{0.7\text{ V}}{R_S} = \frac{0.7\text{ V}}{1\text{ }\Omega} = 0.7\text{ A}$$

$$I_Z = \frac{V_I - V_Z}{R} = \frac{20\text{ V} - 7.5\text{ V}}{6.2\text{ k}\Omega} = 2\text{ mA}$$

Chapter 15

The Zener resistance r_Z @ I_{ZT} = 20 mA is 6 Ω, and since the plot of r_Z versus I_Z is almost linear, then r_Z @ I_Z = 2 mA is approximately 60 Ω, which is computed as follows:

$$r_z = \frac{20\,mA \times 6\,\Omega}{I_Z} = \frac{120\,mV}{2\,mA} = 60\,\Omega$$

$$V_{r\,(p-p)}' = V_{r\,(p-p)}''\left(\frac{r_z}{R+r_z}\right) = 3\,V\left(\frac{60\,\Omega}{6.2\,k\Omega + 60\,\Omega}\right) \cong 28.754\,mV$$

$$V_{r(p-p)} = V_{r\,(p-p)}'\left(\frac{R_2}{R_1}+1\right) = 28.754\,mV\left(\frac{6\,k\Omega}{10\,k\Omega}+1\right) \cong 46\,mV$$

Practice Problem 15-1 — ANALYSIS

Determine the output voltage V_o, output current I_o and $I_{o(max)}$, and output ripple voltage $v_{r(p-p)}$ for the regulator of Figure 15-7, given the following:

$$V_I = 16\,V, \qquad v_{r\,(p-p)}'' = 2\,V(p-p), \qquad R_2 \text{ set at } 7\,k\Omega.$$

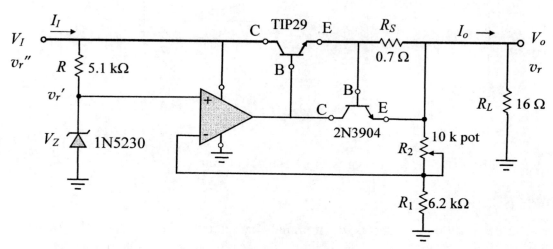

Figure 15-7: Regulator with adjustable output and overload protection

---

Answers: V_o = 10 V, I_o = 0.625 A, $I_{o(max)}$ = 1 A, $v_{r(p-p)}$ = 138 mV

Example 15-2

Adjustable Output — **DESIGN**

Let us now design a regulator with adjustable output and with overload protection for the following specifications:

The output voltage V_o is to be adjustable approximately from 4.7 V to 10 V.

$I_{o(\max)} = 1.5$ A, $v_{r(p\text{-}p)} \leq 0.01\, V_o$, given the following:

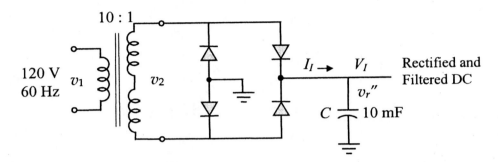

Figure 15-8: Full-wave bridge rectifier with filter capacitor

Solution:

$$v_{2(rms)} = \frac{v_{1(rms)}}{10} = 12\text{ V}$$

$$V_{2(p)} = 1.414 \times v_{2(rms)} \cong 17\text{ V}$$

$$V_{I(p)} = V_{2(p)} - 1.4\text{ V} = 15.6\text{ V}$$

$$I_{I(\max)} = I_{o(\max)} = 1.5\text{ A}$$

$$v_r''{}_{(p\text{-}p)\max} = \frac{I_{I(\max)}}{2fC} = \frac{1.5\text{ A}}{120 \times 10\text{ mF}} = 1.25\text{ V}$$

$$V_{I(DC)} = V_{I(p)} - 0.5\, v_r''{}_{(p\text{-}p)} \cong 15\text{ V}$$

Pick the Zener diode 1N5230: $V_Z = 4.7$ V, $r_Z = 19\ \Omega$ @ $I_{ZT} = 20$ mA

Let $I_Z = 2$ mA and determine the value of R as follows:

$$R = \frac{V_I - V_Z}{I_Z} = \frac{15\text{ V} - 4.7\text{ V}}{2\text{ mA}} = 5.1\text{ k}\Omega$$

$$r_Z \cong 190\ \Omega\ @\ I_Z = 2\text{ mA}$$

$$v_r' = v_r''\left(\frac{r_z}{R + r_z}\right) = 1.25\text{V}\left(\frac{190\ \Omega}{5.1\text{ k}\Omega + 190\ \Omega}\right) = 45\text{ mV}$$

$$V_O = V_Z\left(\frac{R_2}{R_1} + 1\right)$$

Let $R_1 = 82$ kΩ.

$$R_{2(max)} = 82 \text{ k}\Omega \left(\frac{10 \text{ V}}{4.7 \text{ V}} - 1 \right) = 92 \text{ k}\Omega$$

We will use a 100 kΩ potentiometer for R_2 and adjust it for the desired output voltage.

$$R_S = \frac{0.7 \text{ V}}{I_{O(max)}} = \frac{0.7 \text{ V}}{1.5 \text{ A}} = 0.47 \text{ }\Omega$$

$$R_{L(min)} = \frac{V_{O(max)}}{I_{O(max)}} = \frac{10 \text{ V}}{1.5 \text{ A}} = 6.7 \text{ }\Omega$$

$$V_r = V_r' \left(\frac{R_{2(max)}}{R_1} + 1 \right) = 45 \text{ mV} \left(\frac{100 \text{ k}\Omega}{82 \text{ k}\Omega} + 1 \right) = 99.87 \text{ mV}$$

Circuit implementation and diagram:

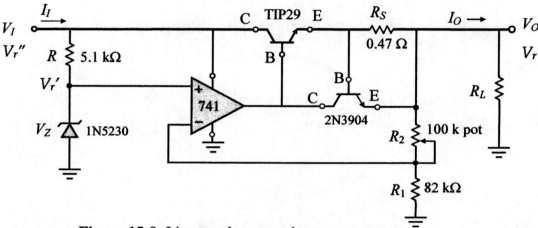

Figure 15-9: Linear voltage regulator
with adjustable output and overload protection

Practice Problem 15-2   *Adjustable Output*   **DESIGN**

Design a regulator with adjustable output and with overload protection for the following specifications: V_O adjustable from approximately 5 V to 15 V,

$I_{O(max)} = 1$ A, $V_r \leq 0.01 V_O$, given that $V_{1(rms)} = 116$ V, $n{:}1 = 8{:}1$, and $C = 8.2$ mF

Answers: Zener is 2N5230, $R = 6.8$ kΩ, $R_1 = 39$ kΩ, $R_2 = 100$ kΩ pot, $V_r \cong 86$ mV

Chapter 15

Three-Terminal IC Regulators

There are also integrated-circuit regulators available with only three terminals (input, output, and ground). These three-terminal regulators are available with fixed and adjustable output voltages. The adjustable version requires two external resistors, preferably one potentiometer and one fixed resistor, in order to adjust the output voltage level. The LM1084 is a 5 A three-terminal regulator from National Semiconductor, which is available with fixed output voltages of 3.3 V, 5 V, and 12 V, and in an adjustable version. It also includes current limiting and thermal protection features. The pin-out and connection diagram of the LM1084 is exhibited in Figure 15-10 below:

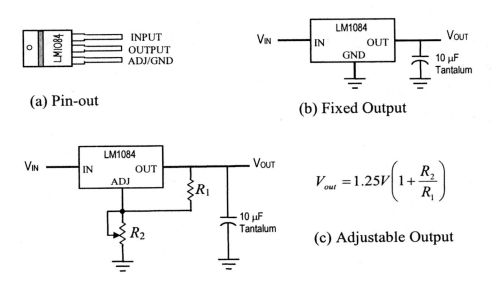

Figure 15-10: Pin-out and connection diagram of the LM1084 3-terminal regulator (Courtesy of National Semiconductor)

The LM1085 is the 3 A version of LM1084 with the same output voltage levels, and LM1086 is the 1.5 A version available with output voltage levels of 2.5 V, 2.85 V, 3.3 V, 3.45 V, 5 V, and adjustable.

The LMS8117A is a 1 A three-terminal regulator from National Semiconductor available with output voltage levels of 1.8 V, 3.3 V, and adjustable.

The LMS1585A and LMS1587 are also 5 A and 3 A three-terminal regulators from National Semiconductor available with output voltage levels of 1.5 V, 3.3 V, and adjustable.

15.3 SWITCHING POWER SUPPLIES

In the traditional power supplies that we have discussed previously, the step-down transformer is a bulky and expensive part of the power supply. The bulkiness of the transformer is due to low line frequency (60 Hz). In addition, the low line frequency requires large filter capacitors to reduce the output ripple, which adds up to the overall cost and physical dimensions of the power supply. Furthermore, the power transistor is operating in the active region and is continuously on, which dissipates considerable power. These three factors combined make the traditional power supplies bulky, more expensive, and less efficient. Switching power supplies provide substantial improvements in all three areas by utilizing a technique called *pulse-width modulation*.

Consider the simple series circuit of Figure 15-11(a), in which a power transistor is switched on and off periodically by a pulse-width modulator, which controls the on and off time of the transistor in one period.

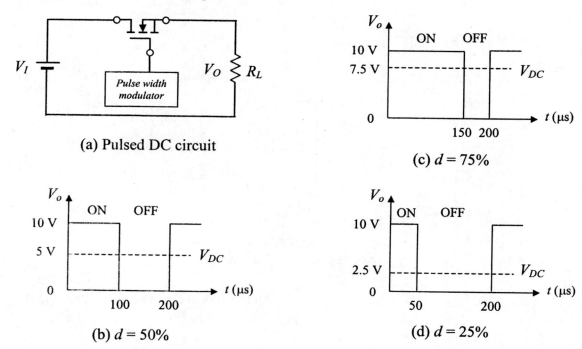

Figure 15-11: Pulse-width modulation

In Figure 15-11(b), the duty-cycle d of the pulse-width modulator is 50%; that is, the transistor is on for 100 μs and off for 100 μs in one period $T = 200$ μs. The resulting pulsed DC output has an average or DC value of 5 V.

In Figure 15-11(c), the transistor is on for 150 μs and off for 50 μs in one period. The resulting pulsed DC output has an average or DC value of 7.5 V.

In Figure 15-11(d), the transistor is on for 50 μs and off for 150 μs in one period. The resulting pulsed DC output has an average or DC value of 2.5 V.

Hence, the larger the duty-cycle D, the higher the average DC voltage is at the output, and *vice versa*. This technique is used to control and regulate the output DC level in switching power supplies. Switching power supplies are also referred to as DC-DC converters, as they take an unregulated DC and convert it very efficiently up or down to a desired level of fixed and regulated DC output voltage.

15.3.1 Buck (Step-Down) DC-DC Converter

The circuit of a buck converter is shown in Figure 15-12, which consists of four principal components: a transistor switch S, which is usually a power MOSFET, a fast switching diode D, an inductor L, and a capacitor C. A pulse-width modulator controls the switch S, which is turned on and off at the rate of the switching frequency $f_s = 1/T$.

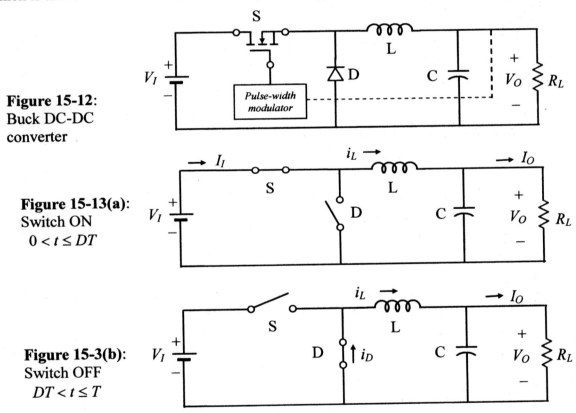

Figure 15-12: Buck DC-DC converter

Figure 15-13(a): Switch ON $0 < t \leq DT$

Figure 15-3(b): Switch OFF $DT < t \leq T$

The transistor switch is turned on and off at the rate of the pulse-width modulator switching frequency f_s with an on duty-cycle D, where

$$f_s = \frac{1}{T}$$

$$D = \frac{t_{on}}{T}$$

The equivalent circuits of the buck converter of Figure 15-12 are shown with the transistor switch on in Figure 15-13(a) and with the switch off in Figure 15-13(b). The corresponding voltage and current waveforms are depicted in Figure 15-14 (a) through (g).

Chapter 15

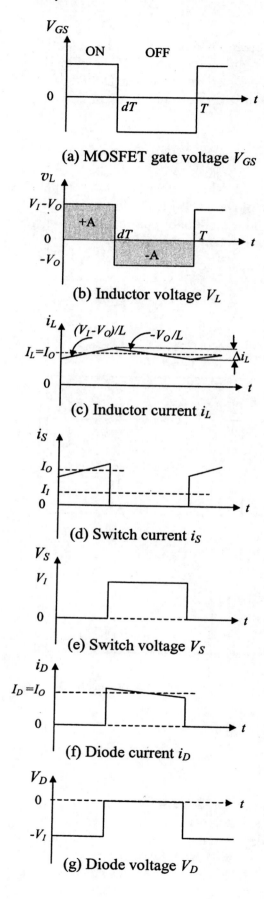

(a) MOSFET gate voltage V_{GS}

(b) Inductor voltage V_L

(c) Inductor current i_L

(d) Switch current i_S

(e) Switch voltage V_S

(f) Diode current i_D

(g) Diode voltage V_D

Assuming that the switch S is turned on at $t = 0$, the diode D is reverse biased and is turned off, as shown in Figure 15-13(a). The voltage across the inductor is $v_L = V_I - V_O$ and the inductor current increases with a slope of $(V_I - V_O)/L$. When the switch is turned off at $t = dT$, the current through the inductor continues flowing through the load and the diode, which has turned on at the instant the switch is turned off. However, the inductor current decreases linearly with a slope of $-V_O/L$. The inductor L and capacitor C form an energy reservoir that maintains the load voltage and current during the time the switch is off.

Figure 15-14: Buck converter voltage and current waveforms

Chapter 15

DC voltage transfer function $M_{V(DC)}$ of the buck converter:

For the steady state, the inductor current at $t = 0$ equals the current at $t = T$, as illustrated in Figure 15-14(c).

$$i_L(0) = i_L(T) \tag{15-9}$$

$$v_L = L\frac{di_L}{dt} \tag{15-10}$$

$$\frac{v_L}{L}dt = di_L \tag{15-11}$$

Integrating both sides over one period results in the following:

$$\frac{1}{L}\int_0^T v_L dt = \int_0^T di_L = [i_L]_0^T = i_L(T) - i_L(0) = 0 \tag{15-12}$$

Therefore, the integral of the inductor voltage over one period equals zero.

$$\int_0^T v_L dt = 0 \tag{15-13}$$

Hence, the average or DC value of the inductor voltage equals zero.

$$v_L(av) = \frac{1}{T}\int_0^T v_L dt = \frac{1}{T}\left(\int_0^{dT} v_L dt + \int_{dT}^T v_L dt\right) = 0 \tag{15-14}$$

$$\int_0^{dT} v_L dt = -\int_{dT}^T v_L dt \tag{15-15}$$

The left side of the above equation is the positive area and the right side is the negative area, as illustrated in Figure 15-14(b); that is,

$$A^+ = A^- \tag{15-16}$$

The above relation is commonly known as the *volt-second balance* principle; that is, the *volt-second* stored equals the *volt-second* released.

$$(V_I - V_O)\, dT = V_O(T - dT) = V_O(1 - d)T \tag{15-17}$$

After some simple algebraic manipulation, the following result is obtained:

$$V_I d = V_O \tag{15-18}$$

$$M_{V(DC)} = \frac{V_O}{V_I} = d \tag{15-19}$$

where $M_{V(DC)}$ is the DC voltage transfer function, and d is the duty-cycle.

For a **lossless** converter the output power equals the input power; that is,

$$P_I = P_O \tag{15-20}$$

$$V_I I_I = V_O I_O \tag{15-21}$$

$$\frac{V_O}{V_I} = \frac{I_I}{I_O} = d \tag{15-22}$$

Chapter 15

For a **lossy** converter the output power does not equal the input power, and as a result, the efficiency is less than 100%.

$$\eta = \frac{P_O}{P_I} = \frac{V_O \cdot I_O}{V_I \cdot I_I} = \frac{V_O}{V_I} \cdot \frac{I_O}{I_I} = \frac{M_{V(DC)}}{d} \quad (15\text{-}23)$$

$$d = \frac{M_{V(DC)}}{\eta} \quad (15\text{-}24)$$

and

$$\frac{I_I}{I_O} = d = \frac{M_{V(DC)}}{\eta} \quad (15\text{-}25)$$

Solving for I_I results in the following equation for the input current:

$$I_I = I_O \frac{M_{V(DC)}}{\eta} \quad (15\text{-}26)$$

Analysis of the buck converter operation

a) Analysis of the converter operation for the time interval with the switch on, $0 \leq t \leq dT$:

$$v_L = V_I - V_O = L\frac{di_L}{dt} \quad (15\text{-}27)$$

$$\frac{v_L}{L}dt = di_L \quad (15\text{-}28)$$

$$i_L = \frac{1}{L}\int_0^t v_L dt + i_L(0) \quad (15\text{-}29)$$

$$i_L = \frac{1}{L}\int_0^{dT}(V_I - V_O)dt + i_L(0) = \left[\frac{V_I - V_O}{L}t\right]_0^{dT} + i_L(0) \quad (15\text{-}30)$$

$$i_L(dT) = \frac{V_I - V_O}{L}(dT) + i_L(0) \quad (15\text{-}31)$$

Then the change in i_L from $t = 0$ to $t = dT$ is the following:

$$\Delta i_L = i_L(dT) - i_L(0) = \frac{V_I - V_O}{L}(dT) + i_L(0) - i_L(0) = \frac{V_I - V_O}{L}(dT) \quad (15\text{-}32)$$

$$\Delta i_L = \frac{(V_I - V_O)dT}{L} = \frac{(V_I - V_O)d}{Lf_s} = \frac{(V_I d - V_O d)}{Lf_s} = \frac{(V_O - V_O d)}{Lf_s} = \frac{V_O(1-d)}{Lf_s} \quad (15\text{-}33)$$

b) Analysis of the converter operation for the time interval with the switch off, $dT \leq t \leq T$:

$$v_L = -V_O = L\frac{di_L}{dt} \quad (15\text{-}34)$$

$$\frac{v_L}{L}dt = di_L \quad (15\text{-}35)$$

$$i_L = \frac{1}{L}\int_0^t v_L dt + i_L(dT) \quad (15\text{-}36)$$

$$i_L = \frac{1}{L}\int_0^t (-V_O)dt + i_L(DT) = \left[\frac{-V_O}{L}t\right]_{DT}^t + i_L(DT) \qquad (15\text{-}37)$$

Evaluation of the above equation at $t = T$ yields the following:

$$i_L(T) = \frac{-V_O}{L}(T - DT) + i_L(DT) \qquad (15\text{-}38)$$

Then the change in i_L from $t = DT$ to $t = T$ is

$$\Delta i_L = i_L(DT) - i_L(T) = i_L(DT) - \left[\frac{-V_O}{L}T(1-D) + i_L(DT)\right] = \frac{V_O}{f_s L}(1-D) \qquad (15\text{-}39)$$

Hence, for both time intervals with the switch on or off, the peak-to-peak ripple current Δi_L of the inductor L is the same and its maximum value is

$$\Delta i_{L(max)} = \frac{V_O(1 - D_{min})}{f_s L_{min}} \qquad (15\text{-}40)$$

The output ripple voltage V_r:

The inductor current i_L has an average or DC value I_L and a ripple or AC component Δi_L, as illustrated in Figure 15-14(c). The DC component I_L equals the output DC load current I_O, which flows through the load R_L, and the AC component is divided between the R_L and the filter capacitor C. However, the impedance of the capacitor is generally much smaller than the load resistor; hence, almost all of the ripple current flows through the capacitor. The impedance of the capacitor comprises a reactance component and an *equivalent series resistance* (*ESR*), which is designated as r_C in the following equations. The reactance component is usually much smaller than the ESR. Hence, the peak-to-peak ripple voltage at the output can be approximated by the following equation:

$$V_r \cong r_C \Delta i_{L(max)} \qquad (15\text{-}41)$$

Substituting for $\Delta i_{L(max)}$ in the above equation results in the following for the output ripple voltage

$$V_r = r_C \frac{V_O(1 - D_{min})}{f_s L} \qquad (15\text{-}42)$$

Apparently, the lower the r_C (*ESR*) and the higher the switching frequency f_s, the lower the output ripple voltage.

Losses and efficiency:

Each individual component of the converter, inductor, capacitor, diode, and the MOSFET, will have its own losses. Following a correct design procedure and selecting the appropriate components can minimize these losses and contribute to the overall efficiency of the converter. Losses in the MOSFET switch are a combination of conduction loss and switching loss. The MOSFET conduction loss P_{con} is proportional to the on duty-cycle D, the on-resistance of the MOSFET r_{DS}, and the switch current $I_{S(rms)}$.

Chapter 15

As illustrated in Figure 15-14(d), $i_S = I_O$ for $0 < t \leq DT$, and $i_S = 0$ for $DT < t \leq T$. Hence, the $I_{S(rms)}^2$ value of the switch current and the corresponding conduction loss is given by the following:

$$I_{S(rms)}^2 = \frac{1}{T}\int_0^T i_S^2 dt = \frac{1}{T}\int_0^{DT} I_O^2 dt = \frac{I_O^2}{T}\int_0^{DT} dt = \frac{I_O^2}{T}[t]_0^{DT} = DI_O^2 \tag{15-43}$$

$$P_{con} = I_{S(rms)}^2 r_{DS} = I_O^2 D r_{DS} \tag{15-44}$$

The MOSFET switching loss P_{sw} is proportional to the switching frequency f_s and the MOSFET output capacitance C_o, and is determined by

$$P_{sw} = f_s C_o V_S^2 = f_s C_o V_I^2 = \frac{f_s C_o V_O^2}{M_{VDC}^2} \tag{15-45}$$

$$P_{FET} = P_{con} + P_{sw}/2 \tag{15-46}$$

Losses in the diode are due to the forward bias voltage V_F and the forward resistance R_F of the non-ideal diode. The loss due to the forward resistance is proportional to R_F of the diode and the diode current $I_{D(rms)}$. As illustrated in Figure 15-14(f), $i_D = 0$ for $0 < t \leq DT$, and $i_D = I_O$ for $DT < t \leq T$. Hence, the $I_{D(rms)}^2$ value of the diode current and the corresponding power loss P_{RF} is determined by

$$I_{D(rms)}^2 = \frac{1}{T}\int_0^T i_D^2 dt = \frac{1}{T}\int_{DT}^T I_O^2 dt = \frac{I_O^2}{T}\int_0^{DT} dt = \frac{I_O^2}{T}[t]_{DT}^T = I_O^2(1-D) \tag{15-47}$$

$$P_{RF} = I_{D(rms)}^2 R_F = I_O^2 (1 - D_{\min})R_F \tag{15-48}$$

The loss due to the forward bias voltage is proportional to the forward voltage drop V_F and the average value of the diode current I_D. The average diode current and the corresponding power loss P_{VF} are determined by

$$I_D = \frac{1}{T}\int_0^T i_D\, dt = \frac{1}{T}\int_{DT}^T I_O\, dt = \frac{I_O}{T}\int_0^{DT} dt = \frac{I_O}{T}[t]_{DT}^T = I_O(1-D) \tag{15-49}$$

$$P_{VF} = I_D V_F = I_O (1 - D_{\min}) V_F \tag{15-50}$$

$$P_D = P_{RF} + P_{VF} \tag{15-51}$$

The loss in the inductor is due to the ESR of the inductor r_L and is determined by

$$P_L = r_L I_L^2 = r_L I_O^2 \tag{15-52}$$

The loss in the filter capacitor is due to the ESR of the capacitor r_C and is determined as follows:

$$P_C = r_C \frac{\Delta i_{L(\max)}^2}{12} \tag{15-53}$$

The total loss in the converter is the sum of all the above losses

$$P_{Loss} = P_{con} + P_{sw} + P_D + P_L + P_C \tag{15-54}$$

The efficiency of the converter is the ratio of the output power to the input power

$$\eta = \frac{P_O}{P_I} = \frac{P_O}{P_O + P_{Loss}} \tag{15-55}$$

Chapter 15

Example 15-3 — Buck Converter ANALYSIS

Let us determine the efficiency and the output ripple voltage of a buck converter with the following specifications:

$V_I = 10$ V $\pm 25\%$, $V_O = 5$ V, $R_L = 5$ Ω, and $f_s = 50$ kHz
MOSFET: $r_{DS} = 0.25$ Ω, $C_o = 100$ pF   Diode: $V_F = 0.5$ V, $R_F = 0.2$ Ω
Inductor: $L = 0.47$ mH, $r_L = 0.3$ Ω Capacitor: $C = 47$ µF, $r_C = 0.25$ Ω

Solution: We will initially assume $\eta = 0.9$.

$$M_{VDC(min)} = \frac{V_O}{V_{I(max)}} = \frac{5\text{ V}}{12.5\text{ V}} = 0.40 \qquad M_{VDC(max)} = \frac{V_O}{V_{I(min)}} = \frac{5\text{ V}}{7.5\text{ V}} = 0.666$$

$$D_{min} = \frac{M_{VDC(min)}}{\eta} = \frac{0.4}{0.9} = 0.444 \qquad D_{max} = \frac{M_{VDC(max)}}{\eta} = \frac{0.666}{0.9} = 0.74$$

$$I_O = \frac{V_O}{R_L} = \frac{5\text{ V}}{5\,\Omega} = 1\text{ A}$$

$$\Delta i_{L(max)} = \frac{V_O(1 - D_{min})}{f_s L} = \frac{5\text{ V}(1 - 0.444)}{50\text{ kHz} \times 0.47\text{ mH}} = 118.3\text{ mA}$$

$$V_r = r_C \,\Delta i_{L(max)} = 0.25\,\Omega \times 118.3\text{ mA} = 29.575\text{ mV}$$

Losses in the MOSFET are determined by Equations 15-44 through 15-46

$$P_{con} = D_{max} I_O^2 r_{DS} = 0.74(1\text{ A})^2(0.25\,\Omega) = 0.185\text{ W}$$

$$P_{sw} = f_s C_o V_{I(max)}^2 = (50\text{ kHz})(100\text{ pF})(12.5\text{ V})^2 = 0.78\text{ mW}$$

$$P_{FET} = P_{con} + P_{sw}/2 = 185\text{ mW} + 0.78\text{ mW}/2 = 185.39\text{ mW}$$

Losses in the diode are determined by Equations 15-48 through 15-51

$$P_{RF} = (1 - D_{min}) R_F I_O^2 = (1-0.444)(0.2\,\Omega)(1\text{ A})^2 = 111.2\text{ mW}$$

$$P_{VF} = I_O (1 - D_{min}) V_F = (1\text{ A})(1 - 0.444)(0.5\text{V}) = 278\text{ mW}$$

$$P_D = P_{RF} + P_{VF} = 111.2\text{ mW} + 278\text{ mW} = 389.2\text{ mW}$$

The loss in the inductor is determined by Equation 15-52

$$P_L = r_L I_O^2 = (0.3\,\Omega)(1\text{ A})^2 = 0.3\text{ W}$$

The loss in the capacitor is determined by Equation 15-53

$$P_C = r_C \frac{\Delta i_{L(max)}^2}{12} = 0.25\,\Omega \frac{(0.01183\text{A})^2}{12} = 0.29\,\mu\text{W}$$

The total loss in the converter is the sum of all the above losses

$$P_{Loss} = P_{con} + P_{sw} + P_D + P_L + P_C$$
$$= 185\text{ mW} + 0.78 + 389.2\text{ mW} + 300\text{ mW} + 0.29\,\mu\text{W} = 875\text{ mW}$$

Chapter 15

$$\eta = \frac{P_O}{P_I} = \frac{P_O}{P_O + P_{Loss}} = \frac{5\,\text{W}}{5.807\,\text{W}} = 86.1\%$$

Since the difference in efficiency is no more than ±5% of the assumed efficiency of 90%, the above answer is fairly close to the actual efficiency. Otherwise, we would assume another value closer to the above found efficiency and repeat the calculations.

Example 15-4 — Buck Converter — DESIGN

Let us now design a buck converter with the following specifications:

$V_I = 10\,\text{V} \pm 20\%$, $V_O = 6\,\text{V}$, $I_{O(min)} = 0.1\,\text{A}$, $I_{O(max)} = 1\,\text{A}$, and $f_s = 150\,\text{kHz}$

Solution: We will initially assume $\eta = 0.85$.

$$M_{VDC(min)} = \frac{V_O}{V_{I(max)}} = \frac{6\,\text{V}}{12\,\text{V}} = 0.50 \qquad M_{VDC(max)} = \frac{V_O}{V_{I(min)}} = \frac{6\,\text{V}}{8\,\text{V}} = 0.75$$

$$D_{min} = \frac{M_{VDC(min)}}{\eta} = \frac{0.5}{0.85} = 0.588 \qquad D_{max} = \frac{M_{VDC(max)}}{\eta} = \frac{0.75}{0.85} = 0.882$$

$$R_{L(max)} = \frac{V_O}{I_{O(min)}} = \frac{6\,\text{V}}{0.1\,\text{A}} = 60\,\Omega \qquad R_{L(min)} = \frac{V_O}{I_{O(max)}} = \frac{6\,\text{V}}{1\,\text{A}} = 6\,\Omega$$

$$L_{min} = \frac{R_{L(max)}(1 - D_{min})}{2 f_s} = \frac{60\,\Omega(1 - 0.588)}{300\,\text{kHz}} = 82.4\,\mu\text{H} \qquad \text{Use } L = 100\,\mu\text{H}.$$

$$\Delta i_{L(max)} = \frac{V_O(1 - D_{min})}{f_S L} = \frac{6\,\text{V}(1 - 0.588)}{150\,\text{kHz} \times 100\,\mu\text{H}} = 165\,\text{mA}$$

$$V_r \leq 0.01 V_O = 60\,\text{mV} = r_C\,\Delta i_{L(max)}$$

$$r_{C(max)} = \frac{V_r}{\Delta i_{L(max)}} = \frac{60\,\text{mV}}{165\,\text{mA}} = 0.36\,\Omega \qquad \text{Let } r_C = 0.15\,\Omega.$$

$$C_{min} = \frac{D_{max}}{2 f_s r_C} = \frac{0.882}{2 \times 150\,\text{kHz} \times 0.15\,\Omega} = 19.6\,\mu\text{F} \qquad \text{Use } C = 27\,\mu\text{F}.$$

$$V_{S(max)} = V_{D(max)} = V_{I(max)} = 12\,\text{V}$$

$$I_{S(max)} = I_{I(max)} = I_{O(max)} \frac{M_{VDC(max)}}{\eta} = 1\,\text{A}\left(\frac{0.882}{0.85}\right) = 1.03\,\text{A}$$

$$I_{D(max)} = I_{O(max)} + \Delta i_{L(max)}/2 = 1\,\text{A} + 0.165\,\text{A} = 1.082\,\text{A}$$

Implementation of the design:

To implement our design of the buck converter, we will utilize National Semiconductor's LM2595J-ADJ regulator, which contains the pulse-width modulator and the MOSFET switch in one *IC* chip. For the inductor we will select Pulse Engineering's PE-53829 (112 µH, 1.26 A, 0.3 Ω), which is designed specifically for National's 150 kHz switching regulators. For the diode we will select Motorola's 1N5817 Schottky barrier diode, and for the output capacitor will select Sprague's 550D276X9020R2 low *ESR* tantalum capacitor (27 μF, 20 V, 0.147 Ω). The output of the LM2595J-ADJ regulator can be adjusted, by means of a potentiometer, to a desired voltage level. The complete circuit implementation is shown in Figure 15-15.

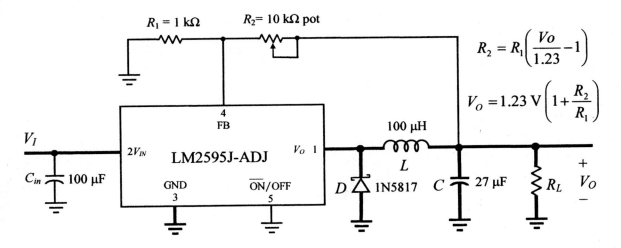

Figure 15-15: Implementation of the buck DC-DC converter

Practice Problem 15-3 — Buck Converter — ANALYSIS

Determine the efficiency and V_r of a buck converter with the following specifications:

$V_I = 10$ V to 15 V, $V_O = 7.5$ V, $R_L = 10$ Ω, and $f_s = 100$ kHz. Initially, assume $\eta = 0.9$.

MOSFET: $r_{DS} = 0.3$ Ω, $C_o = 100$ pF   Diode: $V_F = 0.35$ V, $R_F = 0.25$ Ω

Inductor: $L = 470$ μH, $r_L = 0.3$ Ω   Capacitor: $C = 100$ μF, $r_C = 0.3$ Ω

Answers: $V_r = 21.3$ mV, $\eta = 92\%$

Practice Problem 15-4 — Buck Converter — DESIGN

Design a buck converter with the following specifications:

$V_I = 12$ V to 16 V, $V_O = 7.5$ V, $I_{O(min)} = 0.075$ A, $I_{O(max)} = 2$ A, and $f_s = 150$ kHz

Initially assume $\eta = 0.85$ and let $r_C = 0.25$ Ω.

Answers: $L_{min} = 150$ μH. Use PE-54036, $L = 168$ μH. $C_{min} = 9.8$ μF Use 18 μF.

$V_{S(max)} = V_{D(max)} = 16$ V, $I_{S(max)} = I_{D(max)} \cong 2$ A Use National's LM2595.

Section Summary — Buck Converter — ANALYSIS/DESIGN

Summary of Equations for the Analysis/Design of the Buck Converter

Analysis: Initially assume $\eta = 90\%$.

$$M_{VDC(min)} = \frac{V_O}{V_{I(max)}} \qquad M_{VDC(max)} = \frac{V_O}{V_{I(min)}}$$

$$d_{min} = \frac{M_{VDC(min)}}{\eta} \qquad d_{max} = \frac{M_{VDC(max)}}{\eta}$$

$$\Delta i_{L(max)} = \frac{V_o(1-d_{min})}{f_S L} \qquad v_{r(p-p)} = r_C \Delta i_{L(max)}$$

Losses in the MOSFET: $\quad P_{con} = d_{(max)} I_O^2 r_{DS} \qquad P_{sw} = f_S C_o V_{I(max)}^2$

$$P_{FET} = P_{con} + P_{sw}$$

Losses in the diode: $\quad P_{RF} = (1-d_{min})R_F I_O^2 \qquad P_{VF} = I_O(1-d_{min})V_F$

$$P_D = P_{RF} + P_{VF}$$

Loss in the inductor: $\quad P_L = r_L I_O^2$

Loss in the capacitor: $\quad P_C = r_C \dfrac{\Delta i_{L(max)}^2}{12}$

The total loss: $\quad P_{Loss} = P_{FET} + P_D + P_L + P_C$

Efficiency: $\quad \eta = \dfrac{P_O}{P_I} = \dfrac{P_O}{P_O + P_{Loss}}$

Design: Assume $\eta = 90\%$ and determine $M_{VDC(min)}$, $M_{VDC(max)}$, d_{min}, and d_{max}.

$$R_{L(max)} = \frac{V_O}{I_{O(min)}} \qquad R_{L(min)} = \frac{V_O}{I_{O(max)}}$$

$$L_{(min)} = \frac{R_{L(max)}(1-d_{min})}{2f_S} \qquad \Delta i_{L(max)} = \frac{V_o(1-d_{min})}{f_S L}$$

$$v_{r(p-p)} \leq 0.01 V_O = r_C \Delta i_{L(max)}$$

$$r_{C(max)} = \frac{v_{r(p-p)max}}{\Delta i_{L(max)}} \qquad C_{min} = \frac{d_{max}}{2 f_S r_C}$$

$$V_{S(max)} = V_{D(max)} = V_{I(max)} \qquad I_{S(max)} = I_{I(max)} = I_{O(max)} \frac{M_{VDC(max)}}{\eta}$$

$$I_{S(max)} = I_{D(max)} = I_{O(max)} + \Delta i_{L(max)}/2$$

15.3.2 Boost (Step-Up) DC-DC Converter

The circuit of a boost converter is shown in Figure 15-16, which consists of four components: a transistor switch S, which is usually a power MOSFET, a fast switching diode D, an inductor L, and a capacitor C. A pulse-width modulator driver controls the switch S, which is turned on and off at the rate of the switching frequency $f_s = 1/T$.

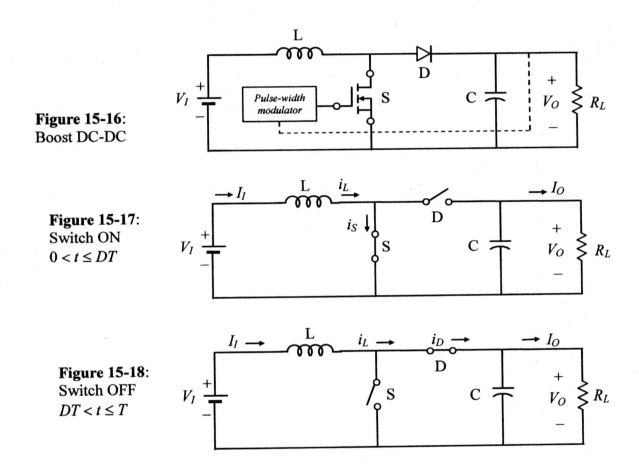

Figure 15-16: Boost DC-DC

Figure 15-17: Switch ON $0 < t \leq DT$

Figure 15-18: Switch OFF $DT < t \leq T$

The transistor switch is turned on and off at the rate of the pulse-width modulator switching frequency f_s with an on duty-cycle D, where

$$f_s = \frac{1}{T}$$

$$D = \frac{t_{on}}{T}$$

The equivalent circuits of the boost converter of Figure 15-16 are shown with the transistor switch on in Figure 15-17, and with the switch off in Figure 15-18. The corresponding voltage and current waveforms are depicted in Figures 15-19(a) through (g).

Chapter 15

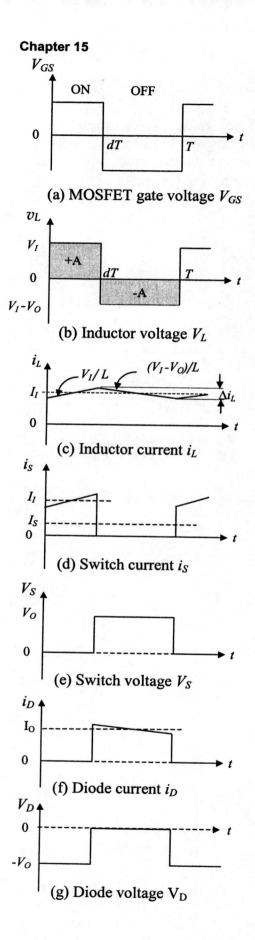

Assuming that the switch S is turned on at $t = 0$, the diode D is reverse-biased and is turned off, as shown in Figure 15-17. The voltage across the inductor is $v_L = V_I$ and the inductor current increases with a slope of V_I/L. When the switch is turned off at $t = dT$, the current through the inductor continues flowing through the load and the diode, which has turned on at the instant the switch is turned off. However, the inductor current decreases linearly with a slope of $(V_I - V_O)/L$. During this time interval, the energy is transferred from the inductor L to the capacitor C and the load resistor R_L.

Figure 15-19: Boost converter voltage and current waveforms

Chapter 15

DC voltage transfer function M_{VDC} of the boost converter:

Applying the *volt-second balance* principle to Figure 15-19(b), the positive area equals the negative area

$$A^+ = A^- \tag{15-56}$$

$$V_I DT = (V_O - V_I)(T - DT) = (V_O - V_I)(1 - D)T \tag{15-57}$$

After some simple algebraic manipulation, the following result is obtained:

$$V_I = V_O(1 - D) \tag{15-58}$$

$$M_{VDC} = \frac{V_O}{V_I} = \frac{1}{1 - D} \tag{15-59}$$

where M_{VDC} is the DC voltage transfer function and D is the duty cycle. For a **lossless** converter the output power equals the input power; that is,

$$P_I = P_O \tag{15-60}$$

$$V_I I_I = V_O I_O \tag{15-61}$$

$$\frac{V_O}{V_I} = \frac{I_I}{I_O} = \frac{1}{1 - D} \tag{15-62}$$

For a **lossy** converter the output power does not equal the input power, and as a result, the efficiency is less than 100%

$$\eta = \frac{P_O}{P_I} = \frac{V_O I_O}{V_I I_I} = M_{VDC}(1 - D) \tag{15-63}$$

$$D = 1 - \frac{\eta}{M_{VDC}} \tag{15-64}$$

$$I_I = \frac{I_O}{\eta} M_{VDC} = \frac{I_O}{1 - D} \tag{15-65}$$

Analysis of the boost converter operation

a) Analysis of the converter operation for the time interval with the switch on, $0 \le t \le DT$.

As depicted in Figure 15-17, the switch is on, the diode is off, and the inductor voltage is

$$v_L = V_I = L \frac{di_L}{dt} \tag{15-66}$$

$$\frac{v_L}{L} dt = di_L \tag{15-67}$$

$$i_L = \frac{1}{L} \int_0^{DT} v_L dt + i_L(0) \tag{15-68}$$

$$i_L = \frac{1}{L} \int_0^{DT} V_I \, dt + i_L(0) = \left[\frac{V_I}{L} t \right]_0^{DT} + i_L(0) \tag{15-69}$$

$$i_L(DT) = \frac{V_I}{L}(DT) + i_L(0) \tag{15-70}$$

Chapter 15

Then the change in i_L from $t = 0$ to $t = DT$ is the following:

$$\Delta i_L = i_L(DT) - i_L(0) = \frac{V_I}{L}(DT) + i_L(0) - i_L(0) = \frac{V_I}{L}(DT) \qquad (15\text{-}71)$$

$$\Delta i_L = \frac{V_I DT}{L} = \frac{V_O(1-D)D}{f_s L}$$

b) Analysis of the converter operation for the time interval with the switch off, $DT \le t \le T$.

As depicted in Figure 15-18, the switch is off, the diode is on, and the inductor voltage v_L equals $V_I - V_O$. Thus,

$$v_L = V_I - V_O = L\frac{di_L}{dt} \qquad (15\text{-}72)$$

$$\frac{v_L}{L}dt = di_L \qquad (15\text{-}73)$$

$$i_L = \frac{1}{L}\int_{DT}^{T} v_L \, dt + i_L(DT) \qquad (15\text{-}74)$$

$$i_L = \frac{1}{L}\int_{dT}^{T}(V_I - V_O)dt + i_L(DT) = \left[\frac{V_I - V_O}{L}t\right]_{DT}^{T} + i_L(DT) \qquad (15\text{-}75)$$

$$i_L(T) = \frac{V_I - V_O}{L}(T - DT) + i_L(DT) \qquad (15\text{-}76)$$

Then the change in i_L from $t = DT$ to $t = T$ is the following:

$$\Delta i_L = i_L(DT) - i_L(T) = i_L(DT) - \left(\frac{V_I - V_O}{L}T(1-D) + i_L(DT)\right) = \frac{V_O - V_I}{f_s L}(1-D) \qquad (15\text{-}77)$$

Substituting for V_I in terms of V_O in the above equation results in the following:

$$\Delta i_L = \frac{V_O - V_O(1-D)}{f_s L}(1-D) = \frac{V_O(1-1+D)}{f_s L}(1-D) = \frac{V_O}{f_s L}D(1-D) \qquad (15\text{-}78)$$

Therefore, for both time intervals with the switch on or off, the peak-to-peak ripple current Δi_L of the inductor L is the same and its maximum value occurs at $D = 0.5$; hence,

$$\Delta i_{L(max)} = \frac{V_O}{4 f_s L} \qquad (15\text{-}79)$$

The output ripple voltage V_r:

The peak-to-peak ripple voltage at the output of the boost converter is $V_r = V_{rC} + V_{C(p\text{-}p)}$. The peak-to-peak ripple voltage across the ESR is

$$V_{rC} \cong r_C I_{D(max)} = \frac{r_C I_{O(max)}}{1 - D_{max}} \qquad (15\text{-}80)$$

and the peak-to-peak ripple voltage across the filter capacitor is

$$V_{C(p\text{-}p)} = \frac{\Delta Q_{max}}{C_{min}} = \frac{I_{O(max)} D_{max} T}{C_{min}} = \frac{V_O D_{max}}{f_s R_{L(min)} C_{min}} \qquad (15\text{-}81)$$

where r_C is the *equivalent series resistance (ESR)* of the filter capacitor C.

Chapter 15

Losses and efficiency:

Each individual component of the converter, inductor, capacitor, diode, and MOSFET will have its own losses. Following a correct design procedure and selecting the appropriate components can minimize these losses and contribute to the overall efficiency of the converter. Losses in the MOSFET switch are a combination of conduction and switching losses. The MOSFET conduction loss P_{con} is proportional to the on-resistance of the MOSFET r_{DS} and the switch current I_S.

As illustrated in Figure 15-19(d), $I_S = I_I = I_O/(1-D)$ for $0 < t \le DT$, and $I_S = 0$ for $DT < t \le T$. Hence, the $I_{S(rms)}^2$ value of the switch current and the corresponding conduction loss are determined as follows:

$$I_{S(rms)}^2 = \frac{1}{T}\int_0^T I_S^2\, dt = \left(\frac{I_O}{1-D}\right)^2 \frac{1}{T}\int_0^{DT} dt = \left(\frac{I_O}{1-D}\right)^2 D \qquad (15\text{-}82)$$

$$P_{con} = I_{S(rms)}^2\, r_{DS} = D\left(\frac{I_O}{(1-D)}\right)^2 r_{DS} \qquad (15\text{-}83)$$

The MOSFET switching loss P_{sw} is proportional to the switching frequency f_s and the MOSFET output capacitance C_o, and is determined with the following equation:

$$P_{sw} = f_s C_o V_S^2 = f_s C_o V_O^2 \qquad (15\text{-}84)$$

$$P_{FET} = P_{con} + P_{sw}/2 \qquad (15\text{-}85)$$

Losses in the diode are due to the forward bias voltage V_F and the forward resistance R_F of the non-ideal diode. The loss due to the forward resistance P_{RF} is proportional to R_F of the diode and the rms value of the diode current. Note that as illustrated in Figure 15-19(f), $I_D = 0$ for $0 < t \le DT$, and $I_D = I_O$ for $DT < t \le T$. Hence, the $I_{D(rms)}^2$ value of the diode current and the corresponding loss in the R_F are the following:

$$I_{D(rms)}^2 = \frac{1}{T}\int_0^T I_D^2\, dt = I_O^2\, \frac{1}{T}\int_{DT}^T dt = I_O^2\, \frac{1}{T}(T-DT) = I_O^2(1-D) \qquad (15\text{-}86)$$

$$P_{RF} = I_{D(rms)}^2 R_F = I_O^2(1-D)R_F \qquad (15\text{-}87)$$

The loss due to forward bias voltage P_{VF} is proportional to the forward voltage drop V_F of the diode and the average diode current I_D, and is determined as follows:

$$P_{VF} = V_F I_D = V_F I_O \qquad (15\text{-}88)$$

$$P_D = P_{RF} + P_{VF} \qquad (15\text{-}89)$$

The loss in the inductor is due to the ESR of the inductor r_L, and is determined as follows:

$$P_L = r_L I_L^2 = r_L I_I^2 = r_L\left(\frac{I_O}{1-D}\right)^2 \qquad (15\text{-}90)$$

The loss in the filter capacitor is due to the ESR of the capacitor r_C and the AC component of the diode current, and is determined as follows:

$$P_C = r_C I_{C(rms)}^2 = r_C I_{D(rms)}^2 = r_C (1-D) I_O^2 \qquad (15\text{-}91)$$

The total loss in the converter is the sum of all the above losses:

$$P_{Loss} = P_{con} + P_{sw} + P_D + P_L + P_C \qquad (15\text{-}92)$$

The efficiency of the converter is the ratio of the output power to the input power

$$\eta = \frac{P_O}{P_I} = \frac{P_O}{P_O + P_{Loss}} \qquad (15\text{-}93)$$

Example 15-5 ANALYSIS

Let us determine the efficiency and the output ripple voltage of a boost converter with the following specifications: $V_I = 8\text{ V} \pm 25\%$, $V_O = 12\text{ V}$, $R_L = 10\text{ }\Omega$, and $f_s = 100\text{ kHz}$
MOSFET: $r_{DS} = 0.15\text{ }\Omega$, $C_o = 100\text{ pF}$ Diode: $V_F = 0.5\text{ V}$, $R_F = 0.3\text{ }\Omega$
Inductor: $L = 470\text{ }\mu\text{H}$, $r_L = 0.2\text{ }\Omega$ Capacitor: $C = 47\text{ }\mu\text{F}$, $r_C = 0.3\text{ }\Omega$

Solution: We will initially assume $\eta = 0.85$.

$$M_{VDC(min)} = \frac{V_O}{V_{I(max)}} = \frac{12\text{ V}}{10\text{ V}} = 1.2 \qquad\qquad M_{VDC(max)} = \frac{V_O}{V_{I(min)}} = \frac{12\text{ V}}{6\text{ V}} = 2$$

$$D_{min} = 1 - \frac{\eta}{M_{VDC(min)}} = 1 - \frac{0.85}{1.2} = 0.29 \qquad D_{max} = 1 - \frac{\eta}{M_{VDC(max)}} = 1 - \frac{0.85}{2} = 0.575$$

$$V_{rC} = r_C I_{D(max)} = \frac{r_C I_{O(max)}}{1 - D_{max}} = \frac{0.3 \times 575\text{ V}}{1 - 0.575} = 0.847\text{ V}$$

$$V_{C(p-p)} = \frac{V_O D_{max}}{f_s R_{L(min)} C_{min}} \quad \frac{V_O = 12 \times 0.575}{150 \times 10^3 \times 10 \times 47 \times 10^{-6}} = 0.0979\text{ V}$$

$$V_r = V_{rC} + V_{C(p-p)} = 0.847 + 0.0979 = 0.9449\text{ V}$$

Losses in the MOSFET are determined from Equations 15-83 through 15-85

$$P_{con} = D_{max}\left(\frac{I_O}{(1 - D_{max})}\right)^2 r_{DS} = 0.575 \times \left(\frac{1.2\text{ A}}{1 - 0.575}\right)^2 \times 0.15\text{ }\Omega = 687.6\text{ mW}$$

$$P_{sw} = f_s C_o V_O^2 = (100\text{ kHz})(100\text{ pF})(12\text{ V})^2 = 1.44\text{ mW}$$

$$P_{FET} = P_{con} + P_{sw}/2 = 687.6\text{ mW} + 1.11\text{ mW}/2 = 688.3\text{ mW}$$

Losses in the diode are determined from Equations 15-87 through 15-89

$$P_{RF} = I_{O(max)}^2 (1 - D_{min}) R_F = 1.2^2 (1 - 0.29)(0.3\text{ }\Omega) = 307\text{ mW}$$

$$P_{VF} = V_F I_O = (0.5\text{ V})(1.2\text{ A}) = 0.6\text{ W}$$

$$P_D = P_{RF} + P_{VF} = 0.307\text{ W} + 0.6\text{ W} = 0.907\text{ W}$$

The loss in the inductor is determined from Equation 15-90

$$P_L = r_L \frac{I_{O(max)}^2}{(1 - D_{max})^2} = 0.2\text{ }\Omega \frac{1.2^2}{(1 - 0.575)^2} = 1.59\text{ W}$$

The loss in the capacitor is determined from Equation 15-91

$$P_C = r_C(1-D_{min})I_O^2 = 0.3\,\Omega(1-0.29)(1.2\text{ A})^2 = 307\text{ mW}$$

The total loss in the converter is sum of all the above losses:

$$P_{Loss} = P_{con} + P_{sw} + P_D + P_L + P_C$$
$$= 0.6876\text{ W} + 0.0014 + 0.907\text{ W} + 1.59\text{ W} + 0.307\text{ W} = 3.493\text{ W}$$
$$\eta = \frac{P_O}{P_I} = \frac{P_O}{P_O + P_{Loss}} = \frac{14.4\text{ W}}{14.4\text{ W} + 3.493\text{ W}} = 80.5\%$$

Since the difference in efficiency is no more than ±5% of the assumed efficiency of 85%, the above answer is fairly close to the actual efficiency. Otherwise, we would assume another value closer to the above found efficiency and repeat the calculations.

Example 15-6 DESIGN

Let us now design a boost converter with the following specifications:

$V_I = 5$ V to 8 V, $V_O = 12$ V, $I_{O(min)} = 30$ mA, $I_{O(max)} = 1$ A, and $f_s = 150$ kHz

Solution: We will initially assume $\eta = 0.85$.

$$M_{VDC(min)} = \frac{V_O}{V_{I(max)}} = \frac{12\text{ V}}{8\text{ V}} = 1.5 \qquad M_{VDC(max)} = \frac{V_O}{V_{I(min)}} = \frac{12\text{ V}}{5\text{ V}} = 2.4$$

$$D_{min} = 1 - \frac{\eta}{M_{VDC(min)}} = 1 - \frac{0.85}{1.5} = 0.4333 \qquad D_{max} = 1 - \frac{\eta}{M_{VDC(max)}} = 1 - \frac{0.85}{2.4} = 0.646$$

$$R_{L(max)} = \frac{V_O}{I_{O(min)}} = \frac{12\text{ V}}{0.030\text{ A}} = 400\,\Omega \qquad R_{L(min)} = \frac{V_O}{I_{O(max)}} = \frac{12\text{ V}}{1\text{ A}} = 12\,\Omega$$

$$L_{min} = \frac{2R_{L(max)}}{27 f_s} = \frac{2\times 400\,\Omega}{27\times 150\text{ kHz}} = 197\,\mu\text{H}$$

$V_r \le 0.01 V_O = 120$ mV $= V_{rc} + V_{C(p-p)}$ Let $V_{rc} = 100$ mV and $V_{C(p-p)} = 20$ mV

$$r_{C(max)} = V_{rc}\frac{1-D_{(max)}}{I_{O(max)}} = 100\text{ mV}\frac{(1-0.646)}{1\text{ A}} = 0.035\,\Omega$$

$$C_{min} = \frac{D_{max}V_O}{f_s R_{L(min)} V_{C(p-p)}} = \frac{0.646\times 12}{150\text{ kHz}\times 12\,\Omega\times 0.02\text{ V}} = 215.3\,\mu\text{F}$$

Use $C = 300\,\mu\text{F}/15\text{ V}/35\text{ m}\Omega$.

$$V_{S(max)} = V_{D(max)} = V_O = 12\text{ V} \qquad I_{S(max)} = I_{D(max)} = I_{I(max)} = \frac{I_{O(max)}}{1-D_{max}} = \frac{1\text{ A}}{1-0.646} = 2.8\text{ A}$$

Implementation of the design:

To implement our design of the boost converter, we will utilize National Semiconductor's LM2586S-ADJ regulator, which contains the pulse-width modulator and the transistor switch in one *IC* chip. For the inductor, we will select Pulse Engineering's PE-53935 with nominal ratings of 250 μH, 1.5 A, and 0.23 Ω. For the diode we will select Motorola's 1N5820 Schottky barrier diode, and for the 100 μF capacitor will select Sprague's 550D107X9020S2 low *ESR* tantalum capacitor ($C = 2 \times 150\ \mu F/15\ V/35\ m\Omega$). The complete circuit implementation is shown in Figure 15-20.

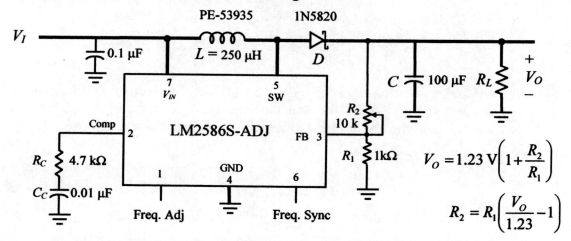

Figure 15-20: Implementation of the boost DC-DC converter

$$R_C \leq \frac{10 V_O^2}{r_C V_{I(min)}} \qquad C_C \geq \frac{40 V_O^2 C_O}{R_C^2 V_{I(min)}}$$

$$R_C \leq \frac{10 \times (12\ V)^2}{0.035\ \Omega \times 5\ V} = 8.2\ k\Omega \quad \text{Use } 4.7\ k\Omega.$$

$$C_C \geq \frac{40 \times 144 \times 100\ \mu F}{(4.7\ k\Omega)^2 \times 5V} = 0.0052\ \mu F \quad \text{Use } 0.01\ \mu F$$

Frequency adjustment:

The switching frequency of the LM2586 regulator can be adjusted from 100 kHz to 200 kHz with an external resistor R_{SET} connected from pin 1 to ground. The resistor values and the corresponding switching frequencies are shown in Table 15-1. The oscillator can also be synchronized with other devices, so that multiple devices can operate at the same switching frequency.

Table 15-1: Frequency Setting Resistor Guide

R_{SET} (kΩ)	Frequency (kHz)
Open	100
200	125
47	150
33	175
22	200

Chapter 15

Section Summary ANALYSIS/DESIGN

Summary of Equations for the Analysis/Design of the Boost Converter

Analysis: Assume $\eta = 80$ to $90\ \%$.

$$M_{VDC(min)} = \frac{V_O}{V_{I(max)}} \qquad M_{VDC(max)} = \frac{V_O}{V_{I(min)}}$$

$$D_{min} = 1 - \frac{\eta}{M_{VDC(min)}} \qquad D_{max} = 1 - \frac{\eta}{M_{VDC(max)}}$$

Losses in the MOSFET: $\quad P_{con} = D_{max}\left[\dfrac{I_O}{(1-D_{max})}\right]^2 r_{DS} \qquad P_{sw} = f_s C_o V_O^2$

$$P_{FET} = P_{con} + P_{sw}/2$$

Losses in the diode: $\quad P_{RF} = I_{O(max)}^2 (1 - D_{min}) R_F \qquad P_{VF} = V_F I_O$

$$P_D = P_{RF} + P_{VF}$$

Loss in the inductor: $\quad P_L = r_L \dfrac{I_{O(max)}^2}{(1-D_{max})^2}$

Loss in the capacitor: $\quad P_C = r_C (1 - D_{min}) I_O^2$

The total loss: $\quad P_{Loss} = P_{con} + P_{sw} + P_D + P_L + P_C$

Efficiency: $\quad \eta = \dfrac{P_O}{P_I} = \dfrac{P_O}{P_O + P_{Loss}}$

Design: Assume $\eta = 80$ to $90\ \%$ and determine $M_{VDC(min)}$, $M_{VDC(max)}$, D_{min}, and D_{max}.

$$R_{L(max)} = \frac{V_O}{I_{O(min)}} \qquad L_{min} = \frac{2 R_{L(max)}}{27 f_s}$$

$$V_r \leq 0.01 V_O = V_{rC} + V_{C(p-p)}$$

Let $V_{rC} \cong 0.8 V_r$ and $V_{C(p-p)} \cong 0.2 V_r$

$$r_{C(max)} = V_{rC} \frac{1 - D_{max}}{I_{O(max)}} \qquad C_{min} = \frac{D_{max} V_O}{f_s R_{L(min)} V_{C(p-p)}}$$

$$V_{S(max)} = V_{D(max)} = V_O \qquad I_{S(max)} = I_{D(max)} = I_{I(max)} = \frac{I_{O(max)}}{1 - D_{max}}$$

677

Chapter 15

Practice Problem 15-5   ANALYSIS

Determine the efficiency and the output ripple voltage of a boost converter with the following specifications:

MOSFET: $r_{DS} = 0.2\ \Omega$, $C_o = 100$ pF   Diode: $V_F = 0.3$ V, $R_F = 0.2\ \Omega$

$V_I = 6$ V to 10 V, $V_O = 16$ V, $I_{O(min)} = 25$ mA, $I_{O(max)} = 1.25$ A, and $f_s = 150$ kHz

Inductor: $L = 680\ \mu$H, $r_L = 0.1\ \Omega$ Capacitor: $C = 100\ \mu$F, $r_C = 0.2\ \Omega$

Answer: $v_{r(p\text{-}p)} = 7.85$ mV, $\eta = 82.4\%$

Practice Problem 15-6   DESIGN

Design a boost converter with the following specifications:

$V_I = 5$ V to 10 V, $V_O = 15$ V, $I_{O(min)} = 30$ mA, $I_{O(max)} = 1.5$ A, and $f_s = 150$ kHz

Answers: $L = 330\ \mu$H, $C = 100\ \mu$F, $V_{S(max)} = V_{D(max)} = 15$ V, $I_{S(max)} = I_{D(max)} \cong 5$ A

15.3.3 Buck-Boost (Inverting Step-Down/Up) DC-DC Converter

The circuit of a buck-boost DC-DC converter is shown in Figure 15-21, which can be utilized as an inverting buck (step-down) or inverting boost (step-up) converter.

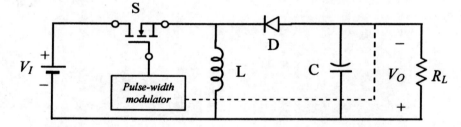

Figure 15-21: Buck-boost DC-DC converter

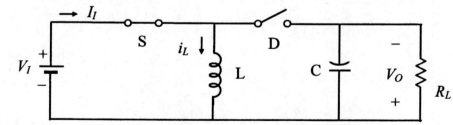

Figure 15-22: Switch ON $0 < t \leq DT$

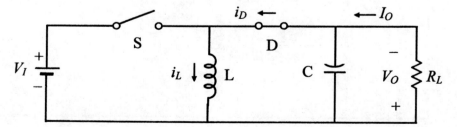

Figure 15-23: Switch OFF $DT < t \leq T$

Chapter 15

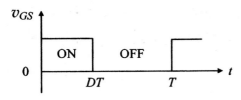

(a) MOSFET gate voltage v_{GS}

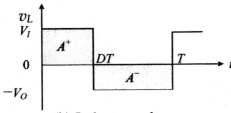

(b) Inductor voltage v_L

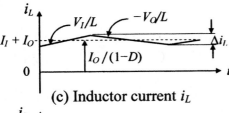

(c) Inductor current i_L

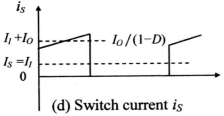

(d) Switch current i_S

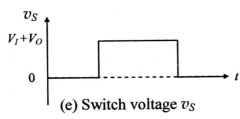

(e) Switch voltage v_S

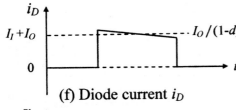

(f) Diode current i_D

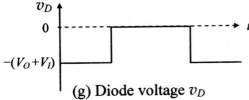

(g) Diode voltage v_D

The transistor switch is turned on and off at the rate of the pulse-width modulator switching frequency f_s with an on duty-cycle D, where

$$f_s = \frac{1}{T}$$

$$D = \frac{t_{on}}{T}$$

The equivalent circuits of the buck-boost converter of Figure 15-21 are shown with the transistor switch on in Figure 15-22, and with the switch off in Figure 15-23. The corresponding voltage and current waveforms are depicted in Figures 15-24 (a) through (g).

Assuming that the switch S is turned on at $t = 0$, the diode D is reverse biased and is turned off, as shown in Figure 15-22. The voltage across the inductor is $v_L = V_I$ and the inductor current increases with a slope of V_I/L. When the switch is turned off at $t = DT$, the current through the inductor continues flowing through the load and the diode, which has turned on at the instant the switch is turned off. However, the inductor current decreases linearly with a slope of $-V_O/L$. During this time interval, the energy is transferred from the inductor L to the capacitor C and the load resistor R_L. The average current I_L through the inductor is the sum of I_I and I_O as shown in Figure 15-24(c).

$$I_L = I_I + I_O$$

Substituting for I_I in terms of I_O in the above equation results in the following equation for I_L:

$$I_L = I_O \frac{D}{1-D} + I_O = I_O \left(\frac{D}{1-D} + 1 \right)$$

$$I_L = \frac{I_O}{1-D}$$

Figure 15-24:
Buck-boost converter
voltage and current waveforms

Chapter 15

DC voltage transfer function M_{VDC} of the boost converter:

Applying the *volt-second balance* principle to Figure 15-24(b), the positive area equals the negative area

$$A^+ = A^- \tag{15-94}$$

$$V_I DT = V_O(T - DT) = V_O(1 - D)T \tag{15-95}$$

After some simple algebraic manipulation, the following result is obtained

$$V_I D = V_O (1 - D) \tag{15-96}$$

$$M_{VDC} = \frac{V_O}{V_I} = \frac{D}{1-D} \tag{15-97}$$

where M_{VDC} is the DC voltage transfer function, and d is the duty cycle.

For a **lossless** converter the output power equals the input power

$$P_I = P_O \tag{15-98}$$

$$V_I I_I = V_O I_O \tag{15-99}$$

$$\frac{V_O}{V_I} = \frac{I_I}{I_O} = \frac{D}{1-D} \tag{15-100}$$

For a **lossy** converter the output power does not equal the input power, and as a result, the efficiency is less than 100%

$$\eta = \frac{P_O}{P_I} = \frac{V_O I_O}{V_I I_I} = \frac{V_O}{V_I} \frac{I_O}{I_I} = M_{VDC}\left(\frac{1-D}{D}\right) \tag{15-101}$$

The voltage transfer function of the lossy converter is determined by solving the above equation for M_{VDC}

$$M_{VDC} = \eta\left(\frac{D}{1-D}\right) \tag{15-102}$$

With some simple algebraic manipulation of the above equation, the duty-cycle D can be determined by

$$D = \frac{M_{VDC}}{M_{VDC} + \eta} \tag{15-103}$$

The input current may also be determined by solving Equation 15-101 for I_I as follows:

$$I_I = I_O \frac{M_{VDC}}{\eta} = I_O\left(\frac{D}{1-D}\right) \tag{15-104}$$

Analysis of the buck-boost converter operation

a) Analysis of the converter operation for the time interval with the switch on, $0 \leq t \leq DT$.

As depicted in Figure 15-22, the switch is on, the diode is off, and the inductor voltage v_L is equal to V_I

$$v_L = V_I = L\frac{di_L}{dt} \tag{15-105}$$

Chapter 15

$$\frac{v_L}{L} dt = di_L \qquad (15\text{-}106)$$

$$i_L = \frac{1}{L} \int_0^{DT} v_L \, dt + i_L(0) \qquad (15\text{-}107)$$

$$i_L = \frac{1}{L} \int_0^{DT} V_I \, dt + i_L(0) = \left[\frac{V_I}{L} t\right]_0^{DT} + i_L(0) \qquad (15\text{-}108)$$

$$i_L(DT) = \frac{V_I}{L}(DT) + i_L(0) \qquad (15\text{-}109)$$

Then the change in i_L from $t = 0$ to $t = DT$ is

$$\Delta i_L = i_L(DT) - i_L(0) = \frac{V_I}{L}(DT) + i_L(0) - i_L(0) = \frac{V_I}{L}(DT) \qquad (15\text{-}110)$$

$$\Delta i_L = \frac{V_I DT}{L} = \frac{V_I D}{f_s L} = \frac{V_O(1-D)}{f_s L} \qquad (15\text{-}111)$$

b) Analysis of the converter operation for the time interval with the switch off, $DT \leq t \leq T$

The switch is off, the diode is on, and the inductor voltage is $v_L = -V_O$. Hence,

$$v_L = -V_O = L \frac{di_L}{dt} \qquad (15\text{-}112)$$

$$\frac{v_L}{L} dt = di_L \qquad (15\text{-}113)$$

$$i_L = \frac{1}{L} \int_{DT}^{T} v_L \, dt + i_L(DT) \qquad (15\text{-}114)$$

$$i_L = \frac{1}{L} \int_{DT}^{T} (-V_O) dt + i_L(DT) = \left[\frac{-V_O}{L} t\right]_{DT}^{T} + i_L(DT) \qquad (15\text{-}115)$$

$$i_L(T) = \frac{-V_O}{L}(T - DT) + i_L(DT) \qquad (15\text{-}116)$$

Then the change in i_L from $t = dT$ to $t = T$ is the following:

$$\Delta i_L = i_L(DT) - i_L(T) = i_L(DT) - \left(\frac{-V_O}{L}T(1-D) + i_L(DT)\right) = \frac{V_O}{f_s L}(1-D) \quad (15\text{-}117)$$

Therefore, for both time intervals with the switch on or off, the peak-to-peak ripple current Δi_L of the inductor L is the same and its maximum value occurs at D_{min}.

$$\Delta i_{L(max)} = \frac{V_O}{f_s L}(1 - D_{min}) \qquad (15\text{-}118)$$

The output ripple voltage V_r:

The maximum peak-to-peak ripple voltage at the output of the buck-boost converter is $V_r = V_{rC} + V_{C(p-p)}$. The peak-to-peak ripple voltage across the capacitor ESR is

$$V_{rC} = r_C I_{D(max)} = \frac{r_C I_{O(max)}}{1 - D_{max}} \qquad (15\text{-}119)$$

Chapter 15

The peak-to-peak ripple voltage across the capacitor is

$$V_{C(p-p)} = \frac{\Delta Q_{max}}{C_{min}} = \frac{I_{O(max)} D_{max} T}{C_{min}} = \frac{D_{max} V_O}{f_s R_{L(min)} C_{min}} \quad (15\text{-}120)$$

where r_C is the *equivalent series resistance* (*ESR*) of the filter capacitor C.

Losses and efficiency:

Each individual component of the converter, inductor, capacitor, diode, and MOSFET will have its own losses. Following a correct design procedure and selecting the appropriate components can minimize these losses and contribute to the overall efficiency of the converter. Losses in the MOSFET switch are a combination of conduction and switching losses. The MOSFET conduction loss P_{con} is proportional to the on-resistance of the MOSFET r_{DS} and the switch current I_S.

As illustrated in Figure 15-19(d), $I_S = I_O/(1-D)$ for $0 < t \leq DT$, and $I_S = 0$ for $DT < t \leq T$. Hence, the $I_{S(rms)}^2$ value of the switch current and the corresponding conduction loss is determined as follows:

$$I_{S(rms)}^2 = \frac{1}{T}\int_0^T I_S^2\, dt = \left(\frac{I_O}{1-d}\right)^2 \frac{1}{T}\int_0^{DT} dt = \left(\frac{I_O}{1-D}\right)^2 D \quad (15\text{-}121)$$

$$P_{con} = I_{S(rms)}^2 r_{DS} = D\left(\frac{I_O}{1-D}\right)^2 r_{DS} \quad (15\text{-}122)$$

The MOSFET switching loss P_{sw} is proportional to the switching frequency f_s and the MOSFET output capacitance C_o, and is determined by

$$P_{sw} = f_s C_o V_S^2 = f_s C_o V_O^2 \quad (15\text{-}123)$$

$$P_{FET} = P_{con} + P_{sw}/2 \quad (15\text{-}124)$$

Losses in the diode are due to the forward bias voltage V_F and the forward resistance R_F of the non-ideal diode. The loss due to the forward resistance (P_{RF}) is proportional to R_F of the diode and the rms value of the diode current. Note that as illustrated in Figure 15-19(f), $I_D = 0$ for $0 < t \leq DT$, and $I_D = I_I + I_O = I_O/(1-D)$ for $DT < t \leq T$. Hence, the $I_{D(rms)}^2$ value of the diode current and the corresponding loss in the R_F is

$$I_{D(rms)}^2 = \frac{1}{T}\int_0^t i_D^2\, dt = \frac{1}{T}\int_{DT}^T \left(\frac{I_O}{1-D}\right)^2 dt = \left(\frac{I_O}{1-D}\right)^2 \frac{1}{T}(T-DT) = \frac{I_O^2}{1-D} \quad (15\text{-}125)$$

$$P_{RF} = I_{D(rms)}^2 R_F = \frac{I_O^2 R_F}{1-D} \quad (15\text{-}126)$$

The loss due to forward bias voltage P_{VF} is proportional to the forward voltage drop V_F of the diode and the average diode current I_D, and is determined by Equation 15-127. Furthermore, the average value of the diode current equals I_O, which is determined as follows:

$$I_D = \frac{1}{T}\int_0^t i_D\, dt = \frac{1}{T}\int_0^T \left(\frac{I_O}{1-D}\right) dt = \frac{1}{T}\int_{DT}^T \left(\frac{I_O}{1-D}\right) dt = \left(\frac{I_O}{1-D}\right)\frac{1}{T}(T-DT) = I_O \quad (15\text{-}127)$$

$$(15\text{-}128)$$

$$P_{VF} = V_F I_D = V_F I_O$$

$$P_D = P_{RF} + P_{VF} \tag{15-129}$$

The loss in the inductor is due to the *ESR* of the inductor r_L, and is determined as follows:

$$P_L = r_L I_{L(rms)}^2 = r_L \left(\frac{I_O}{1-D}\right)^2 \tag{15-130}$$

The loss in the filter capacitor is due to the *ESR* of the capacitor r_C and the AC component of the diode current $I_{D(rms)}$, and is determined as

$$P_C = r_C I_{C(rms)}^2 = r_C \frac{I_O^2}{1-D} \tag{15-131}$$

The total loss in the converter is the sum of all the above losses

$$P_{Loss} = P_{con} + P_{sw} + P_D + P_L + P_C \tag{15-132}$$

The efficiency of the converter is the ratio of the output power to the input power

$$\eta = \frac{P_O}{P_I} = \frac{P_O}{P_O + P_{Loss}} \tag{15-133}$$

Example 15-7   ANALYSIS

Let us determine the efficiency and the output ripple voltage of a boost converter with the following specifications: $V_I = 8\ V \pm 25\%$, $V_O = -12\ V$, $R_L = 10\ \Omega$, and $f_s = 100\ kHz$
MOSFET: $r_{DS} = 0.1\ \Omega$, $C_o = 100\ pF$ Diode: $V_F = 0.3\ V$, $R_F = 0.1\ \Omega$
Inductor: $L = 470\ \mu H$, $r_L = 0.1\ \Omega$ Capacitor: $C = 47\ \mu F$, $r_C = 0.1\ \Omega$

Solution: We will initially assume $\eta = 0.85$.

$$M_{VDC(min)} = \frac{V_O}{V_{I(max)}} = \frac{12\ V}{10\ V} = 1.2 \qquad M_{VDC(max)} = \frac{V_O}{V_{I(min)}} = \frac{12\ V}{6\ V} = 2$$

$$D_{min} = \frac{M_{VDC(min)}}{M_{VDC(min)} + \eta} = \frac{1.2}{1.2 + 0.85} = 0.585 \qquad D_{max} = \frac{M_{VDC(max)}}{M_{VDC(max)} + \eta} = \frac{2}{2 + 0.85} = 0.7$$

$$V_{rC} = \frac{r_C I_{O(max)}}{1 - D_{max}} = \frac{0.1\ \Omega \times 1.2\ A}{1 - 0.7} = 0.4\ V$$

$$V_{C(p-p)} = \frac{V_O D_{min}}{f_s R_{L(min)} C} = \frac{12\ V \times 0.7}{100\ kHz \times 10\ \Omega \times 47\ \mu F} = 0.1787\ V$$

$$V_r = V_{rC} + V_{C(p-p)} = 0.4 + 0.1787 = 0.5787\ V$$

Losses in the MOSFET are determined with Equations 15-122 through 15-124

$$P_{con} = D_{max} \left(\frac{I_O}{1 - D_{max}}\right)^2 r_{DS} = 0.7 \times \left(\frac{1.2\ A}{1 - 0.7}\right)^2 \times 0.1\ \Omega = 1.12\ W$$

$$P_{sw} = f_s C_o V_O^2 = (100\ kHz)(100\ pF)(12\ V)^2 = 1.44\ mW$$

$$P_{FET} = P_{con} + P_{sw}/2 = 1.12\ W + 1.44\ mW/2 = 1.121\ W$$

Losses in the diode are determined by Equations 15-126 through 15-129

$$P_{RF} = \frac{I_O^2 R_F}{1-D_{max}} = \frac{1.2^2}{1-0.7}(0.1\,\Omega) = 0.48\text{ W}$$

$$P_{VF} = V_F I_O = (0.3\text{ V})(1.2\text{ A}) = 0.36\text{ W}$$

$$P_D = P_{RF} + P_{VF} = 0.48\text{ W} + 0.36\text{ W} = 0.84\text{ W}$$

The loss in the inductor is determined by Equation 15-130

$$P_L = r_L \frac{I_{O(max)}^2}{(1-D_{max})^2} = 0.1\,\Omega \frac{1.2^2}{(1-0.7)^2} = 1.6\text{ W}$$

The loss in the capacitor is determined by Equation 15-131

$$P_C = r_C \frac{I_{O(max)}^2}{(1-D_{max})} = 0.1\,\Omega \frac{1.2^2}{(1-0.7)} = 0.48\text{ W}$$

The total loss in the converter is the sum of all the above losses:

$$P_{Loss} = P_{con} + P_{sw} + P_D + P_L + P_C$$
$$= 1.12\text{ W} + 0.00144 + 0.84\text{ W} + 1.6\text{ W} + 0.48\text{ W} = 4.041\text{ W}$$

$$\eta = \frac{P_O}{P_I} = \frac{P_O}{P_O + P_{Loss}} = \frac{14.4\text{ W}}{14.4\text{ W} + 4.041\text{ W}} = 78.09\%$$

Since the difference in efficiency is more than ±5% of the assumed efficiency of 85%, we will assume another value closer to the above found efficiency, such as 75%, and repeat the calculations. Repeating the above calculations with $\eta = 0.75$ yields an efficiency of 75.23%, which is nearly the assumed efficiency.

Example 15-8 DESIGN

Design an inverting step-down buck-boost converter with the following specifications:

$V_I = 10 \pm 2.5$ V, $V_O = -5$ V, $I_{O(min)} = 50$ mA, $I_{O(max)} = 1$ A, and $f_s = 150$ kHz

Solution: We will initially assume $\eta = 0.8$.

$$M_{VDC(min)} = \frac{V_O}{V_{I(max)}} = \frac{5\text{ V}}{12.5\text{ V}} = 0.4 \qquad M_{VDC(max)} = \frac{V_O}{V_{I(min)}} = \frac{5\text{ V}}{7.5\text{ V}} = 0.666$$

$$D_{min} = \frac{M_{VDC(min)}}{M_{VDC(min)} + \eta} = \frac{0.4}{0.4+0.8} = 0.333 \qquad D_{max} = \frac{M_{VDC(max)}}{M_{VDC(max)} + \eta} = \frac{0.666}{0.666+0.8} = 0.454$$

$$R_{L(max)} = \frac{V_O}{I_{O(min)}} = \frac{5\text{ V}}{0.050\text{ A}} = 100\,\Omega \qquad R_{L(min)} = \frac{V_O}{I_{O(max)}} = \frac{5\text{ V}}{1\text{ A}} = 5\,\Omega$$

$$L_{min} = \frac{R_{L(max)}(1-D_{min})^2}{2f_s} = \frac{100\,\Omega(1-0.333)^2}{2\times 150\text{ kHz}} = 148\,\mu\text{H}$$

Use PE-54036, 168 µH, 0.18 Ω

$$\Delta i_{L(max)} = \frac{V_O(1-D_{min})}{f_s L} = \frac{5\text{ V}(1-0.333)}{168\,\mu\text{H} \times 150\text{ kHz}} = 0.132\text{ A}$$

Chapter 15

$$I_{I(max)} = \frac{M_{VDC(max)} I_{O(max)}}{\eta} = \frac{0.666 \times 1\,\text{A}}{0.8} = 0.8325\,\text{A}$$

$$V_{S(max)} = V_{D(max)} = V_{I(max)} = 12.5\,\text{V}$$

$$I_{S(max)} = I_{D(max)} = I_{I(max)} + I_{O(max)} + 0.5\Delta i_{L(max)} = 0.8325 + 1 + 0.132/2 = 1.8985\,\text{A}$$

$$V_r \leq 0.01 V_O = 50\,\text{mV} = V_{rc} + V_{C(p\text{-}p)}$$

Let $V_{rc} = 40\,\text{mV}$ and $V_{C(p\text{-}p)} = 10\,\text{mV}$

$$r_{C(max)} = \frac{V_{rc}}{I_{D(max)}} = \frac{40\,\text{mV}}{1.8985\,\text{A}} = 0.021\,\Omega$$

$$C_{min} = \frac{D_{max} V_O}{f_s R_{L(min)} V_{C(p-p)}} = \frac{0.454 \times 5}{150\,\text{kHz} \times 100\,\Omega \times 0.01\,\text{V}} = 60.5\,\mu\text{F}$$

Pick $C = 100\,\mu\text{F}/10\text{V}/20\,\text{m}\Omega$.

Implementation of the design:

To implement our design of the buck-boost converter, we will utilize National Semiconductor's LM2595J-ADJ regulator, which can be utilized as buck, inverting buck, and inverting boost (buck-boost). The inductor will be Pulse Engineering's PE-54036 with nominal ratings of 168 µH, 1.81 A, and 0.18 Ω. For the diode we will select Motorola's 1N5820 Schottky barrier diode, and for the capacitor will select Intertechnology's 550D107X9010R2 low *ESR*, 100 µF tantalum capacitor ($C = 100\,\mu\text{F}/10\text{V}/20\,\text{m}\Omega$). The complete circuit implementation is shown in Figure 15-25. Note that the positive terminal of the output capacitor is at the ground.

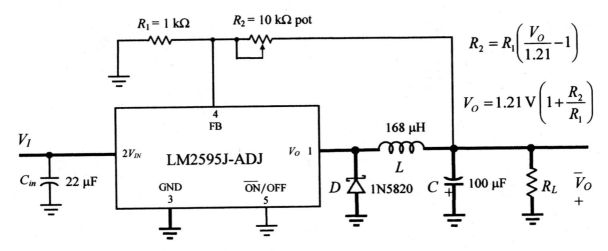

Figure 15-25: Implementation of the inverting buck DC-DC converter

Section Summary — Buck-Boost — ANALYSIS/DESIGN

Summary of Equations for the Analysis/Design of the Buck-Boost Converter

Analysis: Assume $\eta = 80$ to 90%.

$$M_{VDC(min)} = \frac{V_O}{V_{I(max)}} \qquad M_{VDC(max)} = \frac{V_O}{V_{I(min)}}$$

$$D_{min} = \frac{M_{VDC(min)}}{M_{VDC(min)} + \eta} \qquad D_{max} = \frac{M_{VDC(max)}}{M_{VDC(max)} + \eta}$$

Losses in the MOSFET:

$$P_{con} = D_{max}\left(\frac{I_O}{1 - D_{max}}\right)^2 r_{DS} \qquad P_{sw} = f_s C_o V_O^2 \qquad P_{FET} = P_{con} + P_{sw}$$

Losses in the diode:

$$P_{RF} = \frac{I_O^2 R_F}{1 - D_{max}} \qquad P_{VF} = V_F I_O \qquad P_D = P_{RF} + P_{VF}$$

Losses in the inductor and the capacitor:

$$P_L = r_L \frac{I_{O(max)}^2}{(1 - D_{max})^2} \qquad P_C = r_C \frac{I_{O(max)}^2}{(1 - D_{max})}$$

The total loss in the converter is the sum of all the above losses:

$$P_{Loss} = P_{FET} + P_D + P_L + P_C \qquad \eta = \frac{P_O}{P_I} = \frac{P_O}{P_O + P_{Loss}}$$

Design: Assume $\eta = 80$ to 90% and determine $M_{VDC(min)}$, $M_{VDC(max)}$, D_{min}, and D_{max}.

$$R_{L(max)} = \frac{V_O}{I_{O(min)}} = \frac{5\,\text{V}}{0.050\,\text{A}} = 100\,\Omega \qquad R_{L(min)} = \frac{V_O}{I_{O(max)}} = \frac{5\,\text{V}}{1\,\text{A}} = 5\,\Omega$$

$$L_{min} = \frac{R_{L(max)}(1 - D_{min})^2}{2 f_s} \qquad \Delta i_{L(max)} = \frac{V_O(1 - D_{min})}{f_s L}$$

$$I_{I(max)} = M_{VDC(max)} I_{O(max)}$$

$$I_{S(max)} = I_{D(max)} = I_{I(max)} + I_{O(max)} + 0.5 \Delta i_{L(max)} \qquad V_{S(max)} = V_{D(max)} = V_{I(max)}$$

$$V_r \le 0.01 V_O = V_{rc} + V_{C(p\text{-}p)} \qquad V_{rc} \cong 0.8 V_r$$

$$r_{C(max)} = \frac{V_{rc}}{I_{D(max)}} = \frac{V_{rC}(1 - D_{max})}{I_{O(max)}} \qquad C_{min} = \frac{D_{max} V_O}{f_s R_{L(min)} V_{C(p\text{-}p)}}$$

Chapter 15

Practice Problem 15-7 ANALYSIS

Determine the efficiency and the output ripple voltage of a buck-boost converter with the following specifications:

$V_I = 6$ V to 10 V, $V_O = -12$ V, $I_{O(min)} = 25$ mA, $I_{O(max)} = 1.25$ A, and $f_s = 100$ kHz

MOSFET: $r_{DS} = 0.15$ Ω, $C_o = 100$ pF Diode: $V_F = 0.3$ V, $R_F = 0.15$ Ω

Inductor: $L = 680$ μH, $r_L = 0.1$ Ω Capacitor: $C = 100$ μF, $r_C = 0.1$ Ω

Answer: $V_r = 7$ mV(p-p), $\eta = 72.5\%$

Practice Problem 15-8 ANALYSIS

Determine the efficiency and the output ripple voltage of a buck-boost converter with the following specifications:

$V_I = 10$ V to 15 V, $V_O = -8$ V, $I_{O(min)} = 50$ mA, $I_{O(max)} = 2$ A, and $f_s = 200$ kHz

MOSFET: $r_{DS} = 0.15$ Ω, $C_o = 100$ pF Diode: $V_F = 0.3$ V, $R_F = 0.15$ Ω

Inductor: $L = 470$ μH, $r_L = 0.1$ Ω Capacitor: $C = 100$ μF, $r_C = 0.1$ Ω

Answer: $V_r = 5.1$ V(p-p), $\eta = 74.76\%$

Practice Problem 15-9 DESIGN

Design a buck-boost converter with the following specifications:

$V_I = 4$ V to 8 V, $V_O = -10$ V, $I_{O(min)} = 50$ mA, $I_{O(max)} = 1.5$ A, and $f_s = 150$ kHz

Answers: $L_{(min)} = 101$ μH, $C_{min} = 199$ μF, $V_{S(max)} = V_{D(max)} = 8$ V, $I_{S(max)} = I_{D(max)} \cong 6.3$ A

Practice Problem 15-10 DESIGN

Design a buck-boost converter with the following specifications:

$V_I = 10$ V to 15 V, $V_O = -7.5$ V, $I_{O(min)} = 50$ mA, $I_{O(max)} = 1.5$ A, and $f_s = 150$ kHz

Answers: $L_{(min)} = 189$ μH, $C = 322.6$ μF, $I_{S(max)} = I_{D(max)} \cong 1.5$ A

Chapter 15

15.4 SUMMARY

- The primary advantage of using an op-amp as a voltage regulator is that it provides an almost ideal buffer between the rectified DC voltage and the load. In addition, the op-amp can provide gain and thus adjustable output voltage.

- The addition of a power transistor to the basic op-amp regulator can boost the output current. Furthermore, addition of a general-purpose transistor with a current-sense resistor can provide protection against overload and short-circuit.

- The value of the current-sense resistor R_S is chosen such that with $I_I = I_O \leq I_{O(max)}$, the voltage drop across the $R_S = V_{BE(Q2)}$ is less than 0.7 V.

- However, when $I_I = I_O > I_{O(max)}$, which could be due to overload or short-circuit, the voltage drop across the $R_S = V_{BE(Q2)}$ reaches or exceeds the nominal 0.7 V, turning the Q_2 on and Q_1 off, cutting off the output current and dropping the output voltage to zero.

- In the basic traditional power supplies, the step-down transformer is a bulky and expensive part of the power supply. The bulkiness of the transformer is due to low line frequency (60 Hz). In addition, the low line frequency requires large filter capacitors to reduce the output ripple, which adds up to the overall cost and physical dimensions of the power supply. Furthermore, the power transistor is operating in the active region and is continuously on, which dissipates considerable power.

- These three factors combined make the basic traditional power supplies bulky, more expensive, and less efficient. Switching power supplies provide substantial improvements in all three areas by utilizing a technique called *pulse-width modulation*.

- The four principal components in a switching power supply are (1) a transistor switch S, which is usually a power MOSFET; (2) a diode D; (3) an inductor L; and (4) a capacitor C. A pulse-width modulator driver controls the switch S, which is turned on and off at the rate of the switching frequency $f_s = 1/T$.

- Switching power supplies are also referred to as DC-DC converters, as they take an unregulated DC and convert it fairly efficiently up or down to a desired level of fixed and regulated DC output.

Chapter 15

15.5 DEVICE SPECIFICATIONS AND DATA SHEETS

Data sheets of the following devices discussed in this chapter are included for your reference and convenience.

- LM2595 Simple Switcher
  Courtesy of National Semiconductor

- 1N5817 through 1N5819, Schottky Barrier Diodes
  Copyright of Semiconductor Component Industries, LLC. Used by permission.

- List of Inductors designed for National's Simple Switcher
  Courtesy of Pulse Engineering

- Type 550D Solid Tantalum Capacitors
  Courtesy of Vishay Intertechnology

LM2595
SIMPLE SWITCHER® Power Converter 150 kHz 1A Step-Down Voltage Regulator

National Semiconductor

May 1999

General Description

The LM2595 series of regulators are monolithic integrated circuits that provide all the active functions for a step-down (buck) switching regulator, capable of driving a 1A load with excellent line and load regulation. These devices are available in fixed output voltages of 3.3V, 5V, 12V, and an adjustable output version.

Requiring a minimum number of external components, these regulators are simple to use and include internal frequency compensation¹, and a fixed-frequency oscillator.

The LM2595 series operates at a switching frequency of 150 kHz thus allowing smaller sized filter components than what would be needed with lower frequency switching regulators. Available in a standard 5-lead TO-220 package with several different lead bend options, and a 5-lead TO-263 surface mount package. Typically, for output voltages less than 12V, and ambient temperatures less than 50°C, no heat sink is required.

A standard series of inductors are available from several different manufacturers optimized for use with the LM2595 series. This feature greatly simplifies the design of switch-mode power supplies.

Other features include a guaranteed ±4% tolerance on output voltage under specified input voltage and output load conditions, and ±15% on the oscillator frequency. External shutdown is included, featuring typically 85 μA stand-by current. Self protection features include a two stage frequency reducing current limit for the output switch and an over temperature shutdown for complete protection under fault conditions.

Features

- 3.3V, 5V, 12V, and adjustable output versions
- Adjustable version output voltage range, 1.2V to 37V ±4% max over line and load conditions
- Available in TO-220 and TO-263 (surface mount) packages
- Guaranteed 1A output load current
- Input voltage range up to 40V
- Requires only 4 external components
- Excellent line and load regulation specifications
- 150 kHz fixed frequency internal oscillator
- TTL shutdown capability
- Low power standby mode, I_Q typically 85 μA
- High efficiency
- Uses readily available standard inductors
- Thermal shutdown and current limit protection

Applications

- Simple high-efficiency step-down (buck) regulator
- Efficient pre-regulator for linear regulators
- On-card switching regulators
- Positive to negative converter

Note: ¹ Patent Number 5,382,918.

Typical Application (Fixed Output Voltage Versions)

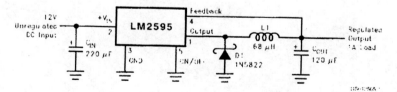

"SIMPLE SWITCHER" and "Switchers Made Simple" are registered trademarks of National Semiconductor Corporation.

© 1999 National Semiconductor Corporation   DS012565   www.national.com

Chapter 15

Connection Diagrams and Ordering Information

Bent and Staggered Leads, Through Hole Package
5–Lead TO-220 (T)

Side View — Pins 1, 3 & 5; Pins 2 & 4

Top View:
- 5 – ON/OFF
- 4 – Feed Back
- 3 – Ground
- 2 – V_{IN}
- 1 – Output

GND

Order Number LM2595T-3.3, LM2595T-5.0, LM2595T-12 or LM2595T-ADJ
See NS Package Number T05D

Surface Mount Package
5-Lead TO-263 (S)

Side View

Top View — Metal Tab GND:
- 5 – ON/OFF
- 4 – Feed Back
- 3 – Ground
- 2 – V_{IN}
- 1 – Output

Order Number LM2595S-3.3, LM2595S-5.0, LM2595S-12 or LM2595S-ADJ
See NS Package Number TS5B

16-Lead Ceramic Dual-in-Line Package (J)

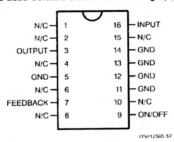

Pin		Pin	
N/C	1	16	INPUT
N/C	2	15	N/C
OUTPUT	3	14	GND
N/C	4	13	GND
GND	5	12	GND
N/C	6	11	GND
FEEDBACK	7	10	N/C
N/C	8	9	ON/OFF

Order Number LM2595J-3.3-QML (5962-9687901QEA),
LM2595J-5.0-QML (5962-9650301QEA),
LM2595J-12-QML (5962-9650201QEA),
or LM2595J-ADJ-QML (5962-9650401QEA)
See NS Package Number J16A
For specifications and information about Military-Aerospace products, please see the Mil-Aero web page at
http://www.national.com/appinfo/milaero/index.html.

Absolute Maximum Ratings (Note 1)

If Military/Aerospace specified devices are required, please contact the National Semiconductor Sales Office/Distributors for availability and specifications.

Maximum Supply Voltage	45V
ON/OFF Pin Input Voltage	$-0.3 \leq V \leq +25V$
Feedback Pin Voltage	$-0.3 \leq V \leq +25V$
Output Voltage to Ground (Steady State)	$-1V$
Power Dissipation	Internally limited
Storage Temperature Range	$-65°C$ to $+150°C$
ESD Susceptibility	
Human Body Model (Note 2)	2 kV
Lead Temperature	
S Package	
Vapor Phase (60 sec.)	$+215°C$
Infrared (10 sec.)	$+245°C$
T Package (Soldering, 10 sec.)	$+260°C$
Maximum Junction Temperature	$+150°C$

Operating Conditions

Temperature Range	$-40°C \leq T_J \leq +125°C$
Supply Voltage	4.5V to 40V

LM2595-3.3 Electrical Characteristics

Specifications with standard type face are for $T_J = 25°C$, and those with **boldface type** apply over **full Operating Temperature Range**.

Symbol	Parameter	Conditions	LM2595-3.3 Typ (Note 3)	LM2595-3.3 Limit (Note 4)	Units (Limits)
SYSTEM PARAMETERS (Note 5) Test Circuit *Figure 1*					
V_{OUT}	Output Voltage	$4.75V \leq V_{IN} \leq 40V$, $0.1A \leq I_{LOAD} \leq 1A$	3.3		V
				3.168/**3.135**	V(min)
				3.432/**3.465**	V(max)
η	Efficiency	$V_{IN} = 12V$, $I_{LOAD} = 1A$	78		%

LM2595-5.0 Electrical Characteristics

Specifications with standard type face are for $T_J = 25°C$, and those with **boldface type** apply over **full Operating Temperature Range**.

Symbol	Parameter	Conditions	LM2595-5.0 Typ (Note 3)	LM2595-5.0 Limit (Note 4)	Units (Limits)
SYSTEM PARAMETERS (Note 5) Test Circuit *Figure 1*					
V_{OUT}	Output Voltage	$7V \leq V_{IN} \leq 40V$, $0.1A \leq I_{LOAD} \leq 1A$	5.0		V
				4.800/**4.750**	V(min)
				5.200/**5.250**	V(max)
η	Efficiency	$V_{IN} = 12V$, $I_{LOAD} = 1A$	82		%

LM2595-12 Electrical Characteristics

Specifications with standard type face are for $T_J = 25°C$, and those with **boldface type** apply over **full Operating Temperature Range**.

Symbol	Parameter	Conditions	LM2595-12 Typ (Note 3)	LM2595-12 Limit (Note 4)	Units (Limits)
SYSTEM PARAMETERS (Note 5) Test Circuit *Figure 1*					
V_{OUT}	Output Voltage	$15V \leq V_{IN} \leq 40V$, $0.1A \leq I_{LOAD} \leq 1A$	12.0		V
				11.52/**11.40**	V(min)
				12.48/**12.60**	V(max)
η	Efficiency	$V_{IN} = 25V$, $I_{LOAD} = 1A$	90		%

Chapter 15

LM2595-ADJ
Electrical Characteristics

Specifications with standard type face are for $T_J = 25°C$, and those with **boldface type** apply over **full Operating Temperature Range**.

Symbol	Parameter	Conditions	LM2595-ADJ Typ (Note 3)	LM2595-ADJ Limit (Note 4)	Units (Limits)
SYSTEM PARAMETERS (Note 5) Test Circuit *Figure 1*					
V_{FB}	Feedback Voltage	$4.5V \leq V_{IN} \leq 40V$, $0.1A \leq I_{LOAD} \leq 1A$ V_{OUT} programmed for 3V. Circuit of *Figure 1*	1.230	1.193/**1.180** 1.267/**1.280**	V V(min) V(max)
η	Efficiency	$V_{IN} = 12V$, $V_{OUT} = 3V$, $I_{LOAD} = 1A$	78		%

All Output Voltage Versions
Electrical Characteristics

Specifications with standard type face are for $T_J = 25°C$, and those with **boldface type** apply over **full Operating Temperature Range**. Unless otherwise specified, $V_{IN} = 12V$ for the 3.3V, 5V, and Adjustable version and $V_{IN} = 24V$ for the 12V version. $I_{LOAD} = 200$ mA.

Symbol	Parameter	Conditions	LM2595-XX Typ (Note 3)	LM2595-XX Limit (Note 4)	Units (Limits)
DEVICE PARAMETERS					
I_b	Feedback Bias Current	Adjustable Version Only, $V_{FB} = 1.3V$	10	50/**100**	nA nA (max)
f_O	Oscillator Frequency	(Note 6)	150	127/**110** 173/**173**	kHz kHz(min) kHz(max)
V_{SAT}	Saturation Voltage	$I_{OUT} = 1A$ (Notes 7, 8)	1	1.2/**1.3**	V V(max)
DC	Max Duty Cycle (ON)	(Note 8)	100		%
	Min Duty Cycle (OFF)	(Note 9)	0		
I_{CL}	Current Limit	Peak Current (Notes 7, 8)	1.5	1.2/**1.15** 2.4/**2.6**	A A(min) A(max)
I_L	Output Leakage Current	Output = 0V (Notes 7, 9) and (Note 10)		50	µA(max)
		Output = −1V	2	15	mA mA(max)
I_Q	Quiescent Current	(Note 9)	5	10	mA mA(max)
I_{STBY}	Standby Quiescent Current	ON/OFF pin = 5V (OFF) (Note 10)	85	200/**250**	µA µA(max)
θ_{JC}	Thermal Resistance	TO-220 or TO-263 Package, Junction to Case	2		°C/W
θ_{JA}		TO-220 Package, Junction to Ambient (Note 11)	50		°C/W
θ_{JA}		TO-263 Package, Junction to Ambient (Note 12)	50		°C/W
θ_{JA}		TO-263 Package, Junction to Ambient (Note 13)	30		°C/W
θ_{JA}		TO-263 Package, Junction to Ambient (Note 14)	20		°C/W
ON/OFF CONTROL Test Circuit *Figure 1*					
V_{IH} V_{IL}	$\overline{ON}$/OFF Pin Logic Input Threshold Voltage	Low (Regulator ON) High (Regulator OFF)	1.3	0.6 2.0	V V(max) V(min)

All Output Voltage Versions
Electrical Characteristics (Continued)

Specifications with standard type face are for $T_J = 25°C$, and those with **boldface type** apply over **full Operating Temperature Range**. Unless otherwise specified, $V_{IN} = 12V$ for the 3.3V, 5V, and Adjustable version and $V_{IN} = 24V$ for the 12V version. $I_{LOAD} = 200$ mA.

Symbol	Parameter	Conditions	LM2595-XX		Units
			Typ (Note 3)	Limit (Note 4)	(Limits)
ON/OFF CONTROL Test Circuit *Figure 1*					
I_H	ON/OFF Pin Input Current	$V_{LOGIC} = 2.5V$ (Regulator OFF)	5		µA
				15	µA(max)
I_I		$V_{LOGIC} = 0.5V$ (Regulator ON)	0.02		µA
				5	µA(max)

Note 1: Absolute Maximum Ratings indicate limits beyond which damage to the device may occur. Operating Ratings indicate conditions for which the device is intended to be functional, but do not guarantee specific performance limits. For guaranteed specifications and test conditions, see the Electrical Characteristics.

Note 2: The human body model is a 100 pF capacitor discharged through a 1.5k resistor into each pin.

Note 3: Typical numbers are at 25°C and represent the most likely norm.

Note 4: All limits guaranteed at room temperature (standard type face) and at temperature extremes (bold type face). All room temperature limits are 100% production tested. All limits at temperature extremes are guaranteed via correlation using standard Statistical Quality Control (SQC) methods. All limits are used to calculate Average Outgoing Quality Level (AOQL).

Note 5: External components such as the catch diode, inductor, input and output capacitors, and voltage programming resistors can affect switching regulator system performance. When the LM2595 is used as shown in the *Figure 1* test circuit, system performance will be as shown in system parameters section of Electrical Characteristics.

Note 6: The switching frequency is reduced when the second stage current limit is activated. The amount of reduction is determined by the severity of current overload.

Note 7: No diode, inductor or capacitor connected to output pin.

Note 8: Feedback pin removed from output and connected to 0V to force the output transistor switch ON.

Note 9: Feedback pin removed from output and connected to 12V for the 3.3V, 5V, and the ADJ. version, and 15V for the 12V version, to force the output transistor switch OFF.

Note 10: $V_{IN} = 40V$.

Note 11: Junction to ambient thermal resistance (no external heat sink) for the TO-220 package mounted vertically, with the leads soldered to a printed circuit board with (1 oz.) copper area of approximately 1 in².

Note 12: Junction to ambient thermal resistance with the TO-263 package tab soldered to a single printed circuit board with 0.5 in² of (1 oz.) copper area.

Note 13: Junction to ambient thermal resistance with the TO-263 package tab soldered to a single sided printed circuit board with 2.5 in² of (1 oz.) copper area.

Note 14: Junction to ambient thermal resistance with the TO-263 package tab soldered to a double sided printed circuit board with 3 in² of (1 oz.) copper area on the LM2595S side of the board, and approximately 16 in² of copper on the other side of the p-c board. See Application Information in this data sheet and the thermal model in *Switchers Made Simple* version 4.3 software.

Typical Performance Characteristics (Circuit of *Figure 1*)

Normalized Output Voltage

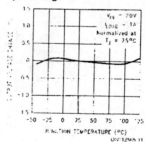

Line Regulation

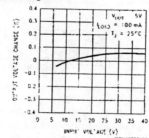

Efficiency

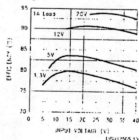

Chapter 15

MOTOROLA
SEMICONDUCTOR TECHNICAL DATA

Order this document by 1N5817/D

Axial Lead Rectifiers

... employing the Schottky Barrier principle in a large area metal–to–silicon power diode. State–of–the–art geometry features chrome barrier metal, epitaxial construction with oxide passivation and metal overlap contact. Ideally suited for use as rectifiers in low–voltage, high–frequency inverters, free wheeling diodes, and polarity protection diodes.

- Extremely Low v_F
- Low Stored Charge, Majority Carrier Conduction
- Low Power Loss/High Efficiency

Mechanical Characteristics
- Case: Epoxy, Molded
- Weight: 0.4 gram (approximately)
- Finish: All External Surfaces Corrosion Resistant and Terminal Leads are Readily Solderable
- Lead and Mounting Surface Temperature for Soldering Purposes: 220°C Max. for 10 Seconds, 1/16" from case
- Shipped in plastic bags, 1000 per bag.
- Available Tape and Reeled, 5000 per reel, by adding a "RL" suffix to the part number
- Polarity: Cathode Indicated by Polarity Band
- Marking: 1N5817, 1N5818, 1N5819

**1N5817
1N5818
1N5819**

1N5817 and 1N5819 are Motorola Preferred Devices

SCHOTTKY BARRIER RECTIFIERS
1 AMPERE
20, 30 and 40 VOLTS

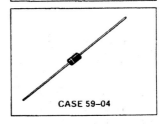

CASE 59–04

MAXIMUM RATINGS

Rating	Symbol	1N5817	1N5818	1N5819	Unit
Peak Repetitive Reverse Voltage Working Peak Reverse Voltage DC Blocking Voltage	V_{RRM} V_{RWM} V_R	20	30	40	V
Non-Repetitive Peak Reverse Voltage	V_{RSM}	24	36	48	V
RMS Reverse Voltage	$V_{R(RMS)}$	14	21	28	V
Average Rectified Forward Current (2) ($V_{R(equiv)} \leq 0.2\ V_R(dc)$, $T_L = 90°C$, $R_{\theta JA} = 80°C/W$, P.C. Board Mounting, see Note 2, $T_A = 55°C$)	I_O	\multicolumn{3}{c}{1.0}	A
Ambient Temperature (Rated $V_R(dc)$, $P_{F(AV)} = 0$, $R_{\theta JA} = 80°C/W$)	T_A	85	80	75	°C
Non-Repetitive Peak Surge Current (Surge applied at rated load conditions, half–wave, single phase 60 Hz, $T_L = 70°C$)	I_{FSM}	\multicolumn{3}{c}{25 (for one cycle)}	A
Operating and Storage Junction Temperature Range (Reverse Voltage applied)	T_J, T_{stg}	\multicolumn{3}{c}{–65 to +125}	°C
Peak Operating Junction Temperature (Forward Current applied)	$T_{J(pk)}$	\multicolumn{3}{c}{150}	°C

THERMAL CHARACTERISTICS (2)

Characteristic	Symbol	Max	Unit
Thermal Resistance, Junction to Ambient	$R_{\theta JA}$	80	°C/W

ELECTRICAL CHARACTERISTICS ($T_L = 25°C$ unless otherwise noted) (2)

Characteristic		Symbol	1N5817	1N5818	1N5819	Unit
Maximum Instantaneous Forward Voltage (1)	($i_F = 0.1$ A) ($i_F = 1.0$ A) ($i_F = 3.0$ A)	v_F	0.32 0.45 0.75	0.33 0.55 0.875	0.34 0.6 0.9	V
Maximum Instantaneous Reverse Current @ Rated dc Voltage (1)	($T_L = 25°C$) ($T_L = 100°C$)	I_R	1.0 10	1.0 10	1.0 10	mA

(1) Pulse Test: Pulse Width = 300 µs, Duty Cycle = 2.0%.
(2) Lead Temperature reference is cathode lead 1/32" from case.

Preferred devices are Motorola recommended choices for future use and best overall value.

Rev 3

Ⓜ **MOTOROLA**

© Motorola, Inc. 1996

1N5817 1N5818 1N5819

NOTE 1 — DETERMINING MAXIMUM RATINGS

Reverse power dissipation and the possibility of thermal runaway must be considered when operating this rectifier at reverse voltages above $0.1\ V_{RWM}$. Proper derating may be accomplished by use of equation (1).

$$T_{A(max)} = T_{J(max)} - R_{\theta JA}P_{F(AV)} - R_{\theta JA}P_{R(AV)} \quad (1)$$

where $T_{A(max)}$ = Maximum allowable ambient temperature
$T_{J(max)}$ = Maximum allowable junction temperature (125°C or the temperature at which thermal runaway occurs, whichever is lowest)
$P_{F(AV)}$ = Average forward power dissipation
$P_{R(AV)}$ = Average reverse power dissipation
$R_{\theta JA}$ = Junction-to-ambient thermal resistance

Figures 1, 2, and 3 permit easier use of equation (1) by taking reverse power dissipation and thermal runaway into consideration. The figures solve for a reference temperature as determined by equation (2).

$$T_R = T_{J(max)} - R_{\theta JA}P_{R(AV)} \quad (2)$$

Substituting equation (2) into equation (1) yields:

$$T_{A(max)} = T_R - R_{\theta JA}P_{F(AV)} \quad (3)$$

Inspection of equations (2) and (3) reveals that T_R is the ambient temperature at which thermal runaway occurs or where $T_J = 125°C$, when forward power is zero. The transition from one boundary condition to the other is evident on the curves of Figures 1, 2, and 3 as a difference in the rate of change of the slope in the vicinity of 115°C. The data of Figures 1, 2, and 3 is based upon dc conditions. For use in common rectifier circuits, Table 1 indicates suggested factors for an equivalent dc voltage to use for conservative design, that is:

$$V_{R(equiv)} = V_{in(PK)} \times F \quad (4)$$

The factor F is derived by considering the properties of the various rectifier circuits and the reverse characteristics of Schottky diodes.

EXAMPLE: Find $T_{A(max)}$ for 1N5818 operated in a 12-volt dc supply using a bridge circuit with capacitive filter such that $I_{DC} = 0.4\ A$ ($I_{F(AV)} = 0.5\ A$), $I_{(FM)}/I_{(AV)} = 10$, Input Voltage = 10 $V_{(rms)}$, $R_{\theta JA} = 80°C/W$.

Step 1. Find $V_{R(equiv)}$. Read F = 0.65 from Table 1.
∴ $V_{R(equiv)} = (1.41)(10)(0.65) = 9.2\ V$.
Step 2. Find T_R from Figure 2. Read $T_R = 109°C$
@ $V_R = 9.2\ V$ and $R_{\theta JA} = 80°C/W$.
Step 3. Find $P_{F(AV)}$ from Figure 4. **Read $P_{F(AV)} = 0.5\ W$

@ $\dfrac{I_{(FM)}}{I_{(AV)}} = 10$ and $I_{F(AV)} = 0.5\ A$.

Step 4. Find $T_{A(max)}$ from equation (3).
$T_{A(max)} = 109 - (80)(0.5) = 69°C$.

**Values given are for the 1N5818. Power is slightly lower for the 1N5817 because of its lower forward voltage, and higher for the 1N5819.

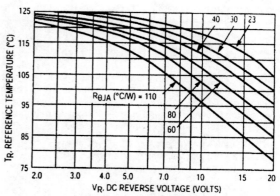

Figure 1. Maximum Reference Temperature 1N5817

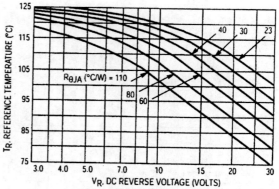

Figure 2. Maximum Reference Temperature 1N5818

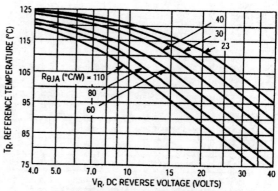

Figure 3. Maximum Reference Temperature 1N5819

Table 1. Values for Factor F

Circuit	Half Wave		Full Wave, Bridge		Full Wave, Center Tapped*†	
Load	Resistive	Capacitive*	Resistive	Capacitive	Resistive	Capacitive
Sine Wave	0.5	1.3	0.5	0.65	1.0	1.3
Square Wave	0.75	1.5	0.75	0.75	1.5	1.5

*Note that $V_{R(PK)} \approx 2.0\ V_{in(PK)}$.   † Use line to center tap voltage for V_{in}.

1N5817 1N5818 1N5819

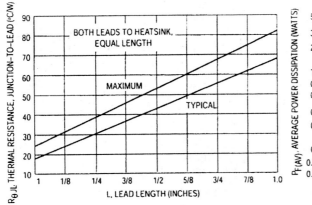

Figure 4. Steady-State Thermal Resistance

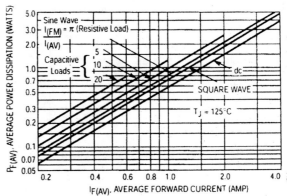

Figure 5. Forward Power Dissipation
1N5817–19

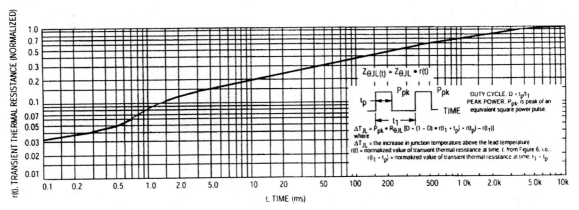

Figure 6. Thermal Response

NOTE 2 — MOUNTING DATA

Data shown for thermal resistance junction–to–ambient ($R_{\theta JA}$) for the mountings shown is to be used as typical guideline values for preliminary engineering, or in case the tie point temperature cannot be measured.

TYPICAL VALUES FOR $R_{\theta JA}$ IN STILL AIR

Mounting Method	Lead Length, L (in)				$R_{\theta JA}$
	1/8	1/4	1/2	3/4	
1	52	65	72	85	°C/W
2	67	80	87	100	°C/W
3	50				°C/W

Mounting Method 1
P.C. Board with 1–1/2" x 1–1/2" copper surface.

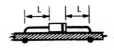

Mounting Method 2

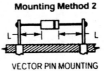

VECTOR PIN MOUNTING

Mounting Method 3
P.C. Board with 1–1/2" x 1–1/2" copper surface.

L = 3/8"

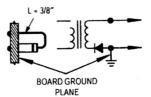

BOARD GROUND PLANE

Rectifier Device Data

Chapter 15

1N5817 1N5818 1N5819

NOTE 3 — THERMAL CIRCUIT MODEL
(For heat conduction through the leads)

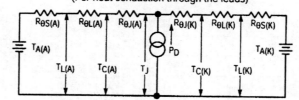

Use of the above model permits junction to lead thermal resistance for any mounting configuration to be found. For a given total lead length, lowest values occur when one side of the rectifier is brought as close as possible to the heatsink. Terms in the model signify:

T_A = Ambient Temperature
T_L = Lead Temperature
$R_{\theta S}$ = Thermal Resistance, Heatsink to Ambient
$R_{\theta L}$ = Thermal Resistance, Lead to Heatsink
$R_{\theta J}$ = Thermal Resistance, Junction to Case
P_D = Power Dissipation

T_C = Case Temperature
T_J = Junction Temperature

(Subscripts A and K refer to anode and cathode sides, respectively.)
Values for thermal resistance components are:
$R_{\theta L}$ = 100°C/W/in typically and 120°C/W/in maximum
$R_{\theta J}$ = 36°C/W typically and 46°C/W maximum.

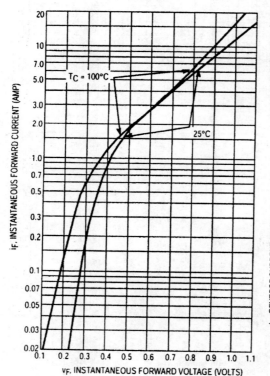

Figure 7. Typical Forward Voltage

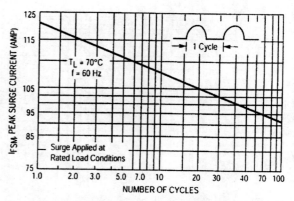

Figure 8. Maximum Non–Repetitive Surge Current

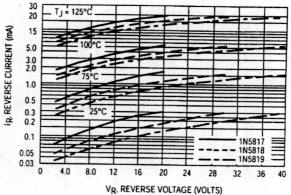

Figure 9. Typical Reverse Current

Chapter 15

Power Products

INDUCTORS DESIGNED FOR NATIONAL'S SIMPLE SWITCHER™

150 kHz

Part Number THT	SMT	Inductance Nominal (µH)	Rated Current (Amps)	Nominal DCR (Ω)	Data Sheet
PE-53801	PE-53801S	259	0.13	3.40	PC503
PE-53802	PE-53802S	178	0.16	2.80	PC503
PE-53803	PE-53803S	118	0.20	1.80	PC503
PE-53804	PE-53804S	79	0.25	1.50	PC503
PE-53805	PE-53805S	55	0.30	1.00	PC503
PE-53806	PE-53806S	39	0.34	0.80	PC503
PE-53807	PE-53807S	26	0.45	0.62	PC503
PE-53808	PE-53808S	374	0.20	2.70	PC503
PE-53809	PE-53809S	256	0.25	2.20	PC503
PE-53810	PE-53810S	176	0.30	1.40	PC503
PE-53811	PE-53811S	118	0.38	1.20	PC503
PE-53812	PE-53812S	78	0.46	0.80	PC503
PE-53813	PE-53813S	55	0.56	0.50	PC503
PE-53814	PE-53814S	39	0.68	0.30	PC503
PE-53815	PE-53815S	26	0.84	0.20	PC503
PE-53816	PE-53816S	17	1.02	0.10	PC503
PE-53817	PE-53817S	375	0.36	1.30	PC503
PE-53818	PE-53818S	252	0.44	0.90	PC503
PE-53819	PE-53819S	173	0.54	0.60	PC503
PE-53820	PE-53820S	115	0.67	0.40	PC503
PE-53821	PE-53821S	78	0.82	0.30	PC503
PE-53822	PE-53822S	54	1.00	0.20	PC503
PE-53823	PE-53823S	38	1.20	0.10	PC503
PE-53824	PE-53824S	26	1.48	0.10	PC503
PE-53825	PE-53825S	18	1.81	0.06	PC503
PE-53826	PE-53826S	377	0.68	1.00	PC503
PE-53827	PE-53827S	248	0.83	0.60	PC503
PE-53828	PE-53828S	168	1.02	0.40	PC503
PE-53829	PE-53829S	112	1.26	0.30	PC503
PE-53830	PE-53830S	77	1.54	0.20	PC503
PE-53831	PE-53831S	53	1.87	0.13	PC503
PE-53932	PE-53932S	37	2.24	0.10	PC503
PE-53933	PE-53933S	24	2.74	0.07	PC503
PE-53934	PE-53934S	17	3.00	0.05	PC503
PE-53935	PE-53935S	250	1.50	0.23	PC503

150 kHz (continued)

Part Number THT	SMT	Inductance Nominal (µH)	Rated Current (Amps)	Nominal DCR (Ω)	Data Sheet
PE-54036	PE-54036S	168	1.81	0.18	PC503
PE-54037	PE-54037S	114	2.22	0.10	PC503
PE-54038	PE-54038S	77	2.70	0.09	PC503
PE-54039	PE-54039S	53	3.00	0.08	PC503
PE-54040	PE-54040S	38	3.00	0.05	PC503
PE-54041	PE-54041S	25	3.00	0.04	PC503
PE-54042	—	167	2.50	0.14	PC503
PE-54043	—	110	3.00	0.09	PC503
PE-54044	PE-54044S	77	3.00	0.08	PC503
PE-53900	—	19	4.50	0.02	PC503

50 kHz

Part Number	Inductance Typical (µH)	I_{DC} (Amps)	Inductance No. DC (µH ±20%)	Package L/W/H (in.)	Data Sheet
PE-53122	2200	0.42	1730	1.440 / .800 / 1.440	PC503
PE-53121	1500	0.62	1150	1.440 / .800 / 1.440	PC503
PE-53114	470	0.64	426	.950 / .600 / 1.000	PC503
PE-52629	680	0.85	657	.710 / .710 / .460	PC503
PE-53146	330	0.90	267	.830 / .450 / .950	PC503
PE-52627	330	0.90	302	.710 / .710 / .460	PC503
PE-53120	1000	0.95	762	1.300 / .700 / 1.400	PC503
PE-53119	680	1.30	562	1.300 / .700 / 1.400	PC503
PE-53145	220	1.40	176	.830 / .450 / .950	PC503
PE-52626	220	1.40	230	.710 / .710 / .460	PC503
PE-53113	150	2.00	130	.950 / .600 / 1.000	PC503
PE-53118	470	2.00	369	1.300 / .700 / 1.400	PC503
PE-53112	47	3.00	38	.650 / .450 / .700	PC503
PE-53115	150	3.00	136	.950 / .600 / 1.000	PC503
PE-53116	220	3.00	167	1.300 / .700 / 1.400	PC503
PE-53117	330	3.00	292	1.300 / .700 / 1.400	PC503
PE-92114K [1]	68	3.00	55	.950 / .600 / 1.000	PC503
PE-92108K [1]	100	3.00	91	.950 / .600 / 1.000	PC503

1. (K) Klipmount package dimensions – see Power Catalog PC503.

CURRENT SENSE INDUCTORS & TRANSFORMERS

Low Profile Self-Leaded Surface Mount Transformers

Part Number*	Secondary Turns	Secondary Inductance	Ipeak (Amps)	Package L/W/H (in.)**	Data Sheet
PE-68210	50	3.8 mH	15.3	.575 / .495 / .280	PC502
PE-68280	100	14.8 mH	16.3	.575 / .495 / .280	PC502
PE-68383	200	59.2 mH	16.4	.575 / .495 / .280	PC502

UL/C-UL Recognized Through Hole Transformers

Part Number	Primary Turns	Primary Inductance	Ipeak (Amps)	Package L/W/H (in.)	Data Sheet
P0581	200	76 mH	34	.810 / .580 / .750	PC502 / PC503
P0582	100	19 mH	35	.810 / .580 / .750	PC502 / PC503
P0583	50	5 mH	36	.810 / .580 / .750	PC502 / PC503

500 kHz Through Hole Inductors & Transformers

Part Number	Secondary Turns	Secondary Inductance	Ipeak (Amps)	Package L/W/H (in.)	Data Sheet
Inductors					
PE-52876	10	150.0 µH	20	.550 / .325 / .580	PC503
PE-52877	20	600.0 µH	20	.550 / .325 / .580	PC503
PE-52878	50	4000.0 µH	20	.550 / .325 / .580	PC503
Transformers					
PE-64976	10	150.0 µH	20	.550 / .325 / .580	PC502
PE-64977	20	600.0 µH	20	.550 / .325 / .580	PC502
PE-64978	50	4000.0 µH	20	.550 / .325 / .580	PC502

VDE Approved Through Hole Inductors & Transformers

Part Number	Secondary Turns	Secondary Inductance	Ipeak (Amps)	Package L/W/H (in.)	Data Sheet
Transformer With 2 - 1T Primaries					
PE-64487	100	20.0 mH	20	.900 / .700 / .700	PC502
PE-64488	200	80.0 mH	20	.900 / .700 / .700	PC502
PE-64517	50CT	5.0 mH	20	.900 / .700 / .700	PC502
PE-64518	100CT	20.0 mH	20	.900 / .700 / .700	PC502
PE-64519	200CT	80.0 mH	20	.900 / .700 / .700	PC502
Transformer With 1T Primary					
PE-63586	50	5.0 mH	20	.900 / .700 / .700	PC502
PE-63587	100	20.0 mH	20	.900 / .700 / .700	PC502
PE-63588	200	80.0 mH	20	.900 / .700 / .700	PC502
PE-63618	100CT	20.0 mH	20	.900 / .700 / .700	PC502
PE-63619	200CT	80.0 mH	20	.900 / .700 / .700	PC502
PE-63691	300CT	180.0 mH	20	.900 / .700 / .700	PC502
PE-67050	50	5.0 mH	35	.750 / .565 / .750	PC502
PE-67100	100	20.0 mH	37	.750 / .565 / .750	PC502
PE-67200	200	80.0 mH	38	.750 / .565 / .750	PC502
PE-67300	300	180.0 mH	37	.750 / .565 / .750	PC502
Inductor Without Primary					
PE-51686	50	5.0 mH	20	.670 / .390 / .800	PC503
PE-51687	100	20.0 mH	20	.670 / .390 / .800	PC503
PE-51688	200	80.0 mH	20	.670 / .390 / .800	PC503
PE-51717	50CT	5.0 mH	20	.670 / .390 / .800	PC503
PE-51718	100CT	20.0 mH	20	.670 / .390 / .800	PC503
PE-51719	200CT	80.0 mH	20	.670 / .390 / .800	PC503

* Add "T" suffix to surface mount part number to indicate Tape & Reel packaging.
** L/W/H is measured on surface mount parts tip to tip (height includes wash area).

SMT - Surface Mount Package   THT - Through Hole Package

550D
Vishay Sprague

Solid-Electrolyte TANTALEX® Capacitors for High Frequency Power Supplies

FEATURES
- Hermetically-sealed, axial-lead solid tantalum capacitors
- Small size and long life
- Exceptional capacitance stability and excellent resistance to severe environmental conditions
- The military equivalent is the CSR21 which is qualified to MIL-C-39003/09

APPLICATIONS
Designed for power supply filtering applications at above 100kHz

At + 85°C: Leakage current shall not exceed 10 times the values listed in the Standard Ratings Tables.

At +125°C: Leakage current shall not exceed 15 times the values listed in the Standard Ratings Tables.

Life Test: Capacitors shall withstand rated DC voltage applied at + 85°C for 2000 hours or derated DC voltage applied at + 125°C for 1000 hours.
Following the life test:
1. DCL shall not exceed 125% of the initial requirement.
2. Dissipation Factor shall meet the initial requirement.
3. Change in capacitance shall not exceed ± 5%.

PERFORMANCE CHARACTERISTICS
Operating Temperature: - 55°C to + 85°C. (To + 125°C with voltage derating.)
Capacitance Tolerance: At 120 Hz, + 25°C. ± 20% and ± 10% standard. ± 5% available as a special.
Dissipation Factor: At 120 Hz, + 25°C. Dissipation factor, as determined from the expression $2\pi fRC$, shall not exceed the values listed in the Standard Ratings Tables.
DC Leakage Current (DCL Max.):
At + 25°C: Leakage current shall not exceed the values listed in the Standard Ratings Tables.

ORDERING INFORMATION

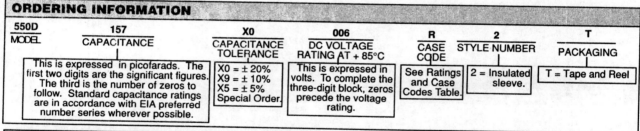

DIMENSIONS in inches [millimeters]

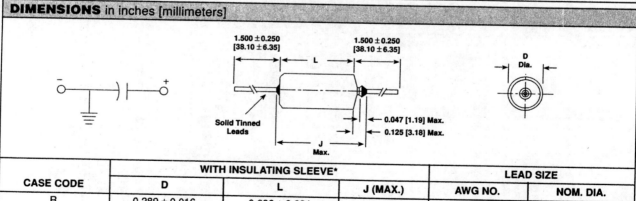

CASE CODE	WITH INSULATING SLEEVE*			LEAD SIZE	
	D	L	J (MAX.)	AWG NO.	NOM. DIA.
R	0.289 ± 0.016 [7.34 ± 0.41]	0.686 ± 0.031 [17.42 ± 0.79]	0.822 [20.88]	22	0.025 [0.64]
S	0.351 ± 0.016 [8.92 ± 0.41]	0.786 ± 0.031 [19.96 ± 0.79]	0.922 [23.42]	22	0.025 [0.64]

*When a shrink-fitted insulation is used, it shall lap over the ends of the capacitor body.

Document Number: 40017
Revision 13-Jun-02

550D

Vishay Sprague

Solid-Electrolyte TANTALEX® Capacitors for High Frequency Power Supplies

STANDARD RATINGS

CAPACITANCE (µF)	CASE CODE	PART NUMBER* CAP. TOL. ± 20%	PART NUMBER* CAP. TOL. ± 10%	MAX. DCL @ +25°C (µA)	MAX. DF @ +25°C 1kHz (%)	MAX. ESR @ +25°C 100kHz (Ohms)
6 WVDC @ +85°C, SURGE = 8 V ... 4 WVDC @ +125°C, SURGE = 5 V							
150	R	550D157X0006R2	550D157X9006R2	9	10	0.065
180	R	550D187X0006R2	550D187X9006R2	11	10	0.060
220	S	550D227X0006S2	550D227X9006S2	12	10	0.055
270	S	550D277X0006S2	550D277X9006S2	13	10	0.050
330	S	550D337X0006S2	550D337X9006S2	15	12	0.045
10 WVDC @ +85°C, SURGE = 13 V ... 7 WVDC @ +125°C, SURGE = 9 V							
82	R	550D826X0010R2	550D826X9010R2	8	8	0.085
100	R	550D107X0010R2	550D107X9010R2	10	8	0.075
120	R	550D127X0010R2	550D127X9010R2	12	8	0.070
150	S	550D157X0010S2	550D157X9010S2	15	8	0.065
180	S	550D187X0010S2	550D187X9010S2	18	8	0.060
220	S	550D227X0010S2	550D227X9010S2	20	10	0.055
15 WVDC @ +85°C, SURGE = 20 V ... 10 WVDC @ +125°C, SURGE = 12 V							
56	R	550D566X0015R2	550D566X9015R2	8	6	0.100
68	R	550D686X0015R2	550D686X9015R2	10	6	0.095
82	S	550D826X0015S2	550D826X9015S2	12	6	0.085
100	S	550D107X0015S2	550D107X9015S2	15	8	0.075
120	S	550D127X0015S2	550D127X9015S2	18	8	0.070
150	S	550D157X0015S2	550D157X9015S2	20	8	0.065
20 WVDC @ +85°C, SURGE = 26 V ... 13 WVDC @ +125°C, SURGE = 16 V							
27	R	550D276X0020R2	550D276X9020R2	5	5	0.145
33	R	550D336X0020R2	550D336X9020R2	7	5	0.130
39	R	550D396X0020R2	550D396X9020R2	8	5	0.120
47	R	550D476X0020R2	550D476X9020R2	9	6	0.110
56	S	550D566X0020S2	550D566X9020S2	11	6	0.100
68	S	550D686X0020S2	550D686X9020S2	14	6	0.095
82	S	550D826X0020S2	550D826X9020S2	16	6	0.085
100	S	550D107X0020S2	550D107X9020S2	20	8	0.075
35 WVDC @ +85°C, SURGE = 46 V ... 23 WVDC @ +125°C, SURGE = 28 V							
8.2	R	550D825X0035R2	550D825X9035R2	3	4	0.250
10	R	550D106X0035R2	550D106X9035R2	4	4	0.230
12	R	550D126X0035R2	550D126X9035R2	4	4	0.210
15	R	550D156X0035R2	550D156X9035R2	5	4	0.190
18	R	550D186X0035R2	550D186X9035R2	6	4	0.175
22	R	550D226X0035R2	550D226X9035R2	8	4	0.160
27	S	550D276X0035S2	550D276X9035S2	9	4	0.145
33	S	550D336X0035S2	550D336X9035S2	11	5	0.130
39	S	550D396X0035S2	550D396X9035S2	14	5	0.120
47	S	550D476X0035S2	550D476X9035S2	16	5	0.110
50 WVDC @ +85°C, SURGE = 65 V ... 33 WVDC @ +125°C, SURGE = 40 V							
5.6	R	550D565X0050R2	550D565X9050R2	4	3	0.300
6.8	R	550D685X0050R2	550D685X9050R2	4	3	0.275
8.2	R	550D825X0050R2	550D825X9050R2	5	3	0.250
10.0	R	550D106X0050R2	550D106X9050R2	5	3	0.230
12.0	R	550D126X0050R2	550D126X9050R2	6	3	0.210
15.0	R	550D156X0050R2	550D156X9050R2	8	3	0.190
18.0	R	550D186X0050R2	550D186X9050R2	9	4	0.175
22.0	S	550D226X0050S2	550D226X9050S2	11	4	0.160

*Insert capacitance tolerance code "X5" for ± 5% units (special order).

Document Number: 40017
Revision 13-Jun-02

PROBLEMS

15.1 Carry out Practice Problem 15-1.

15.2 Determine the output voltage V_O, output current I_O and $I_{O(max)}$, and output ripple voltage $V_{r(p-p)}$ for the linear voltage regulator of Figure 15-2P, given the following:

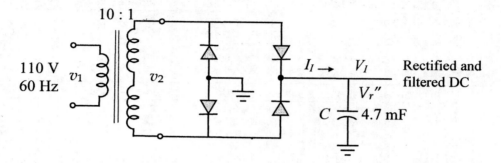

Figure 15-1P: Full-wave bridge rectifier with filter capacitor

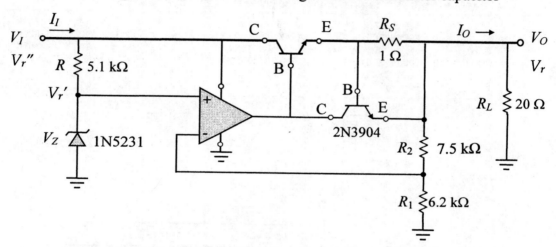

Figure 15-2P: Regulator with adjustable output and overload protection

15.3 Carry out Practice Problem 15-2.

15.4 Design a linear voltage regulator with adjustable output and overload protection for the following specifications: V_O adjustable from approximately 5 V to 15 V. $I_{O(max)} = 0.8$ A, $V_{r(p-p)} \leq 0.01 V_O$, with $V_{1(rms)} = 120$ V, turns ratio = 8:1, $C = 6.8$ mF

15.5 List the advantages of switching power supplies over the traditional basic power supplies.

15.6 Carry out Practice Problem 15-3.

15.7 Determine the efficiency and $v_{r(p-p)}$ of a buck converter with the following specifications: $V_I = 10$ V to 16 V, $V_O = 6$ V, $R_L = 10$ Ω, $f_s = 200$ kHz. Initially assume $\eta = 0.90$.

    MOSFET: $r_{DS} = 0.30$ Ω, $C_o = 100$ pF   Diode: $V_F = 0.35$ V, $R_F = 0.25$ Ω
    Inductor: $L = 470$ μH, $r_L = 0.3$ Ω   Capacitor: $C = 100$ μF, $r_C = 0.3$ Ω

Chapter 15

15.8 Carry out Practice Problem 15-4.

15.9 Design a buck converter with the following specifications:

$V_I = 10$ V $\pm 25\%$, $V_O = 5$ V, $I_{O(min)} = 75$ mA, $I_{O(max)} = 750$ mA, and $f_s = 150$ kHz

15.10 Carry out Practice Problem 15-5.

15.11 Determine the efficiency and the output ripple voltage of a boost converter with the following specifications:

$V_I = 8$ V to 12, $V_O = 15$ V, $R_L = 10$ Ω, and $f_s = 200$ kHz

MOSFET: $r_{DS} = 0.25$ Ω, $C_o = 100$ pF Diode: $V_F = 0.35$ V, $R_F = 0.3$ Ω
Inductor: $L = 470$ μH, $r_L = 0.2$ Ω Capacitor: $C = 47$ μF, $r_C = 0.3$ Ω

15.12 Carry out Practice Problem 15-6.

15.13 Design a boost converter with the following specifications:

$V_I = 10$ V $\pm 25\%$, $V_O = 15$ V, $I_{O(min)} = 75$ mA, $I_{O(max)} = 750$ mA, and $f_s = 150$ kHz

15.14 Carry out Practice Problem 15-7.

15.15 Carry out Practice Problem 15-8.

15.16 Determine the efficiency and the output ripple voltage of a buck-boost converter with the following specifications:

$V_I = 5$ V to 10 V, $V_O = -12.5$ V, $I_{O(min)} = 25$ mA, $I_{O(max)} = 1$ A, and $f_s = 200$ kHz

MOSFET: $r_{DS} = 0.15$ Ω, $C_o = 100$ pF Diode: $V_F = 0.35$ V, $R_F = 0.15$ Ω
Inductor: $L = 680$ μH, $r_L = 0.15$ Ω Capacitor: $C = 100$ μF, $r_C = 0.1$ Ω

15.17 Carry out Practice Problem 15-9.

15.18 Carry out Practice Problem 15-10.

15.19 Design an inverting buck converter with the following specifications:

$V_I = 10 \pm 2$ V, $V_O = -6$ V, $I_{O(min)} = 50$ mA, $I_{O(max)} = 1$ A, and $f_s = 150$ kHz

15.20 Design an inverting boost converter with the following specifications:

$V_I = 7.5 \pm 2.5$ V, $V_O = -12$ V, $I_{O(min)} = 50$ mA, $I_{O(max)} = 1$ A, and $f_s = 150$ kHz

Chapter 16

OSCILLATORS AND WAVEFORM GENERATORS

Analysis & Design

16.1 INTRODUCTION

Another important application of the op-amp is in the area of waveform generation. With the addition of a few external components, the op-amp can be utilized to produce waveforms of desired frequency and amplitude, without any external input. The output waveform may be sinusoidal, rectangular, or triangular. In this chapter, we will present some popular op-amp configurations that serve as oscillators or waveform generators. Furthermore, we will also introduce some other popular integrated circuits (*ICs*) that are specifically designed to be utilized as a function generator or waveform generator, with the addition of some properly selected *R* and *C* components.

16.2 OSCILLATOR PRINCIPLES OF OPERATION

Recall that a negative feedback was used with amplifier circuits to maintain a stable and steady output, which was the amplified version of an input signal. However, an oscillator is generally an amplifier with positive feedback, which produces a periodic alternating output signal or waveform with a certain frequency, without any external input signal. Oscillation is actually a form of instability caused by positive feedback that reinforces and regenerates a signal that would otherwise be lost or die out due to loss of energy, mainly in resistive elements. A block diagram of an oscillator containing the amplifier section (A) and the feedback network (β) is shown in Figure 16-1 below:

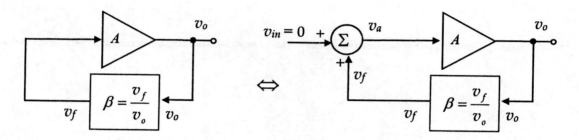

Figure 16-1(a):
Block diagram of an oscillator

Figure 16-1(b):
Equivalent block diagram

From the above two block diagrams we conclude the following:

$$v_a = v_{in} + v_f = 0 + v_f = v_f \tag{16-1}$$

where,

$$v_f = \beta v_o \tag{16-2}$$

$$v_o = A \times v_a = A v_f = A \beta v_o \tag{16-3}$$

Dividing both sides by v_o yields the following:

$$\beta A = 1 \tag{16-4}$$

That is, the overall loop gain of the oscillator must equal 1. Furthermore, since the feedback network (β) may contain resistive and reactive elements, β will be a complex term. Hence, we may want to write Equation 16-4 in a general form, as follows:

$$\beta A = 1 + j\,0 = 1\angle 0° \text{ or } 1\angle \pm 360° \tag{16-5}$$

Note that the above equation requires that the magnitude gain around the loop must equal unity and the overall phase shift must equal 0° or ±360°.

16.3 RC PHASE-SHIFT OSCILLATOR

The *RC* phase-shift oscillator consists of an inverting amplifier for the required gain and a cascade of three *RC* circuits for the feedback network, as shown in Figure 16-2 below:

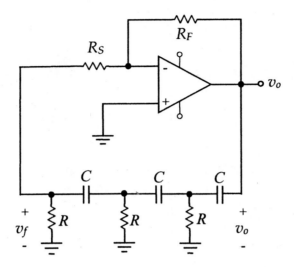

Figure 16-2: *RC* phase-shift oscillator

The inverting amplifier has a phase-shift of 180°; hence, the feedback circuit must provide a phase-shift of another 180° in order to achieve an overall phase-shift of 360°. It can be shown, with some considerable algebraic effort, that the transfer function of the feedback circuit (β) is as follows:

$$\beta = \frac{v_f}{v_o} = \frac{(RCs)^3}{(RCs)^3 + 6(RCs)^2 + 5RCs + 1} \qquad (16\text{-}6)$$

The gain of the inverting amplifier is defined and determined as follows:

$$A = \frac{v_o}{v_f} = -\frac{R_F}{R_S} \qquad (16\text{-}7)$$

Substituting the values for β and A in Equation 16-4 yields the following:

$$\beta A = -\frac{R_F}{R_S} \cdot \frac{(RCs)^3}{(RCs)^3 + 6(RCs)^2 + 5RCs + 1} = 1 \qquad (16\text{-}8)$$

Multiplying both sides of the above equation with the denominator of β results in the following equation:

$$-\frac{R_F}{R_S}(RCs)^3 = (RCs)^3 + 6(RCs)^2 + 5RCs + 1 \qquad (16\text{-}9)$$

We will now substitute $j\omega$ for s in the above equation so that we can solve for the oscillation frequency ω.

$$-\frac{R_F}{R_S}(RCj\omega)^3 = (RCj\omega)^3 + 6(RCj\omega)^2 + 5RCj\omega + 1 \qquad (16\text{-}10)$$

$$j\frac{R_F}{R_S}(RC\omega)^3 = -j(RC\omega)^3 - 6(RC\omega)^2 + 5jRC\omega + 1 \qquad (16\text{-}11)$$

$$j\frac{R_F}{R_S}(RC\omega)^3 + j(RC\omega)^3 + 6(RC\omega)^2 - 5jRC\omega = 1 \qquad (16\text{-}12)$$

$$j\frac{R_F}{R_S}(RC\omega)^3 + j(RC\omega)^3 + 6(RC\omega)^2 - 5jRC\omega = 1 \qquad (16\text{-}13)$$

$$6(\omega RC)^2 + j[\frac{R_F}{R_S}(RC\omega)^3 + (RC\omega)^3 - 5RC\omega] = 1 \qquad (16\text{-}14)$$

Recall that $\beta A = 1 + j\,0$. Equating the real part of the left side of the above equation to the real part of the right side, we obtain the following:

$$6(\omega_o RC)^2 = 1 \qquad (16\text{-}15)$$

$$\omega_o^2 = \frac{1}{6(RC)^2} \qquad (16\text{-}16)$$

$$\omega_o = \frac{1}{\sqrt{6}(RC)} \qquad (16\text{-}17)$$

The ω is the oscillation frequency in rad/sec; however, we can solve for the oscillation frequency (f_o) in Hz by dividing the ω by 2π.

$$f_o = \frac{1}{2\pi\sqrt{6}(RC)} \cong \frac{0.065}{RC} \qquad (16\text{-}18)$$

Equating the imaginary part of the left side of Equation 16-14 to the imaginary part of the right side, we obtain the following:

$$j[\frac{R_F}{R_S}(RC\omega)^3 + (RC\omega)^3 - 5RC\omega] = 0 \qquad (16\text{-}19)$$

$$(RC\omega)^3\left(\frac{R_F}{R_S} + 1\right) - 5RC\omega = 0 \qquad (16\text{-}20)$$

$$(RC\omega)^2\left(\frac{R_F}{R_S} + 1\right) - 5 = 0 \qquad (16\text{-}21)$$

Substituting for ω in the above equation yields the following:

$$(RC)^2 \frac{1}{6(RC)^2}\left(\frac{R_F}{R_S} + 1\right) - 5 = 0 \qquad (16\text{-}22)$$

$$\frac{1}{6}\left(\frac{R_F}{R_S} + 1\right) = 5 \qquad (16\text{-}23)$$

$$\frac{R_F}{R_S} = 30 - 1 = 29 \qquad (16\text{-}24)$$

Example 16-1 — Phase-Shift Oscillator — DESIGN

Let us now design a phase-shift oscillator for an oscillation frequency of 500 Hz.

Solution: Let $C = 0.1\ \mu F$, and solve for the resistor R in Equation 16-18.

$$f_o = \frac{1}{2\pi\sqrt{6}(RC)} \cong \frac{0.065}{RC}$$

$$R = \frac{0.065}{f_o C} = \frac{0.065}{(500\ \text{Hz})(0.1\ \mu F)} = 1.3\ k\Omega, \qquad R = 1.3\ k\Omega$$

Let $R_S = 15\ k\Omega$, $\quad R_F = 29 R_S = 435\ k\Omega$ $\qquad$ Use a 500 $k\Omega$ potentiometer for R_F

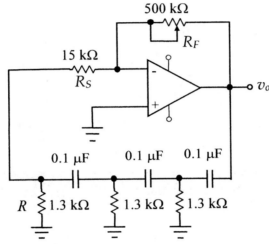

Figure 16-3:
Circuit diagram for the RC phase-shift oscillator of Design Example 16-1

Practice Problem 16-1 — Phase-shift Oscillator — DESIGN

Design a phase-shift oscillator with the oscillation frequency $f_o = 330$ Hz.

Answers: $C = 0.1\ \mu F$, $R = 2\ k\Omega$ $\qquad$ Let $R_S = 10\ k\Omega$, and $R_F = 500\ k\Omega$ pot.

Test Run and Design Verification:

The results of simulation of the above design example, created in Electronics Workbench, are presented in Figures 16-4 and 16-5.

Chapter 16

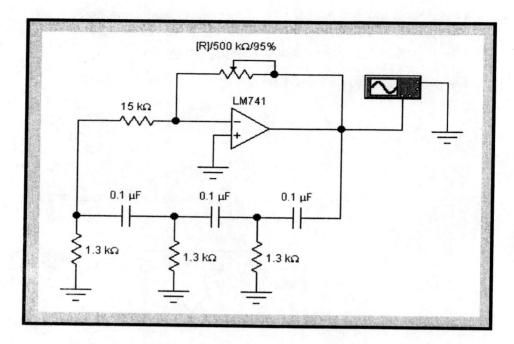

Figure 16-4: Circuit setup for simulation of Design Example 16-1

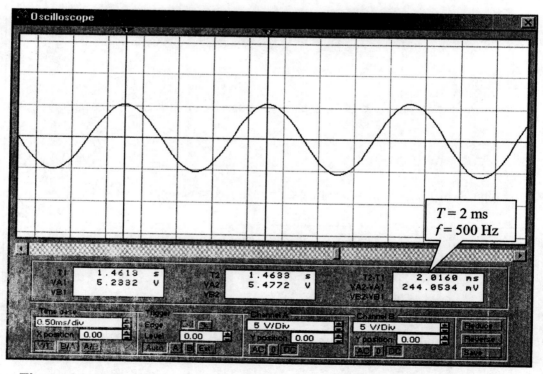

Figure 16-5: The Electronics Workbench oscilloscope showing the output waveform for Design Example 16-1

710

16.4 WIEN-BRIDGE OSCILLATOR

A Wien-bridge oscillator consists of a non-inverting amplifier and a feedback network, which has a series RC circuit in one arm of the Wien-bridge and a parallel RC circuit in the adjoining arm, as shown in Figure 16-6 below:

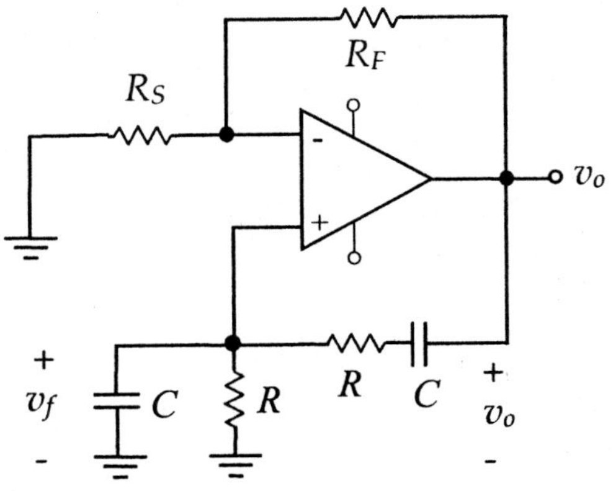

Figure 16-6:
Circuit diagram of a Wien-bridge oscillator

Let $z_1 = R + X_C$ and $z_2 = R \| X_C$, and show both z_1 and z_2 in the s-domain:

$$z_1 = R + \frac{1}{sC} = \frac{sCR+1}{sC} \tag{16-25}$$

$$z_2 = \frac{R \dfrac{1}{sC}}{R + \dfrac{1}{sC}} = \frac{R}{sCR+1} \tag{16-26}$$

Applying the voltage divider rule, we can now solve for v_f, as follows:

$$v_f = v_o \left(\frac{z_2}{z_1 + z_2} \right) \tag{16-27}$$

The transfer function of the feedback network is defined and determined as follows:

$$\beta = \frac{v_f}{v_o} = \frac{z_2}{z_1 + z_2} = \frac{R}{(sCR+1)\left(\dfrac{sCR+1}{sC} + \dfrac{R}{sCR+1}\right)} \tag{16-28}$$

$$\beta = \frac{sCR}{(sCR+1)^2 + sCR} = \frac{sCR}{(sCR)^2 + 2sCR + 1 + sCR} \tag{16-29}$$

$$\beta = \frac{sCR}{(sCR)^2 + 3sCR + 1} \tag{16-30}$$

The gain of the non-inverting amplifier is defined and determined as follows:

$$A = \frac{v_o}{v_f} = \frac{R_F}{R_S} + 1 \tag{16-31}$$

Applying the oscillator principles ($\beta A = 1$) results in the following:

$$\beta A = \left(\frac{R_F}{R_S} + 1\right) \cdot \frac{RCs}{(RCs)^2 + 3RCs + 1} = 1 \qquad (16\text{-}32)$$

We will now substitute $j\omega$ for s in the above equation so that we can solve for the oscillation frequency ω_o.

$$\left(\frac{R_F}{R_S} + 1\right) j\omega RC = -(RC\omega)^2 + 3j\omega RC + 1 \qquad (16\text{-}33)$$

Again, recall that $\beta A = 1 + j\,0$; that is, the real part of the above equation equals 1, and its imaginary part equals zero. Equating the real part of the left side of the above equation to the real part of the right side, we obtain the following:

$$0 = 1 - (RC\omega_o)^2 \qquad (16\text{-}34)$$

$$(RC\omega_o)^2 = 1, \text{ or } RC\omega_o = 1 \qquad (16\text{-}35)$$

$$\omega_o = \frac{1}{RC} \qquad (16\text{-}36)$$

$$f_o = \frac{1}{2\pi RC} \qquad (16\text{-}37)$$

Equating the imaginary part of the left side to the imaginary part of the right side, we obtain the following:

$$\left(\frac{R_F}{R_S} + 1\right) j\omega RC = 3j\omega RC \qquad (16\text{-}38)$$

$$\left(\frac{R_F}{R_S} + 1\right) = 3 \qquad (16\text{-}39)$$

$$\frac{R_F}{R_S} = 2 \qquad (16\text{-}40)$$

The circuit of Figure 16-6, which we have used to study the principles of Wien-bridge oscillator, is rather unstable; any drift can cause the output to saturate or disappear. However, connecting two Zener diodes back-to-back across the R_F, as illustrated in Figure 16-7, can stabilize the operation of the circuit.

Example 16-2 — Wien-Bridge Oscillator — DESIGN

Let us now design a Wien-bridge oscillator for an oscillation frequency of 1 kHz.
Solution: Let $C = 0.1\ \mu F$, and solve for the resistor R in Equation 16-37.

$$f_o = \frac{1}{2\pi RC}$$

$$R = \frac{1}{2\pi f_o C} = \frac{1}{6.28 \times 1\,\text{kHz} \times 0.1\,\mu\text{F}} = 1.59\,\text{k}\Omega \qquad \text{Use } R = 1.6\,\text{k}\Omega$$

Let $R_S = 4.7\,\text{k}\Omega$, $R_F = 2R_S = 9.4\,\text{k}\Omega$ Use a 10 kΩ potentiometer for R_F.

We will also use two 2N5231 Zener diodes for stabilization.

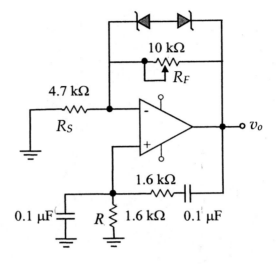

Figure 16-7:
Wien-bridge oscillator with Zener diode adaptive feedback

Test Run and Design Verification:

The results of simulation of the above design example, created in Electronics Workbench, are presented in Figures 16-8 and 16-9:

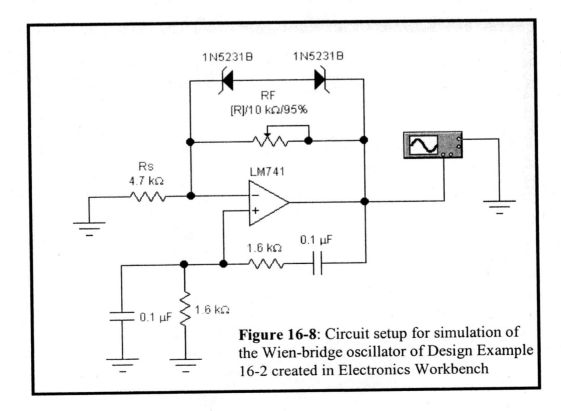

Figure 16-8: Circuit setup for simulation of the Wien-bridge oscillator of Design Example 16-2 created in Electronics Workbench

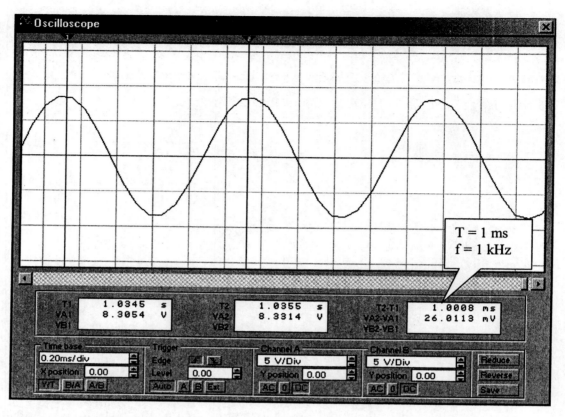

Figure 16-9: The Electronics Workbench oscilloscope showing the output waveform of the Wien-bridge oscillator of Design Example 16-2

Practice Problem 16-2   *Wien-Bridge Oscillator*   **DESIGN**

Design a Wien-bridge oscillator with the oscillation frequency $f_o = 3.3$ kHz.

Answers: $C = 0.01$ µF, $R = 4.8$ kΩ, $R_S = 7.5$ kΩ, $R_F = 20$ kΩ pot.

Practice Problem 16-3   *Wien-Bridge Oscillator*   **ANALYSIS**

Determine the oscillation frequency for a Wien-Bridge oscillator with the following components:

$R = 2$ kΩ, $C = 0.01$ µF, $R_F = 20$ kΩ, $R_S = 10$ kΩ

Answer: $f_o = 8$ kHz

16.5 COLPITTS OSCILLATOR

The Colpitts oscillator, as shown in Figure 16-10, consists of an inverting amplifier and a feedback network, which is a resonant LC network. The oscillation frequency is the resonance frequency of the feedback LC circuit.

Note that at resonance the magnitude of the total inductive reactance of the circuit equals the total capacitive reactance; that is, $X_L = X_{C1} + X_{C2}$.

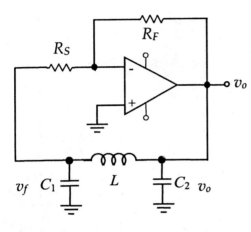

Figure 16-10:
Circuit diagram of the Colpitts oscillator

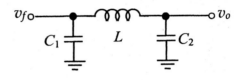

(a) Colpitts oscillator

(b) β network

Applying the voltage divider rule to the feedback circuit, we can solve for v_f, as follows:

$$v_f = v_o \left(\frac{-jX_{C1}}{jX_L - jX_{C1}} \right) = v_o \left(\frac{-jX_{C1}}{j(X_{C2} + X_{C1}) - jX_{C1}} \right) \quad (16\text{-}41)$$

The transfer function of the feedback network is defined and determined as follows:

$$\beta = \frac{-jX_{C1}}{jX_{C2}} = -\frac{X_{C1}}{X_{C2}} = -\frac{C_2}{C_1} \quad (16\text{-}42)$$

The gain of the non-inverting amplifier is defined and determined as follows:

$$A = \frac{v_o}{v_f} = -\frac{R_F}{R_S} \quad (16\text{-}43)$$

Applying the oscillator principles ($\beta A = 1$) results in the following:

$$\beta A = -\frac{R_F}{R_S} \cdot \frac{-C_2}{C_1} = \frac{R_F}{R_S} \cdot \frac{C_2}{C_1} = 1 \quad (16\text{-}44)$$

$$\frac{R_F}{R_S} = \frac{C_1}{C_2} \quad (16\text{-}45)$$

Chapter 16

Since the oscillation occurs at resonance, we must determine the oscillation frequency under resonance conditions. Hence, we will start as follows:

$$X_L = X_{C2} + X_{C1} \tag{16-46}$$

$$\omega L = \frac{1}{\omega C_2} + \frac{1}{\omega C_1} = \frac{\omega C_1 + \omega C_2}{\omega C_1 \cdot \omega C_2} \tag{16-47}$$

Cross multiplication produces the following equation:

$$\omega L \cdot \omega C_1 \cdot \omega C_2 = \omega C_1 + \omega C_2 \tag{16-48}$$

We will now divide both sides by ω, then solve for the oscillation frequency ω_o.

$$\omega_o^2 L \cdot C_1 \cdot C_2 = C_1 + C_2 \tag{16-49}$$

$$\omega_o^2 = \frac{1}{L} \frac{C_1 + C_2}{C_1 \cdot C_2} = \frac{1}{LC_T} \tag{16-50}$$

where C_T is the series combination of C_1 and C_2

$$C_T = \frac{C_1 \cdot C_2}{C_1 + C_2} \tag{16-51}$$

$$\omega_o = \frac{1}{\sqrt{LC_T}} \tag{16-52}$$

$$f_o = \frac{1}{2\pi\sqrt{LC_T}} \tag{16-53}$$

Example 16-3 — Colpitts Oscillator — DESIGN

Let us now design a Colpitts oscillator with an oscillation frequency of 50 kHz.

Solution:

Let $C_1 = C_2 = C = 0.01$ µF, and solve for the inductor L in Equation 16-53.
Having $C_1 = C_2 = C$ makes $C_T = 0.5C = 0.005$ µF, and $C_1/C_2 = 1$. Hence, $R_F/R_S = 1$.

$$f_o^2 = \frac{1}{4\pi^2 LC_T}$$

$$L = \frac{1}{4\pi^2 f_o^2 C_T}$$

$$L = \frac{1}{4\pi^2 (50 \text{ kHz})^2 (0.005 \text{ µF})} = \frac{1}{4\pi^2 (2.5 \times 10^9)(5 \times 10^{-9})} = 2 \text{ mH}$$

One option would be to let $R_F = R_S = 10$ kΩ. However, it would be best to have R_F adjustable, so that we can tune the circuit for the intended oscillation.
Hence, let $R_S = 10$ kΩ and $R_F = 1$ kΩ + 10 kΩ pot.

Chapter 16

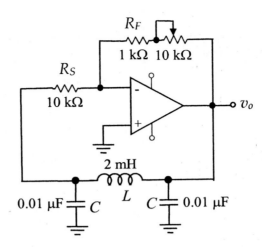

Figure 16-11: Colpitts oscillator of Design Example 16-3

16.6 HARTLEY OSCILLATOR

Like the Colpitts oscillator, the Hartley oscillator consists of an inverting amplifier and a feedback network, which is a resonant LC network. However, the feedback circuit consists of two inductors and a capacitor, as shown in Figure 16-12. At resonance, the capacitive reactance of the circuit equals the total inductive reactance, which is the series combination of X_{L1} and X_{L2}; that is, $X_C = X_{L1} + X_{L2}$

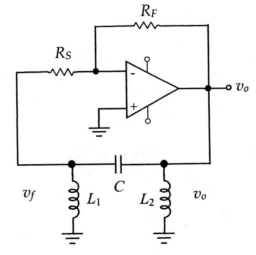

Figure 16-12: Circuit diagram of the Hartley oscillator

The derivations for the Hartley oscillator are similar to those carried out for the Colpitts oscillator. Applying the voltage divider rule to the feedback circuit, we can solve for v_f:

$$v_f = v_o\left(\frac{jX_{L1}}{jX_{L1} - jX_C}\right) = v_o\left(\frac{jX_{L1}}{jX_{L1} - j(X_{L1} + X_{L2})}\right) = v_o\left(\frac{jX_{L1}}{jX_{L2}}\right) \quad (16\text{-}54)$$

The transfer function of the feedback network is defined and determined as follows:

$$\beta = \frac{v_f}{v_o} = \frac{jX_{L1}}{jX_{L2}} = \frac{X_{L1}}{X_{L2}} = \frac{L_1}{L_2} \quad (16\text{-}55)$$

The gain of the non-inverting amplifier is defined and determined as follows:

$$A = \frac{v_o}{v_f} = -\frac{R_F}{R_S} \quad (16\text{-}56)$$

Applying the oscillator principles ($\beta A = 1$) results in the following:

$$\beta A = -\frac{R_F}{R_S} \cdot \frac{L_1}{L_2} = 1 \tag{16-57}$$

$$|A_V| = \frac{R_F}{R_S} = \frac{L_2}{L_1} \tag{16-58}$$

We will now establish the condition required for resonance and determine the oscillation frequency.

$$X_C = X_{L1} + X_{L2} \tag{16-59}$$

$$\frac{1}{\omega C} = \omega L_1 + \omega L_2 = \omega(L_1 + L_2) \tag{16-60}$$

Dividing both sides of the above equation by ω, we get the following:

$$\frac{1}{\omega_o^2 C} = L_1 + L_2 \tag{16-61}$$

$$\omega_o = \frac{1}{\sqrt{CL_T}} \tag{16-62}$$

where, $L_T = L_1 + L_2$

$$f_o = \frac{1}{2\pi\sqrt{CL_T}} \tag{16-63}$$

Example 16-4 — Hartley Oscillator — DESIGN

Let us now design a Hartley oscillator with an oscillation frequency of 100 kHz.

Solution: Let $C = 0.01\ \mu F$, and solve for the inductor L_T in Equation 16-63.

$$f_o^2 = \frac{1}{4\pi^2 CL_T}$$

$$L_T = \frac{1}{4\pi^2 f_o^2 C}$$

$$L_T = \frac{1}{4\pi^2 (100\ \text{kHz})^2 (0.01\ \mu F)} = \frac{1}{4\pi^2 (1\times 10^{10})(1\times 10^{-8})} = 0.25\ \text{mH}$$

Let $L_1 = L_2 = 0.125\ \text{mH} = 125\ \mu H$

For the same reason stated in Example 16-3, let $R_S = 10\ k\Omega$ and $R_F = 1\ k\Omega + 10\ k\Omega$ pot.

Chapter 16

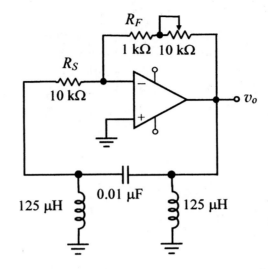

Figure 16-13:
Circuit diagram of the Hartley oscillator for Design Example 16-4

Practice Problem 16-4 Colpitts Oscillator — **DESIGN**

Design a Colpitts oscillator with the oscillation frequency f_o = 100 kHz.

Answers: C = 0.01 µF, L = 0.5 mH, R_S = 10 kΩ, and R_F = 1 kΩ + 10 kΩ pot.

Practice Problem 16-5 Hartley Oscillator — **DESIGN**

Design a Hartley oscillator with the oscillation frequency f_o = 25 kHz.

Answers: C = 0.1 µF, L = 200 µH, R_S = 10 kΩ, and R_F = 1 kΩ + 10 kΩ pot.

Practice Problem 16-6 Colpitts Oscillator — **ANALYSIS**

Determine the oscillation frequency for the oscillator circuit of Figure 16-14.

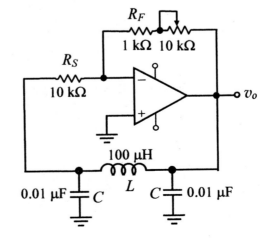

Figure 16-14: Colpitts oscillator

Answer: f_o = 225 kHz

719

16.7 SQUARE-WAVE GENERATOR

Recall that the output of the Schmitt trigger, which was introduced in Chapter 11 as a bi-reference level comparator, is a square wave with $\pm v_{o(p)} = \pm V_{sat}$ of the op-amp. With the addition of a capacitor C and a feedback resistor R, as shown in Figure 16-15(a), the need for an input signal is eliminated and the output frequency can also be controlled by proper selection of the R and C.

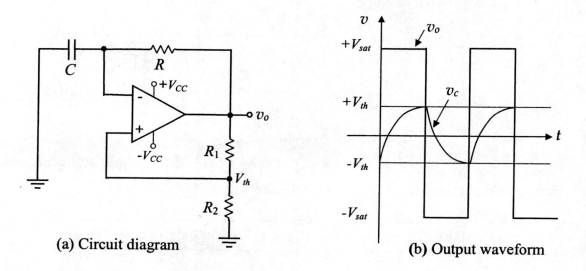

(a) Circuit diagram

(b) Output waveform

Figure 16-15: Square-wave generator

Referring to Equations 11-7 and 11-8, the upper and lower threshold voltages (V_{UT} & V_{LT}) or ($\pm V_{th}$) can be written in one equation as follows:

$$\pm V_{th} = \pm V_{sat} \frac{R_2}{R_1 + R_2} \tag{16-64}$$

It can be shown, with some considerable algebraic effort, that the period of the output waveform is as follows:

$$T = 2RC \ln\left(\frac{2R_2}{R_1} + 1\right) \tag{16-65}$$

$$f_0 = \frac{1}{T} = \frac{1}{2RC \ln(2R_2/R_1 + 1)} \tag{16-66}$$

However, if we select R_1 and R_2 such that $(1 + 2R_2/R_1) = 2.178$ (the natural log base), then $\ln(1 + 2R_2/R_1)$ will equal unity.

$$\frac{2R_2}{R_1} + 1 = 2.718 \tag{16-67}$$

$$2R_2 = 1.718 R_1 \tag{16-68}$$

$$R_2 = 0.859 R_1 \tag{16-69}$$

Chapter 16

Hence, the output frequency is a function of R and C only, and its equation simplifies as follows:

$$f_o = \frac{1}{2RC} \quad (16\text{-}70)$$

Example 16-5 — Square-Wave Generator — DESIGN

Let us now design a square-wave generator with an oscillation frequency of 10 kHz.

Solution: Let $C = 0.01\ \mu F$, and $R_1 = 10\ k\Omega$. Use a 318 op-amp for superior slew rate.

$$R = \frac{1}{2f_o C} = \frac{1}{2(10\ \text{kHz})(0.01\ \mu F)} = \frac{1}{2 \times 10^{-4}} = 5\ k\Omega$$

$$R_2 = 0.859 R_1 = 8.59\ k\Omega$$

Use a 10 kΩ potentiometer for R_2

The output $v_{o(p\text{-}p)}$ can also be controlled by connecting two Zener diodes back-to-back or face-to-face at the output, as shown in Figure 16-16(a).

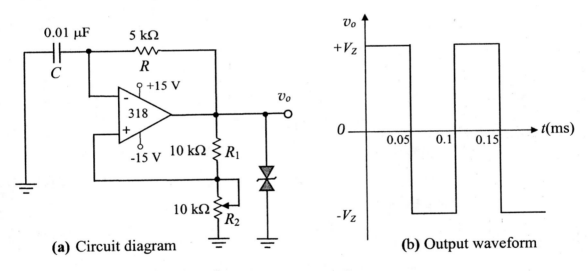

(a) Circuit diagram (b) Output waveform

Figure 16-16: Square-wave generator of Design Example 16-5

Practice Problem 16-7 — Square-Wave Oscillator — DESIGN

Design a square-wave generator with the oscillation frequency $f_o = 5$ kHz.

Answers: $C = 0.01\ \mu F$, $R = 10\ k\Omega$ Let $R_S = 10\ k\Omega$ and $R_F = 10\ k\Omega$ pot.

16.8 THE 555 TIMER

The 555 timer is a popular 8-pin integrated circuit (IC), which may be used in many applications including rectangular waveform generation. Figure 16-17 shows the common configuration of the 555 timer as it is connected to produce a rectangular waveform.

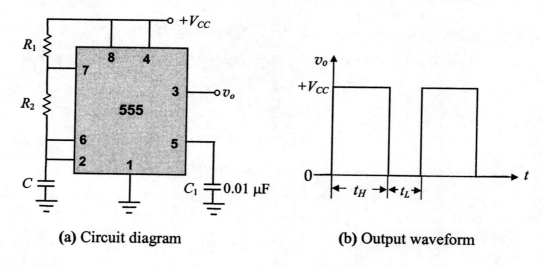

(a) Circuit diagram (b) Output waveform

Figure 16-17: The 555 timer connected as a rectangular waveform generator

The time duration for which the output is high (t_H) is given by the following equation:

$$t_H = 0.69(R_1 + R_2)C \qquad (16\text{-}71)$$

The time duration for which the output is low (t_L) is given by the following equation:

$$t_L = 0.69(R_2)C \qquad (16\text{-}72)$$

Therefore, the period and frequency of the waveform are as follows:

$$T = t_H + t_L = 0.69(R_1 + 2R_2)C \qquad (16\text{-}73)$$

$$f_o = \frac{1}{T} = \frac{1}{0.69(R_1 + 2R_2)C} \qquad (16\text{-}74)$$

For a rectangular waveform, the ratio of the pulse duration (t_H) to the period T is referred to as the *duty cycle* (*d*) of the waveform. A square wave is a rectangular waveform with $d = 0.5$ or 50% duty cycle. Examining the equations for t_H and t_L, we notice that it would not be possible to produce a square wave with the circuit of Figure 16-16. However, there is a simple solution for this problem, and that is to connect a diode across the R_2 and let $R_1 = R_2 = R$, as illustrated in Figure 16-18(a).

Chapter 16

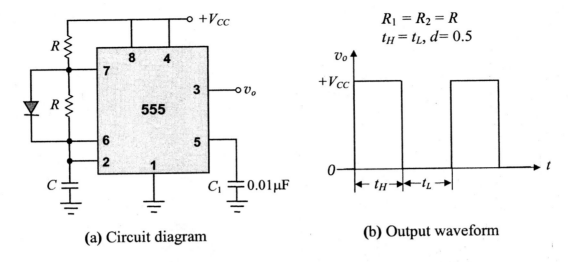

(a) Circuit diagram (b) Output waveform

Figure 16-18: The 555 timer connected as a square-wave generator

When the output is high, the diode is forward-biased, shorting out R_2; hence,

$$t_H = 0.69(R_1)C = 0.69RC \tag{16-75}$$

When the output is low, the diode is unbiased, behaving like an open-circuit; hence,

$$t_L = 0.69(R_2)C = 0.69RC \tag{16-76}$$

$$T = t_H + t_L = 0.69RC + 0.69RC = 1.38RC \tag{16-77}$$

$$f_o = \frac{1}{T} = \frac{1}{1.38RC} \tag{16-78}$$

$$d = \frac{t_H}{T} = \frac{0.69RC}{1.38RC} = 0.5 \tag{16-79}$$

In order to produce a rectangular waveform with a duty cycle less than 50% ($t_H < t_L$), we can pick R_2 larger than R_1, as required. However, the practical solution is to split R_2 into a series combination of a fixed resistor and a potentiometer, so that R_2 can be adjusted for a desired duty cycle.

Example 16-6 — Rectangular Wave — DESIGN

Let us now design a rectangular waveform generator with 25% duty cycle and an oscillation frequency of 100 kHz.

Solution: Let $C = 0.001$ μF. With R_2 shunted by a diode, the equations for the output frequency and duty cycle take the following form:

$$f_o = \frac{1}{T} = \frac{1}{0.69(R_1 + R_2)C} \tag{16-80}$$

$$d = \frac{t_H}{T} = \frac{t_H}{t_H + t_L} = \frac{0.69R_1 C}{0.69(R_1 + R_2)C} = \frac{R_1}{R_1 + R_2} \tag{16-81}$$

Chapter 16

$$d = \frac{R_1}{R_1 + R_2} = 0.25, \qquad R_1 = 0.25(R_1 + R_2), \qquad 0.75R_1 = 0.25R_2, \qquad R_2 = 3R_1$$

$$R_1 + R_2 = \frac{1}{0.69 f_o C} = \frac{1}{0.69(100\text{ kHz})(0.001\ \mu\text{F})} = 14.5\text{ k}\Omega$$

$$R_1 + R_2 = 4R_1 = 14.5\text{ k}\Omega$$

$$R_1 = \frac{14.5\text{ k}\Omega}{4} = 3.625\text{ k}\Omega$$

Let $R_1 = 3.6$ kΩ, $R_2 = 3R_1 = 10.8$ kΩ, and let $R_2 = 10$ kΩ pot $+ 1$ kΩ

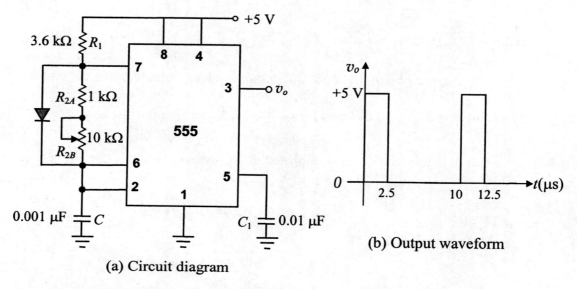

(a) Circuit diagram

(b) Output waveform

Figure 16-19: Rectangular waveform generator of Design Example 16-6

Test Run and Design Verification:

The results of simulation of the above design example, created in Electronics Workbench, are presented in Figures 16-20 and 16-21 below:

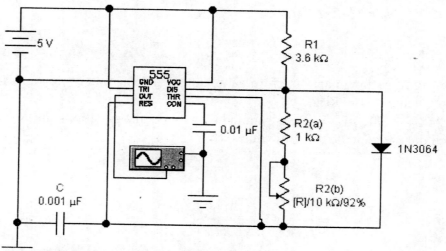

Figure 16-20: Circuit setup for simulation of Design Example 16-6

Chapter 16

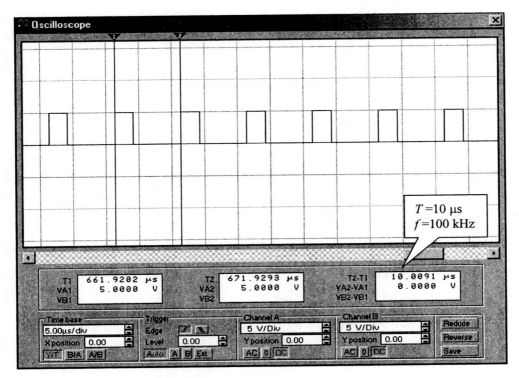

Figure 16-21: The Electronics Workbench oscilloscope showing the output waveform for Design Example 16-6

Practice Problem 16-8   *Square-Wave Generator*   **DESIGN**

Design a square-wave generator with $f_o = 200$ kHz.

Answers: $C = 0.001$ µF, $R_1 = R_2 = 3.6$ kΩ

Practice Problem 16-9   *Rectangular Wave*   **DESIGN**

Design a rectangular waveform generator with $f_o = 200$ kHz and 20% duty-cycle.

Answers: $C = 0.001$ µF, $R_1 = 1.5$ kΩ, $R_2 = 6$ kΩ   Let $R_2 = 2$ kΩ + 5 kΩ pot.

16.9 FUNCTION GENERATOR

A function generator is capable of producing sine wave, square wave, triangular wave, and possibly pulse and ramp outputs. One popular function generator *IC* is the XR-2206, which is a monolithic integrated circuit capable of producing sine, square, triangle, ramp, and pulse waveforms. The output frequency can be selected externally from 0.01 Hz to more than 1Mz. The output signal amplitude can also be controlled with an external potentiometer. The XR-2206 is a 16-pin *IC* that can operate from a dual supply power source (±5 V to ±13 V) and from a single supply power source (10 to 26 V). In addition, it can produce amplitude modulated (AM) and frequency modulated (FM) signals with an external input voltage. We will, however, review some of the setups and arrangements of XR-2206 for generating sine wave, square wave, pulse and ramp outputs. The pin diagram of the XR-2206 is exhibited in Figure 16-22 below:

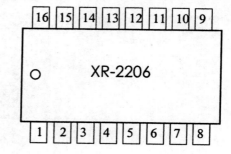

Figure 16-22:
Pin diagram of the XR-2206 function generator *IC*

Sine Wave and Square Wave Generation

The circuit setup for simultaneously generating sine waves and square waves of desired frequency and amplitude is shown in Figure 16-23.

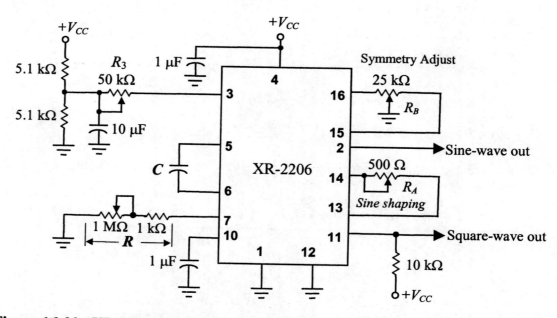

Figure 16-23: XR-2206 circuit setup for generating sinusoidal and square waveforms

Chapter 16

The output frequency is a function of the resistance R connected to pin 7 and the capacitor C connected across pins 5 and 6. This output frequency is determined by the following simple equation:

$$f_o = \frac{1}{RC} \qquad (16\text{-}82)$$

The amplitude of the output waveform can be adjusted with the potentiometer R_3 at pin 3.

Example 16-7 — Sine-Wave Generator — DESIGN

Let us now pick the value of the capacitor C, so that we can produce sinusoidal and square wave outputs with the frequency range from 100 Hz to 100 kHz.
Solution: Let $C = 0.01\ \mu F$. Hence,

$$f_{o(min)} = \frac{1}{R_{max}C} \cong \frac{1}{1\ M\Omega \times 0.01\ \mu F} = 100\ Hz$$

$$f_{o(max)} = \frac{1}{R_{min}C} = \frac{1}{1\ k\Omega \times 0.01\ \mu F} = 100\ kHz$$

Pulse and Ramp Generation

The circuit setup for simultaneously generating pulse train and ramp or sawtooth waveforms of desired frequency and amplitude is shown in Figure 16-24.

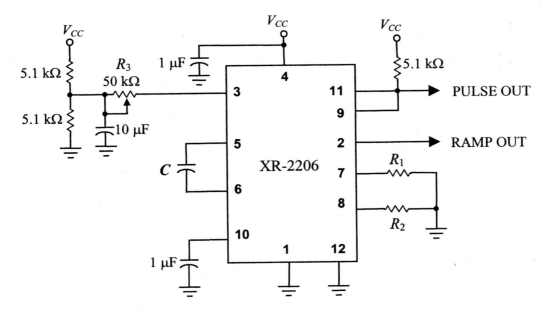

Figure 16-24: XR-2206 circuit setup for pulse and ramp generation (rectangular and sawtooth waveforms)

The output frequency is a function of the resistances R_1 and R_2 connected to pins 7 and 8, and the capacitor C connected across pins 5 and 6, and is determined with the following equation:

$$f_o = \frac{2}{(R_1 + R_2)C} \tag{16-83}$$

The duty cycle of the rectangular waveform is a function of R_1 and R_2 and is defined and determined as follows:

$$duty\ cycle = \frac{t_H}{T} = \frac{R_1}{R_1 + R_2} \tag{16-84}$$

The amplitude of the output waveform can be adjusted with the potentiometer R_3 at pin 3.

Example 16-8 — Pulse & Ramp Generator — DESIGN

Let us now determine the values of C, R_1, and R_2 in order to produce rectangular output with $f_o = 25$ kHz and 25% duty cycle.

Solution: Let $C = 0.01\ \mu F$. Hence,

$$f_o = \frac{2}{(R_1 + R_2)C}$$

$$R_1 + R_2 = \frac{2}{f_o C} = \frac{2}{25\ \text{kHz} \times 0.01\ \mu F} = 8\ k\Omega$$

$$duty\ cycle = \frac{R_1}{R_1 + R_2} = 0.25$$

$$R_1 = 0.25(R_1 + R_2)$$

$$R_2 = 3R_1$$

Let $R_1 = 2\ k\Omega$, and $R_2 = 6\ k\Omega = 3\ k\Omega + 3\ k\Omega$

Practice Problem 16-10 — Pulse Generator — ANALYSIS

Determine the frequency and duty-cycle of the rectangular output of Figure 16-24 with the following components:

$R_1 = 2.7\ k\Omega$, $R_2 = 10\ k\Omega$, and $C = 0.01\ \mu F$

Answers: $f_o = 15.75$ kHz, $d = 21.25\%$

Chapter 16

FSK Generation

FSK stands for Frequency Shift Keying, which is a form of frequency modulation with the modulating signal being binary (logic 0 or 1). That is, the XR2206 can be utilized as a voltage controlled oscillator (VCO) to produce sinusoidal output with a specific frequency (f_0) when the input voltage level is low (logic 0), and produce sinusoidal output with another specific frequency (f_1) when the input voltage level is high (logic 1).

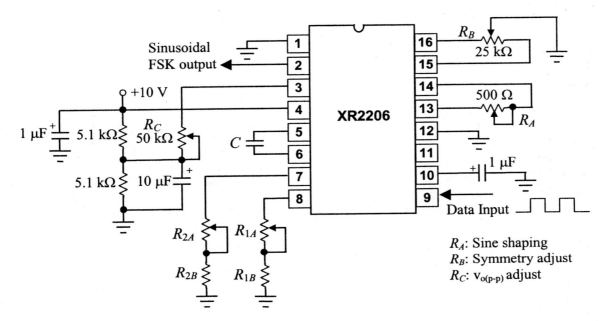

Figure 16-25: Circuit diagram of the XR2206 connected as FSK generator

The output frequencies f_0 and f_1 are achieved by a combination of two resistors R_1 and R_2 connected to pins 7 and 8 and a capacitor C attached across pins 5 and 6, as shown in Figure 16-25.

$$R_1 = \frac{1}{f_0 \times C} \qquad (16\text{-}85)$$

$$R_2 = \frac{1}{f_1 \times C} \qquad (16\text{-}86)$$

Example 16-9 — FSK Generator — DESIGN

Let us determine the values of R_1 and R_2 assuming that f_0 = 1200 Hz, and f_1 = 2200 Hz. Let C = 0.033 µF.

$$R_1 = \frac{1}{f_0 \times C} = \frac{1}{1200 \times 0.033\,\mu F} = 25.25\ k\Omega$$

Let R_1 = 10 kΩ pot + 18 kΩ

$$R_2 = \frac{1}{f_1 \times C} = \frac{1}{2200 \times 0.033\,\mu F} = 13.77\,k\Omega$$

Let $R_2 = 10\,k\Omega$ pot. $+ 5.1\,k\Omega$

Practice Problem 16-11 — FSK Generator — DESIGN

Determine the values of C, R_1 and R_2 for the FSK output with $f_0 = 2025$ Hz and $f_1 = 2225$ Hz.

Answers:

Let $C = 0.033\,\mu F$, $R_1 = 15\,k\Omega = 10\,k\Omega$ pot. $+ 10\,k\Omega$, $R_2 = 13.6\,k\Omega = 10\,k\Omega$ pot. $+ 5.1\,k\Omega$

16.10 SUMMARY

- With the addition of a few external components, an op-amp can be utilized to produce waveforms of desired frequency and amplitude, without any external input. The output waveform may be sinusoidal, rectangular, or triangular.

- An oscillator is generally an amplifier with positive feedback, which produces a periodic alternating output signal with a certain frequency, without any external input signal.

- The oscillator operation principle requires that the loop gain must equal unity; that is, the attenuation in the feedback network must be compensated with the gain of the amplifier. Furthermore, the overall phase shift around the loop must equal 0°.

- The RC phase-shift oscillator, which produces a sinusoidal output, consists of an inverting amplifier for the required gain and a cascade of three RC circuits for the feedback network, as shown in Figure 16-2.

- The Wien-bridge oscillator also produces sinusoidal output. It consists of a non-inverting amplifier and a feedback network, which has a series RC circuit in one arm of the Wien-bridge and a parallel RC circuit in the adjoining arm, as shown in Figure 16-6.

- The feedback network of the Colpitts oscillator is a resonant LC network, which consists of two capacitors and an inductor, as shown in Figure 16-10. The oscillation frequency is the resonance frequency of the LC circuit.

- The feedback network of the Hartley oscillator is also a resonant LC network, which consists of two inductors and a capacitor, as shown in Figure 16-12. The oscillation frequency is the resonance frequency of the LC circuit.

Chapter 16

- All of the above-mentioned oscillators produce sinusoidal waveforms. However, the Schmitt trigger can be utilized as a square-wave generator with the addition of a capacitor C and a feedback resistor R, as shown in Figure 16-15. The output frequency can be controlled by proper selection of R and C.

- The 555 timer is an 8-pin integrated circuit, which may be used in many applications, including rectangular waveform generation. Figure 16-17 shows the common configuration of the 555 timer as it is utilized to produce a rectangular waveform. However, Figures 16-18 and 16-19 show more specific configurations of the 555 timer for producing square wave (50% duty cycle) and rectangular waveform of desired frequency and duty cycle, respectively.

- The XR-2206 is a 16-pin integrated circuit function generator, which is capable of producing sine wave, square wave, triangular wave, pulse, or ramp outputs. The output frequency can be selected externally from 0.01 Hz to more than 1 MHz. The output signal amplitude can also be controlled with an external potentiometer.

Chapter 16

PROBLEMS

16.1   Describe the principles of oscillation.

16.2   Determine the output frequency for the oscillator of Figure 16-1P.

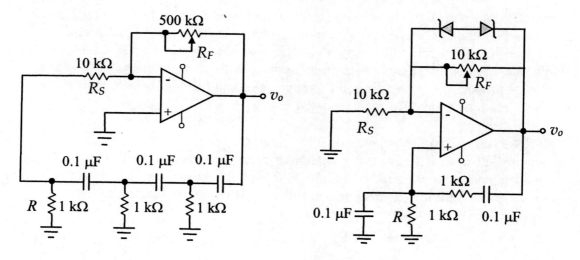

Figure 16-1P: RC phase-shift oscillator

Figure 16-2P: Wien-bridge oscillator

16.3   Carry out Practice Problem 16-1.

16.4   Design a phase-shift oscillator with the oscillation frequency $f_o = 1$ kHz.

16.5   Determine the output frequency for the oscillator of Figure 16-2P.

16.6   Carry out Practice Problem 16-2.

16.7   Carry out Practice Problem 16-3.

16.8   Design a Wien-bridge oscillator with the oscillation frequency $f_o = 2$ kHz.

16.9   Determine the output frequency for the oscillator of Figure 16-3P.

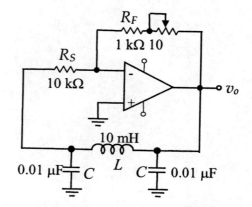

Figure 16-3P: Colpitts oscillator

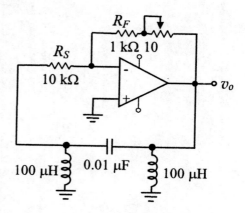

Figure 16-4P: Hartley oscillator

Chapter 16

16.10 Determine the output frequency for the oscillator of Figure 16-4P.

16.11 Carry out Practice Problem 16-4.

16.12 Design a Colpitts oscillator with the oscillation frequency $f_o = 33$ kHz.

16.13 Carry out Practice Problem 16-5.

16.14 Design a Hartley oscillator with the oscillation frequency $f_o = 20$ kHz.

16.15 Carry out Practice Problem 16-6.

16.16 Using an appropriate op-amp, design a square-wave generator with an output frequency $f_o = 25$ kHz and $v_{o(p)} \cong 5$ V.

16.17 Carry out Practice Problem 16-7.

16.18 Using a 555 timer, design a square-wave generator with an output frequency $f_o = 25$ kHz and $v_{o(p)} \cong 5$ V.

16.19 Design a rectangular waveform generator with an output frequency $f_o = 50$ kHz, 25% duty cycle, and $v_{o(p)} \cong 5$ V.

16.20 Carry out Practice Problem 16-8.

16.21 Design a rectangular waveform generator with an output frequency $f_o = 150$ kHz, 20% duty cycle, and $v_{o(p)} \cong 5$ V.

16.22 Carry out Practice Problem 16-9.

16.23 Using XR-2206, design a square-wave generator with an output frequency $f_o = 250$ kHz.

16.24 Using XR-2206, design a sine-wave generator with an output frequency $f_o = 250$ kHz.

16.25 Using XR-2206, design a rectangular waveform generator with an output frequency $f_o = 200$ kHz, 20% duty cycle, and $v_{o(p)} \cong 5$ V.

16.26 Using XR-2206, design a triangular waveform generator with an output frequency $f_o = 100$ kHz.

16.27 Using XR-2206, design a sawtooth waveform (ramp) generator with an output frequency $f_o = 20$ kHz.

16.28 Carry out Practice Problem 16-10.

Chapter 17

DIGITAL-TO-ANALOG AND ANALOG-TO-DIGITAL CONVERSION

Analysis & Design

17.1 INTRODUCTION

Nowadays, the transmission, reception, and processing of information is mainly digital. Although the information might originally be analog such as voice, temperature, pressure, camera recordings, etc., it is digitized before being processed by a computer and/or transmitted over a local area network or the public telephone network. Upon reception, it might be necessary to convert the digital information back to the original analog form after processing. Hence, we will cover the principles of digital-to-analog and analog-to-digital conversions in this chapter. In addition, we will also review some of the integrated circuit devices that are designed to perform these conversions.

17.2 DIGITAL-TO-ANALOG CONVERSION

After the digital systems have received and processed the digital information, it might be necessary to convert the digital information to analog. Consider the voice signal, which is originally analog; however, in order to transmit the voice signal over the telephone network, it is first digitized and then transmitted in the form of 8-bit binary codes. The main reason that voice is transmitted in digital form is that digital transmission is much more immune to noise compared to analog transmission. At the receiver, the received information, which is in the form of 8-bit binary codes, is then converted back to analog. The digital-to-analog converter, also referred to as DAC, converts each binary code to a pulse with amplitude corresponding to the binary value of the code. In general, an n-bit DAC will generate 2^n distinct codes and thus 2^n distinct output voltage levels.

The R-2R Ladder DAC

The R-2R digital-to-analog converter utilizes a series-parallel combination of two resistors with values R and 2R in a ladder network. The circuit diagram of a 4-bit R-2R ladder DAC is illustrated in Figure 17-1. The op-amp buffer isolates the ladder network from the output by providing almost infinite input resistance.

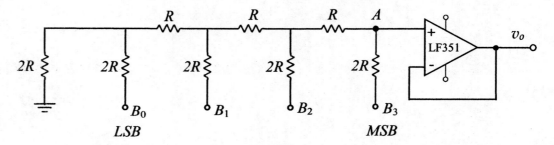

Figure 17-1: Circuit diagram of a 4-bit R-2R ladder DAC

Assuming TTL logic levels for the digital inputs, that is, logic 0 = 0 V, and logic 1 = 5 V, let us determine the output voltage levels for the following input conditions:
(a) $B_0 = B_1 = B_2 = B_3 = 0$ V

All inputs being 0, the non-inverting input voltage of the op-amp $V_A = 0$; hence, $v_o = 0$.
(b) $B_0 = B_1 = B_2 = 0$ V, $B_3 = 5$ V

The equivalent circuit of the DAC with the above input voltage levels is presented in Figure 17-2.

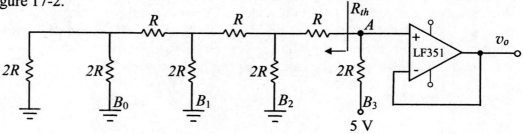

Figure 17-2: Equivalent circuit diagram of the DAC for step (b)

Chapter 17

$R_{th} = R + \{2R || [R + (2R || (R + R))]\}$
$= R + [2R || 2R] = 2R$

$V_A = 5\,V \dfrac{R_{th}}{2R + R_{th}} = 2.5\,V$

$v_o = V_A = 2.5\,V$

Input: $1000 \Rightarrow v_o = 2.5\,V$

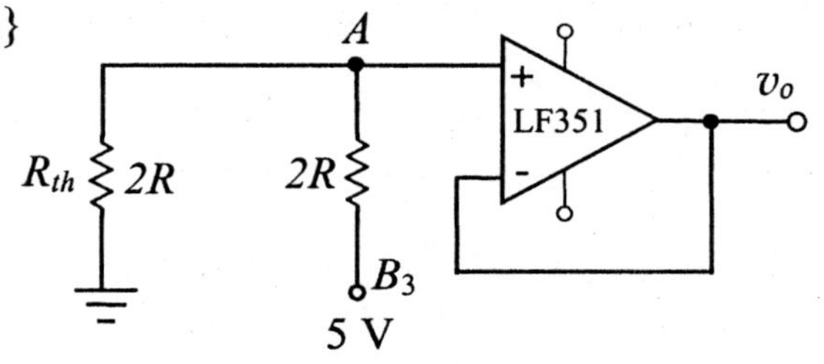

(c) $B_0 = B_1 = B_3 = 0\,V, B_2 = 5\,V$

Figure 17-3: Thevenized equivalent circuit

The equivalent circuit of the DAC with the above input voltage levels is presented in Figure 17-4.

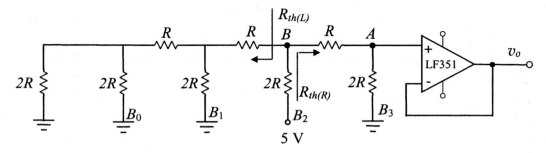

Figure 17-4: Equivalent circuit diagram of the DAC for step (c)

Thevenin's equivalent resistances to the left and to the right of point B are the following:

$R_{th(L)} = R + \{2R || [R + (2R || 2R)]\} = R + [2R || 2R] = 2R$
$R_{th(R)} = R + 2R = 3R$
$R_{th} = R_{th(L)} || R_{th(R)} = 2R || 3R = 1.2R$

$V_B = 5\,V \dfrac{R_{th}}{2R + R_{th}} = 5\,V \dfrac{1.2\,V}{3.2\,V} = 1.875\,V$

$V_A = V_B \dfrac{2R}{R + 2R} = 1.875\,V \dfrac{2}{3} = 1.25\,V$, $v_o = V_A = 1.25\,V$. Input: $0100 \Rightarrow v_o = 1.25\,V$

(d) $B_0 = B_2 = B_3 = 0\,V, B_1 = 5\,V$

The equivalent circuit of the DAC with the above input voltage levels is presented in Figure 17-5.

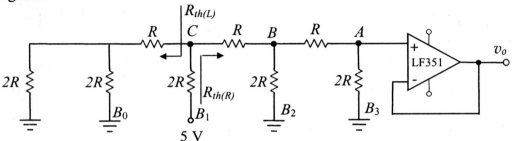

Figure 17-5: Equivalent circuit diagram of the DAC for step (d)

Thevenin's equivalent resistances to the left and to the right of point C are the following:

$R_{th(L)} = R + (2R \| 2R) = R + R = 2R$

$R_{th(R)} = R + [2R \| (2R + R)] = R + 1.2R = 2.2R$

$R_{th} = R_{th(L)} \| R_{th(R)} = 2R \| 2.2R = 1.0476R$

$V_C = 5\,\text{V}\,\dfrac{R_{th}}{2R + R_{th}} = 5\,\text{V} \times \dfrac{1.0476}{3.0476} = 1.71873\,\text{V}$

$V_B = V_C \dfrac{2R}{2R + R} = 1.71873\,\text{V} \times \dfrac{2}{3} = 1.1458\,\text{V}$

$V_A = V_B \dfrac{2R}{2R + R} = 1.1458\,\text{V} \times \dfrac{2}{3} = 0.764\,\text{V}$

$v_o = V_A = 0.764\,\text{V}$ Input: 0010 $\Rightarrow v_o = 0.764\,\text{V}$

(e) $B_1 = B_2 = B_3 = 0\,\text{V},\ B_0 = 5\,\text{V}$

The equivalent circuit of the DAC with the above input voltage levels is depicted in Figure 17-6.

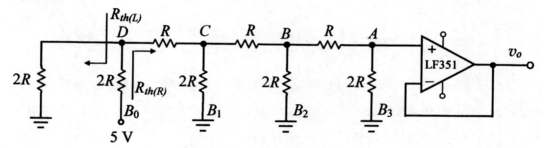

Figure 17-6: Equivalent circuit diagram of the DAC for step (e)

Thevenin's equivalent resistances to the left and to the right of point D are the following:

$R_{th(L)} = 2R$

$R_{th(R)} = R + \{2R \| [R + (2R \| (R + 2R))]\} = R + (2R \| 1.2R) = 2.0475R$

$R_{th} = R_{th(L)} \| R_{th(R)} = 2R \| 2.0475R = 1.011735R$

$V_D = 5\,\text{V}\,\dfrac{R_{th}}{2R + R_{th}} = 5\,\text{V} \times \dfrac{1.011735}{3.011735} = 1.67965\,\text{V}$

$V_C = V_D \dfrac{1.0475R}{R + 1.0475R} = 1.67965\,\text{V} \times \dfrac{1.0475}{2.0475} = 0.8953\,\text{V}$

$V_B = V_C \dfrac{1.2R}{2.2R} = 0.8953\,\text{V} \times \dfrac{1.2}{2.2} = 0.484\,\text{V}$

$V_A = V_B \dfrac{2R}{2R + R} = 0.4884\,\text{V} \times \dfrac{2}{3} = 0.3256\,\text{V}$

$v_o = V_A = 0.3256\,\text{V}$.   Input: 0001 $\Rightarrow v_o = 0.3256\,\text{V}$

Chapter 17

For determining the output for the case when more than one input is high, we can apply the superposition theorem and sum up the corresponding individual outputs. For example, when $B_0 = B_1 = B_2 = B_3 = 5$ V, then the output will be as follows:
$v_o = 0.3125$ V $+ 0.625$ V $+ 1.25$ V $+ 2.5$ V $= 4.6875$ V. Input: 1111 $\Rightarrow v_o = 04.6875$ V

Applying the same procedure, we can determine the outputs for all 16 input bit combinations, as listed in Table 17-1.

Table 17-1:
Input and output of a 4-bit DAC

Input (V) $B_3\ B_2\ B_1\ B_0$	Output Voltage (V)
0 0 0 0	0.0000
0 0 0 5	0.3125
0 0 5 0	0.6250
0 0 5 5	0.9375
0 5 0 0	1.2500
0 5 0 5	1.5625
0 5 5 0	1.8750
0 5 5 5	2.1875
5 0 0 0	2.5000
5 0 0 5	2.8125
5 0 5 0	3.1250
5 0 5 5	3.4375
5 5 0 0	3.7500
5 5 0 5	4.0625
5 5 5 0	4.3750
5 5 5 5	4.6875

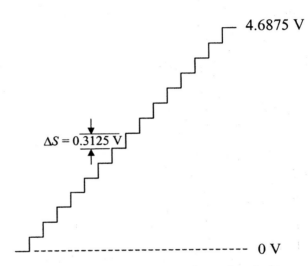

Figure 17-7: 16 output voltage levels of the 4-bit DAC from 0 V to 4.6875 V

Conclusion

From the foregoing analyses we can conclude that for an *n*-bit DAC the step size (ΔS) equals the logic 1 input voltage level (V_H) divided by 2^n.

$$\Delta S = \frac{V_H}{2^n} \qquad (17\text{-}1)$$

Furthermore, the output voltage of the DAC for a given input equals the product of the decimal equivalent of the binary input times ΔS. For example, assuming binary input of 1100 for the above 4-bit DAC, the step size (ΔS) and the output voltage (v_o) may be determined as follows:

$$\Delta S = \frac{5\text{ V}}{2^4} = \frac{5\text{ V}}{16} = 0.3125\text{ V}$$
$$v_o = 12\ \Delta S = 12 \times 0.3125\text{ V} = 3.75\text{ V}$$

Example 17-1 — R-2R DAC — ANALYSIS

Let us determine the step size and the output voltage level for an 8-bit *R-2R* DAC, assuming TTL logic levels.

a) Input bits: 10101010
b) Input bits: 01010101

Solution:

$$\Delta S = \frac{5\text{ V}}{2^8} = \frac{5\text{ V}}{256} = 19.53125 \text{ mV}$$

a) Decimal equivalent $= 2^7 + 2^5 + 2^3 + 2^1 = 170$

$v_o = 170 \Delta S = 170 \times 0.01953125 \text{ V} = 3.32 \text{ V}$

b) Decimal equivalent $= 2^6 + 2^4 + 2^2 + 2^0 = 85$

$v_o = 85 \Delta S = 85 \times 0.01953125 \text{ V} = 1.66 \text{ V}$

Practice Problem 17-1 — R-2R DAC — ANALYSIS

Determine the step size and the output voltage level for a 6-bit *R-2R* DAC, assuming TTL logic levels.

a) Input bits: 110101
b) Input bits: 101011

Answers: $\Delta S = 0.078125$ V   (a) $v_o = 4.14$ V, (b) $v_o = 3.36$ V

Binary-Weighted Resistor DAC

The binary-weighted resistor DAC is simply an inverting summer with $R_n = R/2^n$. The circuit diagram of a 4-bit binary-weighted resistor DAC is shown in Figure 17-8.

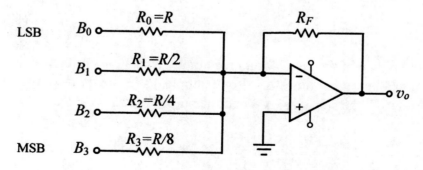

Figure 17-8: Circuit diagram of a 4-bit binary-weighted resistor DAC

$$v_o = -\left(\frac{R_F}{R}B_0 + \frac{R_F}{R/2}B_1 + \frac{R_F}{R/4}B_2 + \frac{R_F}{R/8}B_3\right) \quad (17\text{-}2)$$

$$v_o = -\left(\frac{R_F}{R}B_0 + 2\frac{R_F}{R}B_1 + 4\frac{R_F}{R}B_2 + 8\frac{R_F}{R}B_3\right) \quad (17\text{-}3)$$

$$v_o = -\frac{R_F}{R}(B_0 + 2B_1 + 4B_2 + 8B_3) \quad (17\text{-}4)$$

Assuming $R_F = 1$ kΩ, $R = 10$ kΩ, and TTL logic levels for the digital inputs, that is, logic 0 = 0 V, and logic 1 = 5 V, let us determine the output voltage levels for the following input conditions:

(a) $B_0 = B_1 = B_2 = B_3 = 0$ V. All inputs being 0, the output $v_o = 0$.

(b) $B_1 = B_2 = B_3 = 0$ V, $B_0 = 5$ V. $v_o = -0.1(5$ V$) = -0.5$ V. $0001 \Rightarrow v_o = -0.5$ V

(c) $B_0 = B_2 = B_3 = 0$ V, $B_1 = 5$ V. $v_o = -0.1(10$ V$) = -1$ V. $0010 \Rightarrow v_o = -1$ V

(d) $B_0 = B_1 = B_3 = 0$ V, $B_2 = 5$ V. $v_o = -0.1(20$ V$) = -2$ V. $0100 \Rightarrow v_o = -2$ V

(e) $B_0 = B_1 = B_2 = 0$ V, $B_3 = 5$ V. $v_o = -0.1(40$ V$) = -4$ V. $1000 \Rightarrow v_o = -4$ V

In determining the output for a case in which more than one input is high, we can apply the superposition theorem and sum up the corresponding individual outputs. For example, when $B_0 = B_1 = B_2 = B_3 = 5$ V, then the output will be as follows:

$v_o = -0.5$ V $- 1$ V $- 2$ V $- 4$ V $= -7.5$ V. Input: 1111 $\Rightarrow v_o = -7.5$ V

Practice Problem 17-2   R-2R DAC   ANALYSIS

Determine the output voltage level for the 4-bit binary-weighted resistor DAC of Figure 17-9, assuming TTL logic levels.

a) Input bits: 1101

b) Input bits: 1010

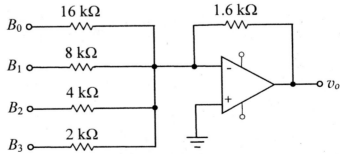

Figure 17-9: A 4-bit binary-weighted resistor DAC

Answers: (a) $v_o = -6.5$ V,   (b) $v_o = -5.5$ V

Monolithic Integrated Circuit DAC

There are integrated circuit digital-to-analog converters available with the number of input bits n ranging from 8 to 24 bits and with current or voltage output. The DAC0808 is a common 8-bit digital-to-analog converter IC chip from National Semiconductor with current output, which can be converted to voltage with an op-amp configuration of I-to-V converter. The DAC0808 uses the R-$2R$ ladder network as the conversion technique. A typical connection diagram of the DAC0808, including the current-to-voltage converter, is displayed in Figure 17-10, where pin 1 is not connected (NC) and normally $R_A = R_B$.

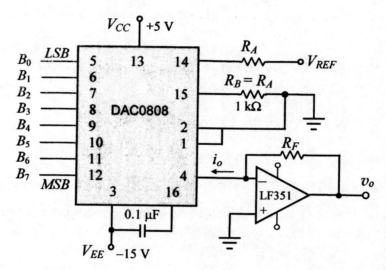

Figure 17-10: Typical connection diagram of DAC0808

$$v_o = V_{REF} \frac{R_F}{R_A}\left(\frac{B_7}{2} + \frac{B_6}{4} + \frac{B_5}{8} + \frac{B_4}{16} + \frac{B_3}{32} + \frac{B_2}{64} + \frac{B_1}{128} + \frac{B_0}{256}\right) \quad (17\text{-}5)$$

For example, for $V_{REF} = 10$ V and $R_F = R_A = 4.7$ kΩ, $I_{RA} \approx V_{REF}/R_A = 10/4.7 = 2.1$ mA,

$$v_o = 10\text{ V}\left(\frac{B_7}{2} + \frac{B_6}{4} + \frac{B_5}{8} + \frac{B_4}{16} + \frac{B_3}{32} + \frac{B_2}{64} + \frac{B_1}{128} + \frac{B_0}{256}\right) \quad (17\text{-}6)$$

Similarly, for $V_{REF} = 2$ V, $R_F = R_A = R_B = 1$ kΩ, $I_{RA} \approx V_{REF}/R_A = 2/1 = 2$ mA, and with all binary inputs being high, the output will be 9.96 V.

The MC1408 is another 8-bit DAC with current output and is manufactured by Motorola. The NE/SE 5018 is an 8-bit DAC with voltage output and is manufactured by Philips Semiconductors, formerly Signetics.

Some of the important parameters that are usually included in manufacturer's data sheets are *resolution, linearity error, offset error,* and *settling time*.

Resolution is determined as $1/2^n$, where n is the number of input bits. For an 8-bit DAC the resolution is $1/2^8 = 1/256 = 0.39\%$.

Nonlinearity or linearity error is the difference between the actual output of the DAC and its ideal straight-line output. The error is normally expressed as a percentage of the full-scale range.

Offset error is the nonzero reading at the output with all binary inputs being zero. This error may be caused by the input offsets of the op-amp and the DAC.

Settling time is the time required for the output of the DAC to reach from zero to full-scale.

17.3 ANALOG-TO-DIGITAL CONVERSION

As mentioned earlier, analog signals are converted to digital form before being processed by a computer. There are several analog-to-digital conversion methods such as *flash* ADC, *tracking* ADC, *integrating* ADC, and *successive approximation*. The one popular method most commonly used is the successive approximation technique. Hence, we will take a brief look at this method of analog-to-digital conversion. A block diagram of a 4-bit successive approximation ADC is exhibited in Figure 17-11.

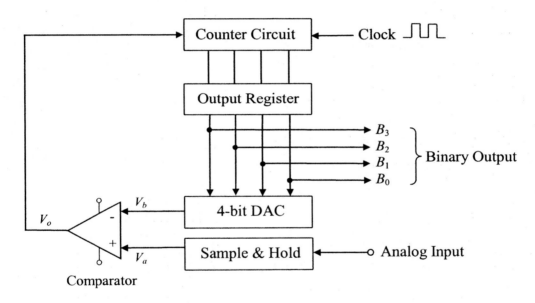

Figure 17-11: Block diagram of a 4-bit successive approximation ADC

The *sample and hold* circuit samples the analog input and holds on to it so that the input voltage level to the comparator (V_a) stays constant during the conversion cycle. The *comparator* compares the output of the sample and hold circuit (V_a) to the output of the DAC (V_b). If $V_a \geq V_b$, the output of the comparator goes high (logic 1); otherwise it goes low (logic 0). For the sake of convenience and simplicity, let us assume that the analog input ranges from 0 V to 15 V, so that it corresponds directly to the 4-bit binary output ranging from 0000 to 1111.

To begin the conversion process, let us assume that $V_a = 10$ V. On the first clock pulse, the MSB of the output register is set high; that is, the output register is loaded with 1000. Hence, the output of the DAC $V_b = 8$ V, which is compared to $V_a = 10$ V. Since $V_a > V_b$,

Chapter 17

the comparator outputs a logic 1, which is the input to the counter circuit. Receiving a logic 1 command, the counter maintains $B_3 = 1$ and sets the next significant bit (B_2) high; that is, the output register is loaded with 1100. Now that $V_b = 12$ V and $V_a = 10$ V, $V_a < V_b$; hence, the comparator outputs a logic 0. Receiving a logic 0 command, the counter resets $B_2 = 0$ and sets the next significant bit (B_1) high; that is, the output register is loaded with 1010. With $V_b = 10$ V and $V_a = 10$ V, $V_a = V_b$; hence, the comparator outputs a logic 1. Receiving a logic 1 command, the counter maintains $B_1 = 1$ and sets the next significant bit (B_0) high; that is, the output register is loaded with 1011. Now that $V_b = 13$ V and $V_a = 10$ V, $V_a < V_b$; hence, the comparator outputs a logic 0. Receiving a logic 0 command, the counter resets $B_0 = 0$ and ends the conversion cycle. Hence, the output register is loaded with 1010 at the end of the conversion cycle, which reads 10 V. Note that for an n-bit DAC, each conversion cycle takes n clock periods; that is, for an 8-bit DAC with clock frequency of 10 MHz, it will take 0.8 μs to complete one conversion cycle.

Monolithic Integrated Circuit ADC

There are a number of integrated circuit analog-to-digital converters available from several semiconductor manufacturers. The TDF8704 is an 8-bit general-purpose analog-to-digital converter *IC* from Philips Semiconductors. It converts the analog input signal to 8-bit binary-coded digital words at a maximum sampling rate of 50 MHz. The full-scale voltage range for the analog input signal is 7 V; that is, $v_{i(p-p)max} = 7$ V.

The pin configuration and a typical connection diagram of the TDF8704 ADC are exhibited in Figure 17-12 and Figure 17-13 below:

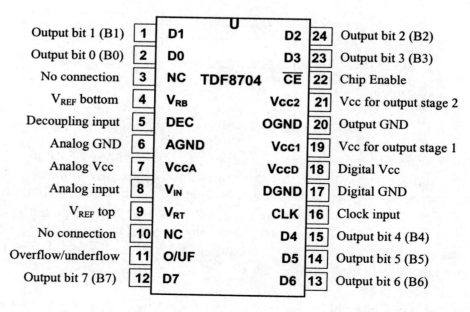

Figure 17-12: Pin configuration of the TDF8704 ADC

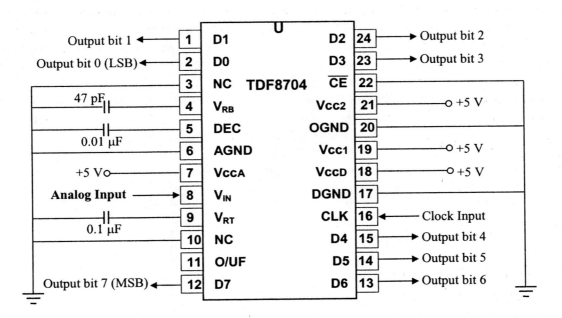

Figure 17-13: Typical connection diagram of the TDF8704 ADC

There are three versions of the TDF8704 analog-to-digital converter: TDF8704/2 with clock frequency of 20 MHz, TDF4704/4 with clock frequency of 40 MHz, and TDF4704/5 with clock frequency of 50 MHz.

The ADC0808 is another 8-bit analog-to-digital converter IC with 8-channel multiplexer and microprocessor compatible control logic.

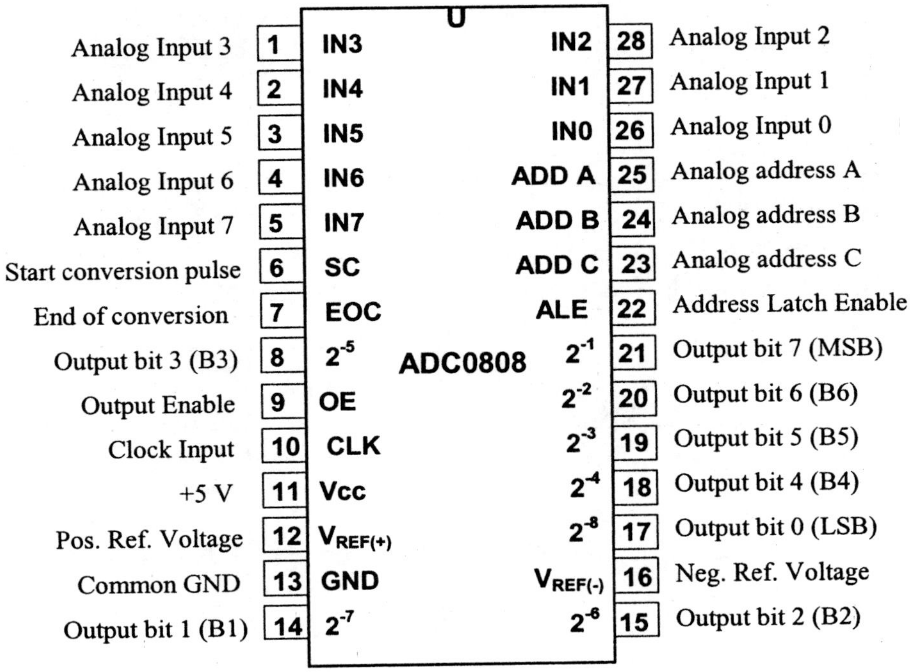

Figure 17-14: Pin configuration of the ADC0808 A/D converter

Chapter 17

The ADC0808 is manufactured by National Semiconductor and uses the successive approximation as the conversion technique. Its typical clock frequency is 640 kHz; however, it can range from a minimum of 10 kHz to a maximum of 1.28 MHz. The 8-channel multiplexer allows up to eight input analog signals to be converted sequentially and outputted digitally over a single medium. The ADC0816 is the 16-channel version of it, which allows up to 16 analog input signals to be converted and multiplexed. The pin diagram of the ADC0808 is exhibited in Figure 17-14, and its typical connection is shown in Figure 17-16(b). The address line logic levels and the corresponding analog input lines of ADC0808 DAC are exhibited in Table 17-2 below:

Table 17-2: Address line logic levels and selected analog inputs

Address line C	Address line B	Address line A	Selected analog input
L	L	L	IN 0
L	L	H	IN 1
L	H	L	IN 2
L	H	H	IN 3
H	L	L	IN 4
H	L	H	IN 5
H	H	L	IN 6
H	H	H	IN 7

Example 17-2 — Mixed Signal — DESIGN

Recall that in Chapter 1 we learned that variations in temperature could have a marked effect on the characteristics of a *p-n* junction. One side effect of this change in the characteristics is that the so-called firing potential (V_F of a diode or V_{BE} of a BJT) of a forward-biased *p-n* junction changes inversely with temperature. That is, as the temperature increases, the firing potential decreases at a rate of approximately 2 mV/ °C.

$$\frac{\Delta V}{\Delta T} = -\frac{2\,\text{mV}}{°\text{C}} \qquad (17\text{-}7)$$

Taking advantage of this change in junction voltage, a diode may be used as a sensor of the temperature. The sensed signal may then be amplified and shifted to a level compatible with an A/D converter whose output may be connected to two 7-segment display units to display the temperature. A block diagram of such a system is given below.

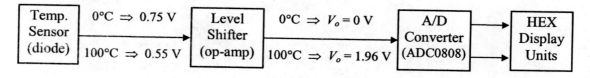

Figure 17-15: Block diagram of the digital thermometer

The thermometer will be calibrated at 0°C (ice water) and 100°C (boiling water).

Chapter 17

<u>Analog-to-digital converter (ADC0808):</u>

Number of bits per code $n = 8$ bits, and full-scale voltage $V_{FS} = 5$ V.
Then the step size is determined with the following equation:

$$\Delta S = \frac{V_{FS}}{2^n - 1} = \frac{5 \text{ V}}{2^8 - 1} = 19.6 \text{ mV} \tag{17-8}$$

Let the change of 1°C in temperature correspond directly to the step size (19.6 mV) of the ADC. Hence, at 1°C/step, $V_{max} = \Delta S \times 100 = 1.96$ V at 100°C, and $V_{min} = 0$ V at 0°C.

<u>Temperature sensor (1N4001):</u>

At 25°C, the nominal $V_F = 0.7$ V. At 0°C, $V_F = 0.7$ V + (2 mV/ °C × 25°C) = 0.75 V.
At 100°C, $V_F = 0.7$ V − (2 mV/ °C × 75°C) = 0.55 V.
As the temperature changes from 0°C to 100°C, $\Delta V = 0.75$ V − 0.55 V = 0.20 V.

<u>Level-shifter (summing amplifier):</u>

The gain needed for the sensed voltage V_F, from the level shifter, is as follows:

$$A_V = -\frac{1.96 \text{ V}}{0.2 \text{ V}} = -9.8 \tag{17-9}$$

At 0°C, $V_F = 0.75$ V and the output of the level-shifter, which is the input to ADC, must equal 0 V. That is, V_F will be amplified by $A_v = -9.8$ producing −7.35 V; hence, a +7.35 V must be added through the second input of the inverting summer in order to produce a net output voltage of 0 V.

$$V_o = -7.35 \text{ V} + 7.35 \text{ V} = 0 \text{ V} \tag{17-10}$$

At 100°C, $V_F = 0.55$ V and the output of the level-shifter must equal 1.96 V. Again, V_F will be amplified by $A_v = -9.8$ producing −5.39 V, which will equal 1.96 V by addition of the same +7.35 V through the second input of the inverting summer.

$$V_o = -5.39 \text{ V} + 7.35 \text{ V} = 1.96 \text{ V} \tag{17-11}$$

<u>Oscillator:</u>

A 555 timer will be utilized to generate the clock signal of 200 kHz with 50% duty cycle for the ADC, as shown in Figure 17-15.
Let $C = 0.001$ μF, then determine R_1 and R_2 with Equations 16-80 and 16-81 as follows:

$$f_o = \frac{1}{0.69(R_1 + R_2)C}$$

$$d = \frac{R_1}{R_1 + R_2} = 0.5 \Rightarrow R_1 = R_2 = R$$

$$R_1 + R_2 = \frac{1}{0.69 f_o C} = \frac{1}{0.69(200 \text{ kHz})(0.001 \text{ μF})} = 7.25 \text{ k}\Omega = 2R$$

$$R_1 = R_2 \cong 3.6 \text{ k}\Omega$$

Let $R_1 = 3.6$ kΩ, and $R_2 = 1$ kΩ pot. + 3 kΩ

The connection and circuit diagrams for all the above mentioned components, including the temperature sensor, level-shifter, clock signal generator, A/D converter, and the HEX display units, are exhibited in Figure 17-16 below:

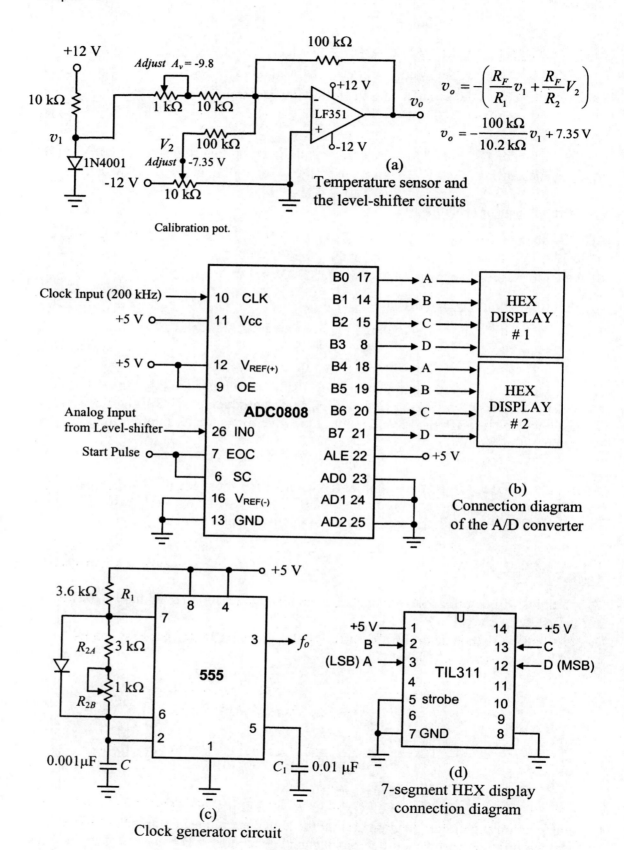

Figure 17-16: Circuit and connection diagrams of the digital thermometer components

Chapter 17

17.4 SUMMARY

- Nowadays, the transmission, reception, and processing of information is mainly digital. Consider voice data, which is originally analog; however, in order to transmit voice over telephone network, it is first digitized and then transmitted in the form of 8-bit binary codes.

- Upon reception, it might be necessary to convert the digital information back to the original analog form after processing.

- The main reason that voice is transmitted in digital form is that digital transmission is much more immune to noise compared to analog transmission.

- The digital-to-analog converter, also referred to as DAC, converts each binary code to a pulse with amplitude corresponding to the binary value of the code.

- In general, an n-bit DAC will generate 2^n distinct codes and thus 2^n distinct output voltage levels.

- The R-2R digital-to-analog converter utilizes a series-parallel combination of two resistors with values R and 2R in a ladder network.

- The binary-weighted resistor DAC is simply an inverting summer with $R_n = R/2^n$.

- There are integrated circuit digital-to-analog converters available with the number of input bits n ranging from 8 to 24 bits, and with current or voltage output.

- The DAC0808 is a common 8-bit DAC *IC* chip from National Semiconductor with current output, which can be converted to voltage with an op-amp I-to-V converter.

- The MC1408 is another 8-bit DAC with current output manufactured by Motorola.

- The NE/SE 5018 is an 8-bit DAC with voltage output manufactured by Philips Semiconductors, formerly Signetics.

- Some of the important parameters that are usually included in manufacturer's data sheets are *resolution*, *linearity error*, *offset error*, and *settling time*.

- Analog signals must be converted to digital form before being processed by a computer.

- There are several analog-to-digital conversion methods such as *flash* ADC, *tracking* ADC, *integrating* ADC, and *successive approximation*. The one popular method most commonly used is the successive approximation technique.

- The successive approximation ADC consists of a sample and hold circuit, a comparator, a successive approximation register (SAR), which is the combination of the counter circuit and the output register, and a digital to analog converter.

- The *sample and hold* circuit samples the analog input and holds on to it so that the input voltage level to the comparator (V_a) stays constant during the conversion cycle.

- The *comparator* compares the output of the sample and hold circuit (V_a) to the output of the DAC (V_b). If $V_a \geq V_b$, the output of the comparator goes high (logic 1), otherwise it goes low (logic 0).

- There are number of integrated circuit analog-to-digital converters available from several semiconductor manufacturers. The TDF8704 is an 8-bit general-purpose analog-to-digital converter *IC* from Philips Semiconductors.

- There are three versions of the TDF8704 analog-to-digital converter: TDF8704/2 with clock frequency of 20 MHz, TDF4704/4 with clock frequency of 40 MHz, and TDF4704/5 with clock frequency of 50 MHz.

- The ADC0808 is another 8-bit analog-to-digital converter *IC* with 8-channel multiplexer and microprocessor compatible control logic.

- The ADC0808 is manufactured by National Semiconductor and uses the successive approximation as the conversion technique. Its typical clock frequency is 640 kHz; however, it can range from a minimum of 10 kHz to a maximum of 1.28 MHz.

- The 8-channel multiplexer allows up to eight input analog signals to be converted sequentially and outputted digitally over a single medium.

- The ADC0816 allows up to 16 analog input signals to be converted and multiplexed.

Chapter 17

PROBLEMS

Section 17.2 Digital-to-Analog Conversion

17.1 Explain why analog signals are usually digitized before being processed by a computer and/or transmitted.

17.2 What is the main advantage of digital transmission over analog transmission?

17.3 Draw the circuit diagram of a basic 6-bit *R-2R* DAC. Assuming TTL logic levels, determine the step size and the output voltage for the following inputs:
   a) 110011
   b) 101101

17.4 Assuming TTL logic levels, determine the step size and the output voltage v_o for an 8-bit *R-2R* DAC with the following inputs:
   a) 11001101
   b) 10110110

17.5 Draw the circuit diagram of a basic 6-bit *binary-weighted resistor* DAC and write its output equation. Let $R_F = 1$ kΩ and $R = 20$ kΩ. Assuming TTL logic levels, determine the output voltage for the following inputs:
   a) 110011
   b) 101101

17.6 Determine the output voltage for the DAC0808 shown in Figure 17-1P with the following inputs:

   a) 11001101
   b) 10110110

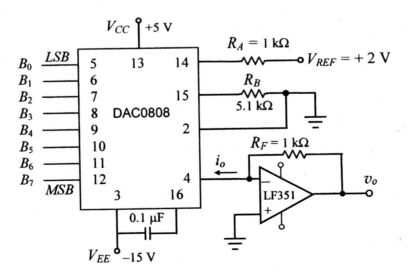

Figure 17-1P: D/A Converter

17.7 Repeat Problem 17.6 with $R_F = 20$ kΩ and $R = 10$ kΩ with the following inputs:
   a) 01010101
   b) 10101010

Chapter 17

Section 17.3 Analog to Digital Conversion

17.8 Explain the function of the sample and hold circuit in an A/D converter.

17.9 Explain the function of the comparator circuit in an A/D converter.

17.10 Assuming that the analog input ranges from 0 V to 15 V, explain the conversion process for the 4-bit ADC of Figure 17-10 with an input voltage level of 6 V.

17.11 Repeat Problem 17.10 with an input voltage level of 8 V.

17.12 Repeat Problem 17.10 with an input voltage level of 12 V.

17.13 Determine the voltage levels at the output of the temperature sensor and the level-shifter in Figure 17-16(a) at 30°C.

17.14 Determine the voltage levels at the output of the temperature sensor and the level-shifter in Figure 17-16(a) at 20°C.

17.15 Determine the voltage levels at the output of the temperature sensor and the level-shifter in Figure 17-16(a) at 50°C.

17.16 Show the numbers displayed on the 7-segment HEX display units of Figure 17-16(d) at the following temperatures:
    a) 25°C
    b) 50°C
    c) 75°C
    d) 100°C

Chapter 18

FOUR-LAYER ELECTRONIC DEVICES AND OPTOELECTRONIC DEVICES

18.1 INTRODUCTION

In addition to the discrete devices that we have covered in Chapters 1 through 9, there are a number of other useful electronic devices designed for specific applications that we would like to discuss in this chapter. These devices belong to two different and independent groups. First, we will discuss the group of electronic devices that are characterized as *four-layer devices* or *thyristors*. Thyristors are commonly used in high-power switching applications to control the amount of power delivered to heavy loads, such as high-current electric motors, heating, and lighting systems. Next, we will discuss the group of electronic devices called *optoelectronic devices*. These devices are light sensitive, whose characteristics can change with the incident light. That is, they emit light when energized, such as *light-emitting diodes*, or are controlled by light, such as *photodiodes* and *phototransistors*.

Chapter 18

18.2 FOUR-LAYER DEVICES

Recall that the BJT comprises three layers of semiconductors, *npn* or *pnp*, that form two *p-n* junctions. Four-layer devices, also called *thyristors*, comprise a category of semiconductor devices whose structure consists of four alternating layers of *p* and *n* materials (*pnpn*) that form three *p-n* junctions. The four-layer devices that are discussed in this chapter are *silicon-controlled rectifiers* (SCRs), *silicon-controlled switches* (SCSs), DIACs, and TRIACs. These devices are commonly used in high current and voltage switching applications, such as: controlling the power delivered to high-power electric motors, heating, and lighting systems.

Silicon-Controlled Rectifier (SCR)

The silicon-controlled rectifier (SCR) is a three-terminal four-layer device, which consists of four alternating layers of *p* and *n* materials as illustrated in Figure 18-1 below. The three terminals of the SCR are referred to as *anode*, *cathode*, and *gate*.

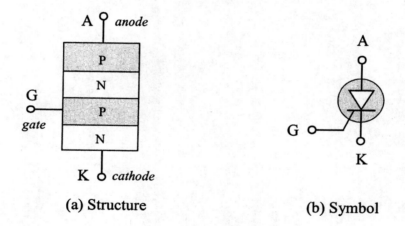

(a) Structure   (b) Symbol

Figure 18-1: SCR structure and schematic symbol

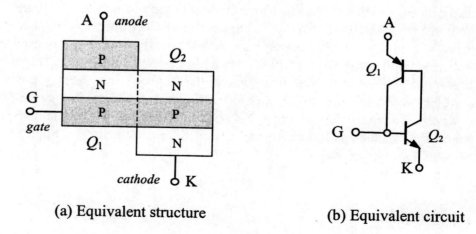

(a) Equivalent structure   (b) Equivalent circuit

Figure 18-2: SCR equivalent structure and circuit

Chapter 18

The silicon-controlled rectifier, as its name implies, is a controlled diode (rectifier), which conducts only when the anode is positive with respect to cathode and the gate-cathode junction is forward-biased, at least momentarily. That is, assuming the anode is positive with respect to the cathode (forward-biased), the application of a positive pulse at the gate turns the SCR on and initiates the conduction.

The analysis of the SCR can be simplified by considering it as a combination of two complementary BJTs (*npn* and *pnp*) with the base of each connected to the collector of the other, as illustrated in Figure 18-2.

Let us first consider the operation of the SCR with the gate terminal open, as illustrated in Figure 18-3 below:

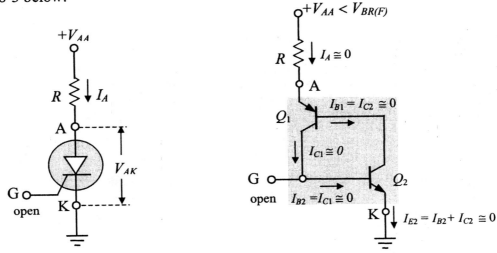

Figure 18-3: The SCR in its *off state* with the gate open and $V_{AA} < V_{BO(F)}$

With the gate open and the voltage V_{AA} less than the *forward breakover voltage* $V_{BO(F)}$, the base currents I_{B2} and I_{B1} are both approximately zero (disregarding the leakage currents), and thus the emitter currents I_{E2} and I_{E1} are both approximately zero, and as a result, both transistors are at the cutoff region and the SCR is in the *off state*. However, if the voltage V_{AA} is larger than or equal to the *forward breakover voltage*, the increased leakage currents drive the transistors into saturation (*on state*), causing a substantial anode-to-cathode current flow. At this point, the SCR is said to be *latched*. The latching action is the result of *regenerative* switching, where one device causes another device to conduct, which, in turn, drives the first device further into conduction. This regenerative process continues until both devices are driven into saturation. The whole process takes no more than a few microseconds. Like the regular diode, the amount of the anode current I_A can be limited with an external resistor R.

A typical plot of the *I-V* characteristics curve of the SCR is exhibited in Figure 18-4 below. Once the SCR is latched, its characteristics curve is similar to that of a silicon diode, and the forward DC voltage drop $V_{AK(on)}$ across the anode-cathode terminals is about 0.9 V to 1 V, which is the sum of the $V_{CE(sat)} + V_{BE}$ of Q_1 and Q_2, respectively. As shown in Figure 18-5, once the SCR is switched on, it remains on provided that the anode current I_A remains larger than the holding current I_H. If the anode current falls below the holding current I_H, either by reducing the supply voltage V_{AA} or by momentarily short-circuiting the anode-cathode terminals, then the SCR reverts back to its *off state* and remains off until the supply voltage V_{AA} is reestablished and the SCR is triggered again at the gate terminal.

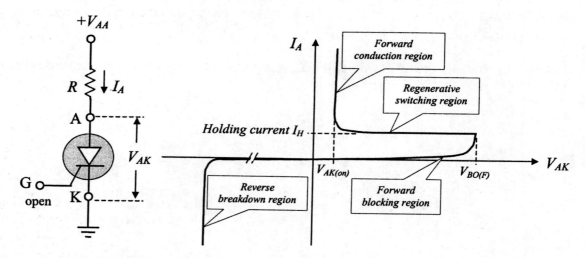

Figure 18-4: The *I-V* characteristics plot of the SCR with the gate open ($I_G = 0$)

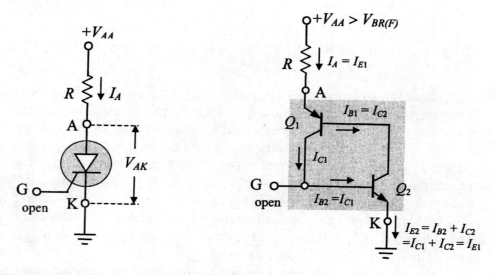

Figure 18-5: The SCR in its *on state* with the gate open and $V_{AA} > V_{BR(F)}$

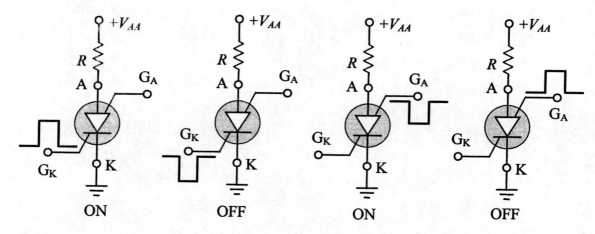

Figure 18-7: Triggering SCS on and off

DIACS and TRIACS

A DIAC is also a four-layer device whose block diagram and equivalent circuit are shown in Figure 18-8 below. The top and bottom layers contain both *p* and *n* materials; the middle layer contains only *n* material, making the device look like a two-sided four-layer structure. The left side of the stack is a *pnpn* structure, which has the characteristics of a gateless SCR. The right side of the stack is an *npnp* structure, which is equivalent to an inverted SCR.

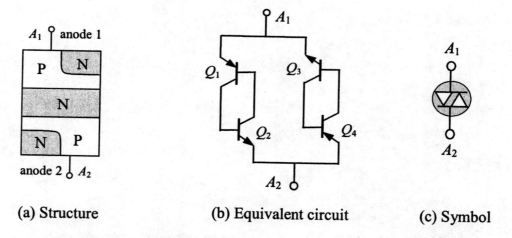

(a) Structure (b) Equivalent circuit (c) Symbol

Figure 18-8: The DIAC structure, equivalent circuit, and symbol

It is obvious from the equivalent circuit that the DIAC will conduct in one direction if A_1 terminal is at a higher potential than A_2, and will conduct in the opposite direction with A_2 terminal being at higher potential than A_1. In other words, a DIAC is a bi-directional diode that can conduct in both directions, one direction at a time.

Chapter 18

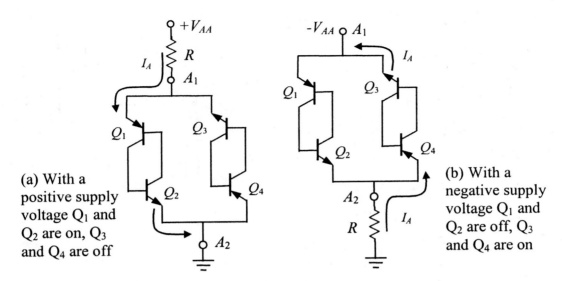

Figure 18-9: DIAC operation and conduction

The *I-V* characteristic of the DIAC is exhibited in Figure 18-10 below, which resembles the characteristics of two SCRs connected in inverse parallel.

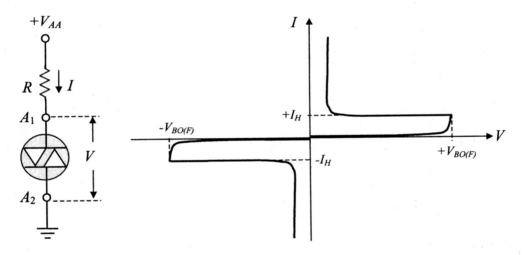

Figure 18-10: The *I-V* characteristics plot of the DIAC

A more common application of the DIAC is as the triggering device for TRIACs, which is demonstrated in Example 18-1 (Figure 18-12).

The TRIAC is equivalent to a DIAC circuit with addition of a gate terminal; it is also equivalent to an inverse parallel combination of two SCRs with the gates tied together forming a single gate terminal. The TRIAC equivalent circuit and its symbol are shown in Figure 18-11 below:

Chapter 18

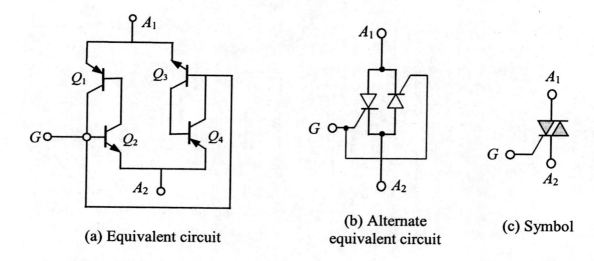

(a) Equivalent circuit

(b) Alternate equivalent circuit

(c) Symbol

Figure 18-11: TRIAC equivalent circuits and symbol

One of the most popular applications of the DIAC and TRIAC combination is utilizing both devices in power control circuits, such as the brightness control of incandescent lights, as depicted in Figure 18-12 below:

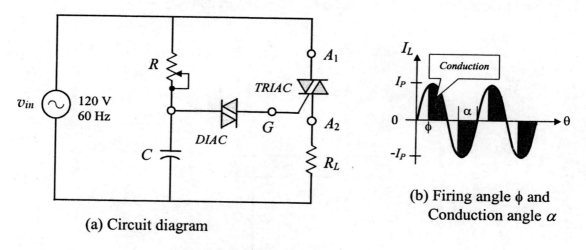

(a) Circuit diagram

(b) Firing angle ϕ and Conduction angle α

Figure 18-12: DIAC-TRIAC power control circuit

The TRIAC in the above circuit is triggered by positive gate current when A_1 is positive with respect to A_2, and is triggered by negative gate current when A_1 is negative with respect to A_2. The positive and negative gate currents are supplied by the DIAC during the positive and negative half-cycles of the input voltage v_{in}. When the voltage across the capacitor reaches beyond the *forward breakover voltage* (V_{BO}) of the DIAC, the DIAC fires and the capacitor discharges rapidly into the gate, triggering the TRIAC into conduction. The same process continues repeatedly for every positive and negative half-cycle of the input voltage v_{in}. The resistor R together with the capacitor C control the time constant for charging the capacitor. The larger the resistor value, the longer it takes to charge the capacitor. In other words, large values of RC delay the charging of the capacitor, increase the firing angle ϕ, and decrease the conduction angle α. The firing angle can be adjusted from approximately 0° to 180°, making it possible for the TRIAC to conduct for the full cycle of the input in one extreme ($\phi = 0°$ and $\alpha = 180°$), and none or for a very small portion of the input cycle in the other extreme ($\phi = 180°$ and $\alpha = 0°$). As a result, the power delivered to the load can be controlled from a minimum of almost zero percent to a maximum of approximately 100%.

The rms current delivered to the load

According to Figure 18-12(b), the rms current delivered to the load is defined as follows:

$$I_{L(rms)} = \sqrt{\frac{1}{\pi}\int_{\phi}^{\pi}(I_p \sin\theta)^2 d\theta} \tag{18-1}$$

$$I_{L(rms)}^2 = \frac{1}{\pi}\int_{\phi}^{\pi}(I_p \sin\theta)^2 d\theta \tag{18-2}$$

$$I_{L(rms)}^2 = \frac{I_p^2}{\pi}\int_{\phi}^{\pi}\frac{1}{2}[1-\cos 2\theta]d\theta \tag{18-3}$$

$$I_{L(rms)}^2 = \frac{I_p^2}{2\pi}\left[\theta - \frac{1}{2}\sin 2\theta\right]_{\phi}^{\pi} \tag{18-4}$$

$$I_{L(rms)}^2 = \frac{I_p^2}{2\pi}\left[(\pi - \frac{1}{2}\sin 2\pi) - (\phi - \frac{1}{2}\sin 2\phi)\right] \tag{18-5}$$

$$I_{L(rms)}^2 = \frac{I_p^2}{2\pi}\left[\pi - \phi + \frac{1}{2}\sin 2\phi\right] \tag{18-6}$$

$$I_{L(rms)}^2 = \frac{I_p^2}{2}\left[\frac{\pi - \phi}{\pi} + \frac{\sin 2\phi}{2\pi}\right] \tag{18-7}$$

$$I_{L(rms)} = \frac{I_p}{\sqrt{2}}\sqrt{\frac{180° - \phi}{180°} + \frac{\sin 2\phi}{2\pi}} \tag{18-8}$$

$$I_{L(rms)} = I_{rms(max)}\sqrt{\frac{\alpha}{180°} + \frac{\sin 2\phi}{2\pi}} \tag{18-9}$$

The firing angle or the time delay for the DIAC to start firing depends on the time required for the voltage across the capacitor V_C to reach the *breakover voltage* V_{BO} of the DIAC, which is a function of the capacitor voltage (magnitude and phase) determined by the voltage divider rule set up by the RC network. Based upon this principle, it can be shown that the firing angle ϕ can be determined by the following equation:

$$\phi = \left|90° - \tan^{-1}\frac{X_C}{R}\right| + \left|\sin^{-1}\frac{V_{BO}}{V_{C(p)}}\right| \tag{18-10}$$

where

Chapter 18

$$V_{C(p)} = I_{C(p)} X_C, \quad I_{C(p)} = \frac{V_p}{Z}, \quad Z = \sqrt{R^2 + X_C^2}, \quad V_p = \sqrt{2} \times V_{rms}, \quad X_C = \frac{1}{2\pi RC}$$

V_{BO} = *breakover voltage* of the DIAC

Example 18-1 — DIAC/TRIAC — ANALYSIS

Let us determine the firing angle, conduction angle, and percent of the power delivered to the load for the power control circuit of Figure 18-12, given that $C = 0.22$ μF, $R = 22$ kΩ, and $V_{BO} = 22$ V. The load consists of six 100 W light bulbs connected in parallel.

Solution:

$v_{rms} = 120$ V, $\quad v_p = v_{rms} \times 1.414 = 170$ V, $\quad P_{L(max)} = 600$ W

$$X_C = \frac{1}{2\pi fC} = \frac{1}{2\pi \times 60 \text{ Hz} \times 0.22 \text{ μF}} = 12 \text{ kΩ}$$

$$Z = \sqrt{R^2 + X_C^2} = \sqrt{22 \text{ kΩ}^2 + 12 \text{ kΩ}^2} = 25 \text{ kΩ}$$

$$I_{C(p)} = \frac{v_p}{Z} = \frac{170 \text{ V}}{25 \text{ kΩ}} = 6.8 \text{ mA}$$

$$V_{C(p)} = I_{C(p)} X_c = 6.8 \text{ mA} \times 12 \text{ kΩ} = 81.6 \text{ V}$$

Having determined the X_C and $V_{C(p)}$, we can now determine the firing angle ϕ with Equation 18-10, as follows:

$$\phi = \left|90° - \tan^{-1}\frac{X_C}{R}\right| + \left|\sin^{-1}\frac{V_{BO}}{V_{C(p)}}\right| = 90° - \tan^{-1}\frac{12 \text{ kΩ}}{22 \text{ kΩ}} + \sin^{-1}\frac{22 \text{ V}}{81.6 \text{ V}} = 46°$$

$$\alpha = 180° - \phi = 180° - 46° = 134°$$

Having determined the firing angle ϕ and the conduction angle α, we will now determine the *peak* and *rms* current delivered to the load, as follows:

$$I_{rms(max)} = \frac{P_{L(max)}}{v_{L(rms)}} = \frac{600 \text{ W}}{120 \text{ V}} = 5 \text{ A} \qquad I_{(p)} = I_{rms(max)} \sqrt{2} = 7.07 \text{ A}$$

$$I_{L(rms)} = I_{rms(max)} \sqrt{\frac{\alpha}{180°} + \frac{\sin 2\phi}{2\pi}} = 5 \text{ A} \sqrt{\frac{134°}{180°} + \frac{\sin(2 \times 46°)}{2\pi}} = 4.75 \text{ A}$$

$$P_L = V_{rms} I_{rms} = 120 \text{ V} \times 4.75 \text{ A} = 570 \text{ W}$$

The *percent power* is determined as the ratio of the actual power delivered (P_L) to the maximum load power ($P_{L(max)}$), as follows:

$$\%P = \frac{P_L}{P_{L(max)}} = \frac{570 \text{ W}}{600 \text{ W}} = 95\%$$

Chapter 18

Practice Problem 18-1 (DIAC/TRIAC) ANALYSIS

Determine the firing angle, conduction angle, and percent of the power delivered to the load for the power control circuit of Figure 18-13.
$V_{BO} = 25$ V, and the load consists of two 150 W light bulbs in parallel.

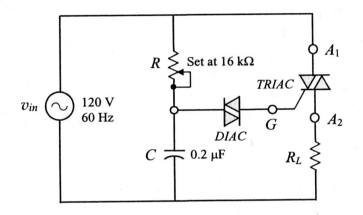

Figure 18-13: DIAC-TRIAC power control circuit

Answers: $\phi = 63.7°$, $\alpha = 116.3°$, $I_{rms} = 2.2$ A, $\%P = 88\%$

18.3 OPTOELECTRONIC DEVICES

Optoelectronic devices are those semiconductor devices whose characteristics change with light, and/or create light when energized such as photoconductive cells, solar cells, photodiodes, phototransistors, light emitting diodes, optocouplers, etc.

Photoconductive Cell / Photoresistor

The photoconductive cell is a semiconductor device whose resistance changes with light. That is, when the photoconductor is exposed to light, its resistance decreases with increased light intensity and vice versa. Photoconductive cells are also referred to as *photoresistors*. The resistance of the typical *photoresistor* ranges from approximately 100 Ω to 100 kΩ, where the latter resistance is called *dark resistance*, which corresponds to no illumination. Photoconductive cells have applications in lighting controls, automatic door openers, and security systems. The symbol for the photoconductive cell or photoresistor is shown in Figure 18-14 below:

Figure 18-14: Schematic symbol for the photoresistor

Example 18-2 — Photoresistor — DEMO

The following example demonstrates the use of a *photoresistor* and an *SCR* in a security system or an automatic door opener in which the relay turns on a motor or activates some other mechanism when energized.

Assuming that the *dark resistance* of the photoresistor used in the circuit of Figure 18-15 is 100 kΩ, and it has a resistance of 10 kΩ when illuminated by a light beam, let us determine if the relay is energized when the light beam is interrupted.

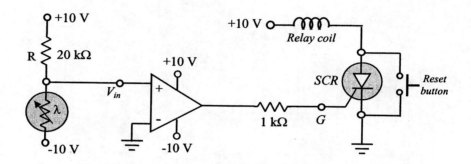

Figure 18-15: Utilizing the photoresistor and SCR in an automatic door opener or a security system

The input voltage to the op-amp comparator V_{in} is determined by the application of the voltage divider rule and superposition theorem as follows:

a) Light beam is on, $R_{on} = 10$ kΩ

$$V_{in} = +10\,V\,\frac{10\,k\Omega}{30\,k\Omega} - 10\,V\,\frac{20\,k\Omega}{30\,k\Omega} = -3.33\,V$$

Since V_{in} is negative, the op-amp output V_o reaches $-V_{sat}$, and so the SCR is off.

b) Light beam is interrupted, $R_{off} = 100$ kΩ

$$V_{in} = +10\,V\,\frac{100\,k\Omega}{120\,k\Omega} - 10\,V\,\frac{20\,k\Omega}{120\,k\Omega} = +6.66\,V$$

Since V_{in} is positive, the op-amp output V_o reaches $+V_{sat}$, the SCR turns on, and the relay is energized.

To reset the system, the reset button is depressed, which short-circuits the SCR momentarily and turns it off.

Photodiodes

A photodiode is a semiconductor device similar to the *p-n* junction diode except that its junction can be exposed to light through a built-in lens or aperture that focuses the incident light at the junction. Like the Zener diode, the photodiode is normally operated under reverse-biased condition. In the absence of light, the amount of current through

Chapter 18

the reverse-biased junction is zero. With sufficient incident light, however, additional electron-hole pairs are generated in the depletion region, causing a substantial amount of reverse current flow. The number of electron-hole pairs generated by the photon energy, and thus, the amount of current flow is proportional to the intensity of the incident light. A typical *I-V* characteristics curve of the photodiode and its schematic symbol are exhibited in Figure 18-16 below:

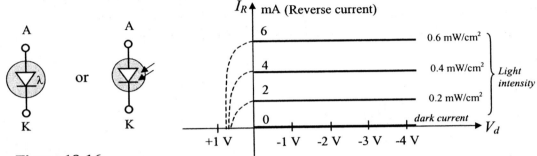

Figure 18-16:
Photodiode symbols and characteristics

Example 18-3 (Photodiode)   DEMO

The following example demonstrates the use of a *photodiode* in a data transfer system utilizing fiber-optic cable as the medium.

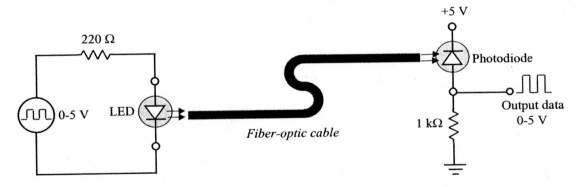

Figure 18-17:
Utilizing the LED and *photodiode* in a fiber-optic data transfer system

The input bit stream has one of two voltage levels, 0 V for logic 0 and 5 V for logic 1, which is a TTL data format. When the input data is a logic 1, the LED turns on and the emitted light travels the length of the fiber-optic cable and turns on the photodiode. Ignoring the voltage drop across the diode, the output will be approximately 5 V, which is a logic 1. When the input data is a logic 0, the LED is off and is emitting no light into the fiber-optic cable. In the absence of light, the photodiode is off and the output voltage is 0 V, which is a logic 0.

Phototransistors

The phototransistor is simply a BJT with a built-in lens that focuses the incident light on the collector-base junction, and usually comes with two terminals, collector and emitter, although there are some available with an additional base terminal, which may be used for external base connection. The collector current I_C increases with the increased light intensity, which is measured in watts per square centimeter (W/cm^2). The output characteristic of a typical phototransistor is exhibited in Figure 18-18 below, which is very similar to that of a conventional BJT. The only difference is that the base current I_B has been replaced with the light intensity. A phototransistor may also be utilized in the above data transfer circuit of Figure 18-17 instead of the photodiode.

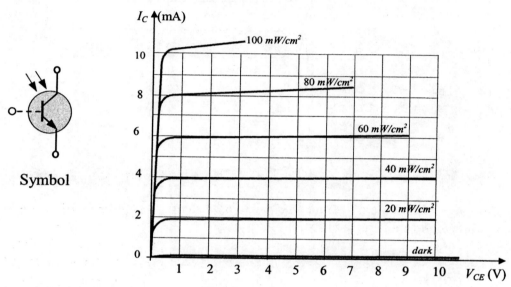

Figure 18-18:
The phototransistor symbol and output characteristics

Optocouplers

An *optocoupler* is basically a combination of an LED and phototransistor in one package. With an input current, the LED lights up and drives the phototransistor. The major advantage of this device is the total electrical isolation of the input circuit from the output circuit. Hence, it is also called an *optoisolator*. The symbol for the optocoupler is shown in Figure 18-19 below:

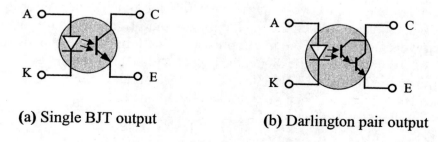

(a) Single BJT output   **(b)** Darlington pair output

Figure 18-19: The optocoupler/optoisolator symbols

Chapter 18

Example 18-4 ANALYSIS

The following example demonstrates one major appliction of the *optocoupler* in a data communication circuit, in which it is utilized to help convert the on/off current pulses in the input *current loop* to a data format known as the RS-232C specification, in which a logic 0 corresponds to a voltage level between +5 V to +15 V, and logic 1 corresponds to a voltage level between –5 V to –15 V.

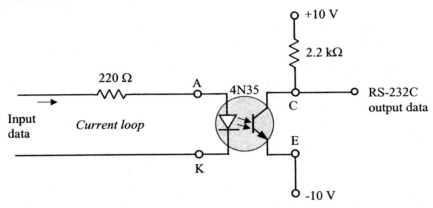

Figure 18-20: Application of the optocoupler in a data conversion circuit

A current pulse at the input, which corresponds to logic 1, turns on the LED at the input side of the optocoupler, which turns on the transistor, driving it to saturation. As a result, the output terminal is at –10 V, which is within the voltage range for logic 1 of the RS-232C data format. With no current pulse at the input loop, which corresponds to logic 0, the LED is off and the transistor is not conducting. Therefore, the output terminal is at +10 V, which is within the voltage range for logic 0 of the RS-232C data format.

Similarly, the optocoupler can be utilized to perform just the opposite of the above conversion; that is, convert the RS-232C data format to *current loop* data format (on/off current pulses), as shown in Figure 18-21 below:

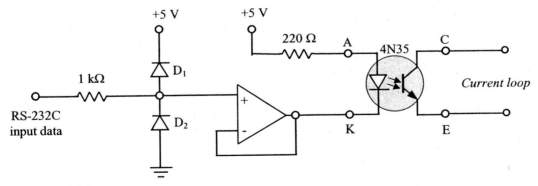

Figure 18-21:
Another application of the optocoupler in a data conversion circuit

Let us assume that the RS-232 voltage level is –10 V for a logic 1, and +10 V for logic 0. With a logic 1 (–10 V) at the input, diode D_1 is reverse-biased and diode D_2 is forward-biased, producing approximately –0.7 V at the input and output of the op-amp buffer. A voltage of –0.7 V at the cathode terminal of the optocoupler turns the LED on and drives the transistor into saturation. Note that saturation can be achieved with proper selection of the current loop resistor and voltage supply level. Hence, a logic 1 at the input is converted to a logic 1 at the output with a different format. With a logic 0 (+10 V) at the input, diode D_2 is reverse-biased and diode D_1 is forward-biased, producing approximately +4.3 V at the input and output of the op-amp buffer. A voltage of 4.3 V at the cathode terminal of the optocoupler turns the LED off and drives the transistor into cutoff. Hence, a logic 0 at the input is converted to a logic 0 at the output. The op-amp buffer functions as an ideal buffer and provides approximately infinite input resistance and zero output resistance; in other words, it behaves as a short circuit for voltage and open circuit for current.

18.4 SUMMARY

- Four-layer devices, also called *thyristors*, comprise a category of semiconductor devices whose structure consists of four alternating layers of *p* and *n* materials (*pnpn*) that form three *p-n* junctions. These devices are commonly used in high current and voltage switching applications, such as controlling the power delivered to high-power electric motors, heating, and lighting systems.

- The *silicon-controlled rectifier* (SCR) is a three-terminal four-layer device, which consists of four alternating layers of *p* and *n* materials. The three terminals of the SCR are referred to as *anode*, *cathode*, and *gate*.

- The SCR is a controlled diode (rectifier), which conducts only when the anode is positive with respect to cathode and the gate-cathode junction is forward-biased, at least momentarily.

- The SCR is turned off only by reducing the anode current I_A to a level less than the holding current I_H, which can be accomplished by switching off the supply voltage or short-circuiting the SCR.

- The *silicon-controlled switch* (SCS) is a four-layer device with four terminals, and its structure is very similar to SCR. The four terminals of the SCS are referred to as *anode*, *cathode*, *cathode gate*, and *anode gate*.

- Unlike the SCR, the *silicon-controlled switch* can be switched on and off through the external gate control. That is, assuming the anode is positive with respect to cathode, the application of a positive pulse at the *cathode gate* turns the SCS on. It can also be turned on by the application of a negative pulse at the *anode gate*. The SCS can be turned off either by application of a negative pulse at the *cathode gate*, or by application of a positive pulse at the *anode gate*. The methods of turning the SCS on or off are illustrated in Figure 18-7.

- A DIAC is a four-layer device whose structure is equivalent to two gateless SCRs connected in inverse-parallel. The more common application of the DIAC is as the triggering device for TRIACs.

- The TRIAC is equivalent to a DIAC circuit with the addition of a gate terminal; it is also equivalent to an inverse parallel combination of two SCRs with the gates tied together forming a single gate terminal.

- One of the most popular applications of the DIAC and TRIAC combination is utilizing both devices in power control circuits, such as brightness control of incandescent lights, as depicted in Figure 18-13.

- Optoelectronic devices are those semiconductor devices whose characteristics change with light, and/or create light when energized, such as photoconductive cells, solar cells, photodiodes, phototransistors, light emitting diodes, optocouplers, etc.

- The *photoconductive cell*, also called *photoresistor*, is a semiconductor device whose resistance changes with light. The resistance of the typical *photoresistor* ranges from approximately 100 Ω to 100 kΩ, where the latter resistance is called *dark resistance*, which corresponds to no illumination.

- A *photodiode* is a semiconductor device similar to the *p-n* junction diode except that its junction can be exposed to light through a built-in lens or aperture that focuses the incident light at the junction. Like the Zener diode, the photodiode is normally operated under reverse-biased condition.

- The *phototransistor* is simply a BJT with a built-in lens that focuses the incident light on the collector-base junction. The collector current I_C increases with the increased light intensity, which is measured in watts per square centimeter (W/cm^2).

- An *optocoupler* is basically a combination of an LED and phototransistor in one package. With an input current, the LED lights up and drives the phototransistor. The major advantage of this device is the total electrical isolation of the input circuit from the output circuit. Hence, it is also called an *optoisolator*.

Chapter 18

18.5 DEVICE SPECIFICATIONS AND DATA SHEETS

Data sheets of the following devices discussed in this chapter are included for your reference and convenience.

- 2N6394 Series Silicon-Controlled Rectifiers
  Courtesy of Semiconductor Component Industries, LLC. Used by permission.
- 2N6071A/B Series TRIACs
  Courtesy of Semiconductor Component Industries, LLC. Used by permission.
- BPW21R Photodiode
  Courtesy of Vishay Intertechnology
- QSE773 Photodiode
  Courtesy of Fairchild Semiconductor
- S289P Darlington Phototransistor
  Courtesy of Vishay Intertechnology
- K817P, K827PH, K847PH Optocouplers
  Courtesy of Vishay Intertechnology
- 4N35, 4N36, 4N37 Optoisolators
  Courtesy of Semiconductor Component Industries, LLC. Used by permission.

Chapter 18

2N6394 Series
Preferred Device

Silicon Controlled Rectifiers
Reverse Blocking Thyristors

Designed primarily for half-wave ac control applications, such as motor controls, heating controls and power supplies.
- Glass Passivated Junctions with Center Gate Geometry for Greater Parameter Uniformity and Stability
- Small, Rugged, Thermowatt Construction for Low Thermal Resistance, High Heat Dissipation and Durability
- Blocking Voltage to 800 Volts
- Device Marking: Logo, Device Type, e.g., 2N6394, Date Code

ON Semiconductor

http://onsemi.com

**SCRs
12 AMPERES RMS
50 thru 800 VOLTS**

TO–220AB
CASE 221A
STYLE 3

*MAXIMUM RATINGS (T_J = 25°C unless otherwise noted)

Rating	Symbol	Value	Unit
Peak Repetitive Off–State Voltage(1) (T_J = –40 to 125°C, Sine Wave, 50 to 60 Hz, Gate Open) 2N6394 2N6395 2N6397 2N6399	V_{DRM}, V_{RRM}	 50 100 400 800	Volts
On-State RMS Current (180° Conduction Angles; T_C = 90°C)	$I_{T(RMS)}$	12	A
Peak Non-Repetitive Surge Current (1/2 Cycle, Sine Wave, 60 Hz, T_J = 125°C)	I_{TSM}	100	A
Circuit Fusing (t = 8.3 ms)	I^2t	40	A^2s
Forward Peak Gate Power (Pulse Width $\leq$ 1.0 µs, T_C = 90°C)	P_{GM}	20	Watts
Forward Average Gate Power (t = 8.3 ms, T_C = 90°C)	$P_{G(AV)}$	0.5	Watts
Forward Peak Gate Current (Pulse Width < 1.0 µs, T_C = 90°C)	I_{GM}	2.0	A
Operating Junction Temperature Range	T_J	–40 to +125	°C
Storage Temperature Range	T_{stg}	–40 to +150	°C

*Indicates JEDEC Registered Data

(1) V_{DRM} and V_{RRM} for all types can be applied on a continuous basis. Ratings apply for zero or negative gate voltage; however, positive gate voltage shall not be applied concurrent with negative potential on the anode. Blocking voltages shall not be tested with a constant current source such that the voltage ratings of the devices are exceeded.

PIN ASSIGNMENT

1	Cathode
2	Anode
3	Gate
4	Anode

ORDERING INFORMATION

Device	Package	Shipping
2N6394	TO220AB	500/Box
2N6395	TO220AB	500/Box
2N6397	TO220AB	500/Box
2N6399	TO220AB	500/Box

Preferred devices are recommended choices for future use and best overall value

2N6394 Series

THERMAL CHARACTERISTICS

Characteristic	Symbol	Max	Unit
Thermal Resistance, Junction to Case	$R_{\theta JC}$	2.0	°C/W
Maximum Lead Temperature for Soldering Purposes 1/8" from Case for 10 Seconds	T_L	260	°C

ELECTRICAL CHARACTERISTICS (T_C = 25°C unless otherwise noted.)

Characteristic	Symbol	Min	Typ	Max	Unit
OFF CHARACTERISTICS					
*Peak Repetitive Forward or Reverse Blocking Current (V_{AK} = Rated V_{DRM} or V_{RRM}, Gate Open) T_J = 25°C T_J = 125°C	I_{DRM}, I_{RRM}	— —	— —	10 2.0	µA mA
ON CHARACTERISTICS					
*Peak Forward On–State Voltage$^{(1)}$ (I_{TM} = 24 A Peak)	V_{TM}	—	1.7	2.2	Volts
*Gate Trigger Current (Continuous dc) (V_D = 12 Vdc, R_L = 100 Ohms)	I_{GT}	—	5.0	30	mA
*Gate Trigger Voltage (Continuous dc) (V_D = 12 Vdc, R_L = 100 Ohms)	V_{GT}	—	0.7	1.5	Volts
Gate Non–Trigger Voltage (V_D = 12 Vdc, R_L = 100 Ohms, T_J = 125°C)	V_{GD}	0.2	—	—	Volts
*Holding Current (V_D = 12 Vdc, Initiating Current = 200 mA, Gate Open)	I_H	—	6.0	50	mA
Turn-On Time (I_{TM} = 12 A, I_{GT} = 40 mAdc, V_D = Rated V_{DRM})	t_{gt}	—	1.0	2.0	µs
Turn-Off Time (V_D = Rated V_{DRM}) (I_{TM} = 12 A, I_R = 12 A) (I_{TM} = 12 A, I_R = 12 A, T_J = 125°C)	t_q		15 35	— —	µs
DYNAMIC CHARACTERISTICS					
Critical Rate–of–Rise of Off-State Voltage Exponential (V_D = Rated V_{DRM}, T_J = 125°C)	dv/dt	—	50	—	V/µs

*Indicates JEDEC Registered Data

(1) Pulse Test: Pulse Width ≤ 300 µsec, Duty Cycle ≤ 2%.

Chapter 18

2N6071A/B Series

Preferred Device

Sensitive Gate Triacs
Silicon Bidirectional Thyristors

Designed primarily for full-wave ac control applications, such as light dimmers, motor controls, heating controls and power supplies; or wherever full-wave silicon gate controlled solid-state devices are needed. Triac type thyristors switch from a blocking to a conducting state for either polarity of applied anode voltage with positive or negative gate triggering.

- Sensitive Gate Triggering Uniquely Compatible for Direct Coupling to TTL, HTL, CMOS and Operational Amplifier Integrated Circuit Logic Functions
- Gate Triggering 4 Mode — 2N6071A,B, 2N6073A,B, 2N6075A,B
- Blocking Voltages to 600 Volts
- All Diffused and Glass Passivated Junctions for Greater Parameter Uniformity and Stability
- Small, Rugged, Thermopad Construction for Low Thermal Resistance, High Heat Dissipation and Durability
- Device Marking: Device Type, e.g., 2N6071A, Date Code

ON Semiconductor

http://onsemi.com

**TRIACS
4 AMPERES RMS
200 thru 600 VOLTS**

TO–225AA
(formerly TO–126)
**CASE 077
STYLE 5**

PIN ASSIGNMENT	
1	Main Terminal 1
2	Main Terminal 2
3	Gate

ORDERING INFORMATION

Device	Package	Shipping
2N6071A	TO225AA	500/Box
2N6071B	TO225AA	500/Box
2N6073A	TO225AA	500/Box
2N6073B	TO225AA	500/Box
2N6075A	TO225AA	500/Box
2N6075B	TO225AA	500/Box

Preferred devices are recommended choices for future use and best overall value.

MAXIMUM RATINGS (T_J = 25°C unless otherwise noted)

Rating	Symbol	Value	Unit
*Peak Repetitive Off-State Voltage(1) (T_J = –40 to 110°C, Sine Wave, 50 to 60 Hz, Gate Open) 2N6071A,B 2N6073A,B 2N6075A,B	V_{DRM}, V_{RRM}	200 400 600	Volts
*On-State RMS Current (T_C = 85°C) Full Cycle Sine Wave 50 to 60 Hz	$I_{T(RMS)}$	4.0	Amps
*Peak Non–repetitive Surge Current (One Full cycle, 60 Hz, T_J = +110°C)	I_{TSM}	30	Amps
Circuit Fusing Considerations (t = 8.3 ms)	I^2t	3.7	A^2s
*Peak Gate Power (Pulse Width ≤ 1.0 µs, T_C = 85°C)	P_{GM}	10	Watts
*Average Gate Power (t = 8.3 ms, T_C = 85°C)	$P_{G(AV)}$	0.5	Watt
*Peak Gate Voltage (Pulse Width ≤ 1.0 µs, T_C = 85°C)	V_{GM}	5.0	Volts
*Operating Junction Temperature Range	T_J	–40 to +110	°C
*Storage Temperature Range	T_{stg}	–40 to +150	°C
Mounting Torque (6-32 Screw)(2)	—	8.0	in. lb.

*Indicates JEDEC Registered Data.
(1) V_{DRM} and V_{RRM} for all types can be applied on a continuous basis. Blocking voltages shall not be tested with a constant current source such that the voltage ratings of the devices are exceeded.
(2) Torque rating applies with use of a compression washer. Mounting torque in excess of 6 in. lb. does not appreciably lower case-to-sink thermal resistance. Main terminal 2 and heatsink contact pad are common.

Chapter 18

2N6071A/B Series

THERMAL CHARACTERISTICS

Characteristic	Symbol	Max	Unit
*Thermal Resistance, Junction to Case	$R_{\theta JC}$	3.5	°C/W
Thermal Resistance, Junction to Ambient	$R_{\theta JA}$	75	°C/W
Maximum Lead Temperature for Soldering Purposes 1/8″ from Case for 10 Seconds	T_L	260	°C

ELECTRICAL CHARACTERISTICS (T_C = 25°C unless otherwise noted; Electricals apply in both directions)

Characteristic		Symbol	Min	Typ	Max	Unit
OFF CHARACTERISTICS						
*Peak Repetitive Blocking Current (V_D = Rated V_{DRM}, V_{RRM}; Gate Open)	T_J = 25°C	I_{DRM}, I_{RRM}	—	—	10	µA
	T_J = 110°C		—	—	2	mA
ON CHARACTERISTICS						
*Peak On-State Voltage(1) (I_{TM} = ±6 A Peak)		V_{TM}	—	—	2	Volts
*Gate Trigger Voltage (Continuous dc) (Main Terminal Voltage = 12 Vdc, R_L = 100 Ohms, T_J = –40°C) All Quadrants		V_{GT}	—	1.4	2.5	Volts
Gate Non–Trigger Voltage (Main Terminal Voltage = 12 Vdc, R_L = 100 Ohms, T_J = 110°C) All Quadrants		V_{GD}	0.2	—	—	Volts
*Holding Current (Main Terminal Voltage = 12 Vdc, Gate Open, Initiating Current = ±1 Adc)	(T_J = –40°C)	I_H	—	—	30	mA
	(T_J = 25°C)		—	—	15	
Turn-On Time (I_{TM} = 14 Adc, I_{GT} = 100 mAdc)		t_{gt}	—	1.5	—	µs

				QUADRANT (Maximum Value)			
Gate Trigger Current (Continuous dc) (Main Terminal Voltage = 12 Vdc, R_L = 100 ohms)	Type	I_{GT} @ T_J	I mA	II mA	III mA	IV mA
	2N6071A 2N6073A 2N6075A	+25°C	5	5	5	10
		–40°C	20	20	20	30
	2N6071B 2N6073B 2N6075B	+25°C	3	3	3	5
		–40°C	15	15	15	20

DYNAMIC CHARACTERISTICS

Characteristic	Symbol	Min	Typ	Max	Unit
Critical Rate of Rise of Commutation Voltage @ V_{DRM}, T_J = 85°C, Gate Open, I_{TM} = 5.7 A, Exponential Waveform, Commutating di/dt = 2.0 A/ms	dv/dt(c)	—	5	—	V/µs

*Indicates JEDEC Registered Data.

(1) Pulse Test: Pulse Width ≤ 2.0 ms, Duty Cycle ≤ 2%.

Chapter 18

BPW21R
Vishay Telefunken

Silicon PN Photodiode

Description

BPW21R is a planar Silicon PN photodiode in a hermetically sealed short TO–5 case, especially designed for high precision linear applications.

Due to its extremely high dark resistance, the short circuit photocurrent is linear over seven decades of illumination level.

On the other hand, there is a strictly logarithmic correlation between open circuit voltage and illumination over the same range.

The device is equipped with a flat glass window with built in color correction filter, giving an approximation to the spectral response of the human eye.

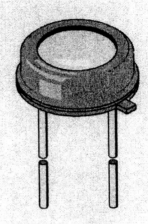

Features

- Hermetically sealed TO–5 case
- Flat glass window with built–in color correction filter for visible radiation
- Cathode connected to case
- Wide viewing angle $\varphi = \pm 50°$
- Large radiant sensitive area (A=7.5 mm^2)
- Suitable for visible radiation
- High sensitivity
- Low dark current
- High shunt resistance
- Excellent linearity
- For photodiode and photovoltaic cell operation

Applications

Sensor in exposure and color measuring purposes

BPW21R
Vishay Telefunken

Absolute Maximum Ratings

T_{amb} = 25°C

Parameter	Test Conditions	Symbol	Value	Unit
Reverse Voltage		V_R	10	V
Power Dissipation	$T_{amb} \leq 50\ °C$	P_V	300	mW
Junction Temperature		T_j	125	°C
Operating Temperature Range		T_{amb}	−55...+125	°C
Storage Temperature Range		T_{stg}	−55...+125	°C
Soldering Temperature	$t \leq 5\ s$	T_{sd}	260	°C
Thermal Resistance Junction/Ambient		R_{thJA}	250	K/W

Basic Characteristics

T_{amb} = 25°C

Parameter	Test Conditions	Symbol	Min	Typ	Max	Unit
Forward Voltage	I_F = 50 mA	V_F		1.0	1.3	V
Breakdown Voltage	I_R = 20 µA, E = 0	$V_{(BR)}$	10			V
Reverse Dark Current	V_R = 5 V, E = 0	I_{ro}		2	30	nA
Diode Capacitance	V_R = 0 V, f = 1 MHz, E = 0	C_D		1.2		nF
	V_R = 5 V, f = 1 MHz, E = 0	C_D		400		pF
Dark Resistance	V_R = 10 mV	R_D		38		GΩ
Open Circuit Voltage	E_A = 1 klx	V_o	280	450		mV
Temp. Coefficient of V_o	E_A = 1 klx	TK_{Vo}		−2		mV/K
Short Circuit Current	E_A = 1 klx	I_k	4.5	9		µA
Temp. Coefficient of I_k	E_A = 1 klx	TK_{Ik}		−0.05		%/K
Reverse Light Current	E_A = 1 klx, V_R = 5 V	I_{ra}	4.5	9		µA
Sensitivity	V_R = 5 V, E_A = $10^{-2}...10^5$ lx	S		9		nA/lx
Angle of Half Sensitivity		φ		±50		deg
Wavelength of Peak Sensitivity		λ_p		565		nm
Range of Spectral Bandwidth		$\lambda_{0.5}$		420...675		nm
Rise Time	V_R = 0 V, R_L = 1k Ω, λ = 660 nm	t_r		3.1		µs
Fall Time	V_R = 0 V, R_L = 1k Ω, λ = 660 nm	t_f		3.0		µs

BPW21R
Vishay Telefunken

Typical Characteristics ($T_{amb} = 25°C$ unless otherwise specified)

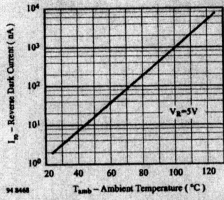

Figure 1. Reverse Dark Current vs. Ambient Temperature

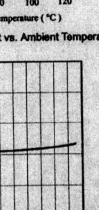

Figure 2. Relative Reverse Light Current vs. Ambient Temperature

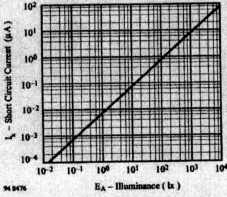

Figure 3. Short Circuit Current vs. Illuminance

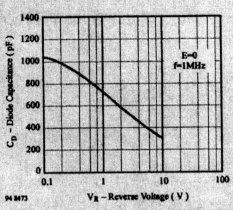

Figure 4. Diode Capacitance vs. Reverse Voltage

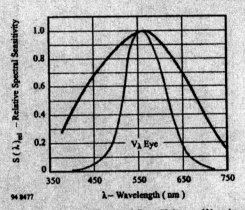

Figure 5. Relative Spectral Sensitivity vs. Wavelength

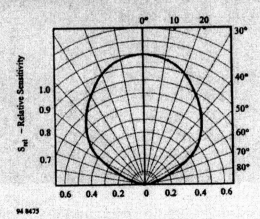

Figure 6. Relative Radiant Sensitivity vs. Angular Displacement

Chapter 18

QSE773
PLASTIC SILICON PIN PHOTODIODE

PACKAGE DIMENSIONS

FEATURES
- Daylight Filter
- Sidelooker Package
- Pin Photodiode
- Wide Reception Angle, 120°
- Chip Size = $.107^2$ inches (2.71^2 mm)

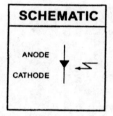

SCHEMATIC

ANODE
CATHODE

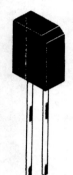

NOTES:
1. Dimensions for all drawings are in inches (mm).
2. Tolerance of ± .010 (.25) on all non-nominal dimensions unless otherwise specified.

1. Derate power dissipation linearly 2.50 mW/°C above 25°C.
2. RMA flux is recommended.
3. Methanol or isopropyl alcohols are recommended as cleaning agents.
4. Soldering iron 1/16" (1.6mm) minimum from housing.
5. As long as leads are not under any stress or spring tension.
6. Light source is an GaAs LED which has a peak emission wavelength of 940 nm.
7. All measuements made under pulse conditions.

ABSOLUTE MAXIMUM RATINGS (T_A = 25°C unless otherwise specified)

Parameter	Symbol	Rating	Unit
Operating Temperature	T_{OPR}	-40 to +85	°C
Storage Temperature	T_{STG}	-40 to +85	°C
Soldering Temperature (Iron)(2,3,4)	T_{SOL-I}	240 for 5 sec	°C
Soldering Temperature (Flow)(2,3)	T_{SOL-F}	260 for 10 sec	°C
Reverse Voltage	V_R	32	V
Power Dissipation(1)	P_D	150	mW

QSE773
PLASTIC SILICON PIN PHOTODIODE

ELECTRICAL / OPTICAL CHARACTERISTICS (T_A =25°C unless otherwise specified)

PARAMETER	TEST CONDITIONS	SYMBOL	MIN	TYP	MAX	UNITS
Reverse Voltage	$I_R = 0.1$ mA	V_R	32	—		V
Dark Reverse Current	$V_R = 10$ V	$I_{R(D)}$	—		30	nA
Peak Sensitivity	$V_R = 5$ V	λ_{PK}		920		nm
Reception Angle @ 1/2 Power		θ		+/-60		Degrees
Photo Current	$E_e = 1.0$ mW/cm^2, $V_{CE} = 5$ V[6]	I_{PH}	30		—	µA
Capacitance	$V_R = 3$ V	C		20		pF
Rise Time	$V_R = 5$ V, $R_L = 1$ K	t_r		50		ns
Fall Time	$V_R = 5$ V, $R_L = 1$ K	t_f		50		ns

TYPICAL PERFORMANCE CURVES

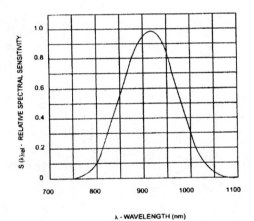

Fig. 1 Relative Spectral Sensitivity vs. Wavelength

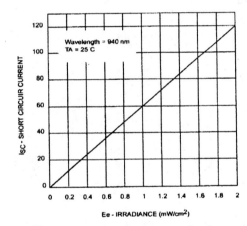

Fig. 2 Short Circuit Current vs. Irradiance

Chapter 18

S289P
Vishay Telefunken

Silicon Darlington Phototransistor

Description

S289P is an extra high sensitive monolithic silicon epitaxial planar Darlington phototransistor in a standard T–1 (ø 3 mm) package.
The epoxy package itself is an IR filter, spectrally matched to GaAs IR emitters with λ_p >850nm. A plastic lens provides a wide viewing angle of ±30°.

Features

- Extra high radiant sensitivity
- Very low temperature drift
- Standard T–1 (ø 3 mm) package with IR filter
- Wide viewing angle φ = ± 30°
- Suitable for near infrared radiation

Applications

Any applications requiring high sensitivity at low light levels e.g.
Direct driving of relays and small motors
Special light barriers and switches

Absolute Maximum Ratings

T_{amb} = 25 °C

Parameter	Test Conditions	Symbol	Value	Unit
Collector Emitter Voltage		V_{CEO}	40	V
Collector Current		I_C	0.1	A
Peak Collector Current	t_p/T = 0.05, t_p ≤ 10 ms	I_{CM}	1	A
Total Power Dissipation	T_{amb} ≤ 25 °C	P_{tot}	185	mW
Junction Temperature		T_j	100	°C
Operating Temperature Range		T_{amb}	–55...+100	°C
Storage Temperature Range		T_{stg}	–55...+100	°C
Soldering Temperature	t ≤ 5 s	T_{sd}	260	°C
Thermal Resistance Junction/Ambient		R_{thJA}	400	K/W

Document Number 81541
Rev. 2, 20-May-99

www.vishay.de • FaxBack +1-408-970-5600

S289P
Vishay Telefunken

Basic Characteristics

$T_{amb} = 25°C$

Parameter	Test Conditions	Symbol	Min.	Typ.	Max.	Unit
Collector Emitter Breakdown Voltage	$I_C = 1$ mA	$V_{(BR)CEO}$	40			V
Collector Dark Current	$V_{CE} = 20$ V, E = 0	I_{CEO}		10	200	nA
Collector Light Current	$E_e = 0.3$ mW/cm², $\lambda = 950$ nm, $V_{CE} = 5$ V	I_{ca}	4	15		mA
Angle of Half Sensitivity		φ		±30		deg
Wavelength of Peak Sensitivity		λ_p		920		nm
Range of Spectral Bandwidth		$\lambda_{0.5}$		830...1000		nm
Collector Emitter Saturation Voltage	$E_e = 0.3$ mW/cm², $\lambda = 950$ nm, $I_C = 1$ mA	V_{CEsat}		0.75	1.1	V
Turn–On Time	$V_S = 5$ V, $I_C = 5$ mA, $R_L = 100$ Ω	t_{on}		40		µs
Turn–Off Time	$V_S = 5$ V, $I_C = 5$ mA, $R_L = 100$ Ω	t_{off}		50		µs

Typical Characteristics ($T_{amb} = 25°C$ unless otherwise specified)

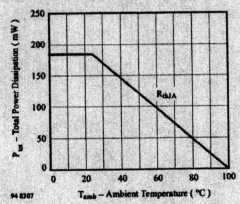

Figure 1. Total Power Dissipation vs. Ambient Temperature

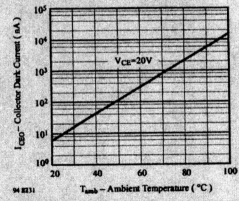

Figure 2. Collector Dark Current vs. Ambient Temperature

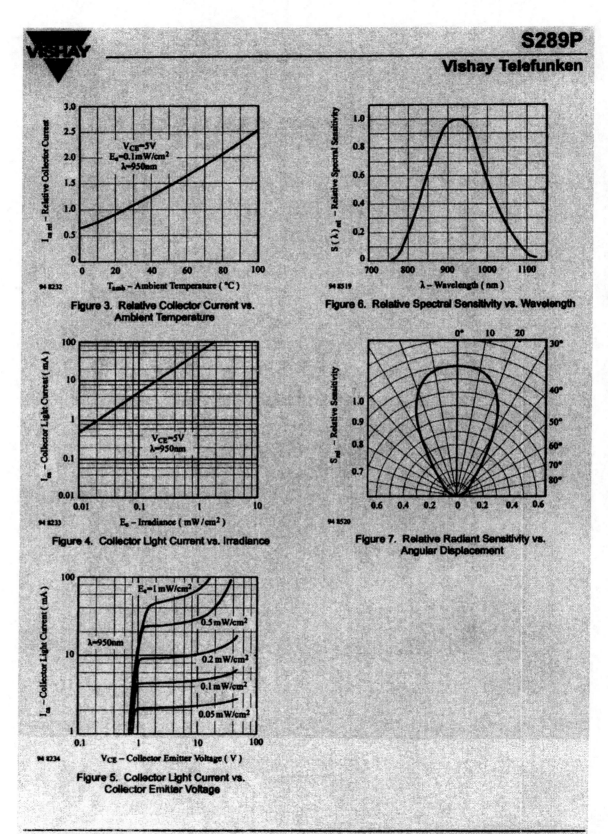

Figure 3. Relative Collector Current vs. Ambient Temperature

Figure 4. Collector Light Current vs. Irradiance

Figure 5. Collector Light Current vs. Collector Emitter Voltage

Figure 6. Relative Spectral Sensitivity vs. Wavelength

Figure 7. Relative Radiant Sensitivity vs. Angular Displacement

Chapter 18

K817P/ K827PH/ K847PH
Vishay Telefunken

Optocoupler with Phototransistor Output

Description
The K817P/ K827PH/ K847PH consist of a phototransistor optically coupled to a gallium arsenide infrared-emitting diode in an 4-lead up to 16-lead plastic dual inline package.
The elements are mounted on one leadframe using a coplanar technique, providing a fixed distance between input and output for highest safety requirements.

Applications
Programmable logic controllers, modems, answering machines, general applications

Features
- Endstackable to 2.54 mm (0.1') spacing
- DC isolation test voltage V_{IO} = 5 kV
- Low coupling capacitance of typical 0.3 pF
- Current Transfer Ratio (CTR) selected into groups
- Low temperature coefficient of CTR
- Wide ambient temperature range
- Underwriters Laboratory (UL) 1577 recognized, file number E-76222
- CSA (C–UL) 1577 recognized, file number E-76222 – Double Protection
- Coupling System U

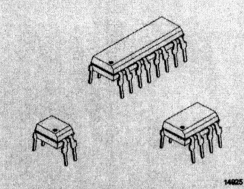

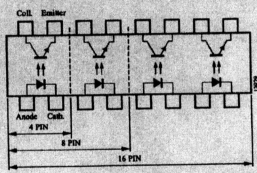

Order Instruction

Ordering Code	CTR Ranking	Remarks
K817P	50 to 600%	4 Pin = Single channel
K827PH	50 to 600%	8 Pin = Dual channel
K847PH	50 to 600%	16 Pin = Quad channel
K817P1	40 to 80%	4 Pin = Single channel
K817P2	63 to 125%	4 Pin = Single channel
K817P3	100 to 200%	4 Pin = Single channel
K817P4	160 to 320%	4 Pin = Single channel
K817P5	50 to 150%	4 Pin = Single channel
K817P6	100 to 300%	4 Pin = Single channel
K817P7	80 to 160%	4 Pin = Single channel
K827P8	130 to 260%	4 Pin = Single channel
K817P9	200 to 400%	4 Pin = Single channel

CTR = Current Transfer Ratio (I_C/I_F)

Rev. A2, 11–Jan–99

K817P/ K827PH/ K847PH
Vishay Telefunken

Absolute Maximum Ratings

Input (Emitter)

Parameter	Test Conditions	Symbol	Value	Unit
Reverse voltage		V_R	6	V
Forward current		I_F	60	mA
Forward surge current	$t_p \leq 10\ \mu s$	I_{FSM}	1.5	A
Power dissipation	$T_{amb} \leq 25°C$	P_V	100	mW
Junction temperature		T_j	125	°C

Output (Detector)

Parameter	Test Conditions	Symbol	Value	Unit
Collector emitter voltage		V_{CEO}	70	V
Emitter collector voltage		V_{ECO}	7	V
Collector current		I_C	50	mA
Peak collector current	$t_p/T = 0.5, t_p \leq 10\ ms$	I_{CM}	100	mA
Power dissipation	$T_{amb} \leq 25°C$	P_V	150	mW
Junction temperature		T_j	125	°C

Coupler

Parameter	Test Conditions	Symbol	Value	Unit
AC isolation test voltage (RMS)	t = 1 min	V_{IO} [1]	5	kV
Total power dissipation	$T_{amb} \leq 25°C$	P_{tot}	250	mW
Operating ambient temperature range		T_{amb}	–40 to +100	°C
Storage temperature range		T_{stg}	–55 to +125	°C
Soldering temperature	2 mm from case, t ≤ 10 s	T_{sd}	260	°C

[1] Related to standard climate 23/50 DIN 50014

Chapter 18

MOTOROLA
SEMICONDUCTOR TECHNICAL DATA

Order this document by 4N35/D

| VDE | UL | CSA | SETI | SEMKO | DEMKO | NEMKO | BABT |

6-Pin DIP Optoisolators
Transistor Output

The 4N35, 4N36 and 4N37 devices consist of a gallium arsenide infrared emitting diode optically coupled to a monolithic silicon phototransistor detector.

- Current Transfer Ratio — 100% Minimum @ Specified Conditions
- Guaranteed Switching Speeds
- Meets or Exceeds all JEDEC Registered Specifications
- *To order devices that are tested and marked per VDE 0884 requirements, the suffix "V" must be included at end of part number. VDE 0884 is a test option.*

Applications

- General Purpose Switching Circuits
- Interfacing and coupling systems of different potentials and impedances
- Regulation Feedback Circuits
- Monitor & Detection Circuits
- Solid State Relays

4N35*
4N36
4N37
[CTR = 100% Min]

*Motorola Preferred Device

STYLE 1 PLASTIC

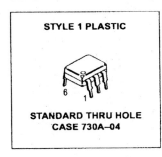

STANDARD THRU HOLE
CASE 730A–04

SCHEMATIC

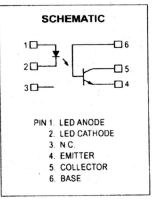

PIN 1. LED ANODE
2. LED CATHODE
3. N.C.
4. EMITTER
5. COLLECTOR
6. BASE

MAXIMUM RATINGS (T_A = 25°C unless otherwise noted)

Rating	Symbol	Value	Unit
INPUT LED			
Reverse Voltage	V_R	6	Volts
Forward Current — Continuous	I_F	60	mA
LED Power Dissipation @ T_A = 25°C with Negligible Power in Output Detector Derate above 25°C	P_D	120 1.41	mW mW/°C
OUTPUT TRANSISTOR			
Collector–Emitter Voltage	V_{CEO}	30	Volts
Emitter–Base Voltage	V_{EBO}	7	Volts
Collector–Base Voltage	V_{CBO}	70	Volts
Collector Current — Continuous	I_C	150	mA
Detector Power Dissipation @ T_A = 25°C with Negligible Power in Input LED Derate above 25°C	P_D	150 1.76	mW mW/°C
TOTAL DEVICE			
Isolation Source Voltage(1) (Peak ac Voltage, 60 Hz, 1 sec Duration)	V_{ISO}	7500	Vac(pk)
Total Device Power Dissipation @ T_A = 25°C Derate above 25°C	P_D	250 2.94	mW mW/°C
Ambient Operating Temperature Range(2)	T_A	–55 to +100	°C
Storage Temperature Range(2)	T_{stg}	–55 to +150	°C
Soldering Temperature (10 sec, 1/16″ from case)	T_L	260	°C

1. Isolation surge voltage is an internal device dielectric breakdown rating.
   For this test, Pins 1 and 2 are common, and Pins 4, 5 and 6 are common.
2. Refer to Quality and Reliability Section in Opto Data Book for information on test conditions.

Preferred devices are Motorola recommended choices for future use and best overall value.
GlobalOptoisolator is a trademark of Motorola, Inc.

4N35 4N36 4N37

ELECTRICAL CHARACTERISTICS (T_A = 25°C unless otherwise noted)[1]

Characteristic		Symbol	Min	Typ[1]	Max	Unit
INPUT LED						
Forward Voltage (I_F = 10 mA)	T_A = 25°C	V_F	0.8	1.15	1.5	V
	T_A = –55°C		0.9	1.3	1.7	
	T_A = 100°C		0.7	1.05	1.4	
Reverse Leakage Current (V_R = 6 V)		I_R	—	—	10	µA
Capacitance (V = 0 V, f = 1 MHz)		C_J	—	18	—	pF
OUTPUT TRANSISTOR						
Collector–Emitter Dark Current (V_{CE} = 10 V, T_A = 25°C)		I_{CEO}	—	1	50	nA
(V_{CE} = 30 V, T_A = 100°C)			—	—	500	µA
Collector–Base Dark Current (V_{CB} = 10 V)	T_A = 25°C	I_{CBO}	—	0.2	20	nA
	T_A = 100°C			100	—	
Collector–Emitter Breakdown Voltage (I_C = 1 mA)		$V_{(BR)CEO}$	30	45	—	V
Collector–Base Breakdown Voltage (I_C = 100 µA)		$V_{(BR)CBO}$	70	100	—	V
Emitter–Base Breakdown Voltage (I_E = 100 µA)		$V_{(BR)EBO}$	7	7.8	—	V
DC Current Gain (I_C = 2 mA, V_{CE} = 5 V)		h_{FE}	—	400	—	—
Collector–Emitter Capacitance (f = 1 MHz, V_{CE} = 0)		C_{CE}	—	7	—	pF
Collector–Base Capacitance (f = 1 MHz, V_{CB} = 0)		C_{CB}	—	19	—	pF
Emitter–Base Capacitance (f = 1 MHz, V_{EB} = 0)		C_{EB}	—	9	—	pF
COUPLED						
Output Collector Current (I_F = 10 mA, V_{CE} = 10 V)	T_A = 25°C	I_C (CTR)[2]	10 (100)	30 (300)	—	mA (%)
	T_A = –55°C		4 (40)	—	—	
	T_A = 100°C		4 (40)	—	—	
Collector–Emitter Saturation Voltage (I_C = 0.5 mA, I_F = 10 mA)		$V_{CE(sat)}$	—	0.14	0.3	V
Turn–On Time	(I_C = 2 mA, V_{CC} = 10 V, R_L = 100 Ω)[3]	t_{on}	—	7.5	10	µs
Turn–Off Time		t_{off}	—	5.7	10	
Rise Time		t_r	—	3.2	—	
Fall Time		t_f	—	4.7	—	
Isolation Voltage (f = 60 Hz, t = 1 sec)		V_{ISO}	7500	—	—	Vac(pk)
Isolation Current[4] (V_{I-O} = 3550 Vpk) 4N35		I_{ISO}	—	—	100	µA
(V_{I-O} = 2500 Vpk) 4N36			—	—	100	
(V_{I-O} = 1500 Vpk) 4N37			—	8	100	
Isolation Resistance (V = 500 V)[4]		R_{ISO}	10^{11}	—	—	Ω
Isolation Capacitance (V = 0 V, f = 1 MHz)[4]		C_{ISO}	—	0.2	2	pF

1. Always design to the specified minimum/maximum electrical limits (where applicable).
2. Current Transfer Ratio (CTR) = I_C/I_F x 100%.
3. For test circuit setup and waveforms, refer to Figure 11.
4. For this test, Pins 1 and 2 are common, and Pins 4, 5 and 6 are common.

Chapter 18

4N35 4N36 4N37

TYPICAL CHARACTERISTICS

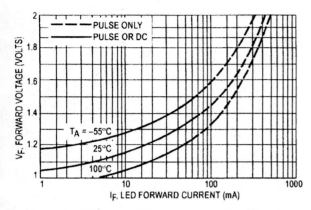

Figure 1. LED Forward Voltage versus Forward Current

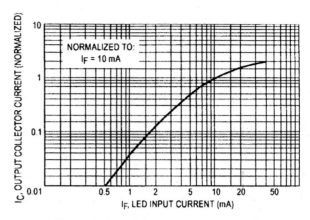

Figure 2. Output Current versus Input Current

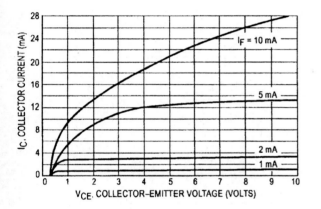

Figure 3. Collector Current versus Collector–Emitter Voltage

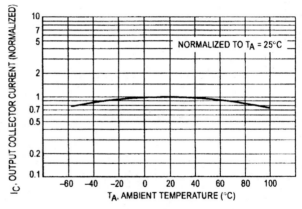

Figure 4. Output Current versus Ambient Temperature

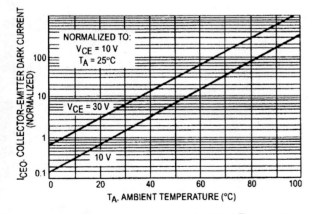

Figure 5. Dark Current versus Ambient Temperature

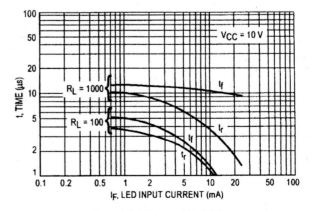

Figure 6. Rise and Fall Times (Typical Values)

Chapter 18

4N35 4N36 4N37

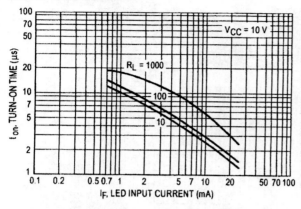

Figure 7. Turn–On Switching Times

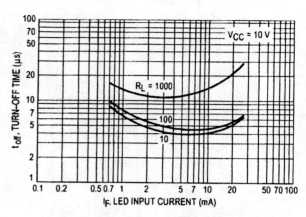

Figure 8. Turn–Off Switching Times

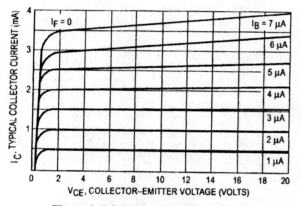

Figure 9. DC Current Gain (Detector Only)

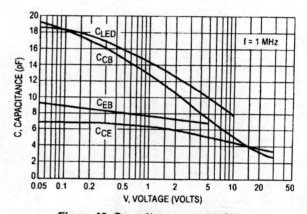

Figure 10. Capacitances versus Voltage

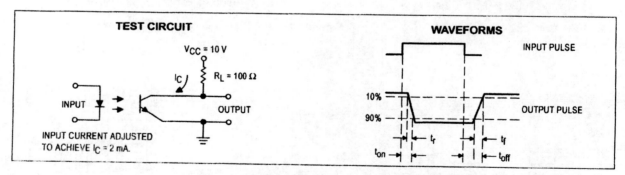

Figure 11. Switching Time Test Circuit and Waveforms

Chapter 18

PROBLEMS

18.1 Sketch the SCR structure, its symbol, and label its terminals.

18.2 Sketch the I-V characteristics of the SCR and label it completely.

18.3 Look up the data sheets for the 2N6394 SCR and determine the following parameters: Current rating (on-state *rms* current), peak forward on-state voltage, gate trigger current, gate trigger voltage, holding current, turn-on time, and turn-off time.

18.4 Describe the two ways in which an SCR can be turned on or latched into conduction.

18.5 Explain the regenerative process in the SCR.

18.6 Describe the two ways in which an SCR can be turned off.

18.7 Sketch the DIAC structure, its symbol, and label its terminals.

18.8 Sketch the I-V characteristics curve of the DIAC and label it completely.

18.9 What is the main application of the DIAC?

18.10 Sketch the TRIAC structure, its symbol, and label its terminals.

18.11 What is the main application of the TRIAC?

18.12 Refer to data sheets for the 2N6071A TRIAC and determine the following parameters: Current rating (on-state *rms* current), peak forward on-state voltage, gate trigger voltage, gate trigger current, holding current, and turn-on time.

18.13 Determine the firing angle ϕ, conduction angle α, and percent power delivered to the load for the circuit of Figure 18-1P below, given that $C = 0.1$ µF, $R = 27$ kΩ, and $V_{BO} = 20$ V. Load consists of five 100 W light bulbs connected in parallel.

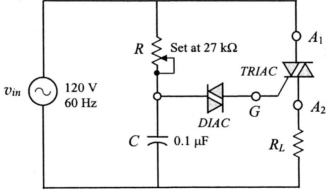

Figure 18-1P:
DIAC-TRIAC power control circuit

18.14 In the DIAC-TRIAC power control circuit of Figure 18-1P, determine the percent power delivered to the load if the firing angle is: (a) 45°, (b) 60°, and (c) 90°.

18.15 Describe the photoresistor and draw its schematic symbol.

18.16 Assuming that the *dark resistance* of the photoresistor used in the circuit of Figure 18-2P is 120 kΩ, and it has a resistance of 20 kΩ when illuminated by a light beam, determine if the relay is energized when the light beam is interrupted and is de-energized when the light beam is on. Show your complete work.

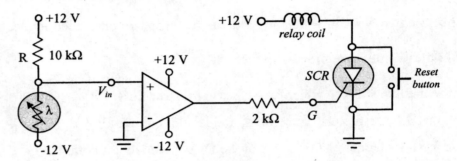

Figure 18-2P: Utilizing the photoresistor and SCR in an automatic door opener or a security system

18.17 Describe the photodiode and draw its schematic symbol.

18.18 Discuss an application of the photodiode.

18.19 Describe the optocoupler and draw its schematic symbol.

18.20 Discuss an application of the optocoupler.

INDEX

A

AC load line, 200, 202, 203, 206
Active filters, 596
    band-pass active filter, 633
    band-stop active filter, 639
    design of active BPF, 637
    design of first-order LPF, 604
    design of first-order HPF, 607
    design of fourth-order LPF, 631
    design of second-order HPF, 621
    design of second-order LPF, 619
    design of third-order LPF, 627
    first-order high-pass, 600
    first-order low-pass, 597
    second-order active filters, 608
        Sallen-Key unity-gain filter, 613
        second-order high-pass, 612
        second-order low-pass, 610
    second-order filter types and parameters, 626
    section summary for the analysis and design of band-pass and band-stop active filters, 643
    section summary for the analysis and design of second-order equal component active filters, 624
    table of higher order filter damping factors (ξ) and low-pass frequency correction factors (k_{LP}), 626
Active region, 232, 263, 266
Amplifier parameters, 159
Analysis of the CE amplifier
    emitter-biased, 163
    voltage-divider biased
        R_E fully bypassed, 167
        R_E partially bypassed, 170
Analysis of the cascode amplifier, 183
Analysis of diff-amp with basic current-mirror, 425
Analysis of diff-amp with Wilson current-mirror, 428
Analog-to-digital conversion, 743
    monolithic IC ADC, 74
Approximate model of BJT, 146

Atomic structure, 2
Atomic weight, 2

B

Bandwidth, 210
Basic differential amplifier, 415
    common mode gain, 417
    design of the basic diff-amp, 419
    differential mode gain, 416
Basic diode circuits, 20
Basic op-amp configurations
    compensation resistor, 476
    current-to-voltage converter, 470, 472
    inverting amplifier, 465
    non-inverting amplifier, 471
    unity-gain buffer, 472
Basic power supply, 43, 62
Basic theory of BJT operation, 68
BJT amplifier fundamentals, 156
BJT amplifier frequency response, 374
BJT cascode amplifier, 380
Bode plots, 351
Boost (step-up) converter, 669
    analysis of, 671
    design of, 675
Breakdown region, 232, 263
Buck (step-down) converter, 659
    analysis of, 665
    design of, 667
Buck-boost DC-DC converter, 678
    analysis of, 680
    design of, 684

C

Cascaded diff-amps, 436, 438
Cascode BJT amplifier
    DC bias analysis, 91
    DC bias design, 98
Cascode MOSFET current-mirror, 410
Common-base amplifier, 89
Common-collector amplifier, 88
Common-emitter characteristics, 71, 73
Conduction band, 6

Colpitts oscillator, 715
Corner frequency, 350
Covalent bonding, 5
CS JFET amplifier, 384
CS MOSFET amplifier, 388
Current boosting, 652
Current gain, 161
Current-mirror current sources
    basic BJT current-mirror, 402
    basic MOSFET current-mirror, 408
    Wilson current-mirror, 404
Cutoff frequency, 350
Cutoff voltage, 80

D
DC bias analysis of BJT amplifier
    base bias, 77
    emitter bias, 81
    voltage-divider bias, 84
DC bias analysis of JFET amplifier
    fixed bias, 235
    self-bias, 237
    voltage-divider bias, 246
DC bias decisions, 213, 219
DC bias design of BJT amplifier, 94
DC bias design of JFET amplifier, 254
DC load line, 78, 201, 202, 203, 206
Depletion-type MOSFET, 261
Design of the cascode amplifier, 219
Design of the CE amplifier, 212
Design of diff-amp with basic current-mirror, 426
Design of diff-amp with Wilson current-mirror, 430
Design of the small-signal JFET and MOSFET amplifiers, 330
Diacs, 758
Diac-triac power control circuit, 760
Diff-amp with current-mirror, 424
Difference amplifier, 486
Differential amplifier, 412
Differentiator, 498
Diffusion current, 7, 8
Digital-to-analog conversion, 736
    binary-weighted resistor DAC, 740
    monolithic IC DAC, 742
    R2R ladder DAC, 736
Diode AC resistance, 17
Diode characteristics, 12
Diode DC resistance, 16
Diode equation, 15
Diode switching circuits, 23
Doping, 7
Drain characteristics
    depletion-type MOSFET, 263
    enhancement-type MOSFET, 266
    JFET, 232

E
Early voltage, 75
Energy bands, 6
Energy gap, 6
Enhancement-type MOSFET, 264
Extrinsic semiconductors, 7

F
Feedback topologies, 526
FET amplifier frequency response, 390
FET cascode amplifier, 391
FET output resistance r_o, 291
Field-effect transistors, 227
Filtering, 53, 55, 58
First-order high-pass RC network, 355
First-order low-pass RC network, 350
Formation of the p-n junction, 9
Forward-bias, 10
Four-layer devices, 754
Frequency response, 210
Frequency response fundamentals, 349
Frequency response of the
    CB BJT amplifier, 377
    CE BJT amplifier, 359
Frequency response plot of the common emitter BJT amplifier, 375
Full-wave bridge rectifier, 50, 55
Full-wave center-tapped rectifier, 47, 55
Function generator, 726
FSK generation, 729

G
General algebraic solution for depletion MOSFET, 264

enhancement MOSFET, 267, 270
self-biased JFET amplifier, 240, 241
voltage-divider biased JFET, 248

H
Half-wave rectifier, 45, 55
Hartley oscillator, 717
Heat sinks and thermal considerations, 585
High cutoff frequency introduced by transistor β, 370
Hints on looking up the FET characteristics, 299
h-parameter model of BJT, 131

I
I_{DSS}, 229
Input bias current (I_{BI}), 539
Input offset current (I_{IO}), 539
Input offset voltage (V_{IO}), 540
Input resistance, 159
Input resistance with FB, 534
Instrumentation amplifier, 490
Integrated circuit regulators, 650
Integrator, 494
Intrinsic semiconductor, 7
Inverting summer, 479
Ionization, 7

J
JFET amplifier, 248
JFET structure, 228
Junction-FET (JFET), 228

L
Large-signal operation, 155
Light-emitting diode, 28
Lower cutoff frequency, 210, 211, 217, 360

M
Maximum output signal swing, 200, 206
Midband gain, 210
Miller's theorem, 366
Mixed signal, 746
MOSFET
 depletion-type MOSFET, 261
 enhancement-type MOSFET, 264
MOSFET differential amplifier, 439
 basic MOSFET diff-amp, 440
 MOSFET diff-amp with cascode current-mirror, 444

N
Negative feedback, 524
Non-inverting summer, 481
N-type impurities, 7
N-type semiconductors, 8

O
Ohmic region, 232, 263, 266
Op-amp characteristics
 AC characteristics, 541
  common mode rejection ratio (CMRR), 541
  gain-bandwidth product, 542
  large signal voltage gain, 541
  slew rate (SR), 542
 DC characteristics, 539
  input bias current (I_{BI}), 539
  input offset current (I_{IO}), 539
  input offset voltage (V_{IO}), 540
  output short-circuit current, 541
  supply voltage rejection ratio (PSRR), 540
Operating point, 78, 200
Operational amplifier, 454
 closed-loop operation, 464
 as LED driver, 458
 characteristics, 455
 comparator, 457
 open-loop operation, 456
 Schmitt trigger, 461
 virtual short and virtual ground, 465
Oscillator principles of operation, 706
Output resistance, 160
Overload and short-circuit protection, 652

P

Photoconductive cell, 763
Photoresistor, 763
Pinch-off voltage, 229, 231, 262
Plotting the transfer characteristics of JFET, 234
p-n junction diode, 12
Power amplifiers, 565
    class A power amplifier, 567
    class AB push-pull power amp, 577
    class B push-pull power amp, 568
    classes of operation, 566
    design of Class AB push-pull power amplifier, 582
    design of push-pull power amp, 575
    heat sinks and thermal considerations, 585
    reducing cross-over distortion, 573
    section summary for analysis and design of class AB power amplifier, 584
    section summary for analysis and design of class B power amplifier, 577
    thermal resistance, 585
Power gain, 161
P-type impurities, 7
P-type semiconductors, 8
Pulse and ramp generator, 727

Q

Q-point, 201

R

RC phase-shift oscillator, 707
Rectangular waveform generator, 724
Regulated power supply, 56, 62
Regulation, 56, 58
Regulator with adjustable output and overload protection, 653
Reverse-bias, 11
Reverse-breakdown, 11
r-parameter model of BJT, 122

S

Sample-and-hold circuit, 483
Saturation current, 82
Saturation voltage, 80
Schmitt trigger, 461
Section summary for the analysis and design of
    boost converter, 677
    buck converter, 668
    buck-boost converter, 686
Section summary for the analysis of
    CE BJT amplifier, 182
    CS JFET or DMOS amplifier, 302
    CD JFET or DMOS amplifier, 309
    CS EMOS amplifier, 322
    CG EMOS amplifier, 326
    current-mirror biased diff-amp, 435
    series-shunt feedback amplifier, 538
Series-shunt feedback, 526
Series-shunt feedback characteristics applied to non-inverting amplifier configuration:
    voltage gain without FB, 533
    voltage gain with FB, 533
    input resistance with FB, 534
    output resistance with FB, 534
    bandwidth with FB, 535
Shockley's equation, 231
Silicon-controlled rectifier (SCR), 754
Silicon-controlled switches, 757
Single-supply operation, 545
Small-signal analysis of the
    CG enhancement MOSFET amplifier, 323
    common-source JFET amplifier, 294
    common-drain JFET amplifier, 303
    CS enhancement MOSFET amplifier, 310
    CS enhancement MOSFET amplifier with partially unbypassed R_S, 313
    FET cascode amplifier, 327
Small-signal circuit model of FET, 290
Small-signal operation of
    BJT amplifier, 155
    FET amplifier, 289
Square-wave generator, 720

Stability comparison of fixed bias circuit versus self-bias circuit, 244
Summing amplifiers, 479
    inverting summer, 479
    non-inverting summer, 481
Switching power supplies, 658
System stability, 525

T

Table of higher order filter damping factors (ξ) and low-pass frequency correction factors (k_{LP}), 626
Temperature effects, 14
The 555 timer, 722
Three-terminal IC regulators, 657
Time-division multiplexer (TDM) circuit, 484
Transconductance g_m, 291
Transformer, 44, 55
Transistor switch
    analysis, 100
    design, 102
Triacs, 758

U

Universal r & h model of BJT, 148
Upper cutoff frequency, 210, 365

V

Valence electrons, 5
Voltage gain, 161
Voltage regulator, 56, 62

W

Wien-bridge oscillator, 711

Z

Zener diode, 24

Answers To Selected Problems

Chapter 1

1.9 (a) $I_D = 6.5$ mA, $V_D = 0.7$ V
 (b) $I_D = 6.818$ mA, $V_D = -60$ V
1.10 (a) $I_D = 9.65$ mA, $V_D = 0.7$ V
 (b) $I_D = 10$ mA, $V_D = -80$ V
1.12 (a) $R_D = 350$ Ω, $r_d = 13$ Ω
 (b) $R_D = 123$ Ω, $r_d = 4.333$ Ω
 (c) $R_D = 76$ Ω, $r_d = 2.6$ Ω
1.13 (a) $R_D = 100$ Ω, $r_d = 8.667$ Ω
 (b) $R_D = 53$ Ω, $r_d = 4.333$ Ω
 (c) $R_D = 33$Ω, $r_d = 2.6$ Ω
1.14 $V_D = 0.25$ V, $I_D = -5.656$ µA
1.15 (a) $V_D = 0.55$ V
 (b) $I_D = -9.05$ µA
1.16 (a) $I_D = 2.7$ mA, $V_D = 0.7$ V
 (b) $I_D = 0$ mA, $V_D = 0.49$ V
1.17 (a) $I_D = 4.924$ mA, $V_D = 0.7$ V
 (b) $I_D = 2.06$ mA, $V_D = 0.7$ V
1.18 (a) $V_O = 9.1$ V, $I_D = 2.545$ mA
 (b) $V_O = 6.7$ mV, $I_D = 2.4545$ mA
1.19 (a) $V_O = 5$ V, $I_D = 4.24$ mA
 (b) $V_O = 2.96$ V, $I_D = 5.7575$ mA
1.24 $I_Z = 10$ mA, $I_{ZM} = 50$ mA,
 $I_{ZK} = 0.25$ mA, $r_z = 17$ Ω @ I_{ZT}
1.25 $V_O = 3.9$ V, $I_Z = 10.1$ mA, $I_{ZM} = 128$ mA
1.26 (a) $V_O = 7.5$ V, $I_Z = 9.75$ mA
 (b) $V_O = 7$ V, $I_D = 0$ mA
1.31 $R = 910$ Ω
1.32 $R = 150$ Ω
1.33 $R = 1$ kΩ
1.34 (a) RLI = 0.2 at $\theta = 40°$ and RLI = 0 at $\theta = 90°$
 (b) RLI = 2 at $I_F = 20$ mA and RLI = 4 at $I_F = 50$ mA
 (c) $V_F = 2$ V at $I_F = 20$ mA and $V_F = 2.1$ V at $I_F = 50$ mA

Chapter 2

2.1 $V_{O(DC)} = 8.98$ V
2.2 $V_{O(DC)} = 8.78$ V for 120 Vrms and $V_{O(DC)} = 8.25$ V for 110 Vrms
2.3 $V_{O(DC)} = 5.73$ V

2.4 $V_{O(DC)} = 5.306$ V
2.5 $V_{O(DC)} = 12$ V
2.6 $V_{O(DC)} = 10.6$ V, $I_{O(DC)} = 106$ mA
2.7 $V_{O(DC)} = 12$ V
2.8 $V_{O(DC)} = 16.45$ V, $I_{O(DC)} = 166.6$ mA, $V_r = 0.42$ V
2.9 $V_{O(DC)} = 16.51$ V, $V_r = 0.2954$ V
2.10 $V_{O(DC)} = 17.7$ V, $V_r = 0.625$ V
2.11(a) $V_O = 7.5$ V, $I_Z = 62.3$ mA, $I_L = 22.7$ mA, $I_S = 85$ mA
   (b) $V_O = 7$ V, $I_Z = 22.3$ mA, $I_L = 22.7$ mA, $I_S = 45$ mA
2.12 $V_O = 5.1$ V, $I_Z = 23.4$ mA, $I_L = 25.5$ mA
2.13 $V_O = 3.3$ V, $I_Z = 0$ mA, $I_L = 33$ mA
2.14 $V_{O(DC)} = 12$ V, $V_r = 0.169$ V, $I_Z = 22$ mA, $I_L = 100$ mA
2.15 $V_{O(DC)} = 5.1$ V, $V_r = 22.7$ mV, $I_Z = 18.58$ mA, $I_L = 68$ mA

Chapter 3

3.7 $\beta = 132$, $I_C = 5.29$ mA, $I_E = 5.32$ mA
3.8(a) $I_C = 8$ mA
   (b) $I_C = 8.2$ mA
   (c) $I_C = 8.4$ mA
   (d) $I_C = 8.5$ mA
3.10 $I_B = 34.25$ μA, $I_C = 4.1$ mA, $V_{CE} = 5.85$ V
3.11(a) $V_{CE} = -5.64$ V
   (b) $I_B = -43.33$ μA
   (c) $I_C = -5.2$ mA
   (d) $V_{CE} = -5.64$ V
3.12(a) $I_B = 39.72$ μA, $I_C = 8$ mA, $V_{CE} = 3$ V
   (b) $I_{BQ} = 40$ μA, $I_{CQ} = 8$ mA, $V_{CEQ} = 3$ V
3.13(a) $I_B = 30.74$ μA
   (b) $I_{BQ} = 30$ μA, $I_{CQ} = 3$ mA, $V_{CEQ} = 5.3$ V
   (c) $\beta = 100$
   (d) $I_{BQ} = 30.74$ μA, $I_{CQ} = 3.075$ mA, $V_{CEQ} = 5.3$ V
3.14 $R_B = 282$ kΩ
3.15 $I_{C(sat)} = 10$ mA, $I_{B(sat)} = 100$ μA, $R_{B(sat)} = 91$ kΩ
3.16 $I_B = 51.36$ μA, $I_C = 5.136$ mA, $V_{CE} = 4.296$ V
3.17(a) $I_B = 23.33$ μA, $I_C = 4.643$ mA, $V_{CE} = 7$ V
3.18 $R_C = 1$ kΩ
3.19 $I_{CQ} = 4.54$ mA, $V_{CEQ} = 7.785$ V
3.20 $R_B = 470$ kΩ
3.21(a) $I_B = 23.347$ μA, $I_C = 4.76$ mA, $V_{CE} = 7.86$ V
   (b) $I_{BQ} = 20$ μA, $I_{CQ} = 4.8$ mA, $V_{CEQ} = 7.8$ V
3.22(a) $I_{BQ} = 15$ μA, $I_{EQ} = 2.4$ mA, $V_{CEQ} = 8.4$ V
   (b) $I_{BQ} = 0$ μA, $I_{CQ} = 2.5$ mA, $V_{CEQ} = 8$ V
3.23(a) $I_{BQ} = 10.73$ μA, $I_{CQ} = 2.146$ mA, $V_{CEQ} = 5.7$ V
   (b) $I_{BQ} = 0$ μA, $I_{CQ} = 2.2$ mA, $V_{CEQ} = 5.44$ V
3.24 $R_C = 4.7$ kΩ

3.25 $I_{CQ} = 1.355$ mA, $V_{CEQ} = 7.8$ V
3.26(a) $I_{BQ} = 0$ μA, $I_{CQ} = 1.416$ mA, $V_{CEQ} = 10.33$ V
    (b) $I_{BQ} = 8.5$ μA, $I_{CQ} = 1.36$ mA, $V_{CEQ} = 10.56$ V
3.27 $I_{BQ} = 31.388$ μA, $I_{CQ} = 6.277$ mA, $V_{CEQ} = 4.46$ V, $V_{BCQ} = -3.767$ V
3.28 $I_{BQ} = 47$ μA, $I_{CQ} = 4.7$ mA, $V_{CEQ} = 6.36$ V, $V_{BCQ} = -5.64$ V
3.29 $V_{CE1} = 4.28$ V, $V_{CE2} = 9.36$ V
3.30 $V_{CE1} = 2.921$ V, $V_{CE2} = 11.502$ V
3.31 $R_E = 910$ Ω, $R_C = 2.7$ kΩ, $R_B = 1.2$ MΩ
3.32 $R_E = 1.1$ kΩ, $R_C = 2.4$ kΩ, $R_1 = 30$ kΩ, $R_2 = 7.5$ kΩ
3.33(a) $R_E = 1$ kΩ, $R_C = 3.9$ kΩ, $R_B = 1.3$ MΩ
    (b) $R_E = 1$ kΩ, $R_C = 3.9$ kΩ, $R_1 = 56$ kΩ, $R_2 = 10$ kΩ
3.34 $R_E = 1.6$ kΩ, $R_C = 5.1$ kΩ, $R_1 = 43$ kΩ, $R_2 = 10$ kΩ
3.35 $R_E = 1.6$ kΩ, $R_C = 5.1$ kΩ, $R_B = 1.3$ MΩ
3.34 $R_E = 1.8$ kΩ, $R_C = 6.2$ kΩ, $R_1 = 51$ kΩ, $R_2 = 11$ kΩ
3.37 $R_E = 3.9$ kΩ, $R_B = 750$ kΩ
3.38 $R_E = 1.3$ kΩ, $R_C = 2.7$ kΩ, $R_1 = 33$ kΩ, $R_2 = 4.3$ kΩ, $R_3 = 10$ kΩ
3.39 $R_E = 1.5$ kΩ, $R_C = 3.6$ kΩ, $R_1 = 39$ kΩ, $R_2 = 5.1$ kΩ, $R_3 = 10$ kΩ
3.40 $R_E = 1.2$ kΩ, $R_C = 3.6$ kΩ, $R_1 = 43$ kΩ, $R_2 = 5.1$ kΩ, $R_3 = 10$ kΩ
3.41 $R_C = 1$ kΩ, $R_B = 82$ kΩ or less
3.42 $R_C = 1.5$ kΩ, $R_B = 33$ kΩ
3.43 $R_C = 1$ kΩ, $R_B = 18$ kΩ
3.44 $t_{on} = 70$ ns, $t_{off} = 250$ ns
3.45 $R_C = 510$ Ω, $R_B = 10$ kΩ
3.46 $R_C = 430$ Ω, $R_B = 18$ kΩ or less

Chapter 4

4.1 $r_b = 1.987$ kΩ, $r_o = 77$ kΩ
4.2 $r_b = 1.72$ kΩ, $r_o = 66.66$ kΩ
4.3 $r_b = 2.2$ kΩ, $r_o = 54.5$ kΩ
4.4 $r_b = 2.437$ kΩ, $r_o = 58.58$ kΩ
4.5 $r_b = 2.47$ kΩ, $r_o = 63.69$ kΩ
4.6 $r_b = 2.8$ kΩ, $r_o = 72$ kΩ
4.7 $r_b = 2.9$ kΩ, $r_o = 72$ kΩ
4.8 $r_b = 2.3$ kΩ, $r_o = 55.55$ kΩ
4.9 $r_e = 12.6$ Ω, $r_o = 48.5$ kΩ, $r_c = 5.825$ MΩ
4.10 $r_b = 3.75$ kΩ, $r_o = 96$ kΩ
4.11 $h_{11} = 4.033$ kΩ, $h_{21} = -2/3$, $h_{12} = 2/3$, $1/h_{22} = 3.3$ kΩ
4.12 $h_{11} = 2.42$ kΩ, $h_{21} = -0.4$, $h_{12} = 0.4$, $1/h_{22} = 5.5$ kΩ
4.13 $h_{11} = 2.42$ kΩ, $h_{21} = -0.4$, $h_{12} = 0.4$, $1/h_{22} = 5.5$ kΩ
4.14 $h_{11} = 160$ Ω, $h_{21} = -1/3$, $h_{12} = 1/2$, $1/h_{22} = 180$ Ω
4.15 $h_{ie} = 2.2$ kΩ, $h_{fe} = 170$, $1/h_{oe} = 50$ kΩ
4.16 $h_{ie} = 2.849$ kΩ, $h_{fe} = 170$, $1/h_{oe} = 50$ kΩ
4.17 $h_{ie} = 3.6$ kΩ, $h_{fe} = 150$, $1/h_{oe} = 94$ kΩ
4.18 $h_{ie} = 2.7$ kΩ, $h_{fe} = 160$, $1/h_{oe} = 65$ kΩ
4.19 $h_{ie} = 2.7$ kΩ, $h_{fe} = 160$, $1/h_{oe} = 65$ kΩ

4.20 $h_{ie} = 3.1$ kΩ, $h_{fe} = 155$, $1/h_{oe} = 78$ kΩ
4.21 $h_{ib} = 22$ Ω, $h_{fb} = 0.9933$, $1/h_{oe} = 86$ kΩ
4.23 $h_{ie} = 3$ kΩ, $h_{fe} = 160$, $1/h_{oe} = 72$ kΩ
4.24 $h_{ie} = 2.3$ kΩ, $h_{fe} = 160$, $1/h_{oe} = 55$ kΩ
4.25 $h_{ie} = r_b = 2.3$ kΩ, $h_{fe} = \beta = 160$, $1/h_{oe} = r_o = 55$ kΩ
4.26 $h_{ie} = r_b = 1.9$ kΩ, $h_{fe} = \beta = 180$, $1/h_{oe} = r_o = 41$ kΩ
4.27 $h_{ib} = r_e = 18.5$ Ω, $h_{fb} = \alpha = 0.9938$, $1/h_{ob} = r_c = 11.4$ MΩ
4.28 $h_{ib} = r_e = 18.5$ Ω, $h_{fb} = \alpha = 1$, $1/h_{ob} = r_c = \infty$
4.29 $h_{ib} = r_e = 12.56$ Ω, $h_{fb} = \alpha = 1$, $1/h_{ob} = r_c = \infty$
4.30 $h_{ie} = r_b = 2.145$ kΩ, $h_{fe} = \beta = 170$, $1/h_{oe} = r_o = \infty$
4.31 $h_{ie} = r_b = 2.717$ kΩ, $h_{fe} = \beta = 170$, $1/h_{oe} = r_o = \infty$
4.32 $h_{ie} = r_b = 1.9$ kΩ, $h_{fe} = \beta = 180$, $1/h_{oe} = r_o = \infty$
4.33 $h_{ie} = r_b = 2.9$ kΩ, $h_{fe} = \beta = 160$, $1/h_{oe} = r_o = \infty$

Chapter 5

5.1 $R_{in} = 1.8$ kΩ, $R_o = 1.1$ kΩ, $A_{v(NL)} = -100$, $A_{v(WL)} = -66.7$, $A_p = 3640$
5.2 $R_{in} = 2.16$ kΩ, $R_o = 680$ Ω, $A_{v(NL)} = -84.6$, $A_{v(WL)} = -64.6$, $A_i = -63.43$, $A_p = 4095.64$
5.3 $R_{in} = 3$ kΩ, $R_o = 1$ kΩ, $A_{v(NL)} = -100$, $A_{v(WL)} = -66.66$, $A_i = -100$, $A_p = 6666$
5.4 $I_{CQ} = 1.65$ mA, $V_{CEQ} = 7$ V, $R_{in} = 2.2$ kΩ, $R_o = 3.15$ kΩ, $A_{v(NL)} = -119.9$, $A_{v(WL)} = -200$, $A_i = -56.17$, $A_p = 6735$, $v_{o(WL)} = -2.4$ V
5.5 $I_{CQ} = 3.24$ mA, $V_{CEQ} = 4.7$ V, $R_{in} = 1.65$ kΩ, $R_o = 3.1$ kΩ, $A_{v(NL)} = -279$, $A_{v(WL)} = -168$, $A_i = -58.97$, $A_p = 9912$, $v_{o(WL)} = -7.59$ V
5.6 $I_{CQ} = 1.84$ mA, $V_{CEQ} = 6.9$ V, $R_{in} = 2.1$ kΩ, $R_o = 3.14$ kΩ, $A_{v(NL)} = -224$, $A_{v(WL)} = -134.28$, $A_i = -60$, $A_p = 8056.8$, $v_{o(WL)} = -6.875$ V
5.7 $I_{CQ} = 1.42$ mA, $V_{CEQ} = 5.63$ V, $R_{in} = 2.13$ kΩ, $R_o = 4.75$ kΩ, $A_{v(NL)} = -259$, $A_{v(WL)} = -129$, $A_i = -58.36$, $A_p = 7541.34$, $v_{o(WL)} = -3.17$ V
5.8 $I_{CQ} = 1.75$ mA, $V_{CEQ} = 7.3$ V, $R_{in} = 1.75$ kΩ, $R_o = 3.15$ kΩ, $A_{v(NL)} = -212$, $A_{v(WL)} = -126.6$, $A_i = -47.14$, $A_p = 5967.7$, $v_{o(WL)} = -5.32$ V
5.9 $I_{CQ} = 1.89$ mA, $V_{CEQ} = 6.68$ V, $R_{in} = 1.54$ kΩ, $R_o = 3.137$ kΩ, $A_{v(NL)} = -228$, $A_{v(WL)} = -136.7$, $A_i = -47.14$, $A_p = 6122.9$, $v_{o(WL)} = -5.35$ V
5.10 $I_{CQ} = 1.53$ mA, $V_{CEQ} = 8.27$ V, $R_{in} = 1.58$ kΩ, $R_o = 3.17$ kΩ, $A_{v(NL)} = -186.5$, $A_{v(WL)} = -111$, $A_i = -37.3$, $A_p = 4351$, $v_{o(WL)} = -4.45$ V
5.11 $I_{CQ} = 1.818$ mA, $V_{CEQ} = 7.96$ V, $R_{in} = 2.7$ kΩ, $R_o = 3.3$ kΩ, $A_{v(NL)} = -91$, $A_{v(WL)} = -53.4$, $A_i = -30.67$, $A_p = 1638$, $v_{o(WL)} = -3$ V
5.12 $I_{CQ} = 1.685$ mA, $V_{CEQ} = 6.35$ V, $R_{in} = 3.3$ kΩ, $R_o = 3.68$ kΩ, $A_{v(NL)} = -76$, $A_{v(WL)} = -43.98$, $A_i = -29$, $A_p = 1275.42$, $v_{o(WL)} = -3.6$ V
5.13 $I_{CQ} = 1.22$ mA, $V_{CEQ} = 8.8$ V, $R_{in} = 5.4$ kΩ, $R_o = 4.12$ kΩ, $A_{v(NL)} = -75.8$, $A_{v(WL)} = -40.5$, $A_i = -46.5$, $A_p = 1782$, $v_{o(WL)} = -2.95$ V
5.14 $I_{CQ} = 1.25$ mA, $V_{CEQ} = 8.7$ V, $R_{in} = 4.75$ kΩ, $R_o = 4.115$ kΩ, $A_{v(NL)} = -76.5$, $A_{v(WL)} = -40.9$, $A_i = -41.34$, $A_p = 1690.7$, $v_{o(WL)} = -2.87$ V
5.15 $I_{CQ} = 1.28$ mA, $V_{CEQ} = 8.4$ V, $R_{in} = 4.4$ kΩ, $R_o = 4.115$ kΩ, $A_{v(NL)} = -76.5$, $A_{v(WL)} = -40.9$, $A_i = -38.2$, $A_p = 1566$, $v_{o(WL)} = -2.811$ V

5.16 $I_{CQ} = 1.22$ mA, $V_{CEQ} = 8.8$ V, $R_{in} = 5.4$ kΩ, $R_o = 4.3$ kΩ, $A_{v(NL)} = -79.19$, $A_{v(WL)} = -41.35$, $A_i = -47.5$, $A_p = 1964.5$, $v_{o(WL)} = -3.017$ V

5.17 $I_{CQ} = 1.09$ mA, $V_{CEQ} = 9.6$ V, $R_{in} = 4.77$ kΩ, $R_o = 4.3$ kΩ, $A_{v(NL)} = -75.63$, $A_{v(WL)} = -39.48$, $A_i = -40.07$, $A_p = 1581.88$, $v_{o(WL)} = -2.78$ V

5.18 $I_{CQ} = 1.685$ mA, $V_{CEQ} = 6.35$ V, $R_{in} = 3.3$ kΩ, $R_o = 3.68$ kΩ, $A_{v(NL)} = -76$, $A_{v(WL)} = -44$, $A_i = -28.5$, $A_p = 1250$, $v_{o(WL)} = -3.025$ V

5.19 $I_{C1Q} = I_{C2Q} = 3.22$ mA, $V_{CE1Q} = 3$ V, $V_{CE2Q} = 3.625$ V, $R_{in} = 1.1$ kΩ, $R_o = 3$ kΩ, $A_{v(NL)} = -375$, $A_{v(WL)} = -228$, $v_{o(WL)} = -2.2$ V

5.20 $I_{C1Q} = I_{C2Q} = 3.225$ mA, $V_{CE1Q} = 2.83$ V, $V_{CE2Q} = 3.625$ V, $R_{in} = 2.65$ kΩ, $R_o = 3$ kΩ, $A_{v(NL)} = -85.7$, $A_{v(WL)} = -52$

Chapter 6

6.1 $V_m = 4.2$ V, $I_m = 1.7$ mA
6.3 $f_L = 86.406$ Hz

Chapter 7

7.1 $I_D = 1.33$ mA
7.2(a) $I_{DSS} = 12$ mA
(b) $V_{GS(off)} = -3$ V
7.5 $V_{GS} = -1$ V, $I_D = 5.33$ mA, $V_{DS} = 5.33$ V
7.6 $V_{GS} = 1.5$ V, $I_D = 3$ mA, $V_{DS} = -9.4$ V
7.7 $V_{GS} = -1.83$ V, $I_D = 1.825$ mA, $V_{DS} = 8.15$ V
7.8 $V_{GS} = -1.83$ V, $I_D = 1.825$ mA, $V_{DS} = 8.15$ V
7.9 $V_{GS} = 1.7$ V, $I_D = 2.25$ mA, $V_{DS} = -8.23$ V
7.10 $V_{GS} = 1.7$ V, $I_D = 2.25$ mA, $V_{DS} = -8.23$ V
7.11 $V_{GS} = -1.817$ V, $I_D = 2.422$ mA, $V_{DS} = 2.19$ V
7.13 $V_{GS} = -1.43$ V, $I_D = 3.275$ mA, $V_{DS} = 5.83$ V
7.14 $V_{GS} = -1.43$ V, $I_D = 3.275$ mA, $V_{DS} = 5.83$ V
7.16 $V_{GS} = -2.2$ V, $I_D = 4.06$ mA, $V_{DS} = 5.065$ V
7.17 $V_{GS} = -1.96$ V, $I_D = 4.133$ mA, $V_{DS} = 5.6$ V
7.18 $V_{GS} = -1.96$ V, $I_D = 4.133$ mA, $V_{DS} = 5.6$ V
7.20 $V_{GS} = 1.96$ V, $I_D = -4.133$ mA, $V_{DS} = -5.58$ V
7.21 $V_{GS} = 1.96$ V, $I_D = -4.133$ mA, $V_{DS} = -5.6$ V
7.23 $R_S = 1$ kΩ, $R_D = 2$ kΩ, $R_1 = 1.3$ MΩ, $R_2 = 150$ kΩ
7.24 $R_S = 1.2$ kΩ, $R_D = 1.6$ kΩ, $R_1 = 1.1$ MΩ, $R_2 = 220$ kΩ
7.25 $R_S = 0.91$ kΩ, $R_D = 2.4$ kΩ, $R_1 = 1.5$ MΩ, $R_2 = 100$ kΩ
7.26 $V_{GS} = -1.5$ V, $I_D = 3.75$ mA, $V_{DS} = 4.75$ V
7.27 $V_{GS} = -1.5$ V, $I_D = 3.75$ mA, $V_{DS} = 4.75$ V
7.28 $V_{GS} = -1.385$ V, $I_D = 4.363$ mA, $V_{DS} = 6.58$ V
7.29 $V_{GS} = -1.385$ V, $I_D = 3.5$ mA, $V_{DS} = 6.58$ V
7.30 $V_{GS} = 3$ V, $I_D = 2$ mA, $V_{DS} = 8.6$ V
7.31 $V_{GS} = 3.1$ V, $I_D = 2.41$ mA, $V_{DS} = 7.712$ V
7.32 $V_{GS} = 2.388$ V, $I_D = 3$ mA, $V_{DS} = 5.8$ V

7.33 $V_{GS} = 2.5$ V, $I_D = 5$ mA, $V_{DS} = 2.5$ V
7.34 $V_{GS} = 2.515$ V, $I_D = 2.65$ mA, $V_{DS} = 2.515$ V
7.35 $V_{GS} = 2.515$ V, $I_D = 2.652$ mA, $V_{DS} = 2.515$ V
7.36 $R_S = 1.5$ kΩ, $R_D = 3.3$ kΩ, $R_1 = 620$ kΩ, $R_2 = 330$ kΩ
7.37 $R_S = 1.2$ kΩ, $R_D = 3$ kΩ, $R_1 = 910$ kΩ, $R_2 = 470$ kΩ
7.38 $V_{GS} = -1.56$ V, $I_D = 3.458$ mA, $V_{DS} = 6.58$ V

Chapter 8

8.1(a) $V_{GSQ} = -1.65$ V, $I_{DQ} = 3.458$ mA, $g_m = 3$ mA/V
   (b) $R_{in} = 192$ kΩ, $R_o = 2.88$ kΩ, $A_{v(NL)} = -8.64$, $A_{v(WL)} = -6.7$, $A_i = -128.6$, $A_p = 862$

8.3(a) $V_{GSQ} = -0.59$ V, $I_{DQ} = 2.49$ mA, $g_m = 12/3$ mA/V
   (b) $R_{in} = 238$ kΩ, $R_o = 2.877$ kΩ, $A_{v(NL)} = -27.06$, $A_{v(WL)} = -34.44$, $A_i = -654.5$, $A_p = 18{,}000$

8.5(a) $V_{GSQ} = -0.122$ V, $I_{DQ} = 1.685$ mA, $g_m = 12.3$ mA/V
   (b) $R_{in} = 238$ kΩ, $R_o = 2.628$ kΩ, $A_{v(NL)} = -16.252$, $A_{v(WL)} = -13.079$, $A_i = -311.28$, $A_p = 4{,}071.23$, $v_{o(WL)} = -4.578$ V

8.6(a) $V_{GSQ} = -2.23$ V, $I_{DQ} = 2.19$ mA, $g_m = 2.7$ mA/V
   (b) $R_{in} = 687$ kΩ, $R_o = 3.31$ kΩ, $A_{v(NL)} = 0.859$, $A_{v(WL)} = 0.866$, $A_i = 59.5$, $A_p = 51.52$, $v_{o(WL)} = 0.7856$V

8.8(a) $V_{GSQ} = -0.726$ V, $I_{DQ} = 2.29$ mA, $g_m = 3$ mA/V
   (b) $R_{in} = 857$ kΩ, $R_o = 3.31$ kΩ, $A_{v(NL)} = 0.9$, $A_{v(WL)} = 0.877$, $A_i = 58.49$, $A_p = 51.3$, $v_{o(WL)} = 0.785$V

8.9(a) $V_{GSQ} = -1.19$ V, $I_{DQ} = 3.28$ mA, $g_m = 8.1$ mA/V
   (b) $R_{in} = 305$ kΩ, $R_o = 2.1$ kΩ, $A_{v(NL)} = -17$, $A_{v(WL)} = -14$, $A_i = -427$, $A_p = 5978$, $v_{o(WL)} = -10.5$V

8.10(a) $V_{GSQ} = -0.88$ V, $I_{DQ} = 3.11$ mA, $g_m = 11.2$ mA/V
   (b) $R_{in} = 283$ kΩ, $R_o = 2.1$ kΩ, $A_{v(NL)} = -23.5$, $A_{v(WL)} = -19.43$, $A_i = -550$, $A_p = 10{,}686$, $v_{o(WL)} = -3.6$V

8.11(a) $V_{GSQ} = 2.5$ V, $I_{DQ} = 3.125$ mA, $g_m = 27.5$ mA/V
   (b) $R_{in} = 103$ kΩ, $R_o = 2.79$ kΩ, $A_{v(NL)} = -76$, $A_{v(WL)} = -60$, $A_i = -618$, $A_p = 37{,}080$, $v_{o(WL)} = -3.04$V

8.12(a) $V_{GSQ} = 3.147$ V, $I_{DQ} = 2.083$ mA, $g_m = 32$ mA/V
   (b) $R_{in} = 77.7$ kΩ, $R_o = 2.85$ kΩ, $A_{v(NL)} = -91$, $A_{v(WL)} = -70$, $A_i = -514$, $A_p = 38{,}080$, $v_{o(WL)} = -3.56$V

8.13(a) $V_{GSQ} = 3.1$ V, $I_{DQ} = 2.04$ mA, $g_m = 50$ mA/V
   (b) $R_{in} = 176$ kΩ, $R_o = 3.25$ kΩ, $A_{v(NL)} = -58.8$, $A_{v(WL)} = -45$, $A_i = -792$, $A_p = 35{,}640$, $v_{o(WL)} = -2.86$V

8.14(a) $V_{GSQ} = 3.1$ V, $I_{DQ} = 2.078$ mA, $g_m = 22$ mA/V
   (b) $R_{in} = 116$ kΩ, $R_o = 3.2$ kΩ, $A_{v(NL)} = -43$, $A_{v(WL)} = -33$, $A_i = -328.8$, $A_p = 12{,}632$, $v_{o(WL)} = 53.7$ mV

8.15(a) $V_{GSQ} = 3.12$ V, $I_{DQ} = 2.38$ mA, $g_m = 60$ mA/V
   (b) $R_{in} = 124$ kΩ, $R_o = 3.2$ kΩ, $A_{v(NL)} = -81$, $A_{v(WL)} = -62$, $A_i = -769$, $A_p = 47678$, $v_{o(WL)} = -3.43$ V

8.16(a) $V_{GSQ} = 2.36$ V, $I_{DQ} = 2.64$ mA, $g_m = 14.4$ mA/V

(b) R_{in} = 65 Ω, R_o = 2.55 kΩ, **9.11** f_{LB} = 6 Hz, f_{LC} = 3.7 Hz, f_{LE} = 33.5 Hz, f_L = 33.5 Hz, $A_{v(WL)}$ = 28.8, A_i = 0.351, A_p = 10, $v_{o(WL)}$ = 1.13 V

8.17(a) V_{GS1Q} = V_{GS2Q} = −1.54 V, I_{DQ} = 2.63 mA, g_m = 3.9 mA/V
   (b) R_{in} = 172 kΩ, R_o = 2.7 kΩ, $A_{v(WL)}$ = −8.287, $v_{o(WL)}$ = −0.783 V
8.18(a) V_{GS1Q} = V_{GS2Q} = −0.633 V, I_{DQ} = 2 mA, g_m = 3.9 mA/V
   (b) R_{in} = 172 kΩ, R_o = 2.7 kΩ, $A_{v(WL)}$ = −23.375, $v_{o(WL)}$ = −0.19 V
8.19 R_{S1} = 51 Ω, R_{S2} = 1.5 kΩ, R_D = 3 kΩ, g_m = 11 mA/V
   R_1 = 2.2 MΩ, R_2 = 360 kΩ
8.20 R_{S1} = 33 Ω, R_{S2} = 1.8 kΩ, R_D = 3 kΩ, g_m = 11.49 mA/V
   R_1 = 1.5 MΩ, R_2 = 330 kΩ
8.21 R_{S1} = 18 Ω, R_{S2} = 1.5 kΩ, R_D = 3 kΩ, g_m = 12.25 mA/V
   R_1 = 1.6 MΩ, R_2 = 300 kΩ
8.22 R_{S1} = 22 Ω, R_{S2} = 1.5 kΩ, R_D = 3.6 kΩ, g_m = 29.7 mA/V
   R_1 = 1.2 MΩ, R_2 = 1 MΩ
8.23 R_{S1} = 51 Ω, R_{S2} = 2.1 kΩ, R_D = 3.6 kΩ, g_m = 31 mA/V
   R_1 = 1.5 MΩ, R_2 = 1 MΩ
8.24 R_{S1} = 24 Ω, R_{S2} = 2.9 kΩ, R_D = 3 kΩ, g_m = 33 mA/V
   R_1 = 2.5 MΩ, R_2 = 1.2 MΩ

Chapter 9

9.1 $|A_v|$ = 42.45 dB
9.2 $|A_v|$ = 47.6 dB
9.3 f_H = 995.2 Hz
9.4 f_H = 2.412 kHz
9.5 f_H = 15.923 kHz
9.6 f_L = 15.923 kHz
9.7 f_L = 5.3 kHz
9.8 f_{LB} = 46.02 Hz, f_{LC} = 19.63 Hz, f_{LE} = 69.2 Hz, f_L = 85.38 Hz
9.9 f_{LB} = 4.723 Hz, f_{LC} = 2.03 Hz, f_{LE} = 70.7373 Hz, f_L = 70.7373 Hz
9.10 f_{LB} = 4.75 Hz, f_{LC} = 8.941 Hz, f_{LE} = 76.151 Hz, f_L = 76.151 Hz
9.11 f_{LB} = 6 Hz, f_{LC} = 3.7 Hz, f_{LE} = 33.5 Hz, f_L = 33.5 Hz
9.12 f_{LB} = 3 Hz, f_{LC} = 1.9 Hz, f_{LE} = 29.48 Hz, f_L = 29.48 Hz, $A_{v(mid)}$ = −41.6, $|v_o(f_L)|$ = 1.794 V(p-p)
9.13 f_{LB} = 6.77 Hz, f_{LC} = 3.8 Hz, f_{LE} = 31.847 Hz, f_L = 31.847 Hz
9.14 $A_{v(mid)}$ = −47, $|v_o(f_L)|$ = 2.402 V(p-p), f_L = 35.6 Hz, f_H = 681 kHz
9.15 $A_{v(mid)}$ = −125.2, $|v_o(f_L)|$ = 2.402 V(p-p), f_{Hi} = 372.59 Hz, f_{Ho} = 21 MHz, f_β = 4.862 MHz
9.16 $A_{v(mid)}$ = −44.34, $|v_o(f_L)|$ = 2.13 V(p-p), f_{Hi} = 838 Hz, f_{Ho} = 19 MHz, f_β = 1.41 MHz
9.17 $A_{v(mid)}$ = −137, f_L = 99 Hz, f_H = 596 kHz
9.18 $A_{v(mid)}$ = −136, f_L = 35 Hz, f_H = 18.7 MHz
9.19 $A_{v(mid)}$ = −112, f_L = 32.5 Hz, f_H = 18.7 MHz
9.20 $A_{v(mid)}$ = −137, f_L = 24 Hz, f_H = 19.9 MHz
9.21 $A_{v(mid)}$ = 1, f_L = 6.932 Hz, f_H = 6.5 MHz

9.22 $A_{v(mid)} = 1$, $f_L = 6.583$ Hz, $f_H = 6.3$ MHz
9.23 $A_{v(mid)} = -138$, $f_L = 92$ Hz, $f_H = 14.74$ MHz
9.24 $A_{v(mid)} = -119.6$, $f_L = 80$ Hz, $f_H = 12.3$ MHz
9.25 $A_{v(mid)} = -91$, $f_L = 67$ Hz, $f_H = 11.6$ MHz
9.26 $A_{v(mid)} = -6.464$, $f_L = 30.412$ Hz, $f_H = 856$ kHz
9.27 $A_{v(mid)} = -7.738$, $f_L = 32.335$ Hz, $f_H = 750$ kHz
9.28 $A_{v(mid)} = -7.745$, $f_L = 32$ Hz, $f_H = 752$ kHz
9.29 $A_{v(mid)} = -32.5$, $f_L = 106.108$ Hz, $f_H = 135.82$ kHz
9.30 $A_{v(mid)} = -34.06$, $f_L = 24.5$ Hz, $f_H = 130.547$ kHz
9.31 $A_{v(mid)} = -8.5$, $f_L = 15.8$ Hz, $f_H = 8.947$ MHz
9.32 $A_{v(mid)} = -21.866$, $f_L = 36.73$ Hz, $f_H = 9.23$ MHz

Chapter 10

10.1(a) $I_o = 3.667$ mA, $R_o = 34$ kΩ
   (b) $R = 6.2$ kΩ, $R_o = 83.333$ kΩ
10.2(a) $I_o = 2.3$ mA, $R_o = 60.869$ kΩ
   (b) $R = 6.5$ kΩ, $R_o = 63.636$ kΩ
10.3 $I_{REF} = I_o = 3.22894$ mA, $R_o = 4.425$ MΩ
10.4(a) $I_o = 4.2$ mA, $R_o = 2.679$ MΩ
   (b) $R = 3.6$ kΩ, $R_o = 3.214$ MΩ
10.5(a) $R = 3.8$ kΩ, $R_o = 40$ kΩ
   (b) $R = 3.6$ kΩ, $R_o = 3.2$ MΩ
10.6 $I_o = 2.43$ mA, $R_o = 26.2$ MΩ
10.7 $I_o = 1.275$ mA, $R_o = 94.11$ kΩ
10.8 $I_o = 2.03$ mA, $R_o = 44$ MΩ
10.9(a) $v_{d2} = 100$ mV(p-p)
   (b) $v_{o1} = 4$ V
   (c) $v_{o2} = -4$ V
   (d) $v_o = 8$ V(p-p)
10.10 $I_E = 1.4$ mA, $V_C = 7.86$ V, $V_{CE} = 8.56$ V
   $A_d = 137.3$, $A_c = -0.5$, CMRR = 274.6 = 48.774 dB
10.11 $R_C = 5.1$ kΩ, $R_E = 4.3$ kΩ, $V_{CC} = V_{EE} = 12$ V
10.12 $I_{REF} = 1.9$ mA, $V_{CE} = 7.2$ V, $A_d = 182.7$, $A_c = -0.0678$, CMRR = 68.575 dB
10.13 $I_{REF} = 2.8$ mA, $I_C = 1.4$ mA, $V_{CE} = 7$ V, $A_d = 167$, $A_c = -0.0578$, CMRR = 69.2 dB
10.14 $I_{REF} = 2.5$ mA, $R_C = 5.2$ kΩ, $R = 4.3$ kΩ, $\pm V_{CC} = \pm 12$ V, $A_c = -0.052$, CMRR = 67.6 dB
10.15 $I_{REF} = 3.32$ mA, $I_C = 1.66$ mA, $R_C = 5.6$ kΩ, $R = 5.1$ kΩ, $V_{CC} = V_{EE} = 18$ V, $A_d = 178.8$, $A_c = -0.001244$, CMRR = 103.15 dB
10.16 $I_{REF} = 3.5$ mA, $I_C = 1.75$ mA, $R_C = 4.3$ kΩ, $R = 3.6$ kΩ, $V_{CE} = 8.175$ V, $A_d = 144.7$, $A_c = -0.006718$, CMRR = 106.66 dB
10.17 $I_{REF} = 2.4$ mA, $V_{CC} = V_{EE} = 11.08$ V, $R_C = 5..1$ kΩ, $R = 4.3$ kΩ, $A_c = -0.0006933$, CMRR = 104.765 dB
10.18 (a) $R_C = 3.6$ kΩ, $R_E = 3$ kΩ, $A_c = -0.6$, CMRR = 44.437 dB, $V_{CC} = V_{EE} = 9.4$ V

(b) $I_{REF} = 3$ mA, $R_C = 3.6$ kΩ, $R = 3$ kΩ, $A_c = -0.0347$, CMRR = 69.2 dB, $V_{CC} = V_{EE} = 9.4$ V

(c) $I_{REF} = 3$ mA, $R_C = 3.6$ kΩ, $R = 2.7$ kΩ, $A_c = -0.00048$, CMRR = 106 dB, $V_{CC} = V_{EE} = 9.4$ V

10.19 $I_{REF} = 2.4$ mA, $V_{CC} = V_{EE} = 20.42$ V, $R_C = 9.1$ kΩ, $R = 7.5$ kΩ, $A_c = -0.00011375$, CMRR = 105 dB

10.20 (a) $R_C = 5.1$ kΩ, $R_E = 4.7$ kΩ, $A_c = -0.5532$, CMRR = 51.16 dB, $V_{CC} = V_{EE} = 19.4$ V

(b) $I_{REF} = 4$ mA, $R_C = 5.1$ kΩ, $R = 4.7$ kΩ, $A_c = -0.08$, CMRR = 68.6 dB, $V_{CC} = V_{EE} = 19.4$ V

(c) $I_{REF} = 4$ mA, $R_C = 5.1$ kΩ, $R = 4.7$ kΩ, $A_c = -0.001$, CMRR = 93.4 dB, $V_{CC} = V_{EE} = 19.4$ V

10.21 (a) $I_{REF} = 2.2$ mA, $I_{C2} = 2.2$ mA, $I_{C4} = 1.1$ mA, $V_{C2} = 4.09$ V, $V_{CE4} = 4.98$ V, $V_{CC} = V_{EE} = 21$ V

(b) $A_{d1} = 139.6$, $A_{d2} = 57.1$, $A_{dm} = 84$ dB

(c) $v_o = 7.97$ V

10.22 $R_E = 3.49$ kΩ. (a) $v_o = 6.377$ V(p-p)(AC only)

(b) $v_o = 6.377$ V(p-p)(AC) + 3.18 V(DC)

10.23 (a) $I_R = I_{REF} = 1.82$ mA, $I_{C1} = I_{C2} = I_{C3} = I_{C4} = I_{C5} = 1.82$ mA, $V_{CE4} = 4$ V

(b) $R_E = 4$ kΩ

(c) (1) $v_o = 10.3$ V(p-p)(AC only), (2) $v_o = 8.25$ V(p-p)(AC) + 4.12 V(DC)

10.24 (a) $I_o = I_{REF} = 2.48$ mA, $I_{D(Q1)} = 1.24$ mA, $V_{DS(Q4)} = 7.95$ V

(b) $A_{d1} = -35$, $A_{d2} = 35$, $A_{dm} = -70$

(c) $A_c = -0.0625$, CMRR = 55 dB

10.25 $I_o = I_{REF} = 1.68$ mA, $I_{D(Q1)} = 0.848$ mA, $V_{DS(Q1)} = 10.84$ V, $A_d = 29.2$, $A_c = -0.0528$, CMRR = 54.851 dB

10.26 (a) $I_o = I_{REF} = 2.03$ mA, $I_{D(Q1)} = 1.015$ mA, $V_{DS(Q4)} = 12.52$ V

(b) $A_{d1} = -20$, $A_{d2} = 35$, $A_{dm} = -40$

(c) $A_c = -1.044 \times 10^{-6}$, CMRR = 145 dB,

10.27 $I_o = I_{REF} = 2$ mA, $I_{D(Q1)} = 1$ mA, $V_{DS(Q1)} = 8$ V, $A_d = 26$, $A_c = -8.6 \times 10^{-4}$, CMRR = 89.6 dB

10.28 $I_o = I_{REF} = 4$ mA, $I_{D(Q1)} = 2$ mA, $R_{D1} = 5.6$ kΩ, $A_d = 25$, $R = 3.6$ kΩ, $A_c = -0.001$, CMRR = 88 dB

Chapter 11

11.6 $I_{LED} = 11.76$ mA

11.8 $I_{RED} = I_G = 17.58$ mA, $I_Y = 15.38$ mA

11.12 $R_F = 100$ kΩ, $R_S = 2$ kΩ

11.13 $A_v = -100$, $R_{in} = 2.2$ kΩ, $R_o = 0$ kΩ, BW = 10 kHz

11.14 $|A_{v(min)}| = 9.1$, $|A_{v(max)}| = 100$, $R_{in} = 1.1$ kΩ, $R_o = 0$, $BW_{min} = 10$ kHz, $BW_{max} = 109.89$ kHz

11.16 $A_v = 101$, $R_{in} = \infty$, $R_o = 0$, BW = 9.9 kHz

11.17 $A_{v(min)} = 51$, $A_{v(max)} = 101$, $R_{in} = \infty$, $R_o = 0$, $BW_{min} = 39.6$ kHz

11.18 $R_F = 100$ kΩ pot, $R_S = 1$ kΩ, $R_o = 0$

11.19 $R_F = 1$ MΩ pot, $R_S = 10$ kΩ, $R_o = 0$
11.20 $v_o = 2.5$ V(DC) − 5 V(p-p)(AC)
11.21 $v_o = -2.5$ V(DC) − 5 V(p-p)(AC)
11.22 $v_o = 2.75$ V(DC) + 10 V(p-p)(AC)
11.23 $v_o = 25(v_1 + v_2 + v_3 + v_4)$
11.26 $v_o = -4$ V(p-p)(AC)
11.27 $v_o = -10$ mV(p-p)(AC)
11.28 (a) $|v_o| = 2$ V(signal) + 0.1 V(noise)
       (b) $|v_o| = 2$ V(signal) + 2 V(noise)
11.29 $|v_o| = 2$ V(signal) + 0.1 V(noise), $R_i = \infty$, $R_o = 0$
11.30 $v_{o(min)} = 720$ mV(rms), $v_{o(max)} = 4.12$ mV(rms)
11.31 $|A_{v(min)}| = 37.36$, $|A_{v(max)}| = 401$, $|v_o| = 1.62$ V(signal) + 25.6 μV(noise)
11.32 $R_F = 80$ kΩ, $R = 4$ kΩ, $C = 0.01$ μF
11.33 $R_F = 160$ kΩ, $R = 10$ kΩ, $C = 0.01$ μF
11.34 $R = 510$ Ω, $R_S = 10$ kΩ, $C = 0.01$ μF, $C_F = 100$ pF
11.35 $R = 20$ kΩ, $R_S = 2$ kΩ, $C = 0.0047$ μF, $C_F = 200$ pF

Chapter 12

12.2 $dA_f/A_f = 0.022\%$, $ISS = 99.8\%$
12.3 $dA_f/A_f = 0.04\%$, $ISS = 99.7\%$
12.4 $dA_f/A_f = 0.0995\%$, $ISS = 99.5\%$
12.5 $A_{vf} = 101$, $BW_f = 9.9$ kHz, $R_{if} = 3.96$ GΩ, $R_{of} = 37.8$ mΩ
12.7 $A_{vf} = 44.99$, $BW_f = 22.22$ kHz, $R_{if} = 8.89$ GΩ, $R_{of} = 16.89$ mΩ
12.8 $A_{vf} = 1$, $BW_f = 1$ MHz, $R_{if} = 400$ GΩ, $R_{of} = 0.375$ mΩ
12.9 $A_{vf} = 34$, $BW_f = 29.4$ kHz, $R_{if} = 11.762$ GΩ, $R_{of} = 12.75$ mΩ
12.11 $V_{IO} = 0.99$ mV
12.12 $PSRR = 5 \times 10^5 = 114$ dB, $V_{IO1} = 0.99$ mV, $V_{IO2} = 1$ mV
12.13 $CMRR = 8,000 = 78$ dB
12.14 $CMRR = 50,000 = 94$ dB
12.16 $SR_{min} = 0.55$ V/μs
12.17 $v_o = 8.18$ V(DC) + 10 V(p-p)(AC), $f_L = 33$ Hz, $f_{o(min)} = 330$ Hz
12.18 $v_o = 8$ V(DC) − 9.1 V(p-p)(AC), $v_{out} = -9.1$ V(p-p)(AC), $f_L = 330$ Hz, $f_H = 22$ kHz, $f_{o(min)} = 3300$ Hz
12.19 $R_F = 200$ kΩ, $R_S = 10$ kΩ, $C_i = 2.2$ μF, $C_o = 10$ μF, $R_1 = R_2 = 100$ kΩ
12.20 (a) $v_o = 8$ V(DC) + 8 V(p-p)(AC), $v_{out} = 8$ V(p-p)(AC)
      (b) $f_L = 22$ Hz, $f_H = 250$ kHz, $f_{o(min)} = 220$ Hz
12.21 $v_o = 6.4$ V(DC) + 5 V(p-p)(AC), $v_{out} = 5$ V(p-p)(AC), $f_L = 10$ Hz, $f_H = 39.6$ kHz, $f_{o(min)} = 100$ Hz
12.22 $R_F = 200$ kΩ, $R_S = 10$ kΩ, $C_1 = 0.47$ μF, $C_2 = 2$ μF, $C_o = 10$ μF, $R_1 = R_2 = 100$ kΩ

Chapter 13

13.2(a) $\eta = 17.36\ \%$, **(b)** $\eta = 11.11\ \%$
13.3(a) $\eta = 19.14\ \%$, **(b)** $\eta = 16\ \%$
13.5(a) $\eta_{maxA} = 25\ \%$, **(b)** $\eta_{maxB} = 78.5\ \%$
13.6(a) $\eta = 65.45\ \%$, $P_D = 1.07$ W **(b)** $\eta = 52.35\ \%$, $P_D = 1.82$ W
13.7 $\eta = 65.5\ \%$, $P_D = 1.65$ W
13.8 $\eta = 58.9\ \%$, $P_D = 7.89$ W
13.9 $V_m = 11$ V, $\eta = 70\ \%$, , $\eta = 69.1\ \%$
13.10 $R_F = 27$ kΩ, $R_S = 3$ kΩ, $R_L = 20$ Ω, $\eta = 71.42\ \%$
13.11 $R_F = 100$ kΩ, $R_S = 2.7$ kΩ, $\eta = 69.1\ \%$
13.12 $R_F = 110$ kΩ, $R_S = 10$ kΩ, $\eta = 67.5\ \%$
13.17 $V_m = 13.5$ V, $\eta = 66.18\ \%$, $I_{DC} = 0.159$ A, $P_O = 3.375$ W, $P_I = 5.1$ W
13.18 $V_m = 13.75$ V, $\eta = 67.5\ \%$, $P_D = 1.82$ W, $P_{D(Q1)} = P_{D(Q2)} = 0.91$ W
13.19 $\pm V_{CC} = \pm 17$ V, $R_F = 27$ kΩ, $R = 3$ kΩ, $R_L = 15$ Ω, $R = 1$ kΩ, $\eta = 69.4\ \%$, $P_D = 3.3$ W
13.20 $\pm V_{CC} = \pm 16$ V, $R_F = 27$ kΩ, $R_S = 3$ kΩ, $I = 27.78$ mA, $R = 510$ Ω, $\eta = 69.44\ \%$
13.21 $P_{D(max)} = 15.49$ W
13.22 $P_{D(max)} = 33.3$ W
13.23 $\theta_{CS} + \theta_{SA} = 2.08°$ C/W
13.24 $\theta_{CS} + \theta_{SA} = 1.333°$ C/W

Chapter 14

14.1 $f_H = 3.38$ kHz, $A_{BP} = 9.46 = 19.52$ dB, -20 dB/dec, $|A_v(f_H)| = 16.52$ dB
14.2 $f_l = 3.38$ kHz, $A_{BP} = 11.93 = 21.53$ dB, 20 dB/dec, $|A_v(f_L)| = 18.53$ dB
14.3 $f_H = 2.2$ kHz, $A_{BP} = 11.9 = 21.51$ dB, -20 dB/dec, $|A_v(f_H)| = 18.51$ dB
14.4 $f_L = 1.56$ kHz, $A_{BP} = 11.91 = 21.52$ dB, 20 dB/dec, $|A_v(f_L)| = 18.52$ dB
14.5 $R_F = 82$ kΩ, $R_S = 9.1$ kΩ, $R = 8.2$ kΩ, $C = 0.001$ μF
14.6 $R_F = 51$ kΩ, $R_S = 4.3$ kΩ, $R = 3.9$ kΩ, $C = 0.01$ μF
14.7 $R_F = 39$ kΩ, $R_S = 4.3$ kΩ, $R = 3.9$ kΩ, $C = 0.01$ μF
14.8 $R_F = 82$ kΩ, $R_S = 9.1$ kΩ, $R = 8.2$ kΩ, $C = 0.001$ μF
14.9 $f_H = 4.82$ kHz, $A_{BP} = 2$, $\xi = 0.5$, $ROR = -40$ dB/dec
14.10 $f_L = 6.63$ kHz, $A_{BP} = 2$, $\xi = 0.5$, $ROR = 40$ dB/dec
14.11 $f_H = 5.3$ kHz, $A_{BP} = 1.667$, $\xi = 0.67$, $ROR = -40$ dB/dec
14.12 $f_L = 1.69$ kHz, $A_{BP} = 1.62, 69$, $ROR = 40$ dB/dec
14.13 $R_F = 12$ kΩ, $R_S = 22$ kΩ, $R = 3.9$ kΩ, $C = 0.01$ μF
14.14 $R_F = 6.2$ kΩ, $R_S = 10$ kΩ, $R = 3.9$ kΩ, $C = 0.01$ μF
14.15 $R_F = 10$ kΩ, $R_S = 18$ kΩ, $R = 3.3$ kΩ, $C = 0.01$ μF
14.16 $R_F = 5.1$ kΩ, $R_S = 9.1$ kΩ, $R = 3.3$ kΩ, $C = 0.01$ μF
14.17 $f_o = 15.9$ kHz, $A_{BP} = 20 = 26$ dB, $\xi = 0.5$, Butterworth filter
14.18 $R_1 = 4.3$ kΩ, $R_{F1} = 22$ kΩ, $R_{S1} = 15$ kΩ, $C = 0.01$ μF, $R_2 = 9.1$ kΩ, $R_{F2} = 300$ kΩ, $R_{S2} = 20$ kΩ
14.19 $R_1 = 3.6$ kΩ, $R_{F1} = 18$ kΩ, $R_{S1} = 12$ kΩ, $C = 0.01$ μF, $R_2 = 7.5$ kΩ, $R_{F1} = 120$ kΩ, $R_{S2} = 18$ kΩ

14.20 $f_o = 21.22$ kHz, $A_{BP} = 25 = 27.96$ dB, $\xi = 0.25$, Chebyshev filter

14.21 $R_1 = 3.9$ kΩ, $R_{F1} = 7.5$ kΩ, $R_{S1} = 7.5$ kΩ, $C = 0.01$ μF, $R_2 = 3.9$ kΩ, $R_{F2} = 24$ kΩ, $R_{S2} = 4.7$ kΩ

14.22 $R_1 = 1.8$ kΩ, $R_{F1} = 4.3$ kΩ, $R_{S1} = 3$ kΩ, $C = 0.0047$ μF, $R_2 = 1.6$ kΩ, $R_{F2} = 7.5$ kΩ, $R_{S2} = 6.2$ kΩ

14.23 $R_1 = 1.6$ kΩ, $R_{F1} = 3.6$ kΩ, $R_{S1} = 24$ kΩ, $C = 0.01$ μF, $R_2 = 1.6$ kΩ, $R_{F2} = 7.5$ kΩ, $R_{S2} = 6.2$ kΩ

14.24 $R_1 = 1.6$ kΩ, $R_{F1} = 1.8$ kΩ, $R_{S1} = 12$ kΩ, $C = 0.01$ μF, $R_2 = 1.6$ kΩ, $R_{F2} = 3.6$ kΩ, $R_{S2} = 3$ kΩ

14.25 $R_1 = 6.8$ kΩ, $R_{F1} = 24$ kΩ, $R_{S1} = 33$ kΩ, $C = 0.01$ μF, $R_2 = 3.6$ kΩ, $R_{F2} = 20$ kΩ, $R_{S2} = 12$ kΩ

14.26 $R_1 = 3.9$ kΩ, $R_{F1} = 13$ kΩ, $R_{S1} = 18$ kΩ, $C = 0.001$ μF, $R_2 = 7.5$ kΩ, $R_{F2} = 20$ kΩ, $R_{S2} = 12$ kΩ

14.27 $R_1 = 3.3$ kΩ, $R_{F1} = 16$ kΩ, $R_{S1} = 12$ kΩ, $C = 0.01$ μF, $R_2 = 3.3$ kΩ, $R_{F2} = 9.1$ kΩ, $R_{S2} = 24$ kΩ, $R_3 = 3.3$ kΩ, $R_{F3} = 20$ kΩ, $R_{S3} = 3.9$ kΩ

14.28 $R_1 = 1.6$ kΩ, $R_{F1} = 3.9$ kΩ, $R_{S1} = 2.7$ kΩ, $C = 0.005$ μF, $R_2 = 2.2$ kΩ, $R_{F2} = 5.6$ kΩ, $R_{S2} = 24$ kΩ, $R_3 = 1.6$ kΩ, $R_{F3} = 10$ kΩ, $R_{S3} = 2$ kΩ

14.29 $R_1 = 8.2$ kΩ, $R_2 = 33$ kΩ, $R_3 = 82$ kΩ, $C = 0.005$ μF, $BW = 2$ kHz

14.30 $R_1 = 8.2$ kΩ, $R_2 = 33$ kΩ, $R_3 = 82$ kΩ, $C = 0.005$ μF, $R_4 = 20$ kΩ, $R_5 = 1$ kΩ, $R_F = 20$ kΩ, $BW = 2$ kHz

14.31 $f_c = 5$ kHz, $Q = 16$, $|A_c| = 10.64$, $BW = 312$ Hz

14.32 $R_1 = 1.6$ kΩ, $R_2 = 33$ kΩ, $R_3 = 91$ kΩ, $C = 0.01$ μF, $R_4 = 10$ kΩ, $R_5 = 1$ kΩ, $R_F = 10$ kΩ, $BW = 1$ kHz

Chapter 15

15.1 $V_O = 10$ V, $I_O = 0.625$ A, $I_{O(max)} = 1$ A, $V_{r(p-p)} = 138$ mV

15.2 $V_O = 11.27$ V, $I_O = 0.5635$ A, $I_{O(max)} = 0.7$ A, $V_{r(p-p)} = 90.6$ mV

15.3 $R = 6.8$ kΩ, $R_1 = 39$ kΩ, $R_2 = 100$ kΩ pot, $R_S = 0.68$ Ω, $R_{L(min)} = 15$ Ω, $V_{r(p-p)} = 86$ mV, Zener diode is 1N5230

15.4 $R = 7.5$ kΩ, $R_1 = 39$ kΩ, $R_2 = 100$ kΩ pot, $R_S = 0.82$ Ω, $R_{L(min)} = 18.75$ Ω, $V_{r(p-p)} = 73.6$ mV

15.5 High efficiency ($\eta = 75$ tom 95 %), high power level (5 W to 50 kW), step-up and step-down voltage conversion, multiple outputs, and small size and weight/W.

15.6 $\eta = 92\%$, $V_{r(p-p)} = 21.3$ mV

15.7 $\eta = 91\%$, $V_{r(p-p)} = 11.1$ mV

15.8 $C = 27$ μF/20 V/0.145 Ω, $L = 168$ μH/1.8 A/0.18 Ω, $R_1 = 1$ kΩ, $R_2 = 10$ kΩ pot, LM2595J-ADJ, 1N5817 Schottky diode, $R_{L(min)} = 3.75$ Ω, $R_{L(max)} = 100$ Ω, $V_{r(p-p)} = 75$ mV, $V_{S(max)} = V_{D(max)} = 16$ V, $I_{S(max)} = I_{D(max)} = 2.067$ A

15.9 $C = 27$ μF/20 V/0.145 Ω, $L = 168$ μH/1.8 A/0.18 Ω, $R_1 = 1$ kΩ, $R_2 = 10$ kΩ pot, LM2595J-ADJ, 1N5817 Schottky diode, $R_{L(min)} = 6.7$ Ω, $R_{L(max)} = 66.7$ Ω, $V_{r(p-p)} = 50$ mV, $V_{S(max)} = V_{D(max)} = 12.5$ V, $I_{S(max)} = I_{D(max)} = 802.5$ mA

15.10 $\eta = 82.3\%$, $V_{r(p-p)} = 0.838$ V

15.11 $\eta = 81.5\%$, $V_{r(p-p)} = 1.078$ V

15.12 $C = 300$ μF/20 V/0.075 Ω, $L = 250$ μH/1.5 A/0.23 Ω, $R_1 = 1$ kΩ, $R_2 = 10$ kΩ pot, LM2586S-ADJ, 1N5820 Schottky diode, $R_C = 4.7$ kΩ, $C_C = 0.01$ μF, $R_{L(min)} = 10$ Ω, $R_{L(max)} = 500$ Ω, $V_{r(p-p)} = 150$ mV, $V_{S(max)} = V_{D(max)} = 15$ V, $I_{S(max)} = I_{D(max)} = 5.28$ A

15.13 $C = 120$ μF/15 V/0.07 Ω, $L = 114$ μH/3 A/0.1 Ω, $R_1 = 1$ kΩ, $R_2 = 10$ kΩ pot, LM2586S-ADJ, 1N5820 Schottky diode, $R_C = 4.7$ kΩ, $C_C = 0.01$ μF, $R_{L(min)} = 20$ Ω, $R_{L(max)} = 200$ Ω, $V_{r(p-p)} = 150$ mV, $V_{S(max)} = V_{D(max)} = 15$ V, $I_{S(max)} = I_{D(max)} = 5.28$ A

15.14 $\eta = 72.5$ %, $V_{r(p-p)} = 75$ mV

15.15 $\eta = 74.76$ %, $V_{r(p-p)} = 450$ mV

15.16 $\eta = 66$ %, $V_{r(p-p)} = 473$ mV

15.17 $C = 68$ μF/15 V/0.095 Ω, $L = 114$ μH/3 A/0.1 Ω, $R_1 = 1$ kΩ, $R_2 = 10$ kΩ pot, LM2595J-ADJ, 1N5820 Schottky diode, $R_C = 4.7$ kΩ, $C_C = 0.01$ μF, $R_{L(min)} = 6.7$ Ω, $R_{L(max)} = 200$ Ω, $V_{r(p-p)} = 100$ mV, $V_{S(max)} = V_{D(max)} = 18$ V, $I_{S(max)} = I_{D(max)} = 6.3$ A

15.18 $C = 360$ μF/10 V/0.06 Ω, $L = 250$ μH/3 A/0.23 Ω, LM2595J-ADJ, 1N5820 Schottky diode, $C_{in} = 22$ μF, $R_{L(min)} = 5$ Ω, $R_{L(max)} = 150$ Ω, $V_{r(p-p)} = 75$ mV, $V_{S(max)} = V_{D(max)} = 22.5$ V, $I_{S(max)} = I_{D(max)} = 1.466$ A

15.19 $C = 360$ μF/10 V/0.06 Ω, $L = 168$ μH/1.8 A/0.18 Ω, LM2595J-ADJ, 1N5820 Schottky diode, $C_{in} = 22$ μF, $R_{L(min)} = 6$ Ω, $R_{L(max)} = 120$ Ω, $V_{r(p-p)} = 60$ mV, $V_{S(max)} = V_{D(max)} = 18$ V, $I_{S(max)} = I_{D(max)} = 2$ A

15.20 $C = 300$ μF/15 V/0.065 Ω, $L = 168$ μH/1.8 A/0.18 Ω, LM2595J-ADJ, 1N5820 Schottky diode, $C_{in} = 22$ μF, $R_{L(min)} = 6$ Ω, $R_{L(max)} = 120$ Ω, $V_{r(p-p)} = 75$ mV, $V_{S(max)} = V_{D(max)} = 22$ V, $I_{S(max)} = I_{D(max)} = 4.1$ A

Chapter 16

16.1 $\beta A = 1$
16.2 $f_o = 650$ Hz
16.3 $C = 0.1$ μH, $R = 2$ kΩ
16.4 $C = 0.1$ μH, $R = 2$ kΩ, $R_S = 10$ kΩ, $R_F = 500$ kΩ pot
16.5 $f_o = 1.6$ kHz
16.6 $C = 0.01$ μH, $R = 4.7$ kΩ + 100 Ω, $R_S = 15$ kΩ, $R_F = 20$ kΩ pot
16.7 $f_o = 8$ kHz
16.6 $C = 0.01$ μH, $R = 4.7$ kΩ, $R_S = 15$ kΩ, $R_F = 20$ kΩ pot
16.7 $f_o = 8$ kHz
16.8 $C = 0.1$ μH, $R = 8.2$ kΩ, $R_S = 7.5$ kΩ, $R_F = 20$ kΩ pot
16.9 $f_o = 22.5$ kHz
16.10 $f_o = 11.25$ kHz
16.11 $C = 0.01$ μH, $L = 0.5$ mH, $R_S = 10$ kΩ, $R_F = 20$ kΩ pot
16.12 $C = 0.01$ μH, $L = 4.65$ mH, $R_S = 10$ kΩ, $R_F = 20$ kΩ pot
16.13 $C = 0.1$ μH, $L = 0.2$ mH, $R_S = 10$ kΩ, $R_F = 20$ kΩ pot
16.14 $C = 0.1$ μH, $L = 0.63$ mH, $R_S = 10$ kΩ, $R_F = 20$ kΩ pot
16.15 $f_o = 225$ kHz
16.16 $C = 0.01$ μH, $R_1 = 10$ kΩ, $R_2 = 10$ kΩ pot
16.17 $C = 0.01$ μH, $R_1 = 10$ kΩ, $R_2 = 10$ kΩ pot
16.18 $C = 0.001$ μH, $R = (27 + 2)$ kΩ, $V_{CC} = 5$ V
16.19 $C = 0.001$ μH, $R_1 = 7.5$ kΩ, $R_2 = 15$ kΩ + 10 kΩ, $V_{CC} = 5$ V

16.20 $C = 0.001$ μH, $R = 2.4$ kΩ, $R_1 = 3.6$ kΩ, $R_2 = 3.6$ kΩ, $\pm V_{CC} = \pm 15$ V
16.21 $C = 0.001$ μH, $C_1 = 20.01$ μF, $R_1 = 2$ kΩ pot, $R_2 = 1$ kΩ + 10 kΩ pot, $V_{CC} = 5$ V
16.22 $C = 0.001$ μH, $C_1 = 20.01$ μF, $R_1 = 2$ kΩ pot, $R_2 = 1$ kΩ + 10 kΩ pot, $V_{CC} = 5$ V
16.23 $C = 0.001$ μH, $R = 1$ kΩ + 10 kΩ pot, $R_A = 0.5$ kΩ, $R_B = 25$ kΩ, $V_{CC} = 15$ V
16.24 $C = 0.001$ μH, $R = 1$ kΩ + 10 kΩ pot, $V_{CC} = 15$ V
16.25 $C = 0.001$ μH, $R_1 = 1$ kΩ, $R_2 = 1$ kΩ + 10 kΩ pot, $V_{CC} = 5$ V
16.26 $C = 0.001$ μH, $R = 1$ kΩ + 10 kΩ pot, $V_{CC} = 5$ V
16.27 $C = 0.01$ μH, $R = 5.1$ kΩ + 10 kΩ pot, $V_{CC} = 5$ V
16.28 $f_o = 15.75$ kHz, $D = 21.26\%$
16.29 $C = 0.033$ μH, $R_1 = 15$ kΩ, $R_2 = 5.1$ kΩ + 10 kΩ pot, $V_{CC} = 5$ V

Chapter 17

17.1 A computer may process them.
17.2 A better signal-to-noise ratio.
17.3(a) $\Delta S = 78.125$ mV, $v_o = 3.984375$ V
   (b) $\Delta S = 78.125$ mV, $v_o = 3.515625$ V
17.4(a) $\Delta S = 19.53125$ mV, $v_o = 3.88$ V
   (b) $\Delta S = 19.53125$ mV, $v_o = 3.51$ V
17.5(a) $v_o = -12.75$ V
   (b) $v_o = -11.25$ V
17.6(a) $v_o = 8$ V
   (b) $v_o = 7.109575$ V
17.7(a) $v_o = 16$ V
   (b) $v_o = 16$ V
17.13 $V_F = 0.69$ V, $V_o = 0.63$ V
17.14 $V_F = 0.71$ V, $V_o = 0.392$ V
17.15 $V_F = 0.65$ V, $V_o = 0.98$ V
17.16(a) 1, 9
   (b) 3, 2
   (c) 4, 8
   (d) 6, 4

Chapter 18

18.3 $I_{T(rms)} = 12$ A, $V_{TM} = 1.7$ V, $I_{GT} = 5$ mA, $t_{on} = 1$ μs, $V_{GT} = 0.7$ V, $I_H = 6$ mA, $t_{off} = 15$ μs
18.4 (a) By making $V_{AA} > V_{BO(F)}$
   (b) By triggering
18.12 $I_{T(rms)} = 4$ A, $V_{TM} = 2$ V, $I_{GT} = 5$ mA, $t_{on} = 1.5$ μs, $V_{GT} = 0.7$ V, $I_H = 30$ mA, $t_{off} = 15$ μs
18.13 $\phi = 55.2°$, $\alpha = 124.8°$, $\%P = 91.6\%$
18.14(a) $\%P = 95.2\%$
   (b) $\%P = 89.5\%$
   (c) $\%P = 70.6\%$
18.16(a) $V_{in} = -4$ V
   (b) $V_{in} = -10.15$ V